Observer Performance Methods for Diagnostic Imaging

IMAGING IN MEDICAL DIAGNOSIS AND THERAPY

Series Editors: Andrew Karellas and Bruce R. Thomadsen

Published titles

Quality and Safety in Radiotherapy
Todd Pawlicki, Peter B. Dunscombe,
Arno J. Mundt, and Pierre Scalliet, Editors
ISBN: 978-1-4398-0436-0

Adaptive Radiation Therapy
X. Allen Li, Editor
ISBN: 978-1-4398-1634-9

Quantitative MRI in Cancer
Thomas E. Yankeelov, David R. Pickens,
and Ronald R. Price, Editors
ISBN: 978-1-4398-2057-5

Informatics in Medical Imaging
George C. Kagadis and Steve G. Langer, Editors
ISBN: 978-1-4398-3124-3

Adaptive Motion Compensation in Radiotherapy
Martin J. Murphy, Editor
ISBN: 978-1-4398-2193-0

Image-Guided Radiation Therapy
Daniel J. Bourland, Editor
ISBN: 978-1-4398-0273-1

Targeted Molecular Imaging
Michael J. Welch and William C. Eckelman,
Editors
ISBN: 978-1-4398-4195-0

Proton and Carbon Ion Therapy
C.-M. Charlie Ma and Tony Lomax, Editors
ISBN: 978-1-4398-1607-3

Physics of Mammographic Imaging
Mia K. Markey, Editor
ISBN: 978-1-4398-7544-5

Physics of Thermal Therapy: Fundamentals and Clinical Applications
Eduardo Moros, Editor
ISBN: 978-1-4398-4890-6

Emerging Imaging Technologies in Medicine
Mark A. Anastasio and Patrick La Riviere, Editors
ISBN: 978-1-4398-8041-8

Cancer Nanotechnology: Principles and Applications in Radiation Oncology
Sang Hyun Cho and Sunil Krishnan, Editors
ISBN: 978-1-4398-7875-0

Image Processing in Radiation Therapy
Kristy Kay Brock, Editor
ISBN: 978-1-4398-3017-8

Informatics in Radiation Oncology
George Starkschall and R. Alfredo C. Siochi,
Editors
ISBN: 978-1-4398-2582-2

Cone Beam Computed Tomography
Chris C. Shaw, Editor
ISBN: 978-1-4398-4626-1

Computer-Aided Detection and Diagnosis in Medical Imaging
Qiang Li and Robert M. Nishikawa, Editors
ISBN: 978-1-4398-7176-8

Cardiovascular and Neurovascular Imaging: Physics and Technology
Carlo Cavedon and Stephen Rudin, Editors
ISBN: 978-1-4398-9056-1

Scintillation Dosimetry
Sam Beddar and Luc Beaulieu, Editors
ISBN: 978-1-4822-0899-3

Handbook of Small Animal Imaging: Preclinical Imaging, Therapy, and Applications
George Kagadis, Nancy L. Ford, Dimitrios N. Karnabatidis, and George K. Loudos Editors
ISBN: 978-1-4665-5568-6

IMAGING IN MEDICAL DIAGNOSIS AND THERAPY

Series Editors: Andrew Karellas and Bruce R. Thomadsen

Published titles

**Comprehensive Brachytherapy:
Physical and Clinical Aspects**
*Jack Venselaar, Dimos Baltas, Peter J. Hoskin,
and Ali Soleimani-Meigooni, Editors*
ISBN: 978-1-4398-4498-4

**Handbook of Radioembolization:
Physics, Biology, Nuclear Medicine,
and Imaging**
*Alexander S. Pasciak, PhD., Yong Bradley, MD.,
and J. Mark McKinney, MD., Editors*
ISBN: 978-1-4987-4201-6

**Monte Carlo Techniques in Radiation
Therapy**
Joao Seco and Frank Verhaegen, Editors
ISBN: 978-1-4665-0792-0

**Stereotactic Radiosurgery and
Stereotactic Body Radiation Therapy**
*Stanley H. Benedict, David J. Schlesinger,
Steven J. Goetsch, and Brian D. Kavanagh,
Editors*
ISBN: 978-1-4398-4197-6

Physics of PET and SPECT Imaging
Magnus Dahlbom, Editor
ISBN: 978-1-4665-6013-0

Tomosynthesis Imaging
Ingrid Reiser and Stephen Glick, Editors
ISBN: 978-1-138-19965-1

Ultrasound Imaging and Therapy
Aaron Fenster and James C. Lacefield, Editors
ISBN: 978-1-4398-6628-3

**Beam's Eye View Imaging in
Radiation Oncology**
Ross I. Berbeco, Ph.D., Editor
ISBN: 978-1-4987-3634-3

**Principles and Practice of
Image-Guided Radiation Therapy
of Lung Cancer**
*Jing Cai, Joe Y. Chang, and Fang-Fang Yin,
Editors*
ISBN: 978-1-4987-3673-2

**Radiochromic Film: Role and
Applications in Radiation Dosimetry**
Indra J. Das, Editor
ISBN: 978-1-4987-7647-9

**Clinical 3D Dosimetry in Modern
Radiation Therapy**
Ben Mijnheer, Editor
ISBN: 978-1-4822-5221-7

**Observer Performance Methods for
Diagnostic Imaging: Foundations,
Modeling, and Applications with
R-Based Examples**
Dev P. Chakraborty, Editor
ISBN: 978-1-4822-1484-0

Observer Performance Methods
for Diagnostic Imaging
Foundations, Modeling, and Applications with
R-Based Examples

Dev P. Chakraborty

CRC Press
Taylor & Francis Group
Boca Raton London New York

CRC Press is an imprint of the
Taylor & Francis Group, an **informa** business

CRC Press
Taylor & Francis Group
6000 Broken Sound Parkway NW, Suite 300
Boca Raton, FL 33487-2742

First issued in paperback 2021

ISBN 13: 978-0-367-78163-7 (pbk)
ISBN 13: 978-1-4822-1484-0 (hbk)

Library of Congress Cataloging-in-Publication Data

Names: Chakraborty, Dev P., author.
Title: Observer performance methods for diagnostic imaging : foundations, modeling, and applications with R-based examples / Dev P. Chakraborty.
Other titles: Imaging in medical diagnosis and therapy ; 29.
Description: Boca Raton, FL : CRC Press, Taylor & Francis Group, [2017] | Series: Imaging in medical diagnosis and therapy ; 29
Identifiers: LCCN 2017031569| ISBN 9781482214840 (hardback ; alk. paper) | ISBN 1482214849 (hardback ; alk. paper)
Subjects: LCSH: Diagnostic imaging–Data processing. | R (Computer program language) | Imaging systems in medicine. | Receiver operating characteristic curves.
Classification: LCC RC78.7.D53 C46 2017 | DDC 616.07/543–dc23
LC record available at https://lccn.loc.gov/2017031569

Visit the Taylor & Francis Web site at
http://www.taylorandfrancis.com

and the CRC Press Web site at
http://www.crcpress.com

Dedication

Dedicated to my paternal grandparents:

Dharani Nath (my "Dadu") and Hiran Bala Devi (my "Thamma")

Contents

Series Preface

Since their inception over a century ago, advances in the science and technology of medical imaging and radiation therapy are more profound and rapid than ever before. Further, the disciplines are increasingly cross-linked as imaging methods become more widely used to plan, guide, monitor, and assess treatments in radiation therapy. Today, the technologies of medical imaging and radiation therapy are so complex and computer-driven that it is difficult for the people (physicians and technologists) responsible for their clinical use to know exactly what is happening at the point of care when a patient is being examined or treated. The people best equipped to understand the technologies and their applications are medical physicists, and these individuals are assuming greater responsibilities in the clinical arena to ensure that what is intended for the patient is actually delivered in a safe and effective manner.

The growing responsibilities of medical physicists in the clinical arenas of medical imaging and radiation therapy are not without their challenges, however. Most medical physicists are knowledgeable in either radiation therapy or medical imaging and expert in one or a small number of areas within their disciplines. They sustain their expertise in these areas by reading scientific articles and attending scientific talks at meetings. In contrast, their responsibilities increasingly extend beyond their specific areas of expertise. To meet these responsibilities, medical physicists must periodically refresh their knowledge of advances in medical imaging or radiation therapy and they must be prepared to function at the intersection of these two fields. To accomplish these objectives is a challenge.

At the 2007 annual meeting of the American Association of Physicists in Medicine in Minneapolis, this challenge was the topic of conversation during a lunch hosted by Taylor & Francis Publishers and involving a group of senior medical physicists (Arthur L. Boyer, Joseph O. Deasy, C.-M. Charlie Ma, Todd A. Pawlicki, Ervin B. Podgorsak, Elke Reitzel, Anthony B. Wolbarst, and Ellen D. Yorke). The conclusion of this discussion was that a book series should be launched under the Taylor & Francis banner, with each volume in the series addressing a rapidly advancing area of medical imaging or radiation therapy of importance to medical physicists. The aim would be for each volume to provide medical physicists with the information needed to understand technologies driving a rapid advance and their applications to safe and effective delivery of patient care.

Each volume in the series is edited by one or more individuals with recognized expertise in the technological area encompassed by the book. The editors are responsible for selecting the authors of individual chapters and ensuring that the chapters are comprehensive and intelligible to someone without such expertise. The enthusiasm of volume editors and chapter authors has been gratifying and reinforces the conclusion of the Minneapolis luncheon that this series of books addresses a major need of medical physicists.

The series Imaging in Medical Diagnosis and Therapy would not have been possible without the encouragement and support of the series manager, Lu Han, of Taylor & Francis Publishers. The editors and authors, and most of all I, are indebted to his steady guidance of the entire project.

William R. Hendee
Founding Series Editor
Rochester, MN

Foreword (Barnes)

Gary T. Barnes, PhD, Professor Emeritus, Department of Radiology, University of Alabama Birmingham

The stated objective of this book is to educate individuals who are interested in evaluating observer performance in diagnostic imaging. It provides insight into the practical methodology of such studies. The author assumes the reader is not a statistician or otherwise an expert in such evaluations, but is persistent and willing to work through the examples given. The author also provides background and references on the evolution of observer performance evaluations from their introduction in the 1940s to the present, as well as insight on common pitfalls and misconceptions.

I have benefited from knowing Dr. Chakraborty for the past thirty-five years. I've been aware of his contributions and also worked with him from 1981 to 1989. His accomplishments during this period were numerous and impressive—the first paper[1] on digital tomosynthesis (1984); development of a digital road mapping system[2] employing an early personal computer, digital frame grabber, and video tape recorder that was used clinically to assess the blood vessel patency in vascular surgery (1985); correction[3] of x-ray image intensifier pin cushion distortion (1985); the first paper[4] to apply FROC analysis to compare two clinical imaging modalities (1986). The breadth of these contributions is impressive and he was the first author on all the resultant papers. There were many other areas where he contributed and was a co-author. Noteworthy is that commercial digital tomosynthesis units are now in clinical use and road mapping is a standard offering on digital fluoroscopy vascular units.

As noted above, Dr. Chakraborty was the first author on the first paper to employ FROC analysis to a clinical comparison of two modalities—conventional screen-film and digital chest units. The digital unit was an early prototype. FROC methodology was chosen because nodule detection was the clinical problem being studied and conventional ROC analysis, or for that matter LROC, new at that time, did not appropriately address the problem of localization and did not (and does not) handle multiple lesions and responses in an image. As a result of this effort, he became aware of the limitations of observer performance studies and subsequently published a number of seminal papers on the subject resulting in a valid figure of merit definition when two or more modalities are compared.

When Dr. Chakraborty has made suggestions to me in the past, I have listened and benefited from listening. His performance in solving a wide breadth of technical and mathematical problems is impressive. This book will benefit individuals interested in observer performance evaluations in diagnostic medical imaging and provide additional insights to those that have worked in the field for many years.

References

1. Chakraborty DP, Yester MV, Barnes GT, Lakshminarayanan AV. Self-masking subtraction tomosynthesis. *Radiology.* 1984;150:225–229.
2. Chakraborty DP, Gupta KL, Barnes GT, Vitek JJ. Digital subtraction angiography apparatus. *Radiology.* 1985;157:547.
3. Chakraborty DP. Image intensifier distortion correction. *Med Phys.* 1987;14(2):249–252.
4. Chakraborty DP, Breatnach ES, Yester MV, Soto B, Barnes GT, Fraser RG. Digital and conventional chest imaging: A modified ROC study of observer performance using simulated nodules. *Radiology.* 1986;158:35–39.

Foreword (Kundel)

Harold L. Kundel, Emeritus Professor, Department of Radiology, Perelman School of Medicine, University of Pennsylvania

"Medicine is, at its core, an uncertain science. Every doctor makes mistakes in diagnosis and treatment."

—*J. Groopman[1]*

Diagnostic imaging is an important component of medical care. It has been estimated that about 10% of physician encounters with patients over age 65 involve imaging.[2] Human observers interpret all of the images and when the diagnostic information in the image is either obscure or ambiguous there is a chance for decision error. Imaging technology also changes, sometimes rapidly; consider the introduction of computed tomography (CT) and magnetic resonance imaging (MRI). Each change in technology requires evaluation to decide whether it improves diagnostic performance and ultimately improves patient care. Radiologists and radiological physicists are well aware of the need for the continuous evaluation of technological advances and long before Evidence Based Medicine became popular, they were measuring both the physical properties of images and the effect of those properties on observer performance.

In fact, the first systematic study of diagnostic performance in radiology was done in the late 1940s, as part of a field study to determine which of four possible chest x-ray techniques was best for tuberculosis screening. Unfortunately, variation in the detection of tuberculosis between observers was so large that it precluded any valid statistical comparison of the relative accuracy of the four x-ray techniques.[3] The large inter-individual variability in performance implies diagnostic error. If two observers disagree, only one can be right. Maybe both are wrong! The biostatistician for the tuberculosis screening study, J. Yerushalmy, identified two major problems in the statistical analysis of the data. First, the true diagnosis of each case was not known, and second, the large variability between and within observers could not be accounted for by the statistical tools available at that time for dealing with sensitivity and specificity, which were considered to be independent components of accuracy.[4]

The statistical toolbox was augmented in the late 1960s when L. Lusted introduced the receiver operating characteristic (ROC) into diagnostic medicine.[5] The ROC curve, which is a plot of true positive responses (sensitivity) against false positive responses (1—specificity), is based on the assumption that sensitivity and specificity are covariates. The detection theory model also implies that sensory variables (ability to detect) and decision variables (bias toward negative or positive decisions) can be separated. The methodology still requires that the investigator (not the observer) knows the true diagnosis, but it produces a single figure of merit for performance, the area under the ROC curve, that is not affected by the observer's decision-making bias. It is the decision bias, the tendency to use either strict or lenient criteria when making a decision about the presence or absence of disease, that is a major source of individual variability.

The ideal ROC study consists of having several observers report on a carefully selected set of cases, normal and abnormal, in which there is a single abnormality, and then report on the cases again after a sufficient interval of time has elapsed to forget previous responses. The ROC area and the measured variability within and between observers can be used in an appropriate statistical model to characterize or compare imaging modalities.[6] An experiment that fits all of the model requirements can be done in the laboratory but in the real world, decision tasks and images are more complicated. Observers may be asked to specify disease location or may be required to classify as well as detect disease, for example, classify a tumor as benign or malignant. Furthermore, some diseases, metastatic lung tumors, for example, may occupy multiple sites.

The ROC methodology has been gradually refined to encompass some of the more realistic situations that are encountered in medical imaging. In particular, software that accounts for case variability and observer variability has been developed and is available on the Internet for general use.[7] The author of this book, Dev Chakraborty, has been particularly instrumental in advancing the analytical capabilities of the ROC methodology for the task of locating abnormalities and for images with multiple abnormalities by developing the free-response ROC (FROC) methodology.

As opposed to most of the books with a primary statistical orientation, this book presents the technology evaluation methodology from the point of view of radiological physics and contrasts the purely physical evaluation of image quality with the determination of diagnostic outcome through the study of observer performance. The reader is taken through the arguments with concrete examples illustrated by code in **R**, an open source statistical language. There is a potential here to make the powerful ROC method of technology evaluation available to a wide community in a format that is readily understandable.

References

1. Groopman J. *How Doctors Think*. New York, NY: Houghton Mifflin Company; 2007.
2. Dodoo MS, Duszak R, Jr., Hughes DR. Trends in the utilization of medical imaging from 2003 to 2011: Clinical encounters offer a complementary patient-centered focus. *J Am Coll Radiol.* 2013;10:507–512.
3. Birkelo CC, Chamberlain WE, Phelps PS, Schools PE, Zacks D, Yerushalmy J. Tuberculosis case finding. *JAMA.* 1947;133:359–365.
4. Yerushalmy J. Statistical problems in assessing methods of medical diagnosis with special reference to x-ray techniques. *Public Health Rep.* 1947;62(40):1432–1449.
5. Lusted LB. *Introduction to Medical Decision Making*. Springfield, IL: Charles C Thomas; 1968.
6. Swets JA, Pickett RM. *Evaluation of Diagnostic Systems: Methods from Signal Detection Theory.* New York, NY: Academic Press; 1982.
7. Hillis SL, Berbaum KS, Metz CE. Recent developments in the Dorfman-Berbaum-Metz procedure for multireader ROC study analysis. *Acad Radiol.* 2008;15:647–661.

Preface

"If you really understand something, you can explain it to your grandmother."

—*Harold L. Kundel, MD*

This book is intended for readers who, like me, start out with almost no knowledge of a field, but who have the patience to learn by doing. I have attempted to explain complex material at a basic level and believe that, even if she did not understand the detailed formula, the final results would make my grandmother say "hmm... this feels right."

People sometimes ask me what exactly it is I do for a living. Here is my answer: radiologists make decisions based on their interpretations of medical images. Contrary to folklore, they do make mistakes—missed lesions and/or false positive findings, some of which have serious consequences. *My work involves modeling their interpretations and, based on these models, objectively estimating their performance.* This sentence is usually met with incredulity among my "hard-core" physicist friends. How can one model and objectively measure something as subjective as human interpretations? Physicists are used to analyzing "hard data" with instruments whose behaviors are well understood. Every physicist knows that the angle a galvanometer moves is proportional to the product of the current and the magnetic field. To put it mildly, the human brain is not as well understood. The story of how I was plunged (unprepared) into this field is described in Chapter 9.

My background is varied, more so than most scientists working in this field that is variously termed observer performance, receiver operating characteristics (ROC) analysis, model observers, psychophysics, and statistical analysis of ROC data. So here is an account of my evolution. This should help the reader appreciate *where I am coming from* and what to expect in my approach to the book.

I did my PhD under the late Prof. Ronald Parks, of superconductivity fame. I worked in an experimental physics laboratory. I got my hands dirty by learning how to use a machine shop to construct measurement apparatus. My PhD thesis was forgettable.[1] I obtained a post-doctoral fellowship at the University of Pennsylvania, under Prof. Alan J. Heeger, a future Nobel laureate, and even had a shared publication.[2] I performed magneto-resistance measurements in the Francis Bitter National Magnet Laboratory at MIT. In spite of my pedigreed background I realized that I was really not good at physics, or, stated alternatively, the competition was too strong (the field was clogged with brilliant folks, and physicists were coming in by the hordes from the Soviet Union). In addition, I had the handicap of a student visa. So, I needed to change fields and go someplace that would help me obtain the proverbial *green-card*. That turned out to be the University of Alabama at Birmingham (UAB), where I moved in 1979.

I worked on hyperthermia—basically heating tumors to about 45° Celsius, to effect preferential cancer-cell kill.[3,4] My boss there was Prof. Ivan Brezovich and we had a close relationship. I still recall our daily noon trips to the UAB Faculty Club, which had the best pecan pie anywhere and which I have since searched for in vain. I had to get my hands really dirty this time: besides machine shop work building various hyperthermia applicators, I helped use them on patients. Since hyperthermia was then considered experimental therapy, we only got patients who had failed all conventional therapies; in short, they were very sick. I recall one who went into shock during the therapy (i.e., heating) session, and Dr. John Durante, then director of the Comprehensive Cancer Center at UAB, had to step in. I also recall an African American patient with melanoma, rare among African Americans. Another patient had cancer of the esophagus and Dr. Brezovich designed a sausage-like tube that contained a concentric electrode and was otherwise filled with saline. The patient endured the incredible task of swallowing this tube so that we could pass current through the electrode and radially outward to the patient's body, thereby preferentially heating the lining of the esophagus. The experiment succeeded but the patient eventually died. These experiences, however painful to watch, were what got me my next job. When Prof. Gary Barnes was looking for a junior faculty member, Ivan told him, "Dev is good with patients." Apparently, that is all that it took to get me my first full-time job!

My work with Prof. Gary Barnes was my entry into diagnostic radiological physics. I was fortunate to have Gary as a boss/mentor, for he was ideal in many respects—his patience and sense of humor is legendary. He still recalls with pride my early tomosynthesis work,[5] which received a Certificate of Merit at RSNA. The images were turned 90° compared to their usual presentation—I did not know the difference! So, the story is each radiologist would step up to see the poster, and turn his or her head 90° to see the images of the inner ear in the orientation they were used to seeing!

A particularly productive research venture, which yielded a technical note[6] in radiology, was the Quantex project. My readers are probably unfamiliar with this device; it was an early fluoroscopic image-processing machine. With its myriad buttons, blinking lights, and IEEE-GPIB interface, allowing every function to be controlled by a computer, it was a dream machine as far as I was concerned. I became aware of work by Gould et al.[7] that used two Quantex video processors, one to perform low-pass filtering and the other to perform high-pass filtering, and subtracting the two image sequences to cancel stationary anatomy, revealing only the iodinated blood vessels. Unfortunately, each Quantex DS-30 box cost about $35,000 and our administrator would not approve buying another one. So, I designed a system that used the only box we had, but added two high-quality VCRs, and built some electronics, all controlled by a Commodore PET computer. I wish I had held on to that computer, as it belongs in a museum. It had no hard drive—the BASIC and machine language code was saved to a cassette drive.

A period of intense collaboration with Dr. Jiri Vitek (an interventional neuroradiologist) resulted. During the evenings, I would wheel my equipment into the angiography suite, place a phantom with which I could mimic iodine being injected into a blood vessel (basically a rubber tube concentric with a large diameter tube filled with water). To prevent damaging the image intensifier I occasionally blocked the x-rays at the tube with a lead plate. Once, after finishing, I forgot to remove the lead plate. To my horror, on the following day, when I was demonstrating my apparatus to Jiri, the technologist looked bewildered when, with a patient on the table, there was no image—I immediately rose to the occasion and nonchalantly removed the lead plate.

This project was a personal success. The first time I showed subtracted images, Jiri said, "This is DSA." DSA is an acronym for digital subtraction angiography. It was also my introduction to the general principle that *to do clinically meaningful research one has to collaborate with clinicians.* Each time I thought the project was ready to be handed over to the technologists, Jiri would come up with one more request. A notable one was *road-mapping.* When iodine is injected the vessels "light up" (actually, they turn dark as iodine contrast material absorbs more x-rays than blood). Jiri asked me if I could capture a static image of the "lit up" vessels and superpose it on the live x-ray image. The static vessel image would then serve as a road map, allowing the radiologist to negotiate the tip of the catheter into the correct blood vessel. To do this I had to learn machine language on the Motorola 6502 processor, the heart of the Commodore PET computer. This sped up a procedure (loading a look-up-table into a part of computer memory) by a factor of at least 100, making the whole idea feasible. At that time, commercial machines did not have the road mapping capability. When a GE digital system was purchased for the angiography suite, I was tasked to roll my cart into the suite to show the GE scientists how it was done. This project started my love affair with computers, both at the software and the hardware levels. To really understand image processing, I had to implement it in software/hardware. *One learns by doing and this philosophy pervades this book.*

The next project was evaluation of Picker International's prototype digital chest machine versus a conventional machine by the same manufacturer. Gary wanted me to do an "ROC" study. I put ROC in quotes, as Gary was deeply suspicious of ROC methodology that simply requires a rating that there are one or more lesions *somewhere* in the image. As Gary would say, "What do you mean by *somewhere*? If you see a lesion, why not point to it?" In fact, radiologists do, but location information cannot be accounted for by the ROC paradigm. Gary was also upset that his requests for NIH funding were denied because they lacked "ROC evaluation." He brought to my attention a publication by Dr. Phillip Bunch and colleagues at Kodak. This often-overlooked paper is described at length in Chapter 12. I conducted a similar study using an anthropomorphic chest phantom with superposed lesion-simulating wax partial spheres. Conducting the study was the easy part. I had radiologists, including the famous Dr. Robert Fraser, interpret the images and mark the locations of perceived lesions on an acrylic overlay. The locations were scored as true positives and false positives according to how close they were to true lesions, and I could construct free-response ROC (FROC) curves and it appeared that in the lung field the two curves (digital and conventional) were similar, while in the mediastinum the digital modality was visibly better. We submitted to radiology and fortunately Prof. Charles E. Metz reviewed the paper. Dr. Metz noted that it was an interesting study, but lacked any statistical analysis. So, I delved into statistical analysis. The only statistics I knew prior to this were limited to the Maxwell-Boltzmann, Fermi-Dirac, and Bose-Einstein distributions and the only statistics book I had in my possession was one by Bevington.[8] I had no idea what a *p*-value meant. In fact, I was in a similar situation as most expected readers of this book. My education began with a book by Swets and Pickett,[9] and I eventually managed to squeeze out confidence intervals and satisfy Charlie, and get the paper published.[10]

A turning point was a letter from Prof. James Sorenson, author of a famous book on Nuclear Medicine,[11] who was also interested in the FROC problem. His letter struck an intellectual nerve and overnight I wrote the FROCFIT program (in FORTRAN) that produced excellent visual fits to our Picker study FROC data. This resulted in a paper in Medical Physics.[12] This was followed soon after by my alternative FROC (AFROC) paper.[13] In those days, getting my papers published was easy!

Due to difficulty getting funding, my FROC work languished for a decade until I obtained funding for an idea (differential ROC or DROC) that eventually turned out to have a fatal flaw. The awful realization signaled a low point in my career, but I did eventually document, for posterity, the reasons why DROC was a bad idea.[14] I used the grant funds to revisit the FROC problem. A key interaction with Prof. Berbaum led to jackknife AFROC (JAFROC) analysis.[15] Another key interaction, with Dr. Claudia Mello-Thoms, resulted in a fuller understanding of the Kundel–Nodine model of diagnostic search yielding the radiological search model,[16,17] a major component of this book.

On a different note, apart from the interactions noted above, doing research in this field has been a solitary experience. I do have a lot of collaborators at the "end-user" level, but not one at the methodology developer level. Instead my FROC work has sparked US-based opposition, which has increased in proportion to its growing acceptance outside the United States.[18,19] With the passing of Swensson, Wagner, Metz, and Dorfman, I have been deprived of the four most qualified scientists who can appreciate my work. Writing this book has also been a solitary experience. It started as an edited book, but I rapidly realized that few of the ROC experts would collaborate with me, which led me to convert it to a sole author book, a decision I have not regretted. In all, I have spent about three years working on this book.

I have had the privilege of learning from some of the best in this field: Prof. Brezovich, who helped me transition from basic physics to medical physics; Prof. Barnes, who helped me transition to imaging physics; Prof. Kundel, who was instrumental in hiring me at the University of Pennsylvania and taught me everything I know about image perception; and Prof. Berbaum, a friend and supporter, who once paid me the compliment that *"you take on difficult projects."* Prof. Metz paid a similar public compliment at the 2001 Airlie Conference Center in Warrenton, VA at the Medical Image Perception Society meeting.

Mr. Xuetong Zhai, currently a graduate student in Bioengineering at the University of Pittsburgh, provided extensive programming and analytical help, especially in connection with the chapters on significance testing and designing a calibrated simulator. On these chapters (9, 10, and 23) he is a co-author. On his own initiative, he developed the RJafroc package, detailed in online chapter 25. This R package forms the software backbone of this book. An essential component of the book is the Online Supplementary material. As with any software, bug fixes and updates will be posted to the BitBucket website listed below. Instructions on how to download material from BitBucket are on www.devchakraborty.com.

Finally, I wish to acknowledge my wife Beatrice for believing in me and for putting up with me for over 26 years and my daughter Rimi, co-founder of minuvida (http://www.minuvida.com), a friend and confidant.

References

1. Chakraborty DP, Parks RD. Resistance anomaly of the order-disorder system Fe3A. *Phys Rev B*. 1978;18:6195–6198.
2. Chakraborty DP, Spal R, Denenstein AM, Lee K-B, Heeger AJ, Azbel MY. Anomalous magnetoresistance of Quasi one-dimensional Hg3-dAsF6. *Phys Rev Lett*. 1979;43:1832–1835.
3. Chakraborty DP, Brezovich IA. Error sources affecting thermocouple thermometry in RF electromagnetic fields. *J Microw Power*. 1982;17(1):17–28.
4. Atkinson WJ, Brezovich IA, Chakraborty DP. Usable frequencies in hyperthermia with thermal seeds. *IEEE Trans Biomed Imaging*. 1984;BME-31(1):70–75.
5. Chakraborty DP, Yester MV, Barnes GT, Lakshminarayanan AV. Self-masking subtraction tomosynthesis. *Radiology*. 1984;150:225–229.
6. Chakraborty DP, Gupta KL, Barnes GT, Vitek JJ. Digital subtraction angiography apparatus. *Radiology*. 1985;157:547.
7. Gould RG, Lipton MJ, Mengers P, Dahlberg R. Digital subtraction fluoroscopic system with tandem video processing units. *Proceedings Volume 0273, Application of Optical Instrumentation in Medicine IX*. Application of Optical Instrumentation in Medicine, San Francisco, 1981. doi:10.1117/12.931794.
8. Bevington PR, Robinson DK. *Data Reduction and Error Analysis*. Boston, MA: McGraw-Hill; 2003.
9. Swets JA, Pickett RM. *Evaluation of Diagnostic Systems: Methods from Signal Detection Theory*. 1st ed. New York, NY: Academic Press; 1982, p. 253.
10. Chakraborty DP, Breatnach ES, Yester MV, Soto B, Barnes GT, Fraser RG. Digital and conventional chest imaging: A modified ROC study of observer performance using simulated nodules. *Radiology*. 1986;158:35–39.
11. Cherry SR, Sorenson JA, Phelps ME. *Physics in Nuclear Medicine*. 4th ed. Orlando, FL: Elsevier, Saunders; 2012, p. 523.
12. Chakraborty DP. Maximum Likelihood analysis of free-response receiver operating characteristic (FROC) data. *Med Phys*. 1989;16(4):561–568.
13. Chakraborty DP, Winter LHL. Free-response methodology: Alternate analysis and a new observer-performance experiment. *Radiology*. 1990;174:873–881.
14. Chakraborty DP. Problems with the differential receiver operating characteristic (DROC) method. Proceedings of the SPIE Medical Imaging 2004: Image Perception, Observer Performance, and Technology Assessment; San Diego, CA; 2004;5372:138–43.
15. Chakraborty DP, Berbaum KS. Observer studies involving detection and localization: Modeling, analysis and validation. *Med Phys*. 2004;31(8):2313–2330.
16. Chakraborty DP. ROC curves predicted by a model of visual search. *Phys Med Biol*. 2006;51:3463–3482.
17. Chakraborty DP. An alternate method for using a visual discrimination model (VDM) to optimize softcopy display image quality. *J Soc Inf Disp*. 2006;14(10):921–926.
18. Hillis SL, Chakraborty DP, Orton CG. ROC or FROC? It depends on the research question. *Med Phys*. 2017;44:1603–1606.
19. Chakraborty DP, Nishikawa RM, Orton CG. Due to potential concerns of bias and conflicts of interest, regulatory bodies should not do evaluation methodology research related to their regulatory missions. *Med Phys*. 2017;44:4403–4406.

Note about Online Supplementary Resources

The following supplementary tools and materials are available online:

- www.devchakraborty.com
- www.expertcadanalytics.com

The online supplementary material, outlined below, is available at

- https://bitbucket.org/dpc10ster/onlinebookk21778, access to which will be granted upon e-mail request to the author at dpc10ster@gmail.com
- https://cran.r-project.org/web/packages/RJafroc/

Chapter 1: Preliminaries

Online Appendix: Introduction to **R/RStudio**: Part I

Chapter 3: Modeling the binary task

Online Appendix 3.A: R code demonstration of sensitivity/specificity concepts
Online Appendix 3.B: Calculating a confidence interval
Online Appendix 3.C: Introduction to R/RStudio: Part II
Online Appendix 3.D: Plotting in R
Online Appendix 3.E: Getting help in R: Part I
Online Appendix 3.F: Getting help in R: Part II
Online Appendix 3.G: What if a package is missing
Online Appendix 3.H: Shaded distributions in R
Online Appendix 3.I: Numerical integration in R

Chapter 4: The ratings paradigm

Online Appendix 4.A: operating points from counts table

Chapter 5: Empirical AUC

Online Appendix 5.A: Calculating the Wilcoxon statistic

Chapter 6: Binormal model

Online Appendix 6.A: Equivalence of two and four parameter models
Online Appendix 6.B: Output of Eng website software
Online Appendix 6.C: Maximizing log likelihood
Online Appendix 6.D: Validating the fitting model
Online Appendix 6.E: Variance of AUC
Online Appendix 6.F: Transformations

Chapter 7: Sources of variability in AUC

Online Appendix 7.A: The bootstrap method in R
Online Appendix 7.B: The jackknife method in R
Online Appendix 7.C: A calibrated simulator for a single dataset
Online Appendix 7.D: Comparison of different methods of estimating variability

Chapter 8: Hypothesis testing

`Six sigma.pdf`

Chapter 9: DBMH analysis

Online Appendix 9.A: The Satterthwaite degree of freedom
Online Appendix 9.B: Demonstration of significance testing formulae
Online Appendix 9.C: Text output file listing
Online Appendix 9.D: Excel ANOVA tables
Online Appendix 9.E: Code for validating DBMH analysis
Online Appendix 9.F: Simulators for validating fixed-reader and fixed-case analyses
Online Appendix 9.G: Code illustrating the meaning of pseudovalues
Online Appendix 9.H: Testing for interactions

Chapter 10: Obuchowski–Rockette–Hillis (ORH) analysis

Online Appendix 10.A: The DeLong method for estimating the covariance matrix
Online Appendix 10.B: Estimation of covariance matrix: single-reader multiple-treatment
Online Appendix 10.C: Comparing DBMH and ORH methods for single-reader multiple-treatment
Online Appendix 10.D: Minimal implementation of ORH method
Online Appendix 10.E: Proof of Eqn. (10.64)
Online Appendix 10.F: Single-treatment multiple-reader analysis

Chapter 11: Sample size estimation

Online Appendix 11.A: Sample size formula using ORH approach
Online Appendix 11.B: ORH fixed-readers random-case (FRRC)
Online Appendix 11.C: ORH random-reader fixed-cases (RRFC)
Online Appendix 11.D: Details on effect-size specification

Chapter 12: The FROC paradigm

Online Appendix 12.A: Code used to generate the FROC plots
Online Appendix 12.B: CAMPI and cross correlation
Online Appendix 12.C: The Bunch transforms

Chapter 13: Empirical operating characteristics possible with FROC data

Online Appendix 13.A: FROC versus AFROC

Chapter 14: Computation and meanings of empirical FROC FOM-statistics and AUC measures

Online Appendix 14.A: Proof of Equivalence theorem for wAFROC
Online Appendix 14.B: Understanding the AFROC and wAFROC
Online Appendix 14.C: Summary of FROC FOMs
Online Appendix 14.D: Numerical demonstrations of FOM-statistic versus AUC equivalences

Chapter 15: Visual search paradigms

Chapter 16: The radiological search model (RSM)

Chapter 17: Predictions of the RSM

Chapter 18: Analyzing FROC data

Chapter 19: Fitting RSM to FROC/ROC data and key findings

Chapter 20: Proper ROC models

Chapter 21: The bivariate binormal model

Chapter 22: Evaluating standalone CAD versus radiologists

Chapter 23: Validating CAD analysis

About the Author

Dev P. Chakraborty received his PhD in physics in 1977 from the University of Rochester, NY. Following postdoctoral fellowships at the University of Pennsylvania (UPENN) and the University of Alabama at Birmingham (UAB), since 1982 he has worked as a clinical diagnostic imaging physicist. He is American Board of Radiology certified in Diagnostic Radiological Physics and Medical Nuclear Physics (1987). He has held faculty positions at UAB (1982–1988), UPENN (1988–2002), and the University of Pittsburgh (2002–2016). At UPENN he supervised hospital imaging equipment quality control, resident physics instruction and conducted independent research. He is an author on 78 peer-reviewed publications, the majority of which are first-authored. He has received research funding from the Whittaker Foundation, the Office of Women's Health, the FDA, the DOD, and has served as principal investigator on several NIH RO1 grants.

His work has covered varied fields: hyperthermia, physical measures of image quality, feasibility of digital tomosynthesis for inner-ear imaging (his RSNA poster received a Certificate of Merit), building a digital subtraction angiography apparatus for interventional neuro-angiography, and computerized analysis of mammography phantom images (CAMPI). He conducted (1986) the first free-response ROC study comparing Picker International's prototype digital chest imaging device to a conventional device. In 1989, he coined the term AFROC to describe the currently widely used operating characteristic for analyzing free-response studies. Since 2004 he has distributed JAFROC software for the analysis of free-response and ROC studies. Over 107 publications have used this software and RJafroc, an enhanced R-version, which are being used in courses and PhD projects worldwide. He is internationally recognized as an expert in observer performance methodology. He recently served as statistical consultant to General Electric on the evaluation of the VolumeRad chest tomosynthesis device.

Dr. Chakraborty's overarching research interest has been measuring image quality, both at physical and at perceptual levels. He showed, via CAMPI, that a widely used mammography QC phantom could be analyzed via an algorithm, achieving far greater precision than radiologic technologists. With the realization that wide variability (about 40%, 1996 study by Beam et al.) affects expert radiologist interpretations, over the past two decades Dr. Chakraborty's research has focused on observer performance measurements, specifically the free-response paradigm, modeling and quantifying visual search performance, and developing associated statistical analysis. He has proposed the radiological search model (RSM) that resolves several basic questions, dating to the 1960s about ROC curves. Recently he has developed an RSM-based ROC curve-fitting method that yields important insights into what is limiting performance. In 2016 Dr. Chakraborty formed Expert CAD Analytics, LLC, to pursue novel ideas to develop expert-level CAD that can be used as a *first* reader; issues with current approaches to CAD are detailed in http://www.expertcadanalytics.com.

Notation

Listed are acronyms and terminology used in the book, their meanings and relevant comments

Acronym / term	Meaning	Comments
	Part A	
Treatment / modality	The imaging chain, excluding the reader	Indexed by $i = 1,2,...,I$
Reader	Radiologist or algorithmic observer	Indexed by $j = 1,2,...,J$
Case	Patient, possibly multiple images and / or modalities	Indexed by $k = 1,2,...,K$
K_1	Total number of non-diseased cases in dataset	Indexed by $k_1 = 1,2,...,K_1$
K_2	Total number of diseased cases in dataset	Indexed by $k_2 = 1,2,...,K_2$
K	Total number of cases in dataset	$K = K_1 + K_2$
TP	True Positive	Correct decision on diseased case
FP	False Positive	Incorrect decision on non-diseased case
TN	True Negative	Correct decision on non-diseased case
FN	False Negative	Incorrect decision on diseased case
D = diagnosis	D = 1 for non-diseased; D = 2 for diseased	
T = truth	T = 1 for non-diseased; T = 2 for diseased	
$P(A\|B)$	Probability of event A given, or conditioned on, event B	Conditional probability
$Se = P(D = 2\|T = 2)$	Sensitivity	
$Se = P(D = 1\|T = 1)$	Specificity	
$P(D), P(\|D) = 1 - P(D)$	Probability of disease presence; probability of disease absence	$P(D) \equiv$ Disease prevalence; ! = Negation
Ac	Accuracy, probability of a correct decision	Eqn. (2.17)
$NPV = P(T = 1\|D = 1)$	Negative predictive value, probability that a non-diseased decision is correct	Eqn. (2.25)
$PPV = P(T = 2\|D = 2)$	Positive predictive value, probability that a diseased decision is correct	Eqn. (2.26)
#TP, #FP, #TN, #FN	Numbers of TP, FP, TN and FN events	
$FPF = \#FP/K_1$	FPF = False Positive Fraction = estimate of $(1 - Sp)$	Abscissa of ROC operating point

(Continued)

Acronym / term	Meaning	Comments
$TPF = \#TP/K_2$	TPF = True Positive Fraction = estimate of Se	Ordinate of ROC operating point
AUC	Area Under Curve	Relevant curve is defined in context
ROC	Receiver Operating Characteristic	One decision per case
ROC-plot	Plot of TPF versus FPF	
ROC-AUC	Area Under ROC Curve	
Empirical AUC	Area under trapezoidal plot, connecting points with lines	
Fitted AUC	AUC predicted by fitting model	
True AUC	Population AUC	
A_z	AUC under binormal model fitted ROC curve	
FOM	Figure of merit; any valid measure of performance, generically denoted θ	
θ_{ij}	Generic FOM for modality i and reader j	$\theta_{i\bullet}$; dot represents average over replaced index
pdf	Probability density function	$pdf(x)dx = P(x < X < x + dx)$
CDF	Cumulative distribution function	$CDF(x) = P(X < x)$
$N(\mu, \sigma^2)$	Normal distribution with mean μ and variance σ^2	$N(0,1)$ is the unit normal distribution
$\phi(z) = \frac{1}{\sqrt{2\pi}} e^{\frac{-z^2}{2}}$	pdf corresponding to $N(0,1)$	Probability density function
$\Phi(y) = \int_{-\infty}^{y} \phi(z)dz$	CDF corresponding to $N(0,1)$	Cumulative distribution function
$\Phi^{-1}(z)$	$\Phi^{-1}(\Phi(z)) = z$	Inverse function; also called quantile function
Z-sample, z	Random decision variable; realized value	
$k_t t$	Case k_t in case-truth state t	
$Z_{k_t t}$	Random Z-sample for case $k_t t$	

(Continued)

Acronym / term	Meaning	Comments
$I(x)$	$I(TRUE) = 1; I(FALSE) = 0$	Indicator function
W	Wilcoxon statistic	Eqn. (5.9)
$\psi(x,y)$	Wilcoxon kernel function	Eqn. (5.10)
(a,b)	Conventional parameters of binormal model	Introduced by Dorfman and Alf
(μ, σ)	Alternate parameterization	$\mu = a/b; \sigma = 1/b$
R_{ROC}	$R_{ROC} \geq 2$	# of ROC bins
ζ_i	$\zeta_i < \zeta_{i+1}; i = 1,2,...,R_{ROC}-1; \zeta_0 = -\infty; \zeta_{R_{ROC}} = \infty$	ROC binning thresholds
ROC binning rule	if $\zeta_{r-1} \leq z < \zeta_r \Rightarrow rating = r; r = 1,2,...,R_{ROC}$	
$O(\zeta) = (FPF(\zeta), TPF(\zeta))$	Operating point defined by threshold ζ	
$O_r \equiv O(\zeta_r); r = 1,2,...,R_{ROC}-1$	O_1 is the uppermost non-trivial point, O_2 is the next lower one	
K_{tr}	Number of counts rated r in truth state t	$r = 1, 2, ..., R_{ROC}; t = 1,2$
$C^2 = \sum\limits_{t=1}^{2}\sum\limits_{r=1}^{R}\dfrac{\left(K_{tr} - \langle K_{tr} \rangle\right)^2}{\langle K_{tr} \rangle}; K_{tr} \geq 5$	Pearson goodness of fit statistic	Also called chi-square statistic
$C^2 \sim \chi^2_{df}$	Expected distribution, assuming model validity	
$p = P\left(X > C^2 \mid X \sim \chi^2_{df}\right)$	Goodness of fit p-value	Fits are generally considered valid if p > 0.01
$df = R_{ROC} - 3$	Degrees of freedom; applies to all 2-parameter ROC models	Need at least 3 points for goodness of fit
$E(X) = \int x \, pdf(x) \, dx$	Expectation of X	Expectation of X
$Cov(X,Y)$	$Cov(X,Y) = E(XY) - E(X)E(Y)$	Covariance of X and Y; E is the expectation

(Continued)

Acronym / term	Meaning	Comments
$Var(X) = Cov(X,X)$	Variance of X	Square root of variance is standard deviation
$d' = \sqrt{2}\Phi^{-1}(A_z)$	Detectability index	
$L = P(data\|parameters)$	Likelihood function	LL = log likelihood
MLE	Maximum likelihood estimate	Parameters maximizing log likelihood function
I	Matrix of second partial derivatives of $-LL$ w.r.t. parameters	Fisher information matrix, Eqn. (6.43)
$Cov(\hat{\theta}) = I^{-1}$	Covariance matrix of parameters	Used to calculate confidence intervals on parameters and functions of parameters
Part B		
$\{c\}$	$c = 1,2,...,C$	Case-set index
$AUC_{\{c\}}; AUC_{\{\bullet\}}$	AUC for case-set index $\{c\}$; average over case-set index	
$Cov_{bs}(AUC)$	Bootstrap estimate of covariance	Eqn. (10.17)
$Cov_{jk}(AUC)$	Jackknife estimate of covariance	Eqn. (10.18)
MRMC	Multiple-reader multiple-case	Reader interprets cases in all modalities
DBMH	Dorfman–Berbaum–Metz–Hillis	Jackknife pseudovalue based significance testing
ORH	Obuchowski–Rockette–Hillis	Figure of merit based significance testing algorithm
$\theta, \theta_{(k_t t)}$	Generic FOM, FOM with case $k_t t$ removed	$\theta_{(k_t t)}$ is the jackknife FOM for case $k_t t$
$\theta_{k_t t} = K\theta - (K-1)\theta_{(k_t t)}$	Jackknife pseudovalue	
$\theta_{1\bullet} = \theta_{2\bullet}; \theta_{1\bullet} \neq \theta_{2\bullet}$	Null hypothesis, two-sided alternative hypothesis	Assuming two treatments
DBM model	$\sigma^2_{YR}, \sigma^2_{YC}, \sigma^2_{YtR}, \sigma^2_{YtC}, \sigma^2_{YRC}, \sigma^2_{YtRC}, \sigma^2_{Y\epsilon}$	Pseudovalue variance components, Y subscript
OR model	$\sigma^2_R, \sigma^2_{tR}, Cov_1, Cov_2, Cov_3, Var$	FOM variance components
$F_{ndf, ddf}$	F-distribution with ndf, ddf degrees of freedom	$F \sim F_{ndf, ddf}$

(Continued)

Acronym / term	Meaning	Comments
$MST, MSTR$, etc.	Mean-square treatment, mean-square treatment reader, etc.	Eqn. 9.12
$\alpha = P(NH\ rejected\|NH\ true)$	Control on Type I error	Size of test
$\beta = P(NH\ not\ rejected\|NH\ false)$	Control on Type II error	$Power = 1 - \beta$
F_{DBMH}	Observed value of DBMH F-statistic	Eqn. 9.23
ddf_H	Hillis denominator degrees of freedom	Eqn. 9.24
$F_{NH} \sim F_{ndf,ddf_H}$	NH distribution of F-statistic	
$F_{1-\alpha,ndf,ddf_H}$	Critical value of F-statistic	$(1-\alpha)$ quantile of $F_{ndf,ddf}$
$p = P(F \geq F_{DBMH} \mid F \sim F_{ndf,ddf_H})$	p-value for rejecting NH	Reject NH if $p < \alpha$
t_{df}	t-distribution with df degrees of freedom	$t \sim t_{df}$
$t_{\alpha/2;df}$	Upper α / 2 quantile of the t-distribution with df degrees of freedom	Eqn. 9.31
$CI_{1-\alpha}$	$1-\alpha$ confidence interval	Generally $\alpha = 0.05$
$F_{ndf,ddf,\Delta}$	Non-central F with non-centrality parameter Δ	
Δ	Non-centrality parameter	$F_{AH} \sim F_{ndf,ddf,\Delta}$
Power	$P(F > F_{crit} \mid F \sim F_{ndf,ddf,\Delta})$	
$z_{\alpha/2}$	Upper α / 2 quantile of the unit normal distribution	
Part C		
LROC; PCL; LROC plot	Location ROC; Probability of correct localization; Plot of PCL versus FPF	One decision and one location per case
ROI	ROI = region of interest; investigator divides image into ROIs	One decision per ROI
FROC	Free-response ROC	Random number of localizations (marks) and decisions (ratings) and per case
FROC-plot	Plot of LLF versus NLF	

(Continued)

Acronym / term	Meaning	Comments
NL	Non-lesion localization	A mark on a non-diseased region
LL	Lesion localization	A mark on a diseased region
$k_t t$	Case k_t in case-truth state t	
$l_s s$	Location l_s in local-truth state s	
$z_{k_t t l_s s}$	Z-sample for localized suspicious region $l_s s$ on case $k_t t$	
L_{k_2}	Number of lesions in case $k_2 2$	
L_T	Total number of lesions in dataset	
L_{max}	Maximum number of lesions per diseased case in dataset	
$NLF = \#NL/K$	Non-lesion localization fraction	Abscissa of FROC operating point
$LLF = \#LL/L_T$	Lesion localization fraction	Ordinate of FROC operating point
Inferred ROC rating	Rating of highest rated mark or minus infinity if case has no marks	
$\vec{f}_L = \{f_L; L = 1, 2, \ldots, l_{max}\}$	Lesion distribution vector	Fraction of diseased cases with L lesions
R_{FROC}	$R_{FROC} \geq 1$	# of FROC bins
ζ_i	$\zeta_i < \zeta_{i+1}; i = 1, 2, \ldots, R_{FROC}; \zeta_0 = -\infty; \zeta_{R_{FROC}+1} = \infty$	FROC binning thresholds
FROC binning rule	if $\zeta_r \leq z < \zeta_{r+1} \Rightarrow rating = r; r = 1, 2, \ldots, R_{FROC}$	
$O_r = (FPF_r, TPF_r)$	Inferred ROC operating point corresponding to threshold ζ_r	Eqn. (13.11), Eqn. (13.15)
$W_{k_2 l_2}$	Lesion weights	Eqn. (13.16)
$wLLF_r$	Weighted LLF corresponding to threshold ζ_r	Eqn. (13.17)
AFROC1	Alternative FROC; NLs on diseased cases not ignored	
wAFROC1	Weighted AFROC; NLs on diseased cases not ignored	

(Continued)

Acronym / term	Meaning	Comments
AFROC	Alternative FROC; NLs on diseased cases ignored	Plot connects uppermost point to (1,1)
AFROC-plot	Plot of LLF versus FPF	
AFROC-AUC	Area under AFROC plot	
wAFROC	Weighted AFROC; NLs on diseased cases ignored	
wAFROC-plot	Plot of wLLF versus FPF	
wAFROC-AUC	Area under wAFROC plot	
EFROC	Exponentially transformed FROC	Section 13.5
Latent NL	A suspicious region that does not correspond to a lesion	
Latent LL	A suspicious region that corresponds to a lesion	
Marked NL or LL	if $z \geq \zeta_1 \Rightarrow$ region is marked	
IDCA	Initial detection and candidate analysis	
RSM	Radiological search model, parameters (μ, λ', ν')	
pSNR	Perceptual signal-to-noise ratio	
$\mu > 0$	Separation of unit variance NL and LL normal distributions	Equal to *pSNR*
$\lambda' > 0$	Poisson parameter of RSM	Average number of latent NLs per case
$\nu'; 0 < \nu' < 1$	Binomial parameter of RSM, constrained to (0,1)	Probability that lesion is found
$\lambda = \lambda' \mu$	Intrinsic parameter corresponding to λ'	Distinction between physical and intrinsic parameters
$\nu' = 1 - exp(-\mu\nu)$	Defines intrinsic parameter ν corresponding to ν'	do:
$Poi(\lambda)$	Poisson distribution with mean λ	
$B(L,\nu)$	Binomial distribution with trial size L and success probability ν	
$NLF_{max} = \lambda'$	Theoretical end-point abscissa of FROC	
$LLF_{max} = \nu'$	Theoretical end-point ordinate of FROC or AFROC	
FPF_{max}	Theoretical end-point abscissa of ROC or AFROC	
TPF_{max}	Theoretical end-point ordinate of ROC	

(Continued)

Acronym / term	Meaning	Comments
$df = R_{FROC} - 3$	RSM goodness of fit degrees of freedom	Need at least 4 points for goodness of fit
$AUC_{RSM}^{ROC}\left(\mu, \lambda, \nu, \vec{f_L}\right)$	RSM predicted AUC	Eqn. (17.25)
S	Search performance; ability to find latent LLs while avoiding latent NLs	Eqn. (17.38)
A_c	Lesion-classification performance; ability to discriminate between latent NLs and LLs	Eqn. (17.39)
Part D		
$N_p\left(\vec{\mu}, \Sigma\right)$	Multivariate normal distribution with mean $\vec{\mu}$ and covariance Σ	Yields p samples per realization; if $p = 1$, p is suppressed
$N_2\left(\vec{\mu}, \Sigma\right)$	Bivariate normal distribution with mean $\vec{\mu}$ and covariance Σ	Yields two samples per realization
$\Sigma = \begin{pmatrix} \sigma_1^2 & \rho_2\sigma_1\sigma_2 \\ \rho_2\sigma_1\sigma_2 & \sigma_2^2 \end{pmatrix}$	Bivariate covariance matrix, showing standard deviations and correlations	
$l(z) = pdf_D(z)/pdf_N(z)$	Likelihood ratio; D = diseased; N = non-diseased	Equals slope of ROC curve
CBM	Contaminated binormal model Parameters (μ_{CBM}, α)	Proper fits to ROC data; Section 20.8
μ_j, α_j	CBM separation and visibility parameters for reader j	Binning related thresholds excluded
$\mu_{jj'}, \alpha_{jj'}, \mu_{f_{jj'}}, \alpha_{f_{jj'}}, \rho_{t_{jj'}}, \rho_{2,jj'}$	CORCBM parameters for reader j-reader j' pairing	Binning related thresholds excluded
PROPROC	A binormal model based proper ROC fitting program with parameters (c, d_a)	Section 20.7
A_{PROP}, A_{CBM}	PROPROC or CBM fitted AUC	Eqn. (20.21) and Eqn. (20.29)
CORROC2	Fits bivariate ROC data to bivariate binormal model	
CORCBM	Fits bivariate ROC data to bivariate contaminated binormal model	

1

Preliminaries

1.1 Introduction

The question addressed by this book is: *How good are radiologists using medical imaging devices at diagnosing disease?* Observer performance measurements, widely used for this purpose, require data collection and analysis methods that fall under the rubric of what is loosely termed *ROC analysis,* where ROC is an abbreviation for Receiver Operating Characteristic.[1] ROC analysis and its extensions form a specialized branch of science encompassing knowledge of diagnostic medical physics, perception of stimuli (commonly studied by psychologists), human observer decision modeling, and statistics. Its importance in medical imaging is due to the evolution of technology and the need to objectively assess advances. The Food and Drug Administration, Center for Devices and Radiological Health (FDA/CDRH), which regulates medical-imaging devices, requires ROC studies as part of its device approval process.* There are, conservatively, at least several hundred publications using ROC studies and a paper[1] by the late Prof. C.E. Metz has been cited over 2025 times. Numerous reviews and tutorial papers have appeared[1-11] and there are books on the statistical analysis[12-14] of ROC data. However, in spite of the numbers of publications and books in this field, basic aspects of it are still misunderstood and lessons from the past have been forgotten, which has seriously held back health care advances—this will be demonstrated in this book.

It is the aim of this book to describe the field in some depth while assuming little statistical background for the reader. That is a tall order. The key to accomplishing this aim is the ability to illustrate abstract statistical concepts and analysis methods with free, cross-platform, open-source software.† **R**, a programming language, and **RStudio**, helper software that makes it much easier to work with **R**, is very popular in the scientific community. (The **PT Mono** font is used to distinguish software specific material from normal text). A brief introduction to **R/RStudio**, while relegated—for organizational reasons—to Online Appendix A, is *essential reading,* preferably running it on a computer, if the reader is to understand subsequent chapters. This advice applies to all chapters in this book—the accompanying online material is essential reading. It is available at https://bitbucket.org/dpc10ster/onlinebookk21778 access to which will be granted upon e-mail request to the author at dpc10ster@gmail.com.

The purpose of this introductory chapter is to lay the groundwork, provide background material and an overview of what is to come. The starting point is *clinical tasks* occurring every day in

---

* See *Statistical Guidance on Reporting Results from Studies Evaluating Diagnostic Tests* available at www.fda.gov/RegulatoryInformation/Guidances.
† If the reader has any doubts about investing in yet another programming language (the author has been through, in historical order, FORTRAN, BASIC, C, Pascal, C++, Visual C, MATLAB, and IDL), the reader may wish to read a summary **R (programming language).pdf** in the online supplementary material directory corresponding to this chapter. It was downloaded from https://en.wikipedia.org/wiki/R(programming_language) on 5/1/17. An eminent scientist who reviewed the original proposal for the book, about three years ago, remarked then, "**R** is a fad." He is currently using it!

imaging centers and hospitals, where images of patients, both symptomatic and asymptomatic, are acquired by radiological technologists and interpreted by radiologists.

1.2 Clinical tasks

In hospital based radiology departments or freestanding imaging centers, imaging studies are conducted to diagnose patients for signs of disease. Examples are chest x-rays, computerized tomography (CT) scans, magnetic resonance imaging (MRI) scans, ultrasound imaging, and so on. A patient does not go directly to a radiology department; rather, the patient first sees a family doctor, internist or general practitioner about an ailment. After a physical examination, perhaps augmented with non-imaging tests (blood tests, electrocardiogram, etc.), the physician may recommend an imaging study. As an example, a patient suffering from a persistent cough yielding mucus and experiencing chills may be referred for chest x-rays to rule out pneumonia. In the imaging suite, a radiologic technician positions the patient with respect to the x-ray beam. Chest x-rays are taken, usually in two projections, back to front (posterior-anterior, or PA-view) and sideways (lateral, or LAT-view).

Each x-ray image is a projection from, ideally, a point source of x-rays, of patient anatomy in the path of the beam, onto a detector, for example, x-ray film or digital detector. Because of differential attenuation, the shadow cast by the x-rays shows anatomical structures within the patient. The technician checks the images for proper positioning and technical quality. A radiologist (a physician who specializes in interpreting imaging studies) interprets them and dictates a report.

Because of the referring physician's report, the radiologist knows why the patient was sent for chest x-rays in the first place, and interprets the images in that context. *At the very outset, one recognizes that images are not interpreted in a vacuum, rather, for a symptomatic patient, the interpretation is done in the context of resolving a specific ailment.* This is an example of a *clinical task* and it should explain why different specialized imaging devices are needed in a radiology department. Radiology departments in the United States are usually organized according to body parts, for example, a chest section, a breast imaging section, an abdominal imaging section, head CT, body CT, cardiac radiology, orthopedic radiology, and so on. Additionally, for a given body part, different means of imaging are generally available. Examples are x-ray mammography, ultrasound, and magnetic resonance imaging of the breast.

1.2.1 Workflow in an imaging study

The workflow in an imaging study can be summarized as follows. The patient's images are acquired. Nowadays almost all images in the United States are acquired digitally, but some of the following concepts are illustrated with analog images; this is not an essential distinction. The digital detector acquired image(s) are processed for optimality and displayed on one or more monitors. These are interpreted by a radiologist in the context of the clinical task implied by the referring physician's notes attached to the imaging request (such as rule out pneumonia). After interpreting the image(s), the radiologist makes a diagnosis, such as, *patient shows no signs of disease* or *patient shows signs of disease*. If signs of disease are found, the radiologist's report will contain a description of the disease and its location, extent, and other characteristics, for example, *diffuse opacity near the bottom of the lungs, consistent with pneumonia*. Alternatively, an unexpected finding can occur, such as *nodular lesion, possibly lung cancer, in the apex of the lungs*. A diseased finding will trigger further imaging, for example, a CT scan, and perhaps biopsy (excision of a small amount of tissue and examination by a pathologist to determine whether it is malignant), to determine the nature of the disease. (In this book, the terms *non-diseased* and *diseased* are used instead of normal and abnormal, or noise and signal plus noise, or target absent and target present, and so on.)

So far, patients with symptoms of disease were considered. Interpreting images of *asymptomatic* patients involves an entirely different clinical task, termed *screening*, described next.

1.2.2 The screening and diagnostic workup tasks

In the United States, women older than 40 years are imaged at yearly intervals using a special x-ray machine designed to optimally image the breast. Here the radiologist's task is to find breast cancer, preferably when it is small and has not had an opportunity to spread, or metastasize, to other organs. Cancers found at an early stage are more likely to be treatable. Fortunately, the incidence of breast cancer is very low, about five per thousand women in the United States, but, because most of the patients are non-diseased, this makes for a difficult task. Figuratively, the radiologist needs to find *five bad needles mixed with 995 good needles in a haystack*. Again, the images are interpreted in context. The family history of the patient is available, the referring physician (the woman's primary care physician and/or gynecologist) has performed a physical examination of the patient, and in some cases, it may be known whether the patient is at high-risk because she has a gene that predisposes her to breast cancer. The interpreting radiologist has to be MQSA-certified (Mammography Quality Standards Act, see Section 1.3.2) to interpret mammograms. If the radiologist finds one or more regions suspicious for breast cancer, the location of each suspicious region is recorded, as it provides a starting point for subsequent patient management. At the author's previous institution, The University of Pittsburgh, the images are electronically marked (annotated) on the digital images. The patient receives a dreaded letter or e-mail, perhaps preceded by a phone call from the imaging center, that she is being recalled for further assessment. When the woman arrives at the imaging center, further imaging, termed a *diagnostic workup*, is conducted. For example, magnification views, centered on the location of the suspicious region(s) found at screening, may be obtained. Magnifying the image reveals more detail. Additional x-ray projections and other types of imaging (e.g., ultrasound, MRI, and perhaps breast CT—still in the research stage) may be used to resolve ambiguity regarding true disease status. If the suspicious region is determined benign, the woman goes home with the good news. This is the most common outcome. If ambiguity remains, a somewhat invasive procedure, termed a *needle biopsy*, is performed whereby a small amount of tissue is extracted from the suspicious region and sent to the pathology laboratory for final determination of malignancy status by a pathologist. Even here, the more common outcome is that the biopsy comes back negative for malignancy. About ten percent of women who are screened by *experts* are recalled for unnecessary diagnostic workups, in the sense that the diagnostic workup and/or biopsy ends up showing no signs of cancer. These unnecessary recalls often cause some physical and much emotional trauma, and result in increased health care costs. About four of every five cancers are detected by experts, that is, about one in five is missed. All of these numbers are for *experts*—there is considerable variability[15] in skill-levels between MQSA-certified radiologists. If cancer is found, radiation, chemotherapy or surgery may be initiated to treat the patient. Further imaging is usually performed to determine the response to therapy (has the tumor shrunk?).

The practice of radiology, and patients served by this discipline, has benefited tremendously from technological innovations. How these innovations are developed and adopted by radiology departments is the next topic.

1.3 Imaging device development and its clinical deployment

Roentgen's 1895 discovery of x-rays found almost immediate clinical applications and started the new discipline of radiology. Initially, two developments were key: optimizing the production of x-rays, as the process is very inefficient, and efficiently detecting the photons that pass through the imaged anatomy: these photons form the radiological image. Consequently, initial developments were in x-ray tube and screen-film detector technologies. Over many decades, these have matured and new modalities have emerged, examples of which are CT in the late 1960s, MRI in the 1970s, and computed radiography and digital imaging in the late 1980s.

1.3.1 Physical measurements

There is a process to imaging device development and deployment into clinical practice. The starting point is to build a *prototype* of the new imaging device. The device is designed in the context of a clinical need and is based on physical principles suggesting that the device, perhaps employing new technology or new ideas, should be an improvement over what is already available, generically termed the *conventional modality*. The prototype is actually the end-point of much research involving engineers, imaging scientists and radiologists.

The design of the prototype is optimized by physical measurements. For example, images are acquired of a block of Lucite™, termed a *phantom*, with thickness equivalent in x-ray penetrability to an average patient. Ideally, the images would be noise free, but x-ray quantum fluctuations and other sources of noise influence the final image and cause them to have noise, termed *radiographic mottle*.[16–18] For conventional x-rays, the kind one might see the doctor putting up on a viewing panel (light box) in old movies, the measurement employs a special instrument called a micro-densitometer, which essentially digitizes narrow strips of the film. The noise is quantified by the standard deviation of the digitized pixel values. This is compared to that expected based on the number of photons used to make the image; the latter number can be calculated from knowledge of the x-ray beam spectrum, intensity and the thickness of the phantom. If the measured noise equals the expected noise (if it is smaller, there is obviously something wrong with the calculation of the expected noise and/or the measurement), image quality is said to be *quantum limited*. Since a fundamental limit, dictated by the underlying imaging physics, has been reached, further noise reduction is only possible by increasing the number of photons. The latter can be accomplished trivially by increasing the exposure time, which, of course, increases radiation dose to the patient. Therefore, as far as image noise is concerned, in this scenario, the system is *ideal* and no further noise optimization is needed. In the author's experience teaching imaging physics to radiology residents, the preceding sentences cause confusion. In particular, the terms *limited* and *ideal* seem to be at odds, but the residents eventually understand it. The point is that if one is up against a fundamental limit, then things are ideal in the sense that they can get no better (physicists do have a sense of humor). In practice this level of perfection is never reached, as the screen-film system introduces its own noise, due to the granularity of the silver halide crystals that form the photographic emulsion and other factors—ever tried digitizing an old slide? Furthermore, there could be engineering limitations preventing attainment of the theoretical limit. Through much iteration, the designer reaches a point at which it is decided that the noise is about as low as it is going to get.

Noise is but one factor limiting image quality. Another factor is *spatial resolution*—the ability of an imaging system to render sharp edges and/or resolve closely spaced small objects. For this measurement, one images an object with a sharp edge, for example, a razor blade. When the resulting image is scanned with a microdensitometer, the trace should show an abrupt transition as one crosses the edge. In practice, the transition is rounded or spread out, resembling a sigmoid function. This is due to several factors. The finite size of the focal spot producing the x-rays produces a penumbra effect, which blurs the edge. The spread of light, within the screen due to its finite thickness, also blurs the edge. [The screen absorbs photons and converts them to visible light to which film is exquisitely sensitive. Without the screen, the exposure would have to increase about thousand-fold.] One can make the screen only so thin, because then it would lack the ability to stop the x-rays that have penetrated the phantom. These photons contain information regarding the imaged anatomy. Ideally, all photons that form the radiological image should be stopped in the detector. Again, an optimization process is involved until the equipment designer is convinced that a fundamental limit has been reached or engineering limitations prevent further improvement.

Another factor affecting image quality is *contrast*—the ability of the imaging system to depict different levels of x-ray penetration. A phantom consisting of a step-wedge, with varying thickness of Lucite™, is imaged and the image scanned with a microdensitometer. The resulting trace should show distinct steps as one crosses the different thickness parts of the step-wedge phantom (this is termed *large area contrast*, to distinguish it from the resolution related blurring occurring at the edges between the steps). The more steps that can be visualized, the better the system. The digital term for this is the *gray-scale*. For example, an 8-bit gray-scale can depict 256 shades of gray. Once again, design considerations and optimization are used to arrive at the design of the prototype.

The preceding is a simplified description of possible physical measurements. In fact, it is usual to measure the spatial frequency dependence of resolution, noise, and overall photon usage efficiency.[19,20] These involve quantities named modulation transfer function (MTF), noise power spectrum (NPS) and detective quantum efficiency (DQE), each of which is a function of spatial frequency (in cycles per mm). The frequency dependence is important in understanding, during the development process, the physical factors limiting image quality.

After an optimized prototype has been made it needs approval from the FDA/CDRH for pre-clinical usage. This involves submitting information about the results of the physical measurements and making a case that the new design is indeed an improvement over existing methods. However, since none of the physical measurements involved radiologists interpreting actual patient images produced by the prototype, observer performance measurements are needed before machines based on the prototype can be marketed. Observer performance measurements, in which the prototype is compared to an existing standard, involve radiologists interpreting a set of patient images acquired on the prototype and on the conventional modality. The truth (is the image of a diseased patient?) is unknown to them but is known to the researcher, that is, the radiologists are blinded to the truth. The radiologists' decisions, classified by the investigator as correct or incorrect, are used to determine the average performance of the radiologists on the prototype and on the existing standard. Specialized statistical analysis is needed to determine whether the difference in performances is in the correct direction and *statistically significant*, that is, unlikely to be due to chance. Observer performance measurements are unique in the sense that the *entire imaging chain* is being evaluated. In order to get a sufficiently large and representative sample of patients and radiologists, such studies are generally performed in a multi-institutional setting.[21] If the prototype's performance equals or exceeds that of the existing standard, it is approved for clinical usage. At this point, the manufacturer can start marketing the device to radiology departments. This is a simplified description of the device approval process. Most imaging companies have experts in this area that help them negotiate a necessarily more complex regulatory process.

1.3.2 Quality control and image quality optimization

Once the imaging device is sold to a radiology department, both routine quality control (QC) and continuous image quality optimization are needed to assure proper utilization of the machine over its life span. The role of QC is to *maintain* image quality at an established standard. Initial QC measurements, termed *acceptance testing*,[22-24] are made to establish base-line QC parameters and a medical physicist establishes a program of systematic checks to monitor them. The QC measurements are relatively simple, typically taking a few hours of technologist time, that look for *changes* in monitored variables. The role of continuous image quality optimization, which is the bread-and-butter of a diagnostic medical physicist, is to resolve site-specific image quality issues. The manufacturer cannot anticipate every issue that may arise when their equipment is used in the field, and it takes a medical physicist, working in collaboration with the equipment manufacturer, technologists and radiologists, to continually optimize the images and solve specific image quality related problems. Sometimes the result is a device that performs better than what the manufacturer was able to achieve. One example, from the author's experience, is the optimization, using special filters and an air-gap technique, of a chest x-ray machine in the 1980s by Prof. Gary T. Barnes, a distinguished medical physicist and the late Prof. Robert Fraser, a famous chest radiologist.[25] The subsequent evaluation of this machine versus a prototype digital chest x-ray machine by the same manufacturer, Picker International, Cleaveland OH, was the author's entry into the field of observer performance.[26]

A good example of QC is the use of the American College of Radiology Mammography Quality Standards Act (ACR-MQSA) phantom to monitor image quality of mammography machines.[27-29] The phantom consists of a (removable) wax insert in an acrylic holder; the latter provides additional absorption and scattering material to more closely match the attenuation and beam hardening* of an average breast. Embedded in the wax insert are *target objects* consisting of six fibrils, five groups of microcalcifications, each containing six specks, and five spherical objects of different sizes, called masses. An image of the phantom, Figure 1.1a is obtained daily, before the first patient is imaged, and is inspected by a technologist, who records the number of target objects of different types that are visible. There is a pass-fail criterion and if the image fails then patients cannot be imaged on that machine until the problem is corrected. At this point, the medical physicist is called in to investigate.

One can perhaps appreciate the subjectivity of the measurement. Since the target locations are known, the technologist can claim to have detected it and the claim cannot be disproved; *unless a claim is falsifiable, it is not science.* While the QC team is trained to achieve repeatable measurements, the author has shown[30-34] that computer analysis of mammography phantom images (CAMPI) can achieve far greater precision and repeatability than human observer readings. Commercial software is currently available from various vendors that perform proprietary analysis of phantom images for various imaging systems (e.g., mammography machines, CT scanners, MRI scanners, ultrasound, etc.).

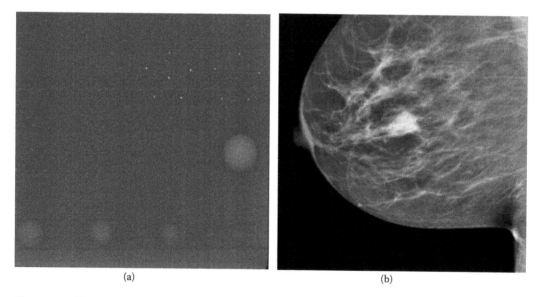

(a)   (b)

Figure 1.1 (a) Image of an American College of Radiology mammography accreditation phantom. The phantom contains *target objects* consisting of six fibrils, five groups of microcalcifications, and five nodule-like objects. An image of the phantom is obtained daily, before the first patient is imaged, and is inspected by a technologist, who records the number of target objects of different types that are visible. On his 27″ iMac monitor, the author sees four fibrils, three speck groups and four masses, which would be graded as a pass. This is a greatly simplified version of the test. The scoring accounts for irregular fibril or partially visible masses, borders, and so on, all of which is intended to get more objectivity out of the measurement. (b) A breast image showing an invasive cancer, located roughly in the middle of the image. Note the lack of similarity between the two images (a) and (b). The breast image is much more complex and there is more information, and therefore more to go wrong than with the phantom image. Moreover, there is variability between patients unlike the fixed image in (a). In the author's clinical experience, the phantom images interpreted visually are a poor predictor of clinical image quality.

---

* Beam hardening is a technical term used to describe the selective removal of low energy photons by an absorber in the path of the beam; i.e., the spectrum of the beam shifts to higher energy, i.e., more penetrating, photons, colloquially termed a *harder* beam.

Figure 1.1b shows a mammogram with a mass-like cancer visible near its center. It is characterized by complex anatomical background, quite unlike the uniform background in the phantom image in Figure 1.1a. In mammography, 30% of retrospectively visible lesions are missed at initial screening and radiologist variability can be as large as 40%.[15] QC machine parameters (e.g., kVp, the kilovoltage accuracy) are usually measured to 1% accuracy. *It is ironic that the weak link, in the sense of greatest variability, is the radiologist but quality control and much effort is primarily focused on measuring/improving the physical parameters of the machine.* This comment is meant to motivate clinical medical physicists, most of who are focused on QC, to become more aware about observer performance methods, where achieving better than 5% accuracy is quite feasible.[35] The author believes there should be greater focus on improving radiologist performance, particularly those with marginal performance. Efforts in this direction, using ROC methods, are underway in the United Kingdom[36,37] by Prof. Alistair Gale and colleagues.

1.4 Image quality versus task performance

In this book, *image quality* is defined as the *fidelity* of the image with respect to some external gold standard of what the ideal image should look like, while *task performance* is how well a radiologist, using the image, accomplishes a given clinical task. For example, if one had an original Rembrandt and a copy, the image quality of the copy is perfect if even an expert appraiser cannot distinguish the copy from the original. The original painting is the gold standard. If an expert can distinguish the copy from the original, its image quality is degraded. The amount of degradation is related to the ease with which the expert can detect the fraud.

A radiological image is the result of x-ray interactions within the patient and the image receptor. Here it is more difficult to define a gold standard. If it exists at all, the gold standard is expected to depend on what the image is being used for, that is, the diagnostic task. An image suitable for soft-tissue disease diagnosis may not be suitable for diagnosis of bone disease. This is the reason why CT scanners have different soft-tissue and bone window/level* settings. With clinical images, a frequently used approach is to have an expert rank-order the images, acquired via different methods, with respect to *clinical appropriateness* or *clinical image quality*. The italics are used to emphasize that these terms are hard to define objectively. In this approach, the gold standard is in the mind of the expert. Since experts have typically interpreted tens of thousands of images in the past, and have lived with the consequences of their decisions, there is considerable merit to using them to judge clinical image quality. However, experts do disagree and biases cannot be ruled out. This is especially true when a new imaging modality is introduced. The initial introduction of computed radiography (CR) was met with some resistance in the United States among technologists, who had to learn a different way of obtaining the images that disrupted their workflow. There was also initial resistance from more experienced radiologists, who were uncomfortable with the appearance of the new images, that is, their gold standard was biased in favor of the modality—plain films—with which they were most familiar. The author is aware of at least one instance where CR had to be imposed by "diktat" from the chairman of the department. Some of us are still more comfortable reading printed material than viewing it on a computer screen, so this type of bias is understandable.

Another source of bias is patient variability, that is, the gold standard depends on the patient. Some patients are easier to image than others in the sense that their images are clearer, that is, they depict anatomical structures that are known to be present more clearly. X-rays pass readily through a relatively slim patient (e.g., an athlete) and there are fewer scattered photons which degrade image quality,[38,39] than when imaging a larger patient (e.g., an NFL linebacker). The image of the former will be clearer; the ribs, the heart shadow, the features of the lungs, and so on, will be better visualized (i.e., closer to what is expected based on the anatomy) than the image of the linebacker. Similar

---

* The window level settings refer to digital manipulations performed on an image in order to optimally display it on a monitor.

differences exist in the ease of imaging women with dense breasts, containing a larger fraction of glandular tissue compared to women with fatty breasts. By imaging appropriately selected patients, one can exploit these facts to make one's favorite imaging system look better. [Prof. Harold Kundel, one of the author's mentors, used to say, "Tell me which modality you want to come out better and I will prepare a set of patient images to help you make your case."]

The approach advocated in this book is to measure performance in imaging tasks staged in a laboratory setting (i.e., under complete control of the researcher) using ROC methodology and its extensions. The approach is objective and eliminates some, but not all, of the biases. Before getting into the details of the ROC approach, one of the alternatives is summarized. Alternatives are needed because ROC measurements are time consuming, although not as time consuming as some of the other approaches, Section 1.8, that measure clinical performance using live interpretations.

1.5 Why physical measures of image quality are not enough

Both high spatial resolution and low noise are desirable characteristics. However, imaging systems do not come unambiguously separated as high spatial resolution and low noise versus low spatial resolution and high noise. There is generally an intrinsic imaging physics dictated tradeoff between spatial resolution and noise. Improving one makes the other worse. For example, if the digital image is smoothed with a spatial filter, then noise will be smaller because of the averaging of neighboring pixels, but the ability to resolve closely spaced structures will be compromised. Therefore, a more typical scenario is deciding whether the decreased noise justifies the accompanying loss of spatial resolution. Clearly the answer to this depends on the clinical task: if the task is detecting relatively large low contrast nodules, then some spatial smoothing may actually be beneficial, but if the task involves detecting small microcalcifications then smoothing will tend to reduce their visibility.

The problem with physical measures of image quality lies in relating them to clinical performance. Phantom images have little resemblance to clinical images, compare Figure 1.1a and b. X-ray machines generally have automatic exposure control: the machines use a brief exposure to automatically sense the thickness of the patient from the detected x-rays. Based on this, the machine chooses the best combinations of technical factors (kVp and tube charge) and image processing. The machine has to be put in a special manual override mode to obtain reasonable images of phantoms, as otherwise the exposure control algorithm, which expects patient anatomy, is misled by the atypical nature of the phantom, compared to typical patient anatomy, into producing very poor images. This type of problem makes it difficult to reproduce problems encountered using clinical images with phantom images. It has been the author's experience that QC failures often lag clinical image quality reported problems: more often than not, clinical image quality problems are reported before QC measurements indicate a problem. This is not surprising since clinical images, for example, Figure 1.1b, are more complex and have more information, both in the clinical and in the information-theoretic sense,[40,41] than the much simpler phantom image shown in Figure 1.1a. So, there is more that can go wrong with clinical images than with phantom images. Manufacturers now design *anthropomorphic* phantoms whose images resemble human x-rays. Often these phantoms provide the option of inserting target objects at random locations; this is done to get more objectivity out of the measurement. Now, if the technologist claims to have found the target, the indicated location can be used to determine if the target was truly detected.

To circumvent the possibility that changes in physical measurements on phantoms may not sensitively track changes in clinical image interpretations by radiologists, one needs to include both the complexity of clinical images and the skill of the radiologists as part of the measurement. *Because of variability in both patient images and radiologist interpretations, such measurements are expected to be more complicated than QC measurements, so to be clear, the author is not advocating observer performance studies as part of QC.* However, they could be built into a continuous quality improvement program, perhaps performed annually. Before giving an overview of the observer performance methods, an alternative modeling-driven approach, that is widely used, is described next.

1.6 Model observers

If one can adequately simulate (or model) the entire imaging process, then one can design mathematical measurements that can be used to decide if a new imaging system is an improvement over a conventional imaging system. Both new and conventional systems are modeled (i.e., reduced to formula that can be evaluated). The field of model observers[42] is based on assuming that this can be done. The FDA/CDRH has a research program called VICTRE: Virtual Imaging Clinical Trials for Regulatory Evaluation. Since everything is done on a computer, the method potentially eliminates time-consuming studies involving radiologists.

A simple example may elucidate the process (for more details one should consult the extensive literature on model observers). Suppose one simulates image noise by sampling a Gaussian random number generator and filling up the pixels in the image with random samples. This simulates a non-diseased image. The number of such images could be quite large, for example, 1000, limited only by one's patience. A second set of simulated diseased images is produced in which one samples a random number generator to create non-diseased images, as before, but this time one adds a small low-contrast but noiseless disk to the center of each image. The procedure yields two sets of images, 1000 with noise only backgrounds and 1000 with different noise backgrounds and the superposed centered low contrast disk. One constructs a *template* whose shape is identical to that of the superposed disk (i.e., one does not simply measure peak contrast at the center of the lesion; rather the shape-dependent contrast of the disk is taken into account). One then calculates the cross-correlation of the template with each of the superposed disks.[30,43] The cross correlation is the sum of the products of pixel values of corresponding pixels, one drawn from the template and the other drawn from the matching position on the disk image. Because of random noise, the cross-correlations from different simulated diseased cases will not be identical, and one averages the 1000 values. Next, one applies the template to the centers of the non-diseased images and computes the cross-correlations as before. Because of the absence of the disk, the values will be smaller (assuming positive disk contrast). The difference between the average of the cross-correlations at disk locations and the average at disk-absent locations is the numerator of a signal-to-noise ratio (SNR) like quantity. The denominator is the standard deviation of the cross-correlations at disk-free locations. To be technical, the procedure yields the SNR of the non-pre-whitening ideal observer.[44] It is an ideal mathematical observer in the sense that for white noise no human observer can surpass this level of performance.[45,46]

Suppose the task is to evaluate two image-processing algorithms. One applies each algorithm to the 2000 images described above and measures SNR for each algorithm. The one yielding the higher SNR, after accounting for variability in the measurements, is the superior algorithm.

Gaussian noise images are not particularly clinical in appearance. If one filters the noise appropriately, one can produce simulated images that are similar to non-diseased backgrounds observed in mammography.[47–49] Other techniques exist for simulating statistically characterized lumpy backgrounds that are a closer approximation to some medical images.[50]

Having outlined one of the alternatives, one is ready for the methods that form the subject matter of this book.

1.7 Measuring observer performance: Four paradigms

Observer performance measurements come in different flavors, types, or *paradigms*. In the current context, a paradigm* is an agreed-upon method for collecting the data. A given paradigm can lend itself to different analyses. In historical order, the paradigms are: (1) the receiver operating

---

* A typical example, or pattern, of something, New Oxford American Dictionary.

characteristic (ROC) paradigm;[1,2,7,51,52] (2) the free-response ROC (FROC) paradigm;[53,54] (3) the location ROC (LROC) paradigm;[55,56] and (4) the region of interest (ROI) paradigm.[57] Each paradigm assumes that the truth is known independently of the modalities to be compared. This implies that one cannot use diagnoses from one of the modalities to define truth—if one did, the measurement would be biased in favor of the modality used to define truth. It is also assumed that the true disease status of the image is known to the researcher but the radiologist is blinded to this information.

In the ROC paradigm, the observer renders a single decision per image. The decision could be communicated using a binary scale (ex. 0 or 1) or declared by use of the terms *negative* or *positive*, abbreviations of *negative for disease* (the radiologist believes the patient is non-diseased) and *positive for disease* (the radiologist believes the patient is diseased), respectively. Alternatively, the radiologist could give an ordered label, termed a *rating*, to each case, where the rating is a number with the property that higher values correspond to greater radiologist's confidence in presence of disease. A suitable ratings scale could be the consecutive integers 1 through 6, where 1 is definitely non-diseased and 6 is definitely diseased.

If data is acquired on a binary scale, then the performance of the radiologist can be plotted as a single *operating point on an ROC plot*. The x-axis of the plot is false positive fraction (FPF), that is, the fraction of non-diseased cases *incorrectly* diagnosed as diseased. The y-axis of the plot is true positive fraction (TPF), that is, the fraction of diseased cases *correctly* diagnosed as diseased. Models have been developed to fit binary or multiple rating datasets. These models predict continuous curves, or *operating characteristics*, along which an operating point can move by varying the radiologist's *reading style*. The reading style is related to the following concept: based on the evidence in the image, how predisposed is a radiologist to declaring a case as diseased. A *lenient*, *lax*, or *liberal* reporting style radiologist is very predisposed even with scant evidence. A *strict* or *conservative* reporting style radiologist requires more evidence before declaring a patient as diseased. This brief introduction to the ROC was given to explain the term *operating characteristic* in ROC. The topic is addressed in more detail in **Chapter 2**.

In the FROC paradigm, the observer marks and rates all regions in the image that are sufficiently suspicious for disease. A mark is the location of the suspicious region and the rating is an ordered label, characterizing the degree of suspicion attached to the suspicious region. In the LROC paradigm, the observer gives an overall ROC-type rating to the image, and indicates the location of the most suspicious region in the image. In the ROI paradigm, the researcher divides each image into a number of adjacent non-overlapping regions of interest (ROIs) that cover the clinical area of interest. The radiologist's task is to evaluate each ROI for presence of disease and give an ROC-type rating to it.

1.7.1 Basic approach to the analysis

The basic approach is to obtain data, according to one of the above paradigms, from a group of radiologists interpreting a common set of images in one or more modalities. The way the data is collected, and the structure of the data, depends on the selected paradigm. The next step is to adopt an objective measure of performance, termed a *figure of merit* (FOM) and a procedure for estimating it for each modality-reader combination. Assuming two modalities, for example, a new modality and the conventional one, one averages FOM over all readers within each modality. If the difference between the two averages (new modality minus the conventional one) is positive, that is an indication of improvement. Next comes the statistical part: is the difference large enough so as to be unlikely to be due to chance? This part of the analysis, termed *significance testing*, yields a probability, or *p-value*, that the observed difference or larger could result from chance even though the modalities have identical performances. If the p-value is small, that it is taken as evidence that the modalities are not identical in performance, and if the difference is in the right direction, the new modality is judged better.

1.7.2 Historical notes

The term *receiver operating characteristic* (ROC) traces its roots to the early 1940s. The receiver in ROC literally denoted a pulsed radar receiver that detected radio waves bounced off objects in the sky, the obvious military application being to detect enemy aircraft. Sometimes the reflections were strong compared to receiver electronic and other sources of noise and the operator could confidently declare that the reflection indicated the presence of aircraft and the operator was correct. This combination of events was termed a *true positive* (TP). At other times, the aircraft was present but due to electronic noise and reflections off clouds, the operator was not confident enough to declare "aircraft present" and this combination of events was termed a *false negative* (FN). Two other types of decisions can be discerned when there was no aircraft in the field of view: (1) the operator mistook reflections from clouds or perhaps a flock of large birds and declared "aircraft present," termed a *false positive* (FP). (2) The operator did not declare "aircraft present" because the reflected image was clear of noise or false reflections and the operator felt confident in a negative decision, termed a *true negative* (TN). Obviously, it was desirable to maximize correct decisions (TPs and TNs) while minimizing incorrect decisions (FNs and FPs). Scientists working on this problem analyzed it as a generic signal detection problem, where the signal was the aircraft reflection and the noise was everything else. A large field called signal detection theory (SDT) emerged.[58] However, even at this early stage, it must have been apparent to the researchers that the problem was incomplete in a key respect: when the operator detects a suspicious signal, there is a location (specifically an azimuth and altitude) associated with it. The operator could be correct in stating "aircraft present" but direct the interceptors to the wrong location. Additionally, there could be multiple enemy aircraft present, but the operator is only allowed the "aircraft present" and "aircraft absent" responses, which fail to allow for multiplicity of suspected aircraft locations. This aspect was not recognized, to the best of the author's knowledge, until Egan coined the term *free-response* in the auditory detection context.[53]

Having briefly introduced the different paradigms, two of which, namely the ROC and the FROC, will be the focus of this book, it is appropriate to see how these measurements fit in with the different types of measurements possible in assessing imaging systems.

1.8 Hierarchy of assessment methods

The methods described in this book need to be placed in context of a six-level hierarchy of assessment methods.[7,59] The cited paper by Fryback and Thornbury titled "The efficacy of diagnostic imaging" is a highly readable account, which also gives a more complete overview of this field, including key contributions by Yerushalmy[60] and Lusted.[61] The term *efficacy* is defined generically as *the probability of benefit to individuals in a defined population from a medical technology applied for a given medical problem under ideal conditions of use.* Demonstration of efficacy at each lower level is a necessary but not sufficient condition to assure efficacy at higher level. The different assessment methods are, in increasing order of *efficacy*:* technical, diagnostic accuracy, diagnostic thinking, therapeutic, patient outcome, and societal, Table 1.1.

The term *clinical relevance* is used rather loosely in the literature. The author is not aware of an accepted definition of clinical relevance apart from its obvious English language meaning. As a working definition, the author has proposed[62] that the clinical relevance of a measurement be defined as its hierarchy-level. A level-5 patient outcome measurement (do patients, on the average, benefit from the imaging study?) is clinically more relevant than a technical measurement of noise on a uniform background phantom or an ROC study. This is because it relates directly to the benefit, or lack thereof, to a group of patients (it is impossible to define outcome efficacy at the

---

* Efficacy is defined as the ability to produce a desired or intended result: e.g., there is little information on the efficacy of this treatment, New Oxford American Dictionary.

Table 1.1 Fryback and Thornbury proposed hierarchy of assessment methods

Efficacy level	Designation	Essential characteristic
1	Technical efficacy	Engineering measures: Resolution, noise, gray-scale range, MTF, NPS, DQE
2	Diagnostic accuracy efficacy	Sensitivity, specificity, ROC area, FROC area
3	Diagnostic thinking efficacy	Positive and negative predictive values
4	Therapeutic efficacy	How treatment or patient management was affected by imaging test
5	Patient outcome efficacy	How result of treatment or management was affected by imaging test
6	Societal efficacy	Cost-benefit from societal viewpoint

Note: Demonstration of efficacy at each lower level is a necessary but not sufficient condition to assure efficacy at higher level. (MTF = modulation transfer function; NPS(f) = noise power spectra as a function of spatial frequency f; DQE(f) = detective quantum efficiency)

individual patient level—at the patient level outcome is a binary random variable, 1 if the outcome was good and 0 if the outcome was bad).

One could make physical measurements but one cannot (yet) predict the average benefit to patients. Successful virtual clinical trials would prove the author wrong. ROC studies are more clinically relevant than physical measurements, and it is more likely that a modality with higher performance will yield better outcomes, but it is not a foregone conclusion. Therefore, higher-level measurements are needed.

However, the time and cost of the measurement increases rapidly with the hierarchy level. Technical efficacy, although requiring sophistical mathematical methods, take relatively little time. ROC and FROC, both of which are level-2 diagnostic accuracy measurements, take more time, often a few months to complete. However, since ROC measurements include the entire imaging chain and the radiologist, they are more clinically relevant than technical measurements, but they do not tell us the effect on diagnostic thinking. After the results of live interpretations are available, for example, patients are diagnosed as diseased or non-diseased, what does the physician do with the information? Does the physician recommend further tests or recommends immediate treatment? This is where the level-3 measurements come in, which measure the effect on diagnostic thinking. Typical level-3 measurements are positive predictive value (PPV) and negative predictive value (NPV). PPV is the probability that the patient is actually diseased when the diagnosis is diseased and NPV is the probability that the patient is actually non-diseased when the diagnosis is non-diseased. These are discussed in more detail in **Chapter 2**.

Unlike level-2 measurements, PPV and NPV depend on disease prevalence. As an example, consider breast cancer, which has low prevalence, about 0.005. Before the image is interpreted and lacking any other history, the mammographer knows only that there is a five in 1000 chance that the woman has breast cancer. After the image is interpreted, the mammographer has more information. If the image was interpreted as diseased, the confidence in presence of cancer increases. For an expert mammographer typical values of sensitivity and specificity are 80% and 90%, respectively (these terms will be explained in the next chapter; for now sensitivity is identical to true positive fraction and specificity is 1-false positive fraction). It will be shown (in **Chapter 2**, Section 2.9.2) that for this example PPV is only 0.04. In other words, even though an expert interpreted the screening mammogram as diseased, the probability that the patient actually has cancer is only 4%. Obviously more tests are needed before one knows for sure whether the patient has cancer—this is the reason for the recall and the subsequent diagnostic workup referred to in Section 1.2.2. The corresponding NPV is 0.999. Negative interpretations by experts are definitely good news for the affected patients and these did not come from an ROC study, or physical measurements, rather they came from actual live clinical interpretations. Actually, NPV and PPV are defined as averages over a group of patients. For example, the 4% chance of cancer following a positive diagnosis is good news, on average. An unlucky patient could be one of the 4-in-100 patients that has cancer following a positive screening diagnosis.

While more relevant than ROC, level-3 measurements like PPV and NPV are more difficult to conduct than ROC studies[18]—they involve following, in real time, a large cohort of patients with images interpreted under actual clinical conditions. Level-4 and higher measurements, namely therapeutic, patient outcome, and societal, are even more difficult and are sometimes politically charged, as they involve cost benefit considerations.

1.9 Overview of the book and how to use it

For the most part, the book follows the historical development; that is, it starts with chapters on ROC methodology, chapters on significance testing, chapters on FROC methodology, chapters on advanced topics and appendices. Not counting **Chapter 1**, the current chapter, the book is organized into five Parts (A–E).

1.9.1 Overview of the book

1.9.1.1 Part A: The receiver operating characteristic (ROC) paradigm

Part A describes the ROC (receiver operating characteristic) paradigm. **Chapter 2** describes the binary decision task. Terminology that is important to master, such as accuracy, sensitivity, specificity, disease prevalence, positive and negative predictive values, is introduced. **Chapter 3** introduces the important concepts of decision variable, the reporting threshold, and how the latter may be manipulated by the researcher, and this chapter introduces the ROC curve. **Chapter 4** reviews the ratings method for acquiring ROC data. **Chapter 5** introduces the binormal model for fitting ratings data. The chapter is heavy on mathematical and computational aspects, as it is intended to take the mystery out of these techniques, which are used in subsequent chapters. *The data fitting method, pioneered by Dorfman and Alf in 1969, is probably one of the most used algorithms in ROC analysis.* **Chapter 7** describes sources of variability affecting any performance measure, and how they can be estimated.

1.9.1.2 Part B: The statistics of ROC analysis

Part B describes the specialized statistical methods needed to analyze ROC data, in particular how to analyze data originating from multiple readers interpreting the same cases in multiple modalities. **Chapter 8** introduces hypothesis-testing methodology, familiar to statisticians, and the two types of errors that the researcher wishes to control, the meaning of the ubiquitous p-value and statistical power. **Chapter 9** focuses on the Dorfman–Berbaum–Metz method, with improvements by Hillis. Relevant formula, mostly from publications by Prof. Steven Hillis, are reproduced without proofs (it is the author's understanding that Dr. Hillis is working on a book on his specialty, which should complement the minimalistic-statistical description approach adopted in this book). **Chapter 10** describes the Obuchowski–Rockette method of analyzing MRMC ROC data, with Hillis' improvements. **Chapter 11** describes sample size estimation in a ROC study, i.e., how to plan a prospective ROC study in order to have a reasonable chance of detecting a true difference.

1.9.1.3 Part C: The FROC paradigm

Part C is unique to this book. Anyone truly wishing to understand human observer visual search performance needs to master the FROC paradigm. The payoff is that the concepts, models and methods described here apply to almost all clinical tasks. **Chapter 17** and **Chapter 1** are particularly important. These were difficult chapters to write and they will take extra effort to comprehend. However, the key findings presented in these chapters and their profound implications should strongly influence future observer performance research. *If the potential of the findings is recognized and used to benefit patients, by even one reader, the author will consider this book a success.*

1.9.1.4 Part D: Advanced topics

Some of the chapters in **Part D** are also unique to this book. **Chapter 20** discusses proper ROC curve fitting and software. The widely used bivariate binormal model, developed around 1980, but never properly documented, is explained in depth, is described in **Chapter 21**. A method for comparing (standalone) CAD to radiologists is described in **Chapter 22**. **Chapter 23** describes validation of the CAD analysis method described in **Chapter 22**.

1.9.1.5 Part E: Appendices

Part E contains two online chapters. **Online Chapter 24** is a description of 14 datasets, all but two of them collected by the author over years of collaborations with researchers who conducted the studies and on which the author helped with analysis and sometimes with manuscript preparation. The datasets provide a means to demonstrate analysis techniques and to validate fitting methods. **Online Chapter 25** is a user-manual for the `RJafroc` package. Since `RJafroc` is used extensively in the book, this is expected to be a useful go-to chapter for the reader. The choice to put these chapters online is to allow the author to update the datasets with new files as they become available and to update the analysis and documentation of `RJafroc` as new features are added.

1.9.2 How to use the book

Each chapter consists of the physical book chapter that one is reading. Additionally, there are good chances that the online directory corresponding to this book will contain two directories, one is called `software` and the other is called `Supplementary Material`.* The `software` directory contains ready-to-run code that is referenced in the book chapter. When one sees such a reference in a chapter, the reader should open the relevant file and run it. Detailed directions are provided in the Online Appendix corresponding to each chapter.

Those new to the field should read the chapters in sequence. It is particularly important to master **Part A**. **Part B** presents the statistical analysis at a level accessible to the expected readers of this book, namely the user community. The only way to really understand this part is to apply the described methods and codes to the online datasets. Understanding the formula in this part, especially those relating to statistical hypothesis testing, requires statistical expertise, which could lead the average reader in unproductive directions. It is best to accept the statisticians' formula and confirm that they work. How to determine if a method works will be described. Readers with prior experience in the field may wish to skim chapters. If they do, it is strongly recommended that they at least run and understand the software examples. This will prepare them for the more complex code in later chapters.

This concludes the introduction of the book. The author trusts the reader will have as much fun reading it as the author had writing it.

---

* The files can be accessed https://bitbucket.org/dpc10ster/onlinebookk21778, access to which will be granted upon e-mail request to the author at dpc10ster@gmail.com. All code can be downloaded free of charge. There are two stipulations: (1) No material can be used commercially without express permission from ExpertCad Analytics, LLC. (2) If material from the website is used in subsequent publications, the book and relevant chapter must be cited.

References

1. Metz CE. ROC methodology in radiologic imaging. *Invest Radiol.* 1986;21(9):720–733.
2. Metz CE. Basic principles of ROC analysis. *Semin Nucl Med.* 1978;8(4):283–298.
3. Metz CE. Some practical issues of experimental design and data analysis in radiological ROC studies. *Invest Radiol.* 1989;24:234–245.
4. Metz C. ROC analysis in medical imaging: A tutorial review of the literature. *Radiol Phys Technol.* 2008;1(1):2–12. doi:10.1007/s12194-007-0002-1.
5. Wagner RF, Beiden SV, Campbell G, Metz CE, Sacks WM. Assessment of medical imaging and computer-assist systems: lessons from recent experience. *Acad Radiol.* 2002;9(11):1264–1277.
6. Wagner RF, Metz CE, Campbell G. Assessment of medical imaging systems and computer aids: A tutorial review. *Acad Radiol.* 2007;14(6):723–748.
7. Kundel HL, Berbaum KS, Dorfman DD, Gur D, Metz CE, Swensson RG. Receiver Operating Characteristic Analysis in Medical Imaging (ICRU Report 79). International Commission on Radiation Units & Measurments, 2008 Contract No.: 1.
8. Zhai X, Chakraborty DP. RJ afroc: Analysis of Data Acquired Using the Receiver Operating Characteristic Paradigm and Its Extensions: R package version 0.1.1, https://cran.r-project.org/web/packages/RJafroc/index.html, https://cran.r-project.org/web/packages/RJafroc/index.html: CRAN; 2015.
9. Chakraborty DP. A brief history of free-response receiver operating characteristic paradigm data analysis. *Acad Radiol.* 2013;20(7):915–919. doi:10.1016/j.acra.2013.03.001.
10. Chakraborty DP. New developments in observer performance methodology in medical imaging. *Semin Nucl Med.* 2011;41(6):401–418.
11. Obuchowski NA. ROC analysis. *Am J Roentgenol.* 2005;184(2):364–372.
12. Zhou X-H, Obuchowski NA, McClish DK. *Statistical Methods in Diagnostic Medicine.* New York, NY: John Wiley & Sons; 2002.
13. Pepe MS. The Statistical Evaluation of Medical Tests for Classification and Prediction. New York, NY: Oxford University Press; 2003.
14. Swets JA, Pickett RM. *Evaluation of Diagnostic Systems: Methods from Signal Detection Theory.* 1st ed. New York, NY: Academic Press; 1982, 253 p.
15. Beam CA, Layde PM, Sullivan DC. Variability in the interpretation of screening mammograms by US radiologists. Findings from a national sample. *Arch Intern Med.* 1996;156(2):209–213.
16. Barnes GT. The dependence of radiographic mottle on beam quality. *Am J Roentgenol.* 1976;127(5):819–824.
17. Barnes GT, Chakraborty DP. Radiographic mottle and patient exposure in mammography. *Radiology.* 1982;145(3):815–821.
18. Barnes GT. Radiographic mottle: A comprehensive theory. *Med Phys.* 1982;9(5):656–667.
19. Cunningham IA, Yao J, Subotic V, eds. Cascaded models and the DQE of flat-panel imagers: Noise aliasing, secondary quantum noise, and reabsorption. Medical Imaging 2002; 2002: International Society for Optics and Photonics. San Diego, CA
20. Wagner RF, Brown DG. Unified SNR analysis of medical imaging systems. *Phys Med Biol.* 1985;30(6):489.
21. Dobbins JT, McAdams HP, Sabol JM, et al. Multi-Institutional evaluation of digital tomosynthesis, dual-energy radiography, and conventional chest radiography for the detection and management of pulmonary nodules. *Radiology.* 2017;282(1):236–250.
22. Chakraborty DP. Acceptance testing, Quality improvement, and dose assessment of fluoroscopy. RSNA Categorical Course in Physics. 1996:81–101.
23. Barnes GT, Frey DG. Mammography Acceptance Testing and Quality control: Documentation and Reports. 1991. Medical Physics Publishing, Madison WI.

24. Chakraborty DP, ed. Routine fluoroscopic quality control. Proceedings of the AAPM Summer School: specification, acceptance testing, and quality assurance of diagnostic x-ray imaging equipment - 1991; 1994; Santa Cruz, CA: American Institute of Physics.
25. Fraser RS, Muller NL, Colman N, Pare, PD. *Fraser and Paré's Diagnosis of Diseases of the Chest.* 4th ed, 4 Vol. Philadephia, PA: Saunders; 1999.
26. Chakraborty DP, Breatnach ES, Yester MV, et al. Digital and conventional chest imaging: A modified ROC study of observer performance using simulated nodules. *Radiology.* 1986;158:35–39.
27. Hendrick RE, Bassett L, Botsco MA, et al. *Mammography Quality Control Manual.* 4th ed. American College of Radiology, Committee on Quality Assurance in Mammography; 1999. Reston, VA.
28. Destouet JM, Bassett LW, Yaffe MJ, Butler PF, Wilcox PA. The ACR's mammography accreditation program: Ten years of experience since MQSA. *J Am Coll Radiol.* 2005;2(7):585–594.
29. McLelland R, Hendrick R, Zinninger MD, Wilcox PA. The American College of Radiology mammography accreditation program. *AJR Am J Roentgenol.* 1991;157(3):473–479.
30. Chakraborty DP. Computer analysis of mammography phantom images (CAMPI): An application to the measurement of microcalcification image quality of directly acquired digital images. *Med Phys.* 1997;24(8):1269–1277.
31. Chakraborty DP, Fatouros PP. Application of computer analyis of mammography phantom images (CAMPI) methodology to the comparison of two digital biopsy machines. Paper presented at: Proceedings of the SPIE Medical Imaging 1998: Physics of Medical Imaging; July 24, 1998; San Diego, CA: SPIE.
32. Chakraborty DP, Sivarudrappa M, Roehrig H, eds. Computerized measurement of mammographic display image quality. Proceedings of the SPIE Medical Imaging 1999: Physics of Medical Imaging; 1999; San Diego, CA: SPIE.
33. Chakraborty DP, ed. Comparison of computer analysis of mammography phantom images (CAMPI) with perceived image quality of phantom targets in the ACR phantom. Paper presented at: Proceedings of the SPIE Medical Imaging 1997: Image Perception; February 26–27, 1997; Newport Beach, CA: SPIE.
34. Chakraborty DP. Computer analysis of mammography phantom images (CAMPI). Paper presented at: Proceedings of the SPIE Medical Imaging 1997: Physics of Medical Imaging. 1997;3032:292–299.
35. Beiden SV, Wagner RF, Campbell G. Components-of variance models and multiple-bootstrap experiments: An alternative method for random-effects, receiver operating characteristic analysis. *Acad Radiol.* 2000;7(5):341–349.
36. Scott HJ, Gale AG. Breast screening: PERFORMS identifies key mammographic training needs. *Br J Radiol.* 2014;79(2), S127–S133. doi:10.1259/bjr/25049149
37. Gale AG, ed. PERFORMS: A self-assessment scheme for radiologists in breast screening. Seminars in Breast Disease; 2003: Elsevier.Citation: GALE, A.G., 2003. PERFORMS - a self assessment scheme for radiologists in breast screening. Seminars in Breast Disease, 6(3), pp. 148–152.
38. Barnes GT. Contrast and scatter in x-ray imaging. *Radiographics.* 1991;11(2):307–323.
39. Curry TS, Dowdey JE, Murry RC. *Christensen's Introduction to the Physics of Diagnostic Radiology.* 3rd ed. Philadelphia, PA: Lea & Febiger; 1984, 515 p.
40. Jaynes ET. Information theory and statistical mechanics. *Phys Rev.* 1957;106(4):620.
41. Claude E. Shannon, Warren Weaver. *The Mathematical Theory of Communication.* University of Illinois Press, 1949. ISBN 0-252-72548-4
42. Barrett HH, Myers K. *Foundations of Image Science.* Hoboken, NJ: John Wiley & Sons; 2003.
43. Chakraborty DP, Eckert MP. Quantitative versus subjective evaluation of mammography accreditation phantom images. *Med Phys.* 1995;22(2):133–143.
44. Tapiovaara MJ, Wagner RF. SNR and noise measurements for medical imaging: I. A practical approach based on statistical decision theory. *Phys Med Biol.* 1993;38:71–92.

45. Burgess AE. The Rose model, revisited. *J Opt Soc Am A.* 1999;16(3):633–646.
46. Burgess AE, Wagner RF, Jennings RJ, Barlow HB. Efficiency of human visual signal discrimination. *Science.* 1981;214(2):93–94.
47. Burgess AE, Jacobson FL, Judy PF. Human observer detection experiments with mammograms and power-law noise. *Med Phys.* 2001;28(4):419–437.
48. Chakraborty DP. An alternate method for using a visual discrimination model (VDM) to optimize softcopy display image quality. *J Soc Inf Display.* 2006;14(10):921–926.
49. Chakraborty DP, Kundel HL. Anomalous nodule visibility effects in mammographic images. Paper presented at: Proceedings of the SPIE Medical Imaging 2001: Image Perception and Performance; 2001; San Diego, CA: SPIE.
50. Bochud FO, Abbey CK, Eckstein MP. Visual signal detection in structured backgrounds. III. Calculation of figures of merit for model observers in statistically nonstationary backgrounds. *J Opt Soc Am A Opt Image Sci Vis.* 2000;17(2):193–205.
51. Metz CE. Receiver operating characteristic analysis: A tool for the quantitative evaluation of observer performance and imaging systems. *J Am Coll Radiol.* 2006;3:413–422.
52. Shiraishi J, Pesce LL, Metz CE, Doi K. Experimental design and data analysis in receiver operating characteristic studies: Lessons learned from reports in radiology from 1997 to 2006. *Radiology.* 2009;253(3):822–830.
53. Egan JP, Greenburg GZ, Schulman AI. Operating characteristics, signal detectability and the method of free-response. *J Acoust Soc Am.* 1961;33:993–1007.
54. Bunch PC, Hamilton JF, Sanderson GK, Simmons AH. A free-response approach to the measurement and characterization of radiographic-observer performance. *J of Appl Photogr Eng.* 1978;4:166–171.
55. Starr SJ, Metz CE, Lusted LB, Goodenough DJ. Visual detection and localization of radiographic images. *Radiology.* 1975;116:533–538.
56. Swensson RG. Unified measurement of observer performance in detecting and localizing target objects on images. *Med Phys.* 1996;23(10):1709–1725.
57. Obuchowski NA, Lieber ML, Powell KA. Data analysis for detection and localization of multiple abnormalities with application to mammography. *Acad Radiol.* 2000;7(7):516–525.
58. Van Trees HL, Bell HL, Tian Z. *Detection Estimation and Modulation Theory, Part I: Detection, Estimation, and Filtering Theory,* 2nd ed., p. 1176. New York, NY: John-Wiley & Sons; 2013.
59. Fryback DG, Thornbury JR. The efficacy of diagnostic imaging. *Med Decis Making.* 1991;11(2):88–94. doi:10.1177/0272989x9101100203.
60. Yerushalmy J. Reliability of chest radiography in the diagnosis of pulmonary lesions. *Am J Surg.* 1955;89(1):231–240.
61. Lusted LB. Introduction to medical decision making. *Am J Phys Med Rehab.* 1970;49(5):322.
62. Chakraborty DP. Clinical relevance of the ROC and free-response paradigms for comparing imaging system efficacies. *Radiat Prot Dosimetry.* 2010;139(1–3):37–41. doi:10.1093/rpd/ncq017.
63. Fenton JJ, Taplin SH, Carney PA, et al. Influence of computer-aided detection on performance of screening mammography. *N Engl J Med.* 2007;356(14):1399–1409. doi:10.1056/NEJMoa066099.
64. Fenton JJ. Is it time to stop paying for computer-aided mammography? *JAMA Intern Med.* 2015;175(11):1837–1838. doi:10.1001/jamainternmed.2015.5319

PART A

The receiver operating characteristic (ROC) paradigm

2

The binary paradigm

2.1 Introduction

In the previous chapter, four observer performance paradigms were introduced: the receiver operating characteristic (ROC), the free-response ROC (FROC), the location ROC (LROC), and the region of interest (ROI). In the chapters comprising this section, i.e., **Chapter 2** through **Chapter 7**, focus is on the ROC paradigm, where each case is rated for confidence in the presence of disease. While a multiple point rating scale is generally used, *in this chapter it is assumed that the ratings are binary*, and the allowed values are 1 versus 2. Equivalently, the ratings could be *non-diseased* versus *diseased*, *negative* versus *positive*, etc. In the literature this method of data acquisition is also termed the *yes/no* paradigm.[1,2] The reason for restricting, for now, to the binary task is that the multiple rating task can be shown to be equivalent to a number of simultaneously conducted binary tasks. So, understanding the simpler method is a good starting point.

Since the truth is also binary, this chapter could be named the *binary-truth binary-decision* task. The starting point is a 2 × 2 table summarizing the outcomes in such studies and useful fractions that can be defined from the counts in this table, the most important ones being true positive fraction (TPF) and false positive fraction (FPF). These are used to construct measures of performance, some of which are desirable from the researcher's point of view, but others are more relevant to radiologists. The concept of disease prevalence is introduced and used to formulate relations between the different types of measures. An **R** example of calculation of these quantities is given that is only slightly more complicated than the demonstration in the online material to the prior chapter.

2.2 Decision versus truth: The fundamental 2 × 2 table of ROC analysis

In this book, the term *case* is used for images obtained, for diagnostic purposes, of a patient. Often multiple images of a patient, sometimes from different modalities, are involved in an interpretation. All images of a single patient, that are used in the interpretation, are collectively referred to as a case. A familiar example is the 4-view presentation used in screening mammography, where two views of each breast are available for viewing.

Let D represent the radiologist's *decision*, with $D = 1$ representing the decision *case is non-diseased* and $D = 2$ representing the decision *case is diseased*. Let T denote the *truth* with $T = 1$ representing *case is actually non-diseased* and $T = 2$ representing *case is actually diseased*. It is assumed that, prior to the interpretation, the radiologist does not know the truth state of the case and the decision is based on information contained in the case. Each decision, one of two values, will be associated with one of two truth states, resulting in an entry in one of four cells arranged in a 2 × 2 layout, termed the *decision versus truth table*, Table 2.1, *which is of fundamental importance in observer performance*. The cells are labeled as follows. The abbreviation *TN*, for true negative, represents a $D=1$ decision on a $T=1$ case. Likewise, *FN*, for false negative, represents a $D=1$ decision on a $T=2$ case (also termed a *miss*). Similarly, *FP*, for false positive, represents a $D=2$ decision on a $T=1$ case (a *false-alarm*) and *TP*, for true positive, represents a $D=2$ decision on a $T=2$ case (a *hit*).

Table 2.2 shows the *numbers* (indicated by the hash symbol prefix) of decisions in each of the four categories shown in Table 2.1. Specifically, #TN is the number of true negative decisions, #FN is the number of false negative decisions, and so on. The last row is the sum of the corresponding columns. The sum of the number of true negative decisions (#TN) and the number of false positive decisions (#FP) must equal the total number of non-diseased cases, denoted K_1. Likewise, the sum of the number of false negative decisions (#FN) and the number of true positive decisions (#TP) must equal the total number of diseased cases, denoted K_2. The last column is the sum of the corresponding rows. The sum of the number of true negative (#TN) and false negative (#FN) decisions is the total number of negative decisions, denoted #N. Likewise, the sum of the number of false positive (#FP) and true positive (#TP) decisions is the total number of positive decisions, denoted #P. Since each case yields a decision, the bottom-right corner cell is #N + #P, which must also equal $K_1 + K_2$, the total number of cases, denoted K. These statements are summarized in Equation 2.1.

$$\left.\begin{aligned}
K_1 &= \#TN + \#FP \\
K_2 &= \#FN + \#TP \\
\#N &= \#TN + \#FN \\
\#P &= \#TP + \#FP \\
K &= K_1 + K_2 = \#N + \#P
\end{aligned}\right\} \qquad (2.1)$$

Table 2.1 The decision versus truth table: The fundamental 2 × 2 table of observer performance, showing the classification of decisions in the binary task

	Case truth *T*	
Radiologist's decision *D*	*Case is actually non-diseased: T = 1*	*Case is actually diseased: T = 2*
Case is diagnosed non-diseased: D = 1	**TN**	**FN (miss)**
Case is diagnosed diseased: D = 2	**FP (false alarm)**	**TP (hit)**

Table 2.2 Decision versus truth table, showing total counts in the different cells

Radiologist's decision: *D*	Case truth: *T*		
	Case is non-diseased: T = 1	*Case is diseased: T = 2*	**Row totals**
Case is non-diseased: D = 1	#TN	#FN	#N = #TN + #FN
Case is diseased: D = 2	#FP	#TP	#P = #FP + #TP
Column totals	$K_1 = \#TN + \#FP$	$K_2 = \#FN + \#TP$	$K_1 + K_2 = \#N + \#P$

Note: The last row/column show the totals of the corresponding columns/rows (# denotes the number of counts in the corresponding cell)

2.3 Sensitivity and specificity

The notation $P(D|T)$ indicates the *probability of diagnosis D given truth state T* (the vertical bar symbol is used to denote a *conditional probability*, that is, what is to the left of the vertical bar depends on the *condition* appearing to the right of the vertical bar being true).

$$P(D|T) = P(\text{patient diagnosis is } D \,|\, \text{patient truth is } T) \tag{2.2}$$

Therefore, the probability that the radiologist will diagnose *case is diseased* when the case is actually diseased is $P(D=2|T=2)$, which is the probability of a true positive $P(TP)$.

$$P(TP) = P(D=2|T=2) \tag{2.3}$$

Likewise, the probability that the radiologist will diagnose *case is non-diseased* when the case is actually diseased is $P(D=1|T=2)$, which is the probability of a false negative $P(FN)$.

$$P(FN) = P(D=1|T=2) \tag{2.4}$$

The corresponding probabilities for non-diseased cases, $P(TN)$ and $P(FP)$, are defined by

$$\left.\begin{aligned} P(TN) &= P(D=1|T=1)\\ P(FP) &= P(D=2|T=1) \end{aligned}\right\} \tag{2.5}$$

Since the diagnosis must be either $D=1$ or $D=2$, for each truth state the probabilities of non-diseased and diseased case diagnoses must sum to unity.

$$\left.\begin{aligned} P(D=1|T=1)+P(D=2|T=1)&=1\\ P(D=1|T=2)+P(D=2|T=2)&=1 \end{aligned}\right\} \tag{2.6}$$

Equivalently, these equations can be written as

$$\left.\begin{aligned} P(TN)+P(FP)&=1\\ P(FN)+P(TP)&=1 \end{aligned}\right\} \tag{2.7}$$

Comments:

An easy way to remember Equation 2.7 is to start by writing down the probability of one of the four probabilities, for example, $P(TN)$, and reversing both terms inside the parentheses, that is, T → F, and N → P. This yields the term $P(FP)$ which, when added to the previous probability, $P(TN)$, yields unity, that is, the first equation in Equation 2.7, and likewise for the second equation.

Because there are two equations in four unknowns, only two of the four probabilities, one per equation, are independent. By tradition these are chosen to be $P(D=1|T=1)$ and $P(D=2|T=2)$, that is, $P(TN)$ and $P(TP)$, which happen to be the probabilities of correct decisions on non-diseased and diseased cases, respectively. The two basic probabilities are so important that they have names: $P(D=2|T=2) = P(TP)$ is termed *sensitivity* (Se) and $P(D=1|T=1) = P(TN)$ is termed *specificity* (Sp).

$$\left.\begin{aligned} Se &= P(TP) = P(D=2|T=2)\\ Sp &= P(TN) = P(D=1|T=1) \end{aligned}\right\} \tag{2.8}$$

The radiologist can be regarded as a diagnostic test yielding a binary decision under the binary truth condition. More generally, any test (e.g., a blood test for HIV) yielding a binary result (positive or negative) under a binary truth condition is said to be *sensitive* if it correctly detects the diseased condition most of the time. The test is said to be *specific* if it correctly detects the non-diseased condition most of the time. Sensitivity is how correct the test is at detecting a diseased condition, and specificity is how correct the test is at detecting a non-diseased condition.

2.4 Reasons for the names sensitivity and specificity

It is important to understand the reason for these names and an analogy may be helpful. Most of us are *sensitive* to temperature, especially if the choice is between ice-cold versus steaming hot. The sense of touch is said to be *sensitive* to temperature. One can imagine some neurological condition rendering a person hypersensitive to temperature, such that the person responds "hot" no matter what is being touched. For such a person, the sense of touch is not very *specific*, as it is unable to distinguish between the two temperatures. This person would be characterized by unit sensitivity (since the response is "hot" to all steaming hot objects) and zero specificity (since the response is never "cold" to ice-cold objects). Likewise, a different neurological condition could render a person hypersensitive to cold, and the response is "cold" no matter what is being touched. Such a person would have zero sensitivity (since the response is never "hot" when touching steaming hot) and unit specificity (since the response is "cold" when touching ice-cold). Already one suspects that there is an inverse relation between sensitivity and specificity.

2.5 Estimating sensitivity and specificity

Sensitivity and specificity are the probabilities of correct decisions, over diseased and non-diseased cases, respectively. The *true* values of these probabilities would require interpreting all diseased and non-diseased cases in the *entire population* of cases. In reality, one has a *finite sample* of cases and the corresponding quantities, calculated from this finite sample, are termed *estimates*. Population values are fixed, and in general unknown, while estimates are random variables. Intuitively, an estimate calculated over a larger number of cases is expected to be closer to the true or population value than an estimate calculated over a smaller number of cases.

Estimates of sensitivity and specificity follow from counting the numbers of TP and TN decisions in Table 2.2 and dividing by the appropriate denominators. For sensitivity, the appropriate denominator is the number of actually diseased cases, namely K_2, and for specificity, the appropriate denominator is the number of actually non-diseased cases, namely K_1. The estimation equations for sensitivity specificity are (estimates are denoted by the circumflex symbol ^):

$$\left. \begin{aligned} \widehat{Se} = \widehat{P(TP)} = \frac{\#TP}{K_2} \\ \widehat{Sp} = \widehat{P(TN)} = \frac{\#TN}{K_1} \end{aligned} \right\} \tag{2.9}$$

The ratio of the number of TP decisions to the number of actually diseased cases is termed *true positive fraction* $\widehat{TPF}$, which is an estimate of sensitivity, or equivalently, an estimate of $P(TP)$. Likewise, the ratio of the number of TN decisions to the number of actually non-diseased cases is termed *true negative fraction* $\widehat{TNF}$, which is an estimate of specificity, or equivalently, an estimate of

Table 2.3 This table shows estimates of two selected probabilities, sensitivity and specificity, in the binary decision task

	Case truth: T	
Radiologist's decision: D	T = 1	T = 2
D = 1	$\widehat{Sp} = \widehat{TNF} = \widehat{P(TN)} = \dfrac{\#TN}{K_1}$	$1 - \widehat{Se} = \widehat{FNF} = \widehat{P(FN)} = \dfrac{\#FN}{K_2}$
D = 2	$1 - \widehat{Sp} = \widehat{FPF} = \widehat{P(FP)} = \dfrac{\#FP}{K_1}$	$\widehat{Se} = \widehat{TPF} = \widehat{P(TP)} = \dfrac{\#TP}{K_2}$

Note: The two other probabilities are the complements of these values. The probabilities follow from dividing the numbers of counts from Table 2.2 by the appropriate denominators.

$P(TN)$. The complements of $\widehat{TPF}$ and $\widehat{TNF}$ are termed *false negative fraction* $\widehat{FNF}$ and *false positive fraction* $\widehat{FPF}$, respectively (Table 2.3).

2.6 Disease prevalence

Disease prevalence, often abbreviated to *prevalence*, is defined as the *actual* or true probability that a randomly sampled case is of a diseased patient, that is, the fraction of the entire population that is diseased. It is denoted, $P(D \mid pop)$ when patients are randomly sampled from the population (*pop*) and otherwise it is denoted $P(D \mid lab)$, where the condition *lab* stands for a laboratory study, where cases may be artificially enriched, and thus not representative of the population value.

$$\left.\begin{aligned} P(D \mid pop) &= P(T = 2 \mid pop) \\ P(D \mid lab) &= P(T = 2 \mid lab) \end{aligned}\right\} \tag{2.10}$$

Since the patients must be either diseased or non-diseased, it follows with either sampling method, that

$$\left.\begin{aligned} P(T = 1 \mid pop) + P(T = 2 \mid pop) &= 1 \\ P(T = 1 \mid lab) + P(T = 2 \mid lab) &= 1 \end{aligned}\right\} \tag{2.11}$$

If a finite number of patients are sampled randomly from the population the fraction of diseased patients in the sample is an estimate of *true* disease prevalence.

$$\widehat{P(D \mid pop)} = \left.\dfrac{K_2}{K_1 + K_2}\right|_{pop} \tag{2.12}$$

It is important to appreciate the distinction between *true* (population) prevalence and *laboratory* prevalence. As an example, true disease prevalence for breast cancer is about five per 1000 patients in the United States, but most mammography studies are conducted with comparable numbers of non-diseased and diseased cases.

$$\left.\begin{aligned} \widehat{P(D \mid pop)} &\sim 0.005 \\ \widehat{P(D \mid lab)} &\sim 0.5 \gg \widehat{P(D \mid pop)} \end{aligned}\right\} \tag{2.13}$$

2.7 Accuracy

Accuracy is defined as the fraction of all decisions that are in fact correct. Denoting accuracy by Ac, one has for the corresponding estimate:

$$Ac = \frac{\#TN + \#TP}{\#TN + \#TP + \#FP + \#FN} \tag{2.14}$$

The numerator is the total number of correct decisions and the denominator is the total number of decisions. An equivalent expression is

$$\widehat{Ac} = \widehat{Sp}\,\widehat{P(!D)} + \widehat{Se} \times \widehat{P(D)} \tag{2.15}$$

The exclamation mark symbol is used to denote the *not* or *negation* operator. For example, $P(!D)$ means the probability that the patient is not diseased. Equation 2.15 applies equally to laboratory or population studies, *provided sensitivity and specificity are estimated consistently*. In other words, one cannot combine a population estimate of prevalence with a laboratory measurement of sensitivity and/or specificity.

Equation 2.15 can be understood from the following argument. $\widehat{Sp}$ is an estimate of the fraction of correct (i.e., negative) decisions on non-diseased cases. Multiplying this by $\widehat{P(!D)}$ yields $\widehat{Sp}\,\widehat{P(!D)}$, the fraction of correct negative decisions on all cases. Similarly, $\widehat{Se} \times \widehat{P(D)}$ is the fraction of correct positive decisions on all cases. Therefore, their sum is the fraction of all (i.e., negative and positive) correct decisions on all cases. A formal mathematical derivation follows. The terms on the right-hand side of Equation 2.9 can be turned around yielding

$$\#TP = K_2\widehat{Se}$$
$$\#TN = K_1\widehat{Sp} \tag{2.16}$$

Therefore

$$\widehat{Ac} = \frac{\#TN + \#TP}{K} = \frac{K_1\widehat{Sp} + K_2\widehat{Se}}{K} = \widehat{Sp}\widehat{P(!D)} + \widehat{Se}\,\widehat{P(D)} \tag{2.17}$$

∎

2.8 Positive and negative predictive values

Sensitivity and specificity have desirable characteristics, insofar as they reward the observer for correct decisions on actually diseased and actually non-diseased cases, respectively, so these quantities are expected to be independent of disease prevalence. Stated alternatively, one is dividing by the relevant denominator, so increased numbers of non-diseased cases are balanced by a corresponding increased number of correct decisions on non-diseased cases, and likewise for diseased cases. However, radiologists interpret cases in a mixed situation where cases could be positive or negative for disease and disease prevalence plays a crucial role in their decision-making—this point will be clarified shortly. Therefore, a measure of performance that is desirable from the researcher's point of view is not necessarily desirable from the radiologist's point of view. It should be obvious that if most cases are non-diseased, that is, disease prevalence is close to zero, specificity, being correct on non-diseased cases, is more important to the radiologist. Otherwise, the radiologist would figuratively be *crying wolf* most of the time. The radiologist who makes too many FPs would discover it from subsequent clinical audits or daily case conferences, which are held in most large imaging departments. There is a cost to unnecessary false positives—the cost of additional imaging and/or needle-biopsy to rule out cancer, not to mention the pain and emotional trauma inflicted on the patients. Conversely, if disease prevalence is high, then sensitivity, being correct on diseased cases, is more important to the radiologist. With intermediate disease prevalence a weighted average of sensitivity and specificity, where the weighting involves disease prevalence, is desirable from the radiologist's point of view.

The radiologist is less interested in the *normalized* probability of a correct decision on non-diseased cases. Rather greater interest is in the probability that a patient diagnosed as non-diseased is actually non-diseased. The reader should notice how the two probability definitions are turned around—more on this below. Likewise, the radiologist is less interested in the *normalized* probability of correct decisions on diseased cases; rather greater interest is in the probability that a patient diagnosed as diseased is actually diseased. These are termed *negative and positive predictive values*, respectively, and denoted NPV and PPV.

Let us start with NPV, defined as the probability, given a non-diseased diagnosis, that the patient is actually non-diseased.

$$NPV = P(T = 1 \mid D = 1) \tag{2.18}$$

Note that this equation is "turned around" from the definition of specificity, in Equation 2.8, repeated below for ease of comparison.

$$Sp = P(D = 1 \mid T = 1) \tag{2.19}$$

To estimate NPV, one divides the number of correct negative decisions (#TN) by the total number of negative decisions (#N). The latter is the sum of the number of correct negative decisions (#TN) and the number of incorrect negative decisions (#FN). Therefore

$$\widehat{NPV} = \frac{\#TN}{\#TN + \#FN} \tag{2.20}$$

Dividing the numerator and denominator by the total number of cases K, one gets:

$$\widehat{NPV} = \frac{\widehat{P_K(TN)}}{\widehat{P_K(TN)} + \widehat{P_K(FN)}} \tag{2.21}$$

The estimate $\widehat{P_K(TN)}$ of the probability of a TN *over all cases* (hence the subscript K) equals the estimate of true negative fraction $(1 - \widehat{FPF})$ multiplied by the estimate that the patient is non-diseased, i.e., $\widehat{P(!D)}$:

$$\widehat{P_K(TN)} = \widehat{P(!D)}(1 - \widehat{FPF}) \tag{2.22}$$

Explanation: A similar logic to that used earlier applies: $(1 - \widehat{FPF})$ is the probability of being correct on non-diseased cases. Multiplying this by the estimate of probability of disease *absence* yields the estimate of $\widehat{P_K(TN)}$.

Likewise, the estimate $\widehat{P_K(FN)}$ of the probability of a FN over all cases equals the estimate of false negative fraction, which is $(1 - \widehat{TPF})$, multiplied by the estimate of the probability that the patient is diseased, i.e., $\widehat{P(D)}$:

$$\widehat{P_K(FN)} = \widehat{P(D)}(1 - \widehat{TPF}) \tag{2.23}$$

Putting this all together, one has

$$\widehat{NPV} = \frac{\widehat{P(!D)}(1 - \widehat{FPF})}{\widehat{P(!D)}(1 - \widehat{FPF}) + \widehat{P(D)}(1 - \widehat{TPF})} \tag{2.24}$$

For the population,

$$NPV = \frac{P(!D)(1-FPF)}{P(!D)(1-FPF)+P(D)(1-TPF)} \tag{2.25}$$

Likewise, it can be shown that *PPV* is given by

$$PPV = \frac{P(D)\times TPF}{P(D)\times TPF + P(!D)\times FPF} \tag{2.26}$$

In words,

$$\text{negative predictive value} = \frac{(1-\text{prevalence})(\text{specificity})}{(1-\text{prevalence})(\text{specificity})+(\text{prevalence})(1-\text{sensitivity})} \tag{2.27}$$

$$\text{positive predictive value} = \frac{(\text{prevalence})(\text{sensitivity})}{(\text{prevalence})(\text{sensitivity})+(1-\text{prevalence})(1-\text{specificity})} \tag{2.28}$$

The equations defining NPV and PPV are actually special cases of Bayes' theorem.[3] The general theorem is

$$P(A\,|\,B) = \frac{P(B\,|\,A)P(A)}{P(B)} = \frac{P(A)P(B\,|\,A)}{P(A)P(B\,|\,A)+P(!A)P(B\,|\,!A)} \tag{2.29}$$

An easy way to remember Equation 2.29 is to start with the numerator, which is the reversed form of the desired probability on the left-hand side, multiplied by an appropriate probability. For example, if the desired probability is $P(A\,|\,B)$, one starts with the reversed form, that is, $P(B\,|\,A)$, multiplied by $P(A)$. This yields the numerator. The denominator is the sum of two probabilities: the probability of B given A, that is, $P(B\,|\,A)$, multiplied by $P(A)$, plus the probability of B given !A, that is, $P(B\,|\,!A)$, multiplied by $P(!A)$.

2.9 Example: Calculation of PPV, NPV, and accuracy

Typical disease prevalence in the United States in screening mammography is 0.005. A typical operating point, for an expert mammographer, is FPF = 0.1, TPF = 0.8. What are NPV and PPV? While this can be done using a hand calculator, since one has **R/RStudio**, why not use it? In the online **software** folder for this chapter, open the **RStudio** project file, always named **software. Rproj** in this book, and use the **Files** menu to open **mainNpvPpv.R**, a listing of which follows.

2.9.1 Code listing

```
# mainNpvPpv.R
rm(list = ls())
# disease prevalence in US screening mammography
prevalence <- 0.005
FPF <- 0.1 # typical operating point
TPF <- 0.8 # do:
specificity <- 1-FPF
sensitivity <- TPF
```

```
NPV <- (1-prevalence)*(specificity)/
   ((1-prevalence)*(specificity) +
     prevalence*(1-sensitivity))
PPV <- prevalence*sensitivity/
   (prevalence*sensitivity +
     (1-prevalence)*(1-specificity))
cat("NPV = ", NPV, "\nPPV = ", PPV, "\n")
accuracy <-(1-prevalence)*
   (specificity)+(prevalence)*(sensitivity)
cat("accuracy = ", accuracy, "\n")
```

Line 4 initializes the variable **prevalence**, the disease prevalence. In other words, **prevalence <- 0.005** causes the value 0.005 to be assigned to the variable **prevalence**. Do not use **prevalence = 0.005** as an assignment statement. It may work some of the time, but it can cause problems when one least expects it. Code that works some of the time is worse, in the author's opinion, than code that never works. In **R**, one does not need to worry about the type of variable—integer, float, double, or declaring variables before using them; this can lead to sloppy programming constructs but for the most part **R** behaves reasonably. Line 5 assigns 0.1 to **FPF** and line 6 assigns 0.8 to **TPF**. Lines 7 and 8 initialize the variables **specificity** and **sensitivity**, respectively.

Line 9–11 calculates **NPV**, using Equation 2.27 and line 12–14 calculates **PPV**, using Equation 2.28. **R** does not use any special continuation character; if a line evaluates to something valid then **R** does not look at the next line, otherwise it does. The user needs to make sure that a line to be continued does not make sense to **R**, and thereby forces **R** to read the next line. Lines 9–11 and 12–14 provide guidance on how to split code onto continuation lines. Line 15 prints the values of **NPV** and **PPV**, with a helpful message. The **cat()** function stands for *concatenate and print the comma-separated components of the argument*. The **cat()** function starts by printing the string variable **"NPV ="**, then it encounters a comma, then the variable name **NPV**, so it prints the value of the variable. Note that a string or character variable is surrounded by quotation marks. Then it encounters another comma, and the string **"PPV ="**, which it prints. Then it encounters another comma and the variable name **PPV**, so it prints the value of this variable. Finally, it encounters the last comma, and the string **"\n"**, which stand for a *newline* character, which sends any subsequent output to the next line; without it any subsequent print statements would appear on the same line, which is usually not the intent. Line 16–17 calculates accuracy, Equation 2.17, and the next line prints it. Click on the **Source** button (in the future this will be abbreviated to **source** the code) on the top-right corner of the source-file window; one gets the following output in the **Console** window [In the first line, > **source(...)**, the ellipsis indicate lines that are context dependent, i.e., the full path name of the file that is being sourced. This line does not contain code output. This convention will be used throughout the book. In general, the > character in the **Console** window means that **R** is waiting for user input. Lines without the > character prefix represent actual code output, which, due to the printed page-width limitation, may spill over onto several lines.]:

2.9.2 Code output

```
> source(...)
NPV = 0.9988846
PPV = 0.03864734
accuracy = 0.8995
```

As stated in Section 1.8 if a woman has a negative diagnosis, chances are very small that she has breast cancer; as shown in this example, the probability that the radiologist is incorrect in the negative diagnosis is 1 − NPV = 0.00111. Even if she has a positive diagnosis, the probability that she actually has cancer is still only 0.039. That is why following a positive screening diagnosis the woman is recalled for further imaging, and if that reveals cause for reasonable suspicion, then additional imaging is performed, perhaps augmented with a needle-biopsy to confirm actual disease status. If the biopsy turns out positive, only then is the woman referred for cancer therapy. Overall, accuracy is 90%, that is, the radiologist is accurate! The numbers in this illustration are for expert radiologists. In practice there is wide variability in radiologist performance.[4]

Consider what happens if the radiologist simply calls every case negative for disease. The radiologist will be correct on all of the actually non-diseased cases but will be incorrect on all of the diseased cases. Since there are 995 non-diseased cases and 5 diseased cases, accuracy will be 0.995, higher than that achieved by the expert in the previous example. *This tells us that accuracy is not a good measure of performance.* If the radiologist responds non-diseased to every case, then both FPF and TPF will be zero. Making these changes, i.e., setting FPF equal to zero and TPF equal to zero, and sourcing the code one gets the following:

2.9.3 Code output

```
> source(...)
NPV = 0.995
PPV = NaN
accuracy = 0.995
```

This confirms our expectation for accuracy. The reason **PPV** is **NaN** (Not a Number) is because one has zero *correct* positive decisions out of a *total* of zero positive decisions, leading to a 0 divided by 0 situation, Equation 2.26.

2.10 PPV and NPV are irrelevant to laboratory tasks

According to the hierarchy of assessment methods described in **Chapter 1**, Table 1.1, PPV and NPV are level-3 measurements, which are calculated from live interpretations. In the clinic, the radiologist adjusts the operating point to achieve a balance between sensitivity and specificity. *The balance depends critically on the known disease prevalence.* Based on geographical location and type of practice, the radiologist, over time, develops an idea of actual disease prevalence, or it can be found in various databases. For example, a breast-imaging clinic that specializes in imaging high-risk women will have higher disease prevalence than the general population and the radiologist is expected to err more on the side of reduced specificity because of the expected benefit of increased sensitivity. However, in the context of a laboratory study, where one uses enriched case sets and retrospective interpretation of cases, the concepts of NPV and PPV are meaningless. For example, it would be rather difficult to perform a laboratory study with 10,000 randomly sampled women, which would ensure about 50 actually diseased patients, which is large enough to get a reasonably precise estimate of sensitivity (estimating specificity is inherently more precise because most women are actually non-diseased). Rather, in a laboratory study one uses enriched data sets where the numbers of diseased-cases is much larger than in the general population, Equation 2.13. *The radiologist cannot interpret these cases pretending that the actual prevalence is very low.* Negative and positive predictive values, while they can be calculated from laboratory data, have very little, if any, clinical meanings, since they have no effect on a radiologist's thinking. As noted in **Chapter 1** the whole purpose of level-3 measurements is to determine the effect on a radiologist's thinking. There are no diagnostic decisions riding on laboratory ROC interpretations of retrospectively acquired patient images. However, PPV and NPV do have clinical meanings when calculated from very large population based live studies.[5-7] For example, the 2011 Fenton et al. study sampled 684,956 women and used the results of live interpretations. In contrast, laboratory ROC studies are typically conducted with 50–100 non-diseased and 50–100 diseased cases. A study using about 300 cases total would be considered a large ROC study.

2.11 Summary

This chapter introduced the terms *sensitivity* (identical to TPF), *specificity* (the complement of FPF), disease prevalence, positive and negative predictive values, and accuracy. It is shown that, due to its strong dependence on disease prevalence, accuracy is a relatively poor measure of performance. Radiologists generally have a good, almost visceral, understanding of positive and negative predictive values, as these terms are relevant in the clinical context, being in effect, their batting averages. A caveat on the use of PPV and NPV calculated from laboratory studies is noted; these quantities only make sense in the context of live clinical interpretations. The next chapter describes a parametric model for the binary task.

References

1. Green DM, Swets JA. *Signal Detection Theory and Psychophysics.* New York, NY: John Wiley & Sons; 1966.
2. Egan JP. *Signal Detection Theory and ROC Analysis.* 1st ed. New York,. NY: Academic Press; 1975.
3. Larsen RJ, Marx ML. *An Introduction to Mathematical Statistics and Its Applications.* 3rd ed. Upper Saddle River, NJ: Prentice-Hall; 2001.
4. Beam CA, Layde PM, Sullivan DC. Variability in the interpretation of screening mammograms by US radiologists. Findings from a national sample. *Arch Intern Med.* 1996;156(2):209–213.
5. Barlow WE, Chi C, Carney PA, et al. Accuracy of screening mammography interpretation by characteristics of radiologists. *J Natl Cancer Inst.* 2004;96(24):1840–1850.
6. Fenton JJ, Taplin SH, Carney PA, et al. Influence of computer-aided detection on performance of screening mammography. *N Engl J Med.* 2007;356(14):1399–1409.
7. Fenton JJ, Abraham L, Taplin SH, et al. Effectiveness of computer-aided detection in community mammography practice. *J Natl Cancer Inst.* 2011;103(15):1152–1161.

3

Modeling the binary task

3.1 Introduction

Chapter 2 introduced measures of performance associated with the binary decision task. Described in this chapter is a 2-parameter statistical model for the binary task. In other words, it shows how one can predict quantities such as sensitivity and specificity based on the values of the parameters of a statistical model. It introduces the fundamental concepts of a *decision variable* and a *decision threshold* (the latter is one of the parameters of the statistical model) that pervade this book, and shows how the decision threshold can be altered by varying experimental conditions. The receiver operating characteristic (ROC) plot is introduced, which shows how the dependence of sensitivity and specificity on the decision threshold is exploited by a measure of performance that is independent of the decision threshold, namely the area AUC under the ROC curve. AUC turns out to be related to the other parameter of the model.

The dependence of variability of the operating point on the numbers of cases is explored, introducing the concept of random sampling and how the results become more stable with larger numbers of cases, or larger sample sizes. These are perhaps intuitively obvious concepts but it is important to see them demonstrated, Online Appendix 3.A. The formula for 95% confidence intervals for estimates of sensitivity and specificity are derived and the calculations are shown explicitly in Online Appendix 3.B.

The final aim of this chapter is to introduce **R** in somewhat greater depth so that the reader can take advantage of it in later chapters. Online Appendix 3.C contains the second part of the **R** tutorial; the first part was in **Chapter 1**. The intent is not to make an **R**-programmer out of the reader; rather it is to get the reader to a level to appreciate its utility in demonstrating abstract formula and concepts. Since little statistical expertise is assumed of the reader, these demonstrations take on added importance. For example, Online Appendix 3.C has detail on the normal distribution, how to sample from it, how to get the probability density function, the quantile function, and so on. Also important is the ability to visualize data, Online Appendix 3.D. *A picture is worth a thousand words*: this cliché is not only true, it is particularly relevant to learning this area of science. A plot summarizes a lot of information into one visual nugget. Considerable emphasis is placed in this book on visualizing data using **R** and almost every displayed plot has a statement in the caption naming the **R** file that generated it. Online Appendix 3.E and 3.F describe how to use the **R** help system. Online Appendix 3.G describes what to do if a package is missing (packages are extensions to **R**; the author's **RJafroc** software is an example of an **R** package). Online Appendix 3.H describes drawing shaded distributions in **R** - anyone working in this field has seen shaded distributions used to illustrate the variation of TPF as a a function of FPF. Finally, the Online Appendix 3.I illustrates numerical integration in **R**.

The starting point is the important concepts of *decision variable* and *decision threshold*.

3.2 Decision variable and decision threshold

The model[1] for the binary task involves three assumptions: (1) the existence of a case-dependent *decision variable* associated with each case, (2) the existence of a case-independent *decision threshold* for reporting individual cases as non-diseased or diseased, and (3) the adequacy of training session(s) in getting the observer to a steady state. In addition, an assumption common to all models is that the observer is blinded to the truth, while the researcher is not.

3.2.1 Existence of a decision variable

Assumption 1: Each case presentation is associated with the occurrence (or realization) of a specific value of a *random scalar sensory variable* yielding an unidirectional measure of *evidence of disease*. The two italicized phrases introduce important terms.

- By *sensory variable* one means something that is sensed internally by the observer (in the cognitive system, associated with the thinking brain) and as such is not directly measurable in the traditional physical sense. A physical measurement, for example, might consist of measuring a voltage difference across two points with a voltmeter. The term *latent* is often used to describe the sensory variable because it turns out that transforming this variable by an arbitrary monotonic non-decreasing transformation has no effect on the ROC—this will become clearer later. Alternative equivalent terms are *psychophysical variable, perceived variable, perceptual variable,* or *confidence level.* The last term is the most common. It is a *subjective* variable since its value is expected to depend on the observer: the same case shown to different observers could evoke different values of the sensory variable. Since one cannot measure it anyway, it would be a very strong assumption to assume that the two sensations are identical. In this book the term *latent decision variable* or simply, *decision variable,* is used, which hopefully gets away from the semantics and focuses instead on what the variable is used for, namely, *making decisions.* The symbol Z will be used for the random decision variable and specific realized values are denoted z; while in general uppercase/lower case are used to differentiate between random and realized values, to avoid confusion the Z in "Z-sample" will always be uppercase. It is *random* in the sense that it varies randomly from case to case; unless the cases are similar in some respect. For example, two variants of the same case under different image processing conditions, or images of twins; in these instances, the corresponding decision variables are expected to be correlated. In the binary paradigm model relevant to this chapter, the decision variables corresponding to different cases are assumed mutually independent.
- The latent decision variable *rank-orders* cases with respect to evidence for *presence* of disease. Unlike a traditional rank-ordering scheme, where 1 is the highest rank, the scale is inverted with larger values corresponding to greater evidence of disease. Without loss of generality, one assumes that the decision variable ranges from $-\infty$ to $+\infty$, with large positive values indicative of strong evidence for presence of disease, and large negative values indicative of strong evidence for absence of disease. The zero value indicates no evidence for presence or absence of disease. (The $-\infty$ to $+\infty$ scale is not an assumption. The decision variable scale could just as well range from a to b, where $a < b$; with appropriate rescaling of the decision variable, there will be no changes in the rank-orderings, and the scale can be made to extend from $-\infty$ to $+\infty$.) Such a decision scale, with increasing values corresponding to increasing evidence of disease, is termed *positive-directed.*

3.2.2 Existence of a decision threshold

Assumption 2: In the binary decision task the radiologist adopts a fixed (i.e., case-independent) *decision threshold* ζ and states, *case is diseased*, if the decision variable is greater than or equal to ζ, that is, $Z \geq \zeta$, and *case is non-diseased* if the decision variable is smaller than ζ, that is, $Z < \zeta$.

- The decision threshold is a *fixed value* used to separate cases reported as diseased from cases reported as non-diseased.
- Unlike the random Z-sample, which varies from case to case, the decision threshold is held fixed for the duration of the study. In some of the older literature[2] the decision threshold is sometimes referred to as *response bias*. The author prefers not to use the term *bias*, which has a negative connotation, whereas, in fact, the choice of decision threshold depends on rational assessment of costs and benefits of different outcomes.
- The choice of decision threshold depends on the *conditions* of the study: perceived or known disease prevalence, cost-benefit considerations, instructions regarding dataset characteristics, personal interpreting style, and so on. There is a transient learning curve during which the observer is assumed to find the optimal threshold and henceforth holds it constant for the duration of the study. The learning-curve is expected to plateau during a sufficiently long training interval.
- Data should only be collected in the fixed threshold state, that is, at the end of the training session.
- If a second study is conducted under different conditions, the observer will determine, after a new training session, the optimal threshold for the new conditions and henceforth hold it constant for the duration of the second study, and so on.

From assumption 2, it follows that

$$1 - Sp = FPF = P(Z \geq \zeta \mid T = 1) \tag{3.1}$$

$$Se = TPF = P(Z \geq \zeta \mid T = 2) \tag{3.2}$$

Explanation: $P(Z \geq \zeta \mid T = 1)$ is the probability that the Z-sample for a non-diseased case is greater than or equal to ζ. According to assumption 2, these cases are incorrectly classified as diseased; that is, they are FP decisions and the corresponding probability is false positive fraction (FPF), which is the complement of specificity (Sp). Likewise, $P(Z \geq \zeta \mid T = 2)$ denotes the probability that the Z-sample for a diseased case is greater than or equal to ζ. These cases are correctly classified as diseased; that is, these are TP decisions and the corresponding probability is true positive fraction (TPF), which is sensitivity (Se).

There are several concepts implicit in Equations 3.1 and 3.2.

- The Z-samples have an associated probability distribution; this is implicit in the notation: $P(Z \geq \zeta \mid T = 1)$. *Diseased-cases are not homogenous.* In some cases, disease is easy to detect, perhaps even obvious. In others, the signs of disease are subtler, and, in some, the disease is almost impossible to detect. *Likewise, non-diseased cases are not homogenous.*
- The probability distributions depend on the truth state T. The distribution of the Z-samples for non-diseased cases is, in general, different from that for the diseased cases. Generally, the distribution for $T = 2$ is shifted to the right of that for $T = 1$ (assuming a positive-directed decision variable scale). Later, specific distributional assumptions will be employed to obtain analytic expressions for the right-hand sides of Equations 3.1 and 3.2.
- Equations 3.1 and 3.2 imply that, via choice of the decision threshold ζ, Se and Sp are under the control of the observer. The lower the decision threshold the higher the sensitivity and the lower the specificity; the converses are also true. Ideally, both sensitivity and specificity should be large, i.e., unity (since they are probabilities they cannot exceed unity). *The tradeoff between sensitivity and specificity says, essentially, that there is no free lunch. In general, the price paid for increased sensitivity is decreased specificity and vice versa.*

3.2.3 Adequacy of the training session

Assumption 3: The observer has complete knowledge of the distributions of actually non-diseased and actually diseased cases and makes rational decision based on this knowledge. Knowledge of the *probabilistic distributions* is completely consistent with not knowing for sure which distribution a specific sample came from, that is, the blindedness assumption common to all observer performance studies.

How an observer can be induced to change the decision threshold is the subject of the following two examples.

3.3 Changing the decision threshold: Example I

Suppose that in the first study a radiologist interprets a set of cases subject to the instructions that it is rather important to identify actually diseased cases and not to worry about misdiagnosing actually non-diseased cases. One way to do this would be to reward the radiologist with $10 for each TP decision but only $1 for each TN decision. For simplicity, assume there is no penalty imposed for incorrect decisions (FPs and FNs) and the case set contains equal numbers of non-diseased and diseased cases, and the radiologist is informed of these facts. It is also assumed that the radiologist is allowed to reach a steady state and responds rationally to the payoff arrangement. Under these circumstances, the radiologist is expected to set the decision threshold at a small value so that even slight evidence of *presence* of disease is enough to result in a *case is diseased* decision. The low decision threshold also implies that considerable evidence of *lack* of disease is needed before a *case is non-diseased* decision is rendered. The radiologist is expected to achieve relatively high sensitivity but specificity will be low. As a concrete example, if there are 100 non-diseased cases and 100 diseased cases, assume the radiologist makes 90 TP decisions; since the threshold for presence of disease is small, this number is close to the maximum possible value, namely 100. Assume further that 10 TN decisions are made; since the implied threshold for evidence of absence of disease is large, this number is close to the minimum possible value, namely 0. Therefore, sensitivity is 90% and specificity is 10%. The radiologist earns 90 × $10 + 10 × $1 = $910 for participating in this study.

Next, suppose the study is repeated with the same cases but this time the payoff is $1 for each TP decision and $10 for each TN decision. Suppose, further, that sufficient time has elapsed between the two study sessions so that memory effects can be neglected. Now the roles of sensitivity and specificity are reversed. The radiologist's incentive is to be correct on actually non-diseased cases without worrying too much about missing actually diseased cases. The radiologist is expected to set the decision threshold at a large value so that considerable evidence of disease-presence is required to result in a *case is diseased* decision, but even slight evidence of absence of disease is enough to result in a *case is non-diseased* decision. This radiologist is expected to achieve relatively low sensitivity, but specificity will be higher. Assume the radiologist makes 90 TN decisions and 10 TP decisions, earning $910 for the second study. The corresponding sensitivity is 10% and specificity is 90%. The numbers in this example are summarized in Table 3.1.

Table 3.1 This table illustrates the dependence of the number of counts in a 2 × 2 table on the payoffs

Decision	TP earns $10, TN earns $1		TP earns $1, TN earns $10	
	$T = 1$	$T = 2$	$T = 1$	$T = 2$
$D = 1$	#TN = 10	#FN = 10	#TN = 90	#FN = 90
$D = 2$	#FP = 90	#TP = 90	#FP = 10	#TP = 10
Se, Sp, Payoff	Se = 0.9, Sp = 0.1, Payoff = $910		Se = 0.1, Sp = 0.9, Payoff = $910	

Note: Reversal of the payoff scheme causes the observer to reverse the roles of sensitivity and specificity to achieve the same payoff.

The incentives in the first study caused the radiologist to accept low specificity in order to achieve high sensitivity; the incentives in the second study caused the radiologist to accept low sensitivity in order to achieve high specificity.

3.4 Changing the decision threshold: Example II

Suppose one asks the same radiologist to interpret a set of cases, but this time the reward for a correct decision is always $1, regardless of the truth state of the case, and as before, there are is no penalty for incorrect decisions. However, the radiologist is told that disease prevalence is only 0.005 and that this is the actual prevalence; that is, the experimenter is not deceiving the radiologist in this regard. (Even if the experimenter attempts to deceive the radiologist, by claiming for example that there are roughly equal numbers of non-diseased and diseased cases, after interpreting a few tens of cases the radiologist will *know* that deception is involved. *Deception is generally not a good idea* as the observer's performance is not being measured in a steady state condition. The observer's decision threshold will change as the observer learns the true disease prevalence.) In other words, only five out of every 1000 cases are actually diseased. This information will cause the radiologist to adopt a high threshold for diagnosing disease-present thereby becoming more reluctant to state that the *case is diseased*. By simply diagnosing all cases as non-diseased, without using any case information, the radiologist will be correct on every disease absent case and earn $995, which is very close to the maximum $1000 the radiologist can earn by using case information to the fullest and being correct on disease-present and disease-absent cases.

In screening mammography, the cost of missing a breast cancer, both in terms of loss of life and a possible malpractice suit, is usually perceived to be higher than the cost of a false positive. This results in a shift toward higher sensitivity at the expense of lower specificity.

If a new study were conducted with a highly enriched set of cases, where the disease prevalence is 0.995 (i.e., only 5 out of every 1000 cases are actually non-diseased), then the radiologist would adopt a low threshold. By simply calling every case non-diseased, the radiologist earns $995.

These examples show that by manipulating the relative costs of correct versus incorrect decisions and/or by varying disease prevalence one can influence the radiologist's decision threshold. *These examples apply to laboratory studies.* Clinical interpretations are subject to different cost-benefit considerations that are generally not under the researcher's control: actual (population) disease prevalence, the reputation of the radiologist, malpractice, and so on.

3.5 The equal variance binormal model

Here is the model for the Z-samples. [Upper-case vs. lower case: upper case Z means a random variable, while lower case z means a realized value.] Using the notation $N(\mu, \sigma^2)$ for the normal (or Gaussian) distribution with mean μ and variance σ^2, it is assumed

1. The Z-samples for non-diseased cases are distributed $N(0,1)$.
2. The Z-samples for diseased cases are distributed $N(\mu, 1); \mu \geq 0$.
3. A case is diagnosed as diseased if its Z-sample $\geq \zeta$, representing the threshold parameter, and non-diseased otherwise.

The constraint $\mu \geq 0$ is needed so that the observer's performance is at least as good as chance. A large negative value for this parameter would imply an observer *so predictably bad that the* *observer is good*; one simply reverses the observer's decision (diseased to non-diseased and vice versa) to get near-perfect performance.*

The model described above is termed the *equal variance binormal model*. (If the common variance is not unity, one can rescale the decision axis to achieve unit variance without changing the predictions of the model.) A more general model, termed the *unequal variance binormal model*, is generally used for modeling human observer data, **Chapter 6**, but for the moment, one does not need that complication. The equal variance binormal model is defined by

$$
\left.\begin{array}{c}
Z_{k_t t} \sim N(\mu_t, 1) \\[2mm]
\mu_1 = 0; \; \mu_2 = \mu
\end{array}\right\}
\tag{3.3}
$$

In Equation 3.3 the subscript t denotes the truth, sometimes referred to as the gold standard, with $t = 1$ denoting a non-diseased case and $t = 2$ denoting a diseased case. The variable $Z_{k_t t}$ denotes the random Z-sample for case $k_t t$, where k_t is the index for cases with truth state t; for example $k_1 1 = 21$ denotes the twenty-first non-diseased case and $k_2 2 = 3$ denotes the third diseased case. To explicate $k_1 1 = 21$ further, the label k_1 indexes the case while the label 1 indicates the truth-state of the case. The label k_t ranges from $1, 2, ..., K_t$, where K_t is the total number of cases with disease state t.

The author departs from usual convention, which labels the cases with a single index k, which ranges from 1 to $(K_1 + K_2)$, and one is left guessing as to the truth-state of each case. Also, the proposed notation extends more readily to the FROC paradigm where two states of truth have to be distinguished, one at the case level and one at the location level, **Chapter 13**.

The first line in Equation 3.3 states that $Z_{k_t t}$ is a random sample from the $N(\mu_t, 1)$ distribution, which has unit variance regardless of the value of t (the reason for naming it the equal variance binormal model). The second line in Equation 3.3 defines μ_1 as zero and μ_2 as μ. Taken together, these equations state that non-diseased case Z-samples are distributed $N(0,1)$ and diseased case Z-samples are distributed $N(\mu,1)$. The name *binormal* arises from the *two* normal distributions underlying this model. It should not be confused with *bivariate*, which means a single distribution yielding two values per sample, where the two values could be correlated see Chapter 21. In the binormal model, the samples from the two distributions are assumed independent of each other.

A few facts concerning the normal (or Gaussian) distribution are summarized next.

3.6 The normal distribution

In probability theory, a *probability density function (pdf)*, or density of a continuous random variable, is a function giving the relative chance that the random variable takes on a given value. For a continuous distribution, the probability of the random variable being *exactly* equal to a given value is zero. The probability of the random variable falling in a range of values is given by the integral of this variable's *pdf* function over that range. For the normal distribution $N(\mu, \sigma^2)$ the *pdf* is denoted $\phi(z \mid \mu, \sigma)$ given by

$$
\phi(z \mid \mu, \sigma) = \frac{1}{\sigma\sqrt{2\pi}} e^{\frac{-(z-\mu)^2}{2\sigma^2}}
\tag{3.4}
$$

By definition,

$$
\phi(z \mid \mu, \sigma)\,dz = P\left(z < Z \le z + dz \mid Z \sim N(\mu, \sigma^2)\right)
\tag{3.5}
$$

---

* In the author's teaching experience this example invariably elicits laughter from the audience. It also reminds the author, in the current (Aug. 2016) political context, of a particular prognosticator (Bill Kristol) whose political predictions are so bad that he is considered good, but not in the way he would like it. Figuratively, he is a Kristol-Ball; see for example https://www.youtube.com/watch?v = UmmGHueOpEs.

The right-hand side of Equation 3.5 is the probability that the random variable Z, sampled from $N(\mu,\sigma^2)$, is between the fixed limits z and $z + dz$. For this reason, $\phi(z\,|\,\mu,\sigma)$ is termed the *probability density function*. The special case $N(0,1)$ is referred to as the *unit normal distribution*; it has zero mean and unit variance and the corresponding notation is $\phi(z)$. The defining equation for the *pdf* of this distribution is

$$\phi(z) = \frac{1}{\sqrt{2\pi}}\,e^{\frac{-z^2}{2}} \tag{3.6}$$

The integral of $\phi(z)$ from negative infinity to z is the probability that a sample from the unit normal distribution is less than or equal to z. Regarded as a function of z, this is termed the *cumulative distribution function* (CDF) and is denoted, in this book, by the symbol Φ. The function $\Phi(z)$, specific to the unit normal distribution, is defined by

$$\Phi(z) \equiv P\big(Z \le z \,|\, Z \sim N(0,1)\big) = \int_{-\infty}^{z} \phi(t)\,dt \tag{3.7}$$

Figure 3.1 shows plots, as functions of z, of the CDF and the *pdf* for the unit normal distribution. Since Z-samples outside ± 3 are unlikely, the plotted range, from -3 to $+3$ includes most of the distribution. The *pdf* is the familiar bell-shaped curve, centered at zero; the corresponding **R** function is **dnorm()**, that is, density of the normal distribution. $\Phi(z)$ increases monotonically from 0 to unity as z increases from $-\infty$ to $+\infty$. It is the sigmoid-shaped curve in Figure 3.1; the corresponding **R** function is **pnorm()**. A related function is the inverse of Equation 3.7. Denoting the left-hand side of Equation 3.7 by p, a probability in the range 0–1, that is,

$$p \equiv \Phi(z) = \int_{-\infty}^{z} \phi(t)\,dt \tag{3.8}$$

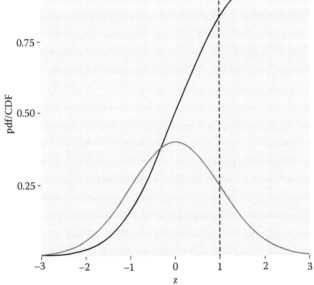

Figure 3.1 The sigmoid shaped curve is the *CDF*, or cumulative distribution function, of the $N(0,1)$ distribution, while the bell-shaped curve is the corresponding *pdf*, or probability density function. The dashed line corresponds to the reporting threshold $\zeta = 1$. The area under the *pdf* to the left of ζ equals the value of *CDF* at the selected ζ, that is, 0.841 (**pnorm(1)** = 0.841). The code for this figure is in **mainUnitNormalPdfCdf.R**.

The inverse of $\Phi(z)$ is that function which when applied to p yields the upper limit z in Equation 3.8, that is,

$$\Phi^{-1}(p) = z \tag{3.9}$$

Since $p \equiv \Phi(z)$ it follows that

$$\Phi^{-1}(\Phi(z)) = z \tag{3.10}$$

This nicely satisfies the property of an inverse function. The inverse function is known in statistical terminology as the *quantile function*, implemented in **R** as the **qnorm()** function. Think of **pnorm()** as a probability and **qnorm()** as a value on the z-axis.

To summarize, **norm** implies the unit normal distribution, **p** denotes a probability distribution function* or *CDF*, **q** denotes a quantile function, and **d** denotes a density function. This convention is used with all distributions in **R**; there is method to the madness.

Open the **software.prj** file corresponding to this chapter. The following **Console** window code snippet demonstrates usage of the functions described above (At the risk of repetition, the > prompt is **R** waiting for a command; the code following the prompt is the command. **R** output is anything not preceded by a > prompt; the reader should type the lines following the > prompts into the **RStudio** Console window and hit Enter, to obtain the listed output):

3.6.1 Code snippet

```
> qnorm(0.025)
[1] -1.959964
> qnorm(1-0.025)
[1] 1.959964
> pnorm(qnorm(0.025))
[1] 0.025
> qnorm(pnorm(-1.96))
[1] -1.96
```

Multiple **R** commands can be placed on a line as long as they are separated by semi-colons and **R** does not need a special continuation character. The first command **qnorm(0.025)** demonstrates the identity:

$$\Phi^{-1}(0.025) = -1.959964 \tag{3.11}$$

qnorm(1-0.025) demonstrates the identity:

$$\Phi^{-1}(1-0.025) = 1.959964 \tag{3.12}$$

---

* In the statistical literature, this is also referred to as the probability distribution function, which unfortunately has the same abbreviation as the probability density function, excepting for a change is case; for clarity, the author will always refer to it as the cumulative distribution function (*CDF*).

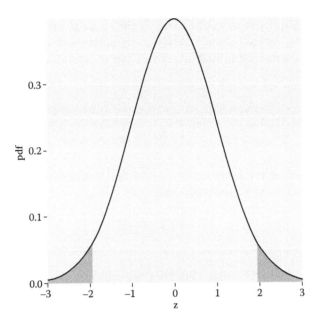

Figure 3.2 This plot illustrates the fact that 95% of the total area under the unit normal *pdf* is contained in the range $|Z| < 1.96$, which can be used to construct a 95% confidence interval for an estimate of a suitably normalized statistic. The area contained in each shaded tail is 2.5%. This figure was generated by sourcing the code in **mainShadedTails.R**.

Equation 3.11 means that the (rounded) value −1.96 is such that the area under the *pdf* to the left of this value is 0.025. Similarly, Equation 3.12 means that the (rounded) value +1.96 is such that the area under the *pdf* to the left of this value is $1 - 0.025 = 0.975$. In other words, −1.96 captures, to its left, the 2.5th percentile of the unit normal distribution, and 1.96 captures, to its left, the 97.5th percentile of the unit normal distribution, Figure 3.2. Since between them they capture 95% of the unit normal *pdf*, these two values can be used to estimate 95% confidence intervals.

> If one knows that a variable is distributed as a unit normal random variable, then the observed value minus 1.96 defines the lower limit of its 95% confidence interval, and the observed value plus 1.96 defines the upper limit of its 95% confidence interval.

The last two commands in Section 3.6.1 demonstrate that **pnorm()** and **qnorm()**, applied in either order, are inverses of each other.

3.6.2 Analytic expressions for specificity and sensitivity

According to assumption #2 in Section 3.3.2, specificity corresponding to threshold ζ is the probability that a Z-sample from a non-diseased case is smaller than ζ. By definition, this is the *CDF* corresponding to the threshold ζ. In other words,

$$Sp(\zeta) = P\left(Z_{k_1 1} < \zeta \mid Z_{k_1 1} \sim N(0,1)\right) = \Phi(\zeta) \tag{3.13}$$

The expression for sensitivity can be derived tediously by starting with the fact that $Z_{k_2 2} \sim N(\mu, 1)$ and using calculus to obtain the probability that a Z-sample for a diseased case equals or exceeds ζ. A quicker way is to consider the random variable obtaining by shifting the origin to μ. A little thought should convince the reader that $Z_{k_2 2} - \mu$ must be distributed as $N(0,1)$. Therefore, the desired probability is (the last step follows from Equation 3.7, with z replaced by $\zeta - \mu$):

$$
\left.
\begin{aligned}
Se(\zeta) &= P\big(Z_{k_2 2} \geq \zeta\big) = P\big((Z_{k_2 2} - \mu) \geq (\zeta - \mu)\big) \\
&= 1 - P\big((Z_{k_2 2} - \mu) < (\zeta - \mu)\big) = 1 - \Phi(\zeta - \mu)
\end{aligned}
\right\}
\tag{3.14}
$$

A little thought (based on the definition of the *CDF* function and the symmetry of the unit normal *pdf* function) should convince the reader that:

$$
1 - \Phi(\zeta) = \Phi(-\zeta)
\tag{3.15}
$$

$$
1 - \Phi(\zeta - \mu) = \Phi(\mu - \zeta)
\tag{3.16}
$$

Instead of carrying the "1 minus" around, one can use more compact notation. Summarizing, the analytical formula for the specificity and sensitivity for the equal variance binormal model are

$$
Sp(\zeta) = \Phi(\zeta)
\tag{3.17}
$$

$$
Se(\zeta) = \Phi(\mu - \zeta)
\tag{3.18}
$$

In these equations, the threshold ζ appears with different signs because specificity is the area under a *pdf* to the *left* of a threshold, while sensitivity is the area to the *right*.

As probabilities, both sensitivity and specificity are restricted to the range 0 to 1. The observer's performance could be characterized by specifying sensitivity *and* specificity, that is, a *pair* of numbers. If both sensitivity and specificity of an imaging system are greater than the corresponding values for another system, then the first system is unambiguously better than the second. But what if sensitivity is greater for the first but specificity is greater for the second? Now the comparison is ambiguous. It is difficult to unambiguously compare two pairs of performance indices. Clearly, a scalar measure is desirable that combines sensitivity and specificity into a single measure of diagnostic performance.

The parameter μ satisfies the requirements of a scalar figure of merit (FOM). Equations 3.17 and 3.18 can be solved for μ as follows. Inverting the equations yields

$$
\zeta = \Phi^{-1}(Sp)
\tag{3.19}
$$

$$
\mu - \zeta = \Phi^{-1}(Se)
\tag{3.20}
$$

Eliminating ζ yields

$$
\mu = \Phi^{-1}(Sp) + \Phi^{-1}(Se)
\tag{3.21}
$$

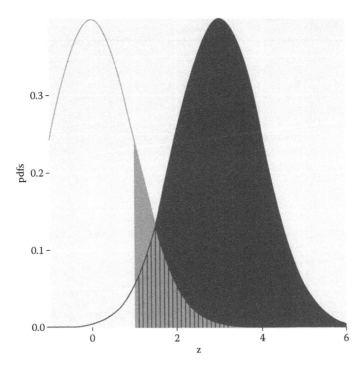

Figure 3.3 The equal variance binormal model for $\mu = 3$ and $\zeta = 1$; the light curve, centered at zero, corresponds to the *pdf* of non-diseased cases and the dark one, centered at $\mu = 3$, corresponds to the *pdf* of diseased cases. The left edge of the light shaded region represents the threshold ζ, set at unity. The dark shaded area, including the common portion with the vertical lines, is sensitivity. The light shaded area including the common portion with the vertical lines is 1-specificity. This figure was generated using **mainShadedPlots.R**, which generates color plots, with "light" replaced by "blue" and "dark" replaced by "red."

Eqn. 3.21 is a useful relation as it converts a *pair* of numbers that is hard to compare between two modalities, in the sense described above, into a *single* FOM. Now it is almost trivial to compare two modalities: the one with the higher μ wins. In reality, the comparison is not trivial since like sensitivity and specificity, μ has to be estimated from a finite dataset and is therefore subject to sampling variability, accounting for which is the subject of Part B of the book.

Figure 3.3 shows the equal variance binormal model for $\mu = 3$ and $\zeta = 1$. The light shaded area, including the common portion with the vertical dark lines, is the probability that a Z-sample from a non-diseased case *exceeds* ζ, which is the *complement* of specificity, i.e., it is false positive fraction, which is $1 - 0.841 = 0.159$. The 0.841 comes from **pnorm(1)** $= 0.841$. The dark shaded area, including the common portion with the vertical dark lines, is the probability that a Z-sample from a diseased case *exceeds* ζ, which is sensitivity or true positive fraction, which is $\Phi(3-1) = \Phi(2) = 0.977$, because **pnorm(2)** $= 0.977$.

Demonstrated next are these concepts using **R** examples.

3.7 Demonstration of the concepts of sensitivity and specificity

The code for this demonstration is **mainBinaryRatings.R** explained in Online Appendix 3.A. **Source** the file. The following output appears in the **Console** window.

3.7.1 Code output

```
> source(...)
seed =  100
K1 =  9
K2 =  11
Specificity =  0.8888889
Sensitivity =  0.9090909
Estimated mu =  2.555818
```

The estimate of μ (2.56) was calculated using Equation 3.21, that is, `mu <- qnorm(Sp) + qnorm(Se)`.

3.7.2 Changing the seed variable: Case-sampling variability

No matter how many times one clicks the **Source** button one always sees the same output shown in Section 3.7.1. This is because at line 5 one sets the **seed** of the random number generator to a fixed value, namely 100. This is like having a perfectly reproducible reader repeatedly interpret the same cases—one always gets the same results. Change **seed** to 101 and click on the **Source** button. One sees:

3.7.2.1 Code output

```
> source(...)
seed =  101
K1 =  9
K2 =  11
Specificity =  0.7777778
Sensitivity =  0.5454545
Estimated mu =  0.878895
```

Changing **seed** is equivalent to sampling a completely new set of patients. *This is an example of case sampling variability.* The effect is quite large (**Se** fell from 0.909 to 0.545 and estimated **mu** fell from 2.56 to 0.88!) because the size of the relevant case set, 11 for sensitivity, is small, leading to large sampling variability.

3.7.3 Increasing the numbers of cases

Increase K_1 and K_2, by a factor of 10 each, and return **seed** to 100. Clicking **Source** yields:

3.7.3.1 Code output

```
> source(...)
seed =   100
K1 =   90
K2 =   110
Specificity =   0.7777778
Sensitivity =   0.8363636
Estimated mu =   1.744332
```

Change **seed** to 101 and click **Source**:

3.7.3.2 Code output

```
> source(...)
seed =   101
K1 =   90
K2 =   110
Specificity =   0.8111111
Sensitivity =   0.7545455
Estimated mu =   1.570862
```

Notice that with increasing sample size that the values are less sensitive to **seed**. Table 3.2 illustrates this trend with ever increasing sample sizes (the reader should confirm the listed values).

As the numbers of cases increase, sensitivity and specificity converge to a common value, around 0.773, and the estimate of the separation parameter converges to the known value. Typing

Table 3.2 Effect of sample size on case-sampling variability of estimates of sensitivity, specificity, and the separation parameter

K_1	K_2	seed	$\left(\widehat{Se}, \widehat{Sp}\right)$	$\hat{\mu}$
9	11	100	(0.889, 0.909)	2.56
		101	(0.778, 0.545)	0.879
90	110	100	(0.778, 0.836)	1.74
		101	(0.811, 0.755)	1.57
900	1100	100	(0.764, 0.761)	1.43
		101	(0.807, 0.759)	1.57
9000	11000	100	(0.774, 0.772)	1.5
		101	(0.771, 0.775)	1.5
∞	∞	NA	(0.773, 0.773)	1.5

$\widehat{Se}$ = estimate of sensitivity, $\widehat{Sp}$ = estimate of specificity, $\hat{\mu}$ = estimate of separation parameter; K_1 and K_2 are the numbers of non-diseased and diseased cases, respectively. Different values of seed generate different case samples. The parameters of the model are $\mu = 1.5$ and $\zeta = \mu/2$. The values for infinite numbers of cases are from the analytical expressions Equations 3.17 and 3.18. NA = not applicable.

`pnorm(0.75)` in the **Console** window yields (the hash-tag symbol comments code; anything to the right of the symbol is ignored by **R**; comments are generally inserted to document code):

3.7.3.3 Code snippet

```
> # example 1
> pnorm(0.75)
[1] 0.7733726
> # example 2
> 2*qnorm(pnorm(zeta))
[1] 1.5
```

Because the threshold is halfway between the two distributions, in this example sensitivity and specificity are identical. In words, with two unit variance distributions separated by 1.5, the area under the diseased distribution (centered at 1.5) *above* 0.75, namely sensitivity, equals the area under the non-diseased distribution (centered at zero) *below* 0.75, namely specificity, and the common value is $\Phi(0.75) = 0.773$, yielding the last row of Table 3.2, and example 1 in the above code snippet. Example 2 in the above code snippet illustrates Equation 3.21: the factor of two arises since in this example sensitivity and specificity are identical.

From Table 3.2, for the same numbers of cases but different seeds, comparing pairs of sensitivity and specificity values is more difficult as four numbers are involved. Comparing $\hat{\mu}$ values is easier, as only two numbers are involved. The tendency of asymptotically becoming independent of case sample is discernible with fewer cases with $\hat{\mu}$, around 90/110 cases, than with sensitivity and specificity pairs. The numbers in the table might appear disheartening in terms of the implied numbers of cases needed to detect a difference in specificity. Even with 200 cases, the difference in specificity for two seed values is 0.081, which is actually a large effect considering that the scale extends from 0 to 1.0. A similar comment applies to differences in sensitivity. The situation is not that bad. One uses an area measure that combines sensitivity and specificity yielding less variability. One uses the ratings paradigm, which is more efficient than the binary one in this chapter. Finally, one takes advantage of correlations that exist between the interpretations in matched-case matched-reader interpretations in two modalities that tend to decrease variability in the AUC-difference even further (most applications of ROC methods involved detecting *differences* in AUCs).

3.8 Inverse variation of sensitivity and specificity and the need for a single FOM

The inverse variation of sensitivity and specificity is modeled in the binormal model by the threshold parameter ζ. From Equation 3.17, specificity at threshold ζ is $\Phi(\zeta)$ and the corresponding expression for sensitivity is $\Phi(\mu - \zeta)$. Since the threshold ζ appears with a *minus* sign, the dependence of sensitivity on ζ will be the *opposite* of the corresponding dependence of specificity. In Figure 3.3, the left edge of the light shaded region represents threshold ζ, set at unity. As ζ is moved towards the left, specificity decreases but sensitivity increases. Specificity decreases because less of the non-diseased distribution lies to the left of the new threshold. In other words, fewer non-diseased cases are correctly diagnosed as non-diseased. Sensitivity increases because more of the diseased distribution lies to the right of the new threshold, in other words more diseased cases are correctly diagnosed as diseased. If an observer has higher sensitivity than another observer, but lower specificity, it is difficult to unambiguously compare them. It is not impossible.[3,4] The unambiguous comparison is difficult for the following reason. Assuming the second observer can be coaxed into adopting a lower threshold, thereby decreasing specificity to match that of the first observer, then it is possible that the second observer's sensitivity, formerly smaller, could now be greater than that of the first observer. A single figure of merit is desirable as compared to sensitivity—specificity analysis. It is possible to leverage the inverse variation of sensitivity and specificity by combing them into a single scalar measure, as was done with the μ parameter in the previous section, Equation 3.21. An equivalent way is by using the area under the ROC plot, discussed next.

3.9 The ROC curve

The receiver operating characteristic (ROC) is defined as the plot of *sensitivity* (*y*-axis) versus 1-*specificity* (*x*-axis). Equivalently, it is the plot of *TPF* (*y*-axis) versus *FPF* (*x*-axis). From Equations 3.14, 3.17, and 3.18 it follows that

$$FPF(\zeta) = 1 - Sp(\zeta) = 1 - \Phi(\zeta) = \Phi(-\zeta) \tag{3.22}$$

$$TPF(\zeta) = Se(\zeta) = \Phi(\mu - \zeta) \tag{3.23}$$

Specifying ζ selects a particular *operating point* on this plot and varying ζ from $+\infty$ to $-\infty$ causes the operating point to trace out the ROC *curve from the origin to (1,1)*. Specifically, as ζ is *decreased* from $+\infty$ to $-\infty$, the operating point *rises* from the *origin (0,0)* to *the end-point (1,1)*. In general, as ζ *increases* the operating point moves *down* the curve, and conversely, as ζ decreases the operating point moves up the curve. The operating point $O(\zeta|\mu)$ for the equal variance binormal model is (the notation assumes the μ parameter is fixed and ζ is varied by the observer in response to interpretation conditions):

$$O(\zeta|\mu) = (\Phi(-\zeta), \Phi(\mu - \zeta)) \tag{3.24}$$

The operating point predicted by the above equation lies *exactly* on the theoretical ROC curve. This condition can only be achieved with very large numbers of cases, so that sampling variability is very small. In practice, with finite datasets, the operating point will almost never be exactly on the theoretical curve.

The ROC curve is the locus of the operating point for fixed μ and variable ζ. Figure 3.4 shows examples of equal variance binormal model ROC curves for different values of μ. Each curve is labeled with the corresponding value of μ. Each has the property that TPF is a monotonically increasing function of FPF and the slope of each ROC curve decreases monotonically as the operating point moves up the curve. As μ increases the curves get progressively *upward-left shifted*, approaching the top-left corner of the ROC plot. In the limit $\mu \rightarrow \infty$ the curve degenerates into two line segments, a vertical one connecting the origin to (0,1) and a horizontal one connecting (0,1) to (1,1)—the ROC plot for a perfect observer.

3.9.1 The chance diagonal

In Figure 3.4 the ROC curve for $\mu = 0$ is the *positive diagonal* of the ROC plot, termed the *chance diagonal*. Along this curve TPF = FPF and the observer's performance is at *chance level*. In the equal variance binormal model, for $\mu = 0$, the *pdf* of the diseased distribution is identical to that of the non-diseased distribution: both are centered at the origin. Therefore, no matter the choice of threshold ζ, TPF = FPF. Setting $\mu = 0$ in Equations 3.22 and 3.23 yields

$$TPF(\zeta) = FPF(\zeta) = \Phi(-\zeta) \tag{3.25}$$

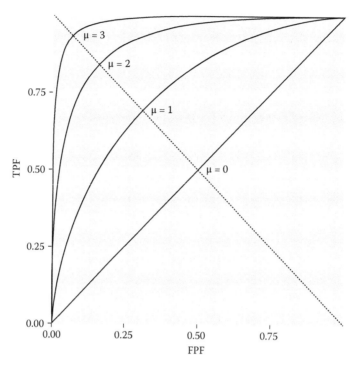

Figure 3.4 ROC plots predicted by the equal variance binormal model for different values of μ. As μ increases the intersection of the curve with the negative diagonal moves closer to the ideal operating point, (0,1) at which sensitivity and specificity are both equal to unity. At the intersection sensitivity equals specificity and $\zeta = \mu/2$; see Section 3.9.2. [The curves were generated using **mainRocCurve EqualVarianceModel.R.**]

In this special case, the dark and light curves in Figure 3.3 coincide. The observer is unable to find any difference between the two distributions. This can happen if the cancers are of such low visibility that diseased cases are indistinguishable from non-diseased ones, or the observer's skill level is so poor that the observer is unable to make use of distinguishing characteristics between diseased and non-diseased cases that do exist, and which experts exploit.

3.9.1.1 The guessing observer

If the cases are indeed impossibly difficult and/or the observer has zero skill at distinguishing between them, the observer has no option but to guess. This rarely happens in the clinic, as too much is at stake, and this paragraph is intended to make the pedagogical point that the observer can move the operating point along the change diagonal. If there is no special incentive, the observer tosses a coin and if the coin lands head up, the observer states that the *case is diseased* and otherwise, states the *case is non-diseased*. When this procedure is averaged over many non-diseased and diseased cases, it will result in the operating point (0.5, 0.5). (Many cases are assumed as otherwise, due to sampling variability, the operating point will not be on the theoretical ROC curve.) To move the operating point downward, for example, to (0.1, 0.1) the observer randomly selects an integer number between 1 and 10, equivalent to a 10-sided coin. Whenever a *one* shows up, the observer states *case is diseased* and otherwise, the observer states *case is non-diseased*. To move the operating point to (0.2, 0.2) whenever a *one or two* shows up, the observer states *case is diseased* and otherwise, the observer states *case is non-diseased*. One can appreciate that simply by changing the probability of stating *case is diseased,* the observer can place the operating point anywhere on the chance diagonal, but wherever the operating point is placed, it will satisfy *TPF = FPF.*

3.9.2 Symmetry with respect to negative diagonal

A characteristic of the ROC curves shown in Figure 3.4 is that they are symmetric with respect to the *negative diagonal*, defined as the straight line joining (0,1) and (1,0) which is shown as the dotted straight line in Figure 3.4. The symmetry property is due to the equal variance nature of the binormal model and is not true for models considered in later chapters. The intersection between the ROC curve and the negative diagonal corresponds to $\zeta = \mu / 2$, in which case the operating point is

$$
\left.\begin{array}{c}
FPF(\zeta) = \Phi\left(-\dfrac{\mu}{2}\right) \\[3mm]
TPF(\zeta) = \Phi\left(\dfrac{\mu}{2}\right)
\end{array}\right\}
\tag{3.26}
$$

The first equation implies

$$
1 - FPF(\zeta) = 1 - \Phi\left(-\frac{\mu}{2}\right) = \Phi\left(\frac{\mu}{2}\right)
\tag{3.27}
$$

Therefore

$$
TPF(\zeta) = 1 - FPF(\zeta)
\tag{3.28}
$$

This equation describes a straight line with unit intercept and slope equal to minus 1, which is the negative diagonal. Since *TPF* = sensitivity and *FPF* = 1-specificity, another way of stating this is that at the intersection with the negative diagonal, sensitivity equals specificity.

3.9.3 Area under the ROC curve

The AUC (abbreviation for *area under curve*) under the ROC curve suggests itself as a measure of performance that is independent of threshold and therefore circumvents the ambiguity issue of comparing sensitivity/specificity pairs, and has other advantages. It is defined by the following integrals:

$$
A_{z;\sigma=1} = \int_0^1 TPF(\zeta) d\left(FPF(\zeta)\right) = \int_0^1 FPF(\zeta) d\left(TPF(\zeta)\right)
\tag{3.29}
$$

Equation 3.29 has the following equivalent interpretations:

1. The first form performs the integration using thin vertical strips, for example, extending from x to $x + dx$, where, for convenience, x is a temporary symbol for FPF. The area can be interpreted as the average *TPF* over all possible values of *FPF*.
2. The second equivalent form performs the integration using thin horizontal strips, for example, extending from y to $y + dy$, where, for convenience, y is a temporary symbol for *TPF*. The area can be interpreted as the average *FPF* over all possible values of *TPF*.

By convention, the symbol A_z is used for the area under the *binormal model predicted* ROC curve. In Equation 3.29, the subscript $\sigma = 1$ is necessary to distinguish it from one corresponding to the unequal variance binormal model to be derived in **Chapter 6**. It can be shown that (the proof is in **Chapter 6**):

$$
A_{z;\sigma=1} = \Phi\left(\frac{\mu}{\sqrt{2}}\right)
\tag{3.30}
$$

Since the ROC curve is bounded by the unit square, AUC must be between zero and one. If μ is non-negative, the area under the ROC curve must be between 0.5 and 1. The chance diagonal, corresponding to $\mu = 0$, yields $A_{z;\sigma=1} = 0.5$, while the perfect ROC curve, corresponding to infinite μ, yields unit area. Since it is a scalar quantity, AUC can be used to less-ambiguously quantify performance in the ROC task than is possible using sensitivity and specificity pairs.

3.9.4 Properties of the equal variance binormal model ROC curve

1. The ROC curve is completely contained within the unit square. This follows from the fact that both axes of the plot are probabilities.
2. The operating point rises monotonically from (0,0) to (1,1).
3. Since μ is positive, the slope of the equal variance binormal model curve at the origin (0,0) is infinite and the slope at (1,1) is zero. The slope along the curve is always non-negative and decreases monotonically as the operating point moves up the curve.
4. AUC is a monotone increasing function of μ. It varies from 0.5 to 1 as μ varies from zero to infinity.

3.9.5 Comments

Property (2): since the operating point coordinates can both be expressed in terms of Φ functions, which are monotone in their arguments, and in each case the argument ζ appears with a negative sign, it follows that as ζ is lowered both *TPF* and *FPF* increase. In other words, the operating point corresponding to $\zeta - d\zeta$ is to the upper right of that corresponding to ζ (assuming $d\zeta > 0$).

Property (3): The slope of the ROC curve can be derived by differentiation with respect to **zeta**, and some algebra:

$$\frac{d(TPF)}{d(FPF)} = \frac{d(\Phi(\mu-\zeta))}{d(\Phi(-\zeta))} = \frac{\phi(\mu-\zeta)}{\phi(-\zeta)} = e^{\mu(\zeta-\mu/2)} \geq 0 \tag{3.31}$$

The above derivation uses the fact that the differential of the *CDF* function yields the *pdf* function, that is,

$$d\Phi(\zeta) = P(\zeta < Z < \zeta + d\zeta) = \phi(\zeta)d\zeta \tag{3.32}$$

Since the slope of the ROC curve can be expressed as a power of e, it is always non-negative. Provided $\mu > 0$, then in the limit $\zeta \to \infty$, the slope at the origin approaches ∞. Equation 3.31 also implies that in the limit $\zeta \to -\infty$ the slope of the ROC curve at the end-point (1,1) approaches zero. For constant $\mu > 0$, the slope is a monotone increasing function of ζ. As ζ *decreases* from $+\infty$ to $-\infty$, the slope decreases monotonically from $+\infty$ to 0.

Figure 3.5 is the ROC curve for the equal variance binormal model for $\mu = 3$. The entire curve is defined by μ. Specifying a particular value of ζ corresponds to specifying a particular point on the ROC curve. In Figure 3.5 the circle corresponds to the operating point (0.159, 0.977) defined by $\zeta = 1$; `pnorm(-1)` = 0.159; `pnorm(3-1)` = 0.977. The operating point lies *exactly* on the curve, as this is a *predicted* operating point.

3.9.6 Physical interpretation of μ

As a historical note, μ is equivalent[2] to a signal detection theory variable denoted d' in the literature (pronounced *dee-prime*). It can be thought of as the *perceptual* signal-to-noise ratio (*pSNR*) of diseased cases relative to non-diseased ones. It is a measure of reader expertise and/or ease of

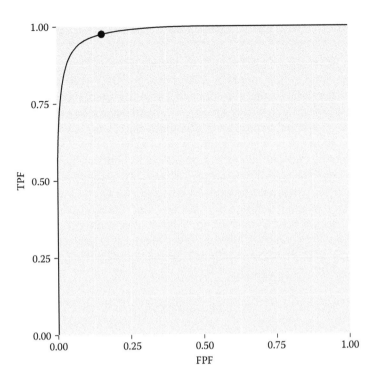

Figure 3.5 ROC curve predicted by equal variance binormal model for $\mu = 3$. The circled operating point corresponds to $\zeta = 1$. The operating point falls exactly on the curve, as these are analytical results. Due to sampling variability, with finite numbers of cases, this is not observed in practice. The code for this plot is in file **mainAnalyticalROC.R**.

detectability of the disease. SNR is a term widely used in engineering, specifically in signal detection theory.[1,5] It dates to the early 1940s when one had the problem[6] of detecting faint radar reflections from a plane against a background of noise. The reader may be aware of the rule-of-thumb that if SNR exceeds three the target is likely to be detected. It will be shown later that the area under the ROC curve is the probability that a diseased case Z-sample is greater than that of a non-diseased one, **Chapter 5**. It is also the probability that a correct choice will be made if a diseased and non-diseased case are shown simultaneously and the observer is asked to pick the diseased case, which is the 2 alternative forced choice (2AFC) paradigm.[1,7] The following code snippet shows that for $\mu = 3$, the probability of detection is 98.3%.

3.9.6.1 Code snippet

```
> pnorm(3/sqrt(2))
[1] 0.9830526
```

For electrical signals, SNR can be measured with instruments but, in the current context of decisions, the *perceptual* SNR is being measured. Physical characteristics that differentiate non-diseased from diseased cases, and how well they are displayed will affect it. In addition, the eye-sight of the observer is an obvious factor; not so obvious is how information is processed by the cognitive system, and the role of the observer's experience in making similar decisions (i.e., expertise). [In the code output shown in Section 3.9.6.1, the output value is preceded by [1]; this means the output is a vector with a single value and the first element is the listed number; other elements of the vector, if present, would have been listed in the same line.]

3.10 Assigning confidence intervals to an operating point

A $(1-\alpha)$ confidence interval (CI) of a statistic is that range* expected to contain the true value of the statistic with probability $(1-\alpha)$. It should be clear that a 99% CI is wider than a 95% CI, and a 90% CI is narrower; in general, the higher the confidence that the interval contains the true value, the wider the range of the CI. Calculation of a *parametric* confidence interval requires a distributional assumption (*non-parametric* estimation methods, which use resampling methods, are described in **Chapter 7**). With a distributional assumption, the method being described now, the parameters of the distribution are estimated, and since the distribution accounts for variability, the needed confidence interval estimate follows. With quantities *TPF* and *FPF*, each of which involves a ratio of two integers, it is convenient to assume a binomial distribution for the following reason: the diagnosis non-diseased versus diseased is a Bernoulli trial, i.e., one whose outcome is binary. A Bernoulli trial is like a coin-toss, a special coin whose probability of landing diseased face up is p, which is not necessarily 0.5 as with a real coin. The limits on p are $0 \le p \le 1$. It is a theorem in statistics[8] that the total number of Bernoulli outcomes of one type, for example, # *FP*, is a binomial-distributed random variable, with success probability $\widehat{FPF}$ and trial size K_1.

$$\# FP \sim B\left(K_1, \widehat{FPF}\right) \tag{3.33}$$

In Equation 3.33, $B(n, p)$ denotes the binomial distribution[8] with success probability p and trial size n:

$$\left.\begin{array}{l} k \sim B(n, p) \\ k = 0, 1, 2, ..., n \end{array}\right\} \tag{3.34}$$

Equation 3.34 states that k is a random sample from the binomial distribution $B(n, p)$. For reference, the *probability mass function (pmf)* of $B(n, p)$ is defined by (the subscript denotes a binomial distribution):

$$pmf_{Bin}(k; n, p) = \binom{n}{k} p^k (1-p)^{n-k} \tag{3.35}$$

For a discrete distribution, one has probability *mass* function; in contrast, for a continuous distribution one has a probability *density* function.

The binomial coefficient $\binom{n}{k}$ appearing in Equation 3.35, to be read as *n pick k*, is defined by

$$\binom{n}{k} = \frac{n!}{k!(n-k)!} \tag{3.36}$$

For large n the binomial distribution $B(n, p)$ asymptotically approaches[8] a normal distribution with mean np and variance $np(1-p)$:

$$B(n, p) \rightarrow N\left(np, np(1-p)\right) \tag{3.37}$$

---

* Since the observed value is a realization of a random variable, the 95% confidence interval is also a random range variable. If the trial were repeated many times, the true value would be included in 95% of the confidence intervals. An individual estimated range is not guaranteed to contain the true value with 95% probability.

Replacing n with K_1 and p with $\widehat{FPF}$, the binomial distribution of $\# FP$ asymptotically approaches a normal distribution with mean $K_1 \widehat{FPF}$ and variance $K_1 \widehat{FPF}(1 - \widehat{FPF})$:

$$\# FP \sim N\left(K_1 \widehat{FPF}, K_1 \widehat{FPF}\left(1 - \widehat{FPF}\right)\right) \tag{3.38}$$

Equation 3.38 implies that for large K_1, FPF follows the normal distribution (the variance has to be divided by the square of the number of non-diseased cases):

$$FPF \sim N\left(\widehat{FPF}, \widehat{FPF}\left(1 - \widehat{FPF}\right) / K_1\right) \tag{3.39}$$

It follows that

$$\left.\begin{aligned} \widehat{\sigma^2_{FPF}} &= \widehat{FPF}\left(1 - \widehat{FPF}\right) / K_1 \\ FPF &\sim N\left(\widehat{FPF}, \widehat{\sigma^2_{FPF}}\right) \end{aligned}\right\} \tag{3.40}$$

Translating the mean to zero and dividing by the square root of the variance, it follows that

$$\frac{FPF - \widehat{FPF},}{\widehat{\sigma}_{FPF}} \sim N(0,1) \tag{3.41}$$

In practice, the normal approximation is adequate if *both* of the following two conditions are *both* met (i.e., $\widehat{FPF}$ is not too close to zero or 1):

$$K_1 \widehat{FPF} > 10 \quad \& \quad K_1\left(1 - \widehat{FPF}\right) > 10 \tag{3.42}$$

From the properties of the normal distribution, it follows that an approximate symmetric $(1 - \alpha) \times 100\%$ confidence interval for FPF is (with this definition a 95% confidence interval corresponds to choosing $\alpha = 0.05$):

$$CI^{FPF}_{1-\alpha} = \left(\widehat{FPF} - z_{\alpha/2} \widehat{\sigma}_{FPF}, \ \widehat{FPF} + z_{\alpha/2} \widehat{\sigma}_{FPF}\right) \tag{3.43}$$

In Equation 3.43 $z_{\alpha/2}$ is the *upper* $\alpha / 2$ quantile of the unit normal distribution, that is, the area to the *right* under the unit normal distribution *pdf* from $z_{\alpha/2}$ to infinity equals $\alpha / 2$. It is the complement of the Φ^{-1} introduced earlier; the difference is that the latter uses the area to the *left*:

$$\left.\begin{aligned} z_{\alpha/2} &= \Phi^{-1}\left(1 - \alpha / 2\right) \\ \alpha / 2 &= \int_{z_{\alpha/2}}^{\infty} \phi(z)dz = 1 - \Phi(z_{\alpha/2}) \end{aligned}\right\} \tag{3.44}$$

The reader should be convinced that the two equations in Equation 3.44 are consistent. Similarly, an approximate symmetric $(1 - \alpha) \times 100\%$ confidence interval for TPF is

$$CI^{TPF}_{1-\alpha} = \left(\widehat{TPF} - z_{\alpha/2} \widehat{\sigma}_{TPF}, \ \widehat{TPF} + z_{\alpha/2} \widehat{\sigma}_{TPF}\right) \tag{3.45}$$

In Equation 3.45,

$$\widehat{\sigma^2_{TPF}} \equiv \frac{1}{K_2} \widehat{TPF}\left(1 - \widehat{TPF}\right) \tag{3.46}$$

The confidence intervals are largest when the probabilities (FPF or TPF) are close to 0.5 and decrease inversely as the square root of the relevant number of cases. The symmetric binomial distribution based estimates can stray outside the allowed range (0–1). Exact confidence intervals[9] that are asymmetric around the central value and which are guaranteed to be in the allowed range can be calculated: it is implemented in **R** in function **binom.test()** and used in **mainConfidenceIntervals.R** in Online Appendix 3.B. Ensure that **mu** is set to 1.5 at line 9. **Source** the code to get the following code output:

3.10.1 Code output

```
> source(...)
alpha =  0.05
K1 =  99
K2 =  111
mu =  1.5
zeta =  0.75
Specificity =  0.777778
Sensitivity =  0.846847
approx 95% CI on Sp =  0.695884 0.859672
Exact 95% CI on Sp =  0.683109 0.855188
approx 95% CI on Se =  0.77985 0.913843
Exact 95% CI on Sp =  0.766148 0.908177
```

The exact and approximate confidence intervals are close to each other. Now change **mu** to five, thereby assuring that both sensitivity and specificity will be close to unity; **source** the code to get the following code output:

3.10.2 Code output

```
> source(...)
alpha =  0.05
K1 =  99
K2 =  111
mu =  5
zeta =  2.5
Specificity =  0.989899
Sensitivity =  0.990991
approx 95% CI on Sp =  0.970202 1.0096
Exact 95% CI on Sp =  0.945003 0.999744
approx 95% CI on Se =  0.973413 1.00857
Exact 95% CI on Sp =  0.950827 0.999772
```

The approximate confidence interval is clearly incorrect as it extends beyond unity. The exact confidence in calculated by avoiding the normal approximation, Equation 3.37. Instead, one numerically calculates a confidence interval $\left(CI_{1-\alpha}^{lower}, CI_{1-\alpha}^{upper}\right)$ such that at most $\alpha/2$ of the binomial distribution *pmf* is below the lower limit $CI_{1-\alpha}^{lower}$ and at most $\alpha/2$ of the binomial distribution *pmf* is above the upper limit $CI_{1-\alpha}^{upper}$.

3.10.3 Exercises

The preceding demonstration should give the reader an idea of the power of **R**, and that one does not have to be a statistician or an expert programmer to benefit from it. Experiment with different values of parameters (e.g., try reducing the numbers of cases, or change the seed), and/or run the code in debug mode, until it makes sense. Change α appropriately to be convinced that a 99% CI is wider than a 95% CI, and a 90% CI is narrower. Finally, as a greater challenge, repeat the code without initializing seed, so that the samples are independent, and confirm that the 95% confidence intervals indeed include the correct analytical value with probability 95%.

3.11 Variability in sensitivity and specificity: The Beam et al. study

In this study[10] fifty accredited mammography centers were randomly sampled in the United States. *Accredited* is a legal/regulatory term implying, among other things, that the radiologists interpreting the breast cases were board-certified by the American Board of Radiology (ABR).[11,12] One hundred eight (108) ABR-certified radiologists from these centers gave blinded interpretations to a common set of 79 randomly selected enriched screening cases containing 45 cases with cancer and the rest normal or with benign lesions. Ground truth for these women had been established either by biopsy or by a two-year follow-up (establishing truth is often the most time-consuming part of conducting a ROC study). The observed range of sensitivity (TPF) was 53% and the range of FPF was 63%; the corresponding range for AUC was 21%, Table 3.3.

In Figure 3.6, a schematic of the data, if one looks at the points labeled (B) and (C) one can mentally construct a smooth ROC curve that starts at (0,0), passes roughly through these points and ends at (1,1). In this sense, the intrinsic performances (i.e., AUCs or, equivalently, the μ parameter) of the two radiologists are similar. The only difference between them is that radiologist (B) is using a lower threshold relative to the radiologist (C). Radiologist (C) is more concerned with minimizing FPs while radiologist (B) is more concerned with maximizing sensitivity. By appropriate feedback, radiologist (C) can perhaps be induced to change the threshold to that of radiologist (B), or they both could be induced to achieve a happy compromise. An example of feedback might be: *you are missing too many cancers and this could get us all into trouble; worry less about reduced specificity and more about increasing your sensitivity.* In contrast, radiologist (A) has an intrinsically greater performance over (B) or (C). No change in threshold is going to get the other two to a similar level of performance as radiologist (A). Extensive training will be needed to bring the underperforming radiologists to the expert level represented by radiologist (A).

Table 3.3 This table illustrates the variability of sample of 108 board-certified radiologists on a common dataset of screening mammograms

Measure	Min%	Max%	Range%
Sensitivity	46.7	100.0	53.3
Specificity	36.3	99.3	63.0
ROC AUC	0.74	0.95	0.21

Note: Reduced variability when one uses AUC, which accounts for variations in reporting thresholds (AUC variability range is 21%, compared to 53% for sensitivity and 63% for specificity).

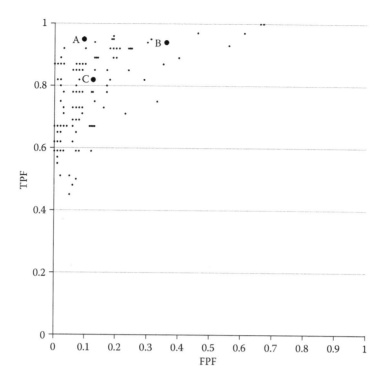

Figure 3.6 Schematic, patterned from the Beam et al. study, showing the ROC operating points of 108 mammographers. Wide variability in sensitivity (40%) and specificity (45%) are evident. Radiologists (B) and (C) appear to be trading sensitivity for specificity and vice versa, while radiologist (A)'s performance is intrinsically superior. The operating points of the three radiologists are emphasized using larger dots.

Figure 3.6 and Table 3.3 illustrate several important principles.

1. Since an operating point is characterized by two values, unless both numbers are higher (e.g., radiologist (A) versus (B) or (C)), it is difficult to unambiguously compare them.
2. While sensitivity and specificity depend on the reporting threshold ζ, the area under the ROC plot is independent of ζ. Using the area under the ROC curve, one can unambiguously compare two readers.
3. Combining sensitivity and the complement of specificity into a single AUC measure yields the additional benefit of lower variability. In Table 3.3, the range for sensitivity is 53% while specificity is 63%. In contrast, the range for AUC is only 21%. This means that much of the observed variations in sensitivity and specificity are due to variations in thresholds, and using AUC eliminates this source of variability. Decreased variability of a measure is a highly desirable characteristic as it implies the measurement is more precise, making it easier to detect genuine changes between readers and/or modalities.

3.12 Discussion

The concepts of sensitivity and specificity are of fundamental importance and are widely used in the medical imaging literature. However, it is important to realize that sensitivity and specificity do not provide a complete picture of diagnostic performance, since they represent performance at a particular threshold. As demonstrated in Figure 3.6, expert observers can and do operate at different points, and the reporting threshold depends on cost-benefit considerations, disease prevalence, and personal reporting styles. If using sensitivity and specificity, the dependence on reporting threshold often makes it difficult to unambiguously compare observers. When one does compare them, there is loss of statistical power (equivalent to loss of precision of the measurement) due to the additional source of variability introduced by the varying thresholds.

The ROC curve is the locus of operating points as the threshold is varied. It and AUC are completely defined by the μ parameter of the equal variance binormal model. Since both are independent of reporting threshold ζ, they overcome the ambiguity inherent in comparing sensitivity/specificity pairs. Both are scalar measures of performance. AUC is widely used in assessing imaging systems. It should impress the reader that a subjective internal sensory perception of disease presence and an equally subjective internal threshold can be translated into an objective performance measure, such as the area under a ROC curve or, equivalently, the μ parameter. The latter has the physical meaning of a perceptual signal-to-noise ratio.

The ROC curve predicted by the equal variance binormal model has a useful property, namely, as the threshold is lowered, its slope decreases monotonically. The predicted curve never crosses the chance diagonal, that is, the predicted ROC curve is *proper*. Unfortunately, as one will see later, most ROC datasets are inconsistent with this model: rather, they are more consistent with a model where the diseased distribution has variance greater than unity. The consequence of this is an *improper* ROC curve, where in a certain range, which may be difficult to see when the data is plotted on a linear scale, the predicted curve actually crosses the chance diagonal and then its slope increases as it hooks up to reach (1,1). The predicted worse than chance performance is unreasonable. Models of ROC curves have been developed that do not have this unreasonable behavior: **Chapter 17**, **Chapter 19** and **Chapter 20**.

The properties of the unit normal distribution and the binomial distribution were used to derive parametric confidence intervals for sensitivity and specificity. These were compared to exact confidence intervals. An important study was reviewed showing wide variability in sensitivity and specificity for radiologists interpreting a common set of cases in screening mammography, but smaller variability in areas under the ROC curve. This is because much of the variability in sensitivity and specificity is due to variation of the reporting threshold, which does not affect the area under the ROC curve. This is an important reason for preferring comparisons based on area under the ROC curve to those based on comparing sensitivity/specificity pairs.

This chapter has demonstrated the equal variance binormal model with **R** examples. These were used to illustrate important concepts of case-sampling variability and its dependence on the numbers of cases. Again, while relegated for organizational reasons to online appendices, *these appendices are essential components of the book*. Most of the techniques demonstrated there will be reused in the remaining chapters. The motivated reader can learn much from studying the online material and running the different main-level functions contained in the software-directory corresponding to this chapter.

References

1. Green DM, Swets JA. *Signal Detection Theory and Psychophysics*. New York, NY: John Wiley & Sons; 1966.
2. Macmillan NA, Creelman CD. *Detection Theory: A User's Guide*. New York, NY: Cambridge University Press; 1991.
3. Skaane P, Bandos AI, Gullien R, et al. Comparison of digital mammography alone and digital mammography plus tomosynthesis in a population-based screening program. *Radiology*. 2013;267(1):47–56.
4. Bandos AI, Rockette HE, Gur D. Use of likelihood ratios for comparisons of binary diagnostic tests: Underlying ROC curves. *Med Phys*. 2010;37(11):5821–5830.

5. Egan JP. *Signal Detection Theory and ROC Analysis.* 1st ed. New York, NY: Academic Press; 1975.
6. Marcum JI. A Statistical Theory of Target Detection by Pulsed Radar. Santa Monica, CA: U.S. Air Force; 1947.
7. Burgess AE. Comparison of receiver operating characteristic and forced choice observer performance measurement methods. *Med Phys.* 1995;22(5):643–655.
8. Larsen RJ, Marx ML. *An Introduction to Mathematical Statistics and Its Applications.* 3rd ed. Upper Saddle River, NJ: Prentice-Hall; 2001.
9. Conover WJ. *Practical Nonparametric Statistics.* New York, NY: John Wiley & Sons; 1971.
10. Beam CA, Layde PM, Sullivan DC. Variability in the interpretation of screening mammograms by US radiologists. Findings from a national sample. *Arch Intern Med.* 1996;156(2):209–213.
11. Hendrick RE, Bassett L, Botsco MA, et al. *Mammography Quality Control Manual.* 4th ed. Reston, VA: American College of Radiology, Committee on Quality Assurance in Mammography; 1999.
12. Barnes GT, Hendrick RE. Mammography accreditation and equipment performance. *Radiographics.* 1994;14(1):129–138.

4

The ratings paradigm

4.1 Introduction

In **Chapter 2** the binary task and associated concepts of sensitivity, specificity, true positive fraction, false positive fraction, positive, and negative predictive values were introduced. **Chapter 3** introduced the concepts of a random scalar decision variable, or Z-sample, for each case, which is compared, by the observer, to a fixed reporting threshold ζ, resulting in two types of decisions, *case is non-diseased* or *case is diseased* depending on whether the realized Z-sample is less than, or greater than, or equal to the reporting threshold. It described a statistical model, for the binary task, characterized by two unit variance normal distributions separated by μ. The concept of an underlying receiver operating characteristic (ROC) curve, with the reporting threshold defining an operating point on the curve, was introduced and the advisability of using the area under the curve as a measure of performance, which is independent of reporting threshold, was stressed.

In this chapter, the more commonly used *ratings* method is described, which yields greater definition to the underlying ROC curve than just one operating point obtained in the binary task, and moreover, is more efficient. In this method, the observer assigns a rating to each case. Described first is a typical ROC counts table and how operating points (i.e., pairs of FPF and TPF values) are calculated from the counts data. A labeling convention for the operating points is introduced. Notation is introduced for the observed integers in the counts table and the rules for calculating operating points are expressed as formula and implemented in **R**. The ratings method is contrasted to the binary method, in terms of efficiency and practicality. A theme occurring repeatedly in this book, that the *ratings are not numerical values but rather they are ordered labels,* is illustrated with an example. A method of collecting ROC data on a 6-point scale is described that has the advantage of yielding an unambiguous single operating point. The forced choice paradigm is described. Two controversies are discussed: one on the utility of discrete (e.g., 1–6) versus quasi-continuous (e.g., 0–100) ratings and the other on the applicability of a clinical screening mammography-reporting scale for ROC analyses. Both of these are important issues and it would be a disservice to the readers of the book if the author did not express his position on them.

4.2 The ROC counts table

In a positive-directed rating scale with five discrete levels, the ratings could be the ordered labels 1: definitely non-diseased, 2: probably non-diseased, 3: could be non-diseased or diseased, 4: probably diseased, 5: definitely diseased. At the conclusion of the ROC study a *ROC counts table* is constructed. This is the generalization to rating studies of the 2×2 decision versus truth table introduced in

Table 4.1 A typical ROC counts table

	Counts in ratings bins				
	Rating = 1	**Rating = 2**	**Rating = 3**	**Rating = 4**	**Rating = 5**
$K_1 = 60$	30	19	8	2	1
$K_2 = 50$	5	6	5	12	22
			Operating points		
	Ratings ≥5	Ratings ≥4	Ratings ≥3	Ratings ≥2	Ratings ≥1
FPF	0.017	0.050	0.183	0.500	1
TPF	0.440	0.680	0.780	0.900	1

Note: Listed in the upper half of the table are the number of cases in specific ratings bins, listed separately for actually non-diseased and actually diseased cases. There are $K_1 = 60$ non-diseased cases and $K_2 = 50$ diseased cases in this dataset. The lower half of the table lists the corresponding FPF and TPF values, that is, the abscissa and ordinate, respectively, of the operating points on the ROC plot.

Chapter 2, Table 2.1. This type of data representation is sometimes called a *frequency table* in the statistical literature, but frequency* means a *rate* of number of events per some unit, so the author prefers the clearer term *counts*.

Table 4.1 is a representative counts table for a 5-rating study that summarizes the collected data. It is the starting point for analysis. The top half of the table lists the number of counts in each ratings bin, listed separately for non-diseased and diseased cases, respectively. The data is from an actual clinical study.[1]

In this example, there are $K_1 = 60$ non-diseased cases and $K_2 = 50$ diseased cases. Of the 60 non-diseased cases, 30 were assigned the 1-rating, 19 were assigned the 2-rating, eight the 3-rating, two the 4-rating, and one received the 5-rating. The distribution of counts is observed to be tilted towards the 1-rating end, but there is some spread and one actually non-diseased case appeared definitely diseased to the observer. In contrast, the distribution of the diseased cases is observed to be tilted towards the 5-rating end. Of the 50 diseased cases, 22 received the 5-rating, 12 the 4-rating, five the 3-rating, six the 2-rating, and five the 1-rating. The spread appears to be more pronounced for the diseased cases, for example, five of the 50 cases appeared to be definitely non-diseased to the observer. A little thought should convince the reader that the observed tilting of the counts, towards the 1-end for actually non-diseased cases, and toward the 5-end for actually diseased cases, is reasonable. However, one is forewarned not to jump to conclusions about the spread of the data being larger for diseased than for non-diseased cases. While it turns out to be true, the ratings are *merely ordered labels*, and modeling is required, described in **Chapter 6**, that uses only the ordering information implicit in the labels, not the actual values, to reach quantitative conclusions.

4.3 Operating points from counts table

It is critical to understand the following example. The bottom half of Table 4.1 illustrates how ROC operating points are calculated from the cell counts. One starts with non-diseased cases that were rated 5 or more (in this example, since 5 is the highest allowed rating, the *or more* clause is superfluous)

---

* frequency |ˈfrēkwənsē| noun (pl. frequencies): The rate at which something occurs or is repeated over a particular period of time or in a given sample: shops have closed with increasing frequency during the period.

- The fact of being frequent or happening often.
- Statistics: The ratio of the number of actual to possible occurrences of an event.
- Statistics: The (relative) number of times something occurs in a given sample.

and divides by the total number of non-diseased cases, $K_1 = 60$. This yields the abscissa of the lowest non-trivial operating point, namely $FPF_{\geq 5} = 1/60 = 0.017$. The subscript on FPF is intended to make explicit which ratings are being cumulated. The corresponding ordinate is obtained by dividing the number of diseased cases rated 5 or more by the total number of diseased cases, $K_2 = 50$, yielding $TPF_{\geq 5} = 22/50 = 0.440$. The coordinates of the lowest operating point are (0.017, 0.44). The abscissa of the next higher operating point is obtained by dividing the number of non-diseased cases that were rated 4 or more by the total number of non-diseased cases, that is, $FPF_{\geq 4} = 3/60 = 0.05$. Similarly, the ordinate of this operating point is obtained by dividing the number of diseased cases that were rated 4 or more by the total number of diseased cases, that is, $TPF_{\geq 4} = 34/50 = 0.680$. The procedure, which at each stage *cumulates* the number of cases equal to or greater (in the sense of increased confidence level for disease presence) than a specified label, is repeated to yield the rest of the operating points listed in Table 4.1. Since they are computed *directly* from the data, without any assumption, they are called *empirical* or *observed* operating points. After this is done once, it would be nice to have a formula implementing the process, one use of which would be to code the procedure. First, one needs appropriate notation for the bin counts.

Let K_{1r} denote the number of non-diseased cases rated r, and K_{2r} denote the number of diseased cases rated r. For convenience, define dummy counts $K_{1(R+1)} = K_{2(R+1)} = 0$, where R is the number of ROC bins. This construct allows inclusion of the origin (0,0) in the formula. *The range of r is* $r = 1, 2, ... (R+1)$. Within each truth-state, the individual bin counts sum to the total number of non-diseased and diseased cases, respectively. The following equations summarize this paragraph:

$$\left.\begin{array}{c} K_1 = \displaystyle\sum_{r=1}^{R+1} K_{1r} \\[2.5em] K_2 = \displaystyle\sum_{r=1}^{R+1} K_{2r} \\[2.5em] K_{1(R+1)} = K_{2(R+1)} = 0 \\[1em] r = 1, 2, ... (R+1) \end{array}\right\} \quad (4.1)$$

To be clear, Table 4.1 is repeated to show the meaning of the counts notation, Table 4.2.

Table 4.2 Explanation of the notation for cell counts for $R = 5$. The upper half of the table, reproduced from Table 4.1, illustrates the notation while the lower half shows the values, or r, which in conjunction with Equation 4.2, determine the operating points.

	Counts in ratings bins				
	Rating = 1	Rating = 2	Rating = 3	Rating = 4	Rating = 5
$K_1 = 60$	$K_{11} = 30$	$K_{12} = 19$	$K_{13} = 8$	$K_{14} = 2$	$K_{15} = 1$
$K_2 = 50$	$K_{21} = 5$	$K_{22} = 6$	$K_{23} = 5$	$K_{24} = 12$	$K_{25} = 22$
			Operating points		
	$r = R = 5$	$r = R - 1 = 4$	$r = R - 2 = 3$	$r = R - 3 = 2$	$r = R - 4 = 1$
	≥ 5	≥ 4	≥ 3	≥ 2	≥ 1
FPF	0.017	0.050	0.183	0.500	1
TPF	0.440	0.680	0.780	0.900	1

The operating points are defined by:

$$FPF_r = \frac{1}{K_1} \sum_{s=r}^{R+1} K_{1s}$$

$$TPF_r = \frac{1}{K_2} \sum_{s=r}^{R+1} K_{2s}$$

$$(4.2)$$

The labeling of the points follows this convention: $r = 1$ corresponds to the upper right corner (1,1) of the ROC plot, a *trivial operating point since it is common to all datasets*. Next, $r = 2$ is the next lowest operating point, etc., and $r = R$ is the lowest non-trivial operating point, and finally $r = R+1$ is the origin (0,0) of the ROC plot, which is also a trivial operating point, because it is common to all datasets. In other words, the operating points are numbered starting with the upper right corner, labeled 1, and working down the curve, each time increasing the label by one.

Applications of Eqn. 4.2: if $r = 1$ one gets the uppermost *trivial* operating point (1,1), then,

$$FPF_1 = \frac{1}{K_1} \sum_{s=1}^{R+1} K_{1s} = \frac{60}{60} = 1$$

$$TPF_1 = \frac{1}{K_2} \sum_{s=1}^{R+1} K_{2s} = \frac{50}{50} = 1$$

$$(4.3)$$

The uppermost non-trivial operating point is obtained for $r = 2$, when

$$FPF_2 = \frac{1}{K_1} \sum_{s=2}^{R+1} K_{1s} = \frac{30}{60} = 0.500$$

$$TPF_2 = \frac{1}{K_2} \sum_{s=2}^{R+1} K_{2s} = \frac{45}{50} = 0.900$$

$$(4.4)$$

The next lowest operating point is obtained for $r = 3$:

$$FPF_3 = \frac{1}{K_1} \sum_{s=3}^{R+1} K_{1s} = \frac{11}{60} = 0.183$$

$$TPF_3 = \frac{1}{K_2} \sum_{s=3}^{R+1} K_{2s} = \frac{39}{50} = 0.780$$

$$(4.5)$$

The next lowest operating point is obtained for $r = 4$:

$$FPF_4 = \frac{1}{K_1} \sum_{s=4}^{R+1} K_{1s} = \frac{3}{60} = 0.050$$

$$TPF_4 = \frac{1}{K_2} \sum_{s=4}^{R+1} K_{2s} = \frac{34}{50} = 0.680$$

$$(4.6)$$

The lowest non-trivial operating point is obtained for $r = 5$:

$$FPF_5 = \frac{1}{K_1} \sum_{s=5}^{R+1} K_{1s} = \frac{1}{60} = 0.017$$

$$TPF_5 = \frac{1}{K_2} \sum_{s=5}^{R+1} K_{2s} = \frac{22}{50} = 0.440$$

(4.7)

The next value $r = 6$ yields the trivial operating point (0,0):

$$FPF_6 = \frac{1}{K_1} \sum_{s=6}^{R+1} K_{1s} = \frac{0}{60} = 0$$

$$TPF_6 = \frac{1}{K_2} \sum_{s=6}^{R+1} K_{2s} = \frac{0}{50} = 0$$

(4.8)

This exercise shows explicitly that an R-rating ROC study can yield, at most, $R-1$ distinct non-trivial operating points; that is, those corresponding to $r = 2, 3, ..., R$.

The modifier *at most* is needed, because if *both counts* (i.e., non-diseased and diseased) for bin r' are zeroes, then that operating point merges with the one immediately below-left of it.

$$FPF_{r'} = \frac{1}{K_1} \sum_{r=r'}^{R+1} K_{1r} = \frac{1}{K_1} \sum_{r=r'+1}^{R+1} K_{1r} = FPF_{r'+1}$$

$$TPF_{r'} = \frac{1}{K_2} \sum_{r=r'}^{R+1} K_{2r} = \frac{1}{K_2} \sum_{r=r'+1}^{R+1} K_{2r} = TPF_{r'+1}$$

(4.9)

Since bin r' is unpopulated, one can relabel the bins to exclude the unpopulated bin, and now the total number of bins is effectively R-1.

Since one is cumulating counts, which can never be negative, the highest non-trivial operating point resulting from cumulating the 2 through 5 ratings has to be to the upper-right of the next adjacent operating point resulting from cumulating the 3 through 5 ratings. This in turn has to be to the upper-right of the operating point resulting from cumulating the 4 through 5 ratings. This in turn has to be to the upper-right of the operating point resulting from the 5 ratings. In other words, as one cumulates ratings bins, the operating point must move monotonically up and to the right, or more accurately, the point cannot move down or to the left. If a particular bin has *zero* counts for non-diseased cases, and *non-zero* counts for diseased cases, the operating point moves vertically *up* when this bin is cumulated; if it has zero counts for diseased cases, and non-zero counts for non-diseased cases, the operating point moves horizontally to the *right* when this bin is cumulated.

It is useful to replace the preceding detailed explanation with a simple algorithm that incorporates all the logic. Online Appendix 4.A describes the **R** code in **MainOpPtsFromCountsTable.R** for calculating operating points from counts for the data in Table 4.1. **Source** the file to get Section 4.3.1 and Figure 4.1a and b.

4.3.1 Code output

```
> source...
FPF
0.01667 0.05 0.1833 0.5
TPF =
```

```
0.44 0.68 0.78 0.9
uppermost point based estimate of mu = 1.282
corresponding estimate of Az = 0.8176
binormal estimate of Az = 0.8696
```

The output lists the values of the arrays FPF and TPF, which correspond to those listed in Table 4.1.

It was shown in **Chapter 3** that in the equal variance binormal model, an operating point determines the parameter μ, Equation 3.21, or equivalently, $A_{z;\sigma=1}$, Equation 3.30. The last three lines of Section 4.3.1 illustrate the application of these formula using the coordinates (0.5, 0.9) of the uppermost non-trivial operating point. It should come as no surprise that the uppermost operating point in Figure 4.1a is *exactly* on the predicted curve: after all, this point was used to calculate $\mu = 1.282$. The corresponding value of ζ can be calculated from Equation 3.17, namely,

$$\Phi^{-1}(Sp) = \zeta \tag{4.10}$$

Alternatively, using Equation 3.18,

$$\mu - \zeta = \Phi^{-1}(Se) \Rightarrow \zeta = \mu - \Phi^{-1}(Se) \tag{4.11}$$

These are coded in Section 4.3.2.

4.3.2 Code snippet

```
> qnorm(1-0.5)
[1] 0
> mu-qnorm(0.9)
[1] 0
```

Either way, one gets the same result: $\zeta = 0$. It should be clear that $\zeta = 0$ makes sense: FPF = 0.5 is consistent with half of the (symmetrical) unit-normal non-diseased distribution being above $\zeta = 0$. The transformed value ζ is a genuine numerical value. *To reiterate, ratings cannot be treated as numerical values, but thresholds, estimated from an appropriate model, can.*

The ROC curve in Figure 4.1a, as determined by the uppermost operating point, passes *exactly* through this point but misses the others. If a different operating point were used to estimate μ and $A_{z;\sigma=1}$, the estimated values would have been different and the new curve would pass exactly through the new selected point. No choice of μ yields a satisfactory visual fit to all of the experimental data points. The reader should confirm these statements with appropriate modifications to the code. This is the reason one needs a modified model, with an extra parameter, namely the unequal variance binormal model, to fit radiologist data (the extra parameter is the ratio of the standard deviations of the two distributions).

Figure 4.1b shows the predicted ROC curve by the unequal variance binormal model, to be introduced in **Chapter 6**. Notice the improved visual quality of the fit. Each observed point is not engraved in stone, rather it is subject to sampling variability. Estimation of confidence intervals for FPF and TPF was addressed in Section 3.10. [A detail: the estimated confidence interval in the preceding chapter was for a *single* operating point; since the multiple operating points

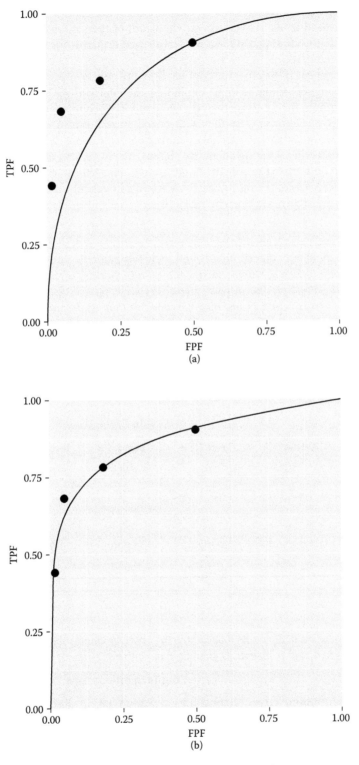

Figure 4.1 (a): Predicted equal-variance binormal model ROC curve for $\mu = 1.282$ superposed on the operating points obtained from the data in Table 4.1. (b): The same data is fitted with a unequal-variance binormal model, with $\mu = 2.17, \sigma = 1.65$, described in **Chapter 6**, in which μ is the separation of the normal distributions and σ is the standard deviation of the diseased distribution; the non-diseased distribution has unit standard deviation. Note the improved visual quality of the fit.

are correlated—some of the counts used to calculate them are common to two or more operating points—the method tends to overestimate the width of the confidence interval. A modeling approach is possible to estimate confidence intervals that accounts for data correlation yielding tighter confidence intervals.]

Consider what happens if the observer does not fully utilize the rating scale. For example, if the observer chooses to ignore the gradations implied by the intermediate ratings 2, 3, and 4, essentially lumping them into the 5-rating, then the observed ROC counts table would be as shown in Table 4.3. Essentially the observer responds with 1s and 5s, so R is effectively equal to two. In this example, the number of *distinct* non-trivial operating points is one, just as in the binary-decision task of **Chapter 2** (note the repeating operating points in the lower half of the table).

What if the observer were to use only the 1 and 2 ratings bins, effectively ignoring the gradations implied by the upper levels of the rating scale? Or stated equivalently, the observer lumps the 3, 4, and 5 rating into the 2-rating, as shown in Table 4.4.

Once again, the number of distinct non-trivial operating points is one, and the coordinates of the operating point (0.500, 0.900) are identical to that in Table 3, where the observer used the 1s and 5s only. This illustrates the intrinsic nature of the ratings as ordered labels. Given binary ratings 1s and 5s versus 1s and 2s, one may not conclude that just because the actual difference

Table 4.3 An atypical ROC counts table where the observer ignores the gradations implied by the intermediate ratings 2, 3, and 4, essentially lumping them into the 5-rating. Otherwise, the structure of this table is identical to Table 4.1.

	Counts in ratings bins				
	Rating = 1	**Rating = 2**	**Rating = 3**	**Rating = 4**	**Rating = 5**
$K_1 = 60$	30	0	0	0	30
$K_2 = 50$	5	0	0	0	45
			Operating points		
	≥5	≥4	≥3	≥2	≥1
FPF	0.500	0.500	0.500	0.500	1
TPF	0.900	0.900	0.900	0.900	1

Note: Even though the operating point (0.500, 0.900) repeats in four columns, there is only one unique value.

Table 4.4 Another atypical ROC counts table where the observer ignores the gradations implied by the intermediate ratings 3, 4, and 5, essentially lumping them into the 2-rating.

	Counts in ratings bins				
	Rating = 1	**Rating = 2**	**Rating = 3**	**Rating = 4**	**Rating = 5**
$K_1 = 60$	30	30	0	0	0
$K_2 = 50$	5	45	0	0	0
			Operating points		
	≥5	≥4	≥3	≥2	≥1
FPF	0	0	0	0.500	1
TPF	0	0	0	0.900	1

between five and one is four times larger than that between two and one, that the discriminabil-ity between the non-diseased and diseased cases is four times larger. As long as the single oper-ating point is unaltered, as between Table 4.3 and Table 4.4, there is no change in performance as quantified by μ or equivalently, by the area $A_{z;\sigma=1}$ under the ROC curve.

4.4 Relation between ratings paradigm and the binary paradigm

Table 4.2 corresponds to $R = 5$. In **Chapter 2** it was shown that the binary task requires a *single* fixed threshold parameter ζ and a decision rule, namely, to give the case a diseased rating of 2 if $Z \geq \zeta$, and a rating of 1 otherwise.

The R-rating task can be viewed as $(R-1)$ simultaneously conducted binary tasks each with its own fixed threshold $\zeta_r, r = 1, 2, ..., R-1$. It is efficient compared to $(R-1)$ sequentially con-ducted binary tasks; however, the onus is on the observer to maintain fixed-multiple thresholds for the duration of the study.

The rating method is a more efficient way of collecting the data compared to running the study repeatedly with appropriate instructions to cause the observer to adopt different fixed thresholds spe-cific to each replication. In the clinical context, such repeated studies would be impractical because it would introduce memory effects, wherein the diagnosis of a case would depend on how many times the case had been seen, along with other cases, in previous sessions. A second reason is that it is dif-ficult for a radiologist to change the operating threshold in response to instructions. To the author's knowledge, repeated use of the binary paradigm has not been used in any clinical ROC study.

How does one model the binning? For convenience, one defines dummy thresholds $\zeta_0 = -\infty$ and $\zeta_R = +\infty$, in which case the thresholds satisfy the ordering requirement $\zeta_{r-1} < \zeta_r$, $r = 1, 2, ..., R$. The *rating rule* is

$$\text{if } \zeta_{r-1} \leq z < \zeta_r \Rightarrow rating = r \tag{4.12}$$

For the dataset in Table 4.1 the *empirical* thresholds (as opposed to *modeled* thresholds via Eqn. (4.10) or Eqn. (4.11); note that empirical thresholds are not true numerical values) are as follows (the super-script E is for empirical):

$$\left.\begin{array}{l} \zeta_r^E = r \\[6pt] r = 1, .., R-1; \\[6pt] \zeta_0^E = -\infty; \ \zeta_R^E = +\infty \end{array}\right\} \tag{4.13}$$

In Table 4.1 the number of bins is $R = 5$. The "simultaneously conducted binary tasks" nature of the rating task can be appreciated from the following examples. Suppose one selects the threshold for the first binary task to be $\zeta_4^E = 4$. Therefore a case rated 5 satisfies $\zeta_4^E \leq 5 < \zeta_5^E$, consistent with Eqn. (4.12). The operating point corresponding to $\zeta_4^E = 4$, obtained by cumulating all cases rated five, yields (0.017, 0.440). In the second binary-task, one selects as threshold $\zeta_3^E = 3$. Therefore, a case rated four satisfies the inequality $\zeta_3^E \leq 4 < \zeta_4^E$, again consistent with Eqn. (4.12). The operating point corresponding to $\zeta_3^E = 3$, obtained by cumulating all cases rated four or five, yields (0.05, 0.680). Similarly, for $\zeta_2^E = 2, \zeta_1^E = 1$ and $\zeta_0^E = -\infty$, which, according to the binning rule Eqn. (4.12), yield counts in bins 3, 2 and 1, respectively. The non-trivial operating points are generated by thresholds ζ_r^E where r = 1, 2, 3 and 4. A five-rating study has four associated thresholds and a corresponding number of equivalent binary studies. In general, an R rating study has R-1 associated thresholds.

4.5 Ratings are not numerical values

The ratings are to be thought of as *ordered labels*, not as numeric values. Arithmetic operations that are allowed on numeric values, such as averaging, are not allowed on ratings. One could have relabeled the ratings in Table 4.2 as A, B, C, D, and E, where A < B, and so on. As long as the counts in the body of the table are unaltered, such relabeling would have no effect on the observed operating points and the fitted curve. Of course, one cannot average the labels A, B, etc. of different cases. The issue with numeric labels is not fundamentally different. At the root is that the difference in thresholds corresponding to the different operating points are not in relation to the difference between their numeric values. There is a way to estimate the underlying thresholds, if one assumes a specific model. For example, the unequal variance binormal model to be described in **Chapter 6**. The thresholds so obtained are genuine numeric values and can be averaged. (Not to hold the reader in suspense, the four thresholds corresponding to the data in Table 4.1 are $\zeta_1 = 0.007676989$, $\zeta_2 = 0.8962713$, $\zeta_3 = 1.515645$, and $\zeta_4 = 2.396711$; see Section 6.4.1. These values would be unchanged if, for example, the labels were doubled, with allowed values 2, 4, 6, 8 and 10, or any of an infinite number of rearrangements that preserves their ordering.)

The temptation to regard confidence levels/ratings as numeric values can be particularly strong when one uses a large number of bins to collect the data. One could use of quasi-continuous ratings scale, implemented for example, by having a slider-bar user interface for selecting the rating. The slider bar typically extends from 0 to 100, and the rating could be recorded as a floating-point number, for example, 63.45. Here too one cannot assume that the difference between a zero-rated case and a 10-rated case is a tenth of the difference between a zero-rated case and a 100-rated case. So, averaging the ratings is not allowed. Additionally, one cannot assume that different observers use the labels in the same way. One observer's 4-rating is not equivalent to another observers 4-rating. *Working directly with the ratings is a bad idea: valid analytical methods use the rankings of the ratings, not their actual values.* The reason for the emphasis is that there are serious misconceptions about ratings. The author is aware of a publication stating, to the effect, that a modality resulted in an increase in average confidence level for diseased cases. Another publication used a specific numerical value of a rating to calculate the single operating point, in the sense explained below, for each observer—this assumes all observers use the rating scale in the same way.

4.6 A single "clinical" operating point from ratings data

The reason for the quotes in the title to this section is that a single operating point on a laboratory ROC plot, no matter how obtained, has little relevance to how radiologists operate in the clinic. However, some consider it useful to quote an operating point from a ROC study. For a 5-rating ROC study, Table 4.1, it is not possible to unambiguously calculate the operating point of the observer in the binary task of discriminating between non-diseased and diseased cases. One possibility would be to use the 3 and above ratings to define the operating point, but one might have chosen 2 and above. A second possibility is to instruct the radiologist that a 4 or higher rating, for example, implies the case would be reported clinically as diseased. However, the radiologist can only pretend that the study, which has no clinical consequences, is somehow a clinical study. If a single laboratory study based operating point is desired,[2] the best strategy*, in the author's opinion, is to obtain the rating via two questions. This method is also illustrated in a book on detection theory,[3] Table 3.1. The first question is: *is the case diseased?* The binary (yes/no) response to this question allows unambiguous calculation of the operating point, as in **Chapter 2**. The second question is: *what is your confidence in your previous decision?*, and allow three responses, namely Low, Medium, and High. The dual-question approach is equivalent to a 6-point rating scale, as shown in Figure 4.2.

---

* The author owes this insight to Prof. Harold Kundel.

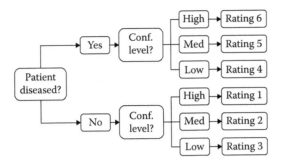

Figure 4.2 This figure describes a method for acquiring ROC data on an effectively 6-point scale that also yields an unambiguous single operating point for declaring patients diseased. The data collection consists of two questions. The answer to the first question, *is the patient diseased?*, allows unambiguous construction of a single operating point for disease presence. The answer to the second question, *what is your confidence level in that decision?*, yields multiple operating points. Note the reversal of the final ratings in the last column in the lower half of the figure.

The ordering of the ratings can be understood as follows. The 4, 5, and 6 ratings are as expected. If the radiologist states the patient is diseased and the confidence level is high, that is clearly the highest end of the scale, that is, 6, and the lower confidence levels, 5 and 4, follow, as shown in Fig. 4.2. If, on the other hand, the radiologist states the patient is non-diseased, and the confidence level is high, then that must be the lowest end of the scale, i.e., 1. The lower confidence levels in a negative decision must be labeled higher than 1, namely 2 and 3, as shown in Fig. 4.2. As expected, the low confidence ratings, namely 3 (non-diseased, low confidence) and 4 (diseased, low confidence), are adjacent to each other. With this method of data-collection, there is no confusion as to what rating defines the single desired operating point as this is determined by the binary response to the first question. The 6-point rating scale is also sufficiently fine to not smooth out the ability of the radiologist to maintain distinct different levels. In the author's experience, using this scale one expects rating noise of about ±½ a rating bin, that is, the same difficult case, shown on different occasions to the same radiologist (with sufficient time lapse or other intervening cases to minimize memory effects) is expected to elicit a 3 or 4, with roughly equal probability.

4.7 The forced choice paradigm

In each of the four paradigms (ROC, FROC, LROC, and ROI) described in **Chapter 1**, patient images are displayed one patient at a time. A fifth paradigm involves presentation of multiple images to the observer, where one image is from a diseased patient, and the rest are from non-diseased patients. The observer's task is to pick the image that is most likely to be from the diseased patient. If the observer is correct, the event is scored as a one and otherwise it is scored as a zero. The process is repeated with other sets of independent patient images, each time satisfying the condition that one patient is diseased and the rest are non-diseased. The sum of the scores divided by the total number of scores is the probability of a correct choice, denoted $P(C)$. If the total number of images presented at the same time is denoted n, then the task is termed n-alternative forced choice ($nAFC$).[4] If only two cases are presented, one diseased and the other non-diseased, then $n = 2$ and the task is 2AFC. In Figure 4.3, in the left image a Gaussian nodule is superposed on a square region extracted from a non-diseased mammogram. The right image is a region extracted from a different non-diseased mammogram (one should not use the same background in the two images—the analysis assumes that different, i.e., independent images, are shown*). If the observer clicks on the left image, a correct choice is recorded. (In some 2AFC-studies, the backgrounds are simulated

---

* The author is aware of a publication using non-diseased cases to simulate both non-diseased cases and as backgrounds to which nodules are added, to simulate diseased cases.

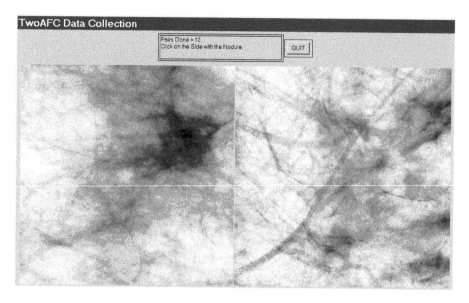

Figure 4.3 Example of image presentation in a 2AFC study. The left image contains, at its center, a positive contrast Gaussian shape disk superposed on a non-diseased mammogram. The right image does not contain a lesion at its center and the background is from a different non-diseased patient. If the observer clicks on the left image it is recorded as a correct choice, otherwise it is recorded as an incorrect choice. The number of correct choices divided by the number of paired presentations is an estimate of the probability of a correct choice, which can be shown to be identical, apart from sampling variability, to the true area under the ROC curve. This is an example of a signal known exactly, location known exactly (SKE-LKE) task widely used by the model observer community.

non-diseased images. They resemble mammograms; the resemblance depends on the expertise of the observer. Expert radiologists can tell they are not true mammograms. They are actually created by filtering the random white noise with a $1/f^3$ spatial filter.[5–12])

The 2AFC paradigm is popular, because its analysis is straightforward, and there exists a theorem[4] that $P(C)$, the probability of a correct choice in the 2AFC task, equals, to within sampling variability, the *true* area under the true (not fitted, not empirical) ROC. Another reason for its popularity is possibly the speed at which data can be collected, sometimes only limited by the speed at which disk-stored images can be displayed on the monitor. While useful for studies into human visual perception on relatively simple images, and the model observer community has performed many studies using this paradigm,[13–20] the author cannot recommend it for clinical studies because it does not resemble any clinical task. In the clinic, radiologists never have to choose the diseased patient out of a pair consisting of one diseased and one non-diseased. Additionally, the forced choice paradigm is wasteful of known-truth images, often a difficult/expensive resource to come by, because better statistics[21] (tighter confidence intervals) are obtained by the ratings ROC method or by utilizing location-specific extensions of the ROC paradigm. (The author is not aware of the 2AFC method being actually used to assess imaging systems using radiologists to perform real clinical tasks on real images.)

4.8 Observer performance studies as laboratory simulations of clinical tasks

Observer performance paradigms (ROC, FROC, LROC, and ROI) should be regarded as experiments conducted in a laboratory (i.e., controlled) setting that are intended to be representative of the actual clinical task. They should not to be confused with performance in a real live clinical setting: there is a known *laboratory effect*.[22-24] For example, in one study radiologists performed better during live clinical interpretations than they did later, on the same cases, in a laboratory ROC study.[22] This is expected because there is more at stake during live interpretations, for example, the patient's health and the radiologist's reputation, than during laboratory ROC studies. The claimed *laboratory effect* has caused some controversy. A paper,[25] titled "*Screening mammography: Test set data can reasonably describe actual clinical reporting, argues against the laboratory effect.*"

Real clinical interpretations happen every day in radiology departments all over the world. In the laboratory, the radiologist is asked to interpret the images as if in a clinical setting and render diagnoses. The laboratory decisions have no clinical consequences, for example, the radiologist will not be sued for mistakes and their ROC study decisions have no impact on the clinical management of the patients. Usually laboratory ROC studies are conducted on retrospectively acquired images. Patients, whose images were used in a ROC study, have already been imaged in the clinic and decisions have already been made on how to manage them.

There is no guarantee that results of the laboratory study are directly applicable to clinical practice. Indeed there is an assumption that the laboratory study correlates with clinical performance. Strict equality is not required, simply that the performance in the laboratory is related monotonically to actual clinical performance. Monotonicity assures preservation of performance orderings, that is, a radiologist has greater performance than another, is expected to be true regardless of how they are measured, in the laboratory or in the clinic. The correlation is taken to be an axiomatic truth by some researchers, when in fact it is an assumption. To the extent that the participating radiologist brings his/her full clinical expertise to bear on each laboratory image interpretation, the assumption is likely to be valid.

This section provoked a strong response from a collaborator. To paraphrase him, *Dear Dev, my friend, I think it is a pity in this book chapter you argue that these studies are simulations. I mean, the reason people perform these studies is because they believe in the results.*

The author also believes in observer performance studies. Otherwise, he would not be writing this book. Distrust of the word **simulation** *seems to be peculiar to this field. Simulations are widely used in hard sciences, for example, they are used in astrophysics to determine conditions dating to 10^{-31} seconds after the big bang. Simulations are not to be taken lightly. Conducting clinical studies is very difficult as there are many factors not under the researcher's control. Observer performance studies of the type described in this book are the closest that one can come to the real thing and the author is a firm believer in them. These studies include key elements of the actual clinical task: the entire imaging system, radiologists (assuming the radiologist take these studies seriously in the sense of bringing their full expertise to bear on each image interpretation), and real clinical images and as such are expected to correlate with real live interpretations. Proving this correlation is going to be difficult as there are many factors that complicated real interpretations. It is not clear to the author that proving or disproving this correlation is ever going to be a settled issue.*

4.9 Discrete versus continuous ratings: The Miller study

There is controversy about the merits of discrete versus continuous ratings.[26-28] Since the late Prof. Charles E. Metz and the late Dr. Robert F. Wagner have both backed the latter (i.e., continuous or quasi-continuous ratings) new ROC study designs sometimes tend to follow their advice. The author's recommendation is to follow the 6-point rating scale as outlined in Figure 4.2. This section provides background for the recommendation.

A widely cited (22,909 citations at the time of writing) 1954 paper by Miller[29] titled "*The Magical Number Seven, Plus or Minus Two: Some Limits on Our Capacity for Processing Information*" is relevant. It is a readable paper, freely downloadable in several languages (www.musanim.com/

miller1956/). In the author's judgment, this paper has not received the attention it should have in the ROC community, and for this reason portions from it are reproduced below. (George Armitage Miller, February 3, 1920—July 22, 2012, was one of the founders of the field of cognitive psychology.)

Miller's first objective was to comment on *absolute judgments of unidimensional stimuli*. Since all (univariate, i.e., single decision per case) ROC models assume a unidimensional decision variable, Miller's work is highly relevant. He comments on two papers by Pollack.[30,31] Pollack asked listeners to identify tones by assigning numerals to them, analogous to the rating task described in this Chapter. The tones differed in frequency, covering the range 100–8000 Hz in equal logarithmic steps. A tone was sounded and the listener responded by giving a numeral (i.e., a rating, with higher values corresponding to higher frequencies). After the listener had made his response, he was told the correct identification of the tone. *When only two or three tones were used, the listeners never confused them. With four different tones, confusions were quite rare, but with five or more tones, confusions were frequent. With fourteen different tones, the listeners made many mistakes.* Since it is so succinct, the entire content of the first (1952) paper by Pollack is reproduced below.

> "In contrast to the extremely acute sensitivity of a human listener to discriminate small differences in the frequency or intensity between two sounds is his relative inability to identify (and name) sounds presented individually. When the frequency of a single tone is varied in equal-logarithmic steps in the range between 100 cps and 8000 cps (and when the level of the tone is randomly adjusted to reduce loudness cues), the amount of information transferred is about 2.3 bits per stimulus presentation. This is equivalent to perfect identification among only 5 tones. The information transferred, under the conditions of measurement employed, is reasonably invariant under wide variations in stimulus conditions."

The term *information* refers to (essentially) the number of levels, measured in *bits* (binary digits), thereby making it independent of the unit of measurement: one bit corresponds to a binary rating scale, two bits to a 4-point rating scale and 2.3 bits to $2^{2.3} = 4.9$, that is, about 5 ratings bins. Based on Pollack's original unpublished data, Miller put an upper limit of 2.5 bits (corresponding to about 6 ratings bins) on the amount of information that is transmitted by listeners who make absolute judgments of auditory pitch. A second paper[31] by Pollack was related to: (1) the frequency range of tones, (2) the utilization of objective reference tones presented with the unknown tone, and (3) the *dimensionality*—the number of independently varying stimulus aspects. Little additional gain in information transmission was associated with the first factor; a moderate gain was associated with the second; and a relatively substantial gain was associated with the third (we return to the dimensionality issue below).

> "Most people are surprised that the number is as small as six. Of course, there is evidence that a musically sophisticated person with absolute pitch can identify accurately any one of 50 or 60 different pitches. Fortunately, I do not have time to discuss these remarkable exceptions. I say it is fortunate because I do not know how to explain their superior performance. So I shall stick to the more pedestrian fact that most of us can identify about one out of only five or six pitches before we begin to get confused."

> "It is interesting to consider that psychologists have been using seven-point rating scales for a long time, on the intuitive basis that trying to rate into finer categories does not really add much to the usefulness of the ratings. Pollack's results indicate that, at least for pitches, this intuition is fairly sound."

As an interesting side-note, Miller states:

"Next you can ask how reproducible this result is. Does it depend on the spacing of the tones or the various conditions of judgment? Pollack varied these conditions in a number of ways. The range of frequencies can be changed by a factor of about 20 without changing the amount of information transmitted more than a small percentage. Different groupings of the pitches decreased the transmission, but the loss was small. For example, if you can discriminate five high-pitched tones in one series and five low-pitched tones in another series, it is reasonable to expect that you could combine all ten into a single series and still tell them all apart without error. When you try it, however, it does not work. The channel capacity for pitch seems to be about six and that is the best you can do."

Miller also quotes work[32] on channel capacities for absolute judgments of *loudness*[32] (2.3 bits), sensation of *saltiness*[33] (1.9 bits), and judgments of *visual position*[34] (3.25 bits).

In contrast to the careful experiments conducted in the psychophysical context to elucidate this issue, the author was unable to find a single study of the number of discrete rating levels that an observer can support in an ROC ratings study. Even lacking such a study, a recommendation has been made to acquire data on a quasi-continuous scale.[27]

There is no question that for multidimensional data, as observed in the second study by Pollack,[31] the observer can support more than 7 ratings bins. To quote Miller:

"You may have noticed that I have been careful to say that this magical number seven applies to one-dimensional judgments. Everyday experience teaches us that we can identify accurately any one of several hundred faces, any one of several thousand words, any one of several thousand objects, etc. The story certainly would not be complete if we stopped at this point. We must have some understanding of why the one-dimensional variables we judge in the laboratory give results so far out of line with what we do constantly in our behavior outside the laboratory. A possible explanation lies in the number of independently variable attributes of the stimuli that are being judged. Objects, faces, words, and the like differ from one another in many ways, whereas the simple stimuli we have considered thus far differ from one another in only one respect."

In the medical imaging context, a trivial way to increase the number of ratings would be to color-code the images using red, green and blue. Now one can assign a red image rated 3, a green image rated 2, and so on, which would be meaningless unless the color encoded relevant diagnostic information. Another ability, quoted in the publication[27] advocating continuous ratings is the ability to recognize faces, again a multidimensional categorization task, as noted by Miller. Also quoted as an argument for continuous ratings is the ability of computer aided detection schemes that calculate many features for each perceived lesion and combine them into a single probability of malignancy, which is on a highly precise floating point 0–1 scale. Radiologists are not computers. Other arguments for a greater number of bins: *it cannot hurt and one should acquire the rating data at greater precision than the noise, especially if the radiologist is able to maintain the finer distinctions.* The author worries that radiologists who are willing to go along with greater precision are over-anxious to co-operate with the experimentalist. In the author's experience, expert radiologists will not modify their reading style and one should be suspicious when overzealous radiologists accede to an investigator's request to interpret images in a style that does not closely resemble the clinic. Radiologists, especially experts, do not like more than about four ratings. The author worked with a famous chest radiologist (the late Dr. Robert Fraser) who refused to use more than four ratings.

Another reason given for using continuous ratings is it reduces instances of data degeneracy. Data is sometimes said to be degenerate if the curve-fitting algorithm, the binormal model and the proper binormal model, cannot fit it. This occurs, for example, if there are no interior points on the ROC plot. Modifying radiologist behavior to accommodate the limitations of analytical methods is inherently dubious. One could simply randomly add or subtract half an integer from the observed ratings, thereby making the rating scale more granular and reduce instances of degeneracy (this is actually done in some ROC software to overcome degeneracy issues). Another possibility is to use the empirical (trapezoidal) area under the ROC curve. This quantity can always be calculated; there are no degeneracy problems with it. Actually, fitting methods now exist that are robust to data degeneracy, such as discussed in **Chapter 19** and **Chapter 20**, so this reason for acquiring continuous data no longer applies.

> The rating task involves a unidimensional scale and the author sees no way of getting around the basic channel-limitation noted by Miller and for this reason recommends a 6-point scale, as in Figure 4.2.

On the other side of the controversy it has been argued that given a large number of allowed ratings levels the observer essentially bins the data into a much smaller number of bins (e.g., 0, 20, 40, 60, 80, 100) and adds a zero-mean noise term to appear to be spreading out the ratings.[35] The author agrees with this reasoning.

4.10 The BI-RADS ratings scale and ROC studies

It is desirable that the rating scale be relevant to the radiologists' daily practice. This assures greater consistency—the fitting algorithms assume that the thresholds are held constant for the duration of the ROC study. Depending on the clinical task, a natural rating scale may already exist. For example, in 1992 the American College of Radiology developed the Breast Imaging Reporting and Data System (BI-RADS) to standardize mammography reporting.[36] There are six assessment categories: category 0 indicates need for additional imaging; category 1 is a negative (clearly non-diseased) interpretation; category 2 is a benign finding; category 3 is probably benign, with short-interval follow-up suggested; category 4 is a suspicious abnormality for which biopsy should be considered; category 5 is highly suggestive of malignancy and appropriate action should be taken. The 4th edition of the BI-RADS manual[37] divides category 4 into three subcategories 4A, 4B, and 4C and adds category 6 for a proven malignancy. The 3-category may be further subdivided into *probably benign with a recommendation for normal or short-term follow-up* and a 3+ category, and *probably benign with a recommendation for immediate follow-up*. Apart from categories 0 and 2, the categories form an ordered set with higher categories representing greater confidence in presence of cancer. How to handle the 0s and the 2s is the subject of some controversy, described next.

4.11 The controversy

Two large clinical studies have been reported in which BI-RADS category data were acquired for > 400,00 screening cases interpreted by many (124 in the first study) radiologists.[38,39] The purpose of the first study was to relate radiologist characteristics to actual performance (e.g., does performance depend on reading volume—the number of cases interpreted per year?), so it could be regarded as a more elaborate version of the Beam et al. study,[40] described in **Chapter 3**. The purpose of the second study was to determine the effectiveness of computer-aided detection (CAD) in screening mammography.

The reported ROC analyses used the BI-RADS assessments labels ordered as follows: 1 < 2 < 3 < 3+ < 0 < 4 < 5. The last column of Table 4.5 shows that with this ordering the numbers of cancer per 1000 patients increases monotonically. The CAD study is discussed later, for now the focus is on the adopted BI-RADS scale ordering that is common to both studies.

The use, specifically the ordering, of the BI-RADS ratings shown in Table 4.5 has been criticized[41] in an editorial, titled "*BI-RADS data should not be used to estimate ROC curves.*" Since BI-RADS is a clinical rating scheme widely used in mammography, the editorial, if correct, implies that ROC analysis of clinical mammography data is impossible. Since the BI-RADS scale was arrived at after considerable deliberation, inability to perform ROC analysis using it would strike at the root of clinical viability of the ROC paradigm. The purpose of this section is to express the reasons why the author has a different take on this controversy.

It is claimed in the editorial[41] that the Barlow et al. method confuses cancer yield with confidence level and that BI-RADS categories 1 and 2 should not be separate entries of the confidence scale, because both indicate no suspicion of cancer.

Table 4.5 The ordering of the BI-RADS ratings shown in the first column correlates with the cancer-rate shown in the last column

BI-RADS assessment	Total number of mammograms	Mammograms in women without breast cancer (%)	Mammograms in women with breast cancer (%)	Cancers per 1,000 screening mammograms
1: Normal	356,030	355,734 (76.2)	296 (12.3)	0.83
2: Benign finding	56,614	56,533 (12.1)	81 (3.4)	1.43
3: Probably benign with a recommendation for normal or short-term follow-up	8,692	8,627 (1.8)	65 (2.7)	7.48
3+: Probably benign with a recommendation for immediate work-up	3,094	3,049 (0.7)	45 (1.9)	14.54
0: Need additional imaging evaluation	42,823	41,442 (8.9)	1,381 (57.5)	32.25
4: Suspicious abnormality, biopsy should be considered	2,022	1,687 (0.4)	335 (13.9)	165.68
5: Highly suggestive of malignancy	237	38 (0.0)	199 (8.3)	839.66

Source: Data from Barlow WE et al., J Natl Cancer Inst., 96, 1840–1850, 2004.

The author agrees with the Barlow et al. ordering of the 2s as more likely to have cancer than the 1s. A category-2 means the radiologist found *something to report*, and the *location* of the finding is part of the clinical report. Even if the radiologist believes the finding is definitely benign, there is a *finite* probability that a category-2 finding is cancer, as evident in the last column of Table 4.5 (1.43 > 0.83). In contrast, there are no findings associated with a category-1 report. (Independent of the Barlow et al. study, a paper,[42] titled *"Benign breast disease and the risk of breast cancer,"* should convince any doubters that benign lesions do have a finite chance of cancer).

The problem with where to put the 0s arises only when one tries to analyze clinical BI-RADS data. In a laboratory study, the radiologist would not be given the category-0 option (in the clinic the zero rating implies additional information is needed before the radiologist will commit to a non-zero rating). In analyzing a clinical study it is incumbent on the study designer to justify the choice of the rating scale adopted. Showing that the proposed ordering agrees with the probability of cancer is justification—and in the author's opinion, given the very large sample size this was accomplished convincingly in the Barlow et al. study. Moreover, the last column of Table 4.5 suggests that any other ordering would violate an important principle, namely, optimal ordering is achieved when each case is rated according to its *likelihood ratio*: defined as the probability of the case being diseased divided by the probability of the case being non-diseased. The likelihood ratio is the betting odds of the case being diseased, which is expected to be monotonic with the empirical probability of the case being diseased, that is, the last column of Table 4.5. Therefore, the ordering adopted in Table 4.5 is equivalent to adopting a likelihood ratio scale and any other ordering would not be monotonic with a likelihood ratio.

The likelihood ratio is described in more detail in **Chapter 20**, which describes ROC fitting methods that yield proper ROC curves, i.e., ones that have monotonically decreasing slope as the operating point moves up the curve from (0,0) to (1,1) and therefore do not (inappropriately) cross the chance diagonal. Key to these fitting methods is adoption of a likelihood ratio scale to rank-order cases, instead of the ratings assumed by the unequal variance binormal model. The proper ROC fitting algorithm implemented in PROPROC software *reorders* confidence levels assumed by the binormal model, **Chapter 20**, Section 20.6. This is analogous to the reordering of the clinical ratings based on cancer rates assumed in Table 4.5. It is illogical to allow reordering of ratings in software but question the same when done in a principled way by a researcher. As expected, the modeled ROC curves in Figure 4.1 of the Barlow publication shows no evidence of improper behavior. This is in contrast to a clinical study (about 50,000 patients spread over 33 hospitals with each mammogram interpreted by two radiologists) using a non-BI-RADS 7-point rating scale which yielded markedly improper ROC curves[43] for the film modality when using ROC ratings (not BI-RADS). This suggests that use of a nonclinical ratings scale for clinical studies, without independent confirmation of the ordering implied by the scale, may be problematical.

The reader might be interested as to reason for the 0-ratings being more predictive of cancer than a 3+ rating, Table 4.5. In the clinic, the 0-rating implies, in effect, *defer decision, incomplete information, additional imaging necessary*. A 0-rating could be due to technical problems with the images: e.g., improper positioning (e.g., missing breast tissue close to the chest wall) or incorrect imaging technique (improper selection of kilovoltage and/or tube charge), making it impossible to properly interpret the images. Since the images are part of the permanent patient record, there are both healthcare and legal reasons why the images need to be optimal. Incorrect technical factors are expected to occur randomly and *not* predictive of cancer. However, if there is a suspicious finding and the image quality is suboptimal, the radiologist may be unable to commit to a decision. They may seek additional imaging, perhaps better compression or a slightly different view angle, to resolve the ambiguity. Such zero ratings are expected with suspicious findings, and therefore are expected to be more predictive of cancer than pure technical reason zero ratings.

(As an aside, the second paper[39] using the ordering shown in Table 4.5 questioned the utility of CAD for breast cancer screening (ca. 2007). This paper was met with a flurry of correspondence[44–51] disputing the methodology (summarized above). The finding regarding lack of utility of CAD has been replicated by more recent studies, again with very large case and reader samples, showing that usage of CAD can actually be detrimental to patient outcome[52] and there has been a call[19] for ending insurance reimbursement for CAD. The current reason given for why CAD does not work is that radiologists do not know how to use it. A recent funding opportunity announcement, https://grants.nih.gov/grants/guide/pa-files/PAR-17-125.html, specifically bars further CAD algorithmic research in favor of determining how to improve utilization of current CAD by radiologists; see http://www.expertcadanalytics.com for full discussion.)

4.12 Discussion

In this chapter, the widely used ratings paradigm was described and illustrated with a sample dataset, Table 4.1. The calculation of ROC operating points from this table was detailed. A formal notation was introduced to describe the counts in this table and the construction of operating points and an **R** example was given. The author does not wish to leave the impression that the ratings paradigm is used only in medical imaging. In fact, the historical reference[3] to the two-question 6-point scale in Figure 4.2, namely Table 3.1 in the book by MacMillan and Creelman, was for a rating study on performance in recognizing odors. The early users of the ROC ratings paradigm were mostly experimental psychologists and psychophysicists interested in studying perception of signals, some in the auditory domain, and some in other sensory domains.

While it is possible to use the equal variance binormal model to obtain a measure of performance, the results depend on the choice of operating point, and evidence was presented for the generally observed fact that most ROC ratings datasets are inconsistent with the equal variance binormal model. This indicates the need for an extended model, to be discussed in **Chapter 6**.

The rating paradigm is a more efficient way of collecting the data compared to repeating the binary paradigm with instructions to cause the observer to adopt different fixed thresholds specific to each repetition. The rating paradigm is also more efficient 2AFC paradigm; more importantly, it is more clinically realistic.

Two controversial but important issues were addressed: the reason for the author's recommendation for adopting a discrete 6-point rating scale and correct usage of clinical BI-RADS ratings in ROC studies. When a clinical scale exists, the empirical disease occurrence rate associated with each rating should be used to order the ratings. Ignoring an existing clinical scale would be a disservice to the radiology community.

The next step is to describe a model for ratings data. Before doing that, it is necessary to introduce an empirical performance measure, namely the area under the empirical or trapezoidal ROC, which does not require any modeling.

References

1. Barnes G, Sabbagh E, Chakraborty D, et al. A comparison of dual-energy digital radiography and screen-film imaging in the detection of subtle interstitial pulmonary disease. *Invest Radiol*. 1989;24(8):585–591.
2. Nishikawa R. Estimating sensitivity and specificity in an ROC experiment. *Breast Imaging*. 2012:690–696.
3. Macmillan NA, Creelman CD. *Detection Theory: A User's Guide*. New York, NY: Cambridge University Press; 1991.
4. Green DM, Swets JA. *Signal Detection Theory and Psychophysics*. New York, NY: John Wiley & Sons; 1966.
5. Burgess AE. Visual perception studies and observer models in medical imaging. *Semin Nucl Med*. 2011;41(6):419–436.
6. Burgess AE. On the noise variance of a digital mammography system. *Med Phys*. 2004;31(7):1987–1995.

7. Burgess AE, Judy PF. Detection in power-law noise: Spectrum exponents and CD diagram slopes. *Proc SPIE*. 2003;5034:57–62.
8. Burgess AE, Jacobson FL, Judy PF. Human observer detection experiments with mammograms and power-law noise. *Med Phys*. 2001;28(4):419–437.
9. Burgess AE. Evaluation of detection model performance in power-law noise. *Proc SPIE*. 2001;4324:419–437.
10. Burgess AE, Jacobson FL, Judy PF. On the detection of lesions in mammographic structure. *SPIE*. 1999;3663(Medical Imaging):304–315.
11. Burgess AE, Chakraborty S. Producing lesions for hybrid mammograms: Extracted tumours and simulated microcalcifications. *Proc SPIE*. 1999;3663:316–321.
12. Chakraborty DP, Kundel HL. Anomalous nodule visibility effects in mammographic images. Paper presented at: Proceedings of the SPIE Medical Imaging 2001: Image Perception and Performance; 2001; San Diego, CA: SPIE.
13. Burgess AE, Li X, Abbey CK. Visual signal detectability with two noise components: Anomalous masking effects. *Journal Opt Soc Am A*. 1997;14(9):2420–2442.
14. Bochud FO, Abbey CK, Eckstein MP. Visual signal detection in structured backgrounds IV, Calculation of figures of merit for model observers in non-stationary backgrounds. *J Opt Soc Am A Opt Image Sci Vis*. 1999;17(2):206–217.
15. Bochud FO, Abbey CK, Eckstein MP. Visual signal detection in structured backgrounds. III. Calculation of figures of merit for model observers in statistically non-stationary backgrounds. *J Opt Soc Am A*. 2000;17(2):193–216.
16. Eckstein M, FO B, Abbey C. Visual signal detection in structured backgrounds IV. Figures of merit for model observers in multiple alternative forced choice with response correlations. *J Opt Soc Am A*. 2000;17:206–217.
17. Eckstein MP, Abbey CK, Bochud FO. A practical guide to model observers for visual detection in synthetic and natural noisy images. In: Kundel H, Beutel J, Van-Metter R, eds. *Handbook of Medical Imaging*. Bellingham, WA: SPIE; 2000:593–628.
18. Abbey CK, Barrett HH. Human- and model-observer performance in ramp-spectrum noise: Effects of regularization and object variability. *J Optical Soc Am A*. 2001;18(3):473–488.
19. Abbey CK, Eckstein MP. Classification image analysis: Estimation and statistical inference for two-alternative forced-choice experiments. *J Vis*. 2002;2(1):66–78.
20. Bochud FO, Abbey CK, Eckstein MP. Search for lesions in mammograms: Statistical characterization of observer responses. *Med Phys*. 2004;31(1):24–36.
21. Burgess AE. Comparison of receiver operating characteristic and forced choice observer performance measurement methods. *Med Phys*. 1995;22(5):643–655.
22. Gur D, Bandos AI, Cohen CS, et al. The "laboratory" effect: Comparing radiologists' performance and variability during prospective clinical and laboratory mammography interpretations. *Radiology*. 2008;249(1):47–53.
23. Gur D, Bandos AI, Fuhrman CR, Klym AH, King JL, Rockette HE. The prevalence effect in a laboratory environment: Changing the confidence ratings. *Acad Radiol*. 2007;14:49–53.
24. Gur D, Rockette HE, Armfield DR, et al. Prevalence effect in a laboratory environment. *Radiology*. 2003;228:10–14.
25. Soh BP, Lee W, McEntee MF, et al. Screening mammography: Test set data can reasonably describe actual clinical reporting. *Radiology*. 2013;268(1):46–53.
26. Rockette HE, Gur D, Metz CE. The use of continous and discrete confidence judgments in receiver operating characteristic studies of diagnostic imaging techniques. *Invest Radiol*. 1992;27:169–172.
27. Wagner RF, Beiden SV, Metz CE. Continuous versus categorical data for ROC analysis: Some quantitative considerations. *Acad Radiol*. 2001;8(4):328–334.
28. Metz CE, Herman BA, Shen J-H. Maximum likelihood estimation of receiver operating characteristic (ROC) curves from continuously-distributed data. *Stat Med*. 1998;17(9):1033–1053.

29. Miller GA. The magical number seven, plus or minus two: Some limits on our capacity for processing information. *Psychol Rev.* 1956;63(2):81–97.

30. Pollack I. The information of elementary auditory displays. *J Acoust Soc Am.* 1952;24(6):745–749.

31. Pollack I. The information of elementary auditory displays. II. *J Acoust Soc Am.* 1953;25(4):765–769.

32. Garner W. An informational analysis of absolute judgments of loudness. *J Exp Psychol.* 1953;46(5):373.

33. Beebe-Center JG, Rogers M, O'connell D. Transmission of information about sucrose and saline solutions through the sense of taste. *J Psychol.* 1955;39(1):157–160.

34. Hake HW, Garner W. The effect of presenting various numbers of discrete steps on scale reading accuracy. *J Exp Psychol.* 1951;42(5):358.

35. Berbaum KS, Dorfman DD, Franken EA, Caldwell RT. An empirical comparison of discrete ratings and subjective probability ratings. *Acad Radiol.* 2002;9(7):756–763.

36. D'Orsi CJ, Bassett LW, Feig SA, et al. *Illustrated Breast Imaging Reporting and Data System.* Reston, VA: American College of Radiology; 1998.

37. D'Orsi CJ, Bassett LW, Berg WA. *ACR BI-RADS-Mammography.* 4th ed. Reston, VA: American College of Radiology; 2003.

38. Barlow WE, Chi C, Carney PA, et al. Accuracy of screening mammography interpretation by characteristics of radiologists. *J Natl Cancer Inst.* 2004;96(24):1840–1850.

39. Fenton JJ, Taplin SH, Carney PA, et al. Influence of computer-aided detection on performance of screening mammography. *N Engl J Med.* 2007;356(14):1399–1409.

40. Beam CA, Layde PM, Sullivan DC. Variability in the interpretation of screening mammograms by US radiologists. Findings from a national sample. *Arch Intern Med.* 1996;156(2):209–213.

41. Jiang Y, Metz CE. BI-RADS data should not be used to estimate ROC curves. *Radiology.* 2010;256(1):29–31.

42. Hartmann LC, Sellers TA, Frost MH, et al. Benign breast disease and the risk of breast cancer. *N Engl J Med.* 2005;353(3):229–237.

43. Pisano ED, Gatsonis C, Hendrick E, et al. Diagnostic performance of digital versus film mammography for breast-cancer screening. *N Engl J Med.* 2005;353(17):1–11.

44. Ciatto S, Houssami N. Computer-aided screening mammography. *N Engl J Med.* 2007;357(1):83–85.

45. Feig SA, Birdwell RL, Linver MN. Computer-aided screening mammography. *N Engl J Med.* 2007;357(1):83–85.

46. Fenton JJ, Barlow WE, Elmore JG. Computer-aided screening mammography. *N Engl J Med.* 2007;357(1):83–85.

47. Gur D. Computer-aided screening mammography. *N Engl J Med.* 2007;357(1):83–85.

48. Nishikawa RM, Schmidt RA, Metz CE. Computer-aided screening mammography. *N Engl J Med.* 2007;357(1):83–85.

49. Ruiz JF. Computer-aided screening mammography. *N Engl J Med.* 2007;357(1):83–85.

50. Berry DA. Computer-assisted detection and screening mammography: Where's the beef? *J Natl Cancer Inst.* 2011;103:1139–1141.

51. Nishikawa RM, Giger ML, Jiang Y, Metz CE. Re: Effectiveness of computer-aided detection in community mammography practice. *J Natl Cancer Inst.* 2012;104(1):77.

52. Philpotts LE. Can computer-aided detection be detrimental to mammographic interpretation? *Radiology.* 2009;253(1):17–22.

5

Empirical AUC

5.1 Introduction

The ROC plot, introduced in **Chapter 3**, is defined as the plot of sensitivity (y-axis) versus 1-specificity (x-axis). Equivalently, it is the plot of TPF (y-axis) versus FPF (x-axis). An equal variance binormal model was introduced which allows an ROC plot to be fitted to a single observed operating point. In **Chapter 4**, the more commonly used ratings paradigm was introduced.

One of the reasons for fitting observed counts data, such as in Table 4.1 in **Chapter 4**, to a parametric model, is to derive analytical expressions for the separation parameter μ of the model or the area AUC under the curve. Other figures of merit, such as the TPF at a specified FPF, or the partial area to the left of a specified FPF, can also be calculated from this model. Each figure of merit can serve as the basis for comparing two readers to determine which one is better. They have the advantage of being single values, as opposed to a pair of sensitivity-specificity values, thereby making it easier to unambiguously compare performances. Additionally, they often yield physical insight into the task, for example, the separation parameter is the perceptual signal-to-noise corresponding to the diagnostic task.

Figure 4.1a and b illustrated that the equal variance binormal model did not fit a clinical dataset and that an unequal variance binormal model yielded a better visual fit. *This turns out to be an almost universal finding.* Before getting into the complexity of the unequal variance binormal model curve fitting, it is appropriate to introduce a simpler *empirical* approach, which is very popular with some researchers. The New Oxford American Dictionary definition of *empirical* is "based on, concerned with, or verifiable by observation or experience rather than theory or pure logic." The method is also termed *nonparametric* as it does not involve any parametric assumptions (specifically normality assumptions). Notation is emphasized for labeling individual cases that is used in subsequent chapters. An important theorem relating the empirical area under the ROC to a formal statistic, known as the Wilcoxon, is described. The importance of the theorem derives from its applications to nonparametric analysis of ROC data.

5.2 The empirical ROC plot

The *empirical* ROC plot is constructed by connecting adjacent observed operating points, including the trivial ones at (0,0) and (1,1), with straight lines. The trapezoidal area under this plot is a nonparametric figure of merit that is threshold independent. Since no parametric assumptions are involved, some prefer it to parametric methods, such as the one to be described in the next chapter. (In the context of AUC, the terms *empirical*, *trapezoidal*, or *nonparametric* all mean the same thing.)

5.2.1 Notation for cases

As in Section 3.5, cases are indexed by $k_t t$ where t indicates the truth-status at the case (i.e., patient) level, with $t = 1$ for non-diseased cases and $t = 2$ for diseased cases. Index k_1 ranges from one to K_1 for non-diseased cases and k_2 ranges from one to K_2 for diseased cases, where K_1 and K_2 are the total number of non-diseased and diseased cases, respectively. In Table 5.1, each case is represented as a label, with labels starting with N for non-diseased and starting with D for diseased cases. There are 11 non-diseased cases, labeled N1—N11, in the upper row of boxes and there are seven diseased cases, labeled D1—D7, in the lower row of boxes.

To address a case, one needs *two* indices. The first index is the column number and the second index is the row number and, moreover, the total number of columns can, and does in this example, depend on the row number. This means the first index has to be t-dependent, that is, k_t, denoting the column index of a case with truth index t and additionally one needs a row number, t, to select the appropriate row of the table. Alternative notation more commonly uses a single index k to label the cases. It reserves the first K_1 positions for non-diseased cases and the rest for diseased cases: that is, $k = 3$ corresponds to the third non-diseased case, $k = K_1 + 5$ corresponds to the fifth diseased case, and so on. Because it extends more easily to more complex data structures, for example, FROC, the author prefers the two-index notation.

5.2.2 An empirical operating point

Let $z_{k_t t}$ represent the (realized) Z-sample of case $k_t t$. For a given reporting threshold ζ, and assuming a *positive-directed* rating scale (i.e., higher values correspond to greater confidence in presence of disease), empirical false positive fraction $FPF(\zeta)$ and empirical true positive fraction $TPF(\zeta)$ are defined by

$$FPF(\zeta) = \frac{1}{K_1} \sum_{k_1=1}^{K_1} I(z_{k_1 1} \geq \zeta) \tag{5.1}$$

$$TPF(\zeta) = \frac{1}{K_2} \sum_{k_2=1}^{K_2} I(z_{k_2 2} \geq \zeta) \tag{5.2}$$

Here, $I(x)$ is the *indicator function* that equals one if x is true and is zero otherwise.

$$\left. \begin{matrix} I(TRUE) = 1 \\ I(FALSE) = 0 \end{matrix} \right\} \tag{5.3}$$

In Equations 5.1 and 5.2, the indicator functions act as *counters*, effectively counting instances where the Z-sample of a case equals or exceeds ζ, and division by the appropriate denominator

Table 5.1 On the need for two indices to label cases in a ROC study. The upper row denotes 11 non-diseased cases, labeled N1—N11, while the lower row denotes seven diseased cases, labeled D1—D7. To address any case one needs two indices: the row number (t = 1 or t = 2) and the column number $k_t t$. Since in general the column number depends on the value of *t*, one needs two indices to specify the column index.

N1	N2	N3	N4	N5	N6	N7	N8	N9	N10	N11
D1	D2	D3	D4	D5	D6	D7				

yields the desired left-hand sides of these equations. The operating point $O(\zeta)$ corresponding to threshold ζ is defined by

$$O(\zeta) = \big(FPF(\zeta), TPF(\zeta)\big) \tag{5.4}$$

The difference between Equations 5.1 and 5.2 versus Equations 3.22 and 3.23 corresponds to that between parametric and nonparametric methods. In **Chapter 3**, an analytical (or parametric, i.e., depending on model parameters) expression for an operating point was obtained. In contrast, in Equations 5.1 and 5.2, one uses the *observed* ratings to calculate the empirical operating point.

5.3 Empirical operating points from ratings data

Consider a ratings *ROC* study with R bins. Describing an R-rating *empirical* ROC plot requires *R-1* ordered *empirical* thresholds, see Equation 4.13, reproduced below (in this chapter the E superscript on each empirical threshold parameter is implicit and therefore not shown):

$$\left. \begin{aligned} \zeta_r &= r \\ r &= 1,.., R-1; \\ \zeta_0 &= -\infty; \; \zeta_R = +\infty \end{aligned} \right\} \tag{5.5}$$

The *discrete* operating point O_r is obtained by replacing ζ with ζ_r in Equation 5.4, where

$$O_r \equiv O(\zeta_r) \equiv (FPF_r, TPF_r) \tag{5.6}$$

Its coordinates are defined by

$$\left. \begin{aligned} FPF_r \equiv FPF(\zeta_r) &= \frac{1}{K_1} \sum_{k_1=1}^{K_1} I\big(z_{k_1 1} \geq \zeta_r\big) \\ TPF_r \equiv TPF(\zeta_r) &= \frac{1}{K_2} \sum_{k_2=1}^{K_2} I\big(z_{k_2 2} \geq \zeta_r\big) \end{aligned} \right\} \tag{5.7}$$

For example,

$$\left. \begin{aligned} FPF_4 &= \frac{1}{K_1} \sum_{k_1=1}^{K_1} I\big(z_{k_1 1} \geq 5\big) = 1/60 = 0.017 \\ TPF_4 &= \frac{1}{K_2} \sum_{k_2=1}^{K_2} I\big(z_{k_2 2} \geq 5\big) = 22/50 = 0.44 \\ O_4 &\equiv (FPF_4, TPF_4) = (0.017, 0.44) \end{aligned} \right\} \tag{5.8}$$

In Table 4.1, a sample clinical ratings data set was introduced. For convenience it is reproduced in Table 5.2. In this example, $R = 5$, corresponding to the 5-ratings bins used to acquire the data.

Shown in Section 5.3.1 is a partial code listing of **mainEmpRocPlot.R** showing implementation of Equation 5.7.

Table 5.2 A typical ROC counts table, reproduced from Chapter 4, Table 4.1.

	Counts in ratings bins				
	Rating = 1	**Rating = 2**	**Rating = 3**	**Rating = 4**	**Rating = 5**
$K_1 = 60$	30	19	8	2	1
$K_2 = 50$	5	6	5	12	22
	Operating points				
	≥ 5	≥ 4	≥ 3	≥ 2	≥ 1
FPF	0.017	0.050	0.183	0.500	1
TPF	0.440	0.680	0.780	0.900	1

5.3.1 Code listing (partial)

```
rm(list = ls()) # mainEmpRocPlot.R
...
K1 <- 60;K2 <- 50
FPF <- c(0, cumsum(rev(c(30, 19, 8, 2, 1)))) / K1)
TPF <- c(0, cumsum(rev(c(5, 6, 5, 12, 22)))) / K2)
...
```

Lines 6-7 construct the counts table shown in Table 5.2 and implements Equation 5.7. Insert a break point at line 9 - i.e., click to the left of the line number. A red dot appears. Source the code. The code pointer - green arrow - should stop at line 9. Exit debug mode - i.e., click on the square red stop symbol. The function **cumsum()** is used to calculate the cumulative sum. The **rev()** function reverses the order of the array supplied as its argument. Copy and paste commands (i.e., the portion of the line appearing after the > symbol) to the **Console** window as shown below, and hit Enter to understand how the code implements Equation 5.7.

5.3.2 Code snippets

```
> c(30, 19, 8, 2, 1)
[1] 30 19 8 2 1
> rev(c(30, 19, 8, 2, 1))
[1] 1 2 8 19 30
> cumsum(rev(c(30, 19, 8, 2, 1)))
[1] 1 3 11 30 60
> c(0, cumsum(rev(c(30, 19, 8, 2, 1))) / K1)
[1] 0.00000 0.01667 0.05000 0.18333 0.50000 1.00000
```

Figure 5.1 is the empirical ROC plot, produced by removing the break point - click on the red dot, it will disappear - and sourcing **mainEmpRocPlot.R**. It illustrates the convention used to label the operating points introduced in Section 4.3 that is, O_1 is the uppermost nontrivial point, and the labels are incremented by unity as one moves down the plot. By convention, not shown are the trivial operating points $O_0 \equiv (FPF_0, TPF_0) \equiv (1,1)$ and $O_R \equiv (FPF_R, TPF_R) \equiv (0,0)$.

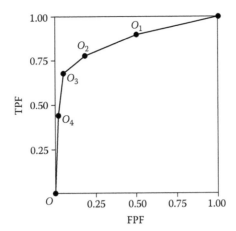

Figure 5.1 Convention: The operating points are numbered starting with the uppermost nontrivial one, which is O_1, and working down the plot. This figure corresponds to the data in Table 5.2. The plot was produced by **mainEmpROCPlot.R**.

5.4 AUC under the empirical ROC plot

Figure 5.2 shows the empirical plot for the data in Table 5.2, and the area under the curve (AUC) is the shaded area. By dropping imaginary vertical lines from the nontrivial operating points onto the x-axis, the shaded area is seen to be the sum of one triangular shaped area and four trapezoids. One may be tempted to write equations to calculate the total area using elementary algebra, but that would be unproductive. There is a theorem (see Section 5.5) that the empirical area is exactly equal to a particular statistic known as the Mann–Whitney–Wilcoxon statistic,[1,2] which, in this book, is abbreviated to the *Wilcoxon* statistic. Calculating this statistic is much simpler than calculating and summing the areas of the triangle and trapezoids or doing planimetry.

5.5 The Wilcoxon statistic

A statistic is any value calculated from observed data. The Wilcoxon statistic is defined in terms of the observed ratings, by

$$W = \frac{1}{K_1 K_2} \sum_{k_1}^{K_1} \sum_{k_2}^{K_2} \psi\left(z_{k_1 1}, z_{k_2 2}\right) \tag{5.9}$$

The function $\psi(x, y)$ is defined by

$$\psi(x, y) = 1 \quad x < y$$
$$\psi(x, y) = 0.5 \quad x = y \tag{5.10}$$
$$\psi(x, y) = 0 \quad x > y$$

ψ is sometimes called the *kernel* function. It is unity if the diseased case is rated higher, 0.5 if the two are rated the same, and zero otherwise. Each evaluation of the kernel function results from a comparison of a case from the non-diseased set with one from the diseased set. In Equation 5.9 the two summations and division by the total number of comparisons yields the observed or *empirical*

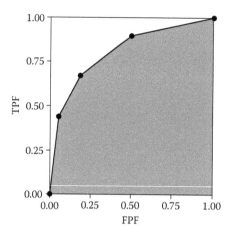

Figure 5.2 The empirical ROC plot corresponding to Table 5.2; the shaded area is the area under this plot, a widely used figure of merit in nonparametric ROC analysis. The plot was produced by the code in **mainEmpiricalAUC.R**.

probability that diseased cases are rated higher than non-diseased ones. Since it is a probability, it can theoretically range from zero to one. However, if the observer has any discrimination ability, one expects diseased cases to be rated equal or greater than non-diseased ones, so in practice one expects:

$$0.5 \leq W \leq 1 \tag{5.11}$$

The limit 0.5 corresponds to a guessing observer, whose operating point moves along the chance diagonal of the ROC plot.

5.6 Bamber's equivalence theorem

Here is the result: the Wilcoxon statistic W equals the area AUC under the *empirical* ROC plot.

$$AUC = W \tag{5.12}$$

Numerical illustration: While hardly a proof, as an illustration of the theorem it is helpful to calculate the sum on the right-hand side of Equation 5.12 and compare it to the result of a direct integration of the area under the empirical ROC curve (i.e., adding the area of a triangle and several trapezoids). **R** provides a function that does just that. It is part of package **caTools** and is called **trapz(x,y)**. It takes two array arguments, **x** and **y**, where in the current case **x** is FPF and **y** is TPF. One has to be careful to include the end-points as otherwise the area will be underestimated. The Wilcoxon W and the empirical area AUC are implemented in Online Appendix 5.A, in file **mainWilcoxon.R**. **Source** the file to get the following output.

5.6.1 Code output

```
> source(,,,)
The wilcoxon statistic is = 0.8606667
direct integration yields AUC = 0.8606667
```

Note the equality of the two estimates.

Proof: The following proof is adapted from a paper by Bamber[3] and while it may appear to be restricted to discrete ratings, the result is, in fact, quite general, that is, it is applicable even if the ratings are acquired on a continuous (floating point variable) scale. The reason is as follows: in an R-rating ROC study the observed Z-samples or ratings take on integer values, 1 through R. *If R is large enough, ordering information present in the continuous data is not lost upon binning (i.e., discretizing)—this is the reason why the proof is quite general.*

In the following it is helpful to keep in mind that one is dealing with *discrete* distributions of the ratings, described by probability *mass* functions as opposed to probability *density* functions, for example, $P(Z_2 = \zeta_i)$ is *not zero*, as would be the case for continuous ratings.

The abscissa of the operating point i is $P(Z_1 \geq \zeta_i)$ and the corresponding ordinate is $P(Z_2 \geq \zeta_i)$. Here Z_1 is a random sample from a non-diseased case and Z_2 is a random sample from a diseased case. The shaded trapezoid, defined by drawing horizontal lines from operating points i (upper) and $i+1$ (lower) to the right edge of the ROC plot (Figure 5.3), created by the code in `MainBamberTheorem.R`, see Online Appendix 5.A, has height:

$$P(Z_2 \geq \zeta_i) - P(Z_2 \geq \zeta_{i+1}) = P(Z_2 = \zeta_i) \tag{5.13}$$

The validity of this equation can perhaps be more easily seen when the first term is written in the form:

$$P(Z_2 \geq \zeta_i) = P(Z_2 = \zeta_i) + P(Z_2 \geq \zeta_{i+1})$$

The lengths of the top and bottom edges of the trapezoid are, respectively,

$$1 - P(Z_1 \geq \zeta_i) = P(Z_1 < \zeta_i)$$

and

$$1 - P(Z_1 \geq \zeta_{i+1}) = P(Z_1 < \zeta_{i+1})$$

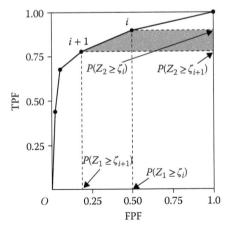

Figure 5.3 Illustration of the derivation of Bamber's equivalence theorem, which shows an empirical ROC plot for R = 5; the shaded area is due to points labeled *i* and *i* + 1. This figure was created by the code in `MainBamberTheorem.R`.

The area A_i of the shaded trapezoid in Figure 5.3 is (the steps are shown explicitly)

$$
\left.
\begin{aligned}
A_i &= \frac{1}{2} P(Z_2 = \zeta_i) \left[P(Z_1 < \zeta_i) + P(Z_1 < \zeta_{i+1}) \right] \\[6pt]
A_i &= P(Z_2 = \zeta_i) \left[\frac{1}{2} P(Z_1 < \zeta_i) + \frac{1}{2} \left(P(Z_1 = \zeta_i) + P(Z_1 < \zeta_i) \right) \right] \\[6pt]
A_i &= P(Z_2 = \zeta_i) \left[\frac{1}{2} P(Z_1 = \zeta_i) + P(Z_1 < \zeta_i) \right]
\end{aligned}
\right\}
\tag{5.14}
$$

In going from the first to the second line of Equation 5.14, use has been made of the last relation, below, derived from Equation 5.13 after replacing the truth subscript 2 with 1 and expressing the two probabilities on the left-hand side of Equation 5.13 in terms of their complementary probabilities.

$$
\left.
\begin{aligned}
1 - P(Z_1 < \zeta_i) - \left(1 - P(Z_1 < \zeta_{i+1}) \right) &= P(Z_1 = \zeta_i) \\[6pt]
P(Z_1 < \zeta_{i+1}) - P(Z_1 < \zeta_i) &= P(Z_1 = \zeta_i) \\[6pt]
P(Z_1 < \zeta_{i+1}) &= P(Z_1 = \zeta_i) + P(Z_1 < \zeta_i)
\end{aligned}
\right\}
\tag{5.15}
$$

Summing all values of i, one gets for the total area under the empirical ROC plot:

$$
AUC = \sum_{i=0}^{R-1} A_i = \frac{1}{2} \sum_{i=0}^{R-1} P(Z_2 = \zeta_i) P(Z_1 = \zeta_i) + \sum_{i=0}^{R-1} P(Z_2 = \zeta_i) P(Z_1 < \zeta_i)
\tag{5.16}
$$

It is shown in Appendix 5.A that the term A_0 corresponds to the triangle at the upper right corner of Figure 5.3. Likewise, the term A_4 corresponds to the horizontal trapezoid defined by the lowest nontrivial operating point.

Equation 5.16 can be restated as

$$
AUC = \frac{1}{2} P(Z_2 = Z_1) + P(Z_1 < Z_2)
\tag{5.17}
$$

The Wilcoxon statistic was defined in Equation 5.9. It can be seen that the comparisons implied by the summations and the weighting implied by the kernel function are estimating the two probabilities in the expression for AUC in Equation 5.17. Therefore,

$$
AUC = W
\tag{5.18}
$$

5.7 The Importance of Bamber's theorem

The equivalence theorem is the starting point for all nonparametric methods of analyzing ROC plots, for example, References 4 and 5. Prior to Bamber's work one knew how to *plot* an empirical operating characteristic and how to *calculate* the Wilcoxon statistic, but their equality had not been proven. This was Bamber's essential contribution. In the absence of this theorem, the Wilcoxon statistic would be *just another statistic*. The theorem is so important that a major paper appeared in *Radiology*[A] devoted to the equivalence. The title of this paper was "*The meaning and use of the area under a receiver operating characteristic (ROC) curve.*" The equivalence theorem literally gives meaning to the empirical area under the ROC.

5.8 Discussion/Summary

In this chapter, a simple method for estimating the area under the ROC plot is described. The empirical AUC is a nonparametric measure of performance. Its simplicity and clear physical interpretation as the AUC under the empirical ROC has spurred much theoretical development. These include the De Long et al. method for estimating the variance of AUC of a single ROC empirical curve, and comparing pairs of ROC empirical curves.[5] A paper by Hanley and Hajian-Tilaki (Ref. 6) is a particularly lucid account of use of the empirical AUC to estimate variability. Bamber's theorem, namely the equivalence between the empirical AUC and the Wilcoxon statistic has been derived and demonstrated. More recently, a first principle approach to analyzing multiple-reader multiple-case (MRMC) datasets, has been described that is based on the empirical AUC.[7,8]

Having described the empirical AUC and its relationship to the Wilcoxon statistic, we return to the subject of parametric modeling of the ROC curve, the subject of the next chapter.

Appendix 5.A Details of Wilcoxon theorem

5.A.1 Upper triangle

For $i = 0$, Equation 5.14 implies (the lowest empirical threshold is unity, the lowest allowed rating, and there are no cases rated lower than one):

$$\left.\begin{aligned} A_0 &= P(Z_2 = 1)\left[\frac{1}{2}P(Z_1 = 1) + P(Z_1 < 1)\right] \\ A_0 &= \frac{1}{2}P(Z_1 = 1)P(Z_2 = 1) \end{aligned}\right\}$$

The base of the triangle is (the point labeled i in Fig. 5.3 corresponds to empirical threshold equal to 2)

$$1 - P(Z_1 \geq 2) = P(Z_1 < 2) = P(Z_1 = 1)$$

The height of the triangle is

$$1 - P(Z_2 \geq 2) = P(Z_2 < 2) = P(Z_2 = 1)$$

5.A.2 Lowest trapezoid

For $i = 4$, Equation 5.14 implies

$$\left.\begin{aligned} A_4 &= P(Z_2 = 5)\left[\frac{1}{2}P(Z_1 = 5) + P(Z_1 < 5)\right] \\ A_4 &= \frac{1}{2}P(Z_2 = 5)\left[P(Z_1 = 5) + 2P(Z_1 < 5)\right] \\ A_4 &= \frac{1}{2}P(Z_2 = 5)\left[P(Z_1 = 5) + P(Z_1 < 5) + P(Z_1 < 5)\right] \\ A_4 &= \frac{1}{2}P(Z_2 = 5)\left[1 + P(Z_1 < 5)\right] \end{aligned}\right\}$$

The upper side of the trapezoid is

$$1 - P(Z_1 \geq 5) = P(Z_1 < 5)$$

The lower side is unity. The average of the two sides is

$$\frac{1 + P(Z_1 < 5)}{2}$$

The height is

$$P(Z_2 \geq 5) = P(Z_2 = 5)$$

Multiplication of the last two expressions yields A_4.

References

1. Wilcoxon F. Individual comparison by ranking methods. *Biometrics*. 1945;1:80–83.
2. Mann HB, Whitney DR. On a test of whether one of two random variables is stochastically larger than the other. *Ann Math Stat*. 1947;18:50–60.
3. Bamber D. The area above the ordinal dominance graph and the area below the receiver operating characteristic graph. *J Math Psychol*. 1975;12(4):387–415.
4. Hanley JA, McNeil BJ. The meaning and use of the area under a receiver operating characteristic (ROC) curve. *Radiology*. 1982;143(1):29–36.
5. DeLong ER, DeLong DM, Clarke-Pearson DL. Comparing the areas under two or more correlated receiver operating characteristic curves: A nonparametric approach. *Biometrics*. 1988;44:837–845.
6. Hanley JA, Hajian-Tilaki KO. Sampling variability of nonparametric estimates of the areas under receiver operating characteristic curves: An update. *Acad Radiol*. 1997;4(1):49–58.
7. Clarkson E, Kupinski MA, Barrett HH. A probabilistic model for the MRMC method, Part 1: Theoretical development. *Acad Radiol*. 2006;13(11):1410–1421.
8. Gallas BD. One-shot estimate of MRMC variance: AUC. *Acad Radiol*. 2006;13(3):353–362.

6

Binormal model

6.1 Introduction

The *equal* variance binormal model was described in **Chapter 3**. The ratings method of acquiring ROC data and calculation of operating points was discussed in **Chapter 4**. It was shown in Figure 4.1a and b, that for a clinical dataset the *unequal* variance binormal model fit the data better than the equal variance binormal model, although how the unequal variance fit was obtained was not discussed. This chapter deals with details of the unequal variance binormal model, often abbreviated to *binormal model*, establishes necessary notation, and derives expressions for sensitivity, specificity, and the area under the predicted ROC curve (due to its complexity, the derivation of area under curve [AUC] appears in (book) Appendix 6.A).

The binormal model is widely used to model *univariate* datasets, in which there is *one* ROC rating per case, as in a single observer interpreting cases, one at a time, in a single modality. By convention the qualifier *univariate* is often omitted, that is, it is implicit. In **Chapter 21** a *bivariate* model will be described where each case yields two ratings, as in a single observer interpreting cases in two modalities, or the homologous problem of two observers interpreting cases in a single modality.

The main aim of this chapter is to take the mystery out of statistical curve fitting. With the passing of Dorfman, Metz, and Swensson, parametric modeling has been neglected. Researchers are instead focusing on nonparametric analysis using the empirical AUC. While useful, empirical AUC yields almost no insight into what is limiting performance. Taking the mystery out of curve fitting will allow the reader to appreciate later chapters that describe more complex fitting methods, which yield important insights into factors limiting performance.

Here is the organization of this chapter. It starts with a description of the binormal model and how it accommodates data binning. An important point, on which there is confusion, namely the invariance of the binormal model to arbitrary monotone transformations of the ratings is explicated with an **R** example. Expressions for sensitivity and specificity are derived. Two notations used to characterize the binormal model are explained. Expressions for the *pdfs* of the binormal model are derived. A simple linear fitting method is illustrated—this used to be the only recourse a researcher had before Dorfman and Alf's seminal publication.[1] The maximum likelihood (ML) method for estimating parameters of the binormal model is detailed. Validation of the fitting method is described; that is, how can one be confident that the fitting method, which makes normality and other assumptions, is valid for a dataset arising from an unknown distribution. Appendix 6.A derives an expression for, the partial area under the ROC curve, defined as the area under the binormal ROC curve from FPF = 0 to FPF = c, where $0 \leq c \leq 1$. As a special case $c = 1$ yields the total area under the binormal ROC.

Online Appendix 6.A describes methods for debugging code and displaying plots. Most of the figures in this book were generated using **ggplot2**, an **R** plotting package (**ggplot2** takes its name from *grammar of graphics*). The interested reader can study the accompanying examples to learn how to use it. Explaining **ggplot2** is outside the scope of this book—there are books on how to use the plotting package[2] (the cited book is just the original book; Amazon has scores of books on **R** and making plots). In the author's view, one learns the minimum to get the job done, and one's understanding grows with usage; the accompanying examples should help in that regard. (For those who wish to learn more there are inexpensive Kindle versions of some books,[3] which are useful to have on your computer. However, more often than not, a Google search beginning with "*ggplot* your question" works.)

This chapter also illustrates usage of a web ROC program that fits ROC ratings data. This is a web implementation of Metz's ROCFIT software. The website output for a sample dataset is in Online Appendix 6.B. As the name implies, ROCFIT fits ROC data. The author has the original Fortran code for this program, but nowadays it is part of software called ROCKIT distributed by the University of Chicago ROC software website (a Google search with the preceding search terms will reveal the correct website address). Unfortunately, the source code for ROCKIT is not available and, at the time of writing, 09/17, the software is no longer supported. This is one reason why the author has chosen to go open-source: this way the software lives on, independent of the vagaries of the NIH funding process.

Online Appendix 6.C explains the author's implementation of ROCFIT, in file **mainRocfitR.R**. The reader can choose to simply execute it by clicking on the **Source** button, or may choose to learn how maximum likelihood estimation works. The results yielded by the program are compared to those from the website implementation. Online Appendix 6.D has details on the goodness of fit implementation. Online Appendix 6.E illustrates computation of the variance of the binormal model AUC. Online Appendix 6.F has various transformations that are useful in ML fitting—they ensure that the parameter optimization function always passes legal values to the function that is being optimized, for example, the standard deviation must always be positive.

The starting point is a familiar dataset.

6.2 The binormal model

For convenience, Table 4.1 from **Chapter 4** is reproduced in Table 6.1.

Table 6.1 A typical ROC counts table copied from **Chapter 4**. Listed in the upper half of the table are the number of cases in specific ratings bins, listed separately for actually non-diseased and actually diseased cases. There are $K_1 = 60$ non-diseased cases and $K_2 = 50$ diseased cases in this dataset. The lower half of the table lists the corresponding FPF and TPF values, that is, the abscissa and ordinate, respectively, of the operating points on the ROC plot.

	Counts in ratings bins				
	Rating = 1	Rating = 2	Rating = 3	Rating = 4	Rating = 5
$K_1 = 60$	30	19	8	2	1
$K_2 = 50$	5	6	5	12	22
			Operating points		
	≥5	≥4	≥3	≥2	≥1
FPF	0.017	0.050	0.183	0.500	1
TPF	0.440	0.680	0.780	0.900	1

In the binormal model the Z-sampling is defined by

$$Z_{k_t t} \sim N\left(\mu_t, \sigma_t^2\right); t = 1, 2 \qquad (6.1)$$

where

$$\left. \begin{array}{l} \mu_1 = 0, \ \mu_2 = \mu \\ \sigma_1^2 = 1, \ \sigma_2^2 = \sigma^2 \end{array} \right\} \qquad (6.2)$$

Equation 6.1 states that the Z-samples for non-diseased cases are distributed as an $N(0,1)$ distribution, that is, the unit normal distribution, while the Z-samples for diseased cases are distributed as an $N\left(\mu, \sigma^2\right)$ distribution, that is, a normal distribution with mean μ and variance σ^2. It is a 2-parameter model.

A more complicated version of this model allows the mean of the non-diseased distribution to be non-zero and its variance to be different from unity. The 4-parameter model is no more general than the 2-parameter model. The reason is that one is free to transform the decision variable, and associated thresholds, by applying arbitrary monotonic increasing function transformations, which do not change the ordering of the ratings and hence do not change the ROC curve. So, if the mean of the noise distribution were non-zero, subtracting this value from all Z-samples would shift the effective mean of the non-diseased distribution to zero (the shifted Z-values are monotonically related to the original values) and the mean of the shifted diseased distribution becomes $\mu_2 - \mu_1$. Next, one divides (division by a positive number is also a monotonic transformation) all Z-samples by σ_1, resulting in the scaled non-diseased distribution having unit variance, and the scaled diseased distribution has mean $(\mu_2 - \mu_1)/\sigma_1$ and variance $(\sigma_2/\sigma_1)^2$. Therefore, if one starts with four parameters then one can, by simple shifting and scaling operations, reduce the model to two parameters, as in Equation 6.1. The two parameters in Equation 6.1 are related to the four parameters by

$$\left. \begin{array}{l} \mu = \left(\mu_2 - \mu_1\right)/\sigma_1 \\ \sigma = \sigma_2/\sigma_1 \end{array} \right\} \qquad (6.3)$$

Online Appendix 6.A has a demonstration of this fact; the purpose is not to illustrate the obvious (however, the author has seen a publication on Bayesian ROC estimation using the 4-parameter model, so perhaps this is not so obvious), rather to demonstrate coding and plotting techniques that will be helpful later on. Sourcing the relevant file, **main4ParameterVs2Parameter.R**, yields two ROC plots, one based on the 2-parameter model, and the other based on the 4-parameter model; one can use the left and right arrow buttons to be convinced that the plots are indeed identical.

6.2.1 Binning the data

In an R-rating ROC study the observed ratings r take on integer values, 1 through R, it being understood that higher ratings correspond to greater confidence for disease. Defining dummy cutoffs $\zeta_0 = -\infty$ and $\zeta_R = +\infty$, the binning rule for a case with a realized Z-sample z is (**Chapter 4**, Equation 4.12):

Binning Rule:

$$if \ \zeta_{r-1} \leq z < \zeta_r \Rightarrow rating = r \qquad (6.4)$$

In the binormal model, the variance σ^2 of the Z-samples for diseased cases is allowed to be different from unity. Most ROC datasets[4] are consistent with $\sigma > 1$. Figure 6.1, generated using **mainRocPdfsWithCutoffs.R**, with $\mu = 1.5, \sigma = 1.5$, illustrates how realized Z-samples are

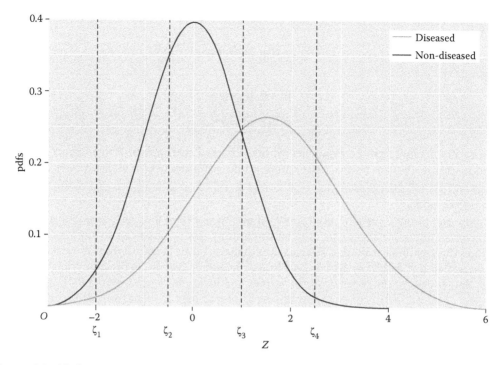

Figure 6.1 This figure shows the usage of two normal distributions and ordered thresholds for modeling data obtained in the ratings paradigm. The example shown is for a 5-rating task, R = 5, with allowed ratings $r = 1,2,...,5$. The thresholds (dashed lines) are at -2.0, -0.5, 1, and 2.5, corresponding to $\zeta_1, \zeta_2, \zeta_3, \zeta_4$, respectively. The dummy thresholds ζ_0 and ζ_5 are at negative and positive infinity, respectively. The decision rule is to label a case with rating r if $\zeta_{r-1} \leq z < \zeta_r$. The plot was generated using `mainRocPdfsWithCutoffs.R`. The non-diseased pdf has zero mean and unit standard deviation, while the diseased distribution has mean 1.5 and standard deviation 1.5.

converted to ratings, that is, it illustrates application of the binning rule. The actual thresholds are at -2.0, -0.5, 1, and 2.5, line 17 in the code file. For example, a case with Z-sample equal to -2.5 would be rated 1, one with Z-sample equal to -1 would be rated 2, cases with Z-samples greater than or equal to 2.5 would be rated 5, and so on.

6.2.2 Invariance of the binormal model to arbitrary monotone transformations

The binormal model is not as restrictive as might appear at first sight. Any monotone increasing transformation* $Y = f(Z)$ applied to the observed Z-samples, and the associated thresholds, will yield the same observed data, e.g., Table 6.1. *This is because such a transformation leaves the ordering of the ratings unaltered and hence results in the same bin counts and operating points.* While the distributions for Y will not be binormal (i.e., two normal distributions), one can safely pretend that one is still dealing with an underlying *latent* binormal model. An alternative way of stating this is that any pair of distributions is allowed as long as they are reducible to a binormal model by a monotonic increasing transformation of Y : e.g., $Z = f^{-1}(Y)$. (If f is a monotone increasing function of its argument, so is f^{-1}.) For this reason the term *latent underlying normal distributions* is sometimes used to describe the binormal model. The robustness of the binormal model has been

---

* A monotone increasing function is defined as one whose value does not decrease as its argument increases, that is, its slope is non-negative (zero slopes in some regions are allowed).

investigated.[5,6] The referenced paper by Dorfman et al. has an excellent discussion of the robustness of the binormal model.

The robustness of the binormal model, that is, the flexibility allowed by the infinite choices of monotonic increasing functions, application of each of which leaves the ordering of the data unaltered, is widely misunderstood. The non-Gaussian appearance of histograms of ratings in ROC studies can lead one to incorrectly conclude that the binormal model is inapplicable to such datasets. To quote a reviewer of one of the author's recent papers,[7] "I have had multiple encounters with statisticians who do not understand this difference.... They show me histograms of data, and tell me that the data is obviously not normal, therefore the binormal model should not be used."

The reviewer is correct. The **R** example below, file **mainLatentTransforms.R**, illustrates the misconception referred to by the reviewer.

6.2.2.1 Code listing

```
# mainLatentTransforms.R
# shows that monontone transformations have no effect on
# AUC even though the pdfs look non-gaussian
# common statistician misconception about ROC analysis

rm( list = ls( all = TRUE ) )
library(RJafroc)
library(ggplot2)
source( 'TrapezoidalArea.R' )
options(digits = 7)

Y <- function(z,mu1,mu2,sigma1,sigma2,f) {
  y <- (1-f)*pnorm((z-mu1)/sigma1)*100
  +f*pnorm((z-mu2)/sigma2)*100
  return( y )
}

fArray <- c(0.1,0.5,0.9)
seedArray <- c(10,11,12)
for (i in 1:3) {
  f <- fArray[i];seed <- seedArray[i];set.seed(seed)
  K1 <- 900;K2 <- 1000
  mu1 <- 30;sigma1 <- 7;mu2 <- 55;sigma2 <- 7
  z1 <- rnorm(K1, mean = mu1, sd = sigma1)
  z1[z1>100] <- 100;z1[z1<0] <- 0
  z2 <- rnorm(K2, mean = mu2, sd = sigma2)
  z2[z2>100] <- 100;z2[z2<0] <- 0
  AUC1 <- TrapezoidalArea(z1, z2)
  Gaussians <- c(z1, z2)
  hist1 <- data.frame(x=Gaussians)
  hist.1 <-  ggplot(data = hist1, mapping = aes(x = x)) +
    geom_histogram(
      binwidth = 1,
      color = "black",
      fill="grey") +
    xlab(label = "Original Rating") +
    ggtitle(label = "Gaussians")
  print(hist.1)

  z <- seq(0.0, 100, 0.1)
  curveData <- data.frame(
```

```
    x = z,
    z =  Y(z,mu1,mu2,sigma1,sigma2,f))
  plot3 <- ggplot(mapping = aes(x = x, y = z)) +
    geom_line(data = curveData) +
    xlab(label = "Original Rating") +
    ylab(label = "Transformed Rating") +
    ggtitle(label = "Monotone Transformation")
  print(plot3)

  y <- Y(c(z1, z2),mu1,mu2,sigma1,sigma2,f)
  y1 <- y[1:K1];y2 <- y[(K1+1):(K1+K2)]
  AUC2 <- TrapezoidalArea( y1, y2)
  hist2 <- data.frame(x=y)
  hist.2 <-  ggplot(data = hist2, mapping = aes(x = x)) +
    geom_histogram(
      binwidth = 1,
      color = "black",
      fill="grey") +
    xlab(label = "Transformed Rating") +
    ggtitle(label = "Latent Gaussians")
  print(hist.2)
  cat("seed =", seed,
      "\nf =", f,
      "\nAUC of actual Gaussians =", AUC1,
      "\nAUC of latent Gaussians =", AUC2, "\n")
}
```

Lines 18, 19 and 21–23 set the parameters of the simulation model. Lines 21–23 simulate continuous ratings in the range 0–100 from a binormal model. Non-diseased cases are sampled from a Gaussian centered at **mu1** = 30 and standard deviation **sigma1** = 7. Diseased cases are sampled from a Gaussian centered at **mu2** = 55 and standard deviation **sigma2** = 7. The variable **f**, which is in the range (0,1), controls the shape of the transformed distribution. If **f** is small, the transformed distribution will be peaked toward 0 and if **f** is unity, it will be peaked at 100. If **f** equals 0.5, the transformed distribution is flat. Insight into the reason for this transformation is in Reference 8, **Chapter 7, ibid**: it has to do with transformations of random variables. The transformation $Y(Z)$ is at lines 12 through 16, in the in-line function **Y**, which implements

$$Y(Z) = \left[(1-f)\Phi\left(\frac{Z-\mu_1}{\sigma_1}\right) + f\Phi\left(\frac{Z-\mu_2}{\sigma_2}\right) \right] \times 100 \qquad (6.5)$$

The multiplication by 100 ensures that the transformed ratings are in the range 0–100. Lines 24–27 realize the random samples and limit them to the range 0 to 100. Line 28 calculates empirical AUC for the binormal samples, saved to variable **AUC1**. Lines 30–38 display the histogram of the binormal ratings. Lines 40–49 display a plot of the transformation function Equation 6.5. Line 51 applies the transformation and saves the results to rating **y**. Line 53 calculates empirical AUC for the transformed rating, **AUC2**, and the remaining lines plot the histogram of the transformed rating. **Source** the code to obtain code output in Section 6.2.2.2 and Figure 6.2a through i. The figure is arranged in a 3 × 3 matrix, where the first column shows histograms corresponding to the binormal ratings Z, the second column shows the transformation $Y(z)$ as a function of z and the third column shows the histogram corresponding to the transformed ratings Y. Each row corresponds to a distinct set of seed and shape parameter (f) values.

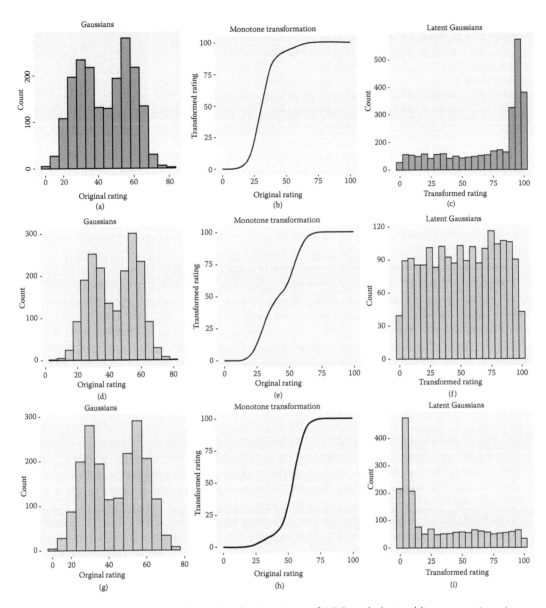

Figure 6.2 (a) through (i): Plots illustrating the invariance of ROC analysis to arbitrary monotone transformations of the ratings. Each of the latent Gaussian plots (c, f and i) appears distinctly not binormal. However, by using the inverse of the monotone transformations shown (b, e and h), they can be transformed to the binormal model histograms (a, d and g). See text for details.

6.2.2.2 Code output

```
> source(...)
seed = 10
f = 0.1
AUC of actual Gaussians = 0.99308
AUC of latent Gaussians = 0.99308
seed = 11
f = 0.5
```

```
AUC of actual Gaussians = 0.9936689
AUC of latent Gaussians = 0.9936689
seed = 12
f = 0.9
AUC of actual Gaussians = 0.9950411
AUC of latent Gaussians = 0.9950411
```

The output lists the values of the seed variable and the value of the shape parameter *f*. *For each value of seed and the shape parameter, the AUCs of the actual Gaussians ratings and the transformed ratings are identical.* Because of the invariance of the AUCs, the transformed ratings qualify as *latent Gaussians.* The values of the binormal parameters were chosen to clearly show two Gaussian peaks illustrating the binormal nature of the plots in the first column, i.e., Figure 6.2a, d, and g. This has the effect of making the AUCs close to unity; with more overlap it is difficult to visualize the two peaks. It is left as an exercise for the reader to plot ROC curves corresponding to actual Gaussian and latent Gaussian variables and show that they are identical.

Figure 6.2b shows the transformation for *f* = 0.1. The steep initial rise of the curve has the effect of flattening the histogram of the transformed ratings at the low end of the rating scale, Figure 6.2c. Conversely, the flatness of the curve near the upper end of the rating range has the effect of causing the histogram of the transformed variable to peak in that range. Figure 6.2e shows the transformation for *f* = 0.5. This time the transformed rating histogram, Figure 6.2f is almost flat over the entire range. Figure 6.2h shows the transformation for *f* = 0.9. This time the transformed rating histogram, Figure 6.2i, is peaked at the low end of the transformed rating scale.

Each histogram in the third column in Figure 6.2 (c, f, and i) appears to be non-Gaussian. The corresponding non-diseased and diseased ratings will fail tests of normality. (Showing this is left as an exercise for the reader.) Nevertheless, the transformed ratings are *latent Gaussians* in the sense that applying the inverses of the transformations shown in Figure 6.2 (b, e, and h) will yield histograms that are strictly binormal. By appropriate changes to the monotone transformation function, the histograms shown in Figure 6.2 (c, f, and i) can be made to resemble a wide variety of shapes. (As another exercise, the reader could modify the transformation function to yield bimodal histograms.) *One concludes that visual examination of the shape of the histogram of ratings yields little, if any, insight into whether the underlying binormal model assumptions are being violated.* Additionally, the flexibility afforded by the latency property implies that the binormal model is likely to be very robust with respect to departures from strict normality. If data binning is employed the strict equality of the AUCs is not expected to hold. Data binning introduces noise, as the ordering information in continuous ratings within a bin is lost.

6.2.3 Expressions for sensitivity and specificity

Let Z_t denote the random Z-sample for truth-state t ($t = 1$ for non-diseased and $t = 2$ for diseased cases). Since the distribution of Z-samples from disease-free cases is $N(0,1)$, the expression for specificity (**Chapter 3**) Equation 3.13, applies. It is reproduced as follows:

$$Sp(\zeta) = P(Z_1 < \zeta) = \Phi(\zeta) \tag{6.6}$$

To obtain an expression for sensitivity, consider that for truth-state $t = 2$, the random variable $\frac{Z_2 - \mu}{\sigma}$ is distributed as $N(0,1)$.

$$\frac{Z_2 - \mu}{\sigma} \sim N(0,1) \tag{6.7}$$

Sensitivity is $P(Z_2 > \zeta)$, which implies, because σ is positive (as in following equation, subtract μ from both sides of the > symbol and divide by σ):

$$Se(\zeta \mid \mu, \sigma) = P(Z_2 > \zeta) = P\left(\frac{Z_2 - \mu}{\sigma} > \frac{\zeta - \mu}{\sigma}\right) \tag{6.8}$$

The right-hand side can be rewritten as follows:

$$Se(\zeta \mid \mu, \sigma) = 1 - P\left(\frac{Z_2 - \mu}{\sigma} \leq \frac{\zeta - \mu}{\sigma}\right)$$
$$= 1 - \Phi\left(\frac{\zeta - \mu}{\sigma}\right) = \Phi\left(\frac{\mu - \zeta}{\sigma}\right) \tag{6.9}$$

Summarizing, the formula for the specificity and sensitivity for the binormal model are

$$Sp(\zeta) = \Phi(\zeta) \tag{6.10}$$

$$Se(\zeta \mid \mu, \sigma) = \Phi\left(\frac{\mu - \zeta}{\sigma}\right) \tag{6.11}$$

The coordinates of the operating point defined by ζ are given by

$$FPF(\zeta) = 1 - Sp(\zeta) = 1 - \Phi(\zeta) = \Phi(-\zeta) \tag{6.12}$$

$$TPF(\zeta \mid \mu, \sigma) = \Phi\left(\frac{\mu - \zeta}{\sigma}\right) \tag{6.13}$$

These expressions allow calculation of the predicted operating point for any ζ. An equation for a curve is usually expressed as $y = f(x)$. An expression of this form for the ROC curve, that is, the y coordinate (*TPF*) expressed as a *function* of the x coordinate (*FPF*), follows upon inversion of the expression for *FPF*, Equation 6.12:

$$\zeta = -\Phi^{-1}(FPF) \tag{6.14}$$

Substitution of Equation 6.14 in the expression for *TPF* in Equation 6.12 yields:

$$TPF = \Phi\left(\frac{\mu + \Phi^{-1}(FPF)}{\sigma}\right) \tag{6.15}$$

This equation gives the desired dependence of TPF on FPF. Discussion of this equation is deferred until it has been recast using conventional notation (see Equation 6.18 below).

6.2.4 Binormal model in conventional notation

The following notation is widely used in the literature[1]:

$$a = \frac{\mu}{\sigma}; b = \frac{1}{\sigma} \tag{6.16}$$

The reason for the (a,b) instead of the (μ,σ) notation is that Dorfman and Alf assumed, in their seminal paper,[1] the diseased distribution (*signal* distribution in signal detection theory) had unit variance, and the non-diseased distribution (*noise*) had standard deviation b ($b > 0$) or variance b^2, and the separation of the two distributions was a, Figure 6.3a. In this example, $a = 1.11$ and $b = 0.556$. Dorfman and Alf's fundamental contribution, namely estimating these parameters from ratings data, to be described below, led to the widespread usage of the (a,b) parameters, estimated by their software (RSCORE), and its newer variants (e.g., RSCORE –II, ROCFIT and ROCKIT).

By dividing the Z-samples by b, the variance of the distribution labeled *Noise* in Fig. 6.3a becomes unity, its mean stays at zero, and the variance of the distribution labeled *Signal* becomes $1/b$, and its mean becomes a/b, as shown in Figure 6.3b. It illustrates that the inverses of Equation 6.16 are

$$\mu = \frac{a}{b}; \sigma = \frac{1}{b} \tag{6.17}$$

Equations 6.16 and 6.17 allow transformation from one notation to the other.

6.2.5 Properties of the binormal model ROC curve

Using the a, b notation, Equation 6.15 for the ROC curve reduces to:

$$TPF = \Phi\left(a + b\Phi^{-1}(FPF)\right) \tag{6.18}$$

Since $\Phi^{-1}(FPF)$ is an increasing function of its argument FPF, and $b > 0$, the argument of the $\Phi(\ )$ function is an increasing function of FPF. Since $\Phi(\ )$ is a monotonically increasing function of *its* argument, TPF is a monotonically increasing function of FPF. This is true regardless of the sign of a. If $FPF = 0$, then $\Phi^{-1}(0) = -\infty$ and $TPF = 0$. If $FPF = 1$, then $\Phi^{-1}(1) = \infty$ and $TPF = 1$. Regardless of the value of a, as long as $b \geq 0$, the ROC curve starts at (0,0) and ends at (1,1), extending monotonically from the origin to (1,1). The requirement $b \geq 0$ is always true since b is a standard deviation, that is, the positive square root of the variance. From Equations 6.12 and 6.13, the expressions for FPF and TPF in terms of model parameters (a,b) are

$$FPF(\zeta) = \Phi(-\zeta) \tag{6.19}$$

$$TPF(\zeta) = \Phi(a - b\zeta) \tag{6.20}$$

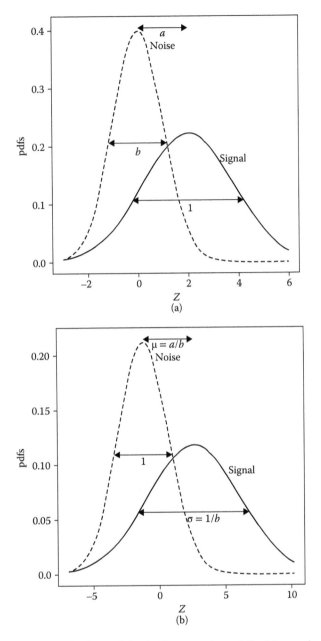

Figure 6.3 Plot (a) shows the definitions of the (a, b) parameters of the binormal model. In plot (b) the x-axis has been rescaled so that the noise distribution has unit variance, thereby showing the relations, Equation 6.16 and 6.17, between (a, b) and the (μ, σ) parameters. In this example, $a = 1.11$, $b = 0.556$, $\sigma = 1.8$ and $\mu = 2$, These figures were generated by the code in **mainPlotAB.R**.

6.2.6 The *pdfs* of the binormal model

According to Equation 6.1, the probability that a Z-sample is smaller than a specified threshold ζ, that is, the *CDF* function, is

$$P\left(Z \leq \zeta \,\middle|\, Z \sim N\left(0, 1\right)\right) = 1 - FPF\left(\zeta\right) = \Phi\left(\zeta\right) \tag{6.21}$$

$$P\left(Z\leq\zeta\middle|Z\sim N\left(\mu,\sigma^2\right)\right)=1-TPF\left(\zeta\right)=\Phi\left(\frac{\zeta-\mu}{\sigma}\right) \tag{6.22}$$

Since the *pdf* is the derivative of the corresponding *CDF* function, it follows that (the subscripts N and D denote non-diseased and diseased cases, respectively):

$$\text{pdf}_N\left(\zeta\right)=\frac{\partial\Phi\left(\zeta\right)}{\partial\zeta}=\phi\left(\zeta\right)\equiv\frac{1}{\sqrt{2\pi}}e^{-\frac{\zeta^2}{2}} \tag{6.23}$$

$$\text{pdf}_D\left(\zeta\right)=\frac{\partial\Phi\left(\frac{\zeta-\mu}{\sigma}\right)}{\partial\zeta}=\frac{1}{\sigma}\phi\left(\frac{\zeta-\mu}{\sigma}\right)=\frac{1}{\sqrt{2\pi}\sigma}e^{-\frac{\left(\frac{\zeta-\mu}{\sigma}\right)^2}{2}} \tag{6.24}$$

The second equation can be written in a,b notation as

$$\text{pdf}_D\left(\zeta\right)=b\phi\left(b\zeta-a\right)=\frac{b}{\sqrt{2\pi}}e^{-\frac{\left(b\zeta-a\right)^2}{2}} \tag{6.25}$$

Generation of *pdfs* for specified values of binormal model parameters μ,σ was illustrated in Figure 6.1 using `mainRocPdfsWithCutoffs.R`.

6.2.7 A JAVA-fitted ROC curve

Described next is a method for fitting data such as in Table 6.1 to the binormal model, that is, determining the parameters (a, b) and the thresholds ζ_r, $r = 1, 2, …, R–1$, to best fit the observed cell counts. The most common method uses an algorithm called *maximum likelihood*, which will be described in the next section. For now, we jump ahead and use a web-based ROC calculator to yield a fitted curve corresponding to the data in Table 6.1. The website is www.rad.jhmi.edu/jeng/javarad/roc/JROCFITi.html. The website may complain that Java is not installed or that Java is not enabled, or that Java is blocked, in which case install it* and/or enable it and/or unblock it. The J in JROCFIT is for Java (not jackknife). A partial screen shot is shown in Figure 6.4. For **Data Format** select Format 3 and enter 5 for the **Number of Ratings Categories** (i.e., the value of R for

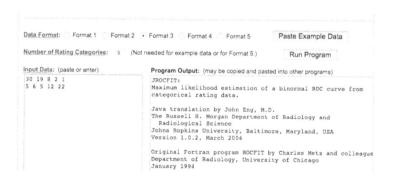

Figure 6.4 Partial screen shot from the website. Note that the data entered on the left-hand side are the numbers in Table 6.1. **Format 3** has been selected and the **Number of Ratings Categories** is set to five. If **Java** is not installed one sees warning messages on this screen. The data is analyzed when one clicks **Run Program**. The entire output is in Online Appendix 6.B.

---

* On the Mac, at the time of writing (8/2014), the author had to use 64-bit Safari, not the then current 32-bit Google Chrome, as the browser. Download and install the 64-bit Java application, and then quit and restart Safari.

the data in Table 6.1). Then enter the numbers, separated by spaces, from Table 6.1 into the panel labeled **Input Data** and click on **Run Program**.

Among the program output, listed in Online Appendix 6.B, are the following results (the estimated thresholds ζ_k are labeled **Z(k)** in the code output; for clarity they have been relabeled below).

6.2.7.1 JAVA output

```
a = 1.3204
b = 0.6075
ζr = 0.0077, 0.9026, 1.6452, 2.1285
```

Note that $b < 1$, which means the diseased distribution is wider than the non-diseased one. As noted earlier, this is the general finding with a wide variety of ROC datasets,[4] quite apart from medical imaging.

Using the fitted parameters determined using the Eng code, one generates the fitted ROC curve and superposes the operating points. Open the file **mainRocCurveFitEng.R**, listed below. Sourcing it yields Figure 6.5 (as usual, code breaking across multiple lines due to page-width restriction counts as a single line, also blank lines are inserted for code readability; they are ignored by **R**).

6.2.7.2 Code listing

```
#mainRocCurveFitEng.R
rm(list = ls())
source("rocY.R")
library(ggplot2);require("grid")
# the famous a and b parameters of the
# Dorfman and Alf binormal model,
# as yielded by the Eng JAVA code

a <- 1.3204; b <- 0.6075

FPF <- seq(0.0, 1, 0.01)
curveData <- data.frame(
  FPF = FPF,
  TPF = rocY(FPF, a, b))
# observed operating points
# Table 6.1
FPF <- c(0.017, 0.050, 0.183, 0.5)
TPF <- c(0.440, 0.680, 0.780, 0.900)
pointsData <- data.frame(
  FPF = FPF, TPF = TPF)
p <- ggplot(mapping = aes(x = FPF, y = TPF)) +
  theme(
    axis.title.y = element_text(size = 25,face="bold"),
    axis.title.x = element_text(size = 30,face="bold")) +
  geom_line(data = curveData, size = 2) +
  geom_point(data = pointsData, size = 5) +
  annotation_custom(
    grob = textGrob(bquote(italic("O")),
                    gp = gpar(fontsize = 32)),
    xmin = -0.03, xmax = -0.03, # adjust the position of "O"
    ymin = -0.03, ymax = -0.03) +
```

```
scale_x_continuous(
    expand = c(0, 0),
    breaks = c(0.25, 0.5, 0.75, 1)) +
  scale_y_continuous(
    expand = c(0, 0),
    breaks = c(0.25, 0.5, 0.75, 1))
print(p)

p <- ggplotGrob(p)
p$layout$clip[p$layout$name=="panel"] <- "off"
grid.draw(p)
```

Line 3 sources file **rocY.R** containing the function **rocY(rocX,a,b),** the *y*-coordinate of the operating point corresponding to abscissa **rocX,** and model parameters **a** and **b**. The implementation corresponds almost literally to Equation 6.18. Line 4 includes the **ggplot2** package (if it has not been installed this line will generate an error; loading a missing package was described in Online Appendix 3.G). Line 9 defines the **(a,b)** parameters; the values were copied from the output of the web-based calculator. Line 11 uses the **seq(0,1,0.01)** function (for sequence) to define an equally spaced array of *FPF* values starting from 0 and ending at 1 with spacing 0.01. These represent abscissa values for the plot that is about to come. Line 12–14 calculates TPF for each value of FPF and combines them into a **data frame** object **curveData** (in **R**, a **data frame** object consists of a list of vectors of the same lengths but possibly different types; think of it as a generalized table). Lines 17 and 18 define the operating points copied from Table 6.1. These are collected into a second data frame object **pointsData** at line 19–20. Lines 21–41 use the **ggplot()** function to create the plot. The last line that is, displays it.

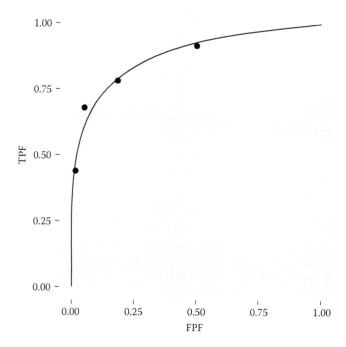

Figure 6.5 Observed operating points from Table 6.1, open circles, and the fitted curve predicted by the web-based calculator, solid line. This figure was generated using **mainRocCurveFitEng.R**. The parameters of the fit are: *a* = 1.3204, *b* = 0.6075.

In this example, **ggplot()** takes as arguments three functions separated by plus signs (with additional plus signs one can keep adding features to the plot). The first one is **mapping = aes(x=FPF,y=TPF)**, which tells **ggplot** to assign the labels FPF and TPF to the x and y axes; the **aes()** function is for aesthetics. Next is **geom_line(data=curveData)**, which tells **ggplot** to plot the densely spaced predicted curve data and connect them with a smooth line. Last is **geom_point(data= pointsData,shape=1,size=3)**, which causes the operating points to be superposed on the plot. The meanings of **shape** and **size** are defined in the help file corresponding to this function.

6.3 Least-squares estimation of parameters

By applying the function Φ^{-1} to both sides of Equation 6.18, one gets (the *inverse* function cancels the *forward* function on the right-hand side):

$$\Phi^{-1}(TPF) = a + b\,\Phi^{-1}(FPF) \tag{6.26}$$

This suggests that a plot of $y = \Phi^{-1}(TPF)$ versus $x = \Phi^{-1}(FPF)$ should follow a straight line with slope b and intercept a. Fitting a straight line to such data is generally performed by the method of least-squares, a capability present in most software packages and even spreadsheets, for example, Excel™. Alternatively, one can simply visually draw the best straight line that fits the points, memorably referred to[8] as *chi-by-eye*. This was the way parameters of the binormal model were estimated prior to Dorfman and Alf's work.[1] The least-squares method is an quantitative/objective way of accomplishing the same aim. If (x_i, y_i) are the data points, one constructs S, the sum of the squared deviations of the observed ordinates, from the predicted values:

$$S = \sum_{i=1}^{R-1} \left(y_i - (a + bx_i) \right)^2 \tag{6.27}$$

The idea is to minimize S with respect to the parameters (a,b). One approach is to differentiate S with respect to a and b and equate each resulting derivate expression to zero. This yields two equations in two unknowns, which are solved for a and b. If the reader has never done this before, one should go through these steps at least once, but it would be smarter in the future to use software that does all this. In **R**, the least-squares fitting function is **lm(y~x)**, which in its simplest form fits a *linear model* $y = ax + b$ using the method of least-squares (in case you are wondering **lm** stands for *linear model*, a whole branch of statistics in itself;[9] in this example one is using its simplest capability). This is illustrated below for the data in Table 6.1. Open the file **mainLeastSquares.R**.

6.3.1 Code listing

```
rm(list = ls()) # mainLeastSquares.R # freeze line numbers
library(ggplot2)
# ML estimates of a and b
# (from Eng JAVA program)
# a <- 1.3204;
# b <- 0.6075
# # these are not used in program;
# just there for comparison

# these are from Table 6.11, last two rows
FPF <- c(0.017, 0.050, 0.183, 0.5)
TPF <- c(0.440, 0.680, 0.780, 0.900)
```

```
# apply the PHI_INV function
phiInvFPF <- qnorm(FPF)
phiInvTPF <- qnorm(TPF)

fit <- lm(phiInvTPF~phiInvFPF)
print(fit)
pointsData <- data.frame(
  phiInvFPF = phiInvFPF,
  phiInvTPF = phiInvTPF)
pointsPlot <- ggplot(
  data = pointsData,
  mapping =
    aes(x = phiInvFPF,
        y = phiInvTPF)) +
  geom_point(size = 5) +
  theme(
    axis.title.y = element_text(size = 25,face="bold"),
    axis.title.x = element_text(size = 30,face="bold")) +
  geom_abline(
    slope = fit$coefficients[2],
    intercept = fit$coefficients[1], size = 2)
print(pointsPlot)
```

At lines 11–12 the operating point data are assigned to arrays FPF and TPF. Lines 15–16 convert the observed FPF and TPF values to $\Phi^{-1}(TPF)$ and $\Phi^{-1}(FPF)$, respectively. Line 18 does the actual least-squares fitting. This line states, in so many words, "perform a least-squares fit with **phiInvFPF** regarded as the independent variable and **phiInvTPF** regarded as the dependent variable, and save the fit to the variable named **fit**." The next line prints the values of the linear fit. The remaining lines display the data points and the superimposed least-squares fitted straight line. **Source** the code yielding the following (partial) output and Figure 6.6.

6.3.2 Code output (partial)

```
...
Call:
lm(formula = ph_inv_TPF ~ ph_inv_FPF)

Coefficients:
(Intercept)   ph_inv_FPF
     1.3288       0.6307
```

The first two lines of the partial output are simply a reminder about the names of the dependent and independent variables. The last line contains the least-squares estimated values, $a = 1.3288$ and $b = 0.6307$. The corresponding maximum likelihood estimates of these parameters, as yielded by the Eng web code, Online Appendix 6.B, are listed in line 5–6 of the main program: $a = 1.3204$ and $b = 0.6075$. The estimates appear to be close, particularly the estimate of a, but there are a few things wrong with the least-squares approach. First, the method assumes that the data points are independent. Because of the manner in which they are constructed, namely by cumulating points, the independence assumption is not valid for ROC operating points. For example, cumulating the 4 and 5 responses constrains the resulting operating point to be above and to the right of the point obtained

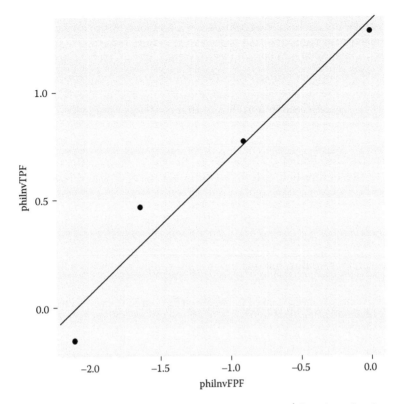

Figure 6.6 Operating points from Table 6.1, transformed by the Φ^{-1} function; the slope of the line is the least-squares estimate of the b-parameter and the intercept is the corresponding a-parameter of the binormal model. The plot was generated by **mainLeastSquares.R**.

by cumulating the 5 responses only. Similarly, cumulating the 3, 4, and 5 responses constrains the resulting operating point to be above and to the right of the point obtained by cumulating the 4 and 5 responses, and so on. The independence assumption of standard least-squares fitting is violated by the data points. The second problem is the linear least-squares method assumes there is no error in measuring x; the only source of error that is accounted for is in the y-coordinate. In fact, both coordinates of a ROC operating point are subject to sampling error.

(A historical note: prior to computers and easy access to statistical functions the analyst had to use a special plotting paper, termed *double probability paper*, that converted probabilities into x and y distances using the inverse function. The complement of the inverse function is sometimes termed the *z-deviate*.[4] Since this term confused the author when he entered this field ca. 1985, and it confuses him even now, he will not use it further.)

6.4 Maximum likelihood estimation (MLE)

The approach taken by Dorfman and Alf[1] was to maximize the likelihood function instead of S. The likelihood function is the probability of the observed data, for example, the cells counts in Table 6.1, given a set of parameter values, that is,

$$L = P(data\,|\,parameters) \tag{6.28}$$

Generally, *data* is suppressed, so likelihood is expressed as a function of the parameters. With reference to Figure 6.1, the probability of a non-diseased case yielding a count in the second bin equals the area under the curve labeled *Non-diseased* bounded by the vertical lines at ζ_1 and ζ_2. In general, the probability of a non-diseased case yielding a count in FP bin r equals the area under the curve labeled *Non-diseased* bounded by the vertical lines at ζ_{r-1} and ζ_r. Since the area to the left of a threshold is the *CDF* corresponding to that threshold, the required probability is $\Phi(\zeta_r) - \Phi(\zeta_{r-1})$:

$$P(\text{count in non-diseased bin } r) = \Phi(\zeta_r) - \Phi(\zeta_{r-1}) \tag{6.29}$$

Similarly, the probability of a diseased case yielding a count in TP bin r equals the area under the curve labeled *Diseased* bounded by the vertical lines at ζ_{r-1} and ζ_r. The area under the diseased distribution to the left of threshold ζ_r is the complement of *TPF* at that threshold, Equation 6.13:

$$1 - \Phi\left(\frac{\mu - \zeta_r}{\sigma}\right) = \Phi\left(\frac{\zeta_r - \mu}{\sigma}\right) \tag{6.30}$$

The area between the two thresholds is

$$P(\text{count in diseased bin } r) = \Phi\left(\frac{\zeta_r - \mu}{\sigma}\right) - \Phi\left(\frac{\zeta_{r-1} - \mu}{\sigma}\right) = \Phi(b\zeta_r - a) - \Phi(b\zeta_{r-1} - a) \tag{6.31}$$

In the above equation, Equation 6.16 was used to express the right hand side in terms of the (a,b) parameters.

Let K_{1r} denote the number of non-diseased cases in the rth bin, and K_{2r} denotes the number of diseased cases in the rth bin. Since the probability of a count in non-diseased case bin r is $\Phi(\zeta_r) - \Phi(\zeta_{r-1})$, the probability of the observed number of counts, *assuming the counts are independent*, is $\left(\Phi(\zeta_r) - \Phi(\zeta_{r-1})\right)^{K_{1r}}$. Similarly, the probability of observing K_{2r} counts in diseased case bin r is $\left(\Phi(b\zeta_r - a) - \Phi(b\zeta_{r-1} - a)\right)^{K_{2r}}$, subject to the same independence assumption. The probability of *simultaneously* observing K_{1r} counts in non-diseased case bin r and K_{2r} counts in diseased case bin r is the product of these individual probabilities (again, an independence assumption is being used—if the cases are different these assumptions are valid):

$$\left(\Phi(\zeta_r) - \Phi(\zeta_{r-1})\right)^{K_{1r}} \left(\Phi(b\zeta_r - a) - \Phi(b\zeta_{r-1} - a)\right)^{K_{2r}} \tag{6.32}$$

Similar expressions apply for all integer values of r ranging from 1 to R. Therefore, the probability of observing the *entire* data set is the *product* of expressions like Equation 6.32, over all values of r:

$$\prod_{r=1}^{R} \left[\begin{array}{l} \left(\Phi(\zeta_r) - \Phi(\zeta_{r-1})\right)^{K_{1r}} \times \\[2mm] \left(\Phi(b\zeta_r - a) - \Phi(b\zeta_{r-1} - a)\right)^{K_{2r}} \end{array} \right] \tag{6.33}$$

We are almost there. A specific combination of $K_{11}, K_{12}, ..., K_{1R}$ counts from K_1 non-diseased cases and $K_{21}, K_{22}, ..., K_{2R}$ counts from K_2 diseased cases can occur the following number of times (given by the multinomial factor shown below):

$$\frac{K_1!}{\displaystyle\prod_{r=1}^{R} K_{1r}!} \frac{K_2!}{\displaystyle\prod_{r=1}^{R} K_{2r}!} \tag{6.34}$$

The likelihood function is the product of Equations 6.33 and 6.34.

$$L\left(a,b,\vec{\zeta}\right) = \left(\frac{K_1!}{\prod\limits_{r=1}^{R} K_{1r}!} \frac{K_2!}{\prod\limits_{r=1}^{R} K_{2r}!} \right) \prod_{r=1}^{R} \left[\begin{array}{c} \left(\Phi(\zeta_r)-\Phi(\zeta_{r-1})\right)^{K_{1r}} \\ \left(\Phi(b\zeta_r-a)-\Phi(b\zeta_{r-1}-a)\right)^{K_{2r}} \end{array} \right] \qquad (6.35)$$

The left-hand side of Equation 6.35 explicitly shows the dependence of the likelihood function on the parameters of the model, namely $a,b,\vec{\zeta}$, where the vector of thresholds is a compact notation for the set of thresholds $\zeta_1,\zeta_2,...,\zeta_R$ (note that since $\zeta_0=-\infty$ and $\zeta_R=+\infty$, only $R-1$ free threshold parameters are involved, and the total number of free parameters in the model is $R+1$). For example, for a 5-rating ROC study, the total number of free parameters is 6, i.e., a, b and 4 thresholds.

Equation 6.35 is forbidding but here comes a simplification. The Φ function is a probability, i.e., in the range 0 to 1, and since ζ_r is greater than ζ_{r-1}, the difference $\Phi(\zeta_r)-\Phi(\zeta_{r-1})$ is positive and less than one. When the difference is raised to the power of K_{1r} (a non-negative integer) a very small number can result. Multiplication of all these small numbers will result in an even smaller number, which may be too small to be represented as a floating-point value, especially as the number of counts increases. To prevent this, one resorts to a "trick." Instead of maximizing the likelihood function $L\left(a,b,\vec{\zeta}\right)$ one chooses to maximize the *logarithm* of the likelihood function (the base of the logarithm is immaterial). The logarithm of the likelihood function is

$$LL\left(a,b,\vec{\zeta}\right) = \log\left(L\left(a,b,\vec{\zeta}\right)\right) \qquad (6.36)$$

Since the logarithm is a monotonically increasing function of its argument, maximizing the logarithm of the likelihood function is equivalent to maximizing the likelihood function. Taking the logarithm converts the product symbols in Equation 6.35 to summations, so instead of multiplying small numbers one is summing them, thereby avoiding underflow errors. Another simplification is that one can ignore the logarithm of the multinomial factor involving the factorials, because these do not depend on the parameters of the model. Putting all this together, one gets the following expression for the logarithm of the likelihood function:

$$\left. \begin{array}{l} LL\left(a,b,\vec{\zeta}\,\middle|\,\overline{K_1},\overline{K_2}\right) \sim \sum\limits_{r=1}^{R} K_{1r} \log\left(\left(\Phi(\zeta_r)-\Phi(\zeta_{r-1})\right)\right) \\[4mm] + \sum\limits_{r=1}^{R} K_{2r} \log\left(\left(\Phi(b\zeta_r-a)-\Phi(b\zeta_{r-1}-a)\right)\right) \end{array} \right\} \qquad (6.37)$$

The symbol ~ is meant to denote that inconsequential terms have been dropped on the right-hand side of Equation 6.37.

If one looks carefully at the left-hand side of Equation 6.37, one sees that it is a function of the model parameters $a,b,\vec{\zeta}$ *and* the observed data, the latter being the counts contained in the vectors $\overline{K_1}$ and $\overline{K_2}$, where the vector notation is used as a compact form for the counts $K_{11},K_{12},...,K_{1R}$ and $K_{21},K_{22},...,K_{2R}$, respectively. The right-hand side of Equation 6.37 is monotonically related to the probability of observing the data given the model parameters (a, b) and $\vec{\zeta}$. If the choice of model parameters is poor, then the probability of observing the data will be small and log-likelihood will be small. With a better choice of model parameters, the probability and log-likelihood will increase. With optimal choice of model parameters, the probability and log-likelihood will be maximized, and the corresponding optimal values of the model parameters are called maximum likelihood estimates (*MLEs*). These are the estimates produced by the programs RSCORE and ROCFIT.

Online Appendix 6.C details the procedure used to maximize the likelihood function and its **R** implementation. **Source** the code in **mainRocfitR.R** to get the following output.

6.4.1 Code output

```
> source(...)
initial parameters =
 1.328 0.6292 0.03703 0.8931 1.506 2.239
final parameters =
 1.32 0.6075 0.007677 0.8963 1.516 2.397
-LL values, initial 141.7
-LL values, final 141.4
covariance matrix =
  0.065  0.025  0.015  0.012  0.007 -0.011
  0.025  0.024  0.005 -0.002 -0.014 -0.043
  0.015  0.005  0.026  0.015  0.011  0.005
  0.012 -0.002  0.015  0.032  0.028  0.029
  0.007 -0.014  0.011  0.028  0.056  0.067
 -0.011 -0.043  0.005  0.029  0.067  0.159

 Az =  0.8705 StdAz =  0.0379
```

The listed values of the parameters, in order, a, b, ζ_1, ζ_3, ζ_3, ζ_4, should be compared to the output of the Eng program: they are quite close. The output also lists the initial and final value of $-LL$ (**R** has functions for *minimizing* a supplied function, so we supply it with the *negative* of the LL function, which is equivalent to *maximizing* the *LL* function): note that the final value of $-LL$ is smaller than the initial value. The covariance matrix, and binormal AUC, by convention denoted A_z, and its standard error are explained below.

6.4.2 Validating the fitting model

Figure 6.5 appears to be a good *visual* fit to the observed operating points, only the operating point corresponding to cumulating ratings 4 and 5 is slightly off the predicted curve. Quantification of the validity of the fitting model is accomplished by calculating the Pearson *goodness of fit* test,[10] also known as the chi-square test, which uses the statistic C^2 defined by[11]

$$C^2 = \sum_{t=1}^{2} \sum_{r=1}^{R} \frac{\left(K_{tr} - \langle K_{tr} \rangle\right)^2}{\langle K_{tr} \rangle}; K_{tr} \geq 5 \tag{6.38}$$

As noted above, the statistic is only valid if the number of counts in each cell is greater than or equal to 5. In the above expression, the $\langle \ \rangle$ symbol denotes an expected value. These are given by

$$\langle K_{1r-1}\rangle = K_1\left(\Phi(\zeta_r)-\Phi(\zeta_{r-1})\right)$$

$$\langle K_{2r-1}\rangle = K_2\left(\Phi(a\zeta_r - b)-\Phi(a\zeta_{r-1}-b)\right)$$

(6.39)

These expressions should make sense: the difference between the two *CDF* functions is the probability of a count in the specified bin, and multiplication by the total number of relevant cases should yield the expected counts (a non-integer).

It can be shown that, under the null hypothesis that the *assumed* probability distribution functions for the counts equals the *true* probability distributions, that is, the model is valid, the statistic C^2 is distributed as

$$C^2 \sim \chi^2_{df}$$

(6.40)

Here χ^2_{df} is the chi-square distribution with degrees of freedom *df* defined by

$$df = (R-1)+(R-1)-(2+R-1) = R-3$$

(6.41)

The right-hand side of the above equation has been written in an expansive form to illustrate the general rule: for R *non-diseased* cells in Table 6.1, the degree of freedom is R–1: this is because when all but one cells are specified, the last is determined, because they must sum to K_1. Similarly, the degree of freedom for the diseased cells is also R–1. Last, we need to subtract the number of free parameters in the model, which is (2+R–1), that is, the a,b parameters and R–1 thresholds. From Equation 6.14 it is evident that if $R = 3$ then $df = 0$. In this situation, there are only two nontrivial operating points and the straight-line fit shown in Figure 6.6 will pass through both of them. With two basic parameters, fitting two points is trivial, and goodness of fit cannot be calculated.

Under the null hypothesis (i.e., model is valid) C^2 is distributed as χ^2_{df}. Therefore, one computes the probability that this statistic is larger than the observed value, i.e., the p-value. If this probability is very small, this means that the deviations of the observed values of the cell counts from the expected values are so large that it is unlikely that the model is correct. The degree of unlikeliness is quantified by the p-value. Poor fits lead to small p-values.

At the 5% significance level, one concludes that the fit is not good if $p < 0.05$. In practice one occasionally accepts smaller values of p, $p > 0.001$ before completely abandoning a model. It is known that adoption of a stricter criterion, e.g., $p > 0.05$, can occasionally lead to rejection of a retrospectively valid model.[8]

Online Appendix 6.D has an **R** implementation of these concepts. Unfortunately, the author just happened to pick a dataset for which the minimum 5-count requirement in each bin is not true, so for the dataset in Table 6.1 the goodness of fit cannot be calculated. This is the reason why these values are not listed in output Section 6.4.1. Bins 3, 4 and 5 would need to be combined to meet the minimum count requirement, reducing the total number of bins to $R = 3$, for which $df = 0$, Equation 6.41. Long story short, the author just happened to pick a bad dataset. There is nothing inherently bad about a dataset, especially a clinical dataset such as Table 6.1, but it cannot be used to show off the prowess of the binormal model. Dorfman and Alf[1] provide examples of goodness of fit calculations using good datasets. A highly readable account of the original Dorfman and Alf paper, and some corrections, can be found in a paper by Grey and Morgan.[12] A point not explicitly made in the ROC curve-fitting literature is that one is free to combine pairs of bins (FP and correspond TP) in arbitrary ways, as long as one calculates the expected values using the same combinations. In particular, the bins do not have to be adjacent.

A simple way around this is to cheat. Uncomment line 19, in **mainRocfit.R** in Online Appendix 6.C, to alter the non-diseased counts as shown below (the cheated values overwrite the correct ones in the preceding line).

6.4.2.1 Code listing (partial)

```
K1 <- c(30,19,8,2,1) # this is the observed data!
K1 <- c(30,19,8,7,5) # this is the cheated data!
K2 <- c(5,6,5,12,22) # this is the observed data!
```

Source the code yielding (The page width limitation causes the covariance matrix output to spill over. The matrix has a column header extending from [,1] to [,6] and each row starts with the index of the first value. For example, 0.029434726 is the [2,1] element of the matrix, i.e., the value at the second row and the first column.):

6.4.2.2 Code output

```
> source(...)
initial parameters =
 1.197 0.8682 -0.1306 0.5217 0.8896 1.505
final parameters =
 1.198 0.8798 -0.1533 0.526 0.8934 1.515
-LL values, initial 167.2
-LL values, final 167.2
covariance matrix =
  0.063  0.029  0.018  0.016  0.014  0.006
  0.029  0.041  0.008 -0.000 -0.007 -0.024
  0.018  0.008  0.023  0.014  0.012  0.007
  0.016 -0.000  0.014  0.023  0.020  0.018
  0.014 -0.007  0.012  0.020  0.027  0.027
  0.006 -0.024  0.007  0.018  0.027  0.050

Az =  0.8158 StdAz =  0.04107
retChisqInitial p-val =  0.663
Chisq =  0.724
Chisq df =  2
Chisq p-val =  0.696
```

This time the output lists goodness of fit statistics. Any p-value greater than 0.01 is considered reasonably good. The actual value of the final chi-square goodness of fit statistic is 0.724, and the number of degrees of freedom is 2 (5 bins minus 3) corresponding to a p-value of 0.696. The author leaves it as an exercise for the reader to plot the data and compare it to the original plot (Figure 6.5). Incrementing the two highest-rated counts for non-diseased cases results, as expected, in reduction of Az from 0.870 to 0.816, a not-significant difference in view of the standard deviation 0.041 (the difference has to be outside twice this value to achieve significance).

Comments on the role of simulations in model validation and testing model robustness are deferred to Section 20.7.3.

6.4.3 Estimating the covariance matrix

We quote, without proofs, some results from estimation theory.[13] The Hessian H of a function $f(\vec{\theta})$ is defined as the partial second derivatives of the function with respect to its parameters; there are $R + 1$ parameters in the model:

$$
H(f(\theta)) = \begin{pmatrix}
\dfrac{\partial^2 f}{\partial^2 \theta_1^2} & \dfrac{\partial^2 f}{\partial\theta_2\,\partial\theta_1} & \cdots & \dfrac{\partial^2 f}{\partial\theta_{R+1}\,\partial\theta_1} \\[2ex]
\dfrac{\partial^2 f}{\partial\theta_1\,\partial\theta_2} & \dfrac{\partial^2 f}{\partial^2 \theta_2^2} & \cdots & \dfrac{\partial^2 f}{\partial\theta_{R+1}\,\partial\theta_2} \\[2ex]
\cdots & \cdots & \cdots & \cdots \\[2ex]
\dfrac{\partial^2 f}{\partial\theta_1\,\partial\theta_{R+1}} & \dfrac{\partial^2 f}{\partial\theta_2\,\partial\theta_{R+1}} & \cdots & \dfrac{\partial^2 f}{\partial^2 \theta_{R+1}^2}
\end{pmatrix} \tag{6.42}
$$

The Hessian is introduced because maximum likelihood estimation software usually calculates the Hessian of the likelihood function (see line 61–65 of **mainRocFitR.R**). The negative of the partial second derivatives of the (natural) logarithm of the likelihood function is the Fisher Information matrix: I.

$$
I = -\begin{pmatrix}
\dfrac{\partial^2 LL}{\partial^2 \theta_1^2} & \dfrac{\partial^2 LL}{\partial\theta_2\,\partial\theta_1} & \cdots & \dfrac{\partial^2 LL}{\partial\theta_{R+1}\,\partial\theta_1} \\[2ex]
\dfrac{\partial^2 LL}{\partial\theta_1\,\partial\theta_2} & \dfrac{\partial^2 LL}{\partial^2 \theta_2^2} & \cdots & \dfrac{\partial^2 LL}{\partial\theta_{R+1}\,\partial\theta_2} \\[2ex]
\cdots & \cdots & \cdots & \cdots \\[2ex]
\dfrac{\partial^2 LL}{\partial\theta_1\,\partial\theta_{R+1}} & \dfrac{\partial^2 LL}{\partial\theta_2\,\partial\theta_{R+1}} & \cdots & \dfrac{\partial^2 LL}{\partial^2 \theta_{R+1}^2}
\end{pmatrix} \tag{6.43}
$$

Finally, the matrix inverse of the Fisher Information matrix is the covariance of the parameters.

$$
Cov(\vec{\theta}) = I^{-1} \tag{6.44}
$$

(The inverse of a matrix is implemented in **R** by the function **solve()**.) Online Appendix 6.C implements these formula and provides a detailed explanation. To put matters in perspective θ is the parameter vector, the diagonal elements of the covariance matrix are the variances of the parameters while the off-diagonal elements are the covariances. Since the parameters are estimated from a single dataset, a fluctuation of a specific parameter, for example, the a-parameter, is likely to be accompanied by corresponding correlated fluctuations in other parameters (b and the thresholds).

6.4.4 Estimating the variance of A_z

The basic idea is this: recall in MLE the parameters are varied until the likelihood function is maximized. By definition, at the maximum the first derivative with respect to each of the parameters is zero. The *sharpness* of the maxima, which is related to the second derivative or *curvature* of the LL function, with respect to a particular parameter, determines the *variance* of the corresponding parameter. The narrower the maximum, the smaller the variance. Conversely, if the maxima is broad, that means there is more uncertainty in the determination of the corresponding parameter estimate; that is, its variance is large. In general, the variance decreases as the number of cases in the dataset increases. The finite number of cases leads to case-sampling variability of the parameters and all quantities derived from them.

The covariance matrix can be used to calculate the variance of *any function* of the parameters. Specifically, one is interested in the variance of the binormal model based estimate of the area under the ROC curve, A_z. We will quote a theorem from statistics. If $g(X,Y)$ is a function of two random variables X and Y, then the variance of $g(X,Y)$ is given by

$$Var\big(g(X,Y)\big) \approx \left(\frac{\partial g}{\partial X}\right)^2 Var(X) + \left(\frac{\partial g}{\partial Y}\right)^2 Var(Y) + 2\frac{\partial g}{\partial X}\frac{\partial g}{\partial Y} Cov(X,Y) \qquad (6.45)$$

From

$$A_z = \Phi\left(\frac{a}{\sqrt{1+b^2}}\right) \qquad (6.46)$$

It follows that

$$\frac{\partial A_z}{\partial a} = \frac{1}{\sqrt{1+b^2}}\phi\left(\frac{a}{\sqrt{1+b^2}}\right)$$

$$\frac{\partial A_z}{\partial b} = \frac{-ab}{\left(1+b^2\right)^{3/2}}\phi\left(\frac{a}{\sqrt{1+b^2}}\right) \qquad (6.47)$$

Online Appendix 6.E has an **R** implementation of these formula. It compares the results to that of the website Java implementation (version 1.0.2, March 2004) of the original Fortran program ROCFIT by the late Prof. Charles E. Metz and colleagues at the Department of Radiology, University of Chicago. The results of the comparison are shown in Table 6.2. For the Eng JAVA implementation, listed are initial values of all parameters, the corresponding final values, and variances. The last column lists the initial and final values of $-LL$. Note that the final value of $-LL$ is smaller than the initial value (maximizing $-LL$ is equivalent to minimizing $-LL$, so the final value is expected to decrease). Listed for the **R** implementation are the same quantities for two runs of

Table 6.2 Comparison of results from website implementation of ROCFIT to the **R** implementation. Listed are initial values, final values and variances of the six parameters needed to fit the data in Table 6.1. For the **R** implementation, the second set of values results from deliberately using incorrect initial values. Note the stability of the results to deliberate de-tuning of the initial values.

Implementation		Quantity	a	b	ζ_1	ζ_2	ζ_3	ζ_4	$-LL$
JAVA		Initial	1.3281	0.6291	0	0.9026	1.6452	2.1285	143.8057
		Final	1.3204	0.6075	0.0077	0.9026	1.6452	2.1285	141.4354
		Variance	0.0656	0.0254	0.0260	0.0317	0.0539	0.1664	NA
R	1st run	Initial	1.3281	0.6292	0.0370	0.8931	1.5061	2.2393	141.6644
		Final	1.3205	0.6075	0.0077	0.8963	1.5156	2.3967	141.4354
		Variance	0.0652	0.0243	0.0259	0.0316	0.0560	0.1585	NA
	2nd run	Initial	1	1	−0.6292	0.3566	1.3424	2.3281	175.8104
		Final	1.3205	0.6075	0.0077	0.8963	1.5156	2.3967	141.4354
		Variance	0.0652	0.0243	0.0259	0.0316	0.0560	0.1585	NA

the program. In the first run good initial values for all parameters were computed as described in Online Appendix 6.C. In the second run the initial values were arbitrarily set to incorrect values, that is, the algorithm was detuned, to check whether the final parameter values and $-LL$ are sensitive to choice of the initial values–they are not, which is a good sign for the algorithm.

Online Appendix 6.F shows parameter transformations that the author finds to be useful in such work, which ensure that illegal values are never passed to the likelihood function. An example of illegal parameters would be $\zeta_2 < \zeta_1$ or $b < 0$. Additionally, **R** has several other minimization algorithms, some of which allow constraints to be placed on parameters.

6.5 Expression for area under ROC curve

Appendix 6.A derives the formula for the partial area under the binormal model.[14] A special case of this formula is the area under the whole ROC curve, reproduced below using both parameterizations of the model:

$$A_z = \Phi\left(\frac{a}{\sqrt{1+b^2}}\right) = \Phi\left(\frac{\mu}{\sqrt{1+\sigma^2}}\right) \tag{6.48}$$

The binormal fitted AUC increases as a increases or as b decreases. Equivalently, it increases as μ increases or as σ decreases. An equivalent d' parameter is defined as the separation of two unit variance normal distributions yielding the same AUC as that predicted by Equation 6.48. It is defined by (the first equation follows from Eqn. 3.30)

$$\left.\begin{array}{l} A_z = \Phi\left(\dfrac{d'}{\sqrt{2}}\right) \\[2mm] \Rightarrow \\[2mm] d' = \sqrt{2}\Phi^{-1}(A_z) \end{array}\right\} \tag{6.49}$$

6.6 Discussion/Summary

This chapter has covered much territory. The binormal model is historically very important and the contribution[1] by Dorfman and Alf was seminal. Prior to their work, there was no valid way of estimating AUC from observed ratings counts as in Table 6.1. Their work and a key paper by Lusted[15] accelerated research using ROC methods. The number of publications using their algorithm, and the more modern versions developed by Metz and colleagues, is probably in excess of 500. Because of its key role, the author has endeavored to take out some of the mystery about how the binormal model parameters are estimated. In particular, a common misunderstanding that the binormal model assumptions are violated by real datasets, when in fact it is quite robust to apparent deviations from normality, is addressed.

While the binormal model has served a key role, it has serious limitations that need to be recognized, especially because alternatives that overcome these limitations are now available, **Chapter 19** and **Chapter 20**.* For example, it is known that the b-parameter of the model is generally observed to be less than one, consistent with the diseased distribution being wider than the non-diseased

---

* The reader may wonder why, if the method is currently obsolete, did the author write a whole chapter on it. The purpose was to introduce the estimation method, as the method carries over to the newer methods. In addition, the binormal model has played a key historical role in the development of this field, and understanding it is important.

one. The ROC literature is largely silent on the reason for this finding: it is accepted as a given. One reason, namely location uncertainty, is presented in **Chapter 17**. If the location of the lesion is unknown, then Z-samples from diseased cases are of two types, samples from the correct lesion location or samples from incorrect locations. The resulting mixture distribution will then appear to have larger variance than samples from non-diseased cases. The mixing need not be restricted to location uncertainty. Even if lesion location is known, if the lesions are non-homogenous (e.g., they contain a range of contrasts and/or sizes) a similar mixture-distribution induced broadening is expected.

If the b-parameter is less than unity, a portion of the curve, near (1,1), crosses the chance-diagonal and hooks upward approaching (1,1) with infinite slope. Usually the hook is not readily visible. For example, in Figure 6.5, one would have to zoom-in on the upper right corner to see it, but the reader should make no mistake about it, the hook is there because $b = 0.6075$ is less than one. In fact, the only instance when the hook is not present occurs if b exactly equals one (if $b > 1$ the hook occurs near the origin). Since b is a continuous variable, the probability is infinitesimal that its estimate is exactly zero. A recent example is Figure 1 appearing in the publication[16] resulting from the Digital Mammographic Imaging Screening Trial (DMIST) clinical trial,[16,17] involving 49,528 asymptomatic women from 33 clinical sites and involving 153 radiologists, where each of the film modality ROC plots crosses the chance diagonal and hooks upwards to (1,1), which as is known, results anytime $b < 1$. An alternative fitting method[18] that avoided this problem was available.

The binormal model is also susceptible to degeneracy problems. If the dataset does not provide any interior operating points (i.e., all observed points lie on the lines defined by FPF = 0 or TPF = 1) then the model fits these points with $b = 0$ and moreover, the fits are not unique.* The resulting straight-line segment fits do not make physical sense and, in essence, lacking other tools, the researcher would have to discard such datasets, always a bad idea, especially when experts are the ones that tend to produce such problem datasets. A newer model, **Chapter 17**, explains the reasons for this tendency of experts to produce problem datasets. Degeneracy problems are addressed by the newer models discussed in **Chapters 19** and **20**.

The next topic to be addressed is the sources of variability affecting AUC estimates and how to estimate them. One source, termed *case sampling*, has already been identified, namely the variability due to the finite nature of a dataset, Section 6.4.4. The next chapter deals with other sources and general methods of estimating them.

Appendix 6.A Expressions for partial and full area under the binormal ROC

This section is based on a cryptic paper by Thompson and Zucchini.[14] In what follows, *FPF* is abbreviates to x and *TPF* to y. Then the equation for the ROC curve is

$$y = \Phi\left(a + b\Phi^{-1}(x)\right) \tag{6.A1}$$

The partial area under the ROC curve from $x = 0$ to $x = c$, where $0 \le c \le 1$, is given by

$$A_{z;c} = \int_0^c y\,dx = \int_0^c dx\left[\Phi\left(a + b\Phi^{-1}(x)\right)\right] \tag{6.A2}$$

Consider the change of variable,

$$x = \Phi(x_1) \tag{6.A3}$$

---

* This is explained in detail in **Chapter 20**, so the reader needs to be patient.

This implies

$$x_1 = \Phi^{-1}(x)$$

$$dx = dx_1\, \phi(x_1)$$

(6.A4)

This yields

$$A_{z;c} = \int_0^c dx\, \Phi(a+bx_1) = \int_{-\infty}^{\Phi^{-1}(c)} dx_1\, \phi(x_1)\, \Phi(a+bx_1)$$

(6.A5)

Employing the definition of the Φ function the integral can be written as the following double integral:

$$A_{z;c} = \int_{x_1=-\infty}^{\Phi^{-1}(c)} \phi(x_1)\, dx_1 \int_{x_2=-\infty}^{a+bx_1} \phi(x_2)\,dx_2$$

(6.A6)

The right hand side of Equation 6.A6 can be expressed as an integral over the bivariate normal distribution by changing variables, from x_1, x_2 to z_1, z_2, as follows:

$$\left.\begin{array}{l} z_2 = x_1 \\[6pt] z_1 = (x_2 - bx_1)\, f \end{array}\right\}$$

(6.A7)

Here f is a quantity to be determined, which will allow one to conveniently complete the transformation to the desired bivariate integral. The second equation in Equation 6.A7 can be written

$$x_2 = \frac{z_1}{f} + bx_1 = \frac{z_1}{f} + bz_2$$

(6.A8)

The Jacobian[19] of the transformation is

$$J = \begin{pmatrix} 0 & 1 \\ \dfrac{1}{f} & b \end{pmatrix}$$

(6.A9)

The magnitude of the determinant of J is

$$\left|\det J\right| = \frac{1}{f}$$

(6.A10)

From a theorem in calculus,[19] the double integral over x_1, x_2 can be expressed in terms of a double integral over z_1, z_2 as follows:

$$A_{z;c} = \frac{1}{f} \int_{z_2=-\infty}^{\Phi^{-1}(c)} \phi(z_2)\,dz_2 \int_{z_1=-\infty}^{z_1^{UL}} \phi\left(\frac{z_1}{f} + bz_2\right) dz_1$$

(6.A11)

The upper limit of the inner integral can be calculated as follows. Using the second equation in Equation 6.A7,

$$z_1^{UL} = \left(x_2^{UL} - bx_1\right) f = (a + bx_1 - bx_1)\, f = af$$

(6.A12)

Then Equation 6.A11 simplifies to

$$A_{z;c} = \frac{1}{f} \int\limits_{z_2=-\infty}^{\Phi^{-1}(c)} \phi(z_2) dz_2 \int\limits_{z_1=-\infty}^{af} \phi\left(\frac{z_1}{f} + bz_2\right) dz_1 \qquad (6.A13)$$

If one performs a change of variable from f to a correlation like quantity ρ (whose magnitude must be less than or equal to unity) defined by

$$f = \sqrt{1-\rho^2} \qquad (6.A14)$$

Then the argument of the right-most ϕ function in Equation 6.A13 simplifies as follows:

$$\frac{z_1}{f} + bz_2 = \frac{z_1 + bz_2\sqrt{1-\rho^2}}{\sqrt{1-\rho^2}} \qquad (6.A15)$$

Recall that we are free to choose f, or equivalently, ρ, as we see fit. We now commit ourselves to a value: we define ρ in terms of the b-parameter to satisfy the following equality:

$$b\sqrt{1-\rho^2} = -\rho \qquad (6.A16)$$

Using this relation the argument of the right-most ϕ function in Equation 6.A13 simplifies as follows:

$$\frac{z_1}{f} + bz_2 = \frac{z_1 + bz_2\sqrt{1-\rho^2}}{\sqrt{1-\rho^2}} = \frac{z_1 - \rho z_2}{\sqrt{1-\rho^2}} \qquad (6.A17)$$

The expression for the partial area under the ROC reduces to

$$A_{z;c} = \frac{1}{\sqrt{1-\rho^2}} \int\limits_{z_2=-\infty}^{\Phi^{-1}(c)} \phi(z_2) dz_2 \int\limits_{z_1=-\infty}^{a\sqrt{1-\rho^2}} \phi\left(\frac{z_1 - \rho z_2}{\sqrt{1-\rho^2}}\right) dz_1 \qquad (6.A18)$$

Equation 6.A16 implies

$$b^2\left(1-\rho^2\right) = \rho^2$$

$$\rho^2\left(1+b^2\right) = b^2 \qquad (6.A19)$$

$$1-\rho^2 = 1 - \frac{b^2}{1+b^2} = \frac{1}{1+b^2}$$

Regardless of the value of b, ρ is assured to have magnitude less than or equal to unity, thereby satisfying the requirement of a correlation coefficient. Therefore,

$$A_{z;c} = \frac{1}{\sqrt{1-\rho^2}} \int\limits_{z_2=-\infty}^{\Phi^{-1}(c)} \phi(z_2) dz_2 \int\limits_{z_1=-\infty}^{\frac{a}{\sqrt{1+b^2}}} \phi\left(\frac{z_1 - \rho z_2}{\sqrt{1-\rho^2}}\right) dz_1 \qquad (6.A20)$$

The integrand is the standard bivariate normal probability density function[20] with correlation coefficient ρ (see Eqn. 26.3.2 in cited reference), denoted $\phi(z_1, z_2; \rho)$. In other words,

$$\phi(z_1, z_2; \rho) = \frac{1}{\sqrt{1-\rho^2}} \phi(z_2) \phi\left(\frac{z_1 - \rho z_2}{\sqrt{1-\rho^2}}\right) \tag{6.A21}$$

Using this function, the expression for the partial area reduces to

$$A_{z;c} = \int_{z_2=-\infty}^{\Phi^{-1}(c)} \int_{z_1=-\infty}^{\frac{a}{\sqrt{1+b^2}}} \phi(z_1, z_2; \rho) \, dz_1 \, dz_2 \tag{6.A22}$$

Appendix 6.A.1 Special case: Total area under the ROC curve

Since c is the upper limit of FPF, setting $c = 1$ yields the total area under the binormal ROC curve:

$$A_z = \int_{z_2=-\infty}^{\infty} \int_{z_1=-\infty}^{\frac{a}{\sqrt{1+b^2}}} \phi(z_1, z_2; \rho) \, dz_1 \, dz_2 = \int_{z_1=-\infty}^{\frac{a}{\sqrt{1+b^2}}} \phi(z_1) \, dz_1 = \Phi\left(\frac{a}{\sqrt{1+b^2}}\right) \tag{6.A23}$$

Since the integral over z_2 is over the entire range, it yields unity, leaving the one-dimensional density function $\phi(z_1)$ inside the integral. The last step follows from the definition of the Φ function. Another equivalent form for the total area under the binormal ROC curve is

$$A_z = \Phi\left(\frac{a}{\sqrt{1+b^2}}\right) = \Phi\left(\frac{\frac{a}{b}}{\sqrt{1+\frac{1}{b^2}}}\right) = \Phi\left(\frac{\mu}{\sqrt{1+\sigma^2}}\right) \tag{6.A24}$$

References

1. Dorfman DD, Alf E. Maximum-likelihood estimation of parameters of signal-detection theory and determination of confidence intervals—Rating-method data. *J Math Psychol.* 1969;6:487–496.
2. Wickham H. *Ggplot2: Elegant Graphics for Data Analysis.* 2nd Ed. AG Switzerland: Springer International Publishing; 2016.
3. Chang W. *R Graphics Cookbook.* Sebastopol, CA: O'Reilly Media; 2012.
4. Green DM, Swets JA. *Signal Detection Theory and Psychophysics.* New York, NY: John Wiley & Sons; 1966.
5. Hanley JA. The robustness of the "binormal" assumptions used in fitting ROC curves. *Med Decis Making.* 1988;8(3):197–203.
6. Dorfman DD, Berbaum KS, Metz CE, Lenth RV, Hanley JA, Abu Dagga H. Proper receiving operating characteristic analysis: The bigamma model. *Acad Radiol.* 1997;4(2):138–149.
7. Zhai X, Chakraborty DP. A bivariate contaminated binormal model for robust fitting of proper ROC curves to a pair of correlated, possibly degenerate, ROC datasets. *Med Phys.* 2017;44(6):2207–2222.

8. Press WH, Teukolsky SA, Vetterling WT, Flannery BP. *Numerical Recipes: The Art of Scientific Computing*. 3rd ed. Cambridge: Cambridge University Press; 2007.

9. McCullagh P, Nelder JA. *Generalized Linear Models*. 2nd ed., Vol 37. Boca Raton, FL: Chapman and Hall/CRC Press; 1989.

10. Pearson K. On the criterion that a given system of deviations from the probable in the case of a correlated system of variables is such that it can be reasonably supposed to have arisen from random sampling. *Breakthroughs in Statistics*. New York, NY: Springer; 1992: 11–28.

11. Larsen RJ, Marx ML. *An Introduction to Mathematical Statistics and Its Applications*. 3rd ed. Upper Saddle River, NJ: Prentice-Hall; 2001.

12. Grey DR, Morgan BJT. Some aspects of ROC curve-fitting: Normal and logistic models. *J Math Psychol*. 1972;9:128–1390.

13. Brandt S. *Data analysis: Statistical and computational methods for scientists and engineers*. 3rd ed,. New York, NY: Springer-Verlag New York Inc.; 1999.

14. Thompson ML, Zucchini W. On the statistical analysis of ROC curves. *Stat Med*. 1989;8(10):1277–1290.

15. Lusted LB. Signal detectability and medical decision making. *Science*. 1971;171:1217–1219.

16. Pisano ED, Gatsonis C, Hendrick E, et al. Diagnostic performance of digital versus film mammography for breast-cancer screening. *N Engl J Med*. 2005;353(17):1–11.

17. Pisano ED, Gatsonis CA, Yaffe MJ, et al. American College of Radiology Imaging Network digital mammographic imaging screening trial: Objectives and methodology. *Radiology*. 2005;236(2):404–412.

18. Dorfman DD, Berbaum KS. A contaminated binormal model for ROC data: Part II. A formal model. *Acad Radiol*. 2000;7(6):427–437.

19. Stein SK, Barcellos A. *Calculus and Analytic Geometry*. 5 ed. New York, NY: McGraw-Hill; 1992.

7

Sources of variability in AUC

7.1 Introduction

In previous chapters, the AUC under the ROC plot was introduced as the preferred way of summarizing performance in the ROC task, as compared to using a pair of sensitivity and specificity values. AUC can be estimated either nonparametrically, as in **Chapter 5**, or parametrically, as in **Chapter 6**, and improved ways of estimating it are described in **Chapters 19** and **20**.

Due to the finite numbers of cases comprising the dataset any estimate of AUC is subject to case sampling variability. This variability can be estimated using the binormal model, from the sharpness of the peak of the likelihood function, Section 6.4.4. This chapter focuses on general sources of variability affecting AUC, regardless of how it is estimated, and other ways of estimating variability.

Here is an outline of this chapter. The starting point is identification of different sources of variability affecting AUC estimates. Considered next is dependence of AUC, however estimated, on a case-set index $\{c\}$, $c = 1,2,...,C$, ignored in the literature. This can lead to confusion among those with less statistical expertise. Considered next is estimating case sampling variability of the empirical estimate of AUC by an analytic method. This is followed by descriptions of two resampling-based methods, namely the bootstrap and the jackknife, both of which have wide applicability (i.e., they are not restricted to estimating AUC variability). The methods are demonstrated using **R** and the implementation of a calibrated simulator is shown and used to demonstrate their validity, that is, showing agreement between the different methods of estimating variability. The dependence of AUC on reader expertise and modality is considered. An important source of variability, namely the radiologist's choice of internal sensory thresholds, is described. Cautionary comments are made regarding indiscriminate usage of empirical AUC as a measure of performance.

Online Appendix 7.A describes **R** implementation of the bootstrap method; Online Appendix 7.B is the corresponding implementation of the jackknife. Online Appendix 7.C describes implementation of the calibrated simulator for single modality single reader ROC datasets. Online Appendix 7.D describes the code that allows comparison of the different methods of estimating case sampling variability.

7.2 Three sources of variability

Statistics deals with variability. Understanding sources of variability affecting AUC is critical to an appreciation of ROC analysis. The author's introduction to this subject (ca. 1984) was through Swets and Pickett's book[1] "*Evaluation of Diagnostic System: Methods from Signal Detection Theory.*" Three sources of variability are identified in that book: *case sampling*, *between-reader*, and *within-reader*.

1. *Consider a single reader interpreting different case-sets.* Case sampling variability arises from the *finite* number of cases comprising each case-set, compared to the potentially very large *population* of cases. (If one could sample every case that exists and have them interpreted by the same reader, there would be no case sampling variability and the poor reader's AUC values [from repeated interpretations of the entire population] would reflect only within reader variability, see #3 below.) Each case-set $\{c\}$, consisting of K_1 non-diseased and K_2 diseased cases, interpreted by the reader yields an AUC value. The notation $\{c\}$ does not mean single cases, rather different *case-sets*. Thus, $\{c\} = \{1\},\{2\}$, and so on, denote different case-sets, each consisting of K_1 non-diseased and K_2 diseased cases; there could be overlap of individual cases between different case-sets, a fact explicitly recognized in the bootstrap method described below.

   There is much data compression in going from individual case ratings to AUC. For a single reader and given case-set $\{c\}$, the ratings can be converted to an $A_{z\{c\}}$ estimate, Equation 6.48. The notation makes explicit the dependence of the AUC measure on the case-set $\{c\}$. One can conceptualize the distribution of $A_{z\{c\}}$'s over different case-sets, each of the same size $K_1 + K_2$, as a normal distribution, that is, $A_{z\{c\}} \sim N\left(A_{z\{\bullet\}}, \sigma^2_{cs+wr}\right)$. The dot notation $\{\bullet\}$ denotes an average over all case-sets. $A_{z\{\bullet\}}$ is the *case sampling mean* of A_z for the single fixed reader and σ^2_{cs+wr} is the *case sampling plus within-reader variance*. The reason for adding the within-reader variance is explained in #3 below. The concept is that the reader interpreting *different case-sets* effectively samples different parts of the population of cases, resulting in variability in measured A_z. Sometimes easier cases are sampled, and sometimes more difficult ones, and so on. This source of variability is expected to decrease with increasing case-set size, that is, increasing $K_1 + K_2$, which is the reason for seeking large numbers of cases in clinical trials. Case sampling (and within-reader) variability also decreases when the cases are more *homogenous*. An example of a more homogenous case sample would be a case-set originating from a small geographical region with limited ethnic variability. This is the reason for seeking multi-institutional clinical trials, because they tend to sample more of the population than patients seen at a single institution.

2. *Consider different readers interpreting a fixed case-set.* Between-reader variability arises from the *finite* number of readers participating in the study compared to the *population* of readers; the population of readers could be all MQSA certified radiologists interpreting screening mammograms in the United States. This time one envisages *different* groups of readers interpreting a fixed case-set $\{1\}$. The different reader's $A_{z;j}$ values (j is the reader index, $j = 1, 2, ..., J$, where J is the total number of readers in the group) are distributed $A_{z;j} \sim N\left(\overline{A_{z;\bullet\{1\}}}, \sigma^2_{br+wr}\right)$, where $\overline{A_{z;\bullet\{1\}}}$ is the reader population mean (the dot symbol replacing the reader index, averages over each group of readers, and the grand average, indicated by the bar symbol, obtains the reader-population mean) for the fixed case-set $\{1\}$ and σ^2_{br+wr} is the between-reader plus within-reader variance. The reason for adding the within-reader variance is explained in #3 below. The concept is that *different groups of readers interpret the same case-set* $\{1\}$, thereby sampling different parts of the reader distribution, causing fluctuations in the measured $A_{z;j\{1\}}$ of the readers. Sometimes better readers are sampled and sometimes not so good ones are sampled. This time there is no data compression—each reader in the sample has an associated $A_{z;j}$. However, variability of the *average* $A_{z;j}$ over the J readers is expected to decrease with increasing J. This is the reason for seeking large reader samples.

3. *Consider a fixed reader, for example, $j = 1$, interpreting a fixed case-set* $\{1\}$. Within-reader variability is due to variability of the ratings for the same case: the same reader interpreting the same case on different occasions will give different ratings to it, causing fluctuations in the measured AUC. This assumes that memory effects are minimized, for example, by sufficient time between successive interpretations. Since this is an intrinsic source of variability (analogous to the internal noise of a voltmeter) affecting each reader's interpretations, *it cannot be separated from case sampling or between-reader variability*; that is, it

cannot be turned off. The last sentence needs further explanation. A measurement of case sampling variability requires a reader, and the reader comes with an intrinsic source of variability that gets added to the case sampling variance, so what is measured is the sum of case sampling and within-reader variances, denoted σ^2_{cs+wr}. Likewise, a measurement of between-reader variability requires a fixed case-set interpreted by different readers, each of whom comes with an intrinsic source of variability that gets added to the between-reader variance, denoted σ^2_{br+wr}. To emphasize this point, an estimate of case sampling variability *always* includes within-reader variability, even if the notation does not show this explicitly. Likewise, an estimate of between-reader variability *always* includes within-reader variability, even if the notation does not show this explicitly.

With this background, the purpose of this chapter is to delve into variability in some detail and, in particular, describe computational methods for estimating them. This chapter introduces the concept of resampling a dataset to estimate variability and the widely used bootstrap and jackknife methods of estimating variance are described. In a later chapter, these are extended to estimating covariance (essentially a scaled version of the correlation) between two random variables.

The starting point is the simplest scenario: a single reader interpreting a case-set.

7.3 Dependence of AUC on the case sample

Suppose a researcher conducts a ROC study with a single reader. One starts by selecting a case-set, that is, a set of proven-truth non-diseased and diseased cases. Another researcher conducting a ROC study at the same institution selects a different case-set, that is, a different set of proven-truth non-diseased and diseased cases. The two case-sets contain the same numbers K_1, K_2 of non-diseased and diseased cases, respectively. Even if the same radiologist interprets the two case-sets, and the reader is perfectly reproducible, the AUC values are expected to be different. Therefore, AUC must depend on a *case sample index*, which is denoted $\{c\}$, where c is an integer: $c = 1, 2, ..., C$, and so on.

$$AUC \rightarrow AUC_{\{c\}} \tag{7.1}$$

Note that $\{c\}$ is not an *individual case* index, rather it is a *case-set* index, that is, different integer values of c denote different *sets*, or *samples*, or *groups*, or *collections of cases*.

What does the dependence of AUC on the c index mean? Different case-sets differ in their *difficulty* levels. A difficult case-set contains a greater fraction of difficult cases than is usual. A difficult diseased case is one where disease is difficult to detect. For example, the lesions could be partly obscured by overlapping normal structures in the patient anatomy; that is, the lesion does not stick out. Alternatively, variants of normal anatomy could mimic a lesion, like a blood vessel viewed end on in a chest radiograph, causing the radiologist to mistake it for a lesion. An easy diseased case is one where the disease is easy to detect. For example, the lesion is projected over smooth background so it sticks out or is more conspicuous.[2] How does difficulty level affect non-diseased cases? A difficult non-diseased case is one where variants of normal anatomy mimic actual lesions and could cause the radiologist to falsely diagnose the patient as diseased. Conversely, an easy non-diseased case is like a textbook illustration of normal anatomy. Every structure in it is clearly visualized and accounted for by the radiologist's knowledge of the patient's non-diseased anatomy, and the radiologist is confident that any abnormal structure, *if present*, would be readily seen. The radiologist is unlikely to falsely diagnose the patient as diseased. Difficult cases tend to be rated in the middle of the rating scale, while easy ones tend to be rated at the ends of the scale.

7.3.1 Case sampling induced variability of AUC

An easy case sample will cause AUC to increase over its average value; interpreting many case-sets and averaging the AUCs determines the average value. Conversely, a difficult case sample will cause the AUC to decrease. Case sampling variability is reflected in variability in the measured AUCs. How does one estimate this essential source of variability? One method, totally impractical in the clinic but easy with simulations, is to have the same radiologist interpret repeated samples of case-sets from the population of cases (i.e., patients), termed *population sampling*, or more viscerally, as the *brute force* method.

Even if one could get a radiologist to interpret different case-sets, it is even more impractical to actually acquire the different case samples of truth-proven cases. Patients do not come conveniently labeled as non-diseased or diseased. Rather, one needs to follow-up on the patients, perhaps do other imaging tests, in order to establish true disease status, or *ground-truth*. In screening mammography, a woman who continues to be diagnosed as non-diseased on successive yearly screening tests in the United States, and has no other symptoms of breast disease, is probably disease-free. Likewise, a woman diagnosed as diseased and the diagnosis is confirmed by biopsy (i.e., the biopsy comes back showing a malignancy in the sampled tissues) is known to be diseased. However, not all patients who are diseased are actually diagnosed as diseased: a typical false negative fraction is 20% in screening mammography.[3] This is where follow-up imaging can help determine true disease status at the initial screen. A false negative mistake is unlikely to be repeated at the next screen. After a year, the tumor may have grown, and is more likely to be detected. Having detected the tumor in the most recent screen, radiologists can go back and retrospectively view it in the initial screen, at which it was missed during the live interpretation. If one knows *where* to look, the cancer is easier to see. The previous screen images would be an example of a difficult diseased case. In unfortunate instances, the patient may die from the previously undetected cancer, which would establish the truth status at the initial screen, too late to do the patient any good. The process of determining actual truth is often referred to as defining the *gold standard*, the *ground truth*, or simply *truthing (the last word, not in the author's dictionary, is used by researchers). One can appreciate from this discussion that acquiring proven cases, particularly diseased ones, is one of the most difficult aspects of conducting an observer performance study.*

There has to be a better way of estimating case sampling variability. With a parametric model, the maximum likelihood procedure provides a means of estimating variability of each of the estimated parameters, which can be used to estimate the variability of A_z, as in **Chapter 6**. The estimate corresponds to case sampling variability (including the inseparable within-reader variability). If unsure about this point, the reader should run the example in **Chapter 6**, specifically mainRocfitR.R, with increased numbers of cases. The estimates of variance of Az will be seen to decrease.

There are other options available for estimating case sampling variance of AUC, and this chapter is not intended to be comprehensive. Three commonly used options are described: the DeLong et al. method, the bootstrap, and the jackknife resampling methods.

7.4 Estimating case sampling variability using the DeLong method

If the figure of merit is the empirical AUC, then a procedure developed by DeLong et al.[4] (henceforth abbreviated to DeLong) that is based on earlier work by Noether[5] and Bamber,[6] is applicable. The author will not go into details of this procedure (implemented in **DeLongVar.R**) but limit to showing that it works. The reader may wish to compare the **R** implementation with the original publication. However, before one can show that it works, one needs to know the true value of the variance of empirical AUC. One cannot use the binormal model maximum likelihood estimation (MLE) estimate of variance as it is an estimate, not to be confused with a true value, and moreover the estimate refers to the fitted AUC, not empirical AUC. [Estimates are realizations of random numbers and are themselves subject to variability, which decreases with increasing case-set size.] Instead, a *brute force* (i.e., simulated population sampling) approach is adopted to determine the true value of the variance of empirical AUC. The simulator provides a means of repeatedly generating case-sets interpreted by the same radiologist, and by sampling it enough times, for example, $C = 10,000$ times, each time calculating empirical AUC, one determines the population mean and standard deviation. The standard deviation determined this way is compared to that yielded by the DeLong method to check if the latter actually works. Open the file **mainDeLongSd.R**; a listing follows.

7.4.1 Code listing

```
rm(list = ls()) # mainDeLongSd.R
source("Wilcoxon.R");source("DeLongVar.R")
seed <- 1;set.seed(seed)
mu <- 1.5;sigma <- 1.3;K1 <- 50;K2 <- 52
cat("seed = ", seed, ", K1 = ", K1, ", K2 = ", K2,
    ", mu = ", mu, ", sigma = ", sigma, "\n")
# brute force method to find the population mean and stdDev. dev.
empAuc <- array(dim = 10000)
for (i in 1:length(empAuc)) {
  zk1 <- rnorm(K1);zk2 <- rnorm(K2, mean = mu, sd = sigma)
  empAuc[i] <- Wilcoxon(zk1, zk2)
}
meanempAuc   <-   mean(empAuc)
stdDevempAuc  <-  sqrt(var(empAuc))
cat("population mean empAuc = ", meanempAuc,
    ", population stdDev empAuc = ", stdDevempAuc, "\n")
# one more trial
zk1 <- rnorm(K1);zk2 <- rnorm(K2, mean = mu, sd = sigma)
empAuc <- Wilcoxon(zk1, zk2)
ret   <- DeLongVar(zk1,zk2)
stdDevDeLong <- sqrt(ret)
cat("1 sample empAuc = ", empAuc,
    ", stdDev DeLong = ", stdDevDeLong, "\n")
```

Line 2 sources the files needed for this code to work: one calculates the Wilcoxon statistic, i.e., the empirical AUC, and the other implements the DeLong method. Line 4 sets the **seed** of the random number generator to 1. The **seed** variable is analogous to the case-set index c. Keeping **seed** fixed realizes the same random numbers each time the program is run. Different values of **seed** result in different, that is, statistically independent, random samples. Line 5 initializes the values needed by the data simulator: the normal distributions are separated by **mu** = 1.5, the standard deviation of the diseased distribution is **sigma** = 1.3, and there are **K1** = 50 non-diseased and **K2** = 52 diseased cases. Lines 9 through 18 implement the brute force method of estimating mean and standard deviation of the population distribution of empirical *AUC* and prints them. The actual data simulation occurs at line 12: the ratings vectors are **zk1** and **zk2**, corresponding to non-diseased and diseased cases, respectively. Line 13 calculates empirical AUC, using function **Wilcoxon()**, and saves it to the array **empAUC**. Lines 15 and 16 calculate the mean and standard deviation of the AUC samples: the latter is the correct value to which the DeLong standard deviation estimate will be compared. Line 21 generates a fresh ROC dataset to which the DeLong method will be applied. Line 22 calculates the new value of the empirical area *AUC* for this dataset and line 23 applies the DeLong method, which returns the variance of the empirical estimate of AUC, whose square root is the standard deviation. Two runs of this code were made, one with the smaller sample size, and the other with

10 times the sample size (the second run takes much longer). A third run was made with the larger sample size but with a different **seed** value. The results follow, as shown in the next section.

7.4.2 Code output

```
> source(...)
seed =  1
K1 =   50
K2 =   52
mu =   1.5
sigma =  1.3
population mean empAuc =   0.819178
population stdDev empAuc =   0.04176683
1 sample empAuc =   0.8626923
stdDev DeLong =   0.03804135
> source(...)
seed =  1
K1 =   500
K2 =   520
mu =   1.5
sigma =  1.3
population mean empAuc =   0.8194576
population stdDev empAuc =   0.01309815
1 sample empAuc =   0.8206962
stdDev DeLong =   0.01309314
> source(...)
seed =  2
K1 =   500
K2 =   520
mu =   1.5
sigma =  1.3
population mean empAuc =   0.8194988
population stdDev empAuc =   0.01300203
1 sample empAuc =   0.8047269
stdDev DeLong =   0.01356696
```

1. An important observation is that as sample-size increases, case sampling variability decreases: 0.0417 for the smaller sample size versus 0.01309 for the larger sample size. The dependence is in proportion to the inverse square root of the numbers of cases, for example, 0.04176683/ sqrt(10) = 0.01320783. This is expected from the central limit theorem[7].
2. With the smaller sample size (K1/K2 = 50/52; the back-slash notation, not to be confused with division, is a convenient way of summarizing the case sample size) the estimated standard deviation (0.038) is within 10% of that estimated by population sampling (0.042). With the larger sample size, (K1/K2 = 500/520) the two are even closer (0.013093 versus 0.01356696—the latter value is for seed = 2).
3. Notice also that the one sample empirical AUC for the smaller case-size is 0.863, which is less than two standard deviations from the population mean 0.819. The two standard deviations come from rounding up 1.96, as in Equation 3.46, where $z_{\alpha/2}$ was defined as the *upper* $\alpha/2$ quantile of the unit normal distribution and $z_{0.025} = 1.96$.
4. To reiterate, with clinical data the DeLong procedure estimates case sampling plus within reader variability. With simulated data as in this example, there is no within-reader variability as the simulator yields identical values for fixed seed.

This demonstration should convince the reader that one does have recourse other than the brute force method, at least when the figure of merit is the empirical area under the ROC. That should come as a relief, as population sampling is impractical in the clinical context. It should also impress the reader, as the DeLong method is able to use information present in a *single* dataset to tease out its *variability*. (This is analogous to the MLE estimate, which is also able to tease out variability based on a parametric fit to a single dataset and examination of the sharpness of the peak of the log-likelihood function, **Chapter 6.**)

Next, two resampling–based methods of estimating case sampling variance of *AUC* are introduced. The word resampling means that the *dataset itself is regarded as containing information regarding its variability*, which can be extracted by sampling from the original data (hence the word resampling). These are general and powerful techniques, applicable to any scalar statistic, not just the empirical AUC, which one might be able to use in other contexts.[7]

Exercises and proposed projects:

1. Exercise the code with different **seed**-values and be convinced that statements in #2 above for the smaller sample size are correct.
    a. Specifically, the differences between the brute force standard deviation **stdDevempAuc**, the true value, and that yielded by the DeLong method **stdDevDeLong** can be accounted for by sampling variability (statistical statement: the differences are not significant at the 5% level).
    b. Note: one does not need to re-establish the true value. With 10,000 samples, variations in **seed** are not expected to alter the true values (Try it! The differences are < 1%). Put lines 21 through 26 inside a **for**-loop, and for each iteration of the for-loop, one saves the value at line 24 in a suitably initialized array **stdDevDeLong**. Finally, compare the empirical 95% confidence interval for **stdDevDeLong** to the true value **stdDevempAuc**.

2. Extend the simulation model to include the effect of binning; the reader may wish to see examples of how binning is easily accomplished by a function in the **RJafroc** package before returning to this. Try binning the data into five or six bins.
3. Include a model for within-reader variability in the simulation model. Does the DeLong method indeed estimate case sampling plus within-reader variability? Hint: this is a modification of Eqn. 6.1 and Eqn. 6.2 to account for replications; r is the replication index; $r = 1$, 2, 3, ...

$$
\left.
\begin{aligned}
& Z_{k_t tr} \sim \mu_t + C_{k_t t} + \varepsilon_{k_t tr} \\
& t = 1, 2; \mu_1 = 0; \mu_2 = 1.5 \\
& C_{k_1 1} \sim N(0, 1^2); C_{k_2 2} \sim N(0, \sigma_{cs}^2); \\
& \varepsilon_{k_t tr} \sim N(0, \sigma_{wr}^2)
\end{aligned}
\right\}
\qquad (7.2)
$$

The variances in Equation 7.2 are for the *Z*-samples, not to be confused with those introduced in the Introduction, which denote AUC variances. In the design of the simulator, the term σ_{wr}^2 represents the ratings variability of a case repeatedly interpreted by the same observer. The term σ_{cs}^2 represents case sampling variability, and to correspond to the code in **mainDeLongSd.R**, it should be set to **sigma**$^2 = (1.3)^2$.

7.5 Estimating case sampling variability of AUC using the bootstrap

The simplest resampling method, at least at the conceptual level, is the bootstrap. It is based on the assumption that it is safe to regard the observed sample as *defining* the population from which it was sampled. Since by definition a population cannot be exhausted, the idea is to resample, with replacement, from the observed dataset. Each resampling step realizes a particular bootstrap data-set denoted $\{b\}$, where $b = 1, 2, ..., B$. The curly brackets emphasize that different integer values of b denote different sets of cases, not individual cases. (In contrast, the notation (k) will be used to denote removing a specific case, k, as in the jackknife procedure to be described shortly. The index b should not be confused with the index c, the case sampling index. The latter denotes repeated sampling from the population, which is impractical in real life. The bootstrap index denotes repeated sampling from a dataset, which is quite feasible.) The procedure is repeated B times, typically B can be as small as 200, but to be safe the author generally uses about 1000–2000 bootstraps. The following example uses Table 4.1 from **Chapter 4**, reproduced in Table 7.1.

For convenience, let us denote cases as follows. The 30 non-diseased cases that received the 1-rating are denoted $k_{1,1}, k_{2,1}, ..., k_{30,1}$. The second index denotes the truth-state of the cases. Likewise, the 19 non-diseased cases that received the 2-rating are denoted $k_{31,1}, k_{32,1}, ..., k_{49,1}$ and so on for the remaining non-diseased cases. The five diseased cases that received the 1 rating are denoted $k_{1,2}, k_{2,2}, ..., k_{5,2}$, the six diseased cases that received the 2-rating are denoted $k_{6,2}, k_{7,2}, ..., k_{11,2}$, and so on. Let us figuratively put all non-diseased cases (think of each case as an index card, with the case notation and rating recorded on it) into one hat (the non-diseased hat) and all the diseased cases into another hat (the diseased hat). Next, one randomly picks one case (card) from the non-diseased hat, records it's rating, and puts the case back in the hat, so it is available to be possibly picked again. This is repeated 60 times for the non-diseased hat resulting in 60 ratings from non-diseased cases. A similar procedure is performed using the diseased hat, resulting in 50 ratings from diseased cases. The author has just described, in painful detail (one might say), the realization of the first bootstrap sample, denoted $\{b = 1\}$. This is used to construct the first bootstrap counts table, shown in Table 7.2.

So, what happened? Consider the 35 non-diseased cases with a 1-rating. If each non-diseased case rated 1 in Table 7.2 were picked one time, the total would have been 30, but it is 35. Therefore, some of the original non-diseased cases rated 1 must have been picked multiple times, but one must also make allowance as there is no guarantee that a specific case was picked at all. Still focusing on the 35 non-diseased cases with a 1-rating in the first bootstrap sample, the picked labels (reordered after the fact, with respect to the first index) might be

$$k_{2,1}, k_{2,1}, k_{4,1}, k_{4,1}, k_{4,1}, k_{6,1}, k_{7,1}, k_{7,1}, k_{9,1}, ..., k_{28,1}, k_{28,1}, k_{30,1}, k_{30,1} \qquad (7.3)$$

Table 7.1 A typical ROC counts table, showing the original data; AUC = 0.870

	Counts in ratings bins				
	Rating = 1	**Rating = 2**	**Rating = 3**	**Rating = 4**	**Rating = 5**
$K_1 = 60$	30	19	8	2	1
$K_2 = 50$	5	6	5	12	22
	Operating points				
	≥5	≥4	≥3	≥2	≥1
FPF	0.017	0.050	0.183	0.500	1
TPF	0.440	0.680	0.780	0.900	1

Table 7.2 The counts table for the first bootstrap dataset; AUC = 0.843

	Counts in ratings bins				
	Rating = 1	Rating = 2	Rating = 3	Rating = 4	Rating = 5
$K_1 = 60$	35	16	9	0	0
$K_2 = 50$	7	9	7	8	19
			Operating points		
	≥ 5	≥ 4	≥ 3	≥ 2	≥ 1
FPF	0.000	0.000	0.150	0.417	1
TPF	0.380	0.540	0.680	0.860	1

In this example, case $k_{1,1}$ was not picked, case $k_{2,1}$ was picked twice, case $k_{3,1}$ was not picked, case $k_{4,1}$ was picked three times, case $k_{5,1}$ was not picked, case $k_{6,1}$ was picked once, and so on. The total number of cases in Equation 7.3 is 35, and similarly for the other cells in this table. Based on the first bootstrapped counts table, one can estimate AUC. Using the website[9] referred to in Chapter 06, one gets $AUC = 0.843$. (It is OK to use a parametric FOM since the bootstrap is a general procedure applicable, in principle, to any *FOM*, not just the empirical *AUC*, as is the DeLong method.) The corresponding value for the original data, Table 7.1, was AUC = 0.870. The first bootstrapped dataset yielded a smaller value than the original dataset because one happened to have picked an unusually difficult bootstrap sample.

(Notice that in the original data there were 6 + 5 = 11 diseased cases that were rated 1 and 2, but in the bootstrapped dataset there are 7 + 9 = 16 diseased cases that were rated 1 and 2. In other words, the number of *incorrect* decisions on diseased cases went up, which would tend to lower AUC. There is also an decrease in number of correct decisions on diseased cases: 8 + 19 = 27 cases rated 4 and 5, as compared to 12 + 22 = 34 in the original dataset. Additionally, there is an increase in the number of *correct* decisions on non-diseased cases, albeit minimally: 35 + 16 = 51 rated 1 and 2 versus 30 + 19 = 49 in the original dataset, and zero counts rated 4 and 5 in the non-diseased versus 2 + 1 = 3 in the diseased. The wordiness of the preceding description illustrates the difficulty, in fact the futility, of correctly predicting the change in performance from an examination of the two ROC counts tables. Too many numbers are changing and, in the above example, one did not even consider the change in counts in the bin labeled 4. Hence, the need for an objective figure of merit, such as the binormal model based *AUC* or the empirical AUC.)

To complete the description of the bootstrap method, one repeats the procedure described in the preceding paragraphs $B = 200$ times, each time running the website calculator (not very practical) and the final result is B values of AUC, denoted:

$$AUC_{\{1\}}, AUC_{\{2\}}, ..., AUC_{\{B\}} \tag{7.4}$$

where $AUC_{\{1\}} = 0.843$, and so on. The bootstrap estimate of the variance of AUC is defined by[7]

$$Var_{bs}\left(AUC\right) = \frac{1}{B-1} \sum_{b=1}^{B} \left(AUC_{\{b\}} - AUC_{\{\bullet\}}\right)^2 \tag{7.5}$$

The right-hand side is the traditional definition of (unbiased) variance. *The dot represents the average over the replaced index.* Of course, running the website code 200 times and recording the outputs is not a productive use of time. The following code implements two methods for estimating AUC, the binormal model estimate of AUC, described in **Chapter 6**, and the empirical AUC, described in **Chapter 5**.

7.5.1 Demonstration of the bootstrap method

Open the project file for this chapter and the file **mainBootstrapSd.R**, Online Appendix 7.A. Make sure the seed variable at line 12 is initialized to 1 and **source** the code yielding the following output. [Think of the seed variable as the case sample index $\{c\}$. Selecting a different seed generates a different case sample.] Since the bootstrap method is applicable to any scalar figure of merit, two options are provided in the code, lines 10 and 11. Currently **FOM <- Az**, which uses the binormal model estimate, but if one reverses the commenting, the empirical *AUC* is used. In about two seconds the author's computer yielded the following output.

7.5.1.1 Code Output for seed = 1

```
> source(...)
FOM =   Az , seed =   1 , B =   200
OrigAUC =   0.8704519 , meanAUC =   0.8671713 , stdAUC =   0.04380523
```

This shows that the AUC of the original data (i.e., before bootstrapping) is 0.870, the mean AUC of the $B = 200$ bootstrapped datasets is 0.867, and the standard deviation of the 200 bootstraps is 0.0438. Now, if one runs the website calculator referenced in the previous chapter on the dataset shown in Table 7.1, one finds that the MLE of the standard deviation of the AUC of the fitted ROC curve is 0.0378. The standard deviation is itself a statistic and there is sampling variability associated with it, that is, there exists such a beast as a standard deviation of a standard deviation; the reader should be satisfied that the bootstrap estimate is near the MLE estimate.

By setting **seed** to different values, one gets an idea of the variability in the estimate of the standard deviation of AUC (to repeat, **seed** is like the case sample index $\{c\}$ and different values correspond to different case-sets). For example, with **seed <- 2**, one gets the following output.

7.5.1.2 Code output for seed = 2

```
> source(...)
FOM =   Az , seed =   2 , B =   200
OrigAUC =   0.8704519 , meanAUC =   0.8673155 , stdAUC =   0.03815402
```

Note that both the mean of the bootstrap samples and the standard deviation have changed, but both are close to the MLE values. One should experiment with other values of seed. Examined next is the dependence of the estimates on B, the number of bootstraps. With **seed <- 1** and **B <- 2000** one gets the following output.

7.5.1.3 Code output for B = 2000

```
> source(...)
#boots =   2000 seed =   1 OrigAUC =   0.8704519 meanAUC =   0.8674622
stdAUC =   0.03833508
```

The estimates are evidently rather insensitive to B, but the computation time was longer, ~13 seconds (running MLE 2000 times in 13 seconds is not bad). It is always a good idea to test the stability of the results to different **B** and **seed** values. Unlike the DeLong method, the bootstrap is broadly applicable to other figures of merit. Specifically, it is not limited to the empirical area under the

ROC. However, it depends on the assumption that the sample itself is representative of the population. [With limited numbers of cases, this could be a bad assumption. [With small sample sizes, it is relatively easy to enumerate the different outcomes of the sampling process and, more importantly, their respective probabilities, leading to what is termed the *exact* bootstrap. It is exact in the sense that it yields results with no **seed** or number of bootstrap **B** dependence.]

7.6 Estimating case sampling variability of AUC using the jackknife

Attention now turns to the second resampling method, termed the *jackknife*, which is computationally less demanding, but as was seen with the bootstrap, with modern personal computers computational limitations are no longer that important, at least for the types of analyses that this book is concerned with.

In this method, the first case is removed, or *jackknifed*, from the set of cases and the MLE (or empirical estimate) is obtained on the resulting dataset, which has one less case. Denote by $AUC_{(1)}$ the resulting value of ROC-AUC. The parentheses around the subscript 1 are meant to emphasize that the AUC value corresponds to that with the first case *removed* from the original dataset. Next, the first case is replaced, and now the second case is removed, the new dataset is analyzed yielding $AUC_{(2)}$, and so on, yielding K (K is the total number of cases; $K = K_1 + K_2$) *jackknife AUC values*.

$$AUC_{(k)}; k = 1, 2, ..., K \qquad (7.6)$$

The corresponding *jackknife pseudovalues* Y_k are defined by

$$Y_k = K\ AUC - (K-1)\ AUC_{(k)} \qquad (7.7)$$

Here AUC denotes the estimate using the entire dataset, that is, not removing any cases. The jackknife pseudovalues will turn out to be of central importance in **Chapter 9**

The jackknife estimate of the variance is defined by[7]

$$Var_{jk}(AUC) = \frac{(K-1)^2}{K} \frac{1}{K-1} \sum_{k=1}^{K} \left(AUC_{(k)} - AUC_{(\bullet)}\right)^2 \qquad (7.8)$$

Since variance of K scalars is defined by

$$Var(x) = \frac{1}{K-1} \sum_{k=1}^{K} \left(x_k - x_{(\bullet)}\right)^2 \qquad (7.9)$$

$$Var_{jk}(AUC) = \frac{(K-1)^2}{K} Var\left(AUC_{(k)}\right) \qquad (7.10)$$

In Equation 7.8 the author has deliberately *not* simplified the right-hand side by cancelling out K-1. The purpose is to show, Equation 7.10, that the usual expression for the variance needs to be multiplied by a *variance inflation factor* $\frac{(K-1)^2}{K}$, which is approximately equal to K, in order to obtain the correct jackknife estimate of variance of AUC. This factor was not necessary for the bootstrap method. That is because the bootstrap samples are more representative of the actual spread in the data. The jackknife samples are more restricted than the bootstrap samples, so the spread of the data is smaller; hence the need for the variance inflation factor.[7]

Source the file **mainJackknifeSd.R**, Online Appendix 7.B, after ensuring that Az is selected as the FOM, yielding the following results

7.6.1 Code output

```
> source(...)
FOM =   Az
OrigAUC =   0.8704519
jackknifeMeanAuc =   0.8704304
stdAUC =   0.03861591
```

Notice that the code does not use a **set.seed()** statement, as no random number generator is needed in the jackknife method (systematically removing and replacing each case, one at a time, is not random sampling, which should further explain the need for the variance inflation factor in Equation 7.10).

7.7 Estimating case sampling variability of AUC using a calibrated simulator

In real life, one does have the luxury of sampling from the population of cases, but with a simulator almost anything is possible. The population sampling method used previously, Section 7.3.2, to compare the DeLong method to a known standard used *arbitrarily* set simulator values (**mu** = 1.5 and **sigma** = 1.3 at line 5 of **mainDeLongSd.R**). There is no guarantee that whether these values are actually representative of real clinical data. In this section, a simple method of implementing population sampling using a *calibrated* simulator is described. The code is in **mainCalSimulator.R** in Online Appendix 7.C. A *calibrated simulator* is one in which parameters are chosen to match those of an actual clinical dataset, so the simulator is calibrated to the specific dataset. Why might one wish to do that? Rather than fish in the dark and set arbitrary values for the simulator parameters, one needs to find realistic values that match an actual clinical dataset. This way one has at least some assurance that the simulator is realistic and therefore its verdict on a proposed method or analysis is more likely to be correct.

As an example, consider a real clinical dataset, such as in Table 7.1. This data set was analyzed by the MLE yielded model parameters, (a, b) and the thresholds $\zeta_1, \zeta_2, \zeta_3, \zeta_4$. The specific values were (in the same order): 1.320453, 0.607497, 0.007675259, 0.8962713, 1.515645, and 2.39671 (listed in 7.7.1, Code Output, below). On each pass through the simulator one samples 60 values from the non-diseased distribution and 50 values from the diseased distribution, implemented in **SimulateRocCountsTable.R**, Online Appendix 7.C, which returns a simulated ROC counts table like Table 7.1. MLE on the ROC counts table yields A_z. The process is repeated $P = 2000$ (p is the population sampling index, ranging from 1 to P) and finally one calculates the mean and standard deviation of the 2000 A_z values. Open the file **mainCalSimulator.R**, Online Appendix 7.C, confirm that FOM is set to Az at line 11 (i.e., it is not commented out) and **source** it. Shown are results of two runs with different values for **seed** (namely, 1 and 2).

7.7.1 Code output

```
> source(...)
seed = 1 , FOM = Az , P = 2000
Calibrated simulator values: a, b, zetas:
 1.32 0.6075 0.007675 0.8963 1.516 2.397
seed = 1
OrigAUC = 0.8705
meanAUC = 0.8677
stdAUC = 0.04033
> source(...)
seed = 2 , FOM = Az , P = 2000
Calibrated simulator values: a, b, zetas:
 1.32 0.6075 0.007675 0.8963 1.516 2.397
seed = 2
OrigAUC = 0.8705
meanAUC = 0.8682
stdAUC = 0.04055
```

The **seed** = 1 estimate of standard deviation of AUC (0.0403) is recorded in **Table 7.3**, row A, sub-row *Population*. The entry for sub-row *MLE* was obtained using the ROCFIT equivalent Eng's JAVA program,[8] Section 6.2.7. The DeLong method entry for row A was obtained using **mainDeLongSd.R** with FOM set to Wilcoxon, as indicated by the asterisk Section 7.3.2. The bootstrap entry was obtained using **mainBootstrapSd.R**, and the jackknife entry was obtained using **mainJackknifeSd.R**. In both cases, **FOM** was set to **Az**. Note that the four estimates are close to each other, around 0.04. This confirms the validity of the different approaches to estimating the case sampling standard deviation, and is a self-consistency check on the calibration process.

Row B repeats the values in row A, except that this time the empirical AUC is being used as the figure of merit. The flexibility afforded by the calibrated simulator is that in using it, one can test various ideas. For example, what happens if the number of cases is increased? One expects the standard deviations in the last column of Table 7.3 to decrease, but by how much? Row B uses datasets generated by the simulator calibrated to the data in Table 7.1. Since the number of cases has not changed, the values are similar to those in row A. In row C the number of cases has been inflated by a factor of 10, and the standard deviations decrease by about a factor of the square root of 10. (Since rows B, C, and D use empirical AUC, MLE estimates are inapplicable.)

Exercise: Use the calibrated simulator to test the effect of changing simulator parameters, particularly mu. As **mu** increases, the standard deviation is expected to decrease, because there is less room for AUC to vary, since it is constrained to be ≤1.

7.8 Dependence of AUC on reader expertise

Suppose one conducts a ROC study with J readers where typically J is about 5 but can be as low as 3 and as high as 20 (the wide variability reflects, in the author's opinion, confusion regarding factors affecting the optimal choice of J and the related issue of statistical power). Each reader interprets the *same case sample*, that is, the same set of cases, but because they have different expertise levels, and for other reasons (see below), the observed ROC counts tables will not be identical. The variance of the observed values of AUC is an empirical estimate of between-reader variance (including

Table 7.3 Comparison of different estimates of the standard deviation of AUC, namely MLE, the DeLong method, bootstrap, jackknife, and population sampling. MLE = maximum likelihood estimate. Shown are results for a real dataset (A, B) and two simulated datasets (C and D) and two methods for estimating AUC: parametric and empirical.

	Dataset	AUC estimate	Var. estimation method	$\sigma(A_z)$
A	$\overline{K_1} = (30,19,8,2,1)$ $\overline{K_2} = (5,6,5,12,22)$	Parametric AUC	MLE	0.0378
			DeLong	*0.0380
			Bootstrap	0.0438
			Jackknife	0.0386
			Population	0.0403
B	$\overline{K_1} = (30,19,8,2,1)$ $\overline{K_2} = (5,6,5,12,22)$	Empirical AUC	MLE	NA
			DeLong	0.0380
			Bootstrap	0.0413
			Jackknife	0.0369
			Population	0.0369
C	Calibrated Simulator $K1 = 60, K2 = 50$		MLE	NA
			DeLong	0.0333
			Bootstrap	0.0366
			Jackknife	0.0335
			Population	0.0359
D	Calibrated Simulator $K1 = 600, K2 = 500$		MLE	NA
			DeLong	0.0113
			Bootstrap	0.0110
			Jackknife	0.0113
			Population	0.0113

* The entry for the DeLong method was obtained using the empirical AUC (P = 2000, B = 2000).

the inseparable within-reader component). File **MainBetweenReaderSd.R** analyzes the Van Dyke[10] dataset, consisting of two modalities and five readers described in **Online Chapter 24**. Source the code file to get the following output.

7.8.1 Code output

```
> source(...)
between-reader variance in modality 1 = 0.003082629
between-reader variance in modality 2 = 0.001304602
avg. between-reader variance in both modalities = 0.002193615
```

Notice that the between-reader (including, as always, within-reader) variance appears to be modality dependent. Determining if the difference is significant requires more analysis. For now, one simply averages the two estimates.

How can one handle between-reader variability in the notation? Each reader's interpretation can be analyzed by MLE to get the corresponding AUC value. The notation for the observed AUC values is

$$AUC_{j\{c\}}, j = 1, 2..., J \tag{7.11}$$

How does one conceptualize reader variability? As stated before, it is due to differences in expertise levels, but there is more to it. Since the single reader is characterized by parameters $\mu, \sigma, \zeta_2, \zeta_3, .., \zeta_{R-1}$ (R is the number of ratings bins; it is assumed that all readers employ the same number of bins, although they may employ it in different ways, that is, the values of the thresholds may be different). While the non-diseased distribution for each reader could have a mean different from 0 and variance different from unity, one can always translate it to zero and scale it to assure that the non-diseased distribution is the unit normal distribution. However, one cannot be assured that the separation and the width of the diseased distribution, and the thresholds, will not depend on the reader. Therefore, the most general way of thinking of reader variability is to put a j subscript on each of the model parameters, yielding $\mu_j, \sigma_j, \zeta_{1,j}, \zeta_{2,j}, .., \zeta_{R-1,j}$. The first two of these define the population ROC curve for reader j, and the corresponding AUC value is (this equation was derived in **Chapter 6**, Equation 6.48):

$$AUC_{j\{c\}} = \Phi\left(\frac{\mu_j}{\sqrt{1 + \sigma_j^2}}\right) \tag{7.12}$$

All else being equal, readers with larger μ_j will perform better because they are better able to separate the non-diseased and diseased cases in z-space than their fellow readers. It is difficult and possibly misleading to try to estimate the differences directly from the observed ROC counts tables, but in general better readers will yield counts more skewed toward the low end of the rating scale on non-diseased cases and more skewed toward the high end of the rating scale for diseased cases. The ideal reader would rate all diseased cases one value (e.g., 5) and all non-diseased cases a smaller fixed value (e.g., 1, 2, 3, or 4), resulting in unit AUC, that is, perfect performance. According to Equation 7.12, a reader with smaller σ_j will also perform better. As noted before, typically the σ parameter is greater than unity. *The reasons for this general finding will be discussed later, but accept the author's word for now that the best the reader can do is to reduce this parameter to unity.* See Summary of **Chapter 6** for reasons for the observation that generally the variance of the diseased distribution is larger than one—it has to do with the inhomogeneity of the distribution of diseased cases and the possibility that a mixture distribution is involved. Regarding thresholds, while the population based performance for a particular reader does not depend on thresholds, the thresholds determine the ROC counts table, so differences in usage of the thresholds will translate to differences in estimates of $AUC_{j\{c\}}$, but this is expected to be a smaller effect compared to the dependence on μ_j & σ_j. To summarize, variability of readers can be attributed to variability in the binormal model parameters and, to a lesser extent, to variability in adopted thresholds.

7.9 Dependence of AUC on modality

Suppose one conducts a ROC study with J (indexed by $j = 1, 2, ... J$) readers but there are I (indexed by $i = 1, 2, ... I$) modalities. This is frequently referred to as the multiple-reader multiple-case (MRMC) paradigm. Each reader interprets the *same case sample*, that is, the same set of cases, in two or more modalities. Here is an example, in file **MainModalityEffect.R**. This file analyzes the same dataset used in Section 7.8.1, **Source** the code file to get the following output:

7.9.1 Code output

```
> source(...)
reader-average FOM in modality 1 = 0.897037
reader-average FOM in modality 2 = 0.9408374
effect size, i.e., FOM modality 1 minus modality 2
= -0.04380032
```

Notice that the second modality has a higher FOM. Determining if the difference is significant requires more analysis as described in **Chapter 9**. The difference between the reader-averaged FOMs is referred to as the *observed effect size*.

How does one handle modality dependence of the FOM in the notation? If K is the total number of cases, the total number of interpretations involved is IJK, each of which results in a rating. MLE analysis yields IJ values for AUC, one for each modality-reader combination. The appropriate notation is

$$AUC_{ij\{c\}} \tag{7.13}$$

The most general way of thinking of reader and modality variability is to put ij subscripts on each of the model parameters, yielding $\mu_{ij}, \sigma_{ij}, \zeta_{2,ij}, \zeta_{3,ij}, ..., \zeta_{R-1,ij}$. For a particular combination of modality and reader, the population ROC curve, as fitted by the binormal model, yields the area under the ROC curve.

$$AUC_{ij\{c\}} = \Phi\left(\frac{\mu_{ij}}{\sqrt{1+\sigma_{ij}^2}}\right) \tag{7.14}$$

Given an MRMC dataset, using MLE one can estimate the parameters $\mu_{ij}, \sigma_{ij}, \zeta_{2,ij}, \zeta_{3,ij}, ..., \zeta_{R-1,ij}$ for each modality-reader combination, and this could be used to design a simulator that is calibrated to the specific clinical dataset, which in turn can be used to illustrate the ideas and to test any proposed method of analyzing the data. However, the problem is more complex; the procedure needs to also account for the correlations arising from the large number of pairings inherent is such a dataset (e.g., reader 1 in modality 1 versus reader 2 in modality 2, since both interpret a common dataset). Designing a MRMC calibrated simulator was, until recently, an unsolved problem. Recent work[11] by the author and Mr. Xuetong Zhai enabled progress in this area, described in **Chapter 23**.

7.10 Effect on empirical AUC of variations in thresholds and numbers of bins

There are actually two effects. (1) The *empirical* AUC will tend to be smaller than the *true* AUC. If there are few operating points and they are clustered together, the difference may be large, Figure 7.1a-d, generated by sourcing **mainEmpVsFit.R**.

7.10.1 Code listing

```
rm( list = ls())#mainEmpVsFit.R
library(RJafroc);library(ggplot2)
seed <- 10;set.seed(seed)
mu <- 2; sigma <- 1.5
cat("Population AUC = ",
    pnorm(mu/sqrt(1+sigma^2)), "\n")
K1 <- 500; K2 <- 500
fp <- rnorm(K1);tp <- rnorm(K2, mu, sigma)
zetas <- c(-Inf, 1.5, 2, 2.5, 3, 4, Inf)
fp1 <- as.numeric(cut(fp, zetas))
tp1 <- as.numeric(cut(tp, zetas))
rocData1 <- Df2RJafrocDataset(fp1, tp1)
plotEmp1 <- PlotEmpiricalOperatingCharacteristics(
    rocData1, 1, 1)
print(plotEmp1$Plot)
empAuc1 <- UtilFigureOfMerit(
    rocData1, fom =
    "Wilcoxon")
cat("Emp. AUC bunched data = ", empAuc1, "\n")
Fit1 <- FitCbmRoc(rocData1)
print(Fit1$fittedPlot)
cat("CBM AUC, bunched data =" , Fit1$AUC,"\n")
zetas <- c(-Inf, -0.5, 0, 1, 1.5, 2, Inf)
fp2 <- as.numeric(cut(fp, zetas))
tp2 <- as.numeric(cut(tp, zetas))
rocData2 <- Df2RJafrocDataset(fp2, tp2)
plotEmp2 <- PlotEmpiricalOperatingCharacteristics(
    rocData2, 1, 1)
print(plotEmp2$Plot)
empAuc2 <- UtilFigureOfMerit(
    rocData2, fom =
    "Wilcoxon")
cat("Emp. AUC well-spaced = ", empAuc2, "\n")
Fit2 <- FitCbmRoc(rocData2)
print(Fit2$fittedPlot)
cat("CBM AUC, well-spaced data =" , Fit2$AUC,"\n")
```

Fig. 7.1a was generated by a binormal model simulator, with thresholds chosen to exaggerate the effect, lines 4 through 7. The population AUC is 0.8664, while the empirical AUC for the bunched data is 0.8030, lines 16–19. However, since interest is in *differences* in AUCs, for example, between two modalities, and the underestimates may tend to cancel, this may not be a serious issue. However, an effect that may be problematical is that the operating points for a given reader may not span the same FPF ranges in the two modalities, in which case the empirical AUCs will be different, as depicted in Figure 7.1a and b. The empirical AUC in Fig. 7.1b, where the operating points span the entire range, is 0.8580, lines 30–33, which is closer to the population value. *Since the usage of the bins is not under the researcher's control, this effect cannot be ruled out.* Fitted AUCs are expected to be less sensitive, but not immune, to this effect. Figure 7.1c is a contaminated binormal model (CBM) fitted curve, line 20, to the same data as in (a), fitted AUC = 0.892, while Figure 7.1d is a CBM fitted curve, line 34, to the same data as in (b), fitted AUC = 0.867. The difference in AUCs between (a) and (b) is −0.055, while that between (c) and (d) is 0.024. The consequences of these effects on the validity of analyses using the empirical AUC have not been studied. (The parameters of the model were $a = 1.33$ and $b = 0.667$, which yields the quoted value of the population AUC. The population value is that predicted by the parameters; it has zero sampling variability. The fitted curves are those predicted by the CBM, discussed in **Chapter 20**.)

(2) The second effect is varying numbers of thresholds or bins between the readers. One could be a radiologist, capable of maintaining at most about 6 bins, and the other an algorithmic observer, such as CAD, capable of maintaining more bins. Moreover, if the radiologist is an expert, the data points will tend to cluster near the initial near vertical part of the ROC (see **Chapter 17** for explanation). This is illustrated using code in file **mainBinVariability.R**. Sourcing this code yields *Figure 7.1e and f.*

7.10.2 Code listing

```
rm( list = ls())#mainBinVariability.R
library(caTools);library(ggplot2);source("rocY.R")

mu <- 2;sigma <- 1.5 # experiment with other values
a <- mu/sigma; b <- 1/sigma # a and b parameters
cat("true AUC = ", pnorm(mu/sqrt(1+sigma^2)), "\n")

x <- seq(0.0, 1, 0.01)
zeta <- c(3, 2.5, 2)
FPF <- pnorm(-zeta);FPF <- c(0,FPF,1)
TPF <- pnorm((mu-zeta)/sigma);TPF <- c(0,TPF,1)
pointsData <- data.frame(FPF = FPF, TPF = TPF)
AUC <- trapz(FPF,TPF)
cat("empirical AUC, sparse points = ", AUC, "\n")
rocPlot1 <- ggplot(mapping = aes(x = FPF, y = TPF)) +
  geom_line(data = pointsData) +
  geom_point(data = pointsData)

print(rocPlot1)

zeta <- seq(3, -2, -0.5)
FPF <- pnorm(-zeta);FPF <- c(0,FPF,1)
TPF <- pnorm((mu-zeta)/sigma);TPF <- c(0,TPF,1)
pointsData <- data.frame(FPF = FPF, TPF = TPF)
AUC <- trapz(FPF,TPF)
cat("empirical AUC, dense point = ", AUC, "\n")
rocPlot2 <- ggplot(mapping = aes(x = FPF, y = TPF)) +
  geom_line(data = pointsData) +
  geom_point(data = pointsData)
print(rocPlot2)
```

In Figure 7.1e and f the effect is dramatic. The expert radiologist trapezoidal AUC is 0.7418, while that for the algorithmic observer is 0.8632; the latter is close to the population value. It is left as an exercise for the reader to demonstrate that, using CBM, one can reduce but not eliminate the severe underestimation of performance that occurs in plot (e).

7.11 Empirical versus fitted AUCs

There is a preference with some researchers to using the empirical AUC as a figure of merit. Its usage enables analyses[12-16] variously referred to as the *probabilistic, mechanistic,* or *first-principles*

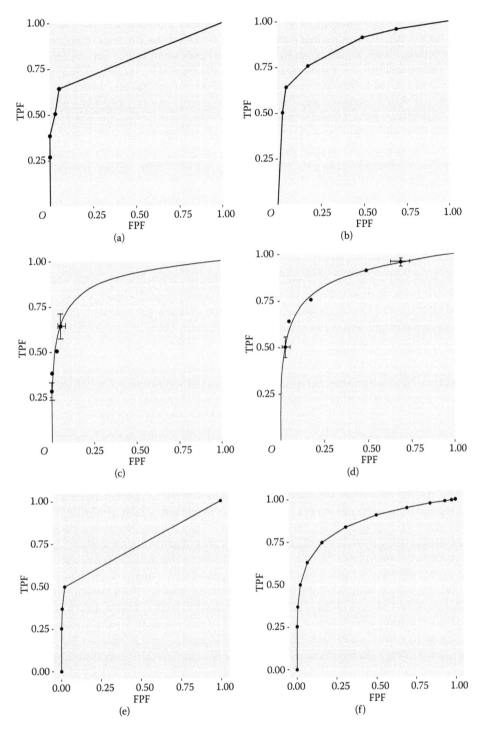

Figure 7.1 Effects on empirical and fitted AUC of threshold placement (a-d) and numbers of thresholds (e-f). Plots (a) and (b) depict empirical plots for two simulated datasets for the same that is, continuous ratings, using different threshold placements. In (a) the thresholds are clustered at low FPF values, while in (b) they are more evenly spaced. Empirical AUCs for the plots are 0.803 for (a) and 0.858 for (b). The clustering in (a) leads to a low estimate of AUC. Plots (c) and (d) are fitted curves corresponding to the same data as in (a) and (b), AUCs equal to 0.892 and 0.867, respectively. For each plot, the population AUC is 0.866. The fitted curves are less sensitive, but not immune, to the data clustering variations. With a large number of evenly spaced points, the empirical AUC is close to that of the fitted curve, as demonstrated in plots (e) and (f). The plots were generated by **mainEmpVsFit.R** (a-d) and **mainBinVariability.R** (e-f).

approach and the *one-sh*ot approach[17] to multiple-reader multiple-case analysis. The author is aware of some statisticians who distrust parametric modeling and the associate normality assumptions (the author hopes that the demonstration of the latency property in Section 6.2.2 may assuage some of their concerns). In addition, empirical AUC frees the researcher from problems with binormal model based fitting, for example, handling degenerate datasets (these problems go away with advanced fitting methods described in later chapters). The fact that the empirical AUC can always be calculated, even, for example, with a single operating point, can make the analyst blissfully unaware of anomalous data conditions. In contrast, the binormal curve-fitting method in **Chapter 6** will complain when the ratings bins are not well populated, for example, by failing to converge. This at least alerts the analyst that conditions are not optimal, and prompt data visualization and consideration of alternate fitting methods.

If empirical AUC is defined by a large number of operating points, such as with continuous ratings obtained with algorithmic observers, then empirical AUC will be nearly equal to the true AUC, to within sampling error. However, with human observers one rarely gets more than about six distinct ratings. The researcher has no control over the internal sensory thresholds used by the radiologist to bin the data, and these could depend on the modality. As demonstrated in the previous section, the empirical AUC is sensitive to the choice of thresholds, especially when the number of thresholds is small, as is usually the case with radiologists, and when the operating points are clustered on the initial near vertical section of the plot, as is also the case with experts.

Chapter 19 describes applications of three advanced methods of fitting ROC curves. The methods were applied to fourteen (14) datasets comprising 43 modalities, 80 readers, and 2012 cases. The binormal model would fail on most of the datasets. The fitted data and the operating points are shown in an online file **RSM Vs. Others.docx** corresponding to the cited chapter. It contains 236 plots, each with three fits and operating point shown. A sampling of these plots for a single dataset, a single modality, and five readers, is shown in Figure 7.2a through e. One can judge from these plots whether threshold variability effects exist. The author believes they cannot be ruled out.

Over-dependence on the empirical AUC can lead to a false-sense of security regarding the validity of the analysis and avoidance of deeper issues affecting radiologist performance. The critique will become clearer when one of the newer fitting methods is described in Chapter 19. Given that methods, to be described, do exist that fit any dataset, the author's advice to users is to consider using them to estimate AUC, instead of indiscriminately using empirical AUC.

A factor arguing in favor of usage of empirical AUC is that some of the newer significance testing procedures cited above, applicable to empirical AUC, might be on firmer theoretical grounds than the ones to be proposed in the following chapters. For example, they explicitly allow for truth-state dependences of some of the variability components, which is reasonable. Careful simulation work needs to be conducted to determine if these advantages counterbalance some of the issues with empirical AUC identified in this chapter. For algorithmic observers yielding detailed many thresholds, use of empirical AUC is entirely appropriate.

7.12 Discussion/Summary

This chapter focused on the factors affecting variability of AUC, namely case sampling and between-reader variability, each of which contain an inseparable within-reader contribution. The only way to estimate within-reader variability is to have the same reader re-interpret the same case-set on multiple occasions (with sufficient time delay to minimize memory effects). This is rarely done and is unnecessary, in the ROC context, to sound experimental design. Some early publications have suggested that such re-interpretations are needed to estimate the within-reader component, but modern analysis, described in the next part of the book, does not require re-interpretations. Indeed, it is a waste of precious reader-time resources. Rather than have the same readers re-interpret the

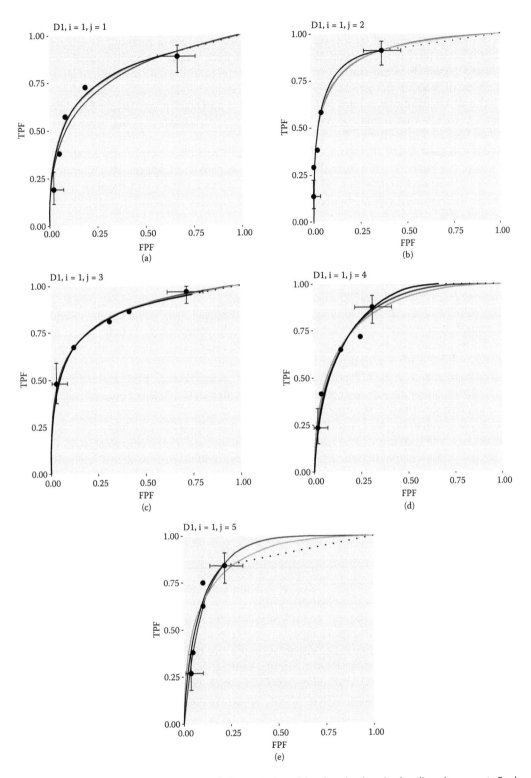

Figure 7.2 (a) through (e) Small sample of the 236 viewable plots in the cited online document. Each panel corresponds to a different reader (the *j*-index in the labels). The modality is the same (the *i*-index) and the dataset is labeled D1. The three curves correspond to different advanced method of fitting ROC data. The interest in this chapter is on the positions of the operating points. Reader (c) traverses more of the FPF range than does reader (e). Empirical AUC may result in a greater error for reader (e) than for reader (c). An explanation of the three fits is deferred to **Chapter 19**.

same case-set on multiple occasions, it makes much more sense to recruit more readers and/or collect more cases, guided by a systematic sample size estimation method. Another reason the author is not in favor of re-interpretations is that the within-reader variance is usually smaller than case sampling and between-reader variances. Re-interpretations would minimize a quantity that is already small, which is not good science.

In the author's judgment, current literature on this topic lacks notational clarity, particularly when it comes to case sampling. An important part of this chapter is the explicit usage of the case-set index $\{c\}$ to describe a key-factor, namely the case sample, on which AUC depends. This index is assumed in the literature, which can lead to confusion, especially when understanding one of the methods used to analyze ROC MRMC data, see **Chapter 10**. Different simulated datasets correspond to different values of $\{c\}$. This indexing leads to a natural, in the author's opinion, understanding of the bootstrap method; one simply replaces $\{c\}$ with $\{b\}$, the bootstrap case-set index.

The bootstrap and jackknife methods described in this chapter have wide applicability. Later they will be extended to estimating the covariance (essentially a scaled correlation) between two random variables. Also described was the DeLong method, applicable to the empirical AUC. Using a real dataset and simulators, all methods were shown to agree with each other, especially when the number of cases is large, Table 7.3 (row D, last column).

The concept of a calibrated simulator was introduced as a way of anchoring a simulator to a real dataset. While relatively easy for a single dataset, the concept has yet to be extended to where it would be useful, namely designing a simulator calibrated to a dataset consisting of interpretations by multiple readers in multiple modalities of a common dataset. Just as a calibrated simulator allowed comparison of the different variance estimation methods to a known standard, obtained by population sampling, a more general calibrated simulator would allow better testing the validity of the analysis described in the next few chapters.

A source of variability not generally considered, namely threshold variability, is introduced, and a cautionary note is struck with respect to indiscriminate usage of the empirical AUC. Finally, the author wishes to reemphasize the importance of viewing ROC plots to detect anomalous conditions that might be otherwise overlooked.

This concludes **Part A** of this book. The next chapter begins **Part B**, namely the statistical analysis of multiple-reader multiple-case (MRMC) ROC datasets.

References

1. Swets JA, Pickett RM. *Evaluation of Diagnostic Systems: Methods from Signal Detection Theory.* 1st ed. New York, NY: Academic Press; 1982.
2. Kundel HL, Revesz G. Lesion conspicuity, structured noise, and film reader error. *Am J Roentgenol.* 1976;126:1233–1238.
3. Beam CA, Layde PM, Sullivan DC. Variability in the interpretation of screening mammograms by US radiologists. Findings from a national sample. *Arch Intern Med.* 1996;156(2):209–213.
4. DeLong ER, DeLong DM, Clarke-Pearson DL. Comparing the areas under two or more correlated receiver operating characteristic curves: A nonparametric approach. *Biometrics.* 1988;44:837–845.
5. Noether GE. *Elements of Nonparametric Statistics.* New York, NY: Wiley & Sons; 1967.
6. Bamber D. The area above the ordinal dominance graph and the area below the receiver operating characteristic graph. *J Math Psychol.* 1975;12(4):387–415.
7. Casella G, Berger RL. *Statistical inference.* Vol 2: Duxbury Pacific Grove, CA; 2002.
8. Efron B, Tibshirani RJ. *An Introduction to the Bootstrap.* Vol. 57. Boca Raton, FL: Chapman & Hall/CRC; 1993.
9. Eng J. ROC analysis: Web-Based Calculator for ROC Curves, http://www.jrocfit.org, 2006.

10. Van Dyke CW, White RD, Obuchowski NA, Geisinger MA, Lorig RJ, Meziane MA. Cine MRI in the diagnosis of thoracic aortic dissection. 79th RSNA Meetings. 1993. Chicago, Il.
11. Zhai X, Chakraborty DP. A bivariate contaminated binormal model for robust fitting of proper ROC curves to a pair of correlated, possibly degenerate, ROC datasets. *Med Phys.* 2017;44:2207–2222.
12. Gallas BD, Pennello Ga, Myers KJ. Multireader multicase variance analysis for binary data. *J Opt Soc Am A Opt Image Sci Vis.* 2007;24(12):70–80.
13. Kupinski MA, Clarkson E, Barrett HH. A probabilistic model for the MRMC method, Part 2: Validation and applications. *Acad Radiol.* 2006;13(11):1422–1430.
14. Clarkson E, Kupinski MA, Barrett HH. A probabilistic model for the MRMC method, Part 1: Theoretical development. *Acad Radiol.* 2006;13(11):1410–1421.
15. Clarkson E, Kupinski MA, Barrett HH. A Model for MRMC AUC Measurements: Theory and Simulations. *Proc SPIE*; Vol. 5749 p 21–31, San Diego, CA 2005.
16. Barrett HH, Kupinski MA, Clarkson E. Probabilistic foundations of the MRMC method. *Prog Biomed Opt Imaging—Proc SPIE.* 2005;5749:21–31.
17. Gallas BD. One-shot estimate of MRMC variance: AUC. *Acad Radiol.* 2006;13(3):353–362.

PART B

Two significance testing methods for the ROC paradigm

8

Hypothesis testing

8.1 Introduction

The problem addressed in this chapter is how to decide whether an estimate of area under curve (AUC) is consistent with a prespecified value. One example of this is when a single reader rates a set of cases in a single modality, from which one estimates AUC, and the question is whether the estimate is statistically consistent with a prespecified value. From a clinical point of view, this is generally not a useful exercise, but its simplicity is conducive to illustrating the broader concepts involved in this and later chapters. The clinically more useful analysis is when multiple readers interpret the same cases in two or more modalities. With two modalities, for example, one obtains an estimate AUC for each reader in each modality, averages the AUC values over all readers within each modality, and computes the inter-modality difference in reader-averaged AUC values. The question forming the main subject of this book is whether the observed difference is consistent with zero.

Each situation outlined above admits a binary (yes/no) answer, which is different from the estimation problem that was dealt with in the maximum likelihood method in **Chapter 6**, where one computed numerical estimates (and confidence intervals) for the parameters of the fitting model.

> Hypothesis testing is the process of dichotomizing the possible outcomes of a statistical study and then using probabilistic arguments to choose one option over the other.

The two competing options, e.g., the AUC estimate is consistent or inconsistent with a prespecified values, are termed the null hypothesis (NH) and the alternative hypothesis (AH), respectively. The hypothesis testing procedure is analogous[1] to the jury trial system in the United States, with 20 instead of 12 jurors, the null hypothesis being the presumption of innocence and the alternative hypothesis being the defendant is guilty, and the decision rule is to assume the defendant is innocent unless all 20 jurors agree the defendant is guilty. If even one juror disagrees, the defendant is deemed innocent (equivalent to choosing an α—defined below—of 0.05, or 1/20).

8.2 Hypothesis testing for a single-modality single-reader ROC study

The binormal model described in **Chapter 6** can be used to generate sets of ratings to illustrate the methods being described in this chapter. To recapitulate, the model is described by

$$\left.\begin{aligned} Z_{k_1 1} &\sim N(0,1) \\ Z_{k_2 2} &\sim N(\mu, \sigma^2) \end{aligned}\right\} \tag{8.1}$$

Set $\mu = 1.5$ and $\sigma = 1.3$ (these values were selected arbitrarily; the reader may wish to experiment with different ones) and simulate $K_1 = 50$ non-diseased cases and $K_1 = 52$ diseased cases. For debugging purposes, the author likes to keep the sizes of the two arrays slightly different; this allows one to quickly check, with a glance at the **Environment** panel, that array dimensions are as expected. The **R**-code in **mainHT1R1M.R** follows the code listing below. (the file name stands for *Hypothesis Testing One-Reader One-Modality*).

8.2.1 Code listing

```
rm(list = ls()) #mainHT1R1M.R
source("Wilcoxon.R")

seed <- 1;set.seed(seed)
mu <- 1.5;sigma <- 1.3;K1 <- 50;K2 <- 52

# cheat to find the population mean and std. dev.
AUC <- array(dim = 10000)
for (i in 1:length(AUC)) {
  zk1 <- rnorm(K1);zk2 <- rnorm(K2, mean = mu, sd = sigma)
  AUC[i] <- Wilcoxon(zk1, zk2)
}
meanAUC    <-   mean(AUC);sigmaAUC   <-   sd(AUC)
cat("pop mean AUC = ", meanAUC,
    "\npop sigma AUC = ", sigmaAUC, "\n")

# one more trial, this is the one we want to compare to meanAUC,
zk1 <- rnorm(K1);zk2 <- rnorm(K2, mean = mu, sd = sigma)
AUC <- Wilcoxon(zk1, zk2)
cat("New AUC = ", AUC, "\n")

z <- (AUC - meanAUC)/sigmaAUC
#z <- qnorm(0.05/2)
cat("z-statistic = ", z, "\n")

# p value for two-sided AH
p2tailed <- pnorm(-abs(z)) + (1-pnorm(abs(z)))
# p value for one-sided AH > 0
p1tailedGT <- 1-pnorm(z)
# p value for one-sided AH < 0
p1tailedLT <- pnorm(z)
alpha   <- 0.05

# critical value for two-sided AH: AUC not equal to meanAUC
z2tailed <- -qnorm(alpha/2)
# critical value for one-sided AH: AUC > meanAUC
z1tailedGT <- qnorm(1-alpha)
# critical value for one-sided AH: AUC < meanAUC
z1tailedLT <- qnorm(alpha)

cat("alpha of test = ", alpha, "\n")
cat("\nTwo-sided AH: AUC not equal to meanAUC", "\n")
cat("Critical value for two-sided AH:", z2tailed, "\n")
cat("p value for two-sided AH:", p2tailed, "\n")
```

```
cat("\nOne-sided AH: AUC > meanAUC", "\n")
cat("Critical value for one-sided AH:", z1tailedGT, "\n")
cat("p value for two-sided AH:", p1tailedGT, "\n")

cat("\nOne-sided AH: AUC < meanAUC", "\n")
cat("Critical value for one-sided AH:", z1tailedLT, "\n")
cat("p value for two-sided AH:", p1tailedLT, "\n")
```

Sourcing the code yields the following output.

8.2.2 Code output

```
> source(...)
pop mean AUC =  0.819178
pop sigma AUC =  0.04176683
New AUC =  0.8626923
z-statistic =  1.04184
alpha of test =  0.05

Two-sided AH: AUC not equal to meanAUC
Critical value for two-sided AH: 1.959964
p value for two-sided AH: 0.297486

One-sided AH: AUC > meanAUC
Critical value for one-sided AH: 1.644854
p value for two-sided AH: 0.148743

One-sided AH: AUC < meanAUC
Critical value for one-sided AH: -1.644854
p value for two-sided AH: 0.851257
```

Lines 7–16 in the code file use the simple (if unimaginative) approach of sampling 10,000 times to estimate the population mean and standard deviation of empirical AUC, denoted below by AUC_{pop} and σ_{AUC}, respectively (similar to that done in **Chapter 7** to validate the different variance estimation methods). Based on the 10,000 simulations, $AUC_{pop} = 0.8192$ and $\sigma_{AUC} = 0.04177$.

Line 20 simulates one more independent ROC study with the same numbers of cases, and the resulting area under the empirical curve is denoted AUC in the code. Is the new value (0.8627) sufficiently different from the population mean 0.8192 to reject the null hypothesis $NH : AUC = AUC_{pop}$? *Note that the answer to this question can be either yes or no: equivocation is not allowed.*

The new value is somewhat close to the population mean, but how does one decide if somewhat close is close enough? Needed is the statistical distribution of the random variable AUC under the hypothesis that the true mean is AUC_{pop}. In the asymptotic limit of a large number of cases (for finite numbers of cases this amounts to an approximation), one can assume that the *pdf* of AUC under the null hypothesis is the normal distribution $N\left(AUC_{pop}, \sigma_{AUC}^2\right)$.

$$pdf_{AUC}\left(AUC \,|\, AUC_{pop}, \sigma_{AUC}\right) = \frac{1}{\sqrt{2\pi}\sigma_{AUC}} \exp\left(-\frac{1}{2}\left(\frac{AUC - AUC_{pop}}{\sigma_{AUC}}\right)^2\right) \tag{8.2}$$

The translated and scaled value is distributed as a unit normal distribution, that is,

$$Z = \frac{AUC - AUC_{pop}}{\sigma_{AUC}} \sim N(0,1) \tag{8.3}$$

(The Z notation here should not be confused with Z-sample, the decision variable or rating of a case in a receiver operating characteristic (ROC) study. The latter, when sampled over a set of non-diseased and diseased cases, yields a realization of AUC. The author trusts the distinction will be clear from the context.) The observed *magnitude* of z is 1.042 (fourth line in code output above).

The ubiquitous p-value is the probability that the *observed magnitude of z, or larger*, occurs under the null hypothesis (NH), that the true mean of Z is zero.

The p-value corresponding to an observed z of 1.042 is given by (as always, uppercase Z is the random variable, while lower case z is a realized value):

$$
\begin{aligned}
P\left(|Z| \geq |z| \,\big|\, Z \sim N(0,1)\right) &= P\left(|Z| \geq 1.042 \,\big|\, Z \sim N(0,1)\right) \\
&= P\left(Z \geq 1.042 \,\big|\, Z \sim N(0,1)\right) + P\left(Z \leq -1.042 \,\big|\, Z \sim N(0,1)\right) \\
&= 2\Phi(-1.042) \\
&= 0.2975
\end{aligned} \tag{8.4}
$$

To recapitulate statistical notation, $P\left(|Z| \geq |z| \,\big|\, Z \sim N(0,1)\right)$ is to be parsed as $P(A|B)$, that is, the probability $|Z| \geq z$ given that $Z \sim N(0,1)$. The last line in Equation 8.4 follows from the symmetry of the unit normal distribution, that is, the area above 1.042 must equal the area below −1.042.

Since z is a continuous variable, the probability that a sampled value will *exactly* equal the observed value is *zero*. Therefore, one must pose the question as stated above, namely, what is the probability that Z is at least as extreme as the observed value (by "extreme" the author means further from zero, in either positive or negative directions)? If the observed was $z = 2.5$ then the corresponding p-value would be $2\Phi(-2.5) = 0.01242$, which is smaller than 0.2975 (**2*pnorm(-2.5)** = 0.01241933). This is cited below as the *second example*.

Under the zero-mean null hypothesis, the larger the magnitude of the observed value of Z, the smaller the p-value and the more unlikely that the data supports the NH. *The p-value can be interpreted as the degree of unlikelihood that the data supports the NH.*

By convention one adopts a *fixed* value of the probability, denoted α and usually $\alpha = 0.05$, which is termed the *size of the test* or the *significance level of the test*, and the decision rule is to reject the null hypothesis if the observed p-*value* $< \alpha$.

$$p < \alpha \Rightarrow \text{Reject NH} \tag{8.5}$$

In the first example, with observed p-value equal to 0.2975, one would not reject the null hypothesis, but in the second example, with observed p-value equal to 0.01242, one would. If the p-value is exactly 0.05 (unlikely with ROC analysis, but one needs to account for it) then one does not reject the NH. In the 20-juror analogy, if one juror insists the defendant is not guilty, then observed P is 0.05, and one does not reject the NH that the defendant is innocent (the double negatives, common in statistics, can be confusing; in plain English, the defendant goes home).

According to the previous discussion, the critical magnitude of z that determines whether to reject the null hypothesis is given by

$$z_{\alpha/2} = -\Phi^{-1}(\alpha/2) \tag{8.6}$$

Note that $z_{\alpha/2} = -\Phi^{-1}(\alpha/2)$ quantile of the unit normal distribution and the reader is reminded that the inverse of the PDF is implemented in **R** as the **qnorm()** function. For $\alpha = 0.05$ this evaluates to 1.95996 (which is sometimes rounded up to two, good enough for government work as the saying goes) and the decision rule is to reject the null hypothesis only if the observed *magnitude* of z is larger than $z_{\alpha/2}$.

The decision rule based on comparing the observed z to a critical value is equivalent to a decision rule based on comparing the observed p-value *to* α. *It is also equivalent, as will be shown later, to a decision rule based on a* $(1-\alpha)$ *confidence interval for the observed statistic. One rejects the NH if the closed confidence interval does not include zero.*

8.3 Type-I errors

Just because one rejects the null hypothesis, as in the second example, does not mean that the null hypothesis is false. Following the decision rule caps, or puts an upper limit on, the probability of incorrectly rejecting the null hypothesis at α. In other words, by agreeing to reject the NH only if $p < \alpha$, one has set an upper limit, namely α, on errors of this type, termed *Type-I* errors. These could be termed *false positives* in the hypothesis testing sense, not to be confused with a false positive occurring on individual case-level decisions. According to the definition of α,

$$P(\text{Type I error}|NH) = \alpha \tag{8.7}$$

To demonstrate the idea, one needs to have a very cooperative reader interpreting new sets of independent cases not just one more time, but 2000 more times (the reason for the 2000 trials will be explained below). The code for this is in file **mainTypeIErrors.R**.

8.3.1 Code listing

```
rm(list = ls()) # mainTypeIErrors.R
source("Wilcoxon.R")

seed <- 1;set.seed(seed)
mu <- 1.5;sigma <- 1.3;K1 <- 50;K2 <- 52

# cheat to find the population mean and std. dev.
AUC <- array(dim = 10000)
for (i in 1:length(AUC)) {
  zk1 <- rnorm(K1);zk2 <- rnorm(K2, mean = mu, sd = sigma)
  AUC[i] <- Wilcoxon(zk1, zk2)
}
sigmaAUC <- sqrt(var(AUC));muAUC <- mean(AUC)

nTrials <- 2000
alpha <- 0.05 # size of test
reject = array(0, dim = nTrials)
```

```
for (trial in 1:length(reject)) {
  zk1 <- rnorm(K1);zk2 <- rnorm(K2, mean = mu, sd = sigma)
  AUC <- Wilcoxon(zk1, zk2)
  z <- (AUC - muAUC)/sigmaAUC
  p <- 2*pnorm(-abs(z)) # p value for individual trial
  if (p < alpha) reject[trial] = 1
}

CI <- c(0,0); width <- -qnorm(alpha/2)
ObsvdTypeIErrRate <- sum(reject)/length(reject)
CI[1] <- ObsvdTypeIErrRate -
  width*sqrt(ObsvdTypeIErrRate*(1-ObsvdTypeIErrRate)/nTrials)
CI[2] <- ObsvdTypeIErrRate
+ width*sqrt(ObsvdTypeIErrRate*(1-ObsvdTypeIErrRate)/nTrials)
cat("alpha = ", alpha, "\n")
cat("ObsvdTypeIErrRate = ",
    ObsvdTypeIErrRate, "\n95% confidence interval = ", CI, "\n")
exact <- binom.test(sum(reject),n = 2000,p = alpha)
cat("exact 95% CI = ", as.numeric(exact$conf.int), "\n")
```

The first 13 lines are identical to the corresponding lines in **mainHT1R1M.R**. Line 15 initializes **NTrials** to 2000 and line 16 initializes α to 0.05. Line 17 initializes all 2000 elements of a vector named **reject** to zeroes. Lines 18 through 24 describes our captive reader interpreting independent sets of cases 2000 times. Each completed interpretation of 102 cases is termed a *trial*. For each trial, line 20 calculates the observed value of *AUC*, the next line calculates the observed *z* statistic, and the next line calculates the observed *p*-value. Line 23 tests the observed *p*-value against the fixed value α and sets the corresponding **reject** flag to unity if $p < \alpha$. In other words, if the trial-specific *p*-value is less than α, one counts an instance of rejection of the null hypothesis. The process is repeated 2000 times.

Line 27 calculates the observed Type-I error rate, denoted **ObsvdTypeIErrRate**, by summing the **reject** array and dividing by the number of trials. Lines 28 and 29 calculate a 95% confidence interval for **ObsvdTypeIErrRate** based on the binomial distribution, as in **Chapter 3**. Finally, lines 30 through 32 print out the relevant results. If one sources the file, one, after a brief delay, gets the following output.

8.3.2 Code output

```
> source(...)
alpha =   0.05
ObsvdTypeIErrRate =   0.049
95% confidence interval =   0.03953934 0.049
exact 95% CI =   0.03995676 0.05939265
```

The second line reminds us that the chosen value for the size of the test is $\alpha = 0.05$. The next line is the observed value of the Type-I error rate (which is a realization of a random variable). The last line is the 95% confidence interval for the observed Type-I error rate (which is also a realization of a random range variable). The fact that this confidence interval includes the chosen value $\alpha = 0.05$ is no coincidence; it shows that the hypothesis testing procedure is working as expected. To distinguish between the selected α (a fixed value) and that observed in a simulation study (a realization of a random variable), the term *empirical α alpha* is used to mean the observed α.

It is a mistake to state that one wishes to minimize the Type-I error probability (the author has seen this comment from a senior researcher, which is the reason for bringing it up). The minimum value of α (a probability) is zero. Run the software with this value of α: *the software will never reject the NH. The downside of minimizing the expected Type-I error rate is that the software will never reject the NH, even when the NH is patently false.* The aim of a valid method of analyzing the data is not minimizing the Type-I error rate, rather, the observed Type-I error rate should equal the specified value of α (0.05 in our example), allowance being made for the inherent variability in its estimate, that is, its confidence interval. This is the reason 2000 trials were chosen for testing the validity of the NH testing procedure. With this choice, the 95% confidence interval, assuming that observed is close to 0.05, is roughly ±0.01 as explained next.

Following analogous reasoning to **Chapter 3**, Equation 3.43, and defining f as the *observed rejection fraction* over T trials, and as usual, F is a random variable and f is a realized value.

$$\sigma_f = \sqrt{f(1-f)/T}$$

$$F \sim N\left(f, \sigma_f^2\right)$$

(8.8)

An approximate $(1-\alpha)$ CI for f is

$$CI_f = \left[f - z_{\alpha/2}\sigma_f, \; f + z_{\alpha/2}\sigma_f\right]$$

(8.9)

If f is close to 0.05, then for 2000 trials, the 0.95 or 95% CI for f is $f \pm 0.01$, that is, `qnorm(alpha/2) * sqrt(.05*(.95)/2000)` = 0.009551683 ~ 0.01.

The only way to reduce the width of the CI, and thereby run a more stringent test of the validity of the analysis, would be to increase the number of trials T. Since the width of the CI depends on the inverse square root of the number of trials, one soon reaches a point of diminishing returns. Usually T = 2000 trials are enough for most statisticians and the author, but examples of more simulations have been published.

8.4 One-sided versus two-sided tests

In the preceding example, the null hypothesis was rejected anytime the *magnitude* of the observed value of z exceeded the cutoff value $z_{\alpha/2} = -\Phi^{-1}(\alpha/2)$. This is a statement of the *alternative hypothesis* $AH: AUC \neq AUC_{pop}$, in other words too high or too low values of z both result in rejection of the null hypothesis. This is referred to as a *two-sided AH* and the resulting p-value is termed a *two-sided* p-value. This is the most common one used in the literature.

Now suppose that the additional interpretation performed by the radiologist, at line 18 in file **mainHT1R1M.R**, was performed following an intervention after which the radiologist's performance is expected to increase. To make matters clearer, assume the interpretations in the 10,000 trials used to estimate AUC_{pop} were performed with the radiologist wearing an old pair of eye-glasses, possibly out of proper strength, and the additional interpretation is performed after the radiologist gets a new set of prescription eye-glasses. Because the radiologist's eyesight has improved, the expectation is that performance should increase. In this situation, it is appropriate to use the one-sided alternative hypothesis $AH: AUC > AUC_{pop}$. Now excessively large values of z result in rejection of the null hypothesis, *but excessively small values of z do not*. The critical value of z is defined by $z_\alpha = \Phi^{-1}(1-\alpha)$, which for $\alpha = 0.05$ is 1.645 (**qnorm (1-alpha)**= 1.644854). Compare 1.64 to the value $z_{\alpha/2} = -\Phi^{-1}(\alpha/2)$ = 1.96 for a two-sided test. If the change is in the expected direction, it is easier to reject the NH with a one-sided test rather than with a two-sided test. The p-value for a one-sided test is given by (the relevant file is **mainHT1R1M.R, line 29**)

$$P(Z \geq 1.042 | NH) = \Phi(-1.042) = 0.1487 \tag{8.10}$$

Notice that this is half the corresponding two-sided test p-value; this is because one is only interested in the area under the unit normal that is above the observed value of z. If the intent is to obtain a significant finding, it is tempting to use one-sided tests. The down side of a one-sided test is that even with a large excursion of the observed z in the other direction, one cannot reject the null hypothesis. So, if the new eye-glasses are so bad as to render the radiologist practically blind (think of a botched cataract surgery) the observed z would be large and negative, but one could not reject the null hypothesis $NH : AUC_{true} = \mu_\theta$.

The one-sided test could be run the other way, with the alternative hypothesis being stated as follows $AH : AUC < AUC_{pop}$. Now large negative excursions of the observed value of AUC cause rejection of the null hypothesis, but large positive excursions do not. The critical value is defined by $z_\alpha = \Phi^{-1}(\alpha)$, which for $\alpha = 0.05$ is -1.645. The p-value is given by (note the reversed sign compared to Equation 8.10)

$$P(Z \leq 1.042 | NH) = \Phi(1.042) = 1 - 0.1487 = 0.8513 \tag{8.11}$$

This is the complement of the value for a one-sided test with the alternative hypothesis going the other way: obviously the probability that Z is smaller than the observed value (1.042) plus the probability that Z is larger than the same value must equal one.

8.5 Statistical power

So far, focus has been on the null hypothesis. The Type-I error probability was introduced, defined as the probability of incorrectly rejecting the null hypothesis, the control, or cap on which is α, usually set to 0.05. What if the null hypothesis is actually false and the study fails to reject it? This is termed a *Type-II* error, the control on which is denoted β, the probability of a Type-II error. The complement of β is called *statistical power*.

Table 8.1 summarizes the two types of errors and the two correct decisions that can occur in hypothesis testing. In the context of hypothesis testing, a Type-II error could be termed a *false negative*, not to be confused with false negatives occurring on individual case-level decisions.

Table 8.1 resembles the 2×2 table encountered in **Chapter 2**, Table 2.1, which led to the concepts of FPF, TPF, and the ROC curve. Indeed, it is possible think of an analogous plot of empirical (i.e., observed) power versus empirical α, which looks like an ROC plot, with empirical α playing the role of FPF and empirical power playing the role of TPF, see below. If $\alpha = 0$, then power = 0; that is, if Type-I errors are minimized all the way to zero, then power is zero and one makes Type-II errors all the time. On the other hand, if $\alpha = 1$ then power = 1 and one makes Type-I errors all the time.

A little history is due at this point. The author's first FROC study, which led to his entry into this field,[2] was published in Radiology in 1986 after a lot of help from a reviewer, who we correctly guessed was the late Prof. Charles E. Metz. Prof. Gary T. Barnes (the author's mentor at that time at the University of Alabama at Birmingham) and the author visited Prof. Charles Metz in Chicago for a day ca. 1986, to figuratively pick Charlie's brain. Prof. Metz referred to the concept outlined in the previous paragraph, as a *"ROC within a ROC."*

This curve does not summarize the result of a single ROC study. Rather it summarizes the probabilistic behavior of the two types of errors that occur when one conducts thousands of such studies, under both NH true and NH false conditions, each time with different values of α, with each trial ending in a decision to reject or not reject the null hypothesis. The long sentence is best explained with an example. Open the file **mainRocWithinRoc.R**.

Table 8.1 This table illustrates the two types of errors and correct decisions that can occur in hypothesis testing

Truth	Decision	
	Fail to reject NH	**Reject NH**
NH is true	$1-\alpha$	α ("FPF")
NH is false	β ("FNF")	Power $= 1 - \beta$

Note: The probability of incorrectly rejecting the NH, a Type-I error, the cap on which is α, is usually set to 0.05. A Type-II error occurs when the NH is false and the study fails to reject it, the control on which is denoted β. The complement of β is statistical power. The two types of errors could be termed *FPF* (declaring a non-existent difference) and *FNF* (failing to declare an existing difference), but these terms should not be confused with the results of a single binary paradigm ROC study.

8.5.1 Metz's ROC within an ROC

8.5.1.1 Code listing

```
rm(list = ls()) # mainRocWithinRoc.R
library(ggplot2)
source("Wilcoxon.R")

seed <- 1;set.seed(seed)
muNH <- 1.5;muAH <- 2.1;sigma <- 1.3
K1 <- 50;K2 <- 52#;K1 <-K1*2;K2 <- K2*2

# cheat to find the population mean and std. dev.
AUC <- array(dim = 10000)
for (i in 1:length(AUC)) {
  zk1 <- rnorm(K1);zk2 <- rnorm(K2, mean = muNH, sd = sigma)
  AUC[i] <- Wilcoxon(zk1, zk2)
}
sigmaAUC <- sqrt(var(AUC));meanAUC <- mean(AUC)

T <- 2000
mu <- c(muNH,muAH)
alphaArr <- seq(0.05, 0.95, length.out = 10)
EmpAlpha <- array(dim = length(alphaArr))
EmpPower <- array(dim =length(alphaArr))
for (a in 1:length(alphaArr)) { # a is index into alpha array
  alpha <- alphaArr[a]
  reject <- array(0, dim = c(2, T))
  for (h in 1:2) {
    for (t in 1:length(reject[h,])) {
      zk1 <- rnorm(K1)
      zk2 <- rnorm(K2, mean = mu[h], sd = sigma)
      AUC <- Wilcoxon(zk1, zk2)
      obsvdZ <- (AUC - meanAUC)/sigmaAUC
```

```
      p <- 2*pnorm(-abs(obsvdZ)) # p value for individual a
      if (p < alpha) reject[h,t] = 1
    }
  }
  EmpAlpha[a] <- sum(reject[1,])/length(reject[1,])
  EmpPower[a] <- sum(reject[2,])/length(reject[2,])
}
# plot the data
EmpAlpha <- c(0,EmpAlpha,1)
EmpPower <- c(0,EmpPower,1)
pointData <- data.frame(EmpAlpha = EmpAlpha, EmpPower = EmpPower)
zetas <- seq(-5, 5, by = 0.01)
muRoc <- 1.8 # found by trial and error
curveData <- data.frame(EmpAlpha = pnorm(-zetas),
                        EmpPower = pnorm(muRoc - zetas))
alphaPowerPlot <- ggplot(mapping = aes(x = EmpAlpha, y = EmpPower)) +
  geom_point(data = pointData, shape = 1, size = 3) +
  geom_line(data = curveData)
print(alphaPowerPlot)
```

Line 6 creates two variables, **muNH** = 1.5 (the binormal model separation parameter under the NH) and **muAH** = 2.1 (the separation parameter under the AH). Under either hypothesis, the same diseased case standard deviation **sigma** = 1.3 and 50 non-diseased and 52 diseased cases are assumed. As before, lines 9 through 15 use the brute force technique to determine population AUC and standard deviation of AUC under the NH condition. Line 17 defines the number of trials **T** = 2000. Line 18 creates a vector **mu** containing the NH and AH values defined at line 6. Line 19 creates **alphaArr**, a sequence of 10 equally spaced values in the range 0.05–0.95, which represent 10 values for α. Lines 20-21 create two arrays of length 10 each, named **EmpAlpha** and **EmpPower**, to hold the values of the observed Type-I error rate, that is, empirical α and the empirical power, respectively. The program will run **T** = 2000 NH and **T** = 2000 AH trials using as α each successive value in **alphaArr** and save the observed Type-I error rates and observed powers to the arrays **EmpAlpha** and **EmpPower**, respectively.

The action begins in line 22, which begins a **for**-loop in **a**, an index into **alphaArr**. Line 23 selects the appropriate value for **alpha** (0.05 on the first pass, 0.15 on the next pass, etc.). Line 24 initializes **reject[2,2000]** with zeroes, to hold the result of each trial; the first index corresponds to hypothesis **h** and the second to trial **t**. Line 25 begins a for-loop in **h**, with **h** = 1 corresponding to the NH and **h** = 2 to the AH. Line 26 begins a for-loop in **t**, the trial index. The code within this block is similar to previous examples. It simulates ratings, computes AUC, calculates the p-value, and saves a rejection of the NH as a one at the appropriate array location **reject[h,t]**. Lines 35 and 36 calculate the empirical α and empirical power for each value of α in **alphaArr**. After padding the ends with zeros and ones (the trivial points), the remaining lines plot the ROC within a ROC. **Source** this code to get Figure 8.1.

Each of the circles in Figure 8.1 corresponds to a specific value of α. For example, the lowest nontrivial circle corresponds to $\alpha = 0.05$, for which the empirical α is 0.049 and the corresponding empirical power is 0.4955. True α increases as the operating point moves up the plot, with empirical α and empirical power increasing correspondingly. The AUC under this curve is determined by the effect size, defined as the difference between the AH and NH values of the separation parameter. If the effect size is zero, then the circles will scatter around the chance diagonal; the scatter will be consistent with the 2000 trials used to generate each coordinate of a point. As the effect size

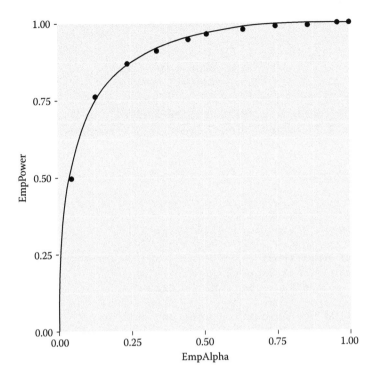

Figure 8.1 Metz's ROC within a ROC: each circle corresponds to a specific value of true α, which increases as the operating point moves up the plot, leading to corresponding increases in empirical alpha (x-axis) and empirical power (y-axis). This figure was generated by the code in **mainRocWithinRoc.R**.

increases, the plot approaches the perfect ROC, that is, approaching the top-left corner. One could use AUC under this ROC as a measure of the incremental performance, the advantage being that it would be totally independent of α, but this would not be practical as it requires replication of the study under NH and AH conditions about 2000 times each and the entire process has to be repeated for several values of α. The purpose of this demonstration was to illustrate the concept behind Metz's profound remark.

It is time to move on to factors affecting statistical power in a single study.

8.5.2 Factors affecting statistical power

Effect size: effect size is defined as the difference in AUC_{pop} values between the alternative hypothesis condition and the null hypothesis condition. Recall that AUC_{pop} is defined as the true value of the empirical ROC-AUC for the relevant hypothesis. One can use the cheat method to estimate it under the alternative hypothesis. The formalism is easier if one assumes it is equal to the asymptotic binormal model predicted value using Equation 6.48. The binormal model yields an *estimate* of the parameters, which only approach the population values in the asymptotic limit of a large number of cases. In the following, it is assumed that the parameters on the right-hand side of Equation 8.12 are the population values,

$$AUC_{pop} = \Phi\left(\frac{\mu}{\sqrt{1+\sigma^2}}\right) \tag{8.12}$$

It follows that effect size (ES) is given by (all quantities on the right-hand side of Equation 8.13 are population values):

$$ES = \Phi\left(\frac{\mu_{AH}}{\sqrt{1+\sigma^2}}\right) - \Phi\left(\frac{\mu_{NH}}{\sqrt{1+\sigma^2}}\right) \tag{8.13}$$

This formula is coded in a function **EffectSize()**. It is called by the code in **mainStatPower.R**. The latter is a stripped-down version of **mainRocWithinRoc.R**. The listing is as follows.

8.5.2.1 Code listing

```
rm(list = ls()) # mainStatPower.R
source("Wilcoxon.R");source("EffectSize.R")

seed <- 1;set.seed(seed)
mu <- 1.5;muAH <- 2.1;sigma <- 1.3
K1 <- 50;K2 <- 52#;K1 <- K1*2;K2 <- K2*2

# cheat to find the population mean and std. dev.
AUC <- array(dim = 10000)
for (i in 1:length(AUC)) {
  zk1 <- rnorm(K1)
  zk2 <- rnorm(K2, mean = mu, sd = sigma)
  AUC[i] <- Wilcoxon(zk1, zk2)
}
sigmaAUC <- sqrt(var(AUC));muAUC <- mean(AUC)

T <- 2000
alpha <- 0.05 # size of test
reject = array(0, dim = T)
for (t in 1:length(reject)) {
  zk1 <- rnorm(K1)
  zk2 <- rnorm(K2, mean = muAH, sd = sigma)
  AUC <- Wilcoxon(zk1, zk2)
  obsvdZ <- (AUC - muAUC)/sigmaAUC
  p <- 2*pnorm(-abs(obsvdZ)) # p value for individual t
  if (p < alpha) reject[t] = 1
}

ObsvdTypeIErrRate <- sum(reject)/length(reject)
CI <- c(0,0);width <- -qnorm(alpha/2)
CI[1] <- ObsvdTypeIErrRate -
  width*sqrt(ObsvdTypeIErrRate*(1-ObsvdTypeIErrRate)/T)
CI[2] <- ObsvdTypeIErrRate +
  width*sqrt(ObsvdTypeIErrRate*(1-ObsvdTypeIErrRate)/T)
cat("alpha = ", alpha, "\n")
cat("#non-diseased images = ", K1,
    "\n#diseased images = ", K2, "\n")
cat("obsvdPower = ", ObsvdTypeIErrRate, "\n")
cat("95% confidence interval = ", CI, "\n")
cat("Effect Size = ", EffectSize(mu, sigma, muAH, sigma), "\n")
```

Ensure that **alpha** at line 18 is set to 0.05. Sourcing **mainStatPower.R** yields the following output.

8.5.2.2 Code output

```
> source(...)
alpha =  0.05
#non-diseased images =   50
#diseased images =   52
obsvdPower =  0.509
95% confidence interval =   0.4870905 0.5309095
Effect Size =  0.08000617
```

The ES for the code above is 0.08 (in AUC units). It should be obvious that if effect size is zero, then power equals α. This is because then there is no distinction between the null and alternative hypotheses conditions (this choice would yield points scattered around the chance diagonal in Figure 8.1). Conversely, as effect size increases, statistical power increases, the limiting value being unity, when every trial results in rejection of the null hypothesis. The reader should experiment with different values of **muAH** to be convinced of the truth of these statements.

Sample size: uncomment the last two statements on line 6, that is, increase the number of cases by a factor of two, that is, $K_1 = 100; K_2 = 104$ and **source** the file. After waiting a short time, one gets the following code output:

8.5.2.3 Code output

```
> source(...)
alpha =  0.05
#non-diseased images =   100
#diseased images =   104
obsvdPower =  0.8435
95% confidence interval =   0.8275767 0.8594233
Effect Size =  0.08000617
```

Doubling the numbers of cases (both non-diseased and diseased) results in statistical power increasing from 0.509 to 0.844. Increasing the numbers of cases decreases σ_{AUC}, the standard deviation of the empirical AUC curve, Equation 8.2. The reader can confirm by looking at the **Environment** panel that the new value of σ_{AUC} is 0.02947, which should be compared to the value 0.04177 for **K1** = 50, **K2** = 52. Recall that σ_{AUC} enters the denominator of the z-statistic, Equation 8.3, so decreasing it will increase the probability of rejecting the null hypothesis.

Alpha: Statistical power depends on α as shown in Figure 8.1: return the sample size to the original values $K_1 = 50; K_2 = 52$. The results below are for two runs of the code, the first with the original value $\alpha = 0.05$, set at line 18, the second with $\alpha = 0.01$.

8.5.2.4 Code output

```
> source(...)
alpha =  0.05
#non-diseased images =   50
#diseased images =   52
obsvdPower =  0.509
95% confidence interval =   0.4870905 0.5309095
Effect Size =  0.08000617
```

```
> source(...)
alpha =  0.01
#non-diseased images =  50
#diseased images =  52
obsvdPower =  0.1915
95% confidence interval =  0.1688365 0.2141635
Effect Size =  0.08000617
```

Statistical power decreases as α decreases (Figure 8.1).

8.6 Comments on the code

The reader may have noticed the Wilcoxon statistic was used to estimate the area under the ROC curve. One could have used **RocfitR.R**, introduced in **Chapter 6**, to obtain maximum likelihood estimates of the area under the binormal model fitted ROC curve. The reasons for choosing the simpler empirical area are as follows. (1) With continuous ratings and 102 data points, the area under the empirical ROC curve is expected to be a close approximation to the fitted area. (2) With maximum likelihood estimation, the code would be more complex. In addition to the fitting routine, one would require a binning routine and that would introduce another variable in the analysis, namely the number of bins. (3) The maximum likelihood fitting code can sometimes fail to converge, while the Wilcoxon method is always guaranteed to yield a result (this is a mixed blessing, see Section 7.10). The non-convergence issue is overcome by modern methods of curve fitting described in later chapters. (4) The aim was to provide an understanding of null hypothesis testing and statistical power without being bogged down in the details of curve fitting.

8.7 Why alpha is chosen to be 5%

One might ask, why is α traditionally chosen to be 5%? It is not a magical number, rather a cost-benefit tradeoff. Choosing too small a value of α would result in greater probability $(1-\alpha)$ of the NH not being rejected, even when it is false, i.e., decreased power, Table 8.1. This would correspond to operating at the low-end of the plot shown in Figure 8.1. Sometimes it is important to detect a true difference between the measured AUC and the postulated value. For example, a new eye-laser surgery procedure is invented and the number of patients is necessarily small as one does not wish to subject a large number of patients to an untried procedure. One seeks some leeway on the Type-I error probability, possibly increasing it to 0.1, in order to have a reasonable chance of success in detecting an improvement in performance due to better eyesight after the surgery. If the NH is rejected and the change is in the right direction, then that is good news for the researcher. One might then consider a larger clinical trial and set alpha at the traditional 0.05, making up the lost statistical power by increasing the number of patients on which the surgery is tried.

*If a whole branch of science hinges on the results of a study, such as discovering the Higg's Boson in particle physics, statistical significance is often expressed in multiples of the standard deviation (σ) of the normal distribution, with the significance threshold set at a much stricter level (e.g., 5σ). This corresponds to an alpha of about 1 in 3.5 million (**1/pnorm(-5)** = 3.5×10^6, a one-sided test of significance). There is an article in Scientific American* (https://blogs.scientificamerican.com/observations/five-sigmawhats-that/) *on the use of* $n\sigma$, *where n is an integer, for example 5, to denote the significance level of a study, and some interesting anecdotes on why such high significance levels (low alpha) are used in some fields of research.*

Similar concerns apply to manufacturing where the cost of a mistake could be the very expensive recall of an entire product line. For background on Six Sigma Performance, see http://www.six-sigma-material.com/Six-Sigma.html. An article downloaded 3/30/17 from https://en.wikipedia.org/wiki/Six_Sigma is included as supplemental material to this chapter (**Six Sigma.pdf**). It has an explanation of why 6σ translates to one defect per 3.4 million opportunities (it has to do with short-term and long-term drifts in a process). In the author's opinion, looking at other fields offers a deeper understanding of this material than simply stating that by tradition one adopts alpha = 5%.

Most observer performance studies, while important in the search for better imaging methods, are not of such earth-shattering importance, and it is somewhat important to detect true differences (AH is true) at a reasonable alpha, so alpha = 5% and beta = 20% represent a good compromise. If one adopted a 5σ criterion, the NH would never be rejected, and progress in image quality optimization would come to a grinding halt. That is not to say that a 5σ criterion cannot be used. Rather, if used, the number of patients needed to detect a reasonable difference (effect size) with 80% probability would be astronomically large. Truth-proven cases are a precious commodity in observer performance studies. Particle physicists working on discovering the Higg's Boson can get away with 5σ criterion because the number of independent observations and/or effect size is much larger than corresponding numbers in observer performance research.

8.8 Discussion/Summary

In most statistical books, the subject of hypothesis testing is demonstrated in different (i.e., non-ROC) contexts. That is to be expected since this field is a small subspecialty of statistics (Prof. Howard E. Rockette, private communication, ca. 2002). Since this book is about ROC analysis, the author decided to use a demonstration using ROC analysis. Using a data simulator, one is allowed to cheat by conducting a very large number of simulations to estimate the population AUC and standard deviation under the null hypothesis. This permitted us to explore the related concepts of Type-I and Type-II errors within the context of ROC analysis. Ideally, both errors should be zero, but the nature of statistics leads one to two compromises. Usually one accepts a Type-I error capped at 5% and a Type-II error capped at 20%. These translate to $\alpha = 0.05$ and statistical power = 80%. The dependence of statistical power on α, the numbers of cases, and the effect size was explored. Statistical power increases with the effect size, it increases with α, and it increases with the sample size (numbers of cases).

In **Chapter 11** sample-size calculations will be described that allow one to estimate the numbers of readers and cases needed to *detect* a specified difference in inter-modality AUCs with an expected statistical power $1-\beta$. The word *detect* in the preceding sentence is shorthand for reject the NH with probability capped at α*while* also rejecting the alternative hypothesis with probability capped at β (see Table 8.1).

This chapter also gave an example of validation of a hypothesis testing method. Statisticians sometimes refer to this as showing a proposed test is a *"5% test."* In other words, one needs to be assured that when the NH is true the probability of NH rejection equals the expected value, namely α, typically chosen to be 5%. Since the observed NH rejection rate over 2000 simulations is a random variable, one does not expect the NH rejection rate to exactly equal 5%, rather the constructed 95% confidence interval (also a random interval variable) should include the NH value with 95% probability.

As noted in the introduction, comparing a single reader's performance to a specified value is not a clinically interesting problem. The next two chapters describe methods for significance testing of multiple-reader multiple-case (MRMC) ROC datasets, consisting of interpretations by a group of readers of a common set of cases in typically two modalities. It turns out that the analyses yield variability estimates that permit sample size calculation. After all, sample size calculation is all about estimation of variability, the denominator of the z-statistic, that is, Equation 8.3, in the context of this chapter. The formula will look more complex, as interest is not in determining the standard deviation of AUC, but in the standard deviation of the inter-modality reader-averaged AUC difference. However, the basic concepts remain the same.

References

1. Larsen RJ, Marx ML. *An Introduction to Mathematical Statistics and Its Applications*. 3rd ed. Upper Saddle River, NJ: Prentice-Hall; 2001.
2. Chakraborty DP, Breatnach ES, Yester MV, Soto B, Barnes GT, Fraser RG. Digital and conventional chest imaging: A modified ROC study of observer performance using simulated nodules. *Radiology*. 1986;158:35–39.

9

Dorfman–Berbaum–Metz–Hillis (DBMH) analysis

9.1 Introduction

In this chapter, *treatment* is used as a generic term for *imaging system, modality,* or *image processing* and *reader* is used as a generic term for *radiologist* or *algorithmic observer,* for example, a computer aided detection (CAD) algorithm. In the context of illustrating hypothesis-testing methods the previous chapter described analysis of a single receiver operating characteristic (ROC) dataset and comparing the observed AUC under the ROC plot to a specified value. Clinically this is not the most interesting problem; rather, interest is usually in comparing performance of a group of readers interpreting a common set of cases in two or more treatments. Such data is termed *multiple-reader multiple-case* (MRMC).* The basic idea is that by sampling a sufficiently large number of readers and a sufficiently large number of cases one might be able to draw conclusions that apply broadly to *other* readers of similar skill levels interpreting *other* similar case-sets in the selected treatments. How one accomplishes this, termed *MRMC analysis,* is the subject of this chapter.

This chapter describes the first truly successful method of analyzing MRMC ROC data, namely the The Dorfman–Berbaum–Metz (DBM) method.[1] The other method, due to Obuchowski and Rockette,[2] is the subject of **Chapter 10**. Both methods have been substantially improved by Hillis.[3–5] Hence the title of this chapter: "Dorfman–Berbaum–Metz–Hillis (DBMH) Analysis." It is not an overstatement that ROC analysis came of age with the methods described in this chapter. Prior to the techniques described here, one knew of the existence of sources of variability affecting a measured AUC value, as discussed in **Chapter 7**, but then-known techniques[6] for estimating the corresponding variances and correlations were impractical.

9.1.1 Historical background

The author was thrown (unprepared) into the methodology field ca. 1985 when, as a junior faculty member, he undertook comparing a prototype digital chest-imaging device (Picker International, ca. 1983) versus an optimized analog chest-imaging device at the University of Alabama at Birmingham. At the outset, a decision was made to use free-response ROC methodology, instead of ROC, as the former accounted for lesion localization. The author and his mentor, Prof. Gary T. Barnes, were influenced in that decision by a publication by Bunch et al.[7] to be described in **Chapter 12**. Therefore, instead of ROC-AUC one had lesion-level sensitivity at a fixed number of

---

* An argument could be made in favor of the term *multiple-treatment multiple-reader,* since *multiple-case* is implicit in any ROC analysis that takes into account correct and incorrect decisions on cases. However, the author will stick with existing terminology.

location-level false positives per case as the figure of merit (FOM). Details of the FOM are not relevant at this time. Suffice to state that methods described in this chapter, which had not been developed in 1983, while developed for analyzing reader-averaged inter-treatment ROC-AUC differences, apply to any scalar FOM. While the author was successful at calculating confidence intervals (this is the heart of what is loosely termed *statistical analysis*) and publishing the work[8] using techniques described in a book[6] titled *Evaluation of Diagnostic Systems: Methods from Signal Detection Theory*, subsequent attempts at applying these methods in a follow-up paper[9] led to negative variance estimates (private communication, Dr. Loren Niklason, ca. 1985). With the benefit of hindsight, negative variance estimates are not that uncommon and the method to be described in this chapter has to deal with that possibility.

The methods[6] described in the cited book involved estimating the different variability components—defined in Chapter 07. Between-reader and within-reader variability (the two cannot be separated) could be estimated from the variance of the AUC values corresponding to the readers interpreting the cases within a treatment and then averaging the variances over all treatments. Estimating case sampling and within-reader variability required splitting the dataset into a few smaller subsets (e.g., a case-set with 60 cases might be split into three subsets of 20 cases each), analyzing each subset to get an AUC estimate, calculating the variance of the resulting AUC values,[6] and scaling the result to the original case size. Because it was based on few values, the estimate was inaccurate and the already case-starved original dataset made it difficult to estimate AUCs for the subsets. Moreover, the division into subsets was at the discretion of the researcher, and therefore unlikely to be reproduced by others. Estimating within-reader variability required re-reading the entire case-set, or at least a part of it. ROC studies have earned a deserved reputation for taking much time to complete, and having to reread a case-set was not a viable option. (Historical note: the author recalls a barroom conversation with Dr. Thomas Mertelmeir after the conclusion of an SPIE meeting ca. 2004, where Dr. Mertelmeir complained mightily, over several beers, about the impracticality of some of the ROC studies required of imaging device manufacturers by the FDA.)

9.1.2 The Wagner analogy

An important objective of modality comparison studies is to estimate the variance of the difference in reader-averaged AUCs between the treatments. For two treatments the variance of the difference is obtained by summing the reader-averaged variance in each treatment and subtracting twice the covariance (a scaled version of the correlation). Therefore, in addition to estimating variances, one needs to estimate correlations. Correlations are present due to the common case-set interpreted by the readers in the different treatments. If the correlation is large, that is, close to unity, then the individual treatment variances tend to cancel, making the constant treatment-induced difference easier to detect. The author recalls a vivid analogy used by the late Dr. Robert F. Wagner to illustrate this point at an SPIE meeting ca. 2008. To paraphrase him, consider measuring from shore the heights of the masts on two adjacent boats in a turbulent ocean. Because of the waves, the heights, as measured from shore, are fluctuating wildly, so the variance of the individual height measurements is large. However, the difference between the two heights is likely to be relatively constant, that is, have small variance. This is because the wave that causes one mast's height to increase also increases the height of the other mast.

9.1.3 The dearth of numbers to analyze and a pivotal breakthrough

The basic issue was that the calculation of AUC reduces the relatively large number of ratings of a set of non-diseased and diseased cases to a *single* number. For example, after completion of a ROC study with five readers and 100 non-diseased and 100 diseased cases interpreted in two treatments, the data is reduced to just 10 numbers, that is, five readers times two treatments. It is difficult to perform statistics with so few numbers. The author recalls a conversation with Prof. Kevin Berbaum at a Medical Image Perception Society meeting in Tucson, Arizona, ca. 1997, in which Dr. Berbaum described the basic idea that forms the subject of this chapter. Namely, *using the jackknife pseudo-values*, Equation 7.7, *as individual case-level figures of merit*. This, of course, greatly increases the amount of data that one can work with; instead of just 10 numbers one now has 2000 pseudovalues ($2 \times 5 \times 200$). If one assumes the pseudovalues behave essentially as case-level data, then they are independent and identically distributed,* and therefore they satisfy the conditions for application of standard analysis of variance (ANOVA) techniques.[10] The relevant paper[1] had already been published in 1992 but other distractions and lack of formal statistical training kept the author from fully appreciating this work until later.

Although methods are available for more complex study designs including partially paired data,[11–13] the author will restrict to fully paired data (i.e., each case is interpreted by all readers in all treatments). There is a long history of how this field has evolved and the author cannot do justice to all methods that are currently available. Some of the methods[14–16] have the advantage that they can handle explanatory variables (termed *covariates*) that could influence performance, for example, years of experience, types of cases, an so on. Other methods are restricted to specific choices of FOM. Specifically, the probabilistic approach[17–22] cited in **Chapter 7,** is restricted to the empirical AUC under the ROC curve, and therefore is not applicable, at the present time, to other FOMs, for example, parametrically fitted ROC AUCs or location specific paradigm FOMs. Instead, the author will focus on methods for which software is readily available (i.e., freely on websites[23,24]), which have been widely used (the method that the author is about to describe has been used in several hundred publications), validated via simulations, apply to any scalar figure of merit, and are therefore widely applicable.

9.1.4 Organization of the chapter

The organization of the chapter is as follows. The concepts of reader and case populations, introduced in **Chapter 7,** are recapitulated. A distinction is made between *fixed* and *random* factors,statistical terms with which one must become familiar. Described next are three types of analysis that are possible with MRMC data, depending on which factors are regarded as random and which as fixed. The general approach to the analysis is described. Two methods of analysis are possible: the jackknife pseudovalue based approach detailed in this chapter and an alternative approach is detailed in **Chapter 10.** The Dorfman–Berbaum–Metz (DBM) model for the jackknife pseudoval-ues is described which incorporates different sources of variability and correlations possible with MRMC data. Calculation of ANOVA-related quantities, termed *mean squares*, from the pseudo-values, are described followed by the significance testing procedure for testing the null hypothesis of no treatment effect. A relevant distribution used in the analysis, namely the F-distribution, is illustrated with **R** examples. The decision rule, that is, whether to reject the NH, calculation of the ubiquitous *p*-value, confidence intervals, and how to handle multiple treatments is illustrated with two datasets. One is an older ROC dataset that has been widely used to demonstrate advances in ROC analysis and the other is a recent dataset involving evaluation of digital chest tomosynthesis versus conventional chest imaging. The approach to validation of DBMH analysis is illustrated with an **R** example. The chapter concludes with a section on the meaning of the pseudovalues. The intent is to explain, at an intuitive level, why the DBM method works, even though use of pseudovalues has been questioned[3] at the conceptual level. For organizational reasons and space limitations, details of the software are relegated to Online Appendices, but they are essential reading, prefer-ably in front of a computer running the software that is part of this book. The author has included

---

* In probability theory and statistics, a sequence or other collection of random variables is independent and identi-cally distributed (IID) if each random variable has the same probability distribution as the others and all are mutually independent.

material in the Online Appendfix that may be obvious to statisticians, for example, an explanation of the Satterthwaite approximation, but are expected to be helpful to others from nonstatistical backgrounds

9.1.5 Datasets

Online Chapter 24 describes a collection of clinical datasets, some of which are used in this chapter to illustrate the methodology. The datasets are embedded in the **RJafroc** package and are named **dataset01, dataset02, …, dataset14**, which are available as soon as **RJafroc** is loaded, i.e., the corresponding **library** command is executed. They are described in function **includedDatasets**. The fourteen (14) datasets comprise 43 modalities, 80 readers, and 2012 cases. Four of these are ROC datasets, one an LROC dataset, and the rest (nine) are free-response ROC (FROC) datasets. Most of the datasets are from the author's international collaborations.

9.2 Random and fixed factors

This paragraph introduces some analysis of variance (ANOVA) terminology. Treatment, reader, and case are *factors* with different numbers of *levels* corresponding to each factor. For a ROC study with two treatments, five readers, and 200 cases, there are two levels of the treatment factor, five levels of the reader factor, and 200 levels of the case factor. *If a factor is regarded as fixed, then the conclusions of the analysis apply only to the specific levels of the factor used in the study. If a factor is regarded as random, the levels of the factor are regarded as random samples from a parent population of the corresponding factor and conclusions regarding specific levels are not allowed. Rather, conclusions apply to the distribution from which the levels are, by assumption, sampled.*

ROC MRMC studies require a sample of cases and interpretations by one or more readers in one or more treatments (in this book the term *multiple* includes, as a special case, *one*). *A study is never conducted on a sample of treatments.* It would be nonsensical to image patients using a sample of all possible treatments known to exist. Every variation of an imaging technique (e.g., different kilovoltage or kVp) or display method (e.g., window-level setting) or image processing technique qualifies as a distinct treatment. The number of possible treatments is very large and, from a practical point of view, most of them are uninteresting. *Rather, interest is in comparing two or more (a few at most) treatments that, based on preliminary studies, are clinically interesting.* One treatment may be computed tomography, the other magnetic resonance imaging, or one may be interested in comparing a standard image processing method to a newly proposed one, or one may be interested in comparing CAD to a group of readers.

This brings out an essential difference between how cases, readers, and treatments have to be regarded in the variability estimation procedure. Cases and readers are usually regarded as *random factors* (there has to be at least one random factor—if not, there are no sources of variability and nothing to apply statistics to!), while treatments are regarded as *fixed factors*. The random factors contribute *stochastic (i.e., random) variability* but the fixed factors do not, rather, they contribute *constant shifts* in performance. The terms *fixed* and *random* factors are used in this specific sense, and are derived, in turn, from ANOVA methods in statistics.[10,25] With two or more treatments, there are shifts in performance of treatments relative to each other, that one seeks to assess the significance of against a background of noise contributed by the random factors. If the shifts are sufficiently large compared to the noise, then one can state, with some certainty, that they are real. Quantifying the last statement uses the methods of hypothesis testing introduced in **Chapter 8**.

9.3 Reader and case populations and data correlations

As discussed in Section 7.2, conceptually there is a reader population, generally modeled as a normal distribution $\theta_j \sim N\left(\overline{\theta_{\bullet\{1\}}}, \sigma^2_{br+wr}\right)$, describing the variation of skill level of readers. The notation closely follows that in the cited section, the only change being that the binormal model estimate A_z has been replaced by a generic FOM, denoted θ. Each reader j is characterized by a different value of θ_j; $j = 1, 2, ..., J$, and one can conceptually think of a bell-shaped curve with variance σ^2_{br+wr} describing between-reader variability of the readers. A large variance implies large spread in reader skill levels.

Likewise, there is a case-population, also modeled as a normal distribution, describing the variations in difficulty levels of the patients. One actually has two unit-variance distributions, one per diseased state, characterized by a separation parameter and conceptually an easy case-set has a larger than usual separation parameter while a difficult case-set has a smaller than usual separation parameter. The effect of the distribution of the separation parameter on the distribution of the FOM can be modeled as a normal distribution $\theta_{\{c\}} \sim N\left(\theta_{\{\bullet\}}, \sigma^2_{cs+wr}\right)$ with variance σ^2_{cs+wr} describing the variations in difficulty levels of different case samples. Note the need for the case-set index, introduced in **Chapter 7**, to specify the separation parameter for a specific case-set (in principle a j-index is also needed as one cannot have an interpretation without a reader, for now, it is suppressed; one can think of the stated equation as applying to the average reader). A small variance σ^2_{cs+wr} implies that different case-sets have similar difficulty levels while a larger variance would imply a larger spread in difficulty levels.

Anytime one has a common random component to two measurements, the measurements are correlated. In the Wagner analogy, the common component is the random height, as a function of time, of a wave, which contributes the same amount to both height measurements (since the boats are adjacent). Since the readers interpret a common case-set in all treatments one needs to account for various types of correlations that are potentially present. These occur due to the different pairings that can occur with MRMC data, where each pairing implies the presence of a common component to the measurements. The pairings include (a) the same reader interpreting the same cases in different treatments, (b) different readers interpreting the same cases in the same treatment, and (c) different readers interpreting the same cases in different treatments. These pairings are more clearly elucidated in **Chapter 10**. The current chapter uses jackknife pseudovalue based analysis to model the variances and the correlations. Hillis has shown that the two approaches are essentially equivalent.[3]

9.4 Three types of analyses

MRMC analysis attempts to draw conclusions regarding the significances of inter-treatment shifts in performance. Ideally a conclusion (i.e., a difference is significant: yes/no; the yes applies if the p-value is less than alpha) should generalize to the respective populations from which the random samples were obtained. In other words, the idea is to generalize from the observed samples to the underlying populations. Three types of analyses are possible depending on which factor(s) one regards as random and which as fixed: *random-reader random-case* (RRRC), *fixed-reader random-case* (FRRC), and *random-reader fixed-case* (RRFC). If a factor is regarded as random, then the conclusion of the study applies to the population from which the levels of the factor were sampled. If a factor is regarded as fixed, then the conclusion applies only to the specific levels of the sampled factor. For example, if the reader is regarded as a random factor, the conclusion generalizes to the reader population from which the readers used in the study were obtained. If a reader is regarded as a fixed factor, then the conclusion applies to the specific readers that participated in the study. Regarding a factor as fixed effectively freezes out the sampling variability of the population and interest then centers only on the specific levels of the factor used in the study. For fixed reader analysis, conclusions about the significances of differences between pairs of readers are allowed; these are not allowed if reader is treated as a random factor. Likewise, treating the case as a fixed factor means the conclusion of the study is specific to the case-set used in the study.

9.5 General approach

This section provides an overview of the steps involved in analysis of MRMC data. Two approaches are described in parallel: a figure of merit (FOM) derived jackknife pseudovalue based approach, detailed in this chapter, and an FOM based approach, detailed in the next chapter. The analysis proceeds as follows:

1. A FOM is selected: *the selection of FOM is the single-most critical aspect of analyzing an observer performance study.* The selected FOM is denoted θ. To keep the notation reasonably compact the usual circumflex hat symbol used previously to denote an estimate is suppressed. The FOM has to be an objective scalar measure of performance with larger values characterizing better performance. [The qualifier *larger* is trivially satisfied; if the figure of merit has the opposite characteristic, a sign change is all that is needed to bring it back to compliance with this requirement.] Examples are empirical AUC, the binormal model-based estimate A_z, other advance method based estimates of AUC, sensitivity at a predefined value of specificity, and so on. An example of a FOM requiring a sign-change is FPF at a specified TPF, where smaller values signify better performance.

2. For each treatment (i) and reader (j) the figure of merit θ_{ij} is estimated from the ratings data. Repeating this over all treatments and readers yields a matrix of observed values θ_{ij}. This is averaged over all readers in each treatment yielding $\theta_{i\bullet}$. The observed effect size, ES_{obs}, is defined as the difference between the reader-averaged FOMs in the two treatments, i.e., $ES_{obs} = \theta_{2\bullet} - \theta_{1\bullet}$. While extensible to more than two treatments, the explanation is more transparent by restricting to two treatments.

3. If the magnitude of ES_{obs} is large one has reason to suspect that there might indeed be a significant difference in *AUCs* between the two treatments, where significant is used in the sense of **Chapter 8**. Quantification of this statement, specifically how large is large, requires the conceptually more complex steps described next.
   a. In the DBMH approach, the subject of this chapter, jackknife pseudovalues are calculated as described in Equation 7.6. A standard ANOVA model with uncorrelated errors is used to model the pseudovalues.
   b. In the ORH approach, the subject of the next chapter, the FOM is modeled directly using a custom ANOVA model with correlated errors.

4. Depending on the selected method of modeling the data (pseudovalue vs. FOM) a statistical model is used which includes parameters modeling the true values in each treatment, and expected variations due to different modeled variability components, e.g., between-reader variability, case sampling variability, *interactions*, (e.g., the possibility that the random effect of a given reader could be treatment dependent) and the presence of *correlations* (between pseudovalues or FOMs) because of the pairings inherent in the interpretations.

5. In RRRC analysis one accounts for randomness in readers and cases. In FRRC analysis one regards the reader as a fixed factor. In RRFC analysis one regards the case as a fixed factor. The statistical model depends on the type of analysis.

6. The parameters of the statistical model are estimated from the observed data.

7. The estimates are used to infer the statistical distribution of the observed effect size, ES_{obs}, regarded as a realization of a random variable, under the null hypothesis (*NH*) the true effect size is zero.
8. Based on this statistical distribution, and assuming a two-sided test, the probability (this is the oft-quoted *p*-value) of obtaining an effect size at least as extreme as that actually observed is calculated, analogous to **Chapter 8**.
9. If the *p*-value is smaller than a preselected value, denoted α, one declares the treatments different at the α- significance level. The quantity α is the control (or cap) on the probability of making a Type I error, defined as rejecting the NH when it is true. It is common to set $\alpha = 0.05$ but depending on the severity of the consequences of a Type I error, as discussed in **Chapter 8**, one might consider choosing a different value. Notice that α is a preselected number while the *p*-value is a realization of a random variable.
10. For a valid statistical analysis, the empirical probability α_{emp} over many (typically 2000) independent NH datasets, that the *p*-value is smaller than α, should equal α to within statistical uncertainty.

9.6 The Dorfman–Berbaum–Metz (DBM) method

The figure of merit has three indices:

1. A treatment index *i*, where *i* runs from 1 to *I*, where *I* is the total number of treatments.
2. A reader index *j*, where *j* runs from 1 to *J*, where *J* is the total number of readers.
3. The often-suppressed case sample index $\{c\}$, where $\{1\}$, i.e., $c = 1$, denotes a set of cases, K_1 non-diseased and K_2 diseased, interpreted by all readers in all treatments, and other integer values of *c* correspond to other sets of cases that, although not interpreted by the readers, could potentially be interpreted using resampling methods such as the bootstrap or the jackknife.

The approach taken by the Dorfman–Berbaum–Metz[1] (DBM) was to use the jackknife resampling method described in **Chapter 7** to calculate FOM *pseudovalues* Y'_{ijk} defined by (the reason for the prime will become clear shortly):

$$Y'_{ijk} = K\theta_{ij} - (K-1)\theta_{ij(k)} \tag{9.1}$$

Here θ_{ij} is the estimate of the figure of merit for reader *j* interpreting all cases in treatment *i* and $\theta_{ij(k)}$ is the corresponding figure of merit with case *k* deleted from the analysis. To adhere to convention and to keep the notation simple the $\{1\}$ index on every figure of merit symbol is suppressed unless it is absolutely necessary for clarity.

Recall from **Chapter 7** that the jackknife is a way of teasing out case-dependence. The left-hand side of Equation 9.1 literally has a case index *k*, with *k* running from 1 to *K*, where *K* is the total number of cases: $K = K_1 + K_2$.

Hillis has proposed a *centering* transformation on the pseudovalues (Hillis calls them *normalized* pseudovalues but to the author *centering* is a more accurate and descriptive term*).

$$Y_{ijk} = Y'_{ijk} + \left(\theta_{ij} - Y'_{ij\bullet}\right) \tag{9.2}$$

---

* Normalize: (In mathematics) multiply (a series, function, or item of data) by a factor that makes the norm or some associated quantity such as an integral equal to a desired value (usually 1). New Oxford American Dictionary, 2016.

The effect of this transformation is that the average of the centered pseudovalues over the case index is identical to the corresponding estimate of the figure of merit.

$$Y_{ij\bullet} = Y'_{ij\bullet} + \left(\theta_{ij} - Y'_{ij\bullet}\right) = \theta_{ij} \tag{9.3}$$

This has the advantage that all confidence intervals are correctly centered. The transformation is unnecessary if one uses the Wilcoxon as the figure of merit, as the pseudovalues calculated using the Wilcoxon as the figure of merit are automatically centered. It is left as an exercise for the reader to show that this statement is true.

*It is understood that, unless explicitly stated otherwise, all calculations from now on will use **centered** pseudovalues.*

Consider N replications of an MRMC study, where a replication means repetition of the study with the same treatments, readers, and case-set $\{1\}$; it is not to be confused with repetitions with different case-sets, that is, the $\{c\}$ index. For N replications per treatment-reader-case combination, the DBM model for the pseudovalues is (n is the replication index, usually $n = 1$, but kept here for now):

$$Y_{n(ijk)} = \mu + \tau_i + R_j + C_k + (\tau R)_{ij} + (\tau C)_{ik} + (RC)_{jk} + (\tau RC)_{ijk} + \varepsilon_{n(ijk)} \tag{9.4}$$

The notation* for the replication index, that is, $n(ijk)$, implies n observations for treatment-reader-case combination ijk.

The basic assumption of the DBM model, Equation 9.4, is that the pseudovalues can be regarded as independent and identically distributed observations. If that is true, the pseudovalue data can be analyzed by standard ANOVA techniques.

9.6.1 Explanation of terms

In Equation 9.4 the term μ is a constant. By definition, the treatment effect τ_i is subject to the constraint

$$\sum_{i=1}^{I} \tau_i = 0 \Rightarrow \tau_\bullet = 0 \tag{9.5}$$

It is shown below, Equation 9.9, that this constraint ensures that μ has the interpretation as the average of the pseudovalues over treatments, readers, cases, and replications, if any.

The right-hand side of Equation 9.4 consists of one fixed and seven random effects. The current analysis assumes readers and cases as random factors (RRRC), so by definition R_j and C_k are random effects, and moreover, any term that includes a random factor is a random effect. For example, $(\tau R)_{ij}$ is a random effect because it includes the R factor. Here is a list of the random terms:

$$R_j, C_k, (\tau R)_{ij}, (\tau C)_{ik}, (RC)_{jk}, (\tau RC)_{ijk}, \varepsilon_{ijk} \tag{9.6}$$

---

* The statistical term is *nested*.

Assumption: Each of the random effects is modeled as a random sample from mutually independent zero-mean normal distributions with variances as specified below:

$$\left.\begin{array}{c} R_j \sim N\left(0,\sigma_R^2\right) \\[2mm] C_k \sim N\left(0,\sigma_C^2\right) \\[2mm] (\tau R)_{ij} \sim N\left(0,\sigma_{\tau R}^2\right) \\[2mm] (\tau C)_{ik} \sim N\left(0,\sigma_{\tau C}^2\right) \\[2mm] (RC)_{jk} \sim N\left(0,\sigma_{RC}^2\right) \\[2mm] (\tau RC)_{ijk} \sim N\left(0,\sigma_{\tau RC}^2\right) \\[2mm] \varepsilon_{n(ijk)} \sim N\left(0,\sigma_\varepsilon^2\right) \end{array}\right\} \qquad (9.7)$$

One could have placed a Y subscript (or superscript) on each of the variances, as they describe fluctuations of the pseudovalues, not FOM values. The latter are the subject of the next chapter. However, this tends to make the notation cumbersome. So here is the **convention**:

Unless explicitly stated otherwise, all variance symbols in this chapter refer to pseudovalues.

Another convention: $(\tau R)_{ij}$ *is not the product of the treatment and reader factors.* Rather, it is a single factor, namely, the treatment-reader factor with IJ levels, subscripted by the index ij and similarly for the other product-like terms in Equation 9.7.

9.6.2 Meanings of variance components in the DBM model

The variances defined in Equation 9.7 are collectively termed *variance components*. Specifically, they are *jackknife pseudovalue variance components*, to be distinguished from figure of merit (FOM) variance components to be introduced in **Chapter 10**. They are in order: $\sigma_R^2, \sigma_C^2, \sigma_{\tau R}^2, \sigma_{\tau C}^2, \sigma_{RC}^2, \sigma_{\tau RC}^2, \sigma_\varepsilon^2$. They have the following meanings (all references to *variance* mean *variance of pseudovalues*).

- The term σ_R^2 is the variance of readers that is independent of treatment or case, which are modeled separately. It is not to be confused with the terms σ_{br+wr}^2 and σ_{cs+wr}^2 used in Section 9.3, which describe the variability of θ measured under specified conditions. (A jackknife pseudovalue is a weighted difference of FOM-like quantities, Equation 9.1. Its meaning will be explored later. For now, a pseudovalue variance is distinct from a FOM variance.)
- The term σ_C^2 is the variance of cases that is independent of treatment or reader. It is not to be confused with the term used in Section 9.3, which describes the variability of a separation parameter.
- The term $\sigma_{\tau R}^2$ is the treatment-dependent variance of readers that was excluded in the definition of σ_R^2. If one were to sample readers and treatments for the same case-set, the net variance would be $\sigma_R^2 + \sigma_{\tau R}^2 + \sigma_\varepsilon^2$.
- The term $\sigma_{\tau C}^2$ is the treatment-dependent variance of cases that were excluded in the definition of σ_C^2. So, if one were to sample cases and treatments for the same readers, the net variance would be $\sigma_C^2 + \sigma_{\tau C}^2 + \sigma_\varepsilon^2$.

- The term σ_{RC}^2 is the treatment-independent variance of readers and cases that was excluded in the definitions of σ_C^2 and σ_R^2. So, if one were to sample readers and cases for the same treatment, the net variance would be $\sigma_R^2 + \sigma_C^2 + \sigma_{RC}^2 + \sigma_\varepsilon^2$.
- The term $\sigma_{\tau RC}^2$ is the variance of treatments, readers, and cases that was excluded in the definitions of all the preceding terms in Equation 9.7. So, if one were to sample treatments, readers and cases the net variance would be $\sigma_R^2 + \sigma_C^2 + \sigma_{\tau R}^2 + \sigma_{\tau C}^2 + \sigma_{RC}^2 + \sigma_{\tau RC}^2 + \sigma_\varepsilon^2$.
- The last term, σ_ε^2 describes the variance arising from different replications of the study using the *same treatments, readers, and cases*. Measuring this variance requires repeating the study several (N) times with the same treatments, readers, and cases, and computing the variance of $Y_{n(ijk)}$, where the additional n-index refers to true replications, $n = 1, 2, ..., N$.

$$\sigma_\varepsilon^2 = \frac{1}{IJK} \sum_{i=1}^{I} \sum_{j=1}^{J} \sum_{k=1}^{K} \frac{1}{N-1} \sum_{n=1}^{N} \left(Y_{n(ijk)} - Y_{\bullet(ijk)} \right)^2 \tag{9.8}$$

The right-hand side of Equation 9.8 is the variance of $Y_{n(ijk)}$, for specific ijk, with respect to the replication index n, averaged over all ijk. In practice, N equals one (i.e., there are no replications) so this variance cannot be estimated (it would imply dividing by zero in Equation 9.8). It has the meaning of *reader inconsistency*, usually termed *within-reader* variability. In the author's experience, rating noise is about half a bin on a 1–6 scale, but this issue has not been studied and rating noise is not to be confused with pseudovalue noise. *As will be shown later, the presence of this inestimable term does not limit one's ability to perform significance testing on the treatment effect without having to replicate the whole study, as implied in earlier work.*[2]

An equation like Equation 9.4 is termed a *linear model* with the left-hand side, the pseudovalue observations, modeled by a sum of fixed and random terms. Specifically, it is a *mixed* model, because the right-hand side has *both* fixed and random effects. Statistical methods have been developed for analysis of linear models.[10,26–28] In essence, one estimates the terms on the right-hand side of Equation 9.4, it being understood that for the random effects, one estimates the *variances* of the zero-mean normal distributions, Equation 9.7, from which the samples are obtained (by assumption).

Estimating the fixed effects is trivial. The term μ is estimated by averaging the left-hand side of Equation 9.4 over all three indices (since $N = 1$).

$$\mu = Y_{1(\bullet\bullet\bullet)} \tag{9.9}$$

Because of the way the treatment effect is defined, Equation 9.5, averaging, which involves summing, over the treatment-index i, yields zero, and all of the remaining random terms yield zero upon averaging, because they are individually sampled from zero-mean normal distributions. To estimate an individual treatment effect, one takes the difference as shown below:

$$\tau_i = Y_{1(i\bullet\bullet)} - \mu \tag{9.10}$$

It is seen that the reader and case averaged difference between two treatments i and i' is estimated by

$$\tau_i - \tau_{i'} = Y_{1(i\bullet\bullet)} - Y_{1(i'\bullet\bullet)} \tag{9.11}$$

Estimating the strengths of the random terms is a little more complicated. It involves methods adapted from least-squares, or maximum likelihood, and more esoteric ways. The author does not care to go into these methods. Instead, results are presented and arguments are made to make them plausible. The starting point is definitions of quantities called *mean squares* and their *expected values*.

9.6.3 Definitions of mean squares

Again, to be clear, one should put a Y subscript (or superscript) on each of the following *definitions*, but that would make the notation unnecessarily cumbersome.

In this chapter, all mean-square quantities are calculated using *pseudovalues*, not figure of merit values. The presence of Y in the following equations should make this clear. Also, the replication index and the nesting notation are suppressed. The notation is abbreviated so MST is the mean square corresponding to the treatment effect, and so on.

$$MST = \frac{JK \sum_{i=1}^{I} (Y_{i\bullet\bullet} - Y_{\bullet\bullet\bullet})^2}{I-1}$$

$$MSR = \frac{IK \sum_{j=1}^{J} (Y_{\bullet j\bullet} - Y_{\bullet\bullet\bullet})^2}{J-1}$$

$$MSTR = \frac{K \sum_{i=1}^{I} \sum_{j=1}^{J} (Y_{ij\bullet} - Y_{ij\bullet} - Y_{\bullet j\bullet} + Y_{\bullet\bullet\bullet})^2}{(I-1)(J-1)} \qquad (9.12)$$

$$MSTC = \frac{J \sum_{i=1}^{I} \sum_{k=1}^{K} (Y_{i\bullet k} - Y_{i\bullet\bullet} - Y_{\bullet\bullet k} + Y_{\bullet\bullet\bullet})^2}{(I-1)(K-1)}$$

$$MSTRC = \frac{\sum_{i=1}^{I} \sum_{j=1}^{J} \sum_{k=1}^{K} (Y_{ijk} - Y_{ij\bullet} - Y_{i\bullet k} - Y_{\bullet jk} + Y_{i\bullet\bullet} + Y_{\bullet j\bullet} + Y_{\bullet\bullet k} - Y_{\bullet\bullet\bullet})^2}{(I-1)(J-1)(K-1)}$$

Note the absence of MSE, corresponding to the epsilon term on the right-hand side of Equation 9.4. With only one observation per treatment-reader-case combination, MSE cannot be estimated; it effectively gets folded into the $MSTRC$ term.

The most general case is presented next, where both readers and cases are regarded as random factors, termed *random-reader random-case* analysis (RRRC).

9.7 Random-reader random-case analysis (RRRC)

The mean squares on the left-hand side of Equation 9.12 can be calculated directly from the pseudovalues. The next step in the analysis is to obtain expressions for their *expected* values in terms of the variances defined in Equation 9.7. Assuming no replications, that is, $N = 1$, the expected mean squares are as follows (Table 9.1). Understanding how this table is derived would lead the author well outside his expertise and the scope of this book. Suffice to say that these are *unconstrained*

Table 9.1 Summary of treatment by reader by case ANOVA expected mean squares

Source	df	E(MS)
T	$(I-1)$	$E(MST) = \sigma_\varepsilon^2 + \sigma_{\tau RC}^2 + K\sigma_{\tau R}^2 + J\sigma_{\tau C}^2 + JK\sigma_\tau^2$
R	$(J-1)$	$E(MSR) = \sigma_\varepsilon^2 + I\sigma_{RC}^2 + IK\sigma_R^2 + K\sigma_{\tau R}^2$
C	$(K-1)$	$E(MSC) = \sigma_\varepsilon^2 + I\sigma_{RC}^2 + IJ\sigma_C^2 + J\sigma_{\tau C}^2$
TR	$(I-1)(J-1)$	$E(MSTR) = \sigma_\varepsilon^2 + \sigma_{\tau RC}^2 + K\sigma_{\tau R}^2$
TC	$(I-1)(K-1)$	$E(MSTC) = \sigma_\varepsilon^2 + \sigma_{\tau RC}^2 + J\sigma_{\tau C}^2$
RC	$(J-1)(K-1)$	$E(MSRC) = \sigma_\varepsilon^2 + I\sigma_{RC}^2$
TRC	$(I-1)(J-1)(K-1)$	$E(MSTRC) = \sigma_\varepsilon^2 + \sigma_{\tau RC}^2$
ε	$N-1=0$	$E(\varepsilon) = \sigma_\varepsilon^2$

Note: E(MS) = expected mean squares; *df* = degrees of freedom of corresponding E(MS).

estimates,[29] which are different from the *constrained estimates* appearing in the original DBM publication[1]. The differences between these two types of estimates is summarized in Ref. 29 and calculation of expected mean square values is detailed in Scheffe's book.[30]

Since treatment is a *fixed* effect, the variance symbol σ_τ^2, which is used for notational consistency in Table 9.1, could cause confusion. It is defined by

$$\sigma_\tau^2 = \frac{1}{I-1}\sum_i^I \left(Y_{i\bullet\bullet} - Y_{\bullet\bullet\bullet}\right)^2 \tag{9.13}$$

The expression on the right-hand side looks like a variance, indeed one that could be calculated for just two treatments ($I = 2$) but, of course, random sampling from a distribution of treatments is not the intent of the notation. Under the null hypothesis,

$$NH : Y_{1\bullet\bullet} = Y_{2\bullet\bullet} \ldots = Y_{I\bullet\bullet} \tag{9.14}$$

This implies that under the NH: $\sigma_\tau^2 = 0$.

The expected mean squares in Table 9.1 are variance-like quantities; specifically, they are weighted linear combinations of the variances appearing in Equation 9.7. For single factors, the column headed degrees of freedom (*df*) is one less than the number of levels of the corresponding factor. Estimating a variance requires first estimating the mean, which imposes a constraint, thereby decreasing *df* by one. For interaction terms, *df* is the product of the degrees of freedom for the individual factors. As an example, the term $(\tau RC)_{ijk}$ contains three individual factors, and therefore $df = (I-1)(J-1)(K-1)$. The number of degrees of freedom can be thought of as the amount of information available in estimating a mean square. As a special case, with no replications, the ε term has zero *df* as $N-1=0$, With only one observation $Y_{n(ijk)}$ there is no information to estimate the variance corresponding to the ε term. To estimate this term, one needs to replicate the study several times—each time the same readers interpret the same cases in all treatments—a very boring task for the reader and totally unnecessary from the researcher's point of view.

9.7.1 Significance testing

If the NH of no treatment effect is true, that is, if $\sigma_\tau^2 = 0$, then according to Table 9.1 the following holds (the last term in the row labeled T in Table 9.1 drops out):

$$E(MST|NH) = \sigma_\varepsilon^2 + \sigma_{\tau RC}^2 + K\sigma_{\tau R}^2 + J\sigma_{\tau C}^2 \tag{9.15}$$

Also, the following linear combination is equal to $E(MST|NH)$:

$$E(MSTR) + E(MSTC) - E(MSTRC)$$

$$= \left(\sigma_\varepsilon^2 + \sigma_{\tau RC}^2 + K\sigma_{\tau R}^2\right) + \left(\sigma_\varepsilon^2 + \sigma_{\tau RC}^2 + J\sigma_{\tau C}^2\right) - \left(\sigma_\varepsilon^2 + \sigma_{\tau RC}^2\right)$$

$$= \sigma_\varepsilon^2 + \sigma_{\tau RC}^2 + J\sigma_{\tau C}^2 + K\sigma_{\tau R}^2 \tag{9.16}$$

$$= E(MST|NH)$$

Therefore, under the NH, the ratio is as follows:

$$\frac{E(MST|NH)}{E(MSTR) + E(MSTC) - E(MSTRC)} = 1 \tag{9.17}$$

In practice, one does not know the *expected* values—that would require averaging each of these quantities, regarded as random variables, over their respective distributions. Therefore, one defines the following statistic, denoted F_{DBM}, using the *observed* values of the mean squares, calculated almost trivially using Equation 9.12:

$$F_{DBM} = \frac{MST}{MSTR + MSTC - MSTRC} \tag{9.18}$$

F_{DBM} is a realization of a random variable. A non-zero treatment effect, that is, $\sigma_\tau^2 > 0$, will cause the ratio to be larger than one, because $E(MST)$ will be larger, see row labeled T in Table 9.1. Therefore, values of $F_{DBM} > 1$ will tend to reject the NH. Drawing on a theorem from statistics,[25] under the NH the ratio of two independent mean squares is distributed as a (central) F-statistic with degrees of freedom corresponding to those of the mean squares forming the numerator and denominator of the ratio (Theorem 12.2.5 in *An Introduction to Mathematical Statistics and Its Applications*, Reference 25). Knowing the distribution of the statistic defined by Equation 9.18 under the NH enables hypothesis testing. This is completely analogous to **Chapter 8** where knowledge of the distribution of AUC under the NH enabled testing the null hypothesis that the observed value of AUC equals a prespecified value.

Under the NH the left-hand side of by (Equation 9.18), that is, F_{DBM}, is distributed according to the F-distribution characterized by two numbers (using the same symbol for a realized values, i.e., the left hand side of Eqn. 9.18, and the corresponding random variable, the left hand side of Eqn. 9.19, should not cause undue confusion, now that it has been acknowledged):

1. A numerator degrees of freedom (*ndf*)—determined by the degrees of freedom of the numerator *MST* of the ratio comprising the F-statistic, i.e., $I - 1$
2. A denominator degrees of freedom (*ddf*)—determined by the degrees of freedom of the denominator of the ratio comprising the F-statistic, to be described below

Summarizing,

$$F_{DBM} \sim F_{ndf,ddf} \tag{9.19}$$

Here,

$$ndf = I - 1 \qquad (9.20)$$

The next topic is estimating *ddf*.

9.7.2 The Satterthwaite approximation

The denominator of the F-ratio is

$$MSTR + MSTC - MSTRC \qquad (9.21)$$

This is not a simple mean square. Rather it is a *linear combination* of mean squares (with coefficients 1, 1, and -1), and the resulting value could even be negative, which is an illegal value for a sample from an F-distribution. In 1941 Satterthwaite[31,32] proposed an approximate degree of freedom for a linear combination of simple mean square quantities. Online Appendix 9.A explains the approximation in more detail. The end result is that the mean square quantity described in Equation 9.21 has an *approximate* degree of freedom defined by (this is called the *Satterthwaite's approximation*):

$$ddf_{Sat} = \frac{\left(MSTR + MSTC - MSTRC\right)^2}{\dfrac{MSTR^2}{(I-1)(J-1)} + \dfrac{MSTC^2}{(I-1)(K-1)} + \dfrac{MSTRC^2}{(I-1)(J-1)(K-1)}} \qquad (9.22)$$

The subscript *Sat* is for Satterthwaite. From Equation 9.22 it should be fairly obvious that, in general, ddf_s is not an integer. To accommodate possible negative estimates of the denominator, Equation 9.21, the original DBM method[1] proposed four expressions for the F-statistic and corresponding expressions for *ddf*. Rather than repeat them here, since they have been superseded by the method described below, the interested reader is referred to Equations 6 and 7 in a 2008 Hillis et al. publication[3], a readable account of the evolution of ideas in this field.

Hillis[4] proposes the following statistic for testing the null hypothesis (the subscript DBMH gives credit to the original formulation by DBM and the subsequent improvements by Hillis):

$$F_{DBMH} = \frac{MST}{MSTR + \max\left(MSTC - MSTRC, 0\right)} \qquad (9.23)$$

Now the denominator cannot be negative. One can think of the F-statistic F_{DBMH} as a signal-to-noise ratio-like quantity, with the difference that both numerator and denominator are variance-like quantities. If the variance represented by the treatment effect is larger than the variance of the noise tending to mask the treatment effect, then F_{DBMH} tends to be large, which makes the observed treatment variance stand out more clearly compared to the noise.

Hillis has shown that the left-hand side of Equation 9.23, regarded as a random variable, is distributed as an F-statistic, with *ndf* defined by Equation 9.20 and denominator degrees of freedom ddf_H defined by

$$ddf_H = \frac{\left[MSTR + \max\left[MSTC - MSTRC, 0\right]\right]^2}{\dfrac{MSTR^2}{(I-1)(J-1)}} \qquad (9.24)$$

Summarizing,

$$F_{DBMH} \sim F_{ndf, ddf_H} \qquad (9.25)$$

Instead of four rules, as in the original DBM method, the Hillis modification involves just one rule, summarized by Equations 9.23 through 9.25. Moreover, the F-statistic is constrained to non-negative values. Using simulation testing Hillis et al.[3] have shown that the DBMH method has better null hypothesis behavior than the original DBM method; the latter tended to be too conservative, typically yielding Type-I error rates smaller than the optimal 5%.

9.7.3 Decision rules, *p*-value, and confidence intervals

The critical value of the F-statistic $F_{1-\alpha,ndf,ddf_H}$ is defined such that fraction $(1-\alpha)$ of the distribution lies to the left of the critical value, in other words, it is the quantile function for the F-distribution.

$$P\left(F \le F_{1-\alpha,ndf,ddf_H} \,\middle|\, F \sim F_{ndf,ddf_H}\right) = 1 - \alpha \tag{9.26}$$

The critical value $F_{1-\alpha,ndf,ddf_H}$ increases as α decreases. The decision rule is that if $F_{DBMH} > F_{1-\alpha,ndf,ddf_H}$, one rejects the NH and, otherwise, one does not. It follows, from the definition of F_{DBMH}, Equation 9.23, that rejection of the NH is more likely if

1. F_{DBMH} is large, Equation 9.23, which occurs if *MST* is large, meaning the treatment effect is large, and/or $MSTR + \max(MSTC - MSTRC, 0)$ is small (see comments following Equation 9.23).
2. α is large: then $F_{1-\alpha,ndf,ddf_H}$ decreases and is more likely to be exceeded by F_{DBMH}.
3. *ndf* is large: the more the number of treatment pairings, the greater the chance that at least one pairing will reject the NH.
4. ddf_H is large: this causes the critical value to decrease, see below, and is more likely to be exceeded by F_{DBMH}.

Playing around with the values in the following code, in file **mainFDist.R**, the reader should be convinced of the truth of the above statements.

9.7.3.1 Code listing

```
rm(list = ls()) #mainFDist.R
require(ggplot2)
alpha   <- 0.05;ndf <- 1;ddf <- 200
fCrit <- qf(1 - alpha, ndf, ddf)
rfCrit <- round(fCrit, 3)
cat("alpha = ", alpha, "ndf = ", ndf, "ddf = ",
    ddf, "fCrit = ", fCrit, "\n")

x <- seq(0, 20, by = 0.01)
pdf <- df(x, ndf, ddf)
plotCurve <- data.frame(x = x, pdf = pdf)
plotArea <- data.frame(
  x = c(0, seq(0, fCrit, by = 0.01), fCrit),
  y = c(0, df(seq(0, fCrit, by = 0.01),
              ndf, ddf), 0))
yMax <- 1
fText <- paste0(
  "F[list(1-",
  as.character(alpha),
  ",", ndf, ",", ddf, ")] == ", rfCrit)
```

```
p <- ggplot() +
  geom _ line(
    data = plotCurve,
    mapping = aes(x = x, y = pdf), size = 2) +
  geom _ polygon(
    data = plotArea,
    mapping = aes(x = x, y = y)) +
  geom _ segment(
    aes(x = fCrit + 3,
        y = yMax / 4,
        xend = fCrit,
        yend = df(fCrit, ndf, ddf)),
    arrow = arrow(
      length = unit(0.5, "cm")), size = 2) +
  annotate("text", x = rfCrit + 4, y = yMax / 4 + 0.04,
           label = fText, size = 7, parse = TRUE) +
  coord _ cartesian(ylim = c(0, yMax))

print(p)
```

After defining **alpha, ndf**, and **ddf**, more or less arbitrarily, in line 3, the critical value **fCrit** is calculated at line 4 using the quantile function **qf()**. The quantile function for the F-distribution **qf()** is the value x such that the area under the F-distribution to the left of x equals the first parameter supplied to the function, which must be a probability in the range 0 to 1. The reader should print out the values of **qf()** for different values of the first parameter; the second parameter is ndf and the third parameter is ddf.

Source this file, using the following parameters values, to get Figure 9.1a through d:

1. Figure 9.1a is for $\alpha = 0.05$, $ndf = 1$, and $ddf = 200$, and the shaded area equals 0.95. The arrow points to the critical value of the F-statistic $F_{1-0.05,1,200} = 3.89$. This is the value that the observed DBMH F-statistic would have to exceed in order for the treatment effect to be significant.
2. Figure 9.1b is for $\alpha = 0.05$, $ndf = 1$, and $ddf = 10$. The arrow points to the critical value of the F-statistic $F_{1-0.05,1,10} = 4.96$. Notice the increase in the critical value of F from 3.89 in (a) to 4.96 in (b), showing that with smaller ddf, the NH is more difficult to reject, because the critical value is larger.
3. Figure 9.1c is for $\alpha = 0.01$, $ndf = 1$, and $ddf = 200$, and now the shaded area equals 0.99. The arrow points to the critical value of the F-statistic $F_{1-0.01,1,200} = 6.76$. Compared to (a), the higher critical corresponds to the smaller α.
4. Figure 9.1d is for $\alpha = 0.05$, $ndf = 5$, and $ddf = 10$. The critical value of the F-statistic is 3.33. This plot should be compared to (b) where $ndf = 1$. Increasing the number of treatments makes it more likely that the NH will be rejected.

The p-value of the test is the probability, under the NH, that an equal or larger value of the F-statistic than F_{DBMH} could occur by chance. In other words, it is the area under the (central) F-distribution F_{ndf,ddf_H} that lies above the observed value F_{DBMH}.

$$p = P\left(F \geq F_{DBMH} \mid F \sim F_{ndf,ddf_H}\right) \tag{9.27}$$

If $p < \alpha$ then the NH that *all treatments are identical* is rejected at significance level α. That informs the researcher that there exists at least one treatment-pair that has a significant

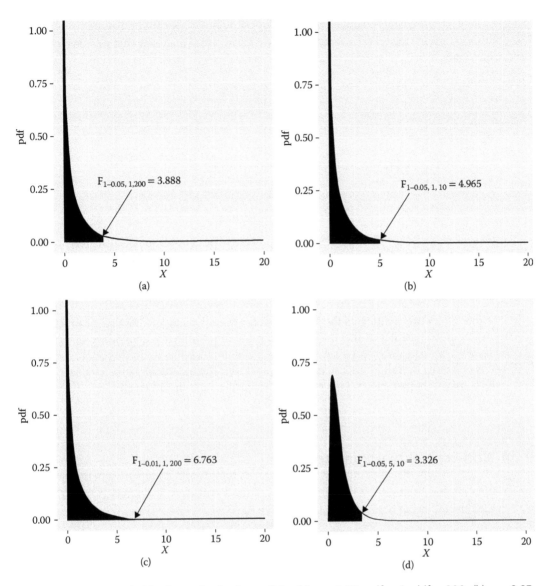

Figure 9.1 (a) through (d): The F-distribution *pdf* for (a) $\alpha = 0.05$, *ndf* = 1, *ddf* = 200; (b) $\alpha = 0.05$, *ndf* = 1, *ddf* = 10; (c) $\alpha = 0.01$, *ndf* = 1, *ddf* = 200; (d) $\alpha = 0.05$, *ndf* = 5, *ddf* = 10. In each case the shaded area is $1 - \alpha$. The code to generate these plots is in **mainFDist.R**.

difference. To identify which pair(s) are different, one calculates confidence intervals for each paired difference. Hillis has shown that the $(1 - \alpha)100$ percent confidence interval for $Y_{i\bullet\bullet} - Y_{i'\bullet\bullet}$ is given by

$$CI_{1-\alpha} = \left(Y_{i\bullet\bullet} - Y_{i'\bullet\bullet}\right) \pm t_{\alpha/2;ddf_H} \sqrt{\frac{2}{JK}\left(MSTR + \max\left(MSTC - MSTRC, 0\right)\right)} \qquad (9.28)$$

Here $t_{\alpha/2;ddf_H}$ is that value such that $\alpha/2$ of the central t-distribution with ddf_H degrees of freedom is contained in the upper tail of the distribution.

$$P\left(T > t_{\alpha/2;ddf_H} \mid T \sim t_{ddf_H}\right) = \alpha/2 \qquad (9.29)$$

In Equation 9.28, the averages indicated by the dot symbols are over the reader and case indices. The notation $\theta_{i\bullet}$ means the reader-averaged FOM for treatment i. Since centered pseudovalues were used, Equation 9.2:

$$\theta_{i\bullet} - \theta_{i'\bullet} = Y_{i\bullet\bullet} - Y_{i'\bullet\bullet} \tag{9.30}$$

Equation 9.28 can be rewritten:

$$CI_{1-\alpha} = \left(\theta_{i\bullet} - \theta_{i'\bullet}\right) \pm t_{\alpha/2;ddf_H} \sqrt{2\frac{MSTR + \max\left(MSTC - MSTRC, 0\right)}{JK}} \tag{9.31}$$

For two treatments, the following equivalent rules could be adopted to reject the NH:

- $F_{DBMH} > F_{1-\alpha, ndf, ddf_H}$
- $p < \alpha$
- $CI_{1-\alpha}$ excludes zero

For more than two treatments the first two rules are equivalent and if a significant difference is found using either of them, then one can use the confidence intervals to determine which treatment pair differences are significantly different from zero. In this book, the first F-test is called the overall F-test and the subsequent tests the treatment pair t-tests. One only conducts treatment pair t-tests if the overall F-test yields a significant result.

9.7.4 Non-centrality parameter

So far attention has been on the NH distribution of the F-statistics. If the AH is true, that is, if $\sigma_\tau^2 \neq 0$, then the following holds instead of Equation 9.17:

$$\left. \begin{aligned} F_{AH} &= \frac{E\left(MST \mid AH\right)}{E\left(MSTR\right) + E\left(MSTC\right) - E\left(MSTRC\right)} \\[2mm] &= \frac{\sigma_\varepsilon^2 + \sigma_{\tau RC}^2 + K\sigma_{\tau R}^2 + J\sigma_{\tau C}^2 + JK\sigma_\tau^2}{\sigma_\varepsilon^2 + \sigma_{\tau RC}^2 + K\sigma_{\tau R}^2 + J\sigma_{\tau C}^2} \\[2mm] &= 1 + \frac{JK\sigma_\tau^2}{\sigma_\varepsilon^2 + \sigma_{\tau RC}^2 + K\sigma_{\tau R}^2 + J\sigma_{\tau C}^2} \end{aligned} \right\} \tag{9.32}$$

The last equation above follows from Equations 9.15 through 9.17. Therefore,

$$F_{AH} = 1 + \Delta \tag{9.33}$$

Δ is known as the *non-centrality parameter*. It is defined by

$$\Delta = \frac{JK\sigma_\tau^2}{\sigma_\varepsilon^2 + \sigma_{\tau RC}^2 + K\sigma_{\tau R}^2 + J\sigma_{\tau C}^2} \tag{9.34}$$

It can be shown that under the AH, F_{AH} is distributed as a *non-central* F-distribution with non-centrality parameter Δ:

$$F_{AH} \sim F_{ndf,ddf,\Delta} \qquad (9.35)$$

The non-central F-distribution will be used for sample size estimation, in **Chapter 11**.

9.8 Fixed-reader random-case analysis

The model is the same as in Equation 9.4 except one puts $\sigma_R^2 = \sigma_{\tau R}^2 = 0$ in Table 9.1. The appropriate test statistic is

$$\frac{E(MST)}{E(MSTC)} = \frac{\sigma_\epsilon^2 + \sigma_{\tau RC}^2 + J\sigma_{\tau C}^2 + JK\sigma_\tau^2}{\sigma_\epsilon^2 + \sigma_{\tau RC}^2 + J\sigma_{\tau C}^2} \qquad (9.36)$$

Under the null hypothesis $\sigma_\tau^2 = 0$:

$$\frac{E(MST|NH)}{E(MSTC)} = 1 \qquad (9.37)$$

Therefore, one defines the F-statistic (replacing *expected* with *observed* values) by

$$F_{DBM|R} = \frac{MST}{MSTC} \qquad (9.38)$$

The observed value $F_{DBM|R}$ (the Roe-Metz notation[33] is used, which indicates that the factor appearing to the right of the vertical bar is regarded as fixed) is distributed as an F-statistic with $ndf = I - 1$ and $ddf = (I - 1)(K - 1)$; the degrees of freedom follow from the rows labeled T and TC in Table 9.1. Therefore, the distribution of the observed value is (no Satterthwaite approximation needed this time as both numerator and denominator are simple mean squares)

$$F_{DBM|R} \sim F_{I-1,(I-1)(K-1)} \qquad (9.39)$$

The null hypothesis is rejected if the observed value of the F-statistic exceeds the critical value.

$$F_{DBM|R} \geq F_{1-\alpha;I-1,(I-1)(K-1)} \qquad (9.40)$$

The *p*-value of the test is the probability that a random sample from the F-distribution Equation 9.39, exceeds the observed value.

$$p = P\left(F \geq F_{DBM|R} \mid F \sim F_{I-1,(I-1)(K-1)}\right) \qquad (9.41)$$

The $(1-\alpha)$ confidence interval for the inter-treatment reader-averaged difference FOM is given by

$$CI_{1-\alpha} = (\theta_{i\bullet} - \theta_{i'\bullet}) \pm t_{\alpha/2;(I-1)(K-1)} \sqrt{2\frac{MSTC}{JK}} \qquad (9.42)$$

9.8.1 Single-reader multiple-treatment analysis

With a single reader interpreting cases in two or more treatments, the reader factor must necessarily be regarded as fixed. The preceding analysis is applicable. One simply puts $J = 1$ in the equations in this Section.

9.8.2 Non-centrality parameter

Analogous to Equation 9.34, the non-centrality parameter is defined by

$$\Delta = \frac{JK\sigma_\tau^2}{\sigma_\varepsilon^2 + \sigma_{\tau RC}^2 + J\sigma_{\tau C}^2} \tag{9.43}$$

Under the AH, the test statistic is distributed as a non-central F-distribution as follows:

$$F_{AH|R} \equiv \frac{MST}{MSTC} \sim F_{I-1,(I-1)(K-1),\Delta} \tag{9.44}$$

These results are used in sample size estimation (**Chapter 11**).

9.9 Random-reader fixed-case (RRFC) analysis

The model is the same as in Equation 9.4 except one puts $\sigma_C^2 = \sigma_{\tau C}^2 = 0$ in Table 9.1. From this table, it follows that

$$\frac{E(MST)}{E(MSTR)} = \frac{\sigma_\varepsilon^2 + \sigma_{\tau RC}^2 + K\sigma_{\tau R}^2 + JK\sigma_\tau^2}{\sigma_\varepsilon^2 + \sigma_{\tau RC}^2 + K\sigma_{\tau R}^2} \tag{9.45}$$

Under the null hypothesis $\sigma_\tau^2 = 0$:

$$\frac{E(MST|NH)}{E(MSTR)} = 1 \tag{9.46}$$

Therefore, one defines the F-statistic (replacing expected values with observed values) by

$$F_{DBM|C} = \frac{MST}{MSTR}$$

The observed value $F_{DBM|C}$ is distributed as an F-statistic with $ndf = I - 1$ and $ddf = (I - 1)(J - 1)$ (see rows labeled T and TR in Table 9.1).

$$F_{DBM|C} \sim F_{I-1,(I-1)(J-1)} \tag{9.47}$$

This is distributed as an F-statistic with $ndf = I - 1$ and $ddf = (I - 1)(J - 1)$.

$$F_{DBM|C} \sim F_{I-1,(I-1)(J-1)} \tag{9.48}$$

The null hypothesis is rejected if the observed value of the F statistic exceeds the critical value.

$$F_{DBM|C} \geq F_{1-\alpha;I-1,(I-1)(J-1)} \qquad (9.49)$$

The p-value of the test is the probability that a random sample from the distribution exceeds the observed value.

$$p = P\left(F \geq F_{DBM|C} \mid F \sim F_{I-1,(I-1)(J-1)}\right) \qquad (9.50)$$

The confidence interval for inter-treatment differences is given by

$$CI_{1-\alpha} = (\theta_{i\bullet} - \theta_{i'\bullet}) \pm t_{\alpha/2;(I-1)(K-1)}\sqrt{2\frac{MSTR}{JK}} \qquad (9.51)$$

9.9.1 Non-centrality parameter

Analogous to Equation 9.34, the non-centrality parameter is defined by

$$\Delta = \frac{JK\sigma_{\tau}^2}{\sigma_{\varepsilon}^2 + \sigma_{\tau RC}^2 + K\sigma_{\tau R}^2} \qquad (9.52)$$

Under the AH, the test statistic is distributed as a non-central F-distribution as follows:

$$F_{AH|C} \equiv \frac{MST}{MSTR} \sim F_{I-1,(I-1)(J-1),\Delta} \qquad (9.53)$$

These results are used in sample size estimation (**Chapter 11**).

9.10 DBMH analysis: Example 1, Van Dyke Data

This example analyzes the Van Dyke dataset, one of the two datasets mentioned in Section 9.1.4 that have been widely used to illustrate advances in ROC analysis. This dataset has two treatments and five readers in each treatment. Two versions of DBMH analysis are given, one is a brief version and the other is a longer version showing more details.

9.10.1 Brief version of code

The code listing Section 9.10.1.1 is the brief version, where all of the analysis is buried in the **R** package **RJafroc**. Since the latter does all the heavy lifting, the listing is blissfully short, but one has no inkling *how* it is done. To see detailed implementation of the formula, one could delve into the **RJafroc** source code (unlike most ROC software, **RJafroc** is open-source and cross-platform, but requires relatively sophisticated programming skills to negotiate) or one could look at Online Appendix 9.B, a longer listing showing straightforward applications of the relevant formula. It is highly recommended that the reader study and run the longer code **mainDBMH.R**. For now, open the file **mainDBMHBrief.R**.

9.10.1.1 Code listing

```
rm(list = ls()) #mainDBMHBrief.R
library(RJafroc)
library(ggplot2)
#fileName <- "Franken1.lrc"
fileName <- "VanDyke.lrc"
UtilOutputReport(
  fileName, format = "MRMC",
  method = "DBMH", fom = "Wilcoxon",
  showWarnings ="FALSE",
  reportFile = "VanDykeOutput.txt")
UtilOutputReport(
  fileName, format = "MRMC", reportFormat = "xlsx",
  method = "DBMH", fom = "Wilcoxon",
  showWarnings = "FALSE",
  reportFile = "VanDykeOutput.xlsx")
rocData <- DfReadDataFile(fileName, format = "MRMC")
plot14 <- PlotEmpiricalOperatingCharacteristics(
  rocData, trts = 1, rdrs = 4)
print(plot14$Plot);print(plot14$Points)
plot24 <- PlotEmpiricalOperatingCharacteristics(
  rocData, trts = 2, rdrs = 4)
print(plot24$Plot);print(plot24$Points)
```

Line 2 loads the **RJafroc** package, documentation for which is available on the CRAN website and **Online Chapter 25**. It encapsulates most of the analysis techniques described in this book, hiding technical detail from the user. In the above code it is used for its capability, via **UtilOutputReport()**, to analyze data and generate output results files in both text (lines 6–10) and Excel™ (lines 11–15) formats, to read ROC data files in different formats, **DfReadDataFile()** line 16, and plot empirical operating characteristic, lines 17 through 22. Line 5 defines **filename** containing the dataset to be analyzed, currently set to **"VanDyke.lrc"**. The name of the data file is passed in line 7 to the function **UtilOutputReport()** with the options **format = MRMC** and **fom = Wilcoxon**, which tells the function to expect a ROC text format dataset and to use the empirical AUC as the figure of merit. The option **reportFile = "VanDykeOutput. txt"** specifies the name of the text format output file. Lines 10 through 12 generate an Excel format output file **reportFormat = "xlsx"**, with name specified by **reportFile = "VanDykeOutput.xlsx"**, described below.

Insert a break point at line 16 and click **Source**. The output is listed below (debugger generated lines are not shown).

9.10.1.2 Code output

```
> debugSource(...)
The report has been saved to VanDykeOutput.txt.
The report has been saved to VanDykeOutput.xlsx.
...
```

The output, in file **VanDykeOutput.txt**, resulting from lines 6 through 10, follows the traditional format of current ROC software available from two websites.[23,24] The full text output is shown in Online Appendix 9.C. A perhaps more readable version is in file **VanDykeOutput.xlsx**, which was created by lines 11 through 15, using the **reportFormat = "xlsx"** option in the call to the function **UtilOutputReport** (). If this option were not specified, as in lines 6 through 10, the output defaults to the text format. The reason for the option **showWarnings = FALSE** is that otherwise, if the output file already exists, which it does in the online directory, the program will stop with a warning displayed in the **Console** window and ask the user **do you wish to overwrite an existing file (y/n)**.

Figure 9.2 shows the organization of the Excel file. It consists of six worksheets named **Summary, FOMs, RRRC, FRRC, RRFC**, and **ANOVA**, which makes it easy, with mouse-clicks, to navigate them. Relatively simple copy and paste operations, and minor reformatting, were used to transfer the output data from worksheets **FOMs, RRRC, FRRC**, and **RRFC** to Tables 9.2 through 9.5 that follow.

1. **Summary**: this lists the analysis date, the input file, the output file, and correspondence between reader and treatment IDs as stated in the input file and as used in the output file. Also listed are the numbers of cases, readers, the selected FOM, the significance testing method (DBMH or ORH), and the variability estimation method (jackknife, bootstrap or DeLong).
2. **FOMs**: this lists the figure of merit for each treatment reader combination, the reader-averaged FOMs for each treatment, and the difference (treatment 1 minus treatment 2, etc.) between the reader-averaged FOMs, which, for two modalities, is the observed effect size (Table 9.2). For more than two modalities the observed effect size is not listed.
3. **RRRC**: results of random-reader random-case analysis (Table 9.3).
4. **FRRC**: results of fixed-reader random-case analysis (Table 9.4).
5. **RRFC**: results of random-reader fixed-case analysis (Table 9.5).
6. **ANOVA**: Analysis of variance tables (Online Appendix 9.D).

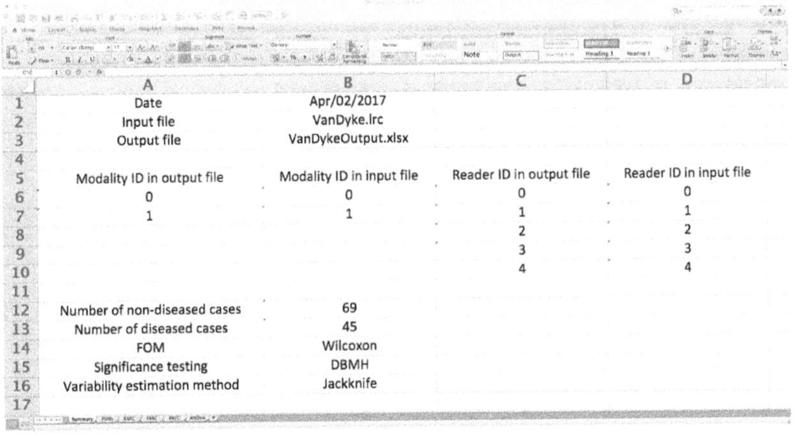

Figure 9.2 Organization of the Excel output file into six worksheets. While difficult to see in the printed version, the worksheet names are **Summary, FOMs, RRRC, FRRC, RRFC**, and **ANOVA**. Shown here is the summary worksheet. The **FOMs** worksheet provides a summary of the FOMs of the readers in the different treatments. The remaining worksheets summarize the results of the different types of analyses and the ANOVA table. The material is similar to that produced by DBM-MRMC software from the University of Iowa ROC website, but organized differently.

9.10.2 Interpreting the output: Van Dyke Data

Table 9.2 shows the figure of merit (from the **FOMs** worksheet) for each reader-treatment combination. The treatments are labeled 1 and 2, and the readers are numbered 1 through 5. For each treatment, the column labeled *Rdr. Avg.* lists the average FOM, over all readers, and the last column lists the observed effect size, that is, the difference of reader-averaged FOMs between treatment 1 and treatment 2 (specifically, 0.8970 – 0.9408 = –0.0438). Readers 1 and 4 have the highest performance, but one does not know at this point whether the differences are statistically significant. It is also evident that all readers had higher FOMs in treatment 2 than in treatment 1. Based on this concordance, chances are reasonable that the difference is significant, but as will be shown shortly, it is not.

Reader 4's very high performance in treatment 2 (0.9994) is interesting. After all, one is using the empirical AUC as the FOM, which is likely an *underestimate* of fitted AUC, so fitted AUC has to be near perfect for this observer. An easy way to check, *and highly recommended before delving into the detailed results*, is to plot the empirical ROC curves. Click on **Continue** to execute the remaining code. One sees the code output below (the column labeled **type** indicates **individual** ROC points, the other choice being **rdrAveraged**; the class column indicates the modality and reader) and Figure 9.3a and b, where 9.3a corresponds to treatment 1 and 9.3b to treatment 2. In the partial code output **genAbscissa** is a generic x-axis, FPF in the current case while **genOrdinate** is a generic y-axis, TPF in the current case. The listed operating points clearly show the steep rise of the ROC and the levelling out near TPF = 1.

9.10.2.1 Code output (partial)

```
   ...
     genAbscissa    genOrdinate      class          type
1  0.00000000       0.0000000    M-0\nR-4    individual
2  0.01449275       0.5111111    M-0\nR-4    individual
3  0.02898551       0.6222222    M-0\nR-4    individual
4  0.15942029       0.6888889    M-0\nR-4    individual
5  0.43478261       0.8444444    M-0\nR-4    individual
6  1.00000000       1.0000000    M-0\nR-4    individual

     genAbscissa    genOrdinate      class          type
1  0.0000000        0.0000000    M-1\nR-4    individual
2  0.0000000        0.5777778    M-1\nR-4    individual
3  0.0000000        0.6666667    M-1\nR-4    individual
4  0.1304348        0.8888889    M-1\nR-4    individual
5  0.6956522        0.9777778    M-1\nR-4    individual
6  1.0000000        1.0000000    M-1\nR-4    individual
```

Table 9.2 FOMs for all combinations of treatments and readers, reader-averaged FOMs for each treatment, and inter-treatment difference in reader-averaged FOMs, which is the observed effect size (treatment 1 minus treatment 2)

Trt	Rdr-1	Rdr-2	Rdr-3	Rdr-4	Rdr-5	Rdr. Avg.	Observed effect size 1–2
1	0.9196	0.8588	0.9039	0.9731	0.8298	0.8970	–0.0438
2	0.9478	0.9053	0.9217	0.9994	0.9300	0.9408	

Note: Notice that all observers performed better in treatment 1 than in treatment 2. The last column lists the observed effect size.

Table 9.3 Results of random-reader random-case (RRRC) analysis, in worksheet RRRC

F statistic	ddf	p-value
4.4563	15.2597	0.0517

95% CI's FOMs, treatment difference

Difference	Estimate	StdErr	df	t	Pr > t	Lower	Upper
1–2	−0.0438	0.0207	15.2597	−2.1110	0.0517	−0.0880	0.0004

95% CI's FOMs, each treatment

Treatment	Estimate	StdErr	df	Lower	Upper
1	0.8970	0.0332	12.7447	0.8252	0.9689
2	0.9408	0.0216	12.7102	0.8941	0.9875

Note: The inter-treatment reader-averaged difference FOM did not reach significance. *ddf* = denominator degrees of freedom of *F*-distribution. *df* = degrees of freedom of t-distribution, *Stderr* = standard error, *CI* = (1 − *alpha*) confidence interval, and *alpha* = 0.05.

For presentation and manuscript preparation purposes, the numeric values in the first two columns of Section 9.10.2.1 can be copied to one's favorite plotting program.

Table 9.3 shows the results of random-reader random-case (RRRC) analysis, in which both reader and cases are regarded as random factors. The first two lines summarize the results of the F-test, Equations 9.23 through 9.27, of the null hypothesis that the two treatments have identical FOMs. The F-statistic is 4.46 with numerator degrees of freedom equal to one (since this is always one less than the number of treatments, it is currently not shown). The denominator degrees of freedom (*ddf*) is 15.26, Equation 9.24, which yields a *p*-value of 0.0517, Equation 9.27, which means one does not reject the null hypothesis (while one came close to rejecting the NH, one either rejects or one does not). It is likely that adding one more reader to the study would have yielded a significant result.

Comments like, *the* p-*value approached significance* or *the* p-*value was slightly greater than 0.05*, that are intended to convey what the experimenter would like to report, represent reporting bias. The study is either significant of not significant.[*]

Table 9.3 lists:

1. The observed difference in reader-averaged FOMs, −0.0438, i.e., the observed effect size (with more than two treatments there would be additional treatment pairings in this part of the output).
2. The standard error[34] of the observed difference in reader-averaged FOMs: the cited reference has a readable account of the distinction between standard error and standard deviation. Basically, the first refers to a population estimate and the second to a sample.
3. The degrees of freedom of the t-distribution used to calculate the confidence interval for the reader-averaged intermodality FOM difference, Equations 9.28 and 9.31.
4. The value of the t-statistic. For two treatments, the F-statistic (4.46) equals the square of the t-statistic (−2.11).
5. The column **Pr** > **t** = 0.0517, is the *p*-value of the t-test, i.e., the probability that the observed magnitude of the t-statistic or larger can occur for identical treatments (this is different, in general, from the p-value of the overall F-test). For two treatments, not coincidentally, the t-test p-value is identical to the *p*-value of the overall F-test. For more than two treatments this equality is not true.

---

[*] The author recalls a dinner at SPIE in Florida, ca. 2000, when Prof. Kevin Berbaum made this point emphatically to the author and Dr. Federica Zanca (then a graduate student).

6. The last two numbers (−0.0880, 0.0004) are the 95% confidence interval for the observed effect size. It straddles zero, consistent with a not significant finding. Metz[35] considers the 95% confidence interval as more informative than a statement of the p-value, with which the author concurs.

The last two rows of Table 9.3 list the individual treatment FOM estimates, the corresponding standard errors, the degrees of freedom[4], and the lower and upper limits of the symmetric confidence intervals (these could stray outside the allowed limits). For example, the confidence interval for the treatment 1 FOM estimate is (0.8252, 0.9689). The cited publication has details about the degrees of freedom of the t-distributions used to calculate the individual treatment CIs.

9.10.2.2 Fixed-reader random-case analysis

Table 9.4 shows the results of fixed-reader random-case (FRRC) analysis. The p-value is 0.0210; that is, the difference is significant at alpha = 0.05. Compared to RRRC analysis, freezing reader variability decreased the p-value resulting in a significant finding. The entries in Table 9.4 are similar in format and meaning to those in Table 9.3, but, as evident from the p-value example, the content is different. Furthermore, since the reader is regarded as a fixed factor, *one can estimate the significances of inter-treatment FOM differences for individual readers*, see last 5 lines in Table 9.4. When reader was regarded as a random factor, as in the prior example, *one cannot compare inter-treatment reader differences, but one can compare inter-treatment differences for readers as a group, that is, the average reader*. When regarded as a fixed effect, reader 4, with treatment difference

Table 9.4 Results of fixed-reader random-case (FRRC) analysis, in worksheet FRRC

F statistic	ddf	p-value
5.4760	113	0.0210

95% CI's FOMs, treatment difference							
Difference	Estimate	StdErr	df	t	Pr > t	Lower	Upper
1–2	−0.0438	0.0187	113.0000	−2.3401	0.0210	−0.0809	−0.0067

95% CI's FOMs, each treatment					
Treatment	Estimate	StdErr	df	Lower	Upper
1	0.897037	0.0243	113	0.8489	0.9452
2	0.9408374	0.0168		0.9076	0.9741

95% CI's FOMs, treatment difference, each reader							
Reader	Treatment	Estimate	StdErr	df	t	Pr > t	Lower	Upper
1	1–2	−0.0282	0.0255	113	−1.1046	0.2717	−0.0787	0.0224
2		−0.0465	0.0263		−1.7694	0.0795	−0.0986	0.0056
3		−0.0179	0.0312		−0.5727	0.5680	−0.0797	0.0440
4		−0.0262	0.0173		−1.5180	0.1318	−0.0605	0.0080
5		−0.1002	0.0441		−2.2734	0.0249	−0.1874	−0.0129

Note: The effect of freezing reader variability is twofold: the p-value is significant, p-value = 0.0210. Additionally, one can compare each reader's performance in the two treatments. Only reader 5 has a significant difference in performance, p-value = 0.0249. Since the overall F-test was significant, at least one reader had to show significant difference in the two treatments, which in this example, happened to be reader 5.

−0.0262 and 95% confidence interval (−0.0605, 0.0080), which straddles zero, the treatment FOM difference for this reader is *not* significant. This does not contradict the finding that the *p*-value for the reader-averaged FOM difference *is* significant. Averaging has its benefits, especially when all readers agree that treatment 2 (cine-MRI) has superior performance for detection of thoracic aortic dissection than spin-echo MRI. Unlike reader 4, the inter-treatment FOM difference for reader 5 is significant: *p*-value = 0.0249, difference = −0.1002, and 95% confidence interval = (−0.1874, −0.0129), which does not straddle zero. Examination of Table 9.4 shows that only reader 5, who has the largest FOM difference, shows a significant inter-treatment difference. At least one reader has to have a significant inter-treatment FOM difference. After all, the result of the overall F-test that all readers haved identical FOMs in both treatments was significant, which means at least one reader, possibly more, has to show a significant inter-treatment FOM difference.

A comment is appropriate regarding reader 4. Recall this reader has the highest performance, Figure 9.3. When performance is this close to perfection it is difficult to show improvement (the difference of two AUCs, each close to unity, will be small), which may explain why this reader did not show a significant difference between the two treatments. A similar situation applies at the low end of performance, that is, AUC = 0.5. If disease is near impossible to see in either treatment, the performance difference is small and it is difficult to show significance (difficult in the sense that more resources, i.e., cases, would be needed to conduct a study with reasonable probability of showing a significant finding). This is the reason for the empirical rule that in the design of the study one should aim for an AUC roughly 0.7–0.8, that is, the cases should not be too easy or too difficult. In the author's judgment, the sweet-spot for AUC has not been well studied and represents an opportunity for research. (As long as the true difference is non-zero, one can always demonstrate a significant difference by increasing the numbers of cases, but then one gets into the issue of whether it is worth it, i.e., the issue of statistical significance vs. clinical significance, as discussed in **Chapter 11**.)

9.10.2.3 Which analysis should one conduct: RRRC or FRRC?

The question as to which analysis to conduct depends on clinical interest and the practical considerations. Ideally one wishes to report random-reader random-case analysis, but for it to be valid the readers and cases should be sampled from different institutions. By convention it is considered

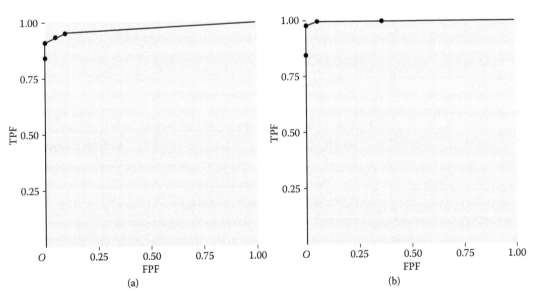

Figure 9.3 Empirical ROC plots for reader 4 in (a) treatment 1 and (b) in treatment 2. These plots are intended to demonstrate the high performance of reader 4, especially in treatment 2. The data in these plots were created by file `mainDBMHBrief.R`.

acceptable, in preliminary studies, to report RRRC results from a single institution. Since readers and cases from a single institution tend to be more homogenous and therefore not representative of true sampling variability, clinical trials must be multi-institutional. By the same token, since the purpose of a multi-institution study is to generalize to the populations of readers and cases, multi-institution studies should not report fixed reader results. A situation where fixed reader analysis is justified is if the readers happen to be special in some way, for example, they are the only ones who can interpret images from a new modality. This could occur in the early stages of the device development cycle. The type of analysis to conduct and report should be decided prior to conducting the study.

9.10.2.4 Random-reader fixed-case analysis

Table 9.5 lists the results of random-reader fixed-case (RRFC) analysis. Since case is regarded as a fixed factor, the results are specific to the dataset that was used. Row 2 column 3 shows that the p-value is 0.0420, which indicates a significant finding. The entries in Table 9.5 are similar in format and meaning to those in Table 9.3, but the content is different. Because reader is regarded as a random factor, results for individual readers are not present. One cannot compare individual levels of a random factor; all one knows is that they are sampled from some distribution. Therefore, one is only allowed conclusions regarding the statistical distribution (mean, variance) of the random factor.

9.10.2.5 When to conduct fixed-case analysis

Regarding case as a fixed factor is only justified (1) if one has very few cases, because they are difficult to acquire, and one is pressed for preliminary results or (2) the case-set is unique in some respect, which makes extrapolation to the population of cases uninteresting or implausible. A good example of the latter is studies conducted with phantoms.[36,37] In the cited studies an anthropomorphic chest phantom (Lungman N1 Multipurpose Chest Phantom, Kyoto Kagaku Company, Japan; https://www.kyotokagaku.com/products/detail03/ph-1.html) representing a 70 Kg male was loaded with simulated nodular lesions of different sizes and contrasts. This phantom allows insertion of lesions inside the chest cavity, so effects of vasculature overlap that mimic lesions are simulated. But there is only one such phantom, manufactured to strict quality control standards assuring that

Table 9.5 Results of random-reader fixed-case (RRFC) analysis, in worksheet RRFC

F statistic	ddf	p-value
8.704	4	0.0420

95% CI's FOMs, treatment difference							
Difference	Estimate	StdErr	df	t	Pr > t	Lower	Upper
1–2	−0.0438	0.0148	4	−2.9503	0.0420	−0.0850	−0.0026

95% CI's FOMs, each treatment					
Treatment	Area	StdErr	df	Lower	Upper
1	0.8970	0.0248	4	0.8281	0.9660
2	0.9408	0.0162		0.8960	0.9857

Note: The result of the overall F-test is significant, p-value = 0.0420. Since the reader is regarded as a random factor, one is not permitted to compare individual readers in the two treatments. However, one can compare the average of the readers in the two treatments and since there is only one pairing possible with two treatments, the resulting p-value is identical to that of the overall F-test.

all copies are almost identical. In this situation, the concept of extrapolating to the populations of phantoms is implausible. Therefore, with phantom studies one should only report random-reader fixed-case analysis.

9.11 DBMH analysis: Example 2, VolumeRad data

The second example uses a data file[38] that illustrates how multiple (i.e., more than one) treatment-pair comparisons are accounted for in the analysis. The code is in file **mainDBMBrief2.R**.

9.11.1 Code listing

```
rm(list = ls()) #mainDBMHBrief2.R
library(RJafroc);library(ggplot2)
fileName <- "CXRinvisible3-20mm.xlsx"
frocData <- DfReadDataFile(fileName, format = "JAFROC")
rocData <- DfFroc2Roc(frocData)
UtilOutputReport(
   dataset = rocData,
   reportFormat = "xlsx",method = "DBMH",
   fom = "Wilcoxon", showWarnings = "FALSE",
   reportFile = "VolumeRadOutput.xlsx")
plotT <- list(1, 2)
plotR <- list(c(1:5), c(1:5))
p <- PlotEmpiricalOperatingCharacteristics(
   rocData, trts = plotT, rdrs = plotR)
print(p$Plot);print(p$Points)
plotT <- list(1, 3)
plotR <- list(c(1:5), c(1:5))
p <- PlotEmpiricalOperatingCharacteristics(
   rocData, trts = plotT, rdrs = plotR)
print(p$Plot);print(p$Points)
```

The data file defined at line 3 actually contains free-response ROC (FROC) data from a recent publication[38] (FROC data, detailed in the next part of the book, consists of zero or more mark-rating pairs per case, where each mark is classified as lesion or non-lesion localization). Line 4 reads the data file and saves the results to a dataset object **frocData** (a *dataset object* is a standardized **RJafroc** variable that can accommodate datasets in any paradigm). To view it, insert a break point at line 6 in the above code, i.e., click to the left of the line number in the Edit panel, a red dot appears, and click **Source** and look at the **Environment** panel, Figure 9.4.

The dataset has four treatments and five readers; the treatments are as follows:

1 = Conventional chest x-rays (two views), abbreviated to CXR.
2 = Conventional chest x-rays plus dual energy views of the chest (dual energy views consist of a "soft-tissue" image plus a "bone" image; the quotes indicate that a clean separation is not possible); dual-energy is abbreviated to DE.
3 = Chest tomosynthesis, i.e., quasi-3D images of the chest, abbreviated to TOMO.
4 = Chest tomosynthesis plus dual energy views of the chest, abbreviated to TOMO+DE.

Since this chapter is about analyzing ROC data, line 5 converts the FROC data to a (highest rating inferred) ROC dataset object, using **DfFroc2Roc()**. Lines 6 through 10 analyze the data and generate the output Excel™ report file named **VolumeRadOutput.xlsx**. Click **Continue** to execute the rest of the code. Table 9.6 is from the worksheet named **RRRC** in the Excel™ output file.

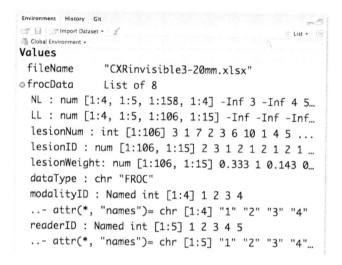

Figure 9.4 The structure of the dataset object **frocData**. It consists of four modalities, five readers, 106 diseased cases containing up to 15 CT-proven lesions per case, 52 non-diseased cases, and the maximum number of NL marks per case is 4. (LL = lesion localization, a correctly localized lesion; NL = non-lesion localization, a marked region that does not correspond to a lesion)

Table 9.6 RRRC worksheet from file **VolumeRad.xlsx**

F statistic	ddf	p-value
13.30457	70.51783	0.00000

95% CI's FOMs, treatment difference							
Difference	Estimate	StdErr	df	t	Pr > t	Lower	Upper
1 – 2	0.00196			0.08800	0.93013	−0.04244	0.04636
1 – 3	−0.10708			−4.80883	0.00001	−0.15148	−0.06267
1 – 4	−0.08806	0.02227	70.51783	−3.95494	0.00018	−0.13247	−0.04366
2 – 3	−0.10903			−4.89682	0.00001	−0.15344	−0.06463
2 – 4	−0.09002			−4.04293	0.00013	−0.13443	−0.04562
3 – 4	0.01901			0.85389	0.39606	−0.02539	0.06342

95% CI's FOMs, each treatment					
Treatment	Estimate	StdErr	df	Lower	Upper
1	0.53828	0.02084	12.64770	0.49314	0.58342
2	0.53632	0.01901	16.11862	0.49605	0.57659
3	0.64536	0.02203	572.70586	0.60209	0.68862
4	0.62634	0.02206	35.00669	0.58155	0.67113

Note: Since this was a true multi-institutional study, RRRC analysis is justified and indeed called for. Note the extremely small p-value using ROC analysis. FROC analysis yielded even smaller p-values.

The first row of the table shows that the overall p-value of the F-test is practically zero (the actual value is 5.6×10^{-7}). This means that at least two of the treatments are significantly different at alpha = 0.05. With four treatments there are six treatment-pairings ($4 \times 3 / 2 = 6$). These are listed in the middle rows of Table 9.6. Looking under column **Pr > t**, one identifies the following treatment pairings as different at the 5% level: 1–3 (meaning reader-averaged FOM of treatment 1 minus that of treatment 3), 1–4, 2–3, and 2–4. Before one can declare a specific treatment pairing as significantly different from zero, two sequential tests must be performed, one of the p-value of the overall F-test and one of the p-value of the t-test of the treatment-pair difference. If the number listed under p-value at row 1 is greater than 0.05, look no further. If the number listed under p-value is ≤ 0.05, then at least one of the treatment-pair differences is bound to be significant, that is, at least one number listed under **Pr > t** in the middle rows will be ≤ 0.05; this much is guaranteed. Only if the p-value of the overall F-test and the p-value of the t-test of the treatment pair difference are both ≤ 0.05, is the treatment-pair to be declared as significantly from zero at the specified alpha.

For a treatment-pairing FOM difference to be considered significant at Type I error probability alpha, two conditions have to be met:
1. The overall p-value of the F-test has to be smaller than alpha.
2. The p-value of the treatment-pairing difference t-test has to be smaller than alpha.
3. No further correction for multiple testing is necessary.

The very small p-value of the overall F-test deserves a few comments. Figure 9.5a shows empirical ROC curves averaged over all readers for conventional CXR (M-1) versus CXR plus dual-energy x-rays (M-2). Lines 11 through 14 generated this plot. The very small visual difference between the two curves is consistent with the values in the row labeled **1–2** in Table 9.6: the small difference in reader-averaged FOM values (0.00196), and the large p-value = 0.93. The result of almost no improvement using dual energy came as a surprise to some of the researchers participating in the study. Figure 9.5b shows empirical ROC curves averaged over all readers for digital chest tomosynthesis CXR (M-3) versus conventional CXR imaging (M-1). Lines 16 through 20 generated this plot. Here the difference is clear. The large difference, among others, is what caused the overall F-test to yield the small p-value.

The close to chance-level performance visible in Figure 9.5a deserves a comment. The lesions were deliberately chosen to be those invisible on chest x-rays; hence the name of the data file **CXRinvisible3-20mm.xlsx**. The lesions were visible on conventional CT but invisible on chest x-rays to the truth panel. Plot b shows the large improvement in performance in detecting these lesions using VolumeRad technology (http://www3.gehealthcare.com/en/products/categories/radiography/advanced_applications/volumerad).

9.12 Validation of DBMH analysis

Having described implementation of DBMH analysis, the next topic is demonstrating its validity. Analysis of MRMC ROC data usually involves testing the null hypothesis (NH) that there is no treatment effect. Application of the software to a dataset yields a binary outcome: the NH is either rejected or it is not. For a valid analysis, the probability of incorrect rejection of the NH, that is, the probability of a Type I error, should equal alpha, typically chosen to be 0.05. Demonstrating this involves simulating many MRMC ROC datasets under NH conditions and determining the observed rejection fraction. The analysis is valid if the observed rejection fraction equals, to within sampling variability, the nominal alpha of the analysis.

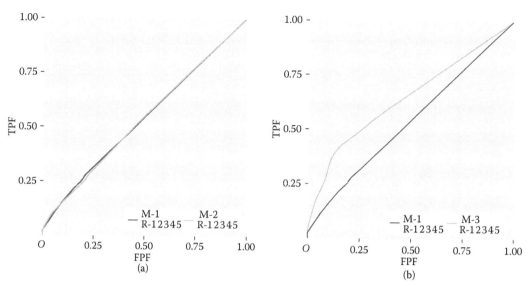

Figure 9.5 (a) shows empirical ROC curves averaged over all readers for conventional CXR (M-1) versus CXR plus dual-energy x-rays (M-2). The very small visual difference between the two treatments is consistent with the row labeled 1–2 in Table 9.6, and the p-value = 0.93; (b) shows empirical ROC curves averaged over all readers for digital chest tomosynthesis CXR (M-3) versus conventional CXR imaging (M-1). Here the difference is clear. This large difference, among others in Table 9.6, is what caused the overall F-test to yield the very small p-value. These plots were generated by **mainDBMBrief2.R**.

To start, one needs an MRMC ROC ratings data simulator with which one can repeatedly simulate independent datasets under the NH condition. Each simulated dataset, consisting of I treatments, J readers, and K cases constitutes a *trial*. The dataset is analyzed by the DBMH method and two things can happen: the NH is rejected, in which case a zero initialized counter variable is incremented by unity, or the NH is not rejected, in which case one does nothing. After $T = 2000$ trials, one divides the total number of rejections of the NH by T. This is the *observed* Type I error rate, α_{obs}. A 95% confidence interval for α_{obs} is

$$CI_{1-\alpha} = \alpha_{obs} \pm z_{\alpha/2} \sqrt{\frac{\alpha_{obs}(1 - \alpha_{obs})}{T}} \tag{9.54}$$

If α_{obs} is approximately 0.05, then the 95% confidence interval is approximately (0.04, 0.06). If the observed confidence interval includes the *nominal* α, the method has passed the validity test. The nominal value of α is that used in Equation 9.26 to calculate the p-value for each trial and one rejects the NH if $p < \alpha$. Obviously, the higher the value of T, the more stringent is the test of NH validity, but because the CI width shrinks as the inverse square root of T, one rapidly reaches a point of diminishing returns. For most statistical work $T = 2000$ is considered sufficient.

It remains to describe the algorithm used to simulate the data. A widely used algorithm was proposed in 1997 by Roe and Metz.[39]

$$Z_{ijkt} = \mu_t^z + \tau_{it}^z + R_{jt}^z + C_{kt}^z + (\tau R)_{ijt}^z + (\tau C)_{ikt}^z + (RC)_{jkt}^z + (\tau RC)_{ijk}^z \tag{9.55}$$

(To be consistent with the literature in this section the single case index notation is used.) The random terms on the right-hand side of Equation 9.55 are sampled from independent zero-mean normal distributions with variances as specified below:

$$
\left.\begin{aligned}
R_{jt}^{z} &\sim N\!\left(0,\left(\sigma_{R}^{z}\right)^{2}\right) \\[4pt]
C_{kt}^{z} &\sim N\!\left(0,\left(\sigma_{C}^{z}\right)^{2}\right) \\[4pt]
(\tau R)_{ijt}^{z} &\sim N\!\left(0,\left(\sigma_{\tau R}^{z}\right)^{2}\right) \\[4pt]
(\tau C)_{ikt}^{z} &\sim N\!\left(0,\left(\sigma_{\tau C}^{z}\right)^{2}\right) \\[4pt]
(RC)_{jkt}^{z} &\sim N\!\left(0,\sigma_{RC}^{z2}\right) \\[4pt]
(\tau RC)_{ijkt}^{z} &\sim N\!\left(0,\left(\sigma_{\tau RC}^{z}\right)^{2}\right)
\end{aligned}\right\}
\tag{9.56}
$$

The variances appearing in the right-hand side of this equation refer to Z-samples, not to pseudovalues, hence the z-superscripts. Equation 9.56 is similar to Equation 9.4 but unlike the latter equation, which models pseudovalues, the Roe and Metz simulator generates Z-samples. Another difference is that the Z-sample variance-components are regarded as known, in fact they are postulated (see below) while the pseudovalue variance components are estimated from the MRMC ratings data. (If one could estimate the Z-sample variance components from MRMC ROC ratings data, one would have a *calibrated simulator*. Currently this is an important unsolved problem. Recent progress is described in **Chapter 23**.) Furthermore, unlike terms on the right-hand side of Equation 9.4, every term on the right-hand side of Equation 9.56 has a truth subscript. This should be fairly obvious as the rating (i.e., Z-sample) of a case depends on its truth-state while the pseudovalues are calculated from FOMs, which crunch the ratings of non-diseased and diseased cases into a single number [a detail: truth state dependence of the pseudovalues is desirable, but not accommodated in the current model]. Finally, the Z-sample variance components appearing on the right-hand side of Equation 9.56 are subject to the following constraint:

$$
\left(\sigma_{C}^{z}\right)^{2}+\left(\sigma_{\tau C}^{z}\right)^{2}+\left(\sigma_{RC}^{z}\right)^{2}+\left(\sigma_{\tau RC}^{z}\right)^{2}=1
\tag{9.57}
$$

In other words, all variance components containing the case factor sum to unit. The constraint ensures that for a given reader and modality, each of the two Gaussian distributions corresponding to non-diseased and diseased cases, respectively, has unit variance. This allows $\mu_{2}^{z}-\mu_{1}^{z}$ to be interpreted as the separation of two-unit variance normal distributions defining the NH modality and $\mu_{2}^{z}-\mu_{1}^{z}+\tau_{22}^{z}$ (for two modalities, all other τ s are zero) defines the separation for the AH modality. The Roe-Metz simulator is an equal-variance binormal model Z-sample simulator.

Apart from the differences just described, the rationale for Equation 9.56 closely follows Equation 9.4. The six variances are collectively referred to as a (Z-sample) *variance structure*. Since both readers and cases are regarded as random factors, the simulator just described generates random-reader random-case data sets. Modifications needed to generate data sets for fixed-reader and fixed-case analyses are described in Online Appendix 9.F.

Roe and Metz[39] have suggested twelve combinations of AUC values and Z-sample variance structures that, based on their experience, are representative of clinical data sets, see Table 1 in Ref. 39. These values are used in the sample code in Online Appendix 9.E (function **RoeMetzVarStr()** in file **RoeMetzZSamples.R**). This is the meaning of the earlier statement that the variances appearing in Equation 9.56 are known or postulated. The combination of variance structure (1 of 12 as defined in Table 1 in Ref. 39), μ_i^z and τ_{it}^z defines the Z-sample simulator.

The code for validating the method is contained in function **mainRejectRate.R** in Online Appendix 9.E, yielding, after about an hour, the following output.

9.12.1 Code output

```
> source(...)
Var Str =  0.011 0.011 0.1 0.1 0.2 0.6 0.702 LH1
NH rejection rate over 2000 simulations:
95% CI lower = 0.04044814
nhRejectRate = 0.055
95% CI upper = 0.05955186
```

The first six numbers in the output are the variance components in the order $\left(\sigma_R^z\right)^2 = 0.011$, $\left(\sigma_{\tau R}^z\right)^2 = 0.011$, $\left(\sigma_C^z\right)^2 = 0.1$, $\left(\sigma_{\tau C}^z\right)^2 = 0.1$, $\left(\sigma_{RC}^z\right)^2 = 0.2$, and $\left(\sigma_{\tau RC}^z\right)^2 = 0.6$, the next two values are AUC = 0.702, and the *name* of the variance structure, LH1, where the first letter L means *low reader variance* and the second letter H means *high data correlation* and 1 specifies AUC = 0.702 (2 would correspond to AUC = 0.855 and 3 would correspond to AUC = 0.962). One should confirm that the variance components satisfy the constraint, Equation 9.57, that is, all terms dependent on the case factor sum to unity.

The third line lists the results of the simulated testing: the observer rejection rate α_{obs} is 0.055 and the 95% CI is (0.040, 0.060), which includes the nominal alpha $\alpha = 0.05$. Therefore, DBMH analysis passes the NH validity test, at least for this simulator. Note that there is a 1 in 20 chance (more generally the specified nominal alpha) that a valid method will fail the NH test: this is because the confidence interval, itself a random interval variable, encompasses the population value with 95% probability. *This means there is a 1 in 20 chance that the CI will not contain the nominal alpha, even though the method is valid. Such is the nature of statistics!* If an analysis never fails the NH validity test, there is something wrong with it. Of course, if the observed fraction is very different from the nominal value, then chances are slim that the method is valid. For example, if the observed rejection rate over 2000 trials is 0.07, the probability of a random excursion from an assumed true value 0.05, which is at least as large as 0.07, is very small: (**pnorm(-abs(0.07-0.05)/sigma)** = 2.031098e-05, where **sigma = sqrt(0.05*(1-0.05)/2000)**).

Of course, waiting an hour for 2000 simulations to complete is not a good use of time. An alternative version of the code, named **mainRejectRateC.R**, is included in the software, where the C stands for *cluster*. A cluster computer is basically a number of single-CPU computers (the jargon is *nodes*), which could be physically inside the same machine, running in parallel. **R** provides a support for parallel computing. Each **for**-loop is executed in parallel. If the number of nodes in the cluster is 20, then each for-loop executes 100 times, and the program keeps track of the NH rejections that occurred in each loop. The result is a reduction in execution time by a factor of about 20. The author's MacBookPro, on which this program was run, has a multiple core CPU (eight, to be exact, function **detectCores()**). It is more difficult to see what is going on in the cluster version of the code, which was the reason for presenting the longer running version first. The following output was obtained in about seven minutes, see following code output (the reader will likely not be able to reproduce the results since the seeds were randomly chosen).

9.12.2 Code output

```
> source(...)
Var Str =  0.011 0.011 0.1 0.1 0.2 0.6 0.702 LH1
Effect size = 0, NH rejection rate over 2000 simulations:
95% CI lower = 0.04775593
empAlpha = 0.058,
95% CI upper = 0.06824407
```

Note the loading of necessary packages needed for parallel computing. If these packages were not downloaded, one needs to **install** them using the methods described in Online Appendix 3.G.

Roe and Metz have published an extensive validation of DBM (the original method, without the Hillis improvements) analysis.[39] The analysis included all twelve combinations of AUC values and Z-sample variance structures and different case-set sizes (e.g., 25/25, 50/50, etc.). More recently a validation of DBMH (with the Hillis improvements) analysis has been published showing improved NH behavior, i.e., observed alphas closer to the nominal value.[40]

9.13 Meaning of pseudovalues

It is shown below that within each truth-state the pseudovalues of cases can be interpreted as individual case-level FOMs, whose fluctuations mirror the variations in the difficulty levels of the cases. Additionally, for the empirical AUC figure of merit, the average of the pseudovalues over non-diseased cases equals the empirical AUC, as does the average of the pseudovalues over diseased cases.

Since the empirical AUC represents performance *averaged over all cases*, the existence of a measure of performance that applies at the *individual case-level* is remarkable. For simplicity, and without losing generality, one considers single-treatment single-reader ROC interpretations of a set of cases. The jackknife pseudovalue $\theta_{k_t t}$ for image $k_t t$ is defined by

$$\theta_{k_t t} = K\theta - (K-1)\theta_{(k_t t)} \tag{9.58}$$

Here K is the total number of images and $\theta_{(k_t t)}$ is the FOM with image $k_t t$ removed and θ is the value when all cases are included. Examination of this equation yields the following insights.

- Removal of an easy case, which could be diseased or non-diseased, leaves behind a slightly more *difficult* case-set, which means that $\theta_{k_t t}$ will decrease, perhaps ever so slightly, relative to the baseline value. Since $\theta_{k_t t}$ appears with a *negative* sign in Eqn. 9.58, the jackknife pseudovalue $\theta_{k_t t}$ will *increase*.
- Likewise, removal of a *difficult* case leaves behind an easier case-set, and the pseudovalue $\theta_{k_t t}$ will *decrease*.
- *Pseudovalue fluctuations mirror fluctuations of difficulty levels of individual cases.*

Source the code **mainMeaningPseudoValues.R** (explained in Online Appendix 9.G) yielding the following output:

9.13.1 Code output

```
> source(...)
data file = VanDyke.lrc , modality 1, reader 1
Number of non-diseased cases = 69
number of diseased cases = = 45
counts table for pseudovalues from non-diseased cases:
-0.092    0.61   0.832   0.887   0.979
    1       2      10       9      47
counts table for ratings from non-diseased cases:
 1  2  3  4  5
47  9 10  2  1
counts table for pseudovalues from diseased cases:
-0.568   0.475   0.828   1.052   1.107
    4       1       2      10      28
counts table for ratings from diseased cases:
 1  2  3  4  5
 4  1  2 10 28
```

9.13.2 Non-diseased cases

Lines 2–10 of the output, corresponding to the non-diseased cases, is summarized in Table 9.7, which shows results for treatment 1, reader 1 for the Van Dyke data set. With five ratings there can occur at most only five unique pseudovalues. Line 6 lists the unique pseudovalues and the next line lists the number of occurrences of each unique value. For example, the unique pseudovalue –0.092 occurs once, 0.610 occurs twice, and so on. Lines 9 and 10 are analogous to lines 6 and 7, except they apply to the ratings. For example, 47 non-diseased cases were rated 1, nine were rated 2, and so on. The sums of the numbers in lines 7 and 10 are each equal to 69, the total number of *non-diseased* cases for this dataset.

Each occurrence of a 1-rating on a non-diseased case is associated with the highest pseudovalue, namely 0.979. This is because a 1 rating is a perfect decision on a non-diseased case, so this case was particularly easy. Since 47 non-diseased cases were rated 1, there are 47 occurrences of the

Table 9.7 Pseudovalue and rating occurrence tables for treatment 1 and reader 1 for *non-diseased cases* in the Van Dyke data set

Pseudovalues	−0.092	0.610	0.832	0.887	0.979	0.91965
# occurrences	1	2	10	9	47	
Ratings	1	2	3	4	5	1.5652
# occurrences	47	9	10	2	1	

Note: With five ratings there can occur at most only five unique pseudovalues. The first row lists the unique pseudovalues and the second row lists the number of occurrences of each unique value. Row 3 and 4 are analogous to rows 1 and 2, except they apply to the ratings. Note the *inverse correspondence* between the ordering of the ratings and the ordering of the pseudovalues.

highest pseudovalue. A 2 rating is the second easiest non-diseased case, and there are nine such occurrences, corresponding to which there are nine occurrences of the second highest pseudovalue (0.887), and so on.

For non-diseased cases, there is an *inverse correspondence* between the ordering of the ratings and the ordering of the pseudovalues: the *lowest* rating corresponds to the *highest* pseudovalue; the second lowest rating corresponds to the second highest pseudovalue, and so on.

It can be shown that[41] the average pseudovalue over non-diseased cases, listed in the last column of Table 9.7, namely 0.91965, equals the *empirical* AUC for this treatment reader combination.

$$\theta = \frac{1}{K_1} \sum_{k_1=1}^{K_1} \theta_{(k_1 1)} \tag{9.59}$$

In other words, multiplying the numbers in the first and second rows of Table 9.7, summing the products and dividing by 69 yields 0.91965, the empirical AUC for treatment 1 and reader 1. Since the average pseudovalue over non-diseased cases equals the FOM the pseudovalues can be interpreted as individual case FOMs. This result may appear non-intuitive because the average over *non-diseased cases* equals a FOM defined over *non-diseased and diseased cases*. However, since each non-diseased pseudovalue is defined using *all* diseased cases and *all but one* non-diseased case, the result is not as surprising as it might seem at first sight. Note that this interpretation is only possible for the empirical AUC, suggesting that the DBMH method may not generalize as well to other FOMs, at least the applicability should not be taken for granted.

A similar calculation using rows 3 and 4 in Table 9.7, corresponding to the ratings, yields a *nonsensical* result (1.565). This is consistent with the point that the author has been hammering away through this book: *the ratings do not have absolute meanings*. The author uses the term *nonsensical* deliberately. If one multiplied each rating by 2 (i.e., the allowed values would be 2, 4, 6, 8, and 10) the value in the last column in Table 9.7 computed using pseudovalues would be unaffected but the value computed using ratings would double.

9.13.3 Diseased cases

Table 9.8 shows results corresponding to Table 9.7 but for *diseased cases*. With five ratings there are again only five unique pseudovalues. However, unlike non-diseased cases, this time there is a *direct correspondence* between the ordering of the rating and the ordering of the pseudovalues:

Table 9.8 Data for treatment 1 and reader 1 for *diseased cases* of the Van Dyke data set: Ordered unique pseudovalues, row 1, the number of occurrences of each distinct pseudovalue, row 2, and corresponding values for ratings, rows 3 and 4

Pseudovalues	−0.568	0.475	0.828	1.052	1.107	0.91965
# occurrences	4	1	2	10	28	
Ratings	1	2	3	4	5	4.2667
# occurrences	4	1	2	10	28	

Note: Note the direct correspondence between the ordering of the ratings and the ordering of the pseudovalues.

the 28 occurrences of the highest pseudovalue (1.107) correspond to the 28 occurrences of the 5 rating, the easiest diseased cases, the 10 occurrences of the second highest pseudovalue (1.052) correspond to the 10 occurrences of the 4 rating, the second most easy diseased cases, and so on. The sum of the number in the second (and fourth) rows is 45, the total number of *diseased* cases. A similar result as Equation 9.59 applies; the average pseudovalue over diseased cases equals the empirical AUC.

$$\theta = \frac{1}{K_2} \sum_{k_2=1}^{K_2} \theta_{(k_2\,2)} \tag{9.60}$$

Since the average pseudovalue over diseased cases equals the FOM, the pseudovalues can be interpreted as individual case FOMs. The same operation with the ratings rows yields a nonsensical value, namely 4.2667.

The interpretation of the pseudovalues as individual case figures of merit makes the fundamental assumption of the DBM model, Equation 9.4, that the pseudovalues can be regarded as independent observations, physically meaningful, at least when the FOM is the empirical AUC. This is the reason why, in the author's opinion, the DBM method works. The ideas described here are not new; they are contained in a 1997 publication by Hanley and Hajian-Tilaki,[42] who refer to pseudovalues as "case-specific measures of case-difficulty." The cited paper gives an excellent account of pseudovalues and their relationship to the DeLong et al. structural components method of estimating variance of empirical AUC.[43]

9.14 Summary

This chapter has detailed analysis of MRMC ROC data using the DBMH method. A reason for the level of detail is that almost all of the material carries over to other data collection paradigms and a thorough understanding of the relatively simple ROC paradigm data is helpful to understanding the more complex ones.

DBMH has been used in several hundred ROC studies (Prof. Kevin Berbaum, private communication ca. 2010). While the method allows generalization of a study finding, for example, rejection of the NH, to the population of readers and cases, the author believes this is sometimes taken too literally. If a study is done at a single hospital, then the radiologists tend to be more homogenous as compared to sampling radiologists from different hospitals. This is because close interactions between radiologists at a hospital tend to homogenize reading styles and performance. A similar issue applies to patient characteristics, which are also expected to vary more between different geographical locations than within a given location served by the hospital. This means is that single hospital study based *p*-values may tend to be biased downwards, declaring differences that may not be replicable if a wider sampling net were cast. The price paid for a wider sampling net is that one must use more readers and cases to achieve the same sensitivity to genuine treatment effects, that is, statistical power (i.e., there is no free lunch).

A third MRMC ROC method, due to Clarkson, Kupinski and Barrett,[19,20] implemented in open-source JAVA software by Gallas and colleagues[22,44] (http://didsr.github.io/iMRMC/) is available on the web. Clarkson et al.[19,20] provide a probabilistic rationale for the DBM model, *provided the figure of merit is the empirical AUC*. The method is elegant but it is only applicable as long as one is using the *empirical* AUC as the figure of merit (FOM). In contrast, the DBMH approach outlined in this chapter, and the approach outlined in the following chapter, are, in principle, applicable to *any* scalar FOM. Broader applicability ensures that significance testing methods described in this, and the following chapter, apply to other ROC FOMs, such as binormal model or other fitted AUCs, and more importantly, to *other* observer performance paradigms, such as free-response ROC paradigm. An advantage of the Clarkson et al. approach is that it predicts truth-state dependence of the variance components. One knows from modeling ROC data that diseased cases tend to have greater variance than non-diseased ones, and there is no reason to suspect that similar differences do not exist between the variance components.

Testing validity of an analysis method via simulation testing is only as good as the simulator used to generate the datasets, and this is where current research is at a bottleneck. The simulator plays a central role in ROC analysis. In the author's opinion this is not widely appreciated. In contrast, simulators are taken very seriously in other disciplines, such as cosmology, high-energy physics, and weather forecasting. The simulator used to validate[3] DBMH was proposed by Roe and Metz[39] in 1997. This simulator has several shortcomings. (1) It assumes that the ratings are distributed like an equal-variance binormal model, which is not true for most clinical datasets (recall that the b-parameter of the binormal model is usually less than one). Work extending this simulator to unequal variance has been published.[3] (2) It does not take into account that some lesions are not visible, which is the basis of the contaminated binormal model (CBM). A CBM model based simulator would use equal variance distributions with the difference that the distribution for diseased cases would be a mixture distribution with two peaks. The radiological search model (RSM) of free-response data, **Chapter 17**, also implies a mixture distribution for diseased cases, and goes farther, as it predicts some case yield no Z-samples, which means they will always be rated in the lowest bin no matter how low the reporting threshold. Both CBM and RSM account for truth dependence by accounting for the underlying perceptual process. (3) The Roe-Metz simulator is out dated; the parameter values are based on datasets then available (prior to 1997). Medical imaging technology has changed substantially in the intervening decades. (4) Finally, the methodology used to arrive at the proposed parameter values is not clearly described. Needed is a more realistic simulator, incorporating knowledge from alternative ROC models and paradigms that is calibrated, by a clearly defined method, to current datasets. At the time of writing this is an open and important research topic.

Since ROC studies in medical imaging have serious health-care related consequences, no method should be used unless it has been thoroughly validated. Much work still remains to be done in proper simulator design, on which the validation is dependent.

Having discussed the DBMH method, it is time to move on to description of an alternate method that seemingly does not require pseudovalues, usage of which has been questioned by Hillis.

References

1. Dorfman DD, Berbaum KS, Metz CE. ROC characteristic rating analysis: Generalization to the population of readers and patients with the Jackknife method. *Invest Radiol.* 1992;27(9):723–731.
2. Obuchowski NA, Rockette HE. Hypothesis testing of the diagnostic accuracy for multiple diagnostic tests: An ANOVA approach with dependent observations. *Commun Stat Simul Comput.* 1995;24:285–308.
3. Hillis SL, Berbaum KS, Metz CE. Recent developments in the Dorfman-Berbaum-Metz procedure for multireader ROC study analysis. *Acad Radiol.* 2008;15(5):647–661.
4. Hillis SL. A comparison of denominator degrees of freedom methods for multiple observer ROC studies. *Stat Med.* 2007;26:596–619.
5. Hillis SL, Obuchowski NA, Schartz KM, Berbaum KS. A comparison of the Dorfman-Berbaum-Metz and Obuchowski-Rockette methods for receiver operating characteristic (ROC) data. *Stat Med.* 2005;24(10):1579–1607.
6. Swets JA, Pickett RM. *Evaluation of Diagnostic Systems: Methods from Signal Detection Theory.* 1st ed. New York, NY: Academic Press; 1982.

7. Bunch PC, Hamilton JF, Sanderson GK, Simmons AH. A free-response approach to the measurement and characterization of radiographic-observer performance. *J Appl Photogr Eng.* 1978;4:166–171.
8. Chakraborty DP, Breatnach ES, Yester MV, Soto B, Barnes GT, Fraser RG. Digital and conventional chest imaging: A modified ROC study of observer performance using simulated nodules. *Radiology.* 1986;158:35–39.
9. Niklason LT, Hickey NM, Chakraborty DP, et al. Simulated pulmonary nodules: Detection with dual-energy digital versus conventional radiography. *Radiology.* 1986;160:589–593.
10. Winer BJ, Brown DR, Michels KM. *Statistical Principles in Experimental Design.* 3 ed. New York, NY: McGraw-Hill; 1991.
11. Metz CE, Herman BA, Roe CE. Statistical comparison of two ROC-curve estimates obtained from partially-paired datasets. *Med Decis Making.* 1998;18(1):110–121.
12. Obuchowski N. Reducing the number of reader interpretations in MRMC studies. *Acad Radiol.* 2009;16:209–217.
13. Hillis SL. A marginal-mean ANOVA approach for analyzing multireader multicase radiological imaging data. *Stat Med.* 2014;33(2):330–360.
14. Toledano AY. Three methods for analyzing correlated ROC curves: A comparison in real data sets. *Stat Med.* 2003;22(18):2919–2933.
15. Ishwaran H, Gatsonis CA. A general class of hierarchical ordinal regression models with applications to correlated ROC analysis. *Can J Stat.* 2000;28(4):731–750.
16. Toledano AY, Gatsonis C. Ordinal regression methodology for ROC curves derived from correlated data. *Stat Med.* 1996;15(16):1807–1826.
17. Barrett HH, Kupinski MA, Clarkson E. Probabilistic foundations of the MRMC method. *Progr Biomed Opt Imaging Proc SPIE.* 2005;5749:21–31.
18. Clarkson E, Kupinski MA, Barrett HH. A model for MRMC AUC measurements: Theory and simulations. *SPIE Proc.* 2005.
19. Clarkson E, Kupinski MA, Barrett HH. A probabilistic model for the MRMC method, part 1: Theoretical development. *Acad Radiol.* 2006;13(11):1410–1421.
20. Kupinski MA, Clarkson E, Barrett HH. A probabilistic model for the MRMC method, part 2: Validation and applications. *Acad Radiol.* 2006;13(11):1422–1430.
21. Gallas BD, Pennello Ga, Myers KJ. Multireader multicase variance analysis for binary data. *J Opt Soc Am A Opt Image Sci Vis.* 2007;24(12):70–80.
22. Gallas BD. One-shot estimate of MRMC variance: AUC. *Acad Radiol.* 2006;13(3):353–362.
23. Berbaum KS, Metz CE, Pesce LL, Schartz KM. *DBM MRMC User's Guide, DBM-MRMC 2.1 Beta Version 2*, http://www-radiology.uchicago.edu/cgi-bin/roc_software.cgi and http://perception.radiology.uiowa.edu, Accessed December 28, 2009.
24. Metz CE. *ROCKIT Beta version*, http://www-radiology.uchicago.edu/krl/ University of Chicago, Accessed February 27, 2007; cited October 2008; 1998.
25. Larsen RJ, Marx ML. *An Introduction to Mathematical Statistics and Its Applications.* 3rd ed. Upper Saddle River, NJ: Prentice-Hall; 2001.
26. Brown H, Prescott R. *Applied Mixed Models in Medicine.* Chichester: John Wiley & Sons, Ltd; 2014.
27. Neter J, Kutner MH, Nachtsheim CJ, Wasserman W. *Applied Linear Statistical Models.* Vol. 4. Chicago, IL: Irwin; 1996.
28. Ravishanker N, Dey DK. *A First Course in Linear Model Theory.* Boca Raton, FL: Chapman & Hall/CRC; 2001.
29. Dorfman DD, Berbaum KS, Lenth RV. Multireader, multicase receiver operating characteristic methodology: A bootstrap analysis. *Acad Radiol.* 1995;2(7):626–633.
30. Scheffe H. *The Analysis of Variance.* Vol. 72. New York, NY:John Wiley & Sons Inc.;1999.
31. Satterthwaite FE. Synthesis of variance. *Psychometrika.* 1941;6(5):309–316.
32. Satterthwaite FE. An approximate distribution of estimates of variance components. *Biometrics.* 1946;2(6):110–114.

33. Roe CA, Metz CE. Variance-component modeling in the analysis of receiver operating characteristic index estimates. *Acad Radiol*. 1997;4(8):587–600.
34. Altman DG, Bland JM. Standard deviations and standard errors. *BMJ*. 2005;331(7521):903.
35. Metz CE. Quantification of failure to demonstrate statistical significance: The usefulness of confidence intervals. *Invest Radiol*. 1993;28:59–63.
36. Thompson JD, Chakraborty DP, Szczepura K, et al. Effect of reconstruction methods and x-ray tube current-time product on nodule detection in an anthropomorphic thorax phantom: A crossed-modality JAFROC observer study. *Med Phys*. 2016;43(3):1265–1274.
37. Thompson JD, Thomas NB, Manning DJ, Hogg P. The impact of grey-scale inversion on nodule detection in an anthropomorphic chest phantom: A free-response observer study. *Br J Radiol*. 2016;89(1064):20160249. doi:10.1259/bjr.20160249.
38. Dobbins JT III, McAdams HP, Sabol JM, et al. Multi-institutional evaluation of digital tomosynthesis, dual-energy radiography, and conventional chest radiography for the detection and management of pulmonary nodules. *Radiology*. 2016;282(1):236–250.
39. Roe CA, Metz CE. Dorfman-Berbaum-Metz method for statistical analysis of multireader, multimodality receiver operating characteristic data: Validation with computer simulation. *Acad Radiol*. 1997;4:298–303.
40. Hillis SL, Berbaum KS. Monte Carlo validation of the Dorfman-Berbaum-Metz method using normalized pseudovalues and less data-based model simplification. *Acad Radiol*. 2005;12(12):1534–1541.
41. Chakraborty DP, Haygood TM, Ryan J, et al. Quantifying the clinical relevance of a laboratory observer performance paradigm. *Br J Radiol*. 2012;57:2873–2904.
42. Hanley JA, Hajian-Tilaki KO. Sampling variability of nonparametric estimates of the areas under receiver operating characteristic curves: An update. *Acad Radiol*. 1997;4(1):49–58.
43. DeLong ER, DeLong DM, Clarke-Pearson DL. Comparing the areas under two or more correlated receiver operating characteristic curves: A nonparametric approach. *Biometrics*. 1988;44:837–845.
44. Gallas BD, Bandos A, Samuelson FW, Wagner RF. A framework for random-effects ROC analysis: Biases with the bootstrap and other variance estimators. *Commun Stat Theory Methods*. 2009;38(15):2586–2603.

Obuchowski–Rockette–Hillis (ORH) analysis

10.1 Introduction

The previous chapter described the Dorfman–Berbaum–Metz (DBM) significance testing procedure[1] for analyzing multiple-reader multiple-case (MRMC) receiver operating characteristic (ROC) data, along with improvements suggested by Hillis. Because the method depends on the assumption that jackknife pseudovalues can be regarded as independent and identically distributed case-level figures of merit, it has been criticized by Hillis who states that the method works but lacks firm statistical foundations.[2–4] The physicist in the author believes that if a method works there must be good reasons why it works and Section 9.13 gave a justification for why the method works, specifically, the empirical area under curve (AUC) pseudovalues qualify as case-level FOM-like quantities; this property was also noted in 1997 by Hanley and Hajian-Tilaki.[5] However, this justification only applies to the empirical AUC, so an alternate approach that does not depend on pseudovalues is desirable.

This chapter presents Hillis' preferred alternative to the Dorfman–Berbaum–Metz–Hillis (DBMH) approach. He has shown that the DBMH method can be regarded as a *working model that gives the right results*, but a method based on an earlier publication[6] by Obuchowski and Rockette, which does not depend on pseudovalues, and predicts more or less the same results, is preferable from a conceptual viewpoint. Since, besides showing the correspondence between the two methods, Hillis has made significant improvements to the original methodology, this chapter is named ORH Analysis, where ORH stands for Obuchowski, Rockette and Hillis. The ORH method has advantages in being able to handle more complex study designs[7] that are outside the scope of this book (the author acknowledges a private communication from Dr. Obuchowski, ca. 2006, that demonstrates the flexibility afforded by the OR approach) and it is possible that applications to other paradigms (e.g., the free-response ROC (FROC) paradigm uses a rather different FOM from empirical ROC-AUC) are better performed with the ORH method.

This chapter starts with a gentle introduction to the Obuchowski and Rockette method. The reason for the gentle introduction is that, in the author's opinion, the method is rather opaque to the user community (as distinct from statisticians). Part of the problem is the notation, namely lack of usage of the case-set index $\{c\}$, which while implicit to statisticians, it's absence can be confusing to those from other disciplines, for example, physics, as in the author's case. The notational issue is highlighted in a key difference of the Obuchowski and Rockette method from the DBM method, namely in how the error term is modeled by a covariance matrix. In this chapter, the structure of the covariance matrix is examined in detail, as it is key to understanding the ORH method.

In the first step of the gentle introduction a single reader interpreting a case-set in multiple treatments is modeled and the results compared to those obtained using the DBMH fixed-reader analysis described in the previous chapter. In the second step, multiple readers interpreting a case-set in multiple treatments is modeled. The two analyses, DBMH and ORH, are compared for the same dataset. The special cases of fixed-reader and fixed-case analyses are described. Single treatment analysis, where interest is in comparing average performance of readers to a fixed value, is described. Three methods of estimating the covariance matrix are described. As before, for organizational reasons illustrative **R** code is relegated to Appendices, but is essential reading.

10.2 Single-reader multiple-treatment model

Consider a single-reader providing ROC interpretations of a common case-set $\{c\}$ in multiple treatments i ($i = 1, 2, ..., I$). Before proceeding, we note that this is *not* homologous (formally equivalent) to multiple readers providing ROC interpretations in a single treatment, Section 10.7. This is because reader is a random factor while treatment is not. In the OR method the figure of merit θ is modeled as

$$\theta_{i\{c\}} = \mu + \tau_i + \varepsilon_{i\{c\}} \tag{10.1}$$

In the Obuchowski and Rockette method[6] one models the figure of merit, *not the pseudovalues*; indeed this is the key difference from the original DBM method.

Recall that $\{c\}$ denotes a set of cases. Equation 10.1 models the observed figure of merit $\theta_{i\{c\}}$ as a constant term μ plus a treatment dependent term τ_i (the treatment-effect) with the constraint,

$$\sum_{i=1}^{I} \tau_i = 0 \tag{10.2}$$

The c-index was introduced in **Chapter 7**. The left-hand side of Equation 10.1 is the figure of merit $\theta_{i\{c\}}$ for treatment i and case-set index $\{c\}$, where $c = 1, 2, ..., C$ denote different independent case-sets sampled from the population, that is, different *collections* of K_1 non-diseased and K_2 diseased cases, *not* individual cases.

This is one place the case-set index is essential for clarity, without it θ_i is a fixed quantity—the figure of merit estimate for treatment i—lacking any index allowing for variability.

Obuchowski and Rockette use a k index, defined as the *kth repetition of the study involving the same diagnostic test, reader and patient (sic)*. In the author's opinion, what is meant is a *case-set* index instead of a *repetition* index. Repeating a study with the same treatment, reader and cases yields *within-reader* variability, which is different from sampling the population of cases with new case-sets, which yields *case sampling plus within-reader* variability. As noted earlier, within-reader variability cannot be turned off and affects the interpretations of all case-sets.

Interest is in extrapolating to the population of cases and the direct way to this end is to sample different case-sets. It is shown below that usage of the case-set index interpretation yields the same results using the DBMH or the ORH methods.

Finally, and this is where newcomers to this field have difficulty understanding what is going on, there is an additive random error term $\varepsilon_{i\{c\}}$ whose sampling behavior is described by a *multivariate normal* distribution with an *I*-dimensional zero mean vector and an $I \times I$ dimensional covariance matrix Σ:

$$\varepsilon_{i\{c\}} \sim N_I\left(\vec{0}, \Sigma\right) \tag{10.3}$$

Here N_I is the *I*-variate normal distribution (i.e., each sample yields *I* random numbers). Obuchowski and Rockette assumed the following structure for the covariance matrix (they describe a more general model, but here one restricts to the simpler one):

$$\Sigma \equiv Cov\left(\varepsilon_{i\{c\}}, \varepsilon_{i'\{c\}}\right) = \begin{cases} Var & i = i' \\ Cov_1 & i \neq i' \end{cases} \tag{10.4}$$

The reason for the subscript 1 in Cov_1 will become clear when one extends this model to multiple readers. The $I \times I$ covariance matrix Σ is

$$\Sigma = \begin{pmatrix} Var & Cov_1 & \dots & Cov_1 & Cov_1 \\ Cov_1 & Var & \dots & Cov_1 & Cov_1 \\ \dots & \dots & \dots & \dots & \dots \\ Cov_1 & Cov_1 & \dots & Var & Cov_1 \\ Cov_1 & Cov_1 & \dots & Cov_1 & Var \end{pmatrix} \tag{10.5}$$

If $I = 2$ then Σ is a symmetric 2×2 matrix for which diagonal terms are the common variances in the two treatments (assumed equal to Var) and off-diagonal terms (each assumed equal to Cov_1) are the co-variances. With $I = 3$ one has a 3×3 symmetric matrix with all diagonal elements equal to Var and all off-diagonal terms are equal to Cov_1, and so on.

> *An important aspect of the Obuchowski and Rockette model is that the variances and co-variances are assumed to be treatment independent. This implies that Var estimates need to be averaged over all treatments. Likewise, Cov_1 estimates need to be averaged over all distinct treatment-treatment pairings.*

A more complex model, with more parameters and therefore more difficult to work with, would allow the variances to be treatment dependent, and the covariances to depend on the specific treatment pairings. For obvious reasons (*Occam's Razor*[*] or the law of parsimony[†]) one wishes to start with the simplest model that, one hopes, captures essential characteristics of the data.

Some elementary statistical results are presented next.

10.2.1 Definitions of covariance and correlation

The covariance of two scalar random variables *X* and *Y* is defined by[8]

$$Cov(X, Y) = E\left[\left(X - E(X)\right)\left(Y - E(Y)\right)\right] = E(XY) - E(X)E(Y) \tag{10.6}$$

---

[*] A scientific and philosophic rule that entities should not be multiplied unnecessarily, which is interpreted as requiring that the simplest of competing theories be preferred to the more complex or that explanations of unknown phenomena be sought first in terms of known quantities (Merriam-Webster dictionary).

[†] The scientific principle that things are usually connected or behave in the simplest or most economical way.

$E(X)$ is the expectation value of X, that is, the integral of x multiplied by its *pdf*:

$$E(X) = \int x \, pdf(x) dx \tag{10.7}$$

The integral is over the range of x. The covariance can be thought of as the common tandem-variation of two random variables. The variance, a special case of covariance, of X is defined by

$$Var(X) \equiv Cov(X,X) = E(X^2) - (E(X))^2 = \sigma_X^2 \tag{10.8}$$

It can be shown using the Cauchy–Schwarz inequality:[9]

$$\left| Cov(X,Y) \right|^2 \leq Var(X) Var(Y) \tag{10.9}$$

A related quantity, the correlation ρ is defined by (the $\sigma's$ are standard deviations):

$$\rho_{xy} \equiv Cor(X,Y) = \frac{Cov(X,Y)}{\sigma_X \sigma_Y} \tag{10.10}$$

According to Equation 10.9 it has the property,

$$\left| \rho_{xy} \right| \leq 1 \tag{10.11}$$

For perfect correlation, ρ_{xy} equals one and for perfect anti-correlation ρ_{xy} equals minus one. For uncorrelated variables, ρ_{xy} equals zero. Statistical independence implies zero correlation but the converse is not true.

10.2.2 Special case applicable to Equation 10.4

Assuming X and Y have the same variance,

$$Var(X) = Var(Y) = Var \equiv \sigma^2 \tag{10.12}$$

A useful theorem applicable to the OR single-reader multiple-treatment model is

$$\left.\begin{aligned} Var(X-Y) &= Var(X) + Var(Y) - 2Cov(X,Y) \\ &= 2(Var - Cov_1) \end{aligned}\right\} \tag{10.13}$$

The first line of the above equation is general, the second line specializes to the OR single-reader multiple-treatment model where the variances are equal and likewise, all covariances in Equation 10.5 are equal. The correlation ρ_1 is defined by (the reason for the subscript 1 on ρ is the same as the reason for the subscript 1 on Cov_1, which will be explained later)

$$\rho_1 = \frac{Cov_1}{Var} \tag{10.14}$$

The $I \times I$ covariance matrix Σ can be written alternatively as (shown below is the matrix for $I = 5$; as the matrix is symmetric, one need only show elements at and above the diagonal)

$$\Sigma = \begin{pmatrix} \sigma^2 & \rho_1\sigma^2 & \rho_1\sigma^2 & \rho_1\sigma^2 & \rho_1\sigma^2 \\ & \sigma^2 & \rho_1\sigma^2 & \rho_1\sigma^2 & \rho_1\sigma^2 \\ & & \sigma^2 & \rho_1\sigma^2 & \rho_1\sigma^2 \\ & & & \sigma^2 & \rho_1\sigma^2 \\ & & & & \sigma^2 \end{pmatrix} \qquad (10.15)$$

10.2.3 Estimation of the covariance matrix

An unbiased estimate of the covariance Equation 10.4 follows from

$$\Sigma_{ii'} = \frac{1}{C-1}\sum_{c=1}^{C}\left(\theta_{i\{c\}} - \theta_{i\{\bullet\}}\right)\left(\theta_{i'\{c\}} - \theta_{i'\{\bullet\}}\right) \qquad (10.16)$$

Sampling different case-sets, as required by Equation 10.16, is unrealistic and in reality, one is stuck with $C = 1$, that is, a single dataset. Therefore, direct application of this formula is impossible. However, as seen when this situation was encountered before in **Chapter 7**, one uses resampling methods to realize, for example, different bootstrap samples, which are resampling-based stand-ins for actual case-sets.[10] If B is the number of bootstraps, then the estimation formula (the subscript bs is for bootstrap)

$$\Sigma_{ii'}\big|_{bs} = \frac{1}{B-1}\sum_{b=1}^{B}\left(\theta_{i\{b\}} - \theta_{i\{\bullet\}}\right)\left(\theta_{i'\{b\}} - \theta_{i'\{\bullet\}}\right) \qquad (10.17)$$

The bootstrap method of estimating the covariance matrix, Equation 10.17, is a direct translation of Equation 10.16. Alternatively, one could have used the jackknife FOM values $\theta_{i(k)}$, that is, the figure of merit with case k removed, to estimate the covariance matrix (the subscript jk is for jackknife):

$$\Sigma_{ii'}\big|_{jk} = \frac{(K-1)^2}{K}\left[\frac{1}{K-1}\sum_{k=1}^{K}\left(\theta_{i(k)} - \theta_{i(\bullet)}\right)\left(\theta_{i'(k)} - \theta_{i'(\bullet)}\right)\right] \qquad (10.18)$$

For simplicity, in this section we depart from the usual two-subscript convention to index each case. So, k ranges from 1 to K, where the first K_1 values represent non-diseased and the following K_2 values represent diseased cases. Jackknife figure of merit values are not to be confused with jackknife pseudovalues. The jackknife FOM value corresponding to a particular case is simply the FOM with the particular case removed. Unlike pseudovalues, jackknife FOM values cannot be regarded as independent and identically distributed. Notice the use of the subscript enclosed in parenthesis (k) to denote the FOM with case k removed, that is, a single case, while in the bootstrap equation, Eqn. 10.17, one uses the curly brackets $\{b\}$ to denote the bth bootstrap case-set, that is, a whole set of K_1 non-diseased and K_2 diseased cases, sampled with replacement from the original dataset. Furthermore, the expression for the jackknife covariance contains a *variance inflation factor*:

$$\frac{(K-1)^2}{K} \qquad (10.19)$$

This factor multiplies the traditional expression for the covariance,[10] shown in square brackets in Equation 10.18. A third method of estimating the covariance, namely the DeLong et al. method,[11] applicable to the empirical AUC, is described later.

10.2.4 Meaning of the covariance matrix in Equation 10.5

Suppose one has the luxury of repeatedly sampling case-sets, each consisting of K cases from the population. A single radiologist interprets these cases in I treatments. Therefore, each case-set $\{c\}$ yields I figures of merit. The final numbers at one's disposal are $\theta_{i\{c\}}$, where $i = 1, 2, ..., I$ and $c = 1, 2, ..., C$. Considering treatment i, the variance of the FOM-values for the different case-sets $c = 1, 2, ..., C$, is an estimate of Var_i for this treatment (as usual, the dot represents an average over the replaced index):

$$\sigma_i^2 \equiv Var_i = \frac{1}{C-1} \sum_{c=1}^{C} \left(\theta_{i\{c\}} - \theta_{i\{\bullet\}} \right) \left(\theta_{i\{c\}} - \theta_{i\{\bullet\}} \right) \tag{10.20}$$

The process is repeated for all treatments and the I-variances are averaged. This is the final estimate of Var appearing in Equation 10.5.

To estimate the covariance matrix, one considers *pairs of FOM values for the same case-set* $\{c\}$ but different treatments, that is, $\theta_{i\{c\}}$ and $\theta_{i'\{c\}}$; *by definition, primed and un-primed indices are different*. Since they are derived from the same case-set, one expects the values to be correlated. For a particularly easy case-set one expects *all I*-estimates to be collectively higher than usual. The process is repeated for different case-sets and one calculates the correlation $\rho_{1;ii'}$ between the two C-length arrays $\theta_{i\{c\}}$ and $\theta_{i'\{c\}}$:

$$\rho_{1;ii'} = \frac{1}{C-1} \frac{\sum_{c=1}^{C} \left(\theta_{i\{c\}} - \theta_{i\{\bullet\}} \right) \left(\theta_{i'\{c\}} - \theta_{i'\{\bullet\}} \right)}{\sigma_i \sigma_{i'}} \tag{10.21}$$

The entire process is repeated for different treatment pairings and the resulting $I(I-1)/2$ distinct values are averaged yielding the final estimate of ρ_1 in Equation 10.15. According to Equation 10.14 one expects the covariance to be smaller than the variance determined as in the previous paragraph.

In most situations, one expects ρ_1 to be positive. There is, perhaps unlikely, a scenario that could lead to anti-correlation and negative ρ_1. This could occur, with complementary treatments, for example, CT versus MRI, where one treatment is good for bone imaging and the other for soft-tissue imaging. In this situation, what constitutes an easy case-set in one treatment could be a difficult case-set in the other treatment. The author is unaware of a practical demonstration of this expectation.

10.2.5 Code illustrating the covariance matrix

As indicated above, the covariance matrix can be estimated using the jackknife or the bootstrap. If the figure of merit is the Wilcoxon statistic, then one can also use the DeLong et al. method.[11] In **Chapter 7**, these methods were described in the context of estimating the variance of AUC. Equations 10.17 and 10.18 extend the bootstrap and the jackknife methods, respectively, to estimating the covariance of AUC (whose diagonal elements are the variances estimated in the earlier chapter). The extension of the DeLong method to covariances is described in Online Appendix 10.A and implemented in file **VarCovMtrxDLStr.R**. It has been confirmed by the author that the implementation of the DeLong method[11] in file **VarCovMtrxDLStr.R** gives *identical* results to those yielded by the SAS macro attributed to DeLong. The file name stands for *variance covariance matrix according to the DeLong structural components method* described in five unnumbered equations following Equation 4 in the cited DeLong et al. reference.

The jackknife, bootstrap, and the DeLong methods are used in file **mainVarCov1.R**, a listing and explanation of which appears in Online Appendix 10.B. **Source** the file yielding the following code output.

10.2.5.1 Code output

```
> source(...)
data file = CXRinvisible3-20mm.xlsx
number of treatments = 4
number of non-diseased cases = 52
number of diseased cases = 106
reader = 1
OR variance components using jackknife
Variance = 0.001614554
Cov1 = 0.0004970402
rho = 0.3078498
OR variance components using bootstrap
Variance = 0.001575106
Cov1 = 0.0005271459
rho = 0.3346733
OR variance components using DeLong method
Variance = 0.001600124
Cov1 = 0.0004926574
rho = 0.3078871
```

The dataset is from a recent study comparing 2D digital chest x-rays, 2D dual energy, and 3D digital chest tomosynthesis,[12] that was used to demonstrate DBMH analysis in Section 9.11. The output shows that while all three estimates of *Var* and *Cov₁* are comparable to within 10%, the DeLong and jackknife estimates are very close (to within 1%; the correlations are even closer). There is seed dependence associated with the bootstrap, but not with the jackknife (estimating sampling variability of a jackknife estimate requires other techniques[10]). For example, running the code with a different **seed** will lead to a different bootstrap estimate, but the jackknife estimate, which does not involve random sampling, rather a systematic leave-one-out procedure, is unaffected.

10.2.6 Significance testing

Why does one go through the trouble of estimating the covariance matrix? The simple reason is that it is needed for significance testing. Define the mean square corresponding to the treatment effect, denoted *MST*, by:

$$MST = \frac{1}{I-1}\sum_{i=1}^{I}(\theta_i - \theta_\bullet)^2 \tag{10.22}$$

Unlike the previous chapter, all mean square quantities defined in this chapter are based on FOMs; specifically, they are not based on pseudovalues. Converting between them is described in Refs. 2–4 and is implemented in the **RJafroc** package.

It can be shown[2] that under the null hypothesis that all treatments have identical performances the test statistic F_{1R} defined below (the 1R subscript denotes single-reader analysis) is distributed approximately as a central F-distribution with $I-1$ numerator degrees of freedom (*ndf*) and infinite denominator degrees of freedom (*ddf*), that is,

$$\left.\begin{aligned}\frac{(I-1)\,MST}{Var - Cov_1} &\sim \chi^2_{I-1} \\[2mm] F_{1R} \equiv \frac{MST}{Var - Cov_1} &\sim F_{I-1,\infty}\end{aligned}\right\} \tag{10.23}$$

(The first form is from Section 3.5 in Reference 2 with two other covariance terms zeroed out because they are multiplied by $J-1=0$. Dividing a χ^2 distributed random variable with $I-1$ degrees of freedom by $I-1$ yields an F-distributed random variable with $ndf = I-1$ and $ddf = \infty$, as in the second form in Equation 10.23. Here is an **R** example: **pf(3.1,4,Inf)** = 0.9853881; **pchisq(3.1*4,4)** = 0.9853881. The software examples show that the CDF of the F-distribution with 4 and infinite degrees of freedom at 3.1 equals the CDF of the χ^2 distribution with 4 degrees of freedom at 3.1 times 4. A little mulling over should convince the reader about the truth of these statements; the example is for $I = 5$ and $MST/(Var - Cov_1) = 3.1$.)

The p-value is the probability that a sample from the $F_{I-1,\infty}$ distribution is greater than or equal to the observed value of the test statistic, namely:

$$p \equiv P\left(F \geq F_{1R} \mid F \sim F_{I-1,\infty}\right) \tag{10.24}$$

The $(1-\alpha)$ confidence interval for the inter-treatment FOM difference is given by

$$CI_{1-\alpha} = \left(\theta_{i\bullet} - \theta_{i'\bullet}\right) \pm t_{\alpha/2;\infty}\sqrt{2\left(Var - Cov_1\right)} \tag{10.25}$$

Comparing Equation 10.25 to Equation 10.13 shows that the term $\sqrt{2\left(Var - Cov_1\right)}$ is the standard error of the inter-treatment FOM difference. This should make intuitive sense; the covariance (i.e., scaled correlation) tends to reduce the variance of the difference (recall the sail boat mast height measurement analogy from the Introduction of **Chapter 9** due to the late Dr. Wagner). The multiplier $t_{\alpha/2;\infty}$ equals 1.96. One has probably encountered the rule that a 95% confidence interval is plus or minus two standard deviations around the central value; the 2 comes from rounding up 1.96, good enough, as they say, for government work.

10.2.7 Comparing DBM to Obuchowski and Rockette for single-reader multiple-treatments

We have shown two methods for analyzing a single reader in multiple treatments and the purpose of this section is to demonstrate that they yield the same results. The DBMH method involved calculating jackknife-derived pseudovalues, Section 9.8 with $J = 1$: with a single reader one must use the fixed reader analysis described in Section 9.8. The ORH method involves estimating the covariance matrix of the error term of the FOM model, Eqn. 10.1, without recourse to calculating pseudovalues. **Source** the file **MainOrDbmh1R.R**, a listing of which appears in Online Appendix 10.C. For convenience, a few relevant lines are shown as follows.

10.2.7.1 Code listing (partial)

```
...
ret1 <- StSignificanceTesting(
   rocData,
   fom = "Wilcoxon",
   method = "DBMH",
   option = "FRRC")
cat("DBMH: F-stat = ",
    ret1$fFRRC,
    "\nddf = ", ret1$ddfFRRC,
    "\nP-val = ", ret1$pFRRC,"\n")

ret2 <- StSignificanceTesting(
   rocData,
```

```
  fom = "Wilcoxon",
  method = "ORH",
  option = "FRRC")
cat("ORH (Jackknife):   F-stat = ",
    ret2$fFRRC, "\nddf = ",
    ret2$ddfFRRC, "\nP-val = ",
    ret2$pFRRC,"\n")

ret3 <- StSignificanceTesting(
  rocData,
  fom = "Wilcoxon",
  method = "ORH",
  option = "FRRC",
  covEstMethod = "DeLong")
cat("ORH (DeLong):   F-stat = ",
    ret3$fFRRC,
    "\nddf = ", ret3$ddfFRRC, "\nP-val = ",
    ret3$pFRRC,"\n")

ret4 <- StSignificanceTesting(
  rocData,
  fom = "Wilcoxon",
  method = "ORH",
  option = "FRRC",
  covEstMethod = "Bootstrap")
cat("ORH (Bootstrap):   F-stat = ",
    ret4$fFRRC, "\nddf = ",
    ret4$ddfFRRC, "\nP-val = ",
    ret4$pFRRC,"\n")
```

The code illustrates different ways of performing the significance testing; Figure 10.1 shows the help-screen for the function **StSignificanceTesting()**. Lines 15 and 16 select reader 1 data for the four treatments in this dataset (the reader should experiment with different choices of selected reader). The first form illustrates usage of the function **SignificanceTesting()** with method specified as **DBMH** and option specified as **FRRC**, for DBMH fixed-reader random-case analysis (with one reader, as always, one must regard the reader as a fixed factor). The second form uses the same function with method specified as **ORH**, which uses the default jackknife method for estimating the covariance matrix. The third form overrides the default with **covEstMethod = DeLong**, which uses the DeLong method for estimating the covariance matrix. The last form uses the **covEstMethod = Bootstrap**, which uses the bootstrap method with 200 bootstraps, the default, which can be overridden, as shown in Figure 10.1, using option **nBoots**.

Sourcing the code yields the following output.

10.2.7.2 Code output

```
> source(...)
data file =   CXRinvisible3-20mm.xlsx
selected reader =   1
DBMH: F-stat =   2.201
ddf =   471
P-val =   0.08719
ORH (Jackknife):   F-stat =   2.201
ddf =   Inf
P-val =   0.08571
```

```
ORH (DeLong):   F-stat =   2.221
ddf =   Inf
P-val =   0.08348
ORH (Bootstrap):   F-stat =   2.18
ddf =   Inf
P-val =   0.08806
```

StSignificanceTesting (RJafroc)   R Documentation

Significance testing

Description

Performs Dorfman-Berbaum-Metz (DBM) or Obuchowski-Rockette (OR) significance testing with Hillis' improvements, for specified dataset; significance testing refers to analysis designed to assign a P-value to reject a null hypothesis (NH); the most common NH is that the reader-averaged figure of merit (FOM) difference between treatments is zero. The results of the analysis are better appreciated in the text or Excel-formatted files produced by UtilOutputReport.

Usage

StSignificanceTesting (dataset, fom = "wJAFROC", alpha = 0.05, method = "DBMH",
covEstMethod = "Jackknife", nBoots = 200, option = "ALL", varCompFlag = TRUE)

Arguments

dataset	The dataset to be analyzed, see RJafroc-package
fom	The figure of merit, default "wJAFROC", see UtilFigureOfMerit
alpha	The significance level of the test of the null hypothesis that all treatment effects are zero; the default alpha is 0.05
method	The significance testing method to be used. There are two options: "DBMH" (the default) or "ORH", representing the Dorfman-Berbaum-Metz and ther Obuchowski-Rockette significance testing methods, respectively
covEstMethod	The method used to estimate the covariance matrix in ORH analysis; it can be "Jackknife", "Bootstrap" or "DeLong", the last assumes fom = "Wilcoxon", otherwise an error results. This parameter is not relevant if the analysis method is "DBMH"
nBoots	The number of bootstraps (default is 200), relevant only if the "Bootstrap" method is used to estimate the covariance matrix in the ORH method
option	Determines which factors are regarded as random vs. fixed: "RRRC" = random-reader random case, "FRRC" = fixed-reader random case, "RRFC" = random-reader fixed case, "ALL" outputs the results of "RRRC", "FRRC" and "RRFC" analyses
varCompFlag	If TRUE, only the appropriate (DBM or OR) variance components (six in all) are returned, default is FALSE

Figure 10.1 This figure shows a screen-shot of the help page on the **StSignificanceTesting()** function. It illustrates different usages of the function illustrated in Sections 10.2.7.1 and 10.2.7.2.

The output lists the results for four methods. Listed first are the results (F-statistic, ddf and p-value) using the DBMH fixed-reader method. The next three lines list the results of the ORH method using different methods of estimating the covariance matrix: the jackknife, the DeLong and the bootstrap. The F-statistics yielded by the DBMH and the ORH/jackknife methods are identical; this is because the jackknife method was used to estimate the covariance matrix needed for the OR method and the DBMH approach always uses jackknife-derived pseudovalues. However, the degrees of freedom are different. The 471 in DBMH comes from $(I-1)(K-1) = 3 \times 157$ whereas in the OR method it is infinite; see Equations 22 and 23 in Ref. 2. Because K is generally a large number the effect of the difference in degrees of freedom on p-values is minimal. As expected, the DeLong method gives results very similar to the jackknife method, while the bootstrap yields slightly different results, due to the different method used to estimate the covariance matrix.

The demonstration should convince one that the replication index in the original Obuchowski and Rockette publication[4] is being interpreted correctly, namely as in Equation 10.1, a case-set index is needed, not a replication index.

10.3 Multiple-reader multiple-treatment ORH model

The previous sections served as a gentle introduction to the single-reader multiple-treatment Obuchowski and Rockette method. This section extends it to multiple readers interpreting a common case-set in multiple treatments (MRMC). The extension is, in principle, fairly straightforward. Compared to Equation 10.1, one needs an additional j index for readers, and additional random terms to model reader and treatment-reader variability, and the error term needs to be modified appropriately to account for the additional reader factor.

The general Obuchowski and Rockette model for fully paired multiple-reader multiple-treatment interpretations is

$$\theta_{ij\{c\}} = \mu + \tau_i + R_j + (\tau R)_{ij} + \varepsilon_{ij\{c\}} \qquad (10.26)$$

The fixed treatment effect τ_i is subject to the usual constraint, Equation 10.2. The first two terms on the right-hand side of Equation 10.26 have their usual meanings: a constant term μ representing performance averaged over treatments and readers, and a treatment effect τ_i ($i = 1, 2, ..., I$). The next two terms are, by assumption, mutually independent random samples specified as follows: R_j denotes the random treatment-independent contribution to the figure of merit of reader j ($j = 1, 2, ..., J$), modeled as a sample from a zero-mean normal distribution with variance σ_R^2; $(\tau R)_{ij}$ denotes the treatment-dependent random contribution of reader j in treatment i, modeled as a sample from a zero-mean normal distribution with variance $\sigma_{\tau R}^2$. There is a notational clash with similar variance component terms defined for the DBMH model—except in the DBMH case they applied to pseudovalues. The meaning should be clear from the context. Summarizing,

$$\left. \begin{array}{c} R_j \sim N\left(0, \sigma_R^2\right) \\[2mm] (\tau R)_{ij} \sim N\left(0, \sigma_{\tau R}^2\right) \end{array} \right\} \qquad (10.27)$$

For a single dataset $c = 1$. An estimate of μ follows from averaging over the i and j indices (the averages over the random terms yield zeroes):

$$\mu = \theta_{\bullet\bullet\{1\}} \qquad (10.28)$$

As before the dot subscript denotes an average over the replaced index. Averaging over the j index and subtracting yields an estimate of τ_i:

$$\tau_i = \theta_{i\bullet\{1\}} - \theta_{\bullet\bullet\{1\}} \qquad (10.29)$$

The τ_i estimates obey Equation 10.2. For example, with two treatments, the values of τ_i must be the negatives of each other.

The error term on the right-hand side of Equation 10.26 is more complex than the corresponding DBM model error term. Obuchowski and Rockette model this term with a multivariate normal distribution with a length $(I \times J)$ zero-mean vector and a $(I \times J) \times (I \times J)$ covariance matrix Σ. In other words,

$$\varepsilon_{ij\{c\}} \sim N_{I \times J}\left(\vec{0}, \Sigma\right) \qquad (10.30)$$

Here $N_{I \times J}$ is the $I \times J$ variate normal distribution. The covariance matrix Σ is defined by four parameters, Var, Cov_1, Cov_2, Cov_3, defined as follows:

$$Cov\left(\varepsilon_{ij\{c\}}, \varepsilon_{i'j'\{c\}}\right) = \begin{cases} Var & i = i', j = j' \\[2mm] Cov_1 & i \neq i', j = j' \\[2mm] Cov_2 & i = i', j \neq j' \\[2mm] Cov_3 & i \neq i', j \neq j' \end{cases} \qquad (10.31)$$

Apart from fixed effects, the model in Equation 10.31 contains six parameters:

$$\sigma_R^2, \sigma_{\tau R}^2, Var, Cov_1, Cov_2, Cov_3 \qquad (10.32)$$

This is the same number of variance component parameters as in the DBMH model, which should not be a surprise since one is modeling the data with equivalent models. The Obuchowski and Rockette model Equation 10.26 looks simpler because four covariance terms are hidden in the ε term. As with the singe-reader multiple-treatment model, the covariance matrix is assumed to be independent of treatment or reader, as allowing treatment and reader dependencies would greatly increase the number of parameters that would need to be estimated.

It is implicit in the Obuchowski–Rockette model that the Var, Cov_1, Cov_2, Cov_3 estimates need to be averaged over all applicable treatment-reader combinations.

10.3.1 Structure of the covariance matrix

To understand the structure of this matrix, recall that the diagonal elements of a square covariance matrix are variances and the off-diagonal elements are covariances. With two indices ij one can still imagine a square matrix where *each dimension is labeled by a pair of indices ij*. One ij pair corresponds to the horizontal direction, and the other ij pair corresponds to the vertical direction. To visualize this, consider the simpler situation of two treatments ($I = 2$) and three readers ($J = 3$). The resulting 6×6 covariance matrix would look like this:

$$\Sigma = \begin{pmatrix} (11,11) & (12,11) & (13,11) & (21,11) & (22,11) & (23,11) \\ (11,12) & (12,12) & (13,12) & (21,12) & (22,12) & (23,12) \\ (11,13) & (12,13) & (13,13) & (21,13) & (22,13) & (23,13) \\ (11,21) & (12,21) & (13,21) & (21,21) & (22,21) & (23,21) \\ (11,22) & (12,22) & (13,22) & (21,22) & (22,22) & (23,22) \\ (11,23) & (12,23) & (13,23) & (21,23) & (22,23) & (23,23) \end{pmatrix} \qquad (10.33)$$

Shown in each cell of the matrix is a pair of *ij*-values, serving as *column* indices, followed by a pair of *ij*-values serving as *row* indices, and a comma separates the pairs. For example, the first column is labeled by (11,xx), where xx depends on the row. The second column is labeled (12,xx), the third column is labeled (13,xx), and the remaining columns are successively labeled (21,xx), (22,xx), and (23,xx). Likewise, the first row is labeled by (yy,11), where yy depends on the column. The following rows are labeled (yy,12), (yy,13), (yy,21), (yy,22), and (yy,23). Note that the reader index increments faster than the treatment index.

The diagonal elements are evidently those cells where the row and column index-pairs are equal. These are (11,11), (12,12), (13,13), (21,21), (22,22), and (23,23). According to Equation 10.31 the entries in these cells would be *Var*.

$$\Sigma = \begin{pmatrix} Var & (12,11) & (13,11) & (21,11) & (22,11) & (23,11) \\ & Var & (13,12) & (21,12) & (22,12) & (23,12) \\ & & Var & (21,13) & (22,13) & (23,13) \\ & & & Var & (22,21) & (23,21) \\ & & & & Var & (23,22) \\ & & & & & Var \end{pmatrix} \tag{10.34}$$

According to Equation 10.31 the entries in cells with different treatment index pairs but identical reader index pairs would be Cov_1 (as an example, the cell (21,11) has the same reader index, namely 1, but different treatment indices, 2 and 1, so it is replaced by Cov_1):

$$\Sigma = \begin{pmatrix} Var & (12,11) & (13,11) & Cov_1 & (22,11) & (23,11) \\ & Var & (13,12) & (21,12) & Cov_1 & (23,12) \\ & & Var & (21,13) & (22,13) & Cov_1 \\ & & & Var & (22,21) & (23,21) \\ & & & & Var & (23,22) \\ & & & & & Var \end{pmatrix} \tag{10.35}$$

Similarly, the entries in cells with identical treatment index pairs but different reader index pairs would be Cov_2:

$$\Sigma = \begin{pmatrix} Var & Cov_2 & Cov_2 & Cov_1 & (22,11) & (23,11) \\ & Var & Cov_2 & (21,12) & Cov_1 & (23,12) \\ & & Var & (21,13) & (22,13) & Cov_1 \\ & & & Var & Cov_2 & Cov_2 \\ & & & & Var & Cov_2 \\ & & & & & Var \end{pmatrix} \tag{10.36}$$

Finally, the entries in cells with different treatment index pairs and different reader index pairs would be Cov_3:

$$\Sigma = \begin{pmatrix} Var & Cov_2 & Cov_2 & Cov_1 & Cov_3 & Cov_3 \\ & Var & Cov_2 & Cov_3 & Cov_1 & Cov_3 \\ & & Var & Cov_3 & Cov_3 & Cov_1 \\ & & & Var & Cov_2 & Cov_2 \\ & & & & Var & Cov_2 \\ & & & & & Var \end{pmatrix} \tag{10.37}$$

To understand these terms, consider how they might be estimated. Suppose one had the luxury of repeating the study with different case-sets, $c = 1, 2, ..., C$. Then the variance term Var can be estimated as follows:

$$Var = \left\langle \frac{1}{C-1} \sum_{c=1}^{C} \left(\theta_{ij\{c\}} - \theta_{ij\{\bullet\}} \right) \left(\theta_{ij\{c\}} - \theta_{ij\{\bullet\}} \right) \right\rangle_{ij} \tag{10.38}$$

Of course, in practice one would use the bootstrap or the jackknife as a stand-in for the c-index, but for pedagogical purposes one maintains the fiction that one has a large number of case-sets at one's disposal (not to mention the time spent by the readers interpreting them). Notice that the left-hand side of Equation 10.38 does not have treatment or reader indices. This is because implicit in the notation is averaging the observed variances over all treatments and readers, as implied by $\langle \ \rangle_{ij}$. Likewise, the covariance terms are estimated as follows:

$$Cov_1 = \left\langle \frac{1}{C-1} \sum_{c=1}^{C} \left(\theta_{ij\{c\}} - \theta_{ij\{\bullet\}} \right) \left(\theta_{i'j\{c\}} - \theta_{i'j\{\bullet\}} \right) \right\rangle_{ii'jj}$$

$$Cov_2 = \left\langle \frac{1}{C-1} \sum_{c=1}^{C} \left(\theta_{ij\{c\}} - \theta_{ij\{\bullet\}} \right) \left(\theta_{ij'\{c\}} - \theta_{ij'\{\bullet\}} \right) \right\rangle_{iijj'} \tag{10.39}$$

$$Cov_3 = \left\langle \frac{1}{C-1} \sum_{c=1}^{C} \left(\theta_{ij\{c\}} - \theta_{ij\{\bullet\}} \right) \left(\theta_{i'j'\{c\}} - \theta_{i'j'\{\bullet\}} \right) \right\rangle_{ii'jj'}$$

In Equation 10.39 the convention is that primed and unprimed variables are *always* different.

Since there are no treatment and reader dependencies on the left-hand sides of the above equations, one averages the estimates as follows:

(i) For Cov_1, one averages over all combinations of *different* treatments and *same* readers, as denoted by $\langle \ \rangle_{ii'jj}$.
(ii) For Cov_2, one averages over all combinations of *same* treatment and *different* readers, as denoted by $\langle \ \rangle_{iijj'}$.
(iii) For Cov_3, one averages over all combinations of *different* treatments and *different* readers, as denoted by $\langle \ \rangle_{ii'jj'}$.

10.3.2 Physical meanings of covariance terms

The meanings of the different terms follow similarly to those given in Section 10.2.4. The diagonal term Var of the covariance matrix Σ is the variance of the figure of merit values obtained when reader j interprets different case-sets in treatment i. Each case-set yields a number $\theta_{ij\{c\}}$ and the variance of the C numbers, averaged over the $I \times J$ treatments and readers, is Var. It captures the total variability due to varying difficulty levels of the case-sets and within-reader variability.

$\rho_{1;ii'jj}$ is the correlation of the figure of merit values obtained when the same reader j interprets a case-set in different treatment i,i'. Each case-set, starting with $c = 1$, yields two numbers $\theta_{ij\{1\}}$ and

$\theta_{i'j\{1\}}$; the process is repeated for C case-sets. The correlation of the two C-length arrays, averaged over all pairings of different treatments and same readers, is ρ_1. Because of the common contributions due to the shared reader and cases, ρ_1 will be non-zero. For large common variation, the two arrays become almost perfectly correlated, and ρ_1 will approach unity. For zero common variation, the two arrays become independent, and ρ_1 equals zero. Translating to covariances, one has $Cov_1 < Var$.

$\rho_{2;iij'}$ is the correlation of the figure of merit values obtained when different readers j, j' interpret the same case-set in the same treatment i. As before, this yields two numbers, and upon repeating over C case-sets, one has two C-length arrays, whose correlation, upon averaging over all distinct pairings of same treatments and different readers, yields ρ_2. If one assumes that common variation between different-reader same-treatment FOMs is smaller than the common variation between same-reader different-treatment FOMs, then ρ_2 will be smaller than ρ_1 *This is equivalent to stating that readers agree more with themselves on different treatments than they do with other readers on the same treatment.* Translating to covariances, one has $Cov_2 < Cov_1 < Var$.

$\rho_{3;ii'jj'}$ is the correlation of the figure of merit values obtained when different readers j, j' interpret the same case set in different treatments i, i', and so on, yielding ρ_3. This is expected to yield the smallest correlation.

Summarizing, one expects the following ordering for the terms in the covariance matrix:

$$Var \geq Cov_1 \geq Cov_2 \geq Cov_3 \qquad (10.40)$$

10.3.3 ORH random-reader random-case analysis

A model such as Equation 10.26 cannot be analyzed by standard analysis of variance (ANOVA) techniques. Because of the correlated structure of the error term a customized ANOVA is needed (in standard ANOVA models, such as those used in DBMH, the covariance matrix of the error term is diagonal with all diagonal elements equal to a common variance, represented by the epsilon term in the DBM model, Eqn. 9.4).

One starts with the null hypothesis (NH) that the true figures of merit of all treatments are identical, that is,

$$NH : \tau_i = 0 \qquad (i = 1, 2, ..., I) \qquad (10.41)$$

The analysis described next considers both readers and cases as random effects. Because of the special nature of the covariance matrix, a modified F-statistic is needed,[2-4,7] denoted F_{ORH}, defined by

$$F_{ORH} = \frac{MST}{MSTR + J \max(Cov_2 - Cov_3, 0)} \qquad (10.42)$$

Equation 10.42 incorporates Hillis' modification, which ensures that the constraint $Cov_2 \geq Cov_3$ is always obeyed and avoids negative denominators. The mean square (MS) terms are defined by (*these are calculated directly using FOM values, not pseudovalues*)

$$\left. \begin{aligned} MST &= \frac{J}{I-1} \sum_{i=1}^{I} (\theta_{i\bullet} - \theta_{\bullet\bullet})^2 \\ MSTR &= \frac{1}{(I-1)(J-1)} \sum_{i=1}^{I} \sum_{j=1}^{J} (\theta_{ij} - \theta_{i\bullet} - \theta_{\bullet j} + \theta_{\bullet\bullet})^2 \end{aligned} \right\} \qquad (10.43)$$

In their original paper[6] Obuchowski and Rockette stated that their proposed test statistic F (basically Equation 10.42 without the constraint implied by the *max* function) is distributed as an F-statistic with numerator degree of freedom $ndf = (I-1)$ and denominator degree of freedom $ddf = (I-1)(J-1)$. It turns out that the test is unduly conservative, meaning it is unusually reluctant to reject the null hypothesis.

In this connection, the author has two historical anecdotes. The late Dr. Robert F. Wagner once stated to the author (ca. 2001) that the sample-size tables published by Obuchowski,[13,14] using the unmodified version of Equation 10.42, predicted such a high number of readers and cases that he was doubtful about the chances of anyone conducting a practical ROC study.

The second story is that the author once conducted NH simulations using the Roe-Metz simulator described in the preceding chapter and the significance testing as described in the Obuchowski–Rockette paper, Ref. 6. The method did not reject the null hypothesis even once in 2000 trials! Recall that with $\alpha = 0.05$ a valid test should reject the null hypothesis about 100 ± 20 times in 2000 trials. The author recalls (ca. 2004) telling Dr. Steve Hillis about this issue, and he suggested a different value for the denominator degrees of freedom (ddf), substitution of which magically solved the problem, that is, the simulations rejected the null hypothesis about 5% of the time. The new ddf value is defined below (ndf is unchanged), with the subscript H denoting the Hillis modification:

$$ndf = I - 1 \tag{10.44}$$

$$ddf_H = \frac{\left[MSTR + \max\left(J\left(Cov_2 - Cov_3 \right), 0 \right) \right]^2}{\dfrac{[MSTR]^2}{(I-1)(J-1)}} \tag{10.45}$$

[If $Cov_2 \leq Cov_3$, which would violate the expected ordering Equation 10.40, this reduces to the expression originally suggested by Obuchowski and Rockette.] With Hillis' changes, under the null hypothesis, the observed statistic F_{ORH}, defined in Equation 10.42, is distributed as an F-statistic with $ndf = (I-1)$ and $ddf = ddf_H$ degrees of freedom:[2-4]

$$F_{ORH} \sim F_{ndf,ddf_H} \tag{10.46}$$

10.3.4 Decision rule, *p*-value, and confidence interval

The critical value of the F-statistic for rejection of the null hypothesis is $F_{1-\alpha,ndf,ddf_H}$, that is, the critical value such that fraction $(1-\alpha)$ of the area under the distribution lies to its left. From the definition of F_{ORH}, rejection of the NH is more likely if $MS(T)$ increases (the treatment effect is larger), $MS(TR)$ decreases (there is less contamination of the treatment effect by treatment-reader variability), the greater of Cov_2 or Cov_3 decreases (there is less contamination of the treatment effect by between-reader and treatment-reader variability), α increases (allowing a greater probability of Type I errors), ndf increases (the more the number of treatment pairings, the greater the chance that at least one pair will reject the NH), or ddf_H increases (this lowers the critical value of the F-statistic).

The *p*-value of the test is the probability, under the NH, that an equal or larger value of the F-statistic than F_{DBMH} could be observed by chance. In other words, it is the area under the F-distribution F_{ndf,ddf_H} that lies above the observed value F_{DBMH}:

$$p = P\left(F \geq F_{ORH} \mid F \sim F_{ndf,ddf_H} \right) \tag{10.47}$$

The $(1-\alpha)100$ percent confidence interval for $(\theta_{i\bullet} - \theta_{i'\bullet})$ is given by (the average is over the reader index; the case-set index $\{1\}$ is suppressed):

$$CI_{1-\alpha} = (\theta_{i\bullet} - \theta_{i'\bullet}) \pm t_{\alpha/2;ddf_H} \sqrt{\frac{2}{J}\left(MSTR + J \max\left(Cov_2 - Cov_3, 0\right)\right)} \qquad (10.48)$$

The next section describes special cases of ORH analysis.

10.4 Special cases

The following extends the analysis to fixed-reader, fixed-case, and single treatment analyses. The relevant results from Hillis papers[2-4] are quoted below.

10.4.1 Fixed-reader random-case (FRRC) analysis

Using the vertical bar notation $|R$ to denote that the reader is regarded as a fixed effect,[15] the appropriate F-statistic for testing the null hypothesis $NH : \tau_i = 0 \ (i = 1, 2..., I)$ is[3]

$$F_{ORH|R} = \frac{MST}{\left[Var - Cov_1 + (J-1)\max\left(Cov_2 - Cov_3, 0\right)\right]} \qquad (10.49)$$

$F_{ORH|R}$, a realization of a random variable, is distributed as an F-statistic with

$$\left.\begin{array}{l} ndf = I-1 \\[6pt] ddf = \infty \end{array}\right\} \qquad (10.50)$$

$$F_{ORH|R} \sim F_{I-1,\infty} \qquad (10.51)$$

Alternatively, as with Equation 10.23,

$$(I-1) F_{ORH|R} \sim \chi^2_{I-1} \qquad (10.52)$$

For $J = 1$, Equation 10.49 reduces to Equation 10.23.

The critical value of the statistic is $F_{1-\alpha;I-1,\infty}$ which is that value such that fraction $(1-\alpha)$ of the area under the distribution lies to the left of the critical value. The null hypothesis is rejected if the observed value of the F-statistic exceeds the critical value:

$$F_{ORH|R} > F_{1-\alpha;I-1,\infty} \qquad (10.53)$$

The p-value of the test is the probability that a random sample from the distribution $F_{I-1,\infty}$ exceeds the observed value of the F-statistic defined in Equation 10.49:

$$p = P\left(F \geq F_{ORH|R} \mid F \sim F_{I-1,\infty}\right) \qquad (10.54)$$

The $(1-\alpha)$ symmetric confidence interval for the difference figure of merit between modalities i and i is given by

$$CI_{1-\alpha} = (\theta_{i\bullet} - \theta_{i'\bullet}) \pm t_{\alpha/2;\infty} \sqrt{\frac{2}{J}\left[Var - Cov_1 + (J-1)\max\left(Cov_2 - Cov_3, 0\right)\right]} \qquad (10.55)$$

One can think of the numerator terms inside the square root, on the right-hand side of Equation 10.55, as the variance of the inter-treatment FOM difference per reader, and the division by J is needed as the readers, as a group, have smaller variance in inverse proportion to their numbers.

The NH is rejected if any of the following equivalent conditions are met:

- The observed value of the F-statistic exceeds the critical value $F_{1-\alpha;I-1,\infty}$.
- The p-value defined by Equation 10.54 is less than α.

Notice that for $J = 1$, Equation 10.55 reduces to Equation 10.25.

10.4.2 Random-reader fixed-case (RRFC) analysis

When case is treated as a fixed factor, the appropriate F-statistic for testing the null hypothesis $NH : \tau_i = 0 \; (i = 1, 2..., I)$ is

$$F_{ORH|C} = \frac{MST}{MSTR} \tag{10.56}$$

$F_{OR\,H|C}$ is distributed as an F-statistic with

$$\left.\begin{array}{l} ndf = I - 1 \\ ddf = (I-1)(J-1) \end{array}\right\} \tag{10.57}$$

In other words,

$$F_{ORH|C} \sim F_{I-1,(I-1)(J-1)} \tag{10.58}$$

The critical value of the statistic is $F_{1-\alpha;I-1,(I-1)(J-1)}$, which is that value such that fraction $(1-\alpha)$ of the distribution lies to the left of the critical value. The null hypothesis is rejected if the observed value of the F statistic exceeds the critical value:

$$F_{ORH|C} > F_{1-\alpha;I-1,(I-1)(J-1)} \tag{10.59}$$

The p-value of the test is the probability that a random sample from the distribution exceeds the observed value:

$$p = P\left(F > F_{ORH|C} \mid F \sim F_{(I-1),(I-1)(J-1)}\right) \tag{10.60}$$

The $(1-\alpha)$ confidence interval is given by

$$CI_{1-\alpha} = (\theta_{i\bullet} - \theta_{i'\bullet}) \pm t_{\alpha/2;(I-1)(J-1)} \sqrt{\frac{2}{J} MSTR} \tag{10.61}$$

It is time to reinforce the formula with examples.

10.5 Example of ORH analysis

A minimal version of ORH analysis, but which shows all of the steps, is implemented in file **mainORH.R** listed and explained in Online Appendix 10.D (the **RJafroc** package[16] has the full implementation with more detailed output, Figure 10.2). **Source** this file to get the following partial output and Figure 10.2.

10.5.1 Code output (partial)

```
> source(...)
alpha = 0.05
data file =  CXRinvisible3-20mm.xlsx
number of treatments =  4
number of readers =  5
number of non-diseased cases =  52
number of diseased cases =  =  106

Random reader random case analysis
Hillis ddfH =  70.52
F statistic is  13.3 and critical value of F is  2.735
pvalue =  5.645e-07
For pairing 1-2
mean diff is  0.001959
and 95% CI is  -0.04244 0.04636
For pairing 1-3
mean diff is  -0.1071
and 95% CI is  -0.1515 -0.06267
For pairing 2-3
mean diff is  -0.109
and 95% CI is  -0.1534 -0.06463
For pairing 1-4
mean diff is  -0.08806
and 95% CI is  -0.1325 -0.04366
For pairing 2-4
mean diff is  -0.09002
and 95% CI is  -0.1344 -0.04562
For pairing 3-4
mean diff is  0.01901
and 95% CI is  -0.02539 0.06342
```

After listing relevant details about the data file, the numbers of treatments, readers, and cases, the results are presented in three parts: random-reader random-case, fixed-reader random-case and random-reader fixed-case. In the above partial listing only results for random-reader random-case are shown for each part, listed are *ddf* (*ndf* = 3 is implicit, as there are four treatments), the observed F-statistic, the critical value of the F-statistic, the *p*-value for testing the NH that all treatments have identical FOMs, the observed inter-treatment FOM differences, and corresponding 95% confidence intervals. As in **Chapter 9**, when more than two treatments are involved, one needs to make two sequential tests before declaring a specific inter-treatment difference as significant: the overall F-statistic has to be significant and the $(1-\alpha)$ confidence interval for the specific inter-treatment difference must not include zero. The *p*-value for the inter-treatment difference, which is calculated by **RJafroc**, is not included in the minimal implementation.

As described in Section 9.11, for this dataset[12] treatment M-1 refers to 2-view digital chest x-rays (CXR) with a flat-panel detector. Treatment M-2 refers to CXR + dual energy images (DE), treatment M-3 refers to chest tomosynthesis images (TOMO) with the GE VolumeRad device, and treatment M-4 to TOMO + DE.

For RRRC the *p*-value is 5.6×10^{-7} but for FRRC is 6.3×10^{-7}. Usually one expects freezing a component of variability to decrease the *p*-value, but when the *p*-value is already so small, rounding and other sources of variability can distort this expectation (recall that quantities calculated in the analysis, e.g., the covariance matrix, are *estimates*, each subject to a sampling error). The reason for

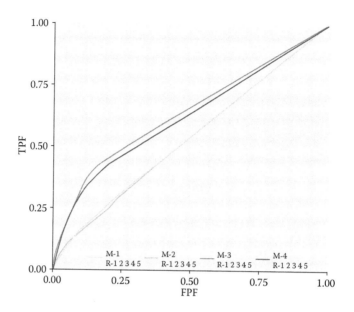

Figure 10.2 Empirical ROC plots averaged over five radiologists for the four treatments. Since the nodules were invisible on chest x-rays, performance for both M-1 (CXR) and M-2 (CXR+DE) are close to chance level. Tomosynthesis resulted in a highly significant improvement, but tomosynthesis plus dual energy yielded a slightly lower ROC compared to M-3. Differences between modalities M-3 or M-4 and M-1 or M-2 were significant. (M-1 = CXR = 2view chest x-rays, M-2 = DE = dual energy + CXR, M-3 = TOMO = GE VolumeRad tomosynthesis, and M-4 = TOMO + DE. **MainOrh.R** was used to generate this figure.)

such low p-values was described in Section 9.11. As usual, plotting is always useful to gain insights into the data. Figure 10.2 shows empirical ROC curves, averaged over readers, for the four treatments. The differences are visually obvious. With such large differences, between modalities (3 or 4) and (1 or 2), a very small p-value is expected. The plots also show that modalities 3 and 4 are not that different and the same applies to modalities 1 and 2.

The following code snippet lists the reader-averaged empirical AUCs for the four treatments (highlight **trtMeans**, line 67, and click **Run**).

10.5.2 Code snippet

```
> trtMeans
[1] 0.5382801 0.5363208 0.6453556 0.6263425
```

Since this analysis was restricted to lesions not visible on CXR (the full dataset contained lesions of different sizes 3–20 mm, all visible on CT images to independent expert radiologist truthers*), for treatment M1, performance is close to chance-level (empirical AUC = 0.538). Adding dual-energy (M2) had minimal effect (empirical AUC = 0.536, an insignificant decrease). The corresponding confidence intervals for the difference M1-M2, listed in Section 10.5.1, include zero for all three ORH analyses. The ROC plot for chest tomosynthesis is clearly above the chance diagonal

---

* Truthers is the word used by all collaborators on the project; some may cringe but this is how new words are ultimately adopted into the English Language.

(empirical AUC = 0.645), and the confidence intervals for the differences M1-M3 and M2-M3 do not include zero (they are both on the negative side of zero, because a larger FOM is being subtracted from a smaller one). Adding dual energy to chest tomosynthesis had minimal effect (empirical AUC = 0.626), so differences M1-M4 and M2-M4 are significant but difference M3-M4 is not (the reader should confirm these statements with the listed confidence intervals). Figure 10.2 was generated by the last few lines in **mainORH.R**.

10.6 Comparison of ORH and DBMH methods

Source the code in **MainOrhDbmh.R**. This code is brief since practically everything occurs inside **RJafroc**.

10.6.1 Code listing

```
rm(list = ls()) #mainOrhDbmh.R
library(RJafroc)
ROC <- FALSE
if (ROC) {
  #fileName <- "Franken1.lrc"
  fileName <- "VanDyke.lrc"
  rocData <- DfReadDataFile(
    fileName,
    format = "MRMC")
} else {
  fileName <- "CXRinvisible3-20mm.xlsx"
  frocData <- DfReadDataFile(
    fileName,
    format = "JAFROC")
  rocData <- DfFroc2Roc(frocData)
  rm(frocData)
}

UtilOutputReport(
  dataset = rocData,
  fom = "Wilcoxon",
  method = "DBMH",
  reportFormat = "xlsx",
  reportFile = "DBMH.xlsx",
  showWarnings = FALSE)
UtilOutputReport(
  dataset = rocData,
  fom = "Wilcoxon",
  method = "ORH",
  reportFormat = "xlsx",
  reportFile = "ORH.xlsx",
  showWarnings = FALSE)
```

After reading the relevant data file, currently set to **CXRinvisible3-20mm.xlsx**, and converting it to a highest rating ROC dataset object named **rocData**, the code uses the **RJafroc** package function **UtilOutputReport()** to analyze the dataset and generate a report using in the first call the DBMH method and in the second call the ORH method; see **method** option in lines 22 and 29. **Source** this file to get the following:

10.6.2 Code output

```
> source(...)
The report has been saved to DBMH.xlsx.
The report has been saved to ORH.xlsx.
```

The output is in two Excel files: **DBMH.xlsx** and **ORH.xlsx**. The reader should compare these files and be convinced that, except for minor differences in fixed-reader p-values due to the differences in degrees of freedom, they are identical. The reader should also confirm that information in **ORH.xlsx** agrees with that in Section 10.5.1, where the details of the analysis are not hidden inside **RJafroc**.

10.7 Single-treatment multiple-reader analysis

Suppose that one has data in a single treatment and multiple readers. One wishes to determine whether the performance of the readers as a group equals some specified value.

In a Section 10.2 *single-reader multiple-treatment* analysis was described. Attention now turns to *single-treatment multiple-reader* analysis and they are not the same! After all, treatment is a fixed factor while reader is a random factor, so one cannot simply use the previous analysis with reader and treatment interchanged (my graduate student tried to do just that, and he is quite smart, hence the reason for this warning; one can use the previous analysis *if* reader is regarded as a fixed factor, and a function in **RJafroc** called **StSignificanceTestingSingleFixedFactor()** does just that).

In the analysis described in this section, reader is regarded as a random factor. The average performance of the readers is estimated and compared to a specified value. Hillis[2,3,7] has described the appropriate modification of the OR model when all readers interpret the same cases in a single treatment. Two approaches are described in the cited papers, one using the DBM pseudovalue based model and the other based on the OR model with appropriate modification. The second approach is summarized below.

For single-treatment multiple-reader ORH analysis, the figure of merit model is (contrast the following equation to Equation 10.1 noting the absence of an i index; if multiple modalities are present the current analysis is applicable to data in each treatment analyzed one at a time):

$$\theta_{j\{c\}} = \mu + R_j + \varepsilon_{j\{c\}} \tag{10.62}$$

One wishes to test the NH $\mu = \mu_0$ where μ_0 is some pre-specified value. It follows from the previous equation that (since $c = 1$, in the interest of brevity, one can suppress it)

$$\theta_{\bullet} = \mu \tag{10.63}$$

The variance of the reader-averaged FOM can be shown[6] by (the reference is to the original OR publication, specifically Equation 2.3 in Ref. 6)

$$\sigma_{\theta_{\bullet}}^2 = \frac{1}{J}\left(\sigma_R^2 + Var + (J-1)Cov_2\right) \tag{10.64}$$

Connection to existing literature: rather than attempt to derive the preceding equation, it is shown how it follows from the existing literature.[6] For convenience Equation 2.3 in Ref. 6 is reproduced below.

$$Var\left(\hat{\theta}_{i\bullet\bullet}\right)=(1/J)\left[\sigma_b^2+\sigma_{ab}^2+\left(\sigma_w^2/K\right)+\sigma_c^2\left(1+[J-1]r_2\right)\right] \tag{10.65}$$

In the OR notation, the FOM has three indices, θ_{ijk}. One deletes the i index as one is dealing with a single treatment and one can drop the average over the k index, as one is dealing with a single dataset; σ_b^2 in the OR notation is what we are calling σ_R^2; for single-treatment the treatment-reader interaction term σ_{ab}^2 is absent; and for single replication the term σ_w^2/K (in OR notation K is the number of replications) is absent, or, more accurately, the within-reader variance σ_w^2 is absorbed into the case sampling variance σ_c^2 as the two are inseparable); the term σ_c^2 is what we are calling Var; and $\sigma_c^2 r_2$ in OR paper is what we are calling Cov_2.

An alternative first principles derivation, due to Mr. Xuetong Zhai, is in Online Appendix 10.E.

One needs to replace σ_R^2 in Equation 10.64 with an expected value. Again, rather than attempt to derive the following equation, it is shown how it follows from the existing literature.[7] We start with Table I in Ref. 7. This lists expected means squares for the OR model, analogous to Table 9.1 in **Chapter 9**, for the DBM model. For a single treatment (in the notation of the cited reference, $t = 1$ and the treatment-reader variance component goes away and the term σ_ε^2 is what we are calling Var), it follows that

$$E(MSR)=\sigma_R^2+Var-Cov_2 \tag{10.66}$$

Substituting Equation 10.66 in Equation 10.64 yields,

$$\sigma_{\theta\bullet}^2=\frac{1}{J}\left(E(MSR)+JCov_2\right) \tag{10.67}$$

An estimate of MSR is given by (from here on it is understood that MSR is an *estimate*, i.e., the circumflex notation is suppressed; the same is true for Cov_2)

$$MSR\equiv\widehat{MSR}=\frac{1}{J-1}\sum_{j=1}^{J}\left(\theta_j-\theta_\bullet\right)^2 \tag{10.68}$$

Replacing the expected mean-square value with the estimate and avoiding negative covariance, which could lead to a negative variance estimate, one has

$$\sigma_{\theta\bullet}^2=\frac{1}{J}\left(MSR+J\max\left(Cov_2,0\right)\right) \tag{10.69}$$

The observed value of the t-statistic (the subscript emphasizes that this statistic applies to the single treatment analysis) for testing the NH is

$$t_{I=1}\equiv\frac{\mu-\mu_0}{\sigma_{\theta\bullet}}=\left(\theta_\bullet-\mu_0\right)\sqrt{\frac{J}{\left(MSR+J\max\left(Cov_2,0\right)\right)}} \tag{10.70}$$

This is distributed as a t-statistic with $df_H^{I=1}$ degrees of freedom:

$$t_{I=1}=\frac{\mu-\mu_0}{\sigma_{\theta\bullet}}\sim t_{df_H^{I=1}} \tag{10.71}$$

In the above equation, Hillis single-treatment degree of freedom $df_H^{I=1}$ is defined by[7]

$$df_H^{I=1} = \left[\frac{MSR + J\max\left(\widehat{JCov_2},0\right)}{MSR}\right]^2 (J-1) \tag{10.72}$$

The p-value of the test is the probability that a random sample from the specified t-distribution exceeds the magnitude of the observed value:

$$p = P\left(t > |t_{I=1}| \,|\, t \sim t_{df_H^{I=1}}\right) \tag{10.73}$$

Therefore, a $100(1-\alpha)\%$ confidence interval for $\theta_g - \mu_0$ is

$$(\theta_\bullet - \mu_0) \pm t_{\alpha/2;df_H^{I=1}} \sigma_{\theta_{i\bullet}} = (\theta_\bullet - \mu_0) \pm t_{\alpha/2;df_H^{I=1}} \sqrt{\frac{1}{J}\left(MSR + \max\left(JCov_2,0\right)\right)} \tag{10.74}$$

The single treatment method is implemented in **mainSingleTreatment.R**. The relevant code is listed in Online Appendix 10.F. **Source** the code to get the following output.

10.7.1 Code output

```
> source(...)
data file =  CXRinvisible3-20mm.xlsx
The NH is that thetaDot = mu0, where thetaDot=  0.5383
and mu0 =  0.5834
The mean FOM for the anal2zed treatment is: 0.5383
The 95 % CI for the preceding value is: ( 0.4931 , 0.5834 )
The t-statistic to test
H0: (analyzed treatment = standard) is: -2.166
and the and p-value is  0.05
The difference in rdr.avg  minus standard =  -0.04514
The 95 % CI of the preceding value is ( -0.09028 , 8.362e-07 )
```

For this dataset, the NH that reader-averaged AUC in treatment 1 equals 0.583422 (the latter value was deliberately chosen to demonstrate that when the p-value is 0.05 the 95% confidence interval touches zero) is (just) not rejected and the p-value is (just) greater than 0.05. Change the comparison value at line line 32 to 0.6 and **Source** the code. Now the NH is rejected with p-value = 0.0113 and the 95% confidence interval for the difference FOM is (−0.107, −0.017).

An application of this method to comparing the performance of a group of radiologists to stand-alone computer-aided diagnosis (CAD) on the same set of images is presented in **Chapter 22**.

10.8 Discussion/Summary

This chapter described the Obuchowski–Rockette method as modified by Hillis. It has the same number of parameters as the DBMH method described in the preceding chapter, but the model Equation 10.26 appears simpler as some terms are hidden in the structure of the error term. In this chapter the NH condition was considered. Extension to the alternative hypothesis, i.e., estimating statistical power, is deferred to Online Appendix 11. The extension is a little simpler with the DBMH model as it is a standard ANOVA model. For example, the expressions for the DBMH non-centrality parameter was readily defined in **Chapter 9**, for example, Section 9.7.4. Hillis has derived expressions allowing transformation between quantities in the two methods, and this is the approach adopted in this book and implemented in Online Appendix 11.

Online Appendix 10.A describes **R** implementation of the DeLong method for estimating the covariance matrix for empirical AUC. Since the main difficulty understanding the original OR method is conceptualizing the covariance matrix, the author has explained this at an elementary level, using a case-set index which is implicit in the original OR paper.[6] This was the reason for the gentle introduction analyzing performance of a single reader in multiple treatments. The jackknife, bootstrap, and the DeLong methods, all implemented in Online Appendix 10.B, should reinforce understanding of the covariance matrix. The DBM and ORH methods are compared for this special case in Online Appendix 10.C. A minimal implementation of the ORH method for MRMC data is given in Online Appendix 10.D, which is a literal implementation of the relevant formula. The special case of multiple readers in a single treatment is coded in Online Appendix 10.F. This will be used in **Chapter 22** where standalone CAD performance is compared to a group of radiologists interpreting the same cases.

The original publication by Dorfman, Berbaum, and Metz[1] and the subsequent publication by Obuchowski and Rockette[6] were major advances. Hillis' work showing their equivalence unified the two apparently disparate analyses, and this was a major advance. The Hillis papers, while difficult reads, are revisited by the author often.

This concludes two methods used to analyze ROC MRMC datasets. A third method, restricted to the empirical AUC, is also available.[17-20] As noted earlier, the author prefers methods that are applicable to other estimates of AUC, not just the empirical ROC area, and to other data collection paradigms.

The next chapter takes on the subject of sample size estimation using either DBMH or the ORH method.

References

1. Dorfman DD, Berbaum KS, Metz CE. ROC characteristic rating analysis: Generalization to the population of readers and patients with the Jackknife method. *Invest Radiol.* 1992;27(9):723–731.
2. Hillis SL, Obuchowski NA, Schartz KM, Berbaum KS. A comparison of the Dorfman-Berbaum-Metz and Obuchowski-Rockette methods for receiver operating characteristic (ROC) data. *Stat Med.* 2005;24(10):1579–1607.
3. Hillis SL. A comparison of denominator degrees of freedom methods for multiple observer ROC studies. *Stat Med.* 2007;26:596–619.
4. Hillis SL, Berbaum KS, Metz CE. Recent developments in the Dorfman-Berbaum-Metz procedure for multireader ROC study analysis. *Acad Radiol.* 2008;15(5):647–661.
5. Hajian-Tilaki KO, Hanley JA, Joseph L, Collet JP. Extension of receiver operating characteristic analysis to data concerning multiple signal detection tasks. *Acad Radiol.* 1997;4:222–229.
6. Obuchowski NA, Rockette HE. Hypothesis testing of the diagnostic accuracy for multiple diagnostic tests: An ANOVA approach with dependent observations. *Commun Stat Simul Comput.* 1995;24:285–308.
7. Hillis SL. A marginal-mean ANOVA approach for analyzing multireader multicase radiological imaging data. *Stat Med.* 2014;33(2):330–360.
8. Larsen RJ, Marx ML. *An Introduction to Mathematical Statistics and Its Applications.* 3rd ed. Upper Saddle River, NJ: Prentice-Hall; 2001.
9. Strang G. *Linear Algebra and Its Applications.* 4 ed. Stamford, CT: Cengage Learning; 2005.
10. Efron B, Tibshirani RJ. *An Introduction to the Bootstrap.* Vol. 57. Boca Raton, FL: Chapman & Hall/CRC; 1993.
11. DeLong ER, DeLong DM, Clarke-Pearson DL. Comparing the areas under two or more correlated receiver operating characteristic curves: A nonparametric approach. *Biometrics.* 1988;44:837–845.

12. Dobbins JT, McAdams HP, Sabol JM, et al. Multi-Institutional evaluation of digital tomosynthesis, dual-energy radiography, and conventional chest radiography for the detection and management of pulmonary nodules. *Radiology*. 2016;282(1);236–250.

13. Obuchowski NA. Sample size calculations in studies of test accuracy. *Stat Methods Med Res*. 1998;7(4):371–392.

14. Obuchowski NA. Sample size tables for receiver operating characteristic studies. *Am J Roentgenol*. 2000;175(3):603–608.

15. Roe CA, Metz CE. Variance-component modeling in the analysis of receiver operating characteristic index estimates. *Acad Radiol*. 1997;4(8):587–600.

16. Zhai X, Chakraborty D. RJafroc: Analysis of data acquired using the receiver operating characteristic paradigm and its extensions. 2015, https://cran.r-project.org/web/packages/RJafroc/.

17. Clarkson E, Kupinski MA, Barrett HH. A probabilistic model for the MRMC method, part 1: Theoretical development. *Acad Radiol*. 2006;13(11):1410–1421.

18. Kupinski MA, Clarkson E, Barrett HH. A probabilistic model for the MRMC method, part 2: Validation and applications. *Acad Radiol*. 2006;13(11):1422–1430.

19. Gallas BD. One-shot estimate of MRMC variance: AUC. *Acad Radiol*. 2006;13(3):353–362.

20. Gallas BD, Bandos A, Samuelson FW, Wagner RF. A framework for random-effects ROC analysis: Biases with the bootstrap and other variance estimators. *Commun Stat Theory Methods*. 2009;38(15):2586–2603.

11

Sample size estimation

11.1 Introduction

The question addressed in this chapter is: *how many readers and cases* (usually abbreviated to *sample size) should one employ to conduct a "well-planned" receiver operating characteristic (ROC) study?* The reasons for the quotes around well-planned will become clear below. If cost were no concern, the reply would be: *as many readers and cases as one can get.* There are other causes affecting sample-size, for example, the data collection paradigm and analysis. However, this chapter is restricted to the MRMC ROC data collection paradigm, with data analyzed by the DBMH or the substantially equivalent ORH methods described in previous chapters. *It turns out that, provided one can specify comparable effect sizes between different paradigms (i.e., in the same units, ideally AUCs under the ROC curves), the methods described in this chapter are applicable to other paradigms;* see **Chapter 19** for sample size estimation for free-response ROC (FROC) studies. *For this reason, it is important to understand the concepts of sample-size estimation in the simpler ROC context.*

For simplicity and practicality, this chapter is restricted to analysis of 2-treatment data ($I = 2$). The purpose of most imaging system assessment studies is to determine, for a given diagnostic task, whether radiologists perform better using a *new* treatment over the *conventional* treatment, and whether the difference is statistically significant. Therefore, the 2-treatment case is most commonly encountered. While it is possible to extend the methods to more than two treatments, these extensions are not clinically interesting.

Assume the figure of merit (FOM) θ is chosen to be the area AUC under the ROC curve (Empirical or fitted is immaterial, as far as the formula are concerned, so long as one sticks with a choice. However, the choice of fitted versus empirical will affect statistical power. The dependence has not been systematically studied). The statistical analysis determines the significance level of the study, that is, the probability or p-value for *incorrectly* rejecting the null hypothesis (NH) that the two θs are equal: $NH : \theta_1 = \theta_2$, where the subscripts refer to the two treatments. If the p-value is smaller than a pre-specified α, typically set at 5%, one rejects the NH and declares the treatments different at the α significance level. Statistical power is the probability of *correctly* rejecting the null hypothesis when the alternative hypothesis $AH : \theta_1 \neq \theta_2$ is true, **Chapter 8**.

The value of the *true* difference between the treatments, that is, the *true* effect size is, of course, unknown. If it were known, there would be no need to conduct an ROC study. One would simply adopt the treatment with the higher θ. Sample-size estimation involves making an *educated guess* regarding the true effect size (*ES*), called the *anticipated* ES, and denoted d. To quote:[1] "any calculation of power amounts to specification of the anticipated effect size." Increasing the anticipated effect size d will increase statistical power but may represent an unrealistic expectation of the true difference between the treatments in the sense that it overestimates the ability of technology/ radiologists to achieve this much improvement. An unduly small d might be clinically insignificant, besides requiring a very large sample-size to achieve sufficient power.

Statistical power depends on the *magnitude* of d divided by the standard deviation $\sigma(d)$ of d, that is, $D = |d| / \sigma(d)$. The sign is relevant as it determines whether the project is worth pursuing at all (see Section 11.8.4). The ratio is termed[2] Cohen's D. When this signal-to-noise-ratio-like quantity is large, statistical power is large. Reader and case variability and data correlations determine $\sigma(d)$. No matter how small the anticipated d, as long as it is finite, using sufficiently large numbers of readers and cases $\sigma(d)$ can be made sufficiently small to achieve any desired statistical power. Of course, a very small effect size may not be clinically significant. There is a difference between *statistical significance* and *clinical significance*.

What determines clinical significance? A seemingly small effect size, for example, 0.01 ROC AUC units, could be clinically significant if it applies to a large population, where the small benefit in correct decision rate is amplified by the number of patients benefiting from the new treatment/ modality. In contrast, for an orphan disease, that is, one with very low prevalence, an effect size of 0.05 might not be enough to justify the additional cost of the new treatment. The improvement might have to be 0.1 before it is worth it for a new treatment to be brought to market. One hates to monetize life and death issues, but there is no getting away from it as cost/benefit issues determine clinical significance. The arbiters of clinical significance are engineers, imaging scientists, clinicians, epidemiologists, insurance companies, and those who set government health care policies. The engineers and imaging scientists determine whether the effect size the clinicians would like is feasible from technical and scientific viewpoints. The clinician determines, based on incidence of disease and other considerations, for example, altruistic, malpractice, cost of the new device, and insurance reimbursement, what effect size is justifiable. Cohen has suggested that d values of 0.2, 0.5, and 0.8 be considered small, medium, and large, respectively, but he has also argued against their indiscriminate usage. However, after a study is completed, clinicians often find that an effect size that biostatisticians label as small may, in certain circumstances, be clinically significant and an effect size that they label as large may in other circumstances be clinically insignificant. Clearly, this is an unsettled issue. Some suggestions on choosing a clinically significant effect size are made in Section 11.12.

Having developed a new imaging modality, the R&D team wishes to compare it to the existing standard with the short-term goal of making a submission to the FDA to allow them to perform pre-market testing of the device. The long-term goal is to commercialize the device. Assume the R&D team has optimized the device based on physical measurements, **Chapter 1**, perhaps supplemented with anecdotal feedback from clinicians based on a few images. Needed at this point is a *pilot* study. A pilot study, conducted with a relatively small and practical sample size, is intended to provide estimates of different sources of variability and correlations. It also provides an estimate of the effect size, termed the *observed effect size, d*. Based on results from the pilot study, the sample-size tools described in this chapter permit estimation of the numbers of readers and cases that will reduce $\sigma(d)$ sufficiently to achieve the desired power for the larger pivotal study. (A distinction could be made in the notation between observed and anticipated effect sizes, but it will be clear from the context. Later, it will be shown how one can make an educated guess about the anticipated effect size from an observed effect size.)

This chapter is concerned with MRMC studies that follow the *fully crossed factorial design* meaning that each reader interprets a common case-set in all treatments. Since the resulting pairings (i.e., correlations) tend to decrease $\sigma(d)$ (since the variations occur in tandem, they tend to cancel out in the difference; see the Introduction in **Chapter 9** for Dr. Robert Wagner's sailboat analogy), it yields more statistical power compared to an unpaired design, and consequently this design is frequently used. Two sample-size estimation procedures for MRMC are based on the Hillis-Berbaum method[3] and the Obuchowski–Rockette[4-6] method. With recent work by Hillis, the two methods have been shown to be substantially equivalent.

Table 11.1 Outcome of a ROC study: Two types of errors and correct decisions at the end of a ROC study

	Decision	
Truth	Fail to reject NH	Reject NH
NH is true	$1-\alpha$	α
NH is false	β	Power $= 1-\beta$

Note: The close parallel to the fundamental 2×2 table of ROC analysis.

This chapter will focus on the DBMH approach. Since it is based on a standard ANOVA model, it is easier to extend the NH testing procedure described in **Chapter 9** to the alternative hypothesis, which is relevant for sample size estimation. Online Appendix 11.A shows how to translate the DBMH formula to the ORH method.[7] Online Appendix 11.B describes fixed-reader and Online Appendix 11.C describes fixed-case analysis in the ORH framework.

Given an effect size, and choosing this wisely is the most difficult part of the process, the method described in this chapter uses pseudovalue variance components estimated by the DBMH method to predict sample-sizes (i.e., different combinations of numbers of readers and cases) necessary to achieve a desired power.

The starting point is a brief recapitulation of the 2×2 table of hypothesis testing.

11.2 Statistical power

The concept of statistical power was introduced in **Chapter 8** but it is worth repeating. There are two possible decisions following a test of a null hypothesis (NH): reject or fail to reject the NH. Each decision is associated with a probability on an erroneous conclusion, Table 11.1. If the NH is true and one rejects it, the probability of the ensuing Type-I error is α. If the NH is false and one fails to reject it, the probability of the ensuing Type-II error is β. Statistical power is the complement of β, that is,

$$\text{Power} = 1-\beta \tag{11.1}$$

Thus, statistical power is defined as the probability of (correctly) rejecting the null hypothesis when the null hypothesis is false. Typically, one aims for $\beta = 0.2$ or less, that is, a statistical power of 80% or more. Again, like alpha = 0.05, this is a convention, and more nuanced cost-benefit considerations may cause the researcher to adopt a different value in the study design.

11.3 Sample size estimation for random-reader random-cases

The procedure is illustrated using the DBMH significance testing method. For convenience, the DBMH model is repeated below with the case-set index suppressed:

$$Y_{n(ijk)} = \mu + \tau_i + R_j + C_k + (\tau R)_{ij} + (\tau C)_{ik} + (RC)_{jk} + (\tau RC)_{ijk} + \varepsilon_{n(ijk)} \tag{11.2}$$

As usual, the treatment effects τ_i are subject to the constraint that they sum to zero. Assuming two modalities, the observed effect size is defined by

$$d = \theta_{1\bullet} - \theta_{2\bullet} \tag{11.3}$$

It is a realization of a random variable, so one has some leeway in the choice of anticipated effect size. In the significance-testing procedure described in **Chapter 9** interest was in the distribution of the F-statistic when the NH is true. For sample size estimation, one needs to know the distribution of the statistic when the NH is *false*. It was shown in Chapter 09 that then the observed F-statistic, Equation 9.35, is distributed as a non-central F-distribution $F_{ndf,ddf,\Delta}$ with non-centrality parameter Δ:

$$F_{DBMH}|AH \sim F_{ndf,ddf,\Delta} \tag{11.4}$$

The non-centrality parameter Δ was defined, Equation 9.34, by

$$\Delta = \frac{JK\sigma_{Y;\tau}^2}{\left(\sigma_{Y;\varepsilon}^2 + \sigma_{Y;\tau RC}^2\right) + K\sigma_{Y;\tau R}^2 + J\sigma_{Y;\tau C}^2} \tag{11.5}$$

To minimize confusion, this equation has been rewritten here using the subscript Y to explicitly denote pseudo-value derived quantities (in **Chapter 9** this subscript was suppressed; it is needed here because this chapter considers both DBMH and ORH methods, albeit the latter is in the Online Appendix corresponding to this chapter).

To avoid a negative denominator, Hillis suggests the following modification:

$$\Delta = \frac{JK\sigma_{Y;\tau}^2}{\left(\sigma_{Y;\varepsilon}^2 + \sigma_{Y;\tau RC}^2\right) + K\sigma_{Y;\tau R}^2 + J\max\left(\sigma_{Y;\tau C}^2, 0\right)} \tag{11.6}$$

This expression depends on three variance components, $\left(\sigma_{Y;\varepsilon}^2 + \sigma_{Y;\tau RC}^2\right)$, the two terms are inseparable, $\sigma_{Y;\tau R}^2$ and $\sigma_{Y;\tau C}^2$. The *ddf* term appearing in Equation 11.4 was defined by Equation 9.24 (this quantity does not change between NH and AH):

$$ddf \equiv ddf_H = \frac{\left[MSTR + \max\left[MSTC - MSTRC, 0\right]\right]^2}{\dfrac{MSTR^2}{(I-1)(J-1)}} \tag{11.7}$$

The mean squares in this expression can be expressed in terms of the three variance components appearing in Equation 11.6. Hillis and Berbaum[3] have derived these expressions and they will not be repeated here (Equation 4 in reference 3). **RJafroc** implements a function to calculate the mean squares, **UtilMeanSquares()**, and then *ddf* is calculated using Equation 11.7. The sample size functions in this package need only the three variance components appearing in Eqn. 11.7 (the computation of *ddf* is implemented internally).

For two treatments, since the individual treatment effects must be the negatives of each other (because they sum to zero), it is easily shown that

$$\sigma_{Y;\tau}^2 = d^2/2 \tag{11.8}$$

Therefore, for two treatments the numerator of the expression for Δ becomes $JKd^2/2$. [For three treatments, $I = 3$, assuming the reader-averaged FOMs are spaced at intervals of d, it is easily shown that the numerator of the expression for Δ becomes $2JKd^2$, which is four times the value for two treatments, meaning that power is expected to increase dramatically. For the rest of the chapter it is assumed that one is limited to two treatments.]

11.3.1 Observed versus anticipated effect size

Assuming no other similar studies have already been conducted with the treatments in question, the observed effect size, although "merely an estimate," is the best information available at the end of the pilot study regarding the value of the true effect size. From the two previous chapters one knows that the significance testing software will report not only the observed effect size, but also a 95% confidence interval for it. It will be shown later how one can use this information to make an educated guess regarding the value of the anticipated effect size.

11.3.2 A software-break from the formula

At this point, it is helpful to see what is meant using numbers and plots instead of abstract formula. Instead of a coffee-break, think of this as a software-break. Open file **mainNonCentralF.R**, a listing of which follows. Ensure that **ddf** at line 5 is set to 10.

11.3.2.1 Code listing

```
rm(list = ls()) #mainNonCentralF.R
library(ggplot2)
alpha   <- 0.05
ndf <- 1
ddf <- 10
ncpArr <- c(0,2,5,10)
fCrit <- qf(1-alpha, ndf,ddf)
rfCrit <- round(fCrit, 3)
cat("critical value for rejecting NH:", fCrit,"\n")
x <- seq(0, 20, 0.1)
for (i in 1:length(ncpArr))
{
  pdf <- df(x,ndf,ddf,ncp=ncpArr[i])
  cat("ndf = ", ndf, ", ddf = ", ddf,
      "\nncp = ", ncpArr[i], "\nprob > fCrit = ",
      1-pf(fCrit, ndf, ddf, ncp = ncpArr[i]), "\n")
  plotCurve <- data.frame(x = x, pdf = pdf)
  yMax <- 1
  fText <- paste0("F[list(1-",as.character(alpha),
                  ",", ndf, ",", ddf, ")] == ", rfCrit)
  p <- ggplot() +
    geom_line(
      data = plotCurve,
      mapping = aes(x = x, y = pdf), size = 1) +
    xlab("z")
  print(p)
  next
}
```

The relevant function is **df()**, the *pdf* or density function of the F-distribution for specified values of *ndf*, *ddf*, and *ncp* (the non-centrality parameter, i.e., Δ). This is called at line 13 and the result is saved to array *pdf*; the next 3 lines print the relevant values and lines 17–26 plot the *pdf* versus *x*. **Source** this code to get the plots, Figure 11.1a through d, and the following output.

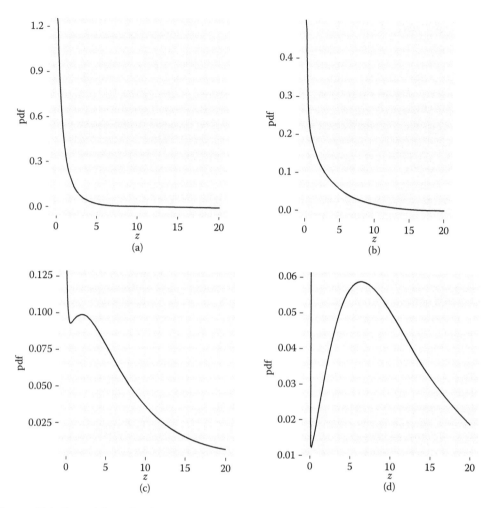

Figure 11.1 Plots of the F-distribution *pdf* for $ndf = 1$, $ddf = 10$ and different values of the non-centrality parameter *ncp*. (a): $ncp = 0$; (b): $ncp = 2$; (c): $ncp = 5$; (d): $ncp = 10$. For $ncp = 0$ the distribution is identical to the usual (i.e., central) F-distribution; as *ncp* increases the distribution shifts increasingly to the right, thereby making it more likely that a random sample will exceed a specified value. In other words, the NH is more likely to be rejected. Not visible on all plots except (c), for $ndf = 1$, the pdf at $z = 0$ is infinite.

11.3.2.2 Code output for *ddf* = 10

```
> source(...)
critical value for rejecting NH: 4.964603
ndf = 1 , ddf = 10
ncp = 0
prob > fCrit = 0.05
ndf = 1 , ddf = 10
ncp = 2
prob > fCrit = 0.2490518
ndf = 1 , ddf = 10
ncp = 5
prob > fCrit = 0.5238753
ndf = 1 , ddf = 10
ncp = 10
prob > fCrit = 0.8128068
```

Another factor to consider is the effect of *ddf*. Increasing *ddf* from 10 to 100 changes the output of **mainNonCentralF.R** to the following.

11.3.2.3 Code output for *ddf* = 100

```
> source(...)
critical value for rejecting NH: 3.936143
ndf =   1 , ddf =   100
ncp =   0
prob > fCrit =   0.05
ndf =   1 , ddf =   100
ncp =   2
prob > fCrit =   0.2883607
ndf =   1 , ddf =   100
ncp =   5
prob > fCrit =   0.6004962
ndf =   1 , ddf =   100
ncp =   10
prob > fCrit =   0.8793619
```

Increasing *ddf* by a factor of 10 decreases **fCrit** from 4.96 to 3.94, thereby *increasing* the probability that a sample from the non-central F-distribution will exceed the critical value. Therefore, power values increase compared to those shown in 11.3.2.2, but the effect is not as dramatic as that of increasing Δ.

Figure 11.1a corresponds to *ncp* = 0, which yields the usual (i.e., *central*) F-distribution that was used in the previous two chapters. The integral under this distribution is unity (this is true for all plots in Figure 11.1). The critical value for rejecting the NH denoted $F_{1-\alpha, ndf, ddf}$, **fCrit** in the code, is that value of *x* such that the probability of exceeding *x* is $\alpha = 0.05$, **fCrit** = 4.965, line 9. Notice the use of the quantile function **qf()** at line 7 to determine this value. Also, the default value of *ncp*, namely zero, is used. Specifically, line 7 does not pass a fourth argument to **qf()**. The decision rule for rejecting the NH uses the NH distribution of the F-statistic; that is, the decision rule is to reject the NH if the observed value of the F-statistic exceeds the critical value. The third line of the output lists **prob > fCrit** = 0.05 because this is how **fCrit** was defined. Plot (b) corresponds to *ncp* = 2. The distribution is noticeably shifted to the right, thereby making it more likely that the observed value of the F-statistic will exceed the critical value. This time **prob > fCrit** = 0.25, which is statistical power. Plot (c) corresponds to *ncp* = 5, **prob > fCrit** = 0.52, and statistical power is 0.52. Plot (d) corresponds to *ncp* = 10, **prob > fCrit** = 0.81, and statistical power is 0.81. The effect of the shift of the distribution is obvious in plots (c) and (d). Considering a vertical line at *x* = 4.965, 81.2% the distribution in plot (d) lies to the right of this line. Of course, what causes the shift is the large value of the non-centrality parameter. The larger that large value of the non-centrality parameter the greater the shift to the right.

11.4 Dependence of statistical power on estimates of model parameters

Examination of the expression for Δ, Equation 11.6, shows that statistical power increases if:

- *The numerator is large.* This occurs if: (a) the anticipated effect size *d* is large. Since effect size enters as the square, Equation 11.8, it has a particularly strong effect, or (b) if $J \times K$ is large.

Both of these results should be intuitive, as a large effect size and a large sample size result in increased probability of rejecting the NH.

- *The denominator is small.* The first term in the denominator is $\sigma_{Y;\varepsilon}^2 + \sigma_{Y;\tau RC}^2$. The two terms cannot be separated. This is the residual variability of the jackknife pseudovalues. It makes sense that the smaller the variability, the larger the non-centrality parameter and the statistical power.

- *The next term in the denominator is* $K\sigma_{Y;\tau R}^2$, the treatment-reader variance component multiplied by the total number of cases. The reader variance $\sigma_{Y;R}^2$ has no effect on statistical power, because it has an equal effect on both treatments and cancels in the difference. Rather, the treatment-reader component $\sigma_{Y;\tau R}^2$ contributes noise, confounding the estimate of the effect size.

- *The variance components estimated by the ANOVA procedure are realizations of random variables and, as such, subject to noise.* The presence of the K term, usually large, can amplify the effect of noise in the estimate of $\sigma_{Y;\tau R}^2$, making the sample size estimation procedure less accurate. Since the estimate of the treatment-reader variance is based on fewer samples (typically number of readers << number of cases), it is particularly susceptible to noise.

- *The final term in the denominator is* $J\sigma_{Y;\tau C}^2$. The variance $\sigma_{Y;C}^2$ has no impact on statistical power, as it cancels the difference. The treatment-case variance component introduces noise into the estimate of the effect size, thereby decreasing power. Since it is multiplied by J, the number of readers, and typically $J << K$, the error amplification is not as bad as with the treatment-reader variance component.

11.5 Formula for random-reader random-case (RRRC) sample size estimation

Having performed a *pilot* study and planning to perform a *pivotal* study, sample size estimation follows the following procedure, which assumes that both the reader and case are treated as random factors. Different formula, described later, apply when either the reader or case is treated as a fixed factor.

- Perform DBMH analysis on the pilot data. This yields the observed effect size as well as estimates of all relevant variance components and mean squares appearing in Equations 11.6 and 11.7.

- This is the difficult but critical part: make an educated guess regarding the effect size, d, that one is interested in detecting (i.e., hoping to reject the NH with probability $1 - \beta$). The author prefers the term *anticipated* effect size to *true* effect size (the latter implies knowledge of the true difference between the modalities which would obviate the need for a pivotal study). Two scenarios are considered below. In the first scenario, the effect size is assumed equal to that observed in the pilot study, that is, $d = d_{obs}$. In the second, so-called best-case scenario, one assumes that the anticipate value of d is the observed value plus two-sigma of the confidence interval, in the correct direction that is, $d = |d_{obs}| + 2\sigma$. Here σ is 1/4 the width of the 95% confidence interval for d_{obs}. Anticipating more than 2σ greater than the observed effect size would be overly optimistic. These points will become clearer when example datasets are analyzed below.

- Calculate statistical power, using the distribution implied by Equation 11.4, to calculate the probability that a random value of the relevant F-statistic will exceed the critical value, as in Section 11.3.2.

- If power is below the desired or target power, one tries successively larger value of J *and/or* K until the target power is reached.

11.6 Formula for fixed-reader random-case (FRRC) sample size estimation

It was shown in Section 9.8.2 that for fixed-reader analysis the non-centrality parameter is defined by

$$\Delta = \frac{JK\sigma_{Y;\tau}^2}{\sigma_{Y;\varepsilon}^2 + \sigma_{Y;\tau RC}^2 + J\sigma_{Y;\tau C}^2} \tag{11.9}$$

The sampling distribution of the F-statistic under the AH is

$$F_{AH|R} \equiv \frac{MST}{MSTC} \sim F_{1-1,(I-1)(K-1),\Delta} \tag{11.10}$$

11.7 Formula for random-reader fixed-case (RRFC) sample size estimation

It was shown in Section 9.9.1 that for fixed-case analysis the non-centrality parameter is defined by

$$\Delta = \frac{JK\sigma_{Y;\tau}^2}{\sigma_{Y;\varepsilon}^2 + \sigma_{Y;\tau RC}^2 + K\sigma_{Y;\tau R}^2} \tag{11.11}$$

Under the AH, the test statistic is distributed as a non-central F-distribution as follows:

$$F_{AH|C} \equiv \frac{MST}{MSTR} \sim F_{1-1,(I-1)(J-1),\Delta} \tag{11.12}$$

11.8 Example 1

The starting point, as always, is a pilot study. In the first example, the Van Dyke dataset[8] is regarded as a pilot study. Two implementations are shown. The first is an open implementation, which shows direct application of the relevant formula, including usage of the mean squares, which in principle can be calculated from the three variance components. The second implementation uses the relevant **RJafroc** function, which only needs as input the three variance components. Shown first is the open implementation. The relevant file is **mainSsDbmhOpen.R**, where **SSDbmh** means *sample size using the DBMH method*. The file listing follows.

11.8.1 Code listing

```
rm(list = ls()) #mainSsDbmhOpen.R
library(RJafroc)
alpha <- 0.05;cat("alpha = ", alpha, "\n")
fileName <- "VanDyke.lrc"
#fileName <- "Franken1.lrc"
rocData <- DfReadDataFile(
    fileName,
    format = "MRMC")
retDbm <- StSignificanceTesting(
```

```
      dataset = rocData,
      fom = "Wilcoxon",
      method = "DBMH")
varYTR <- retDbm$varComp$varComp[3]
varYTC <- retDbm$varComp$varComp[4]
varYEps <- retDbm$varComp$varComp[6]
effectSize <- retDbm$ciDiffTrtRRRC$Estimate
cat("effect size = ", effectSize, "\n")
#RRRC
J <- 10; K <- 163
ncp <- (0.5*J*K*(effectSize)^2)/
  (K*varYTR+max(J*varYTC,0)+varYEps)
MS <- UtilMeanSquares(rocData,
                      fom = "Wilcoxon",
                      method = "DBMH")
ddf <- (MS$msTR+max(MS$msTC-MS$msTRC,0))^2/
  (MS$msTR^2)*(J-1)
FCrit <- qf(1 - alpha, 1, ddf)
Power <- 1-pf(FCrit, 1, ddf, ncp = ncp)
cat("J =", J, "\nK =", K,"\nFCrit =",
    FCrit, "\nddf =", ddf, "\nncp =",
    ncp, "\nRRRC power =", Power, "\n")


#FRRC
J <- 10; K <- 133
ncp <- (0.5*J*K*(effectSize)^2)/
  (max(J*varYTC,0)+varYEps)
ddf <- (K-1)
FCrit <- qf(1 - alpha, 1, ddf)
Power <- 1-pf(FCrit, 1, ddf, ncp = ncp)
cat("J =", J, "\nK =", K,"\nFCrit =", FCrit,
    "\nddf =", ddf, "\nncp =",
    ncp, "\nFRRC power =", Power, "\n")


#RRFC
J <- 10; K <- 53
ncp <- (0.5*J*K*(effectSize)^2)/(K*varYTR+varYEps)
ddf <- (J-1)
FCrit <- qf(1 - alpha, 1, ddf)
Power <- 1-pf(FCrit, 1, ddf, ncp = ncp)
cat("J =", J, "\nK =", K,"\nFCrit =", FCrit,
    "\nddf =", ddf, "\nncp =",
    ncp, "\nRRFC power =", Power, "\n")
ddf <- (J-1)
```

Line 3 sets alpha to 0.05, line 4 selects the data file for analysis (**VanDyke.lrc**). At lines 6–8 the function **DfReadDataFile()** reads the data file and saves the result to an object called **rocData**. The **format** statement tells the function to expect the **.lrc** text file format used by University of Iowa and University of Chicago ROC website software. Lines 9–12 perform **StSignificanceTesting()** using the DBMH method with empirical AUC as the figure of merit. Lines 13–15 extract the variance components from the object returned in line 9. Insert a break point at line 16 and click **Source**. Use the following code snippet to reveal the names and array indexing of the pseudovalue variance components.

```
> retDbm$varComp
                         varComp
Var(R)                   0.0015349993
Var(C)                   0.0272492343
Var(T*R)                 0.0002004025
Var(T*C)                 0.0119752962
Var(R*C)                 0.0122647286
Var(T*R*C) + Var(Error)  0.0399716032
```

For example, the treatment-reader pseudovalue variance component is the third element of **retDbm$varComp**, and so on. Line 16 extracts the observed effect size, which, in this example, is used as the anticipated effect size. Lines 18 through 31 perform RRRC analysis and prints the results. The non-centrality parameter **ncp** is calculated at lines 20–21. The mean squares are extracted at lines 22–24 and used to calculate **ddf** at lines 25–26. Lines 29–31 prints the RRRC results for 10 readers and 163 cases in the pivotal study. Lines 33 through 42 perform FRRC analysis and the remaining code performs RRFC analysis. For each analysis, the numbers of readers and cases were chosen with the foreknowledge (Section 11.8.4) that the expected power will be 80%. Click **Continue** to execute the rest of the code.

11.8.2 Code output

```
> source(...)
alpha =  0.05
effect size =  -0.04380032
J = 10
K = 163
FCrit = 4.127057
ddf = 34.33427
ncp = 8.126982
RRRC power = 0.7911126
J = 10
K = 133
FCrit = 3.912875
ddf = 132
ncp = 7.987384
FRRC power = 0.8011167
J = 10
K = 53
FCrit = 5.117355
ddf = 9
ncp = 10.04872
RRFC power = 0.8049666
```

For 10 readers, the numbers of cases needed for 80% power is largest (163) for RRRC and least for RRFC (53). For all three analyses, the expectation of 80% power is met. Intermediate quantities such as the critical value of the F-statistic, **ddf** and **ncp** are shown. The reader should confirm that the code does in fact implement the relevant formula. Shown next is the **RJafroc** implementation. The relevant file is **mainSsDbmh.R**, a listing of which follows.

11.8.3 Code listing

```
rm(list = ls()) #mainSsDbmh.R
library(RJafroc)

fileName <- "VanDyke.lrc"
#fileName <- "Franken1.lrc"
OptimisticScenario <-  FALSE
alpha <- 0.01
cat("alpha = ", alpha, "\n")
rocData <- DfReadDataFile(
  fileName, format = "MRMC")
retDbm <- StSignificanceTesting(
  dataset = rocData,
  fom = "Wilcoxon",
  method = "DBMH")
varYTR <- retDbm$varComp$varComp[3]
varYTC <- retDbm$varComp$varComp[4]
varYEps <- retDbm$varComp$varComp[6]
effectSize <- retDbm$ciDiffTrtRRRC$Estimate
cat("effect size =", effectSize, "\n")
effectSize <- abs(effectSize)
sigma <- (retDbm$ciDiffTrtRRRC$`CI Upper`
          -retDbm$ciDiffTrtRRRC$`CI Lower`)/4
if (OptimisticScenario == TRUE) {
  effectSize <- effectSize + 2*sigma
}

cat("p-value = ", retDbm$pRRRC,
    "\nanticipated effectSize = ", effectSize,
    "\nCI Lower =", retDbm$ciDiffTrtRRRC$`CI Lower`,
    "\nCI Upper =", retDbm$ciDiffTrtRRRC$`CI Upper`, "\n")
powTab <- SsPowerTable(
  alpha = alpha,
  effectSize = effectSize,
  desiredPower = 0.8,
  method = "DBMH",
  option = "ALL",
  varYTR, varYTC, varYEps)
print(powTab)
```

At line 6 the flag **OptimisiticScenario** is set to **FALSE**, which means the *anticipated* effect size, that is, the value entering the formula stated above, is set equal to the *observed* effect size. If this flag were set to **TRUE**, the anticipated effect size would be two-sigma greater than the observed value in the correct direction, implemented at line 24. Lines 21–22 calculate sigma as 1/4 the width of the 95% confidence interval. Lines 31–37 use the function **SsPowerTable()** to generate the power table and line 38 prints it. Notice **option** = *ALL* requesting RRRC, FRRC, and RRFC analyses. Also, **effectSize = effectSize** is not a tautology; on the left-hand **effectSize** is a *case-sensitive* option understood by the function that is being called, while the right-hand **effectSize** is the value defined at lines 18 and 24. **Source** the file **mainSsDbmh.R** to get the following partial output.

11.8.4 Code output (partial)

```
> source(...)
alpha =  0.05
effect size = -0.04380032
p-value =  0.05166569
anticipated effectSize =  0.04380032
CI Lower = -0.0879595
CI Upper = 0.0003588544
$powerTableRRRC
   numReaders numCases power
2            3     >2000   <NA>
3            4      1089    0.8
4            5       344  0.801
5            6       251  0.801
6            7       211  0.801
7            8       188  0.801
8            9       173  0.801
9           10       163  0.802
...


$powerTableFRRF
   numReaders numCases power
2            3       209  0.800
3            4       182  0.801
4            5       166  0.802
5            6       155  0.801
6            7       147  0.801
7            8       141  0.801
8            9       137  0.802
9           10       133  0.801
...


$powerTableRRFC
   numReaders numCases power
2            3     >2000   <NA>
3            4     >2000   <NA>
4            5       289    0.8
5            6       151  0.801
6            7       102    0.8
7            8        78  0.803
8            9        63  0.804
9           10        53  0.805
...
```

The power table lists the different combinations of numbers of readers and cases that yield a *target power* of 80% at an alpha of 5%. The listing is in three groups, corresponding to RRRC, FRRC, and RRFC generalizations. For example, with four readers, RRRC analysis, 1089 cases are needed, clearly impractical, with five one needs 344 cases, with six one needs 251 cases, and so on. Based on this table, one might reasonably select six readers and about 250 cases. (With less than four readers the program predicts greater than 2000 cases, and besides, one can hardly claim random-reader generalization with less than four readers.)

Table 11.2 Numbers in the shaded part of the table are the numbers of cases *K* needed to achieve 80% power for the indicated number of readers *J*

Data	Optimisitic scenario	d	Gen.	# readers *J*						
				4	5	6	7	8	9	10
VD	FALSE	−0.0438	RRRC	1089	344	251	211	188	173	163
			FRRC	182	166	155	147	141	137	133
			RRFC	>2000	289	151	102	78	63	53
	TRUE	−0.0659	RRRC	62	50	44	40	38	36	35
			FRRC	47	43	40	38	37	36	35
			RRFC	62	35	24	<20	<20	<20	<20
	NA	−0.05	RRRC	361	213	170	148	134	125	119
			FRRC	140	128	119	113	109	106	103
			RRFC	529	166	99	70	55	45	38
FR	FALSE	0.0109	RRRC	>2000	>2000	>2000	>2000	1666	1419	1237
			FRRC	>2000	1958	1632	1400	1225	1089	980
			RRFC	>2000	>2000	>2000	>2000	1666	1419	1237
	TRUE	0.0188	RRRC	926	579	421	332	274	233	203
			FRRC	404	323	270	232	203	181	163
			RRFC	926	579	421	332	274	233	203
	NA	+0.03	RRRC	739	462	336	265	219	186	162
			FRRC	323	259	216	185	162	145	130
			RRFC	739	462	336	265	219	186	162

Note: VD = VanDyke, FR = Franken dataset; d = anticipated effect size; Gen. = generalization; RRRC = random-reader random-case; FRRC = fixed-reader random-case; RRFC = random-reader fixed-case.

The sample-size values shown above are summarized in Table 11.2, which lists results for both datasets (VD = VanDyke, FR = Franken), both scenarios, and three generalizations. The numbers in the last seven columns of the table are the numbers of cases needed to achieve 80% power for the indicated number of readers *J*. Estimates are shown for three cases: flag `OptimisiticScenario` set to **FALSE, TRUE**, effect size arbitrarily set to −0.05 for the Van Dyke dataset, and 0.03 for the Fran-ken dataset.

It is clear from the table that the Van Dyke dataset is easier in the sense that fewer resources are needed to plan a study with 80% power.

First question: does one even need to perform a pivotal study? If the pilot study returns a significant difference, one has rejected the NH and that is all there is to it. There is no need to perform the pivotal study, unless one tweaks the new treatment and/or casts a wider sampling net to make a stronger argument, perhaps to the FDA, that the treatments are indeed generalizable, and that the difference is in the right direction (new treatment FOM > conventional treatment FOM). If a *significant difference is observed in the opposite direction* (e.g., new treatment FOM < conventional treatment FOM) one cannot justify a pivotal study with an expected effect size in the opposite direction. Since the Van Dyke pilot study came close to rejecting the NH and the observed effect size (see below) is not too small, a pivotal study is justified.

Second question: what results did the pilot study return? The results of the analysis are contained in the variable `retDbm` (line 11) and the FOMs are contained in `retDbm$fomArray`. Use the code below to list the FOMs, the *F*-statistic, the *ddf,* and the *p*-value (notice how **RStudio** gives helpful hints as one starts typing in a variable name; also, available **list** members are shown when one enters **$** following a **list** variable).

11.8.5 Code snippet

```
> retDbm$fomArray
            Rdr - 0     Rdr - 1     Rdr - 2     Rdr - 3     Rdr - 4
Trt - 0 0.9196457 0.8587762 0.9038647 0.9731079 0.8297907
Trt - 1 0.9478261 0.9053140 0.9217391 0.9993559 0.9299517
> retDbm$fRRRC;retDbm$ddfRRRC;retDbm$pRRRC
[1] 4.456319
[1] 15.25967
[1] 0.05166569
> effectSize
[1] 0.04380032
> retDbm$fFRRC;retDbm$ddfFRRC;retDbm$pFRRC
[1] 5.475953
[1] 113
[1] 0.02103497
> retDbm$fRRFC;retDbm$ddfRRFC;retDbm$pRRFC
[1] 8.704
[1] 4
[1] 0.04195875
```

For RRRC analysis the study did not reach significance, p-value = 0.0517. *The observed F-statistic is 4.46 with degrees of freedom *ndf* = 1 (this is always # treatments minus 1), *ddf* = 15.3, and* **effectSize** = −0.0438.

> *This is the difference (treatment 1 minus treatment 2) between the reader-averaged FOMs in the two treatments. Lacking any other information, the observed effect size is the best estimate of the effect size to be anticipated. Assuming treatment 2 is the new treatment, the difference is going the right way, and because the pilot study did not reach significance, there is reason to conduct a pivotal study. A reasonable choice would be six readers and 251 cases, Section 11.8.4.*

To print the upper and lower 95% confidence intervals, type in **retOrh$ciDiffTrtRRRC$** and select the appropriate choice (in the code snippet below it is not necessary to type in the single quote, and so on; just use **RStudio**'s prompting abilities to the fullest) and repeat with the other appropriate choice. Observe that these values are centered on the **effectSize** value printed above.

11.8.6 Code snippet

```
> retDbm$ciDiffTrtRRRC$`CI Lower`
[1] -0.0879595
> retDbm$ciDiffTrtRRRC$`CI Upper`
[1] 0.0003588544
```

The observed effect size is a realization of a random variable. The mean of the confidence interval (CI) is −0.044. One could use this as a reasonable anticipated value and calculate sample size as was done above with the `OptimisticScenario` flag set to **FALSE** (the sample size does not depend on the sign of the effect size, but the decision to perform a pivotal study most definitely does). *CI's generated like this, with independent sets of data, are expected to encompass the true value with 95% probability.* The lower end (greatest magnitude of the difference) of the confidence interval is −0.088, which is the optimistic estimate of the anticipated effect size obtained with the `OptimisticScenario` flag set to **TRUE**, and yields small sample sizes (Table 11.2). For example, the number of cases is 50 for five readers and RRRC generalization. With this low number one would be justified in anticipating a smaller effect size, for example, −0.05. After all, −0.05 is only slightly greater in magnitude than the observer effect size −0.044. A 2004 publication[3] used this effect size to illustrate the methodology for the Van Dyke dataset. The results for this effect size are in the Van Dyke section of the table, in the row marked NA, because the `OptimisticScenario` flag is irrelevant when effect size is overridden.

11.9 Example 2

In the second example, the Franken dataset[9] is considered the pilot study. Modality 1 is digital imaging of neonatal chest anatomy and modality 2 is conventional analog imaging. Reverse the commenting on lines 4 and 5 of file **mainSsDbmh.R** to analyze this dataset. Sourcing it with `OptimisticScenario` flag set to **FALSE** yields the output summarized in Table 11.2 in the rows labeled FALSE. The large numbers of cases indicate that, based on cost considerations, this dataset may not justify a pivotal study. Why are so many cases needed? Either the effect size is too small and/or the variances are too large. Actually, both are true. *The observed effect size (0.011) is a factor of 4 smaller than that for the Van Dyke data.* The Franken dataset was acquired in the early days of digital imaging technology, and scores of papers were published using ROC analysis to determine the pixel size requirements to match conventional analog imaging. Nowadays, digital technology has matured and practically replaced analog in all areas of medical imaging in the US. Results for the `OptimisticScenario` flag set to **TRUE** are also summarized in Table 11.2. This corresponds to $d = 0.0188$. Also shown are results for an anticipated effect size of +0.03, which is a factor of 1.6 larger than the best-case scenario, and too optimistic considering the status of then-existing digital technology. Even with this effect size, the Franken dataset is a more difficult one, in the sense of requiring greater resources to achieve 80% power. All of the Franken dataset sample size numbers in the shaded part of Table 11.2 are larger than the corresponding Van Dyke values.

11.9.1 Changing default alpha and power

Look at the help page for **SsPowerTable()**. (To find the help page for any **R** package, see **Chapter 3**, Online Appendix 3.E and Online Appendix 3.F.) Under **Usage** one sees:

```
SsPowerTable(alpha = 0.05, effectSize = 0.05, desiredPower = 0.8,
method = "DBMH", option = "ALL", ...)
```

The default effect size (0.05) is already being overridden at line 33 by **effectSize = effectSize**. To override the default alpha to 0.01 change line 7 of file **mainSsDbmh.R** to 0.01. With the Van Dyke dataset selected and `OptimisticScenario` flag set to **FALSE, source** the file, yielding the following output.

11.9.2 Code output (partial)

```
alpha =   0.01
effect size = -0.04380032
p-value =  0.05166569
anticipated effectSize =   0.04380032
CI Lower = -0.0879595
CI Upper = 0.0003588544
$powerTableRRRC
    numReaders numCases power
1              3      >2000  <NA>
2              4      >2000  <NA>
3              5      >2000  <NA>
4              6        962   0.8
5              7        482   0.8
6              8        371 0.801
7              9        317   0.8
8             10        285   0.8
```

If one wishes to control the probability of a Type I error to less than 1%, the price paid is a greatly increased sample size. For six readers, instead of 251 cases at alpha = 5%, one needs 962 cases at alpha = 1%. The reason for this should be clear from **Chapter 8**. A smaller alpha implies a larger critical value for the F-statistic.

11.10 Cautionary notes: The Kraemer et al. paper

A paper by Kraemer et al.,[10] titled "Caution regarding the use of pilot studies to guide power calculations for study proposals," is informative. Everything hinges on the choice of effect size. Once it is specified, everything else follows from the sample size formula (with due specification of the two types of errors and the desired generalization). There are three strategies for dealing with how to choose the effect size quandary:

1. In the first strategy, a convenient sample size is set, and whatever statistical power results are accepted. This is common in this field, and the author is aware of comments to the effect: *use six readers and about 100 cases, equally split between non-diseased and diseased, and one should be fine.* Indeed, many studies cluster around these values (even smaller sample sizes have been used in some publications).
2. In the second strategy, researchers set the effect size based on their experience with other ROC studies, sometimes without consideration of the specific clinical context of the proposed study.
3. *In the third strategy, the one forming the basis of this chapter, the researchers propose (or institutional review boards, IRBs, insist) that a small pilot study be conducted to estimate the effect size.* After the pilot study is completed the researchers (or the IRB) makes a post-hoc (i.e., after the fact) decision *whether the observed effect size is clinically significant.* If it does not meet the criterion, the study would not be proposed at all or, if proposed, would not be approved or funded (i.e., the study would be aborted). On the other hand, if the pilot study observed effect size is clinically significant, power calculations for the main study are conducted based on the observed effect size.

Kraemer et al.[10] show that the most likely outcomes from the third strategy are: (1) studies worth performing are aborted and (2) studies that are not aborted are underpowered. This is a sobering thought and is something the researcher should keep in mind to guard against prematurely giving up on a new idea or, at the other extreme, being excessively optimistic about a new idea. The Kraemer et al. paper is actually quite an interesting read and highly recommended.

11.11 Prediction accuracy of sample size estimation method

The observed power is a realization of a random variable: there is no guarantee that using the predicted sample size will actually achieve 80% power. Checking prediction accuracy of a sample size estimating method requires simulations similar to those used to test NH validity, Section 9.12. (If one studies the code used for checking NH validity in **Chapter 9**, in file `mainRejectRate.R` at line 11, one sees `tau22 <- 0.0*mu2`. The zero multiplier specifies the NH condition. If a nonzero multiplier is used, then the AH condition is true and the observed NH rejection fraction, over 2000 simulations, is an estimate of *true power*. The reason for using a multiplier, rather than an additive term, will become clear in Section 11.12.)

The author has published a paper where the prediction accuracy of the sample size estimation method described in this chapter was assessed.[11] Two thousand NH pilot datasets were generated using two Roe and Metz ratings simulators[12] with complementary characteristics. Each data set consisted of five readers interpreting 50 normal and 50 abnormal cases in two modalities. The baseline NH area under the ROC curve was AUC = 0.855. Each simulated pilot dataset was analyzed by DBMH to get the necessary pseudovalue variance components. Then the sample size estimation procedure was applied to predict the number of cases K_i (i = 1, 2, …, 2000), assuming 10 readers in the pivotal study, to achieve 80% power for an anticipated effect size of 0.06 in AUC units. For each predicted K_i, one could conduct 2000 additional simulations under AH conditions to estimate empirical power, but in practice a simpler interpolation procedure was used to reduce the computational burden. Figure 11.2a and b shows typical interpolation curves for the three generalization conditions for the two simulators studied, one with high reader variance and low case variance (Figure 11.2a) and the other with the opposite characteristics (Figure 11.2b).

These plots illustrate several points. An obvious one is that power increases with the number of cases. A less obvious point is that for a given number of cases RRRC generalization yields the least power (solid lines). Focusing on Figure 11.2a, a subtle point is that for the high reader variability simulator, freezing reader variability (labeled random-cases, dashed line) yields the greatest increase in power as compared to freezing case-variability (labeled random-readers, dotted line). Because Figure 11.2b is a low reader-variability high case-variability simulator, the opposite trends

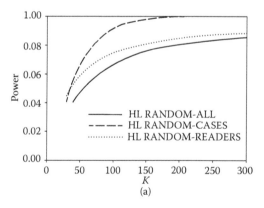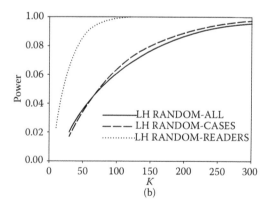

Figure 11.2 (a) Interpolation curves for the high reader variance low case variance simulator. Random-all (i.e., RRRC) yields the lowest curve implying more cases needed for 80% power. The random-case (i.e., FRRC) curve is considerably upward-left shifted relative to RRRC, the result of freezing the large reader variability, while the RRFC curve is less upward-left shifted. FRRC generalization will require fewer cases relative to RRFC generalization to achieve a desired power. (b) Similar to (a), low reader variance, high case variance simulator. The FRRC curve is slightly upward-left shifted relative to RRRC, implying a small reduction in number of cases for 80% power, but the RRFC curve is considerably upward-left shifted, implying a large reduction in needed number of cases for desired power. RANDOM-ALL = RRRC; RANDOM-CASES = FRRC; RANDOM-READERS = RRFC. (Reproduced from Chakraborty DP, *Acad Radiol.*, 17, 628–638, 2010. With permission.)

are observed. A final point: by reader and case variability, one means the treatment-reader $\sigma^2_{Y;\tau R}$ and treatment-case $\sigma^2_{Y;\tau C}$ pseudovalue variance components, respectively, and of course, the error term $\sigma^2_{Y;\varepsilon}$, which involves both reader and treatment variability. The treatment-independent terms, $\sigma^2_{Y;R}$ and case variability $\sigma^2_{Y;C}$, do not affect sample size, as these contribute *constant* shifts in performances between the two treatments.

Sometimes the sample size method failed to find number of cases below 2000, referred to as *clipping*. A pilot study yielding >2000 cases would need to be reanalyzed some other way, for example, changing the method of estimating AUC, increasing the effect size, or number of readers, to avoid clipping. Using the interpolation procedure, for example, Figure 11.2, the number of cases $K_{0.75}$ and $K_{0.90}$ were determined, which corresponded to 0.75 and 0.90 powers. The fraction of the 2000 pilot study simulations where K_i was included in the range $K_{0.75}$ and $K_{0.90}$ was defined as $Q_{0.75,0.90}$, the *quality or prediction-accuracy of the sample size method* (the rationale for the asymmetric interval around 80% is that a slight overestimate is desirable to an underestimate), that is, $Q_{0.75,0.90} = \mathrm{Prob}(K_{0.75} < K_i < K_{0.90})$. The final Hillis-Berbaum (HB) prediction, K_{HB}, was defined as the median of K_i over the trials where $K_i < 2000$ and the corresponding power P_{HB} was determined from the appropriate interpolation curve.

Figure 11.3a and b show normalized histograms of K_i ($i = 1, 2, ..., 2000$) for the low Figure 11.3a and high Figure 11.3b reader variability simulators, respectively, under the random-reader random-case condition. In Figure 11.3a the area under the histogram between the lines labeled $K_{0.75}$ and $K_{0.75}$ is the prediction-accuracy $Q_{0.70,0.95} = 38\%$ and $K_{HB} = 225$ (the HB prediction for required number of cases) and $K_{0.80} = 162$ (true value for required number of cases). An overestimate of the required number of cases is not necessarily a bad thing. The peak at 2000 cases representing the clipped predictions contributes 13% to the area. In Figure 11.3b, $Q_{0.70,0.95} = 26\%$, $K_{HB} = 159$, and $K_{0.80} = 190$, so the HB method is underestimating the number of cases. The area contributed by the peak at 2000 cases is 39%. Prediction-accuracy was generally higher under low reader variability conditions than under high reader variability condition, which is consistent with comments in Section 11.4.

Appendix B of the referenced paper has a discussion of why the method is not as good for large reader variability. The reason has to do with the higher variability of the modality-reader variance component $\sigma^2_{\tau R}$, Table 1, Ref. 11. Moreover, the variability of this variance component is larger as the number of modality-reader combinations, over which it is averaged, is relatively small. Since, the number of cases, which could be in the hundreds, multiplies it, the net effect on variability is amplified. An overestimate of $\sigma^2_{Y;\tau R}$ tends to decrease the power and the HB-method compensates by increasing the number of cases. A sufficiently large overestimate leads to clipping.

11.12 On the unit for effect size: A proposal

Effect size in ROC analysis is almost invariably specified in AUC units. Typically, 0.05 is considered a small to moderate effect size.[13] Because AUC is restricted to the range (0,1), this creates the problem that *the meaning of an effect size depends on the baseline NH value of AUC to which it is added*. An effect size of 0.05 might qualify as a small effect size if the baseline value is 0.7, since the improvement is from 0.70 to 0.75. On the other hand, if the baseline value is 0.95, then the same effect size implies the new treatment has AUC = 1.00. In other words, performance is perfect, implying an infinite effect size (see Figure 11.4).

In the author's judgment, it makes more sense to specify effect size in d' units, where baseline or NH d' is defined as follows (d' is not to be confused with Cohen's D):

$$d' = \sqrt{2}\Phi^{-1}(AUC) \tag{11.13}$$

This equation is derived from Equation 3.30, that is, one imposes the equal variance binormal model, and calculates the separation parameter yielding the observed baseline AUC. One can now specify small, medium, and large effect sizes, on the d' scale, as incremental *multiples* of the baseline d' value, say 0.2, 0.4, and 0.6, patterned after Cohen's suggested values. For example, with the multiplier equal to 0.2, the AH d' would be $1.2 \times d'$.

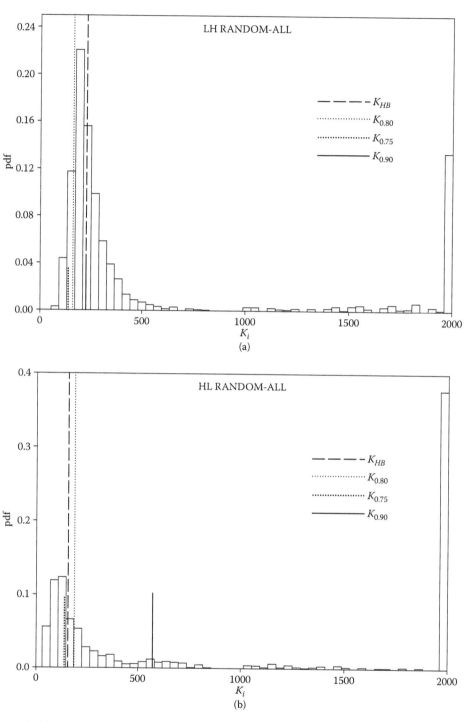

Figure 11.3 (a) Normalized histogram of K_i (i = 1, 2, ..., 2000) for the LH (low reader variability high case variability) simulator under the random-all condition. Each value of i corresponds to an independent pilot data set. The area under the histogram between the lines labeled $K_{0.75}$ and $K_{0.90}$ is the prediction-accuracy $Q_{0.75, 0.90} = 38\%$ and $K_{HB} = 225$ and $K_{0.80} = 162$. The peak at 2000 cases representing the clipped predictions contributed 13% to the area. (b) is similar to (a) except it applies to the HL simulator (high reader variability low case variability) under the random-all condition: $Q_{0.75, 0.90} = 26\%$, $K_{HB} = 159$ and $K_{0.80} = 190$. Note the large reduction in the prediction-accuracy performance index. The area contributed by the peak at 2000 cases is 39%. [RANDOM-ALL = random-reader random=case] (Reproduced from Chakraborty DP, *Acad Radiol.*, 17, 628–638, 2010. With permission.)

Figure 11.4a shows, for different baseline or NH values of AUC, the effect size on the d'-multiple scale. The U-shaped dark curve is for an effect size = 0.05 on the AUC scale, the light line is for the d'-multiple scale. As expected, near about 0.95, the effect size on the d'-multiple scale approaches a very large value, as one is demanding an improvement from 0.95 to 1.00. The y-axis of the plot, labeled **DpMultipler**, is the fraction by which d' would have to increase to give the desired AUC effect size of 0.05. For example, a value of one means the separation parameter d' would need to double. The increase in the dark curve at the low end, near AUC = 0.5, is because there the baseline d' approaches 0, so any increase in separation would be an infinite multiplier compared to the baseline. The code generating these figures, **mainEffectSizeFixedAucEs.R**, is explained in Online Appendix 11.D.

Figure 11.4b, generated by **mainEffectSizeFixedDpMultiple.R**, explained in Online Appendix 11.D, shows the effect size on the AUC scale, the dark curve, for a fixed effect size expressed as a d' multiplier equal to 0.2, the light straight line, as a function of baseline AUC. This could be viewed as the other side of the story told by Figure 11.4a. If one keeps the d' multiplier

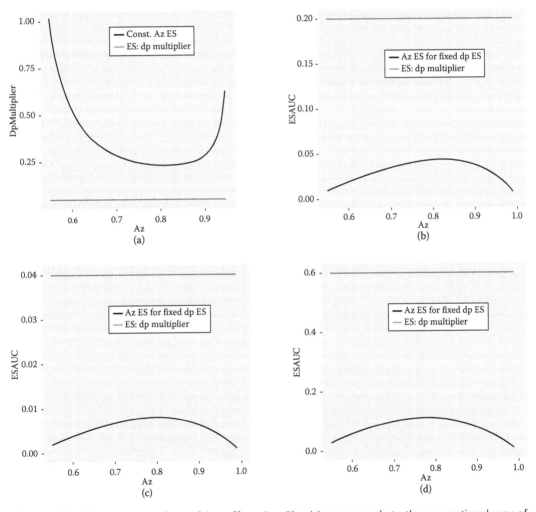

Figure 11.4 Different ways of specifying effect size. Plot (a) corresponds to the conventional way of using a fixed difference in AUC, which implies an infinite relative increase in separation parameter at 0.5 and 1. Plots (b) through (d) correspond to the preferred way of specifying effect size, namely as a multiple of the baseline d' value, which avoids the infinities. Plot (b) represents a small effect size, expressed as a constant multiple, 0.2, of d'. Plot (c) represents a medium effect size as the multiplier is 0.4. Plot (d) represents a large effect size as the multiplier is 0.6. The code generating these figures is in **mainEffectSizeFixedAucEs.R** and **mainEffectSizeFixedDpMultiple.R**.

equal to 0.2, then effect size in AUC units must decrease at the two ends of the plot where the corresponding Figure 11.4a plot increased. Note that with this way (i.e., plot b) of specifying the effect size the AH AUC is always in the valid range. Because in the plateau near the center, the effect size in AUC units is close to 0.04, the author proposes a d' multiplier equal to 0.2 as a small effect size. Figure 11.4c shows the corresponding plot for a d' multiplier equal to 0.4, which the author proposes as a medium effect size, since the plateau AUC effect size is about 0.08. Finally, Figure 11.4d shows the plot for a d' multiplier equal to 0.6, which the author proposes as a large effect size, since the plateau AUC effect size is about 0.12.

The reader may wonder why the d' based effect size is specified as a *multiplier* instead of as an *additive* effect. Consider the task of detecting small blood vessel stenosis in cranial x-rays. The diagnostic medical physicist is probably familiar with cross-table views used to acquire these images in patients with cranial aneurisms or stroke. (The author constructed a digital subtraction angiography (DSA) apparatus for this very purpose.)[14] Just prior to the imaging, with images acquired rapidly at about one per second in those days, the patient is injected with iodine contrast directly into an internal carotid artery. The cranial vessels light up and one uses DSA to subtract the non-vessel background and thereby improve visualization of the blood vessels. So, and this is the key point relevant to this section: the SNR of vessels is *proportional* to the iodine concentration and proportional (roughly) to the square of the diameter of the vessel, as the latter determines the volume of iodine in unit length of the vessel. A completely blocked vessel, where iodine cannot penetrate, will have SNR = 0 (i.e., no iodine signal) and a large vessel with greater volume of iodine per unit length will have greater SNR. This suggests the following model for d':

$$d' = d'_0 ES_{d'}$$

(11.14)

Here, d'_0 is the baseline NH perceptual SNR of the vessel and $ES_{d'}$ is effect size expressed as a multiplier. If the baseline value is zero, the vessel has zero diameter and no amount of iodine can bring it out. On the other hand, if baseline d'_0 is large, the vessel has a finite diameter and the effect of the iodine will be proportionately larger.

One last example before moving on: the author implemented a method called computer analysis of mammography phantom images (CAMPI), referred to in **Chapter 1**, as a way of quantifying mammography phantom image quality. Figure 11.5 is an example from one of the published CAMPI studies.[15]

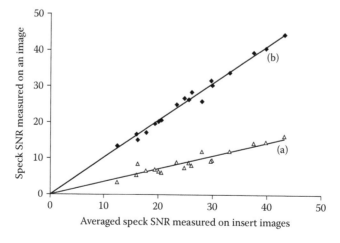

Figure 11.5 (a) Plot of individual speck SNRs as measured on a test image versus averaged individual speck SNRs measured on the insert images. There are 18 points in all, corresponding to the 18 specks in the first 3-microcalcification groups in the ACR phantom. (b) Similar plot to (a) except an insert image has been plotted along the y-axis; as expected, its slope is close to unity.

The target objects (i.e., meant to be detected) are three speck groups, each containing six specks, meant to simulate microcalcifications, see Figure 1.1a. A brief background on CAMPI[16-20] is necessary. There are two types of images involved in CAMPI, *insert* images and *test* images. Insert images are contact-radiographs of a thin wax plate embedded in which are the target objects. A number of insert images are obtained under low kVp conditions to yield high quality images of the specks (low kVp increases subject contrast and the thin plate means that there is minimal scatter degradation of the images). The test images, on the other hand, are obtained under normal conditions with the insert inside a 4.5 cm thick Lucite block. Higher kVp is needed to penetrate the thicker phantom, and both contrast and scatter degrade the image compared to images of the insert. All images were digitized using a Lumisys Model LS-100 digitizer (Lumisys Inc., Sunnyvale, CA), with a pixel size of 50 µm × 50 µm and a gray level resolution of 12 bits. All insert images were digitally aligned and averaged to further reduce noise.

Figure 11.5, plot labeled a, shows test image SNR (signal-to-noise ratio) for each speck on a test image plotted against corresponding averaged insert image SNRs. It follows a straight line through the origin, but the slope is less than unity, because the test image SNRs are *proportionately* smaller in SNR than the corresponding insert image SNRs. Figure 11.5, plot labeled b, shows SNR of one of the insert images versus corresponding averaged insert image SNRs. The slope of this line is close to unity. Note that one can get from (a) to (b) not by adding a constant value. Rather, one needs to *multiply* the slope by a constant value. (At the risk of stating the obvious, in this analogy, the insert images represent one treatment and the test image represents another treatment, and SNR is a unit normal distribution separation measure completely analogous to d'.)

11.13 Discussion/Summary

In the author's experience, the topic of sample-size estimation evokes some trepidation in non-statistician investigators engaging in ROC studies. Statisticians who understand the specialized techniques that have been developed for ROC studies may not be readily available, or when available, may lack sufficient clinical background to interact knowledgeably with the researcher. Lacking this resource, the investigator looks to the literature for similar studies and follows precedents. It is not surprising that published studies tend to cluster around similar numbers, for example, about five readers and 50–100 cases. Sample-size methodology is a valuable tool since it potentially allows non-experts to plan reasonably powered ROC studies to answer questions like: *is an image processing method better than another?* However, proper usage of these tools requires understanding of how they work, or at the very least, to be able to cogently discuss the issues involved with expert statisticians.

Sample size estimation methodology for MRMC ROC studies has gone through several iterations. It started with work by Obuchowski,[4-6,21,22] which, as noted earlier in Chapter 10, Section 10.3.3, led to some consternation, because it predicted excessively large sample sizes and also suggested that, in addition to conducting a pilot study, one needed to estimate within-reader variability by repeating the interpretations with the same readers and same cases. As noted by Hillis et al., the within reader component of variance does not need to be separately estimated.[7] Two updates by Hillis and colleagues followed, one[3] in 2004 and the other in 2011.[7] The latest paper, titled "Power estimation for multireader ROC methods: An updated and unified approach," still lacks fixed-reader and fixed-case analyses and it has some errors, for example, page 135 in Ref. 7, expression for df_2. Given the number of parameters entering the computations, numerical errors are almost inevitable in a step-by-step description. This is the reason the author prefers the software approach; software is consistent. If it is wrong, it is consistently wrong, and it will consistently remember that $J-1$ is supposed to be 7 and not 4, as in the cited example.

It is the author's preference, as in this chapter, not to mix the statistically well-known quantities mean squares with variance components in sample size formula. Mean squares are not as portable as variance components. The latter are intrinsic quantities that can be exchanged between different researchers without having to know the details of how they were calculated. With mean squares the connection depends on the sample size over which the mean squares were calculated, see Table 1 in Ref. 7. The intermixing of mean squares and variance components in publications, while of little consequence to statisticians, can make the formula unnecessarily dense for the rest of us.

The main issue is the selection of the effect size, and since it appears as the square, it can lead to widely divergent sample sizes. This is where a pilot study is helpful, but the results need to be interpreted with caution. The fundamental problem is this: a *clinically meaningful and technologically achievable* effect size, consistent with the pilot study, needs to be chosen. Good communication between the researcher and those familiar with the clinical picture is essential. Additionally, the author believes that the current practice of specifying effect size in AUC units is not appropriate. A method of specifying effect size as small, medium, or large is proposed, which uses multiples of the separation parameter and makes clinical sense. This can be used in situations where the pilot study is believed not to provide good information. Rather than repeat the whole pilot study, one can, with some justification, propose an effect size based on a multiple of the observed separation of the normal distributions describing the ROC. In addition, the chosen effect size should not greatly exceed that revealed by the pilot study.

For all its tortuous progress, current sample size methodology provides a principled approach to planning a ROC study. The approach, recommended in some circles, of using magic numbers like six readers and 50 cases, with no justification for their selection other than the reputation of the person proposing it, is one the author cannot condone.

This concludes **Part B** of the book. It is time to move on to **Part C**, namely the FROC paradigm.

References

1. ICRU. Statistical analysis and power estimation. *J ICRU*. 2008;8:37–40.
2. Cohen J. *Statistical power analysis for the behavioral sciences*. 2nd Ed. Lawrence Erlbaum Associates, Publishers, Hillsdale NJ 1988.
3. Hillis SL, Berbaum KS. Power estimation for the Dorfman-Berbaum-Metz method. *Acad Radiol*. 2004;11(11):1260–1273.
4. Obuchowski NA. Multireader, multimodality receiver operating characteristic curve studies: Hypothesis testing and sample size estimation using an analysis of variance approach with dependent observations. *Acad Radiol*. 1995;2:S22–S29.
5. Obuchowski NA. Sample size calculations in studies of test accuracy. *Stat Methods Med Res*. 1998;7(4):371–392.
6. Obuchowski NA. Sample size tables for receiver operating characteristic studies. *Am J Roentgenol*. 2000;175(3):603–608.
7. Hillis SL, Obuchowski NA, Berbaum KS. Power estimation for multireader ROC methods: An updated and unified approach. *Acad Radiol*. 2011;18(2):129–142.
8. Van Dyke CW, White RD, Obuchowski NA, Geisinger MA, Lorig RJ, Meziane MA. Cine MRI in the diagnosis of thoracic aortic dissection; 79th RSNA Meetings; 1993; Chicago, IL.
9. Franken EA, Jr., Berbaum KS, Marley SM, et al. Evaluation of a digital workstation for interpreting neonatal examinations: A receiver operating characteristic study. *Invest Radiol*. 1992;27(9):732–737.
10. Kreaemer HC, Mintz J, Noda A, Tinklenberg J, Yesavage JA. Caution regarding the use of pilot studies to guide power calculations for study proposals. *Arch Gen Psychiatry*. 2006;63:484–489.
11. Chakraborty DP. Prediction accuracy of a sample-size estimation method for ROC studies. *Acad Radiol*. 2010;17:628–638.
12. Roe CA, Metz CE. Dorfman-Berbaum-Metz method for statistical analysis of multireader, multimodality receiver operating characteristic data: Validation with computer simulation. *Acad Radiol*. 1997;4:298–303.

13. Beiden SV, Wagner RF, Campbell G. Components-of variance models and multiple-bootstrap experiments: An alternative method for random-effects, receiver operating characteristic analysis. *Acad Radiol.* 2000;7(5):341–349.

14. Chakraborty DP, Gupta KL, Barnes GT, Vitek JJ. Digital subtraction angiography apparatus. *Radiology.* 1985;157:547.

15. Chakraborty DP. Physical measures of image quality in mammography. Paper presented at: Proc. SPIE 2708, Medical Imaging 1996: Physics of Medical Imaging1996; Newport Beach CA.

16. Chakraborty DP, Sivarudrappa M, Roehrig H. Computerized measurement of mammographic display image quality. Paper presented at: Proceedings of the SPIE Medical Imaging 1999: Physics of Medical Imaging; 1999; San Diego, CA.

17. Chakraborty DP, Fatouros PP. Application of computer analyis of mammography phantom images (CAMPI) methodology to the comparison of two digital biopsy machines. Paper presented at: Proceedings of the SPIE Medical Imaging 1998: Physics of Medical Imaging; 24 July 1998. Proceedings Volume 3336, Medical Imaging 1998: Physics of Medical Imaging; (1998); doi: 10.1117/12.317066 Event: Medical Imaging '98, 1998, San Diego, CA, United States.

18. Chakraborty DP. Comparison of computer analysis of mammography phantom images (CAMPI) with perceived image quality of phantom targets in the ACR phantom. Paper presented at: Proceedings of the SPIE Medical Imaging 1997: Image Perception; 26–27 February 1997; Newport Beach, CA.

19. Chakraborty DP. Computer analysis of mammography phantom images (CAMPI). *Proc SPIE Med Imaging 1997 Phys Med Imaging.* 1997;3032:292–299. Proceedings Volume 3032, Medical Imaging 1997: Physics of Medical Imaging; (1997); doi: 10.1117/12.273996 Event: Medical Imaging 1997, 1997, Newport Beach, CA, United States.

20. Chakraborty DP. Computer analysis of mammography phantom images (CAMPI): An application to the measurement of microcalcification image quality of directly acquired digital images. *Med Phys.* 1997;24(8):1269–1277.

21. Obuchowski NA. Computing sample size for receiver operating characteristic studies. *Invest Radiol.* 1994;29(2):238–243.

22. Obuchowski NA, McLish DK. Sample size determination for diagnostic accuracy studies involving binormal ROC curve indices. *Stat Med.* 1997;16:1529–1542.

PART C

The free-response ROC (FROC) paradigm

12

The FROC paradigm

12.1 Introduction

Until now focus has been on the receiver operating characteristic (ROC) paradigm. For diffuse interstitial lung disease,* and diseases like it, where disease location is implicit (by definition *diffuse interstitial lung disease* is spread through and confined to lung tissues) this is an appropriate paradigm in the sense that possibly essential information is not being lost by limiting the radiologist's response in the ROC study to a single rating. The *extent* of the disease, that is, how far it has spread within the lungs, is an example of essential information that is still lost.[1] Anytime essential information is not accounted for in the analysis, as a physicist, the author sees a red flag. There is room for improvement in basic ROC methodology by modifying it to account for extent of disease. However, this is not the direction taken in this book. Instead, the direction taken is accounting for *location* of disease.

In clinical practice it is not only important to identify whether the patient is diseased, but also to offer further guidance to subsequent care-givers regarding other characteristics (such as location, size, extent) of the disease. In most clinical tasks if the radiologist believes the patient may be diseased, there is a location (or more than one location) associated with the manifestation of the suspected disease. Physicians have a term for this: *focal disease*, defined as *a disease located at a specific and distinct area.*

For focal disease, the ROC paradigm restricts the collected information to a single rating representing the confidence level that there is disease *somewhere* in the patient's imaged anatomy. The emphasis on *somewhere* is because it begs the question: if the radiologist believes the disease is *somewhere*, why not have them to point to it? In fact, they do point to it in the sense that they record the location(s) of suspect regions in their clinical report, but the ROC paradigm cannot use this information. *Neglect of location information leads to loss of statistical power as compared to paradigms that account for location information.* One way of compensating for reduced statistical power is to increase the sample size, which increases the cost of the study and is also unethical, because one is subjecting more patients to imaging procedures[2]

---

* Diffuse interstitial lung disease refers to disease within both lungs that affects the interstitium or connective tissue that forms the support structure of the lungs' air sacs or alveoli. When one inhales, the alveoli fill with air and pass oxygen to the blood stream. When one exhales, carbon dioxide passes from the blood into the alveoli and is expelled from the body. When interstitial disease is present, the interstitium becomes inflamed and stiff, preventing the alveoli from fully expanding. This limits both the delivery of oxygen to the blood stream and the removal of carbon dioxide from the body. As the disease progresses, the interstitium scars with thickening of the walls of the alveoli, which further hampers lung function.

and not using the optimal paradigm/analysis. This is the *practical* reason for accounting for location information in the analysis. The *scientific* reason is that including location information yields a wealth of insight into what is limiting performance; these are discussed in **Chapter 16** and **Chapter 19**. This knowledge could have significant implications—currently widely unrecognized and unrealized—for how radiologists and algorithmic observers are designed, trained and evaluated. There are other scientific reasons for accounting for location, namely it accounts for unexplained features of ROC curves. Clinicians have long recognized problems with ignoring location[1,3] but, with one exception,[4] much of the observer performance experts have yet to grasp it.

This part of the book, the subject of which has been the author's prime research interest over the past three decades, starts with an overview of the FROC paradigm introduced briefly in **Chapter 1**. Practical details regarding how to conduct and analyze an FROC study are deferred to **Chapter 18**. The following is an outline of this chapter. Four observer performance paradigms are compared using a visual schematic as to the kinds of information collected. An essential characteristic of the FROC paradigm, namely search, is introduced. Terminology to describe the FROC paradigm and its historical context is described. A pioneering FROC study using phantom images is described. Key differences between FROC and ROC data are noted. The FROC plot is introduced and illustrated with **R** examples. The dependence of population and empirical FROC plots on perceptual signal-to-noise ratio *(pSNR)* is shown. The expected dependence of the FROC curve on *pSNR* is illustrated with a solar analogy—understanding this is key to obtaining a good intuitive feel for this paradigm. The finite extent of the FROC curve, characterized by an end-point, is emphasized. Two sources of radiologist expertise in a search task are identified: search and lesion-classification expertise, and it is shown that an inverse correlation between them is expected.

The starting point is a comparison of four current observer performance paradigms.

12.2 Location specific paradigms

Location-specific paradigms take into account, to varying degrees, information regarding the locations of perceived lesions, so they are sometimes referred to as *lesion-specific* (or lesion-level[5]) paradigms. Usage of this term is discouraged. In this book, the term *lesion* is reserved for true malignant* lesions[†] (distinct from *perceived lesions* or *suspicious regions* that may not be true lesions).

All observer performance methods involve detecting the presence of true lesions. So, ROC methodology is, in this sense, also lesion-specific. On the other hand, location is a *characteristic* of true and perceived focal lesions, and methods that account for location are better termed *location-specific* than lesion-specific.

There are three location-specific paradigms: the free-response ROC (FROC),[6,7-11] the location ROC (LROC),[12-16] and the region of interest (ROI).[17,18]

---

* Benign lesions are simply normal tissue variants that resemble a malignancy, but are not malignant.
† Lesion: a region in an organ or tissue that has suffered damage through injury or disease, such as a wound, ulcer, abscess, tumor, and so on.

Figure 12.1 shows a mammogram as it might be interpreted according to current paradigms—these are not actual interpretations, just schematics to illustrate essential differences
between the paradigms. The arrows point to two real lesions (as determined by subsequent
follow-up of the patient) and the three lightly shaded crosses indicate perceived lesions or
suspicious regions. From now on, for brevity, the author will use the term *suspicious region*.

The numbers and locations of suspicious regions depend on the case and the observer's skill
level. Some images are so obviously non-diseased that the radiologist sees nothing suspicious in
them, or they are so obviously diseased that the suspicious regions are conspicuous. Then there is
the gray area where one radiologist's suspicious region may not correspond to another radiologist's
suspicious region.

In Figure 12.1, evidently the radiologist found one of the lesions (the lightly shaded cross near the
left-most arrow), missed the other one (pointed to by the second arrow), and mistook two normal
structures for lesions (the two lightly shaded crosses that are relatively far from the true lesions).
To repeat, the term *lesion* is always a true or real lesion. The prefix *true* or *real is implicit*. The term
suspicious region is reserved for any region that, as far as the observer is concerned, has lesion-like
characteristics, but may not be a true lesion.

1. In the ROC paradigm, Figure 12.1 (top left), the radiologist assigns a single rating indicating the confidence level that there is at least one lesion *somewhere* in the image.* Assuming
   a 1 through 5 positive directed integer rating scale, if the left-most lightly shaded cross is a
   highly suspicious region then the ROC rating might be 5 (highest confidence for presence of
   disease).
2. In the FROC paradigm, Figure 12.1 (top right), the dark shaded crosses indicate *suspicious
   regions that were marked or reported in the clinical report,* and the adjacent numbers are the
   corresponding ratings, which apply to specific regions in the image, unlike ROC, where the
   rating applies to the whole image. Assuming the allowed positive-directed FROC ratings
   are 1 through 4, two marks are shown, one rated FROC-4, which is close to a true lesion,
   and the other rated FROC-1, which is not close to any true lesion. The third suspicious
   region, indicated by the lightly shaded cross, was not marked, implying its confidence level
   did not exceed the lowest reporting threshold. The marked region rated FROC-4 (highest
   FROC confidence) is likely what caused the radiologist to assign the ROC-5 rating to this
   image in the top-left figure. (For clarity the rating is specified alongside the applicable
   paradigm.)
3. In the LROC paradigm, Figure 12.1 (bottom-left), the radiologist provides a rating summarizing confidence that there is at least one lesion somewhere in the image (as in the ROC paradigm) *and* marks the *most suspicious region* in the image. In this example, the rating might
   be LROC-5, the 5 rating being the same as in the ROC paradigm, and the mark may be the
   suspicious region rated FROC-4 in the FROC paradigm, and, since it is close to a true lesion,
   in LROC terminology it would be recorded as a *correct localization*. If the mark were not near
   a lesion it would be recorded as an *incorrect localization*. Only one mark is *allowed* in this
   paradigm, and in fact one mark is *required* on every image, even if the observer does not find

---

* The author's imaging physics mentor, Prof. Gary T. Barnes, had a way of emphasizing the word "somewhere" when he
  spoke about the neglect of localization in ROC methodology, as in, "what do you mean the lesion is *somewhere* in the
  image? If you can see it you should point to it." Some of his grant applications were turned down because they did not
  include ROC studies, yet he was deeply suspicious of the ROC method because it neglected localization information.
  Around 1983 he guided the author toward a publication by Bunch et al., to be discussed in Section 12.4 and that started
  the author's career in this field.

any suspicious region to report. The forced mark has caused confusion in the interpretation of this paradigm and its usage. The late Prof. "Dick" Swensson has been the prime contributor to this paradigm.

4. In the ROI paradigm, the researcher segments the image into a number of ROIs and the radiologist rates each ROI for presence of at least one suspicious region somewhere within the ROI. The rating is similar to the ROC rating, except it applies to the segmented ROI, not the whole image. Assuming a 1 through 5 positive-directed integer rating scale, in Figure 12.1 (bottom-right) there are four ROIs and the ROI at ~9 o'clock might be rated ROI-5 as it contains the most suspicious light cross, the one at ~11 o'clock might be rated ROI-1 as it does not contain any light crosses, the one at ~3 o'clock might be rated ROI-2 or 3 (the light crosses would tend to increase the confidence level), and the one at ~7 o'clock might be rated ROI-1. When different views of the same patient anatomy are available, it is assumed that all images are segmented consistently, and the rating for each ROI takes into account all views of that ROI in the different views. In the example shown in Figure 12.1 (bottom-right), each case yields four ratings. The segmentation shown in the figure is a schematic. In fact, the ROIs could be clinically driven descriptors of location, such as *apex of lung* or *mediastinum*, and the image does not have to have lines showing the ROIs (which would be distracting to the radiologist). The number of ROIs per image can be at the researcher's discretion and there is no requirement that every case have a fixed number of ROIs. Prof. Obuchowski has been the principal contributor to this paradigm.

The rest of the book focuses on the FROC paradigm. It is the most general paradigm, special cases of which accommodate other paradigms. As an example, for diffuse interstitial lung disease, clearly a candidate for the ROC paradigm, the radiologist is implicitly pointing to the lung when disease is seen.

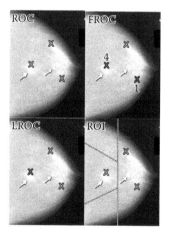

Figure 12.1 A mammogram interpreted according to current observer performance paradigms. The arrows indicate two real lesions and the three light crosses indicate suspicious regions. Evidently the radiologist saw one of the lesions, missed the other lesion, and mistook two normal structures for lesions. ROC (top-left): the radiologist assigns a single confidence level that *somewhere* in the image there is at least one lesion. FROC (top-right): the dark crosses indicate suspicious regions that are *marked* and the accompanying numerals are the FROC ratings. LROC (bottom-left): the radiologist provides a single rating that somewhere in the image there is at least one lesion and marks the most suspicious region. ROI (bottom-right): the image is divided into a number of regions of interest (by the researcher) and the radiologist rates each ROI for presence of at least one lesion somewhere within the ROI.

12.3 The FROC paradigm as a search task

The FROC paradigm is equivalent to a *search* task. Any search task has two components: (1) *finding* something and (2) *acting* on it. An example of a search task is looking for lost car keys or a milk carton in the refrigerator. Success in a search task is finding the object. Acting on it could be driving to work or drinking milk from the carton. There is *search-expertise* associated with any search task. Husbands are notoriously bad at finding the milk carton in the refrigerator (the author owes this analogy to Dr. Elizabeth Krupinski). Like anything else, *search expertise is honed by experience,* that is, lots of practice. While the author is not good at finding the milk carton in the refrigerator, he is good at finding files in his computer.

Likewise, a medical imaging search task has two components (1) finding suspicious regions and (2) acting on each finding (*finding*, used as a noun, is the actual term used by clinicians in their reports), that is, determining the relevance of each finding to the health of the patient, and whether to report it in the official clinical report. A general feature of a medical imaging search task is that the radiologist does not know a priori whether the patient is diseased and, if diseased, how many lesions are present. In the breast-screening context, it is known a priori that about five out of 1000 cases have cancers, so 99.5% of the time, odds are that the case has no malignant lesions (the probability of finding benign suspicious regions is much higher,[19] about 13% for women aged 40–45). The radiologist searches the images for lesions. If a suspicious region is found, and provided it is sufficiently suspicious, the relevant location is *marked* and *rated* for confidence in being a lesion. The process is repeated for each suspicious region found in the case. A radiology report consists of a listing of search-related actions specific to each patient. To summarize:

> Free-response data = variable number (≥ 0) of mark-rating pairs per case. It is a record of the search process involved in finding disease and acting on each finding.

12.3.1 Proximity criterion and scoring the data

In the first two clinical applications of the FROC paradigm,[9,20] the marks and ratings were indicated by a grease pencil on an acrylic overlay aligned, in a reproducible way, to the CRT displayed chest image. Credit for a correct detection and localization, termed a lesion-localization or *LL*-event,[*] was given only if a mark was sufficiently close to an actual diseased region. Otherwise, the observer's mark-rating pair was scored as a non-lesion localization or *NL*-event.

> *The use of ROC terminology, such as true positives or false positives to describe FROC data, seen in the literature on this subject, including the author's earlier papers[6], is not conducive to clarity, and is strongly discouraged.*

The classification of each mark as either an LL or an NL is referred to as *scoring* the marks.

> Definition:
> NL = non-lesion localization, that is, a mark that is not close to any lesion.
> LL = lesion localization, that is, a mark that is close to a lesion.

---

[*] The proper terminology for this paradigm has evolved. Older publications and some newer ones refer to these as true positive (TP) event, thereby confusing a ROC-related term that does not involve search with one that does.

What is meant by *sufficiently close*? One adopts an *acceptance radius* (for spherical lesions) or *proximity criterion* (the more general case). What constitutes close enough is a clinical decision, the answer to which depends on the application.[21–23] This source of arbitrariness in the FROC paradigm, which has been used to question its usage,[24] is more in the mind of some researchers than in the clinic. It is not necessary for two radiologists to point to the same pixel in order for them to agree that they are seeing the same suspicious region. Likewise, two physicians (e.g., the radiologist finding the lesion on an x-ray and the surgeon responsible for resecting it) do not have to agree on the exact center of a lesion in order to appropriately assess and treat it. More often than not, clinical common sense can be used to determine whether a mark actually localized the real lesion. When in doubt, the researcher should ask an independent radiologist (i.e., not one of the participating readers) how to score ambiguous marks.

For roughly spherical nodules a simple rule can be used. If a circular lesion is 10 mm in diameter, one can use the touching-coins analogy to determine the criterion for a mark to be classified as lesion localization. Each coin is 10 mm in diameter, so if they touch, their centers are separated by 10 mm, and the rule is to classify any mark within 10 mm of an actual lesion center as an LL mark, and if the separation is greater, the mark is classified as an NL mark. A recent paper[25] using FROC analysis gives more details on appropriate proximity criteria in the clinical context. Generally, the proximity criterion is more stringent for smaller lesions than for larger ones. However, for very small lesions allowance is made so that the criterion does not penalize the radiologist for normal marking jitter. For 3D images, the proximity criteria is different in the x-y plane versus the slice thickness axis.

For clinical datasets, a rigid definition of the proximity criterion should *not* be used; deference should be paid to the judgment of an independent expert.

12.3.2 Multiple marks in the same vicinity

Multiple marks near the same vicinity are rarely encountered with radiologists, especially if the perceived lesion is mass-like (the exception would be if the perceived lesions were speck-like objects in a mammogram, and even here radiologists tend to broadly outline the region containing perceived specks—in the author's experience they do not spend their valuable clinical time marking individual specks with great precision). However, algorithmic readers, such as a CAD algorithm, are not radiologists and do tend to find multiple regions in the same area. Therefore, algorithm designers generally incorporate a clustering step[26] to reduce overlapping regions to a single region and assign to it the highest rating (i.e., *the rating of the highest rated mark, not the rating of the closest mark*). The reason for using the highest rating is that this gives full and deserved credit for the localization. Other marks in the same vicinity with lower ratings need to be discarded from the analysis. Specifically, they should not be classified as NLs, because each mark has successfully located the true lesion to within the clinically acceptable criterion, that is, any one of them is a good decision because it would result in a patient recall and point further diagnostics.

12.3.3 Historical context

The term *free-response* was coined in 1961 by Egan et al.[7] to describe a task involving the detection of brief audio tone(s) against a background of white-noise (white-noise is what one hears if an FM tuner is set to an unused frequency). The tone(s) could occur at any instant within an active listening interval, defined by an indicator light bulb that is turned on. The listener's task was to respond by pressing a button at the specific instant(s) when a tone(s) was perceived (heard). The listener was uncertain how many true tones could occur in an active listening interval and when they might occur. Therefore, the number of responses (button presses) per active interval was a priori unpredictable: it could be zero, one, or more. The Egan et al. study did not require the listener to rate each button press, but apart from this difference and with a two-dimensional image replacing the one-dimensional listening interval, the acoustic signal detection study is similar to a common task in medical imaging, namely, prior to interpreting a screening case for possible breast cancer, the radiologist does not know how many diseased regions are actually present and, if present, where they are located. Consequently, the case (all four views and possibly prior images) is *searched* for regions that appear to be suspicious for cancer. If one or more suspicious regions are found, and the level of suspicion of at least one of them exceeds the radiologists' minimum reporting threshold, the radiologist reports the region(s). At the author's former institution (University of Pittsburgh, Department of Radiology) the radiologists digitally outline and annotate (describe) suspicious region(s) that are found. As one would expect from the low prevalence of breast cancer, in the screening context United States, and assuming expert-level radiologist interpretations, about 90% of breast cases do not generate any marks, implying case-level specificity of 90%. About 10% of cases generate one or more marks and are recalled for further comprehensive imaging (termed *diagnostic workup*). Of marked cases about 90% generate one mark, about 10% generate two marks, and a rare case generates three or more marks. Conceptually, a mammography screening report consists of the locations of regions that exceed the threshold and the corresponding levels of suspicion, reported as a Breast Imaging Reporting and Data System (BI-RADS) rating.[27,28] This type of information defines the free-response paradigm as it applies to breast screening. Free-response is a clinical paradigm. *It is a misconception that the paradigm forces the observer to keep marking and rating many suspicious regions per case*—as the mammography example shows, this is not the case. The very name of the paradigm, *free-response*, implies, in plain English, *no forcing*.

Described next is the first medical imaging application of this paradigm.

12.4 A pioneering FROC study in medical imaging

This section details a FROC paradigm phantom study with x-ray images conducted in 1978 that is often overlooked. With the obvious substitution of clinical images for the phantom images, this study is a template for how a FROC experiment should ideally be conducted. A detailed description of it is provided to set up the paradigm, the terminology used to describe it, and concludes with the FROC plot, which is still widely (and *incorrectly*, see **Chapter 17**) used as the basis for summarizing performance in this paradigm.

12.4.1 Image preparation

Bunch et al.[3] conducted a free-response paradigm study using simulated lesions. They drilled 10–20 small holes (the simulated lesions) at random locations in ten 5 cm x 5 cm x 1.6 mm Teflon™ sheets. A Lucite™ plastic block 5 cm thick was placed on top of each Teflon™ sheet to decrease contrast and increase scatter, thereby appropriately reducing visibility of the holes (otherwise the hole detection task would be too easy; as in ROC, it is important that the task not be too easy or too difficult). Imaging conditions (kVp, mAs) were chosen such that, in preliminary studies, approximately 50% of the simulated lesions were correctly located at the observer's lowest confidence level. To minimize memory effects, the sheets were rotated, flipped or replaced between exposures. Six radiographs of four adjacent Teflon sheets, arranged in a 10 cm x 10 cm square, were obtained. Of these six radiographs, one was used for training purposes and the remaining five for data collection. Contact radiographs (i.e., with high visibility of the simulated lesions, similar in concept to the insert images of computerized analysis of mammography phantom images [CAMPI] described in Section 11.12 and Online Appendix 12.B; the cited online appendix provides a detailed description of the calculation of SNR in CAMPI) of the sheets were obtained to establish the true lesion locations. Observers were told that each sheet contained from zero to 30 simulated lesions. A mark had to be within about 1 mm to count as a correct localization; *a rigid definition was deemed unnecessary* (the emphasis is because this simple and practical advice is ignored, not by the user community, but by ROC methodology experts). Once images had been prepared, observers interpreted them. The following is how Bunch et al. conducted the image interpretation part of their experiment.

12.4.2 Image interpretation and the 1-rating

Observers viewed each film and *marked* and *rated* any visible holes with a felt-tip pen on a transparent overlay taped to the film at one edge (this allowed the observer to view the film directly without the distracting effect of previously made marks—in digital interfaces it is important to implement a show/hide feature in the user interface).

The observers used a 4-point ordered rating scale with 4 representing *most likely a simulated lesion* to 1 representing *least likely a simulated lesion*. Note the meaning of the 1-rating: least likely a simulated lesion. There is confusion with some using the FROC-1 rating to mean *definitely not a lesion*. If that were the observer's understanding, then logically the observer would fill up the entire image, especially parts outside the patient anatomy, with 1s, as each of these regions is *definitely not a lesion*. Since the observer did not behave in this unreasonable way, the meaning of the FROC-1 rating, as they interpreted it, or were told, must have been: *I am done with this image, I have nothing more to report on this image, show me the next one*.

When correctly used, the 1-rating means there is some finite, small, probability that the marked region is a lesion. In this sense, the free-response rating scale is *asymmetric*. Compare the 5-rating ROC scale, where ROC-1 = *patient is definitely not diseased* and ROC-5 = *patient is definitely diseased*. This is a *symmetric* confidence level scale. In contrast, the free-response confidence level scale labels different degrees of *positivity* in presence of disease. Table 12.1 compares the ROC 5-rating study to a FROC 4-rating study.

The FROC rating is one less than the corresponding ROC rating because the ROC-1 rating is not used by the observer; the observer indicates such images by the simple expedient of *not* marking them.

12.4.3 Scoring the data

Scoring the data was defined (Section 12.3.1) as the process of classifying each mark-rating pair as NL or LL, that is, as an incorrect or a correct decision, respectively. In the Bunch et al. study, after each case was read the person running the study (i.e., Phil Bunch) compared the marks on the overlay to the true lesion locations on the contact radiographs and scored the marks as lesion localizations (LLs: lesions correctly localized to within about 1 mm radius) or non-lesion localizations (NLs: all other marks). Bunch et al. actually used the terms *true positive* and *false positive* to describe these events. This practice, still used in publications in this field, is confusing because there is ambiguity about whether these terms, commonly used in the ROC paradigm, are being applied to the case as a whole or to specific regions in the case.

Table 12.1 Comparison of ROC and FROC rating scales: Note the FROC rating is one less than the corresponding ROC rating and that there is no rating corresponding to ROC-1. The observer's way of indicating definitely non-diseased images is by simply not marking them.

ROC paradigm		FROC paradigm	
Rating	Observer's categorization	Rating	Observer's categorization
1	Definitely not-diseased	NA	Image is not marked
2	...	1	Just possible it is a lesion
3		2	...
4		3	
5	Definitely diseased	4	Definitely a lesion

Note: NA = not available.

12.4.4 The free-response receiver operating characteristic (FROC) plot

The free-response receiver operating characteristic (FROC) plot was introduced, also in an auditory detection task, by Miller[29] as a way of visualizing performance in the free-response auditory tone detection task. In the medical imaging context, assume the marks have been classified as NLs (non-lesion localizations) or LLs (lesion localizations), along with their associated ratings. Non-lesion localization fraction (*NLF*) is defined as the total number of NLs at or above a threshold rating divided by the total number of cases. Lesion localization fraction (*LLF*) is defined as the total number of LLs at or above the same threshold rating divided by the total number of lesions. The FROC plot is defined as that of LLF (ordinate) versus NLF, as the threshold is varied. While the ordinate *LLF* is a proper fraction, for example, 30/40 assuming 30 LLs and 40 true lesions, the abscissa is an *improper* fraction that can exceed unity, for example, 35/21 assuming 35 NLs on 21 cases. The NLF notation is not ideal; it is used for notational symmetry and compactness.

> **Definitions:**
> - NLF = cumulated NL counts at or above threshold rating divided by total number of cases.
> - LLF = cumulated LL counts at or above threshold rating divided by total number of lesions.
> - The FROC curve is the plot of LLF (ordinate) versus NLF.
> - The upper-right most operating point is termed the *end-point* and its coordinates are denoted (NLF_{max}, LLF_{max}).

Following Miller's suggestion, Bunch et al.[8,30] plotted lesion localization fraction (*LLF*) along the ordinate versus non-lesion localization fraction (*NLF*) along the abscissa. Corresponding to the different threshold ratings, pairs of (*NLF, LLF*) values, or operating points on the FROC, were plotted. For example, in a positive directed 4-rating FROC study, such as employed by Bunch et al., four FROC operating points result: those corresponding to marks rated 4s; those corresponding to marks rated 4s or 3s; the 4s, 3s, or 2s; and finally, the 4s, 3s, 2s, or 1s. An R-rating (integer R > 0) FROC study yields at most R operating points. So, Bunch et al. were able to plot only four operating points per reader, Figure 6 in Ref. 8.* Lacking a method of fitting a continuous FROC curve to the operating points, they did the best they could, and manually French-curved fitted curves. In 1986, the author followed the same practice in his first paper on this topic.[9] In 1989 the author described[6] a method for fitting such operating points, and developed software called FROCFIT, but the fitting method is obsolete as the underlying statistical model has been superseded; see **Chapter 16**. Moreover, it is now known, see **Chapter 17,** that the FROC plot is a poor visual descriptor of performance.

If continuous ratings are used, the procedure is to start with a high threshold so none of the ratings exceed the threshold, and gradually lower the threshold. Every time the threshold crosses the rating of a mark, or possibly multiple marks, the total count of LLs and NLs exceeding the threshold is divided by the appropriate denominators yielding the raw FROC plot. For example, when an LL rating just exceeds the threshold, the operating point jumps up by

---

* Figure 7 *ibid* has about 12 operating points as it includes three separate interpretations by the same observer. Moreover, the area scaling implicit in the paper assumes a homogenous and isotropic image, that is, the probability of a NL is proportional to the image area over which it is calculated, which is valid for a uniform background phantom. Clinical images are not homogenous and isotropic and therefore not scalable in the Bunch et al. sense.

1/(total number of lesions), and if two LLs simultaneously just exceed the threshold, the operating point jumps up by 2/(total number of lesions). If an NL rating just exceeds the threshold, the operating point jumps to the right by 1/(total number of cases). If an LL rating and an NL rating simultaneously just exceed the threshold, the operating point moves diagonally, up by 1/(total number of lesions) and to the right by 1/(total number of cases). The reader should get the general idea by now and recognize that the cumulating procedure is very similar to the manner in which ROC operating points were calculated, the only differences being in the quantities being cumulated and the relevant denominators.

Having seen how a binned data FROC study is conducted and scored, and the results French-curved as a FROC plot, typical simulated plots, generated under controlled conditions, are shown next, both for continuous ratings data and for binned rating data. Such demonstrations, illustrating trends, are impossible using real datasets. The reader should take the author's word for it (for now) that the simulator used is the simplest one possible that incorporates key elements of the search process. Details of the simulator are given in **Chapter 16**, but for now the following summary should suffice.

The simulator is characterized by three parameters μ, λ, and ν. The ν parameter characterizes the *ability of the observer to find lesions*. The λ parameter characterizes the *ability of the observer to avoid finding non-lesions*. The μ parameter characterizes the *ability of the observer to correctly classify a found suspicious region as a true lesion or a non-lesion*. The reader should think of μ as a perceptual signal-to-noise ratio (*pSNR*) or conspicuity of the lesion, similar to the separation parameter of the binormal model, that separates two normal distributions describing the sampling of ratings of NLs and LLs. The simulator also needs to know the number of lesions per diseased case, as this determines the number of possible LLs on each case. Finally, there is a threshold parameter ζ_1 that determines whether a found suspicious region is actually marked. If ζ_1 is negative infinity, then all found suspicious regions are marked and conversely, as ζ_1 increases, only those suspicious regions whose confidence levels exceed ζ_1 are marked. The concept of *pSNR* is clarified in Section 12.5.2.

12.5 Population and binned FROC plots

Figure 12.2a through c shows simulated population FROC plots when the ratings are not binned, generated by file **mainFrocCurvePop.R** described in Appendix 12.A. FROC data from 20,000 cases, half of them non-diseased, are generated (the code takes a while to finish). The very large number of cases minimizes sampling variability, hence the term, *population curves*. Additionally, the reporting threshold ζ_1 was set to negative infinity to ensure that all suspicious regions are marked. With higher thresholds, suspicious regions with confidence levels below the threshold would not be marked and the rightward and upward traverses of the curves would be truncated. Plots in Figure 12.2a through c correspond to μ equal to 0.5, 1, and 2, respectively. Plots in Figure 12.2d through f correspond to 5-rating binned data for 50 non-diseased and 50 diseased cases, and the same values of μ; the relevant file is **mainFrocCurveBinned.R**.

1. Plots in Figure 12.2a through c show quasi-continuous plots while Figure 12.2 d through f show operating points, five per plot, connected by straight line segments, so they are termed *empirical FROC curves*, analogous to the empirical ROC curves encountered in previous chapters. At

a microscopic level plots (a) through (c) are also discrete, but one would need to zoom in to see the discrete behavior (upward and rightward jumps) as each rating crosses a sliding threshold.

2. The empirical plots in the bottom row (d) through (f) of Figure 12.2 are subject to sampling variability and will not, in general, match the population plots. The reader should try different values of the **seed** variable in the code.

3. *In general, FROC plots do not extend indefinitely to the right.* Figure 5 in the Bunch et al. paper is incorrect in implying, with the arrows, that the plots extend indefinitely to the right. (Notation differences: In Bunch et al., *P(TP)* or v is equivalent to the author's *LLF*. The variable Bunch et al. call λ is equivalent to *NLF* in this book.)

4. Like a ROC plot, the population FROC curve rises monotonically from the origin, initially with infinite slope (this may not be evident for Figure 12.2a, but it is true; see code snippet 12.5.1). If all suspicious regions are marked, that is, $\zeta_1 = -\infty$, the plot reaches its upper rightmost limit, termed the *end-point*, with zero slope (again, this may not be evident for (a), but it is true [see code snippet below]; here *x* and *y* are arrays containing NLF and LLF, respectively). In general, these characteristics, that is, initial infinite slope and zero final slope, are not true for empirical plots Figure 12.2d through f.

12.5.1 Code snippet

```
> mu
[1] 0.5
> (y[2]-y[1])/(x[2]-x[1]) # slope at origin
[1] Inf
> (y[10000]-y[10000-1])/(x[10000]-x[10000-1]) # slope at end-point
[1] 0
```

5. *Assuming all suspicious regions are marked, the end-point* (NLF_{max}, LLF_{max}) *represents a literal end of the extent of the population FROC curve. This will become clearer in following chapters, but for now it should suffice to note that the region of the population FROC plot to the upper right of the end-point is inaccessible to the observer. [If sampling variability is taken into account it is possible for the observed end-point to extend into this inaccessible space.]*

6. *There is an inverse correlation between* LLF_{max} *and* NLF_{max} *analogous to that between sensitivity and specificity in ROC analysis. The end-point* (NLF_{max}, LLF_{max}) *of the FROC tends to approach the point (0,1) as the perceptual SNR of the lesions approaches infinity. As* μ *decreases the FROC curve approaches the x-axis and extends to large values along the abscissa, as in Figure 12.2b. This is the chance-level FROC, where the reader detects few lesions, and makes many NL marks.*

7. The slope of the population FROC decreases monotonically as the operating point moves up the curve, always staying non-negative, and it approaches zero, flattening out at an ordinate *less than unity*. Some publications[31] (Figure 3 *ibid.*) and Reference [32] (Figure 1 *ibid.*) incorrectly show *LLF* reaching unity. This is generally not the case unless the lesions are particularly conspicuous. This is well known to CAD researchers and to anyone who has conducted FROC studies with radiologists. *LLF* reaches unity for large μ, which can be confirmed by setting μ to a large value, for example, 10, Figure 12.3a. On the unit variance normal distribution scale, a value of 10, equivalent to 10 standard deviations, is effectively infinite.

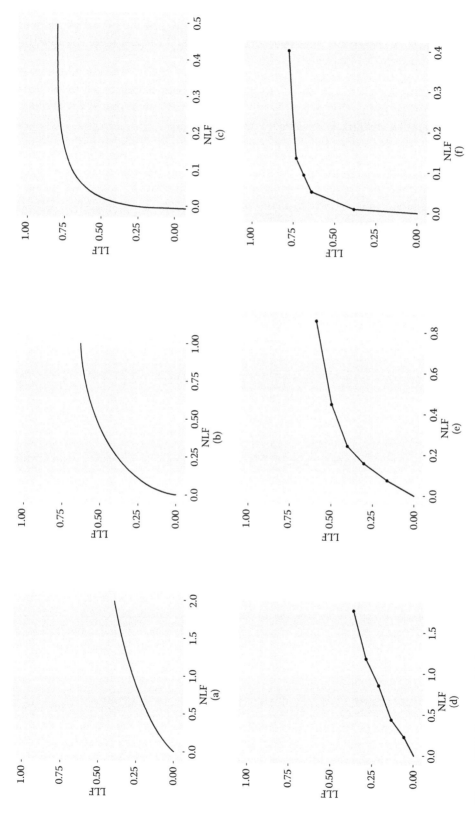

Figure 12.2 Top row, (a) through (c): Population FROC plots for $\mu = 0.5$, 1, 2; the other parameters are $\lambda = 1$, $\nu = 1$, $\zeta_1 = -\infty$, and $L_{max} = 2$ is the maximum number of lesions per case in the dataset. The plots in the bottom row (d) through (f) correspond to 50 non-diseased and 70 diseased cases, where the data was binned into five bins, and other parameters are unchanged. As μ increases, the uppermost point moves upwards and to the left. The top row of images was produced by **MainFrocCurvePop.R** while the bottom row by **mainFrocCurveBinned.R**.

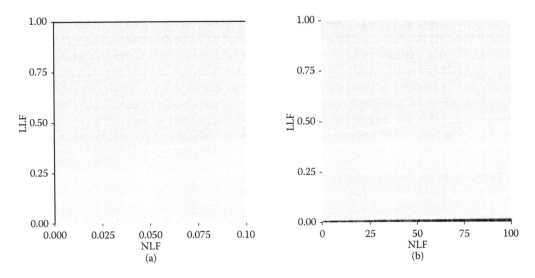

Figure 12.3 (a) FROC plot for $\mu = 10$ in code file **mainFrocCurvePop.R**. Note the small range of the NLF axis (it extends to 0.1). In this limit the ordinate reaches unity but the abscissa is limited to a small value; see solar analogy Section 12.6 for explanation. (b) This plot corresponds to $\mu = 0.01$, depicting near *chance-level* performance. Note the greatly increased traverse in the x-directions and the slight upturn in the plot near NLF = 100.

12.5.2 Perceptual SNR

The shape and extent of the FROC plot is, to a large extent, determined by the *perceptual** SNR of the lesions, *pSNR*, modeled by parameter μ. Perceptual SNR is the ratio of perceptual signal to perceptual noise. To get to perceptual variables one needs a model of the eye-brain system that transforms physical image brightness variations to corresponding perceived brightness variations, and such models exist.[33–35] For uniform background images, like the phantom images used by Bunch et al., a physical signal can be measured by a *template* function that has the same attenuation profile as the true lesion; an overview of this concept was given in Section 1.6. Assuming the template is aligned with the lesion the *cross-correlation* between the template function and the image pixel values is related to the numerator of SNR. The cross-correlation is defined as the summed product of template function pixel values times the corresponding pixel values in the actual image. Next, one calculates the cross-correlation between the template function and the pixel values in the image when the template is centered over regions known to be *lesion free*. Subtracting the mean of these values (over several lesion free regions) from the centered value gives the numerator of SNR. The denominator is the standard deviation of the cross-correlation values in the lesion free areas. Appendix 12.B has details on calculating *physical* SNR, which derives from the author's CAMPI (computer analysis of mammography phantom images) work[36–40]. To calculate *perceptual* SNR, one repeats these measurements but the visual process, or some model of it (e.g., the Sarnoff JNDMetrix™ visual discrimination model[35,41,42]), is used to filter the image prior to calculation of the cross-correlations.

An analogy may be helpful at this point. *Finding the sun in the sky is a search task, so it can be used to illustrate important concepts.*

---

* Since humans make the decisions, it would be incorrect to label these as *physical* signal-to-noise ratios; that is the reason for qualifying them as *perceptual* SNRs.

12.6 The solar analogy: Search versus classification performance

Consider the sun, regarded as a lesion to be detected, with two daily observations spaced 12 hours apart, so that at least one observation period is bound to have the sun somewhere up there. Furthermore, the observer is assumed to know their GPS coordinates and have a watch that gives accurate local time, from which an accurate location of the sun can be deduced. Assuming clear skies and no obstructions to the view, the sun will always be correctly located and no reasonable observer will ever generate a non-lesion localization or *NL*, that is, no region of the sky will be erroneously marked.

FROC curve implications of this analogy are as follows:

- Each 24-hour day corresponds to two trials in the Egan et al. sense,[6] or two cases—one diseased and one non-diseased—in the medical imaging context.
- The denominator for calculating *LLF* is the total number of AM days, and the denominator for calculating *NLF* is twice the total number of 24-hour days.
- *Most important, $LLF_{max} = 1$ and $NLF_{max} = 0$.*

In fact, even when the sun is not directly visible due to heavy cloud cover, since the actual location of the sun can be deduced from the local time and GPS coordinates, the rational observer will still mark the correct location of the sun and not make any false sun localizations or non-lesion localizations, NLs, in the context of FROC terminology. Consequently, even in this example $LLF_{max} = 1$ and $NLF_{max} = 0$.

The conclusion is that in a task where a target is known to be present in the field of view and its location is known, the observer will always reach $LLF_{max} = 1$ and $NLF_{max} = 0$. Why are *LLF* and NLF subscripted *max*? By randomly *not* marking the position of the sun even though it is visible, for example, using a coin toss to decide whether or not to mark the sun, the observer can "walk down" the *y*-axis of the FROC plot, reaching $LLF = 0$ and $NLF = 0$.* Alternatively, the observer uses a very large threshold for reporting the sun, and as this threshold is lowered the operating point "walks down" the curve. The reason for allowing the observer to walk down the vertical is simply to demonstrate that a continuous FROC curve from the origin to the highest point (0,1) can in fact be realized.

Now consider a fictitious otherwise earth-like planet where the sun can be at *random* positions, rendering GPS coordinates and the local time useless. All one knows is that the sun is somewhere, in the upper or lower hemispheres, subtended by the sky. If there are no clouds and consequently one can see the sun clearly during daytime, a reasonable observer would still correctly locate the sun while not marking the sky with any incorrect sightings, so $LLF_{max} = 1$ and $NLF_{max} = 0$. This is because, in spite of the fact that the expected location is unknown, the high contrast sun is enough to trigger the peripheral vision system, so that even if the observer did not start out looking in the correct direction, peripheral vision will *drag* the observer's gaze to the correct location for foveal viewing.

The implication of this is that two fundamentally different mechanisms from that considered in conventional observer performance methodology, namely *search* and *lesion-classification*, are involved. Search describes the process of *finding* the lesion while not finding non-lesions. Classification describes the process, once a possible sun location has been found, of recognizing that it is indeed the sun and marking it. Recall that search involves two steps: finding the object of the search and acting on it. Search and lesion-classification performances quantify the abilities of an observer to perform these steps.

---

* The logic is very similar to that used in Section 3.9.1 to describe how the ROC observer can "walk along" the chance diagonal of the ROC curve.

Think of the eye as two cameras: a low-resolution camera (peripheral vision) with a wide field-of-view, plus a high-resolution camera (foveal vision) with a narrow field-of-view. If one were limited to viewing with the high-resolution camera one would spend so much time steering the high-resolution narrow field-of-view camera from spot to spot that one would have a hard time finding the desired stellar object. Having a single high-resolution narrow field-of-view would also have negative evolutionary consequences as one would spend so much time scanning and processing the surroundings with the narrow field of view vision that one would miss dangers or opportunities. Nature has equipped us with essentially two cameras. The first low-resolution camera is able to digest large areas of the surroundings and process it rapidly so that if danger (or opportunity) is sensed, then the eye-brain system rapidly steers the second high-resolution camera to the location of the danger (or opportunity). This is nature's way of optimally using the finite resources of the eye-brain system. For a similar reason, astronomical telescopes come with a wide field-of-view lower resolution spotter scope.

Since the large field-of-view low-resolution peripheral vision system has complementary properties to the small field-of-view high-resolution foveal vision system, one expects an inverse correlation between search and lesion-classification performances. Stated generally, search involves two complementary processes: *finding* the suspicious regions and *deciding* whether the found region is actually a lesion, and an inverse correlation between performance in the two tasks is expected, (see **Chapter 19**).

When cloud cover completely blocks the fictitious random-position sun there is no stimulus to trigger the peripheral vision system to guide the fovea to the correct location. Lacking any stimulus, the observer is reduced to guessing and is led to different conclusions depending upon the benefits and costs involved. If, for example, the guessing observer earns a dollar for each LL and is fined a dollar for each NL, then the observer will likely not make any marks as the chance of winning a dollar is much smaller than losing many dollars. For this observer $LLF_{max} = 0$ and $NLF_{max} = 0$, and the operating point is "stuck" at the origin. If, on the other hand, the observer is told every LL is worth a dollar and there is no penalty to NLs, then with no risk of losing, the observer will fill up the sky with marks. In the second situation, the locations of the marks will lie on a grid determined by the ratio of the 4π solid angle (subtended by the spherical sky) and the solid angle Ω subtended by the sun. By marking every possible grid location, the observer is guaranteed to detect the sun and earn a dollar irrespective of its random location and reach $LLF = 1$, but now the observer will generate lots of non-lesion localizations, so maximum NLF will be large:

$$NLF_{max} = \frac{4\pi}{\Omega} \tag{12.1}$$

The FROC plot for this guessing observer is the straight line joining $(0,0)$ to $(NLF_{max}, 1)$. For example, if the observer fills up half the sky, then the operating point, averaged over many trials, is

$$\left(0.5\,NLF_{max}, 0.5\right) \tag{12.2}$$

Radiologists do not guess—there is much riding on their decisions—so in the clinical situation, if the lesion is not seen, the radiologist will not mark the image at random.

The analogy is not restricted to the sun, which one might argue is an almost infinite SNR object and therefore atypical. As another example, consider finding stars or planets. In clear skies, if one knows the constellations, herein lies the role of expertise, one can still locate bright stars and planets like Venus or Jupiter. With fewer bright stars and/or obscuring clouds, there will be false sightings and the FROC plot could approach a flat horizontal line at ordinate equal to zero, but the astronomer will not fill up the sky with false sightings of a desired star.

False sightings of objects in astronomy do occur. Finding a new astronomical object is a search task, with two outcomes: correct localization (LL) or incorrect localization (NL). At the time of writing there is a hunt for a new planet, possibly a gas giant,* that is much further away than even the newly demoted Pluto. There is an astronomer in Australia† who is particularly good at finding super novae (an exploding star; one has to be looking in the right region of the sky at the right time to see the relatively brief explosion). His equipment is primitive by comparison to the huge telescope at Mt. Palomar, but his advantage is that he can rapidly point his 15 "telescope at a new region of the sky and thereby cover a lot more sky in a given unit of time than is possible with the 200" Mt. Palomar telescope. His *search expertise* is particularly good. Once correctly localized or pointed to, the Mt. Palomar telescope will reveal a lot more detail about the object than is possible with the smaller telescope, that is, it has high lesion-classification accuracy. In the medical imaging context this detail (the shape of the lesion, its edge characteristics, presence of other abnormal features, etc.) allows the radiologist to diagnose whether the lesion is malignant or benign. Once again one sees that there should be an inverse correlation between search and lesion-classification performances.

Prof. Jeremy Wolfe of Harvard University is an expert in visual search and the interested reader is referred to his many publications.[43,44] As noted by him, rare items are often missed. To paraphrase him, *things that are not seen often are often not seen.*[45] So the problem faced by an astronomer looking for supernova events, a terminal security agency baggage inspector looking for explosives, and the radiologist interpreting a screening mammogram for rare cancers, are similar at a fundamental level: all of these are low prevalence search tasks.

12.7 Discussion/Summary

This chapter has introduced the FROC paradigm, the terminology used to describe it, and a common operating characteristic associated with it, namely the FROC. In the author's experience this paradigm is widely misunderstood. The following rules are intended to reduce the confusion:

- Avoid using the term *lesion-specific* to describe *location-specific* paradigms.
- Avoid using the term *lesion* when one means a *suspicious region* that may not be a true lesion.
- Avoid using ROC-specific terms, such as true positive and false positive, that apply to the whole case, to describe location-specific terms such as lesion and non-lesion localization, that apply to localized regions of the image.
- Avoid using the FROC-1 rating to mean in effect "*I see no more signs of disease in this image,*" when in fact it should be used as the lowest level of a reportable suspicious region. The former usage amounts to "wasting" a confidence level.
- Do not show FROC curves as reaching the unit ordinate, as this is the exception rather than the rule.
- Do not conceptualize FROC curves as extending to arbitrarily large values to the right.
- Arbitrariness of the proximity criterion and multiple marks in the same region are not clinical constraints. Interactions with clinicians will allow selection of an appropriate proximity criterion for the task at hand and the second problem (multiple marks in the same region) only occurs with algorithmic observers and is readily fixed.

Additional points made in this chapter are: There is an inverse correlation between LLF_{max} and NLF_{max} analogous to sensitivity and specificity in ROC analysis. The end-point (NLF_{max}, LLF_{max}) of the FROC curve tends to approach the point (0,1) as the perceptual SNR of the lesions approaches

---

* https://en.wikipedia.org/wiki/Tyche_(hypothetical_planet)
† https://en.wikipedia.org/wiki/Robert_Evans_(astronomer)

infinity. The solar analogy is highly relevant to understanding the search task. In search tasks, two types of expertise are at work: search and lesion-classification performances, and an inverse correlation between them is expected.

Online Appendix 12.A describes, and explains in detail, the code used to generate the population FROC curves shown in Figure 12.2a through c. Online Appendix 11.B details how one calculates *physical* signal-to-noise ratio (SNR) for an object on a uniform noise background. This is useful in understanding the concept of *perceptual* signal-to-noise ratio denoted μ. Online Appendix 12.C is for those who wish to understand the Bunch et al. paper[8] in more depth. This paper has certain transformations, sometimes referred to as the *Bunch transforms*, which relate a ROC plot to a FROC plot and vise-versa. *It is not a model of FROC data.* The reason for including it is that this important paper is much overlooked, and if the author does not write it, no one else will.

The FROC plot is the first proposed way of visually summarizing FROC data. The next chapter deals with different empirical operating characteristics that can be defined from an FROC dataset.

References

1. Black WC. Anatomic extent of disease: A critical variable in reports of diagnostic accuracy. *Radiology*. 2000;217(2):319–320.
2. Halpern SD, Karlawish JH, Berlin JA. The continuing unethical conduct of underpowered clinical trials. *JAMA*. 2002;288(3):358–362.
3. Black WC, Dwyer AJ. Local versus global measures of accuracy: An important distinction for diagnostic imaging. *Med Decis Making*. 1990;10(4):266–273.
4. Obuchowski NA, Mazzone PJ, Dachman AH. Bias, underestimation of risk, and loss of statistical power in patient-level analyses of lesion detection. *Eur Radiol*. 2010;20:584–594.
5. Alberdi E, Povyakalo AA, Strigini L, Ayton P, Given-Wilson R. CAD in mammography: Lesion-level versus case-level analysis of the effects of prompts on human decisions. *Int J Comput Assist Radiol Surg*. 2008;3(1):115–122.
6. Chakraborty DP. Maximum likelihood analysis of free-response receiver operating characteristic (FROC) data. *Med Phys*. 1989;16(4):561–568.
7. Egan JP, Greenburg GZ, Schulman AI. Operating characteristics, signal detectability and the method of free-response. *J Acoust Soc Am*. 1961;33:993–1007.
8. Bunch PC, Hamilton JF, Sanderson GK, Simmons AH. A free-response approach to the measurement and characterization of radiographic-observer performance. *J Appl Photogr Eng*. 1978;4:166–171.
9. Chakraborty DP, Breatnach ES, Yester MV, Soto B, Barnes GT, Fraser RG. Digital and conventional chest imaging: A modified ROC study of observer performance using simulated nodules. *Radiology*. 1986;158:35–39.
10. Chakraborty DP, Winter LHL. Free-response methodology: Alternate analysis and a new observer-performance experiment. *Radiology*. 1990;174:873–881.
11. Chakraborty DP, Berbaum KS. Observer studies involving detection and localization: Modeling, analysis and validation. *Med Phys*. 2004;31(8):2313–2330.
12. Starr SJ, Metz CE, Lusted LB, Goodenough DJ. Visual detection and localization of radiographic images. *Radiology*. 1975;116:533–538.
13. Starr SJ, Metz CE, Lusted LB. Comments on generalization of Receiver Operating Characteristic analysis to detection and localization tasks. *Phys Med Biol*. 1977;22:376–379.
14. Swensson RG. Unified measurement of observer performance in detecting and localizing target objects on images. *Med Phys*. 1996;23(10):1709–1725.
15. Judy PF, Swensson RG. Lesion detection and signal-to-noise ratio in CT images. *Med Phys*. 1981;8(1):13–23.
16. Swensson RG, Judy PF. Detection of noisy visual targets: Models for the effects of spatial uncertainty and signal-to-noise ratio. *Percept Psychophys*. 1981;29(6):521–534.

17. Obuchowski NA, Lieber ML, Powell KA. Data analysis for detection and localization of multiple abnormalities with application to mammography. *Acad Radiol.* 2000;7(7):516–525.

18. Rutter CM. Bootstrap estimation of diagnostic accuracy with patient-clustered data. *Acad Radiol.* 2000;7(6):413–419.

19. Ernster VL. The epidemiology of benign breast disease. *Epidemiol Rev.* 1981;3(1):184–202.

20. Niklason LT, Hickey NM, Chakraborty DP, et al. Simulated pulmonary nodules: Detection with dual-energy digital versus conventional radiography. *Radiology.* 1986;160:589–593.

21. Haygood TM, Ryan J, Brennan PC, et al. On the choice of acceptance radius in free-response observer performance studies. *BJR.* 2012;86(1021): 42313554.

22. Chakraborty DP, Yoon HJ, Mello-Thoms C. Spatial localization accuracy of radiologists in free-response studies: Inferring perceptual FROC curves from mark-rating data. *Acad Radiol.* 2007;14:4–18.

23. Kallergi M, Carney GM, Gaviria J. Evaluating the performance of detection algorithms in digital mammography. *Med Phys.* 1999;26(2):267–275.

24. Gur D, Rockette HE. Performance assessment of diagnostic systems under the FROC paradigm: Experimental, analytical, and results interpretation issues. *Acad Radiol.* 2008;15:1312–1315.

25. Dobbins JT III, McAdams HP, Sabol JM, et al. Multi-institutional evaluation of digital tomosynthesis, dual-energy radiography, and conventional chest radiography for the detection and management of pulmonary nodules. *Radiology.* 2016;282(1):236–250.

26. Hartigan JA, Wong MA. Algorithm AS 136: A k-means clustering algorithm. *J R Stat Soc Ser C Appl Stat.* 1979;28(1):100–108.

27. D'Orsi CJ, Bassett LW, Feig SA, et al. *Illustrated Breast Imaging Reporting and Data System.* Reston, VA: American College of Radiology; 1998.

28. D'Orsi CJ, Bassetty LW, Berg WA. *ACR BI-RADS-Mammography.* 4th ed. Reston, VA: American College of Radiology; 2003.

29. Miller H. The FROC curve: A representation of the observer's performance for the method of free-response. *J Acoust Soc Am.* 1969;46(6):1473–1476.

30. Bunch PC, Hamilton JF, Sanderson GK, Simmons AH. A free-response approach to the measurement and characterization of radiographic-observer performance. *Proc SPIE.* 1977;127:124–135. Boston, MA.

31. Popescu LM. Model for the detection of signals in images with multiple suspicious locations. *Med Phys.* 2008;35(12):5565–5574.

32. Popescu LM. Nonparametric signal detectability evaluation using an exponential transformation of the FROC curve. *Med Phys.* 2011;38(10):5690–5702.

33. Van den Branden Lambrecht CJ, Verscheure O. Perceptual quality measure using a spatiotemporal model of the human visual system. SPIE Proceedings Volume 2668, Digital Video Compression: Algorithms and Technologies. Event: Electronic Imaging: Science and Technology, 1996, San Jose, CA; 1996. doi: 10.1117/12.235440.

34. Daly SJ. Visible differences predictor: An algorithm for the assessment of image fidelity. *Digital images and human vision* 4 (1993): 124–125. SPIE/IS&T 1992 Symposium on Electronic Imaging: Science and Technology; 1992; San Jose, CA.

35. Lubin J. A visual discrimination model for imaging system design and evaluation. In: Peli E, ed. *Vision Models for Target Detection and Recognition.* Vol. 2, pp. 245–357. Singapore: World Scientific; 1995.

36. Chakraborty DP, Sivarudrappa M, Roehrig H. Computerized measurement of mammographic display image quality. Paper presented at: Proc SPIE Medical Imaging 1999: Physics of Medical Imaging; 1999; San Diego, CA.

37. Chakraborty DP, Fatouros PP. Application of computer analyis of mammography phantom images (CAMPI) methodology to the comparison of two digital biopsy machines. Paper presented at: Proc SPIE Medical Imaging 1998: Physics of Medical Imaging; 24 July 1998, 1998.

38. Chakraborty DP. Comparison of computer analysis of mammography phantom images (CAMPI) with perceived image quality of phantom targets in the ACR phantom. Paper presented at: Proc. SPIE Medical Imaging 1997: Image Perception; 26–27 February 1997; Newport Beach, CA.
39. Chakraborty DP. Computer analysis of mammography phantom images (CAMPI). *Proc SPIE Med Imaging 1997 Phys Med Imaging*. 1997;3032:292–299.
40. Chakraborty DP. Computer analysis of mammography phantom images (CAMPI): An application to the measurement of microcalcification image quality of directly acquired digital images. *Med Phys*. 1997;24(8):1269–1277.
41. Siddiqui KM, Johnson JP, Reiner BI, Siegel EL. Discrete cosine transform JPEG compression vs. 2D JPEG2000 compression: JNDmetrix visual discrimination model image quality analysis. Paper presented at: Medical Imaging; 2005. SPIE Proceedings Volume 5748, Medical Imaging 2005: PACS and Imaging Informatics; doi: 10.1117/12.596146, San Diego, CA.
42. Chakraborty DP. An alternate method for using a visual discrimination model (VDM) to optimize softcopy display image quality. *J Soc Inf Display* 2006;14(10):921–926.
43. Wolfe JM. Guided search 2.0: A revised model of visual search. *Psychonomic Bull Rev*. 1994;1(2):202–238.
44. Wolfe JM. Visual search. In: Pashler H, ed. *Attention*. London: University College London Press; 1998.
45. Wolfe JM, Horowitz TS, Kenner NM. Rare items often missed in visual searches. *Nature*. 2005;435(26):439.

13

Empirical operating characteristics possible with FROC data

13.1 Introduction

Operating characteristics are visual depicters of performance. Quantities derived from operating characteristics can serve as figures of merit (FOMs), that is, quantitative measures of performance. For example, the area under an empirical ROC is a widely used FOM in receiver operating characteristic (ROC) analysis. This chapter defines empirical operating characteristics possible with FROC data.

Here is the organization of this chapter. A distinction between latent* and actual marks is made followed by a summary of free-response ROC (FROC) notation applicable to a single dataset where modality and reader indices are not needed. This is a key table, which will be referred to in later chapters. Following this, the chapter is organized into two main parts: formalism and examples. The formalism sections, Sections 13.3 through 13.9, give formula for calculating different empirical operating characteristics. While dry reading, it is essential to master, and the concepts are not that difficult. The notation may appear dense because the FROC paradigm allows an a priori unknown number of marks and ratings per case, but deeper inspection should convince the reader that the apparent complexity is needed. When applied to the FROC plot the formalism is used to demonstrate an important fact, namely the *semi-constrained* property of the observed end-point, unlike the *constrained* ROC end-point, whose upper limit is (1,1).

The second part, Sections 13.10 through 13.14, consists of coded examples of operating characteristics. Section 13.15 is devoted to clearing up confusion, in a clinical journal, about "location-level true negatives," traceable in large part to misapplication of ROC terminology to location-specific tasks. Unlike other chapters, in this chapter most of the code is not relegated to online appendices. This is because the concepts are most clearly demonstrated at the code level. The FROC data structure is examined in some detail. Raw and binned FROC, AFROC and ROC plots are coded under controlled conditions. Emphasized is the fact that unmarked non-diseased regions, confusingly termed "location level true negatives," are unmeasurable events that should not be used in analysis. A simulated algorithmic observer and a simulated expert radiologist are compared using both FROC and AFROC curves, showing that the latter is preferable. The code for this is in an online appendix. The chapter

---

* In previous publications the author has termed these possible or potential NLs or LLs; going by the dictionary definition of latent (that is, of a quality or state) existing but not yet developed or manifest, the present usage seems more appropriate. The latent mark should not be confused with the latency property of the decision variable, that is, the invariance of operating points to arbitrary monotone increasing functions of the decision variable.

concludes with recommendations on which operating characteristics to use and which to avoid. In particular, the alternative free-response operating characteristic (AFROC) has desirable properties that make it the preferred way of summarizing performance. An interesting example is given where AFROC-AUC = 0.5 can occur, and indicates better-than-chance level performance.

The starting point is the distinction between latent and actual marks and FROC notation.

13.2 Latent versus actual marks

From **Chapter 12**, FROC data consists of mark-rating pairs. Each mark indicates the location of a region suspicious enough to warrant reporting and the rating is the associated confidence level. A mark is recorded as lesion localization (LL) if it is sufficiently close to a true lesion according to the adopted proximity criterion; otherwise, it is recorded as non-lesion localization (NL).

- To distinguish between perceived suspicious regions and regions that were actually marked, it is necessary to introduce the distinction between *latent* marks and *actual* marks. A latent mark is defined as a suspicious region, regardless of whether it was marked. A latent mark becomes an actual mark if it is marked.
- A latent mark is a latent *LL* if it is close to a true lesion and otherwise it is a latent *NL*. A non-diseased case can only have latent *NLs*. A diseased case can have latent *NLs* and latent *LLs*.

13.2.1 FROC notation

Recall from Section 3.2, that the ROC paradigm requires the existence of a *case-dependent* decision variable (Z-sample) and a *case-independent* decision threshold ζ, and the rule that if $z \geq \zeta$ the case is diagnosed as diseased and otherwise the case is diagnosed as non-diseased as usual, upper case Z vs. lower case z denotes the difference between a random variable and a realized value. Analogously, FROC data requires the existence of a *case* and *location-dependent* Z-sample associated with each latent mark and a *case-independent* reporting threshold ζ and the rule that a latent mark is marked if $z \geq \zeta$. One needs to account, in the notation, for case and locations dependences of z and the distinction between case-level and location-level ground truth. For example, a diseased case can have many localized regions that are non-diseased and a few diseased regions (the lesions).

Clear notation is vital to understanding this paradigm. FROC notation is summarized in Table 13.1 and it is important to bookmark this table, as it will be needed to understand the subsequent development of this subject. For ease of referencing, the table is organized into three columns: the first column is the row number, the second column has the symbol(s), and the third column has the meaning(s) of the symbol(s).

Row 1: The *case-truth* index t refers to the case (or patient), with $t = 1$ for non-diseased and $t = 2$ for diseased cases.

Row 2: Two indices $k_t t$ (row 2) are needed to select case k_t in truth-state t (recall the need for two case-level indices in ROC analysis, Table 5.1).

Row 3 and 4: For a similar reason, two more indices $l_s s$ are needed to select latent mark l_s in *local* truth-state s, where $s = 1$ corresponds to a latent NL and $s = 2$ corresponds to a latent LL. One can think of l_s as indexing the *locations* of different latent marks with local truth-state s.

Row 5: The realized Z-sample for case $k_t t$ and latent NL mark $l_1 1$ is denoted $z_{k_t t l_1 1}$. Latent NL marks are possible on non-diseased and diseased cases (i.e., both values of t are allowed). The range

Table 13.1 This table summarizes FROC notation. See Section 13.2.1 for details

Row #	Symbols	Meanings
1	t	Case-level truth-state: $t = 1$ for non-diseased and $t = 2$ for diseased case
2	$k_t t; k_t = 1, 2, ..., K_t$	Case k_t in case-level truth-state t; K_t is the total number of cases in truth-state t
3	s	Mark-level truth-state: $s = 1$ for *NL* and $s = 2$ for *LL* marks
4	$l_s s$	Latent mark l_s in mark-level truth-state s
5	$z_{k_t t h_1 1}; z_{k_2 2 l_2 2}$	Z-sample for case $k_t t$ and latent mark $l_1 1; -\infty < z_{k_t t h_1 1} < \infty$ provided $l_1 \neq \varnothing$; otherwise it is an unobservable event; Z-sample for case $k_2 2$ and latent mark $l_2 2$; unmarked lesions are assigned negative infinity ratings
6	ζ_1	Lowest reporting threshold; latent mark is marked only if $z_{k_t t l_s s} \geq \zeta_1$
7	$\zeta_r; r = 1, 2, ..., R_{FROC}$	If $\zeta_r \leq z_{k_t t l_s s} < \zeta_{r+1}$ mark is assigned rating r; dummy thresholds are $\zeta_0 = -\infty$, and $\zeta_{R_{froc}+1} = \infty$; R_{FROC} is the number of FROC bins
8	$N_{k_t t} \geq 0, N_T$	Numbers of latent NLs on case $k_t t$; N_T is the total number of marked NLs in the dataset
9	$L_{k_2}, L_T = \sum\limits_{k_2=1}^{K_2} L_{k_2}$	Number of lesions in diseased case $k_2 2$; total number of lesions in dataset
10	l_1, l_2	Indexing latent marks: $l_1 = \{\varnothing\} \oplus \{1, 2, ..., N_{k_t t}\}; l_2 = \{1, 2, ..., L_{k_2}\}$

of a Z-sample is $-\infty < z_{k_t t l_1 1} < \infty$, provided $l_1 \neq \varnothing$; otherwise, it is an unobservable event (see text box below). The Z-sample of a latent LL is $z_{k_2 2 l_2 2}$. Unmarked lesions are assigned null set labels and negative infinity ratings; this is the meaning of $\left(z_{k_2 2 l_2 2} | l_2 = \varnothing \right) = -\infty$.

Row 6 and 7: A latent mark is actually marked if $z_{k_t t l_s s} \geq \zeta_1$, where ζ_1 is the lowest reporting threshold adopted by the observer. Additional thresholds $(\zeta_2, \zeta_3, ...)$ are needed to accommodate greater than one FROC bins. If marked, a latent *NL* is recorded as an actual *NL*, and likewise if marked, a latent *LL* is recorded as an actual *LL*.

- If not marked, a latent NL is an unobservable event: more on this in Section 13.15. This is a major source of confusion among some researchers familiar with the ROC paradigm who use the highly misleading term *location-level true negative* for unmarked latent NLs.
- In contrast, unmarked lesions are *observable* events—one knows (trivially) which lesions were not marked. In the analyses, unmarked lesions are assigned $-\infty$ ratings, guaranteed to be smaller than any rating used by the observer.

Row 8: $N_{k_t t} \geq 0$ is the total number of latent *NL* marks on case $k_t t$. N_T is the total number of latent NLs in the dataset.

It is an a priori unknown *modality-reader-case dependent non-negative random integer*. It is incorrect to estimate it by dividing the image area by the lesion area because not all regions of the image are equally likely to have lesions, lesions do not have the same size, and most important, clinicians don't work that way. The best insight into the number of latent *NLs* per case is obtained from eye-tracking studies,[1] **Chapter 15**, and even here information is incomplete, as eye-tracking studies can only measure foveal gaze and cannot track lesions found by peripheral vision. Based on the author's experience, in screening mammography, clinical considerations limit the number of regions per case (4-views) that an expert will consider for marking to relatively small numbers, typically less than about three. About 80% of non-diseased cases have no marks. The obvious reason is that because of the low disease prevalence, about 0.5%, marking too many cases would result in unacceptably high recall rates.

Row 9: $L_{k_2} > 0$ is the number of lesions in diseased case $k_2 2$. Since lesions can only occur on diseased cases, a second case-truth subscript, as in $L_{k_2 2}$, is superfluous. L_T is the total number of lesions in the dataset.

Row 10: The label $l_1 = \{1, 2, ..., N_{k_t t}\}$ indexes latent *NL* marks, provided the case has at least one NL mark, in which case and otherwise $N_{k_t t} = 0$ and $l_1 = \varnothing$, the *null set*. The possible values of l_1 are $l_1 = \{\varnothing\} \oplus \{1, 2, ..., N_{k_t t}\}$. The null set applies when the case has no latent NL marks and $\oplus$ is the *exclusive-or* symbol (*exclusive-or* is used in the English sense: *one or the other, but not neither nor both*). In other words, l_1 can *either* be the null set *or* take on positive integer values. Likewise, $l_2 = \{1, 2, ..., L_{k_2}\}$ indexes latent *LL* marks. Unmarked LLs are assigned negative infinity ratings, see row 5. The null set notation is *not* needed for latent LLs.

Having covered notation, attention turns to the empirical plots possible with FROC data. The historical starting point is the FROC plot.

13.3 Formalism: The empirical FROC plot

In **Chapter 12**, the FROC was defined as the plot of *LLF* (along the ordinate) versus *NLF*. Using the notation of Table 13.1, and assuming binned data,* then, corresponding to the operating point determined by rating *r*, the coordinates are $NLF(\zeta_r)$, the total number of NLs rated $\geq$ threshold ζ_r divided by the total number of cases, and $LLF(\zeta_r)$, the total number of LLs rated $\geq$ threshold ζ_r divided by the total number of lesions:

$$NLF_r \equiv NLF(\zeta_r) = \frac{\# NLs\ rated \geq \zeta_r}{\# cases} \tag{13.1}$$

$$LLF_r \equiv LLF(\zeta_r) = \frac{\# LLs\ rated \geq \zeta_r}{\# lesions} \tag{13.2}$$

The empirical FROC plot connects adjacent operating points (NLF_r, LLF_r), including the origin (0,0) *but not (1,1)*, with straight lines. Equation 13.1 is equivalent to

$$NLF_r = \left(\frac{1}{K_1 + K_2}\right) \sum_{t=1}^{2} \sum_{k_t=1}^{K_t} \sum_{l_1}^{N_{k_t t}} I\left(z_{k_t t l_1 1} \geq \zeta_r \middle| l_1 \neq \varnothing\right) \tag{13.3}$$

The *indicator function*, Equation 5.3, yields unity if its argument is true and zero otherwise, so it acts like a "counter." The conditioning $l_1 \neq \varnothing$ and $z_{k_t t l_1 1} \geq \zeta_r$ in Equation 13.3 ensure that only *marked* non-diseased regions contribute to *NLF*. The summations yield the total number of NLs

---

* This is not a limiting assumption: if the data is continuous, for finite numbers of cases, no ordering information is lost if the number of ratings is chosen large enough. This is analogous to Bamber's theorem in **Chapter 5**, where a proof, although given for binned data, is applicable to continuous data.

in the dataset with Z-samples $\geq \zeta_r$ and dividing by the total number of cases yields NLF_r. Equation 13.3 also shows explicitly that NLs on *both non*-diseased ($t = 1$) *and* diseased cases ($t = 2$) contribute to NLF.

Likewise, Equation 13.2 is equivalent to

$$LLF_r = \frac{1}{L_T} \sum_{k_2=1}^{K_2} \sum_{l_2=1}^{L_{k_2}} I\left(z_{k_2 2 l_2 2} \geq \zeta_r\right) \tag{13.4}$$

Because unmarked lesions are assigned the $-\infty$ rating, Equation 13.4 need not be conditioned on $l_2 \neq \varnothing$. For obvious reasons, a third summation over t is not needed, and both truth-state indices on the right-hand side of Equation 13.4 are $t = s = 2$. Unlike NLF, only diseased cases and LLs contribute to LLF. The denominator is the total number of lesions in the dataset (see row 9 in Table 13.1).

So far, the implicit assumption has been that each case or patient is represented by one image. When a case has multiple images or *views*, the above definitions are referred to as *case-based* scoring. A *view-based* scoring of the data is also possible, in which the denominator in Equation 13.1 is the total number of views. Furthermore, in view-based scoring multiple lesions on different views of the same case are counted as different lesions, even though they may correspond to the same physical lesion.[2] The total number of lesion localizations is divided by the total number of lesions visible to the truth panel in all views, which is the counterpart of Equation 13.4. When each case has a single image, the two definitions are equivalent. With four views per patient in screening mammography, case-based NLF is four times larger than view-based NLF. Since a superior system tends to have smaller NLF values, the tendency among researchers is to report view-based FROC curves, because, frankly, it makes their systems "look better" (this is an actual private comment from a prominent CAD researcher).

13.3.1 The semi-constrained property of the observed end-point of the FROC plot

The term *semi-constrained* means that while the observed end-point *ordinate* is *constrained* to the range (0,1) the corresponding *abscissa* is not. Similar to the ROC (Figure 5.1) the operating points are labeled by r, with $r = 1$ corresponding to the uppermost observed point, $r = 2$ is the next lower operating point, and $r = R_{FROC}$ corresponds to the operating point closest to the origin. The number of thresholds equals the number of FROC bins—note the difference from the ROC paradigm, where the number of thresholds was one less than the number of ROC bins. Here is another critical difference:

While $r = R_{FROC} + 1$ yields the trivial operating point (0,0), $r = 0$ *does not yield a defined point.*

To understand this important statement, consider the expression, using Equation 13.3, for NLF_0:

$$NLF_0 = \left(\frac{1}{K_1 + K_2}\right) \sum_{t=1}^{2} \sum_{k_t=1}^{K_t} \sum_{l_1}^{N_{k_t t}} I\left(z_{k_t t l_1 1} \geq \left(\zeta_0 = -\infty\right)\Big| l_1 \neq \varnothing\right) \tag{13.5}$$

The right-hand side can be separated into two terms, the contribution of latent NL marks with Z-samples in the range $z \geq \zeta_1$ and those in the range $-\infty \leq z < \zeta_1$. The first term equals the abscissa of the uppermost *observed* operating point, NLF_1:

$$NLF_1 = \left(\frac{1}{K_1 + K_2}\right) \sum_{t=1}^{2} \sum_{k_t=1}^{K_t} \sum_{l_1}^{N_{k_t t}} I\left(z_{k_t t l_1 1} \geq \zeta_1\right) = \frac{N_T}{K_1 + K_2} \tag{13.6}$$

In the above equation, N_T is the total number of actual NL marks in the dataset (row 8 in Table 13.1). Since each case could have zero or more *NLs*, NLF_1 is unconstrained and, in particular, can exceed one.

Unlike the ROC plot, which is completely contained in the unit square, the FROC plot is not.

The second term is

$$\left(\frac{1}{K_1+K_2}\right)\sum_{t=1}^{2}\sum_{k_t=1}^{K_t}\sum_{l_1}^{N_{k_t t}}I\left(-\infty \leq z_{k_t tl_1 1} < \zeta_1\right) = \frac{??}{K_1+K_2} \tag{13.7}$$

It represents the contribution of unmarked NLs (i.e., latent NLs with Z-samples below ζ_1). It determines how much further to the right the observer's NLF would have moved, relative to NLF_1, if one could get the observer to lower the reporting criterion to $-\infty$. Since in practice the observer will not oblige, this term cannot be evaluated.

Another way of stating this important point is that unmarked NLs, as indicated by the question marks in the numerator of the right-hand side of Equation 13.7, represent *unobservable events*.

Turning attention to LLF_0,

$$LLF_0 = \frac{\sum_{k_2=1}^{K_2}\sum_{l_2=1}^{L_{k_2}}I\left(z_{k_2 2l_2 2} \geq -\infty\right)}{L_T} = 1 \tag{13.8}$$

Unlike unmarked latent NLs, unmarked lesions can safely be assigned the $-\infty$ rating. *Such an assignment is allowed because an unmarked lesion is an observable event.* The right-hand side of Equation 13.8 can be evaluated and indeed, it evaluates to unity. However, since the corresponding abscissa NLF_0 is undefined, one cannot plot this point. A trivial but important statement: a plotted point requires two coordinates. This should not be construed to mean that an ordinate of unity is potentially achievable, if only one could find the appropriate x-coordinate to assign to it. In most clinical studies, the observer who marks every suspicious region does not reach unit ordinate. *Taken together, it follows that the observed end-point is semi-constrained in the sense that its abscissa is not limited to the range (0,1).*

The next lowest value of LLF can be plotted:

$$LLF_1 = \frac{\sum_{k_2=1}^{K_2}\sum_{l_2=1}^{L_{k_2}}I\left(z_{k_2 2l_2 2} \geq \zeta_1\right)}{L_T} \leq 1 \tag{13.9}$$

The numerator is the total number of lesions that were actually marked. The ratio is the fraction of lesions that are marked. The above expression is the ordinate of the observed end-point. Since the corresponding abscissa is defined, Eqn. 13.6, this point can be plotted.

The formalism should not obscure the fact that Equations 13.6 and 13.9 are obvious conclusions about the observed end-point of the FROC, namely the ordinate is constrained to $\leq$ unity while the abscissa is unconstrained and one does not know how far to the right it might extend were the observer to report every suspicious region.

13.4 Formalism: The alternative FROC (AFROC) plot

In **Chapter 12**, work by Bunch et al.[3] was discussed. Figure 4 *ibid.* anticipated another way of visualizing FROC data*. The author subsequently termed this the *alternative* FROC (AFROC) plot.[4] The AFROC is defined as the plot of $LLF(\zeta)$ along the ordinate versus $FPF(\zeta)$ along the abscissa. So how does one get FPF, an *ROC* paradigm quantity, from *FROC* data?

13.4.1 Inferred ROC rating

By adopting a rule for converting the zero or more mark-rating data per case to a single rating per case, and most commonly the highest rating assumption is used, it is possible to *infer* ROC data points from mark-rating data. The rating of the highest rated mark on a case, or $-\infty$ if the case has no marks, is defined as the *inferred ROC* rating for the case. Other rules to obtain a single rating from a variable number of ratings on a case, such as the average rating or a stochastically dominant rating, have been described,[5] but the highest rating method is by far the simplest and most intuitive.

Definition:
The rating of the highest rated mark on a case, or $-\infty$ if the case has no marks, is defined as its *inferred* ROC rating.

Inferred ROC ratings on non-diseased cases are referred to as inferred FP ratings and those on diseased cases as inferred TP ratings. When there is little possibility for confusion, the prefix *inferred* is suppressed. Using the by now familiar cumulation procedure, *FP* counts are cumulated to calculate *FPF* and likewise, *TP* counts are cumulated to calculate *TPF*.

Definitions:
- FPF = cumulated inferred FP counts ≥ threshold divided by total number of non-diseased cases.
- TPF = cumulated inferred TP counts ≥ threshold divided by total number of diseased cases.

As will become clearer later, the AFROC plot includes an important extension from the observed end-point to (1,1).

Definition of AFROC plot
- The alternative free-response operating characteristic (AFROC) is the plot of LLF versus inferred FPF.
- The plot includes an extension from the observed end-point to (1,1).

The mathematical definition of the AFROC follows.

---

* The late Prof. Richard Swensson did not like the author's choice of the word *alternative* in naming this operating characteristic. The author had no idea in 1989 how important this operating characteristic would later turn out to be, otherwise a more meaningful name would have been proposed.

13.4.2 The AFROC plot and AUC

The highest Z-sample ROC false positive (FP) rating for non-diseased case $k_1 1$ is defined by

$$FP_{k_1 1} \triangleq max_{l_1} \left(z_{k_1 1 l_1 1} \big| z_{k_1 1 l_1 1} \geq \zeta_1 \right) \, \big\| -\infty \big| l_1 = \varnothing \tag{13.10}$$

The single vertical bar | is the *conditioning* operator; for example, *A|B is event A assuming condition B is true.* It ensures that only marked regions enter the calculation. The double vertical bars ‖ denote the logical OR operator. The basic idea is simple: $FP_{k_1 1}$ is the maximum over all marked Z-samples occurring on non-diseased case $k_1 1$, or $-\infty$ if the case has no marks. Assignment of the $-\infty$ rating is allowed because an unmarked non-diseased case is an observable event. The corresponding false positive fraction FPF_r is defined by

$$FPF_r \equiv FPF(\zeta_r) = \frac{1}{K_1} \sum_{k_1=1}^{K_1} I \left(FP_{k_1 1} \geq \zeta_r \right) = \frac{1}{K_1} \sum_{k_1=1}^{K_1} I \left(max_{l_1} \left(z_{k_1 1 l_1 1} \big| z_{k_1 1 l_1 1} \geq \zeta_1 \right) \geq \zeta_r \right) \tag{13.11}$$

The indicator function is a "counter." If $max_{l_1}(...)$ is greater than or equal to ζ_r, it yields unit count, and otherwise, it yields zero. The maximum is taken over all marked NLs. Lesion localization fraction, LLF_r, is defined, as before, by Equation 13.4. The empirical AFROC plot connects adjacent operating points (FPF_r, LLF_r), including the origin $(0,0)$, with straight lines *plus* a straight-line segment connecting the observed end-point to $(1,1)$.

> *The area under this plot is defined as the empirical AFROC AUC.* A computational formula for it will be given in the next chapter.

13.4.2.1 Constrained property of the observed end-point of the AFROC

$r = R_{FROC} + 1$ yields the trivial operating point $(0,0)$ and $r = 0$ yields the trivial point $(1,1)$:

$$FPF_0 = \frac{\sum_{k_1=1}^{K_1} I \left(FP_{k_1 1} \geq -\infty \right)}{K_1} = 1 \tag{13.12}$$

Because every non-diseased case is assigned a rating, and therefore counted, the right-hand side evaluates to unity. This is obvious for marked cases. Since each unmarked case also gets a rating, albeit a $-\infty$ rating, it is counted (the argument of the indicator function in Equation 13.12 is true even when the inferred FP rating is $-\infty$).

Since the value of LLF_0 is unity, Equation 13.8, and this time the corresponding value FPF_0 exists, Equation 13.12, *one may plot it.* The empirical AFROC plot is obtained by adjacent operating points, including the trivial ones, with straight lines.

Key points:
- The ordinates LLF of the FROC and AFROC are identical; unlike the empirical FROC, whose observed end-point has the semi-constrained property, the AFROC end-point is constrained.
- Anticipating what is to come, the AFROC plot, especially a weighted[6] version of it, is of fundamental importance in the analysis of FROC data[7] and the FROC plot is a poor summary of performance.

- While the AFROC plot was anticipated by Bunch et al.[3] in 1978, they labeled the FROC plot as the *preferred form*, Figure 5 *ibid.*, when in fact it is the other way around. Also, their AFROC plots should end at (1,1) and not plateau at lower values, as shown in Figure 4 *ibid.*

13.4.2.2 Chance level FROC and AFROC

The chance level FROC was addressed in the previous chapter; it is a *flat-liner*, hugging the x-axis, except for a slight upturn at large NLF.

The AFROC of a guessing observer is not the line connecting (0,0) to (1,1). This is a misconception.[9] A guessing observer will also generate a *flat-liner*, but this time the plot ends at FPF = 1, and the straight-line extension will be a vertical line connecting this point to (1,1). In the limit $\mu \rightarrow 0$, AFROC-AUC of a guessing observer tends to zero.

Figure 13.1 shows *near guessing* FROC and AFROC plots. These plots were generated by the code in **mainOCsRaw.R** with **mu** = 0.1, **K1** = 50, **K2** = 70, $\zeta_1 = -1$ and other parameters as in code listing in Section 13.10.1. One does not expect to observe curves like Figure 13.1 with radiologists as they rarely guess in the clinic—there is too much at stake.

To summarize, AFROC AUC of a guessing observer is zero. On the other hand, suppose an expert radiologist views screening images and the lesions on diseased cases are very difficult, even for the expert, and the radiologist does not find any of them. Being an expert, the radiologist successfully screens out non-diseased cases and sees nothing suspicious in any of them—this is one measure of the expertise of the radiologist, i.e., not mistaking variants of normal anatomy for false lesions on non-diseased cases. Accordingly, the expert radiologist does not report anything, and the operating point is stuck at the origin. Even in this unusual situation, one would be justified in connecting the origin to (1,1) and claiming area under AFROC is 0.5. The extension gives the radiologist credit for not marking any non-diseased cases; of course, the radiologist does not get any credit for marking any of the lesions. An even better radiologist, who finds and marks some of the lesions, will score higher, and AFROC-AUC will then exceed 0.5. See Section 17.7.4 for a software demonstration of this unusual but instructive situation.

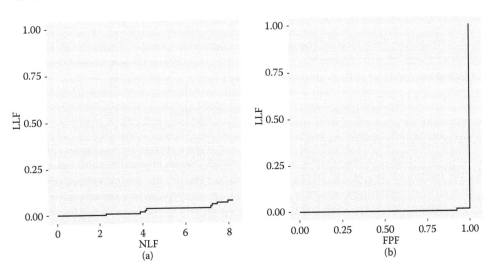

Figure 13.1 (a) Near guessing observer's FROC and (b) AFROC plots generated by the code in **mainOCsRaw.R** with **mu** = 0.1, **K1** = 50, **K2** = 70, $\zeta_1 = -1$ and other parameters as in code listing in Section 13.10.1.

13.5 The EFROC plot

An exponentially transformed FROC (EFROC) plot has been proposed[10] that, like the AFROC, is contained within the unit square. The EFROC inferred FPF is defined by (this is yet another way of inferring ROC data, albeit only FPF, from FROC data):

$$FPF_r^{EFROC} \equiv FPF^{EFROC}(\zeta_r) \triangleq 1 - \exp(-NLF(\zeta_r)) \tag{13.13}$$

In other words, one computes NLF_r using NLs rated $\geq \zeta_r$ on all cases, Equation 13.3, and then transforms it to FPF_r^{EFROC} according to Equation 13.13. Note that FPF so defined is in the range (0,1). The empirical EFROC plot connects adjacent coordinates (FPF_r^{EFROC}, LLF_r), including the origin (0,0), with straight lines *plus* a straight-line segment connecting the observed end-point to (1,1). The area under the empirical EFROC has been proposed as a figure of merit for FROC data. It has the advantage, compared to the FROC, of being contained in the unit square. It has the advantage over the AFROC of using *all* NL ratings, not just the highest rated ones, but this is a mixed blessing. The effect on statistical power compared to the AFROC has not been studied, but the author expects the advantage to be minimal (because the highest rated NL contains more information than a randomly selected NL mark). A disadvantage is that cases with more LLs get more importance in the analysis; this can be corrected by replacing LLF with wLLF (see Equation 13.17). Another disadvantage is that inclusion of NLs on diseased cases causes the EFROC plot to depend on diseased prevalence. In addition, as with several papers in this field, there are misconceptions: the cited publication shows the EFROC as smoothly approaching (1,1). In fact, Figure 1 *ibid.*, resembles the ROC curve predicted by the equal variance binormal model. The author expects the EFROC to resemble the AFROC curves shown below, for example, Figure 13.2k. Furthermore, the statement in Section C *ibid.* "*By operating under the free-response conditions, the observer will mark and score all suspicious locations*" (emphasis added) repeats serious misconceptions in this field. Not all suspicious regions are reported; even CAD reports a small fraction of the suspicious regions that it finds. In spite of these concerns, the EFROC represents the first recognition by someone other than the author, of significant limitations of the FROC curve, and that an operating characteristic for FROC data that is completely contained within the unit square is highly desirable. The empirical EFROC-AUC FOM is implemented in **RJafroc** software.

13.6 Formalism for the inferred ROC plot

The inferred true positive (TP) Z-sample for diseased case $k_2 2$ is defined by

$$TP_{k_2 2} \triangleq \left(max_{l_1 l_2}\left(\left(z_{k_2 2 l_1 1}, z_{k_2 2 l_2 2}\right)\Big| l_1 \neq \varnothing\right)\right) \Big\| \left(max_{l_2}\left(z_{k_2 2 l_2 2}\right)\Big| l_1 = \varnothing\right) \tag{13.14}$$

The right hand of the logical OR clause ensures that if the case has no NL marks, the maximum is over the LL marks. On the left side of the logical OR clause, applicable when the case has NL marks, the maximum is over all *marked* NLs and all LLs on the case (to reiterate, an *unmarked* NL is an *unobservable* event; the evaluation shown in Equation 13.14 involves observable events only).

The formula appears complex, but the basic idea is simple: $TP_{k_2 2}$ is the maximum over all ratings, NLs *and* LLs, whichever is higher, occurring on diseased case $k_2 2$, or $-\infty$ if the case has no marks. The $-\infty$ assignment is justified because an unmarked diseased case is an observable event. The highest Z-sample inferred true positive fraction TPF_r is defined by

$$TPF_r \equiv TPF(\zeta_r) = \frac{1}{K_2} \sum_{k_2=1}^{K_2} I\left(TP_{k_2 2} \geq \zeta_r\right) \tag{13.15}$$

The definition of FPF_r is the same as before, that is, Equation 13.11. The inferred ROC plot connects adjacent coordinates (FPF_r, TPF_r), including the origin $(0,0)$, with straight lines *plus* a straight-line segment connecting the observed end-point to $(1,1)$.

13.7 Formalism for the weighted AFROC (wAFROC) plot

The AFROC ordinate defined in Equation 13.4 gives equal importance to every *lesion* on a case. Therefore, *a case with more lesions will have more influence on the AFROC* (this is explained in depth in **Chapter 14**). This is undesirable since each case (i.e., patient) should get equal importance in the analysis. As with ROC analysis, one wishes to draw conclusions about the population of cases and each case is regarded as an equally valid sample from the population. In particular, one does not want the analysis to be skewed toward cases with a greater-than-average number of lesions. (Historical note: the author became aware of how serious this issue could be when a researcher contacted him about using FROC methodology for nuclear medicine bone scan images, where the number of lesions on diseased cases can vary from a few to a hundred!)

Another issue is that the AFROC assigns equal clinical importance to each lesion in a case. Lesion weights were introduced[7] to allow for the possibility that the clinical importance of finding a lesion might be lesion-dependent[11] (the referenced conference proceeding paper, included in the Online Supplemental material corresponding to this chapter, should be of interest to the more advanced reader). For example, it is possible that an easy-to-find lesion is less clinically important than a harder-to-find one, therefore the figure of merit should give more importance to the harder-to-find lesion. Clinical importance in this context could be the mortality associated with the specific lesion type, which can be obtained from epidemiological studies.[12] Let $W_{k_2 l_2} \geq 0$ denote the weight (i.e., clinical importance) of lesion l_2 in diseased case k_2 (since weights are only applicable to diseased cases, one can, without ambiguity, drop the case-level and location-level truth subscripts, i.e., $W_{k_2 2 l_2 2}$ would be superfluous). For each diseased case $k_2 2$ the weights are subject to the constraint,

$$\sum_{l_2=1}^{L_{k_2}} W_{k_2 l_2} = 1 \tag{13.16}$$

The constraint assures that each diseased case exerts equal importance in determining the weighted AFROC (wAFROC) operating characteristic, regardless of the number of lesions in it (this is explained in depth in **Chapter 14**).

The weighted lesion localization fraction $wLLF_r$ is defined by[13]

$$wLLF_r \equiv wLLF(\zeta_r) = \frac{1}{K_2} \sum_{k_2=1}^{K_2} \sum_{l_2=1}^{L_{k_2}} W_{k_2 l_2} I\left(z_{k_2 2 l_2 2} \geq \zeta_r\right) \tag{13.17}$$

(The conditioning operator is not needed because every lesion gets a rating.)

The empirical *wAFROC* plot connects adjacent operating points $(FPF_r, wLLF_r)$, including the origin $(0,0)$, with straight lines *plus* a straight-line segment connecting the observed end-point to $(1,1)$. The area under this plot is the empirical weighted AFROC AUC.

13.8 Formalism for the AFROC1 plot

Historically[8], the AFROC originally used a different definition of FPF, which is retrospectively termed the AFROC1 plot. Since NLs can occur on diseased cases, it is possible to define an inferred "FP" rating on a *diseased* case as the maximum of all NL ratings on the case, or $-\infty$ if the case has no

NLs. The quotes emphasize that this is non-standard usage of ROC terminology. In a ROC study, an FP can only occur on a *non-diseased* case. Since both case-level truth-states are allowed, the highest false positive Z-sample for case $k_t t$ is (the superscript 1 is necessary to distinguish it from Equation 13.10):

$$FP_{k_t t}^1 \triangleq max_{l_1} \left(z_{k_t t l_1 1} \middle| l_1 \neq \varnothing \right) \middle\| -\infty \middle| l_1 = \varnothing \tag{13.18}$$

$FP_{k_t t}^1$ is the maximum over all marked NL Z-samples, labeled by the location index l_1, occurring on case $k_t t$, or $-\infty$ if $l_1 = \varnothing$. One is allowed to assign the $-\infty$ rating because a case with no NL marks is an observable event. The corresponding false positive fraction $FPF_r^1 (\zeta_r)$ is defined by (the superscript 1 is necessary to distinguish it from Equation 13.11):

$$FPF_r^1 \equiv FPF_r^1 (\zeta_r) = \frac{1}{(K_1 + K_2)} \sum_{t=1}^{2} \sum_{k_t=1}^{K_t} I\left(FP_{k_t t}^1 \geq \zeta_r \right) \tag{13.19}$$

Note the subtle differences between Equation 13.11 and Equation 13.19. The latter counts FPs on non-diseased *and* diseased cases while Equation 13.11 counts FPs *only* on non-diseased cases. Accordingly, the denominators in the two equations are different. The advisability of allowing a diseased case to be both a TP and a FP is questionable from both clinical and statistical considerations. However, allowing this possibility leads to the following definition: the empirical alternative FROC1 (AFROC1) plot connects adjacent operating points $\left(FPF_r^1, LLF_r \right)$, including the origin (0,0), with straight lines *plus* a straight-line segment connecting the observed end-point to (1,1). The only difference between it and the AFROC plot is in the x-axis.

Based on considerations of statistical power alone, tested with a simulator that did not include asymmetry effects between NLs on diseased and non-diseased cases, the author made a recommendation to use the AFROC1 curve as the basis of analysis.[14] This recommendation was a mistake, not the author's only mistake,[15–17] and was subsequently corrected.[18] [The rationale for pointing to one's own mistakes is the author's opinion that, even in retrospect, the mistakes are instructive. This is especially true with Ref. 15, which is included in the Online Supplemental material.]

13.9 Formalism: The weighted AFROC1 (wAFROC1) plot

The empirical weighted AFROC1 (wAFROC1) plot connects adjacent operating points $\left(FPF_r^1, wLLF_r \right)$, including the origin (0,0), with straight lines *plus* a straight-line segment connecting the observed end-point to (1,1). The only difference between it and the wAFROC plot is in the x-axis. *Usage of the wAFROC1 plot as the basis of analysis is currently recommended for datasets with only diseased cases.*

So far, the description has been limited to abstract definitions of various operating characteristics possible with FROC data. Now it is time to put numbers into the formula and see actual plots. The starting point is the FROC plot.

13.10 Example: Raw FROC plots

The FROC plots shown below were generated using the data simulator introduced in **Chapter 12**. The examples are similar to the population FROC curves shown in that chapter, Figure 12.2a–c, but the emphasis here is on understanding the FROC data structure. To this end, smaller numbers of cases, not 20,000 as in the previous chapter, are used. Examples are given using continuous ratings, termed *raw data*, and binned data, for a smaller dataset and for a larger dataset. With a smaller dataset, the logic of constructing the plot is more transparent but the operating points are more susceptible to sampling variability. The examples illustrate key points distinguishing the free-response paradigm from ROC. The author believes a good understanding of this relatively complex paradigm is obtained from a detailed examination at the coding level.

The file **mainOCsRaw.R** (*OCs* stands for generic *operating characteristics*, which can be FROC, AFROC, or inferred ROC, etc.) utilizes the **RJafroc** package. The functions **SimulateFrocData()**—encountered in the previous chapter—and **PlotEmpiricaOperatingCharacteristics()** are included in the package. As their names suggest, they simulate FROC data and plot *empirical* operating characteristics, respectively. A listing follows.

13.10.1 Code listing

```
rm(list = ls()) # mainOCsRaw.R
library(RJafroc);library(ggplot2)

seed <- 1;set.seed(seed)
mu <- 1;lambda <- 1;nu <- 1
zeta1 <- -1;K1 <- 5;K2 <- 7
Lmax <- 2;Lk2 <- floor(runif(K2, 1, Lmax + 1))

frocDataRaw <- SimulateFrocDataset(
  mu = mu, lambda = lambda, nu = nu, I = 1, J = 1,
  K1 = K1, K2 = K2, lesionNum = Lk2, zeta1 = zeta1)

plotFROC <- PlotEmpiricalOperatingCharacteristics(
  dataset = frocDataRaw,
  trts= 1,
  rdrs = 1,
  opChType = "FROC")
p <- plotFROC$Plot +
  theme(legend.position="none") +
  theme(axis.title.y = element_text(size = 25,face="bold"),
        axis.title.x = element_text(size = 30,face="bold"))
p$layers[[1]]$aes_params$size <- 2 # line
p$layers[[2]]$aes_params$size <- 5 # points
print(p)

plotAFROC <- PlotEmpiricalOperatingCharacteristics(
  dataset = frocDataRaw,
  trts= 1,
  rdrs = 1,
  opChType = "AFROC"
)
p <- plotAFROC$Plot +
  theme(legend.position="none") +
  theme(axis.title.y = element_text(size = 25,face="bold"),
        axis.title.x = element_text(size = 30,face="bold"))
p$layers[[1]]$aes_params$size <- 2 # line
p$layers[[2]]$aes_params$size <- 5 # points
print(p)

plotROC <- PlotEmpiricalOperatingCharacteristics(
  dataset = frocDataRaw,
```

```
    trts= 1,
    rdrs = 1,
    opChType = "ROC")
p <- plotROC$Plot +
  theme(legend.position="none") +
  theme(axis.title.y = element_text(size = 25,face="bold"),
        axis.title.x = element_text(size = 30,face="bold"))
p$layers[[1]]$aes_params$size <- 2 # line
p$layers[[2]]$aes_params$size <- 5 # points
print(p)

retRocRaw <- DfFroc2Roc(frocDataRaw)
```

The parameters **mu**, **lambda** and **nu** are each set to 1 for the ensuing 12 plots, Fig. 13.2 (a–l). Ensure that at line 6, **K1** is set to 5, **K2** to 7 and **zeta1** is set to −1. The code actually generates FROC, AFROC and ROC plots, lines 13–24, 26–38, and 40–51. In order not to be overwhelmed with plots, insert a break point at line 26, which has the effect of suppressing the AFROC and ROC plots, and **source** the code yielding Figure 13.2a. The code should be familiar from the previous chapter and the explanation in Online Appendix 12.A. The discreteness, that is, the relatively big jumps between data points, is due to the small numbers of cases. Exit debug mode (click square **stop** button), increase the numbers of cases to **K1 <- 50;K2 <- 70** and **source** the code again, yielding Figure 13.2b for the new FROC plot. The fact that Figure 13.2a does not match (b), especially near *NLF* = 0.25, is not an aberration; plot (a), with only 12 cases, is subject to more sampling variability than plot (b), with 120 cases. Try different **seed** values to be satisfied on this point (this is case-sampling at work!).

In Figure 13.2, the first column corresponds to five non-diseased, seven diseased cases, and reporting threshold equal to −1. The second column corresponds to 50 non-diseased, 70 diseased cases, and reporting threshold equal to −1. The third column corresponds to 50 non-diseased, 70 diseased cases, and reporting threshold equal to +1. The discreteness (jumps) in the plots in the first column is due to the small number of cases. The discreteness is less visible in the second and third columns. If the number of cases is increased further, the plots will approach continuous plots, like those shown in Chapter 12, Fig. 12.2 (a)–(c). In the current figure, in plot (c), note the smaller traverse of the FROC plot. It is actually a replica of plot (b) truncated at a smaller value of NLF. With a higher reporting threshold, fewer NL/LL events exceed the marking threshold. Plot (d) shows a binned FROC plot corresponding to five non-diseased and seven diseased cases. Plot (e) shows a binned FROC plot corresponding to 50 non-diseased and 70 diseased cases for reporting threshold set to −1. Plot (f) shows the binned FROC plot for reporting threshold set to +1. The resulting plot has a smaller traverse than (e) and is *not* identical to a truncated version of (e). This is because binning was performed after truncating the raw data. Plots (g–l) have the same structure as plots (a–f) but show AFROC curves. All AFROC plots are contained within the unit square, and unlike the semi-constrained property of the observed FROC end-point, the observed AFROC end-point is constrained to lie within the unit square. This has important consequences in terms of defining a valid figure of merit. All plots were generated by alternately sourcing **mainOCsRaw.R** or **mainOCsBinned.R** under the stated conditions.

13.10.2 Simulation parameters and effect of reporting threshold

The simulator generates a non-negative random integer for each non-diseased case, representing *latent NLs*, and two non-negative random integers for each diseased case, representing *latent NLs* and/or latent *LLs*. For each latent *NL* or *LL*, the simulator samples an appropriate normal distribution to generate raw *Z*-samples (i.e., floating point numbers). The unit variance normal

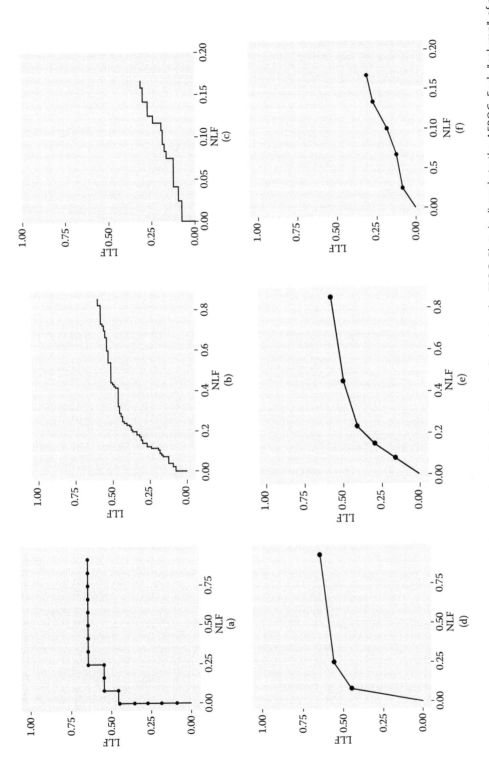

Figure 13.2 Examples of simulated raw and binned FROC and AFROC plots. Plots (a–f) apply to the FROC. Plots (g–l) apply to the AFROC. Each "column" of plots corresponds to the same simulated data, while the rows alternate between raw and binned data. See text for further details. (*Continued*)

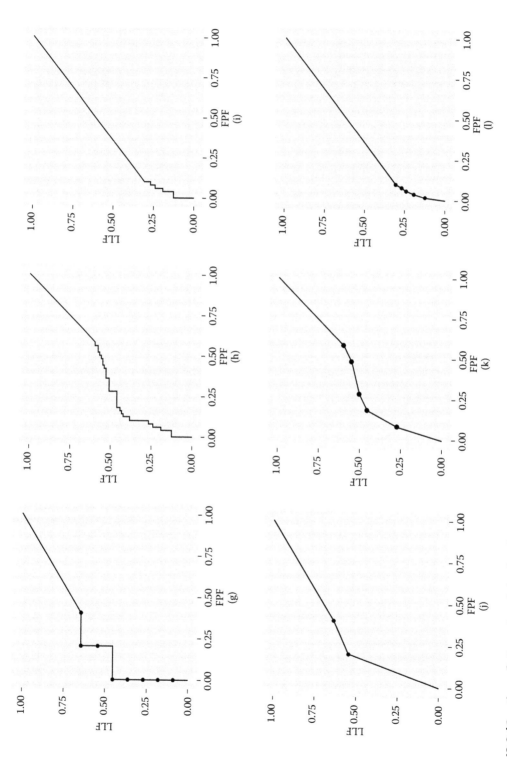

Figure 13.2 (Continued) Examples of simulated raw and binned FROC and AFROC plots. Plots (a–f) apply to the FROC. Plots (g–l) apply to the AFROC. Each "column" of plots corresponds to the same simulated data, while the rows alternate between raw and binned data. See text for further details.

distributions for sampling latent *NL* and/or *LL* ratings are separated by μ, the perceptual signal-to-noise ratio introduced in the previous chapter. The simulation parameter **zeta1** (representing the lowest reporting threshold, denoted ζ_1 in Table 13.1) determines whether the latent mark is actually marked; only locations generating ratings ≥ **zeta1** are actually marked. *The simulator returns actual (not latent) marks and their ratings.* Increasing **zeta1** will yield fewer marked *NLs* and *LLs* per case and the FROC plot will have a shorter upward and rightward traverse, as demonstrated next.

Use the following code snippet, for the dataset corresponding to Fig. 13.2 (b), to determine the coordinates of the end-point in Figure 13.2b (**genAbscissa** stands for a generic abscissa and **genOrdinate** stands for a generic ordinate; the specific value meant will be clear from the context).

13.10.2.1 Code snippets

```
Browse[2]> max(plotFROC$Points$genAbscissa)
[1] 0.85
Browse[2]> max(plotFROC$Points$genOrdinate)
[1] 0.5961538
```

Note that these values agree with the visual estimates from Figure 13.2b. Exit debug mode (click the red square button), increase **zeta1** to +1 and click **source**. Confirm the lower values with the following code snippets.

13.10.2.2 Code snippets

```
Browse[2]> max(plotFROC$Points$genAbscissa)
[1] 0.1666667
Browse[2]> max(plotFROC$Points$genOrdinate)
[1] 0.3076923
```

Note the new FROC curve in the **Plots** window, Figure 13.2c. Increasing **zeta1** resulted in fewer ratings exceeding the new value, so the end-point has moved down and to the left compared to Figure 13.2b. Plot (c) is actually a replica of plot (b) truncated at a smaller value of NLF.

13.10.3 Number of lesions per case

Exit debug mode, return **zeta1** to −1, reduce the numbers of cases to 5/7 and **source** the code. At line 7, **Lmax** is the maximum number of lesions per case, in the dataset, two in this example. The array **Lk2[1:7]**, generated by the uniform random number generator **runif()**, is guaranteed to be in the range 1 to **Lmax**. It contains the number of lesions in each diseased case whose contents are shown below.

13.10.3.1 Code snippets

```
> Lk2;sum(Lk2)
[1] 1 1 2 2 1 2 2
[1] 11
> max(floor(runif(1000, 1, Lmax + 1)))
[1] 2
```

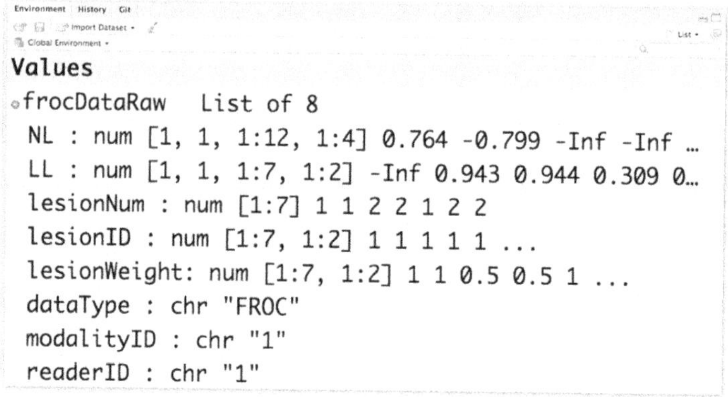

Figure 13.3 The structure of a FROC dataset object. It is a list variable containing, in order, the NL array, the LL array, and so on. See details in **RJafroc** documentation files accessible from **RStudio** help.

The second line shows that the first two diseased cases have one lesion each, the third and fourth have two lesions each, and so on. The next line shows that the total number of lesions in the dataset is 11. The last line shows that, even with a thousand simulations, the number of lesions per diseased case is indeed limited to two.

13.10.4 FROC data structure

At lines 9 through 11 **SimulateFrocData set** (), with appropriate parameters, returns the simulated data, which is saved to **frocDataRaw**, which, as its name suggests, represents the raw FROC data. Figure 13.3, a screenshot of the **Environment** panel, shows the structure of **frocDataRaw** for the parameters that generated Fig. 13.2(a). It is a **list** of eight variables, the first two of which are **NL[1,1,1:12,1;4]** and **LL[1,1,1:7,1:2]**, representing *NL* and *LL* ratings, respectively. The first two dimensions are needed to accommodate the more general situation with multiple modalities (the first dimension) and multiple readers (the second dimension). The third dimension accommodates the case index and the fourth dimension accommodates the location index; it is needed because a case can generate zero or more marks.* The **list** member **lesionNum[1:7]** corresponds to the number of lesions per diseased case. **lesionID[1:7,1:2]** labels the lesions in the dataset; the second dimension is needed to accommodate multiple lesions on the same case, compare Figure 13.3 to code snippet Section 13.10.4.1. Diseased case 1 has one lesion, labeled 1, and the **−Inf** means that a second lesion on this case does not exist. Diseased case 3 has two lesions, labeled 1 and 2. Lesion labels are needed because one needs to keep track of which lesion receives which rating. Just as different cases need unique labels, think of different lesions within a case as *mini-cases*, each of which requires a unique label.

13.10.4.1 Code snippets

```
> str(frocDataRaw$lesionID)
 num [1:70, 1:2] 1 1 1 1 1 1 1 1 1 1 ...
> frocDataRaw$lesionID[1,]
[1]   1 -Inf
> frocDataRaw$lesionID[3,]
[1] 1 2
```

---

* The structure of **frocDataRaw** accommodates ROC, FROC, and ROI paradigms. In the special case of ROC data, the length of the fourth dimension would be one.

Here are some key differences from the ROC paradigm:

- In the ROC paradigm, each non-diseased case generates one FP and each diseased case generates one TP.
- In a FROC study, each non-diseased case can generate zero or more NLs and each diseased case can generate zero or more NLs *and zero* or more LLs.
- The number of lesions in a case limits the number of LLs possible in that case.

For the NL array, the case dimension has 12 values. This is because it needs to accommodate NLs on five non-diseased *and* seven diseased cases. The following is a list of are *more differences between ROC and FROC:*

- In a ROC study, each case generates exactly one rating.
- In a FROC study, each case can generate zero or more (0, 1, 2, ...) mark-rating pairs.
  - The number of marks per case is a random variable as is the rating of each mark.
  - Each mark corresponds to a distinct location on the image and associated with it is a rating, that is, confidence level in presence of disease at the region indicated by the mark.

Is it any wonder that some[19-22] have issues with this paradigm? It is more complicated than ROC, no doubt about it. Clinical data does not fit neatly into the 2 × 2 truth table of ROC analysis.

13.10.5 Dimensioning of the NL and LL arrays

Section 13.10.5.1 shows the contents of the **NL** and **LL** arrays for the dataset corresponding to Fig. 13.2a (for the NL array the entries numbered 6–12 indicate NLs on diseased cases).

13.10.5.1 Code snippets

```
> frocDataRaw$NL[1,1,,]
         [,1]     [,2]     [,3]    [,4]
 [1,]   0.764   -Inf    -Inf   -Inf
 [2,]  -0.799   -Inf    -Inf   -Inf
 [3,]   -Inf    -Inf    -Inf   -Inf
 [4,]   -Inf    -Inf    -Inf   -Inf
 [5,]   -Inf    -Inf    -Inf   -Inf
 [6,]  -1.148   -Inf    -Inf   -Inf
 [7,]  -0.289   -Inf    -Inf   -Inf
 [8,]  -0.299  -0.412   -Inf   -Inf
 [9,]   0.252   -Inf    -Inf   -Inf
[10,]  -0.892   -Inf    -Inf   -Inf
[11,]   0.436   0.377  -0.224  -1.24
[12,]   0.133   -Inf    -Inf   -Inf
> frocDataRaw$LL[1,1,,]
        [,1]    [,2]
 [1,]  -Inf   -Inf
 [2,]  0.943  -Inf
 [3,]  0.944  -Inf
 [4,]  0.309  -Inf
 [5,]  0.522  -Inf
 [6,]  0.764  -Inf
 [7,]  1.388  0.897
```

Copy and paste the command lines into the **Console** window to reproduce these values. An explanation of how **R** prints out arrays may be helpful. **frocDataRaw\$NL[1,1,,]** is a 2×2 array, specifically **[12,5]**, i.e., 12 rows and 5 columns. The output shows the row and column indices of the printed values. For example, -0.4115108 is at row 8 and column 2. The first five rows (since **K1** $= 5$) refer to non-diseased cases and the rest refer to diseased cases. The number of columns of **frocDataRaw\$NL[1,1,,]** is just large enough to accommodate the case(s) generating the most NLs. For the sixth diseased case, that is, the eleventh sequentially numbered case, the simulator actually generated four latent NLs—this fact can be confirmed by going into the simulator function in debug mode. Only three values are listed ($0.4356833, 0.3773956, -0.2242679$) because the fourth Z-sample fell below **zeta1** and is therefore shown as **-Inf**. *Recall the simulator returns actual marks, not latent marks.** For the LL array, the third dimension has seven values (since **K2** $= 7$) and the fourth dimension has two values because at least one diseased case had two lesions.

13.11 Example: Binned FROC plots

In the preceding example, continuous ratings data was available and data binning was not employed. Shown next are FROC plots when the data is binned. The code is in **mainOCsBinned.R**. Insert a breakpoint at line 27, ensure that the parameters are set to those corresponding to Fig. 13.2a and **source** the file yielding Figure 13.2d. Next, exit debug mode, increase the sample size to 50/70 and **source** the code again, yielding Figure 13.2e. Set the lowest reporting threshold to +1 and source the file yielding Figure 13.2f.

13.11.1 Code Listing: Binned data

```
rm(list = ls()) # mainOCsBinned.R
library(RJafroc);library(ggplot2)

seed <- 1;set.seed(seed)
mu <- 1;lambda <- 1;nu <- 1
zeta1 <- -1;K1 <- 5;K2 <- 7
Lmax <- 2;Lk2 <- floor(runif(K2, 1, Lmax + 1))

frocDataRaw <- SimulateFrocDataset(
  mu = mu, lambda = lambda, nu = nu, I = 1, J = 1,
  K1 = K1, K2 = K2, lesionNum = Lk2, zeta1 = zeta1)

frocDataBin <- DfBinDataset(frocDataRaw, desiredNumBins = 5)
plotFROC <- PlotEmpiricalOperatingCharacteristics(
  dataset = frocDataBin,
  trts= 1,
  rdrs = 1,
  opChType = "FROC")
p <- plotFROC$Plot +
  theme(legend.position="none") +
  theme(axis.title.y = element_text(size = 25,face="bold"),
        axis.title.x = element_text(size = 30,face="bold"))
p$layers[[1]]$aes_params$size <- 2 # line
p$layers[[2]]$aes_params$size <- 5 # points
print(p)
```

---

* This is not a science issue; with extra programming, one could have shrunk the fourth dimension so that extra negative infinities do not show up. In most cases one does not really need to know the details of what happened inside the simulator.

```
afrocDataRaw <- DfFroc2Afroc(frocDataRaw)
afrocDataBin <- DfBinDataset(afrocDataRaw, desiredNumBins = 5)
plotAFROC <- PlotEmpiricalOperatingCharacteristics(
  dataset = afrocDataBin,
  trts= 1,
  rdrs = 1,
  opChType = "AFROC"
)
p <- plotAFROC$Plot +
  theme(legend.position="none") +
  theme(axis.title.y = element_text(size = 25,face="bold"),
        axis.title.x = element_text(size = 30,face="bold"))
p$layers[[1]]$aes_params$size <- 2 # line
p$layers[[2]]$aes_params$size <- 5 # points
print(p)

rocDataRaw <- DfFroc2Roc(frocDataRaw)
rocDataBin <- DfBinDataset(rocDataRaw, desiredNumBins = 5)
plotROC <- PlotEmpiricalOperatingCharacteristics(
  dataset = rocDataBin,
  trts= 1,
  rdrs = 1,
  opChType = "ROC")
print(plotROC$Plot)
```

Line 13 uses the function **DfBinDataSet()** to bin the data. A bin is defined as having at least one count in both NL and LL arrays. This function returns a **dataset** object that contains the binned **NL** and **LL** arrays. With only 12 cases, the data only supports three bins, Figure 13.2d. Increasing the number of cases by a factor of 10 each yields Figure 13.2e, which supports five bins. Examine the structure of the data for the 50/70 dataset by copying and pasting commands to the **Console** Window to reproduce the following output (The **Browse [2]>** is coming from the code debugger or browser. Anything coming after the > symbol is a command and anything lacking a preceding > symbol is output).

13.11.2 Code snippets

```
> str(frocDataBinned$NL)
 num [1, 1, 1:120, 1:4] -Inf 2 -Inf -Inf 1 ...
> str(frocDataBinned$LL)
 num [1, 1, 1:70, 1:2] -Inf 3 4 2 1 ...
> table(frocDataBinned$NL)

-Inf    1    2    3    4    5
 378   49   24   10   10    9

> sum(as.numeric(table(frocDataBinned$NL)))

[1] 480
> table(frocDataBinned$LL)

-Inf    1    2    3    4    5
```

```
   78    10    10    10    15    17
> sum(as.numeric(table(frocDataBinned$LL)))
[1] 140

> sum(Lk2)
[1] 104

> sum(Lk2) - sum(as.vector(table(frocDataBinned$LL)))[2:6])
[1] 42
```

In Section 13.11.2, the six command lines are those preceded by a > symbol. The rest are output produced by program. The reader should copy and paste the commands into the **Console** window and hit **Enter** to confirm the output values. The **table()** function converts an array into a counts table. In the first usage, there are $120 \times 4 = 480$ elements in the array: see confirmatory commands/output in Section 13.11.2. From the output of **table(frocDataBin$NL)** one sees that there are 378 entries in the NL array that equal **−Inf**, 49 entries that equal 1, and so on. These sum to 480. Because the fourth dimension of the NL array is determined by cases with the most NLs, therefore, on cases with fewer NLs, this dimension is padded with **−Inf**s. One does not know how many of the 378 **−Inf**s are latent NLs. The actual number of latent NLs could be considerably smaller, and the number of marked NLs even smaller (as this is determined by ζ_1). The last three statements are important to understand and will be further explicated below.

The LL array contains $70 \times 2 = 140$ values. From the output of **table(frocDataBin$1L)** one sees that there are 78 entries in the LL array that equal **−Inf**, 10 entries that equal 1, and so on. These sum to 140. Since the total number of lesions is 104 (last four lines in Section 13.11.2), *the number of unmarked lesions is known*. Specifically, summing the LL counts in bins 1 through 5 (corresponding to indices 2–6, since index 1 applies to the minus infinities) and subtracting from the total number of lesions one gets: $104 − (10 + 10 + 10 + 15 + 17) = 104 – 62 = 42$; see the last line of Section 13.11.2. Therefore, the number of unmarked lesions is 42. The listed value 78, listed a few lines before, is an overestimate because it includes the **−Inf** counts from the fourth dimension **−Inf** padding of the *LL* array. This happens because some other diseased case had lesions in those location-holders.

Use Section 13.11.3 to view the FROC operating points.

13.11.3 Code snippets

```
> plotFROC$Points$genAbscissa
[1]  0.000 0.075 0.158 0.242 0.442 0.850
> plotFROC$Points$genOrdinate
[1]  0.000 0.163 0.308 0.404 0.500 0.596
```

Table 13.2 illustrates the calculation of FROC operating points. The method is similar to that used to construct ROC operating points shown in Table 4.1. The data corresponds to 50 non-diseased and 70 diseased cases containing 104 lesions. The denominator for LLF is 104, and that for NLF is 120. The counts in the upper half of Table 13.2, showing binned FROC data in tabular form, were constructed using the values in Section 13.11.2. *Note the unknown number of unmarked NLs*. The corresponding FROC operating points, constructed using the cumulation process, are shown in the lower half of Table 13.2.

Attention now turns to the AFROC plot.

Table 13.2 The upper half of this table shows the binned FROC data constructed using the values in Section 13.11.2. The corresponding FROC operating points, from section 13.11.3, are shown in the lower half of the table.

	Unmarked	Bins and counts				
		Bin 1	Bin 2	Bin 3	Bin 4	Bin 5
NL	Unknown	49	24	10	10	9
LL	42	10	10	10	15	17
		FROC operating points				
		Bins ≥ 5	Bins ≥ 4	Bins ≥ 3	Bins ≥ 2	Bins ≥ 1
	NLF	0.075	0.158	0.242	0.442	0.850
	LLF	0.163	0.308	0.404	0.500	0.596

Note: There is an unknown number of unmarked NLs. The corresponding FROC operating points (Section 13.11.3) are shown in the lower half of the table. The data corresponds to 50 non-diseased and 70 diseased cases containing 104 lesions, corresponding to Fig. 13.2(E). The denominator for LLF is 104 and that for NLF is 120.

13.12 Example: Raw AFROC plots

The code for the AFROC is in **mainOCsRaw.R**, Section 13.10.1, specifically lines 26 through 38. Delete all plots (broom symbol in lower right panel under **Plots**), insert a break point at line 40, and **source** the code for the conditions corresponding to Fig. 13.2(a). The resulting second plot shown in Figure 13.2g is the raw AFROC. The code snippets, Section 13.12.1, illustrate conversion from FROC to inferred ROC ratings. Since the relevant highest rating code is internal to the plotting function, line 53 demonstrates this externally, using the function **DfFroc2Roc()**, which converts a FROC dataset object to an inferred ROC dataset object and saves it to **retRocRaw**. Insert the cursor anywhere on line 53 and click **Run**. The resulting ROC data structure is shown in Figure 13.4.

False positive ROC ratings are stored in the first **K1** positions of the **$NL** array,* and true positive ratings are stored in the **$LL** array, which has length **K2**. The fourth dimension of either array has unit length, as there is no need to accommodate multiple decisions per case. Copy and paste

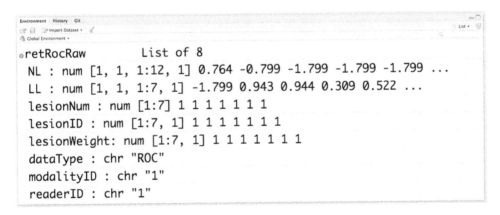

Figure 13.4 Structure of the inferred ROC dataset object. Compare to Figure 13.3 for the FROC dataset object. Unlike the former, the **lesionNum, lesionID,** and **lesionWeight** arrays are filled with ones.

---

* For programming ease, it is necessary to use NL to include FP as a special case and similarly for LL.

commands into the **Console** window to produce Section 13.12.1. The conversion from the NL and LL arrays to **retRocRaw$NL** and **retRocRaw$LL** arrays should be transparent. In this example, the LL-ratings are highest, but this is not always true. The reader should experiment with the parameters on lines 5 and 6 (try increasing **lambda** to 10 to generate 10 times more NLs, on average, then chances that one of them is the highest rating on a diseased case are larger) and/or seed values to be convinced that the conversion to highest rating always works. (The –1.799 is used as a numeric stand-in for negative infinity; line 28; it was chosen to be one less than the lowest explicit rating in the dataset. Since ordering is preserved, any number lower than the lowest explicit rating would have sufficed. Explicit ratings are those excluding negative infinities. See documentation for **DfFroc2Roc** for details.)

13.12.1 Code snippets

```
> frocDataRaw$NL[1,1,1:K1,]
        [,1]  [,2] [,3] [,4]
[1,]   0.764 -Inf -Inf -Inf
[2,]  -0.799 -Inf -Inf -Inf
[3,]    -Inf -Inf -Inf -Inf
[4,]    -Inf -Inf -Inf -Inf
[5,]    -Inf -Inf -Inf -Inf

> retRocRaw$NL[1,1,1:K1,1]
[1]   0.764 -0.799 -1.799 -1.799 -1.799

> frocDataRaw$NL[1,1,(K1+1):(K1+K2),]
        [,1]     [,2]    [,3] [,4]
[1,]    -Inf    -Inf    -Inf -Inf
[2,]  -0.289    -Inf    -Inf -Inf
[3,]  -0.299  -0.412    -Inf -Inf
[4,]   0.252    -Inf    -Inf -Inf
[5,]  -0.892    -Inf    -Inf -Inf
[6,]   0.436   0.377  -0.224 -Inf
[7,]   0.133    -Inf    -Inf -Inf

> frocDataRaw$LL[1,1,,1]
[1]   -Inf 0.943 0.944 0.309 0.522 0.764 1.388

> retRocRaw$LL[1,1,1:K2,1]
[1] -1.799  0.943  0.944  0.309  0.522  0.764  1.388
```

Figure 13.2h results from sourcing **mainOCsRaw.R** with the same parameter values as in Fig. 13.2(b) and Figure 13.2i results from sourcing **mainOCsRaw.R** with the same parameter values as in Fig. 13.2(c). Shown next are binned AFROC plots.

13.13 Example: Binned AFROC plots

The AFROC plot is produced by the code in **mainOCsBinned.R**, Section 13.11.1, specifically lines 27–41. The AFROC plot is shown in Figure 13.2j for the the same parameter values as in Fig. 13.2(a) and in Figure 13.2k for the the same parameter values as in Fig. 13.2(b). Figure 13.2l is the AFROC plot for the same parameter values as in Fig. 13.2(c). To create the counts table, one needs the relevant binned counts and cumulated fractions. These can be calculated for the

the same parameter values as in Fig. 13.2(b), as shown in Section 13.13.1 (be sure to **source** the entire file with no breakpoints and appropriately set parameters; then copy and paste the relevant lines into the **Console** window).

13.13.1 Code snippets

```
> table(afrocDataBin$NL[1,1,1:K1,1])

-Inf    1    2    3    4    5
  21    5   10    5    5    4
> table(afrocDataBin$LL[1,1,1:K2,])

-Inf    1    2    3    4    5
  78    5    5    5   19   28
> plotAFROC$Points$genAbscissa
[1] 0.00 0.08 0.18 0.28 0.48 0.58 1.00
> plotAFROC$Points$genOrdinate
[1] 0.0000000 0.2692308 0.4519231 0.5000000 0.5480769 0.5961538
    1.0000000
> sum(Lk2) - sum(c(5, 5, 5, 19, 28))
[1] 42
```

The reader is encouraged to study this code carefully to understand the entries in Table 13.3. There is a subtle difference in the second **table** command, as one needs to count over the fourth index for LLs. The entry under **-Inf** includes locations without lesions, leading to an overestimate, namely 78, for unmarked lesions. The correct number of unmarked lesions is obtained in the final line. The entries under bins 1 through 5 in Table 13.3 are calculated in lines 1–6. The operating points are calculated in lines 7–10. The entry under unmarked FP is calculated in line 3 and that under LL is calculated in the last line. Table 13.3 shows the binning process used to cumulate counts to calculate the AFROC operating points. The denominator for LLF is 104, the total number of lesions, and that for FPF is 50, the total number of non-diseased cases. The unmarked lesion count, 42, is listed in the column under "Unmarked." This time the corresponding count for FP events is *known*, since it equals the known number of unmarked non-diseased cases.

Table 13.3 The upper half of this table shows the binned AFROC data in tabular form, constructed using the values in Section 13.13.1

	Unmarked	Bin 1	Bin 2	Bin 3	Bin 4	Bin 5
FP	21	5	10	5	5	4
LL	42	5	5	5	19	28
		\multicolumn{5}{AFROC operating points}				
		Bins ≥ 5	Bins ≥ 4	Bins ≥ 3	Bins ≥ 2	Bins ≥ 1
FPF		0.08	0.18	0.28	0.48	0.58
LLF		0.269	0.452	0.5	0.548	0.596

Note: The corresponding AFROC operating points are shown in the lower half of the table. The data corresponds to 50 non-diseased and 70 diseased cases containing 104 lesions. The denominator for LLF is 104 and that for FPF is 50.

13.14 Example: Binned FROC/AFROC/ROC plots

Figure 13.2a through l showed raw and binned FROC and AFROC plots for three datasets, a 5/7 dataset with threshold at −1, a 50/70 dataset with threshold at −1 and a 50/70 dataset with threshold at +1, and in all cases, the **lambda** parameter was set to unity. The purpose of this section is to compare binned FROC, AFROC, and ROC plots for the first two datasets with one change, namely, the **lambda** parameter is set to two; this increases the number of NLs on the average by a factor of two. The inferred ROC is the third plot produced by the code in **mainOCsBinned.R**, Section 13.11.1, specifically lines 43–56. Remove any breakpoints in the code and **source** it twice, once with the 5/7 dataset and once with the 50/70 dataset, with threshold set to −1, and with the appropriate change to **lambda**. Figure 13.5a through f shows FROC, AFROC, and ROC plots. The top row (a) through (c) corresponds to five non-diseased and seven diseased cases, and the bottom row (d) through (f) to 50 non-diseased and 70 diseased cases. [It is observed, in the first row of plots, that the number of points decreases from left to right: the FROC has the most ratings; the AFROC reduces each non-diseased cases to one rating while the ROC further reduces each diseased case to one rating.]

Examination of these plots, particularly the lower row, where sampling variability is lower, reveals the following characteristics. *AFROC and ROC plots are contained within the unit square, but the FROC plot is not (the last statement was not true for the smaller value of lambda, hence the change to demonstrate it). The ROC plot lies above the AFROC plot; compare (b) to (c) and (f) to (e). This is because sometimes an NL rating on a diseased case exceeds the LL rating, and therefore counts as the TP rating. Since the ordinate of the AFROC is defined by marked lesions, the ratings of NLs on the same cases are irrelevant.* All plots have a steep slope near the origin, but not infinity, as these are empirical plots (to confirm infinite slope switch to raw curves and increase the number of cases, as in Figure 12.2a through c. With the exception of the AFROC, the slope decreases monotonically as one moves up the plot. The slope of the AFROC decreases as one moves up the plot until one reaches the observed end-point generated by cumulating all ratings. The AFROC plot literally ends there, analogous to the observed FROC end-point but unlike the FROC, where one does not know what comes next. With the AFROC, the researcher is justified in connecting the observed end-point to the upper right corner of the unit square. This line makes an important contribution to free-response performance, to be shown later in **Chapter 14**.

Use the following code snippets to extract the counts and operating points necessary to construct Table 13.4 for the 50/70 dataset.

Table 13.4 The upper half of this table shows the binned inferred ROC data, constructed using the values in Section 13.14.1

	Unmarked	Bin 1	Bin 2	Bin 3	Bin 4
FP	21	11	9	5	5
TP	9	5	5	19	42
		Inferred ROC Operating Points			
		Bins ≥ 4	Bins ≥ 3	Bins ≥ 2	Bins ≥ 1
FPF		0.08	0.18	0.36	0.58
TPF		0.457	0.729	0.800	0.871

Note: The corresponding ROC operating points are shown in the lower half of the table. The data corresponds to 50 non-diseased and 70 diseased cases containing 104 lesions. The denominator for TPF is 70 and that for FPF is 50. Including the counts in the unmarked columns yields the trivial (1,1) point.

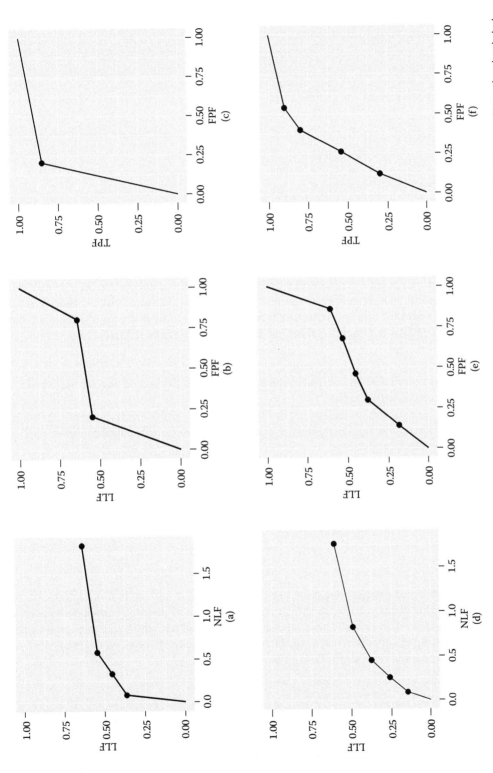

Figure 13.5 From left to right, FROC, AFROC, and ROC plots: the top row corresponds to 5/7 cases; the lower row to 50/70 cases. Note that lambda has been increased to 2 to show explicitly that the observed FROC end-point is semi-constrained.

13.14.1 Code snippets

```
> table(rocDataBin$NL[1,1,1:K1,])

  1  2  3  4  5
 21 11  9  5  4
> table(rocDataBin$LL[1,1,1:K2,])

  1  2  3  4  5
  9  5  5 19 32
> plotROC$Points$genAbscissa
[1] 0.00 0.08 0.18 0.36 0.58 1.00
> plotROC$Points$genOrdinate
[1] 0.000 0.457 0.729 0.800 0.871 1.000
```

13.15 Misconceptions about location-level "true negatives"

The quotes around *true negatives* are intended to illustrate the misconception that results when one inappropriately applies ROC terminology to the FROC paradigm. Section 13.10.5.1 shows that for the 5/7 dataset, **lambda** = one and reporting threshold set to –1, the first non-diseased case has one NL rated 0.7635935. The remaining three entries for this case are each **–Inf**.

What really happened is only known internally to the simulator. To the data analyst, the following possibilities are indistinguishable:

- Four latent NLs, one of whose ratings exceeded ζ_1, that is, three location-level true negatives occurred on this case.
- Three latent NLs, one of whose ratings exceeded ζ_1, that is, two location-level true negatives occurred on this case.
- Two latent NLs, one of whose ratings exceeded ζ_1, that is, one location-level true negative occurred on this case.
- One latent NL, whose rating exceeded ζ_1, that is, 0 location-level true negatives occurred on this case.

The second non-diseased case has one NL mark rated –0.7990092 and similar ambiguities occur regarding the number of latent NLs. The third, fourth, and fifth non-diseased cases have no marks. All four location-holders on each of these cases are filled with **–Inf**, which indicates unassigned values corresponding to either *absence of any latent NL or presence of one or more latent NLs that did not exceed* **zeta1** *and therefore did not get marked*.

To summarize:

Absence of an actual *NL* mark, indicated by a –∞ rating, could be due to either (1) non-occurrence of the corresponding latent *NL* or (2) occurrence of the latent *NL* but its rating did not exceed ζ_1. One cannot distinguish between the two possibilities.

For those who insist on using ROC terminology to describe FROC data, and there are some, the second possibility would be termed a location-level true negative (TN). Their logic is as follows: there was the possibility of an NL mark, which they term an FP, but the observer did not make it. Since the complement of an FP event is a TN event, this was a TN event. However, as just shown, one cannot determine whether it was a TN event or whether there was no latent event in the first place. Here is the conclusion:

There is no place in the FROC lexicon for a location-level TN. This fact has been misunderstood/ignored. There is even a recommendation[23] stating: "Tip: In a lesion-level analysis, be sure to explain how you calculate the number of true-negative findings." As explained in the Introduction in Chapter 12, the term *lesion-level* is ambiguous. The author believes the editors meant *location-specific*. The next part of the recommendation, "be sure to explain how you calculate the number of true-negative findings," sets up an impossible task.

The author's response to comments by a reviewer questioning the validity of analysis based on the AFROC is included in a document "**OnTrueNegatives.pdf**" in the online supplemental material for this chapter. It illustrates location-level true negative confusion in the mind of an expert statistician who is familiar with ROC but not with FROC. The paper in question was eventually published.[24]

If **zeta1 = -Inf** then all latent marks are actually marked and the ambiguities mentioned above disappear. Make this change to confirm, for the dataset used in Fig. 13.2(a) and using code file **mainOCsRaw.R**, that there were actually four latent NLs on the sixth diseased case (the eleventh sequential case), (Section 13.15.1), but the one rated −1.237538 fell below the previous value $\zeta_1 = -1$ and was consequently not marked.

13.15.1 Code snippets

```
> frocDataRaw$NL[1,1,,]
          [,1]    [,2]    [,3]    [,4]
 [1,]    0.764   -Inf    -Inf   -Inf
 [2,]   -0.799   -Inf    -Inf   -Inf
 [3,]    -Inf    -Inf    -Inf   -Inf
 [4,]    -Inf    -Inf    -Inf   -Inf
 [5,]    -Inf    -Inf    -Inf   -Inf
 [6,]   -1.148   -Inf    -Inf   -Inf
 [7,]   -0.289   -Inf    -Inf   -Inf
 [8,]   -0.299  -0.412   -Inf   -Inf
 [9,]    0.252   -Inf    -Inf   -Inf
[10,]   -0.892   -Inf    -Inf   -Inf
[11,]    0.436   0.377  -0.224  -1.24
[12,]    0.133   -Inf    -Inf   -Inf
> frocDataRaw$LL[1,1,,]
          [,1]    [,2]
 [1,]    -Inf    -Inf
 [2,]    0.943   -Inf
 [3,]    0.944   -Inf
 [4,]    0.309   -Inf
 [5,]    0.522   -Inf
 [6,]    0.764   -Inf
 [7,]    1.388   0.897
```

So, one might wonder, why not ask the radiologists to report everything they see, no matter how low the confidence level? Unfortunately, that would be contrary to their clinical task, where there is a price to pay for excessive NLs. It would also be contrary to a principle of good experimental design; one should keep interference with actual clinical practice, where the interference is designed to make the data easier to analyze, to a minimum.

A limited study in screening mammography was conducted* where radiologists were asked, after completing their usual screening interpretation, if they had considered any other regions in the case as possibly positive for malignant lesions, no matter how low the confidence level. The author's understanding of the results of this unpublished study is that they reported very few additional locations. Nodine and Kundel have shown via eye-movement recordings performed on radiologists that the latter are sometimes not consciously aware of regions that were fixated long enough to qualify as a latent mark, so the jury on this is still out, that is, it is not clear that the radiologists actually considered very few additional locations.

13.16 Comments and recommendations

13.16.1 Why not use NLs on diseased cases?

The original definition of the AFROC[4,8] but missing the number 1 currently appended to the acronym, see Section 13.8, was introduced in 1989. It used the maximum rated NL on every case to define the FPF-axis. The paper by Bunch et al.[3] suggested the same procedure. At that time, it seemed a good idea to include all available information and not discard any highest rated NLs. The author recalls a discussion around 2000 at SPIE Medical Imaging with Dr. Berkman Sahiner who argued for not including highest rated NLs on diseased cases in the AFROC—the author does not recall the reasoning. At that time, the author stated that ignoring this data was, on general principles, not a good idea. In retrospect, the author was wrong. Usage of the AFROC1 as the basis of analysis is not recommended. The only exception is when the case-set contains only diseased cases; it is not clear to the author why anyone would wish to conduct an observer performance study with diseased cases only.

The reason for excluding highest rated NLs on diseased cases is that they play a fundamentally different role in the clinic from those on non-diseased cases. A recall due to a highest rated NL on a diseased case where the lesion was not seen is actually not that bad. It would have been better if the recall were for the right reason, that is, the lesion was seen, but with a recall for the wrong reason at least the doctors get a second chance to find the lesion. On the other hand, a recall resulting from a highest rated NL on a non-diseased case is unequivocally bad. The patient is unnecessarily subjected to further imaging and perhaps invasive procedures, like needle-biopsy, in order to rule out cancer that she does not have.

Another subtler reason is that including highest rated NLs makes the AFROC1 curve disease-prevalence dependent (this issue was mentioned earlier in connection with the EFROC). Two investigators sampling from the same population, but one using a low-prevalence dataset while the other uses an enriched high-prevalence dataset, will obtain different AFROC1 curves for the same observer. This is because observers are generally less likely to mark NLs on diseased cases. This could be the satisfaction of search effect[25] where it is known that diseased cases are less likely to generate NL marks than non-diseased ones. It is as if finding a lesion satisfies the radiologist's need to find something in the patient's image that is explanatory of the patient's symptoms. Also, from the clinical perspective, finding a lesion is enough to trigger more extensive imaging, so it is not necessary to find *every other reportable suspicious region* in the image, because the radiologist knows that a more extensive workup is in the works for this patient. Suffice to say the author has datasets showing strong dependence of the number of NLs per case on disease state. More commonly, the number of NLs per case (the abscissa of the uppermost operating point on the FROC) is

* Prof. David Gur, private communication, ca. 2007.

larger if calculated over non-diseased cases than over diseased cases. So, the observed FROC and the AFROC1 will be disease prevalence dependent. If disease prevalence is very low, the curves will approach one limit, extending to larger NLF_{max} and FPF_{max}, and if disease prevalence is high, the curve will approach a different limit, extending to smaller NLF_{max} and FPF_{max}.

13.16.2 Recommendations

Table 13.5 summarizes the different operating characteristic possible with FROC data.

The recommendations are based on the author's experience with simulation testing and many clinical datasets. They involve a compromise between statistical power (the ability to discriminate between modalities that are actually different) and reliability of the analysis (i.e., it yields the right p-value).

AFROC1 versus AFROC: Unlike the AFROC1 figures of merit, the AFROC figures-of-merit do not use non-lesion localization data on diseased cases, so there is loss of statistical power with using the AFROC FOM. However, AFROC analyses are more likely to be reliable. The AFROC1 figures of merit involve two types of comparisons: (1) those between LL-ratings and NL-ratings on non-diseased cases and (2) those between LL-ratings and NL-ratings on diseased cases. The comparisons have different clinical implications, and mixing them does not appear to be desirable. The problem is avoided if one does not use the second type of comparison. This requires further study, but the issue does not arise if the dataset contains only diseased cases when the AFROC1 figures of merit should be used.

Weighted versus non-weighted: Weighting (i.e., using wAFROC or wAFROC1 FOM) assures that all diseased cases get equal importance, regardless of the number of lesions on them, a desirable

Table 13.5 Summary of operating characteristics possible with FROC data and recommendations. In most cases the AUC under the wAFROC is the desirable operating characteristic.

Operating characteristic	Abscissa	Ordinate	Comments	Recommended FOM?
ROC	FPF	TPF	Highest rating used to infer both FPF and TPF	Yes, if overall sensitivity and specificity are desired
FROC	NLF	LLF	Defined by marks; unmarked cases do not contribute	No
AFROC	FPF	LLF	Highest rating used to infer FPF	Yes, AUC, if number of lesions per case is less than four and lesion weighting is not relevant
AFROC1	FPF[1]	LLF	Maximum NL ratings over *every* case contribute to FPF[1]	Yes, AUC, only when there are zero non-diseased cases and if lesion weighting is not relevant
wAFROC	FPF	wLLF	Weights, which sum to unity, affect ordinate only	Yes, AUC
wAFROC1	FPF[1]	wLLF	Weights affect ordinate only; maximum NL rating over *every* case contributes to FPF[1]	Yes, AUC, only when there are zero non-diseased cases

statistical characteristic, so weighted analysis is recommended. Based on the author's experience, there is little difference between the two analyses when the number of lesions varies from one to three. There is some loss of statistical power in using weighted over non-weighted figures of merit, but the benefits versus ROC analysis are largely retained. Unless there are clinical reasons for doing otherwise, equal weighting is recommended.

The (highest rating inferred) ROC curve is sometimes desirable to get case-level sensitivity and specificity, as these quantities have well understood meanings to clinicians. For example, the highest non-trivial point in Figure 13.5f, defined by counting all highest rated marks, yields a relatively stable estimate of sensitivity and specificity, as described in a recent publication.[26]

The validity of the highest rating assumption has been questioned.[27] Two other methods of inferring ROC data from FROC data that have been suggested[5] are implemented in **RJafroc**: the average rating and the stochastically dominant rating. The author has applied both methods of inferring ROC data, in addition to the highest rating method, to the data used in References [27]. The results are insensitive to the choice of inferring method. So, if the highest rating method is not valid, neither are any of the other proposed methods.* A paper supporting the validity of the highest rating assumption has since appeared.[28] The highest rating assumption has a long history. See, for example, Swensson's LROC paper[29] and other papers published by Swensson and Judy. It is intuitive. If an observer sees a highly suspicious region and a less suspicious region, why would the observer want to dilute the severity of the condition by averaging the ratings? The highest rating captures the rating of the most significant clinical finding on the case, which is usually the reason for further clinical follow-up. Much of the confusion[22] regarding this issue is due to a misunderstanding of the meaning of the term *lesion*.

The AFROC and wAFROC are contained within the unit square and provide valid area measures for comparing two treatments. Except in special cases, this is not possible with the FROC.

The reason for the recommendation against the FROC follows.

13.16.3 FROC versus AFROC

Figure 13.6a shows FROC plots and 13.6b shows AFROC plots for two simulated observers, a simulated CAD observer and a RAD (expert radiologist) observer. The code to generate these plots, and explanations, are in Online Appendix 13.A.1, file **mainFrocVsAfroc.R**. Parameters that do not change between the two observers are $v = 1$, $\lambda = 1$, $K_1 = 500$, and $K_2 = 700$. The large numbers of cases were used to minimize sampling variability and allow clearer conclusions. In the plots CAD corresponds to $\mu = 1$ and $\zeta_1 = -1$ while RAD corresponds to $\mu = 1.5$ and $\zeta_1 = 1.5$. Increasing μ, which results in increased separation of the two-unit variance normal distributions, increases performance; recall the solar analogy in **Chapter 12**. Changing ζ_1 does not alter performance, rather, it decreases the observed range of the curve. If ζ_1 is larger, as in RAD, fewer NL and LL ratings exceed the higher threshold, and therefore less of the latent curve is observed. [The situation is analogous to the equal–variance binormal model, where a separation parameter determined AUC for the ROC curve, while ζ determines the operating point on the curve. As ζ increases, the operating point moves down the ROC curve, so the observed part, extending from the origin to the end-point, shrinks.]

The code also prints the AUCs under the two AFROC plots in Figure 13.6b. For CAD it is 0.608 and for RAD it is 0.674. The RAD observer has larger AUC, consistent with the visual impression

---

* The author, who conceived the study and did almost all of the analyses, withdrew his name from the paper following disagreement over whether to include results using other methods of inferring ROC data.

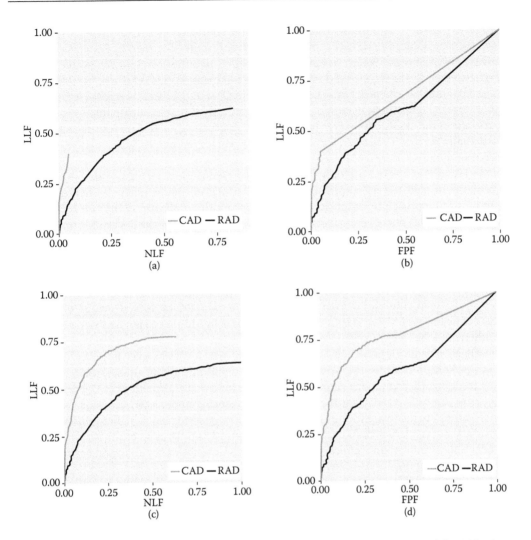

Figure 13.6 (a) FROC curves for the CAD observer (dark line), $\mu = 1$ and $\zeta_1 = -1$, and the RAD observer, $\mu = 1.5$ and $\zeta_1 = 1.5$, (light line). Note the much steeper rise and shorter horizontal traverse of the RAD observer, suggesting superior performance, which is difficult to quantify from the FROC curves, as a universal AUC measure cannot be defined and, if defined over the common NLF range, would ignore most NLs from the CAD observer. (b) This plot shows corresponding AFROC curves for CAD observer (dark line) and the RAD observer (light line). The AUC under the RAD observer is clearly greater than that for the CAD observer, even though the AUC estimate is biased downward against RAD. AUCs under the two AFROC plots are 0.608 for CAD and 0.674 for RAD. (c) FROC curves for CAD observer and the RAD observer for $\zeta_1 = -\infty$, which is impractical with radiologist observers but almost possible with CAD. (d) AFROC curves for CAD observer and the RAD observer for $\zeta_1 = -\infty$. AUCs under the two AFROC plots are 0.601 for CAD and 0.778 for RAD. The code to generate these plots is in file **mainFrocVsAfroc.R**.

in Figure 13.6b and as expected from the larger μ. The reader should try different **seed** values (currently set to one) to be convinced that the higher performance of RAD is not a sampling artifact.

From the plot in Figure 13.6a one intuitively suspects the RAD observer is performing better than CAD. The intuition is based on the much steeper rise and much shorter traverse along the NLF axis for the RAD observer as compared to CAD. The RAD observer is better at finding lesions and producing fewer NLs, both of which are desirable characteristics. One suspects that if this observer could be induced to relax the threshold and report more NLs, then LLF would exceed that of the CAD observer while NLF_{max} would remain smaller than the corresponding value for CAD. However, based on the FROC curve alone, it is difficult to quantify this intuition. [Ability to fit the FROC curve would have helped, but as will be shown in Chapter 17, this is not possible for the radiologist observer.] Figure 13.6c corresponds to 13.6a, the only difference being that $\zeta_1 = -\infty$, so the entire traverse of the FROC curves are visible for both observers. This confirms the expectation that RAD is actually the better observer. Figure 13.6d shows corresponding AFROC curves for $\zeta_1 = -\infty$, the corresponding AUCs are 0.601 and 0.778.

Plots in Figure 13.6c and d are only possible with a simulator. In practice, one cannot get the RAD observer to report every suspicious region. Therefore, one is restricted to analyzing Figure 13.6a and b. One approach to quantification is to compare the AUCs under the two curves in the *common* range of NLF where *both* curves contribute LLF values. Since the RAD observer generates the smaller NLF_{max}, denoted NLF_{max}^{RAD}; this means one can compare the AUCs under the two FROC curves in the common range $NLF = 0$ to NLF_{max}^{RAD}. It is obvious from Figure 13.6a that in this range the RAD observer yields the larger AUC (imagine dropping a vertical line from the end-point of the RAD curve to the x-axis; the relevant areas under the two curves are to the left of this line). However, this would entail a big price in terms of ignored data, namely all marks of the CAD observer defined by $NLF^{CAD} > NLF_{max}^{RAD}$ are ignored. This is in addition to ignored unmarked cases, that is inherent to the FROC curve.

The AFROC plot in Figure 13.6b shows clearly that the RAD observer is performing better than the CAD observer. Since the AFROC is contained within the unit square, there is no question how to extend the curve: one simply connects the observed end-point to (1,1) with a straight line. Actually, the AFROC AUC for the RAD observer is being underestimated. Had the observer relaxed the criterion, the straight-line extension would have started from a higher value of the ordinate, yielding an even larger difference, see plot in Figure 13.6d. In spite of the underestimation, which affects RAD more than it affects CAD, the AFROC still shows superior performance for the RAD observer in plot (b).

The following code snippets are provided to show how to extract the coordinates of the end-point (CAD threshold set to –1 and RAD threshold set to 1.5). [As before, **genAbscissa** is a generic abscissa, NLF in the current context, while **genOrdinate** is a generic ordinate, LLF in the current context]

13.16.3.1 Code snippets

```
> max(froc$Points$genAbscissa[froc$Points$class
+                             == levels(froc$Points$class)[1]])
[1] 0.8275
> max(froc$Points$genAbscissa[froc$Points$class
+                             == levels(froc$Points$class)[2]])
[1] 0.04916667
> max(froc$Points$genOrdinate[froc$Points$class
+                             == levels(froc$Points$class)[1]])
[1] 0.6192308
> max(froc$Points$genOrdinate[froc$Points$class
+                             == levels(froc$Points$class)[2]])
[1] 0.3980769
```

This looks like gobbledygook, the main complication is that there are two plots contained in one object; the first two lines extract from the NLF array those with class 1, which corresponds to CAD, and takes the maximum of the returned values, printed at line 3. The next two lines obtains the corresponding maximum for RAD. Lines 8–9 and lines 11–12 repeat these for the LLF arrays. One can use the **str()** command to unravel what is going on, as shown below (with multiple plots per graph, **class** is used as a factor variable to distinguish between the different plots; for example **class** = 1 is the CAD plot and **class** = 2 is the RAD plot).

13.16.3.2 Code snippets

```
> str(froc$Points$genAbscissa[froc$Points$class
+                           == levels(froc$Points$class)[1]])
 num [1:1638] 0 0 0 0 0.000833 ...
> str(froc$Points$genOrdinate[froc$Points$class
+                           == levels(froc$Points$class)[1]])
 num [1:1638] 0 0.000962 0.001923 0.002885 0.002885 ...
> str(froc$Points$genAbscissa[froc$Points$class
+                           == levels(froc$Points$class)[2]])
 num [1:474] 0 0 0 0 0 0 0 0 0 ...
> str(froc$Points$genOrdinate[froc$Points$class
+                           == levels(froc$Points$class)[2]])
 num [1:474] 0 0.000962 0.001923 0.002885 0.003846 ...
```

Bottom line: the end-point coordinates are (0.828, 0.619) for CAD and (0.049, 0.398) for RAD. These values confirm the visual estimates from the plots in Figure 13.6a.

Figure 13.7a exaggerates the difference between CAD and RAD. The CAD parameters are the same as in Figure 13.6a but the RAD parameters are **mu** = 2 and **zeta1** = +2. Doubling the separation parameter over that of CAD has a huge effect on performance. The end-point coordinates for RAD are (0.015, 0.421). This time AUC under the common region defined by NLF = zero to NLF = 0.015 would exclude almost all marks made by CAD. The AFROCs in plot (b) show the markedly greater performance of RAD compared to CAD (the AUCs are 0.608 for CAD and 0.708 for RAD). The difference is larger, in spite of the downward bias working against the RAD observer; as described earlier, the straight line part would have begun from a higher starting point had the observer used a lower threshold. Figure 13.7c and d are the FROC and AFROC plots for the two observers when **zeta1** is set to **-Inf**. The corresponding AUCs are 0.601 and 0.872.

The final example, Figure 13.8, shows that when there is a small difference in performance, then there is less loss of information from using the FROC as a basis for measuring performance. The CAD parameters are the same as in Figure 13.6a but the RAD parameters are **mu** = 1.1 and **zeta1** = –1. This time there is much more common overlap in the plot in Figure 13.8a and the area measure is counting most of the marks for both readers (but still not accounting for unmarked non-diseased cases). The superior AFROC-based performance of RAD is also apparent in the plot in Figure 13.8b.

A misconception exists that using the rating of only one NL mark, as in AFROC, sacrifices statistical power. In fact, the chosen mark is a special one, namely the highest rated NL mark on a non-diseased case, which carries more information than a randomly chosen NL mark. If the sampling distribution of the Z-sample were uniform, then the highest sample is a *sufficient statistic*, meaning that it carries all the information in the samples. The highest rated Z-sample from a normal

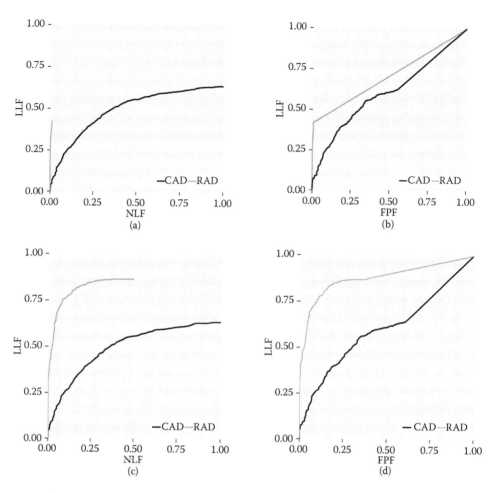

Figure 13.7 (a) FROC curves for CAD observer (dark line) and the RAD observer (light line). The CAD observer is identical to that shown in Figure 13.6a. The RAD observer is characterized by **mu** = 2 and **zeta1** = 2. This time it is impossible to compare the two FROC curves, as the common range is very small. However, the AFROCs clearly show the expected superiority of the RAD observer, in spite of the severe underestimate of the AUC of the RAD observer. AUCs under the two AFROC plots are 0.608 for CAD and 0.708 for RAD. Plots (c) and (d) correspond to (a) and (b), respectively, with **zeta1** = **-Inf** for both observers. AUCs under the two AFROC plots in (d) are 0.601 for CAD and 0.872 for RAD.

distribution is not a sufficient statistic, so there is some loss of information, but not as much as would occur with a randomly picked Z-sample.

13.16.4 Other issues with the FROC

Loss of statistical power is not the only issue with the FROC. Because it counts NLs on both diseased and non-diseased cases, the curve depends on disease prevalence in the dataset. Because the numbers of LLs per case is variable, the curve gives undue importance to those diseased cases with unusually large numbers of lesions. As noted in 13.16.2, the clinical importance of an NL on a non-diseased case differs from that on a diseased case. The FROC curve ignores this distinction. (See Section 17.10 for examination of AUCs under FROC and AFROC and whether they agree with the area under the ROC curve, the latter being regarded as the gold standard. It is shown there that the AFROC AUC agrees with ROC AUC but the FROC AUC does not, except in very special cases.)

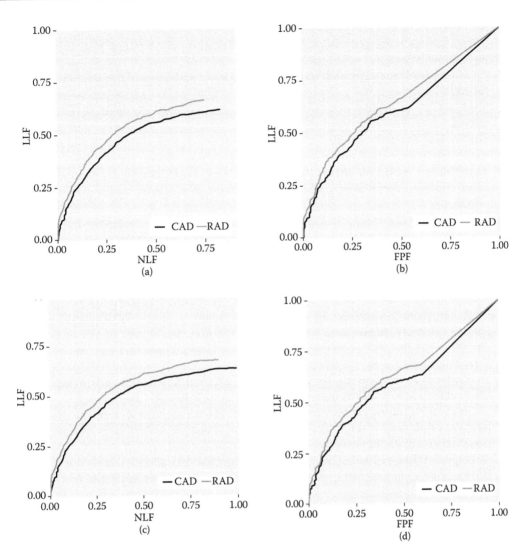

Figure 13.8 (a) and (b) FROC/AFROC curves for CAD and RAD observers. The CAD observer is identical to that shown in Figure 13.7a. The RAD observer is characterized by **mu** = 1.1 and **zeta1** = −1. This time it is possible to compare the two FROC curves, as the common NLF range is large. Both FROC and AFROC show the expected slight superiority of the RAD observer. AUCs under the two AFROC plots are 0.608 for CAD and 0.634 for RAD. Plots (c) and (d) correspond to (a) and (b), respectively, with **zeta1** = **−Inf** for both observers. Since **zeta1** in (a) and (b) is already quite small, lowering it to −∞ does not pick up too many additional marks. AUCs under the two AFROC plots in (d) are 0.601 for CAD and 0.624 for RAD.

13.17 Discussion/Summary

This chapter started with the difference between latent and actual marks and the notation to describe FROC data. The notation is exploited in deriving formula for FROC, AFROC, and inferred ROC operating characteristics obtainable from FROC data. Coded examples are given of FROC, AFROC, and ROC curves, for both raw and binned data, using a FROC data simulator. These allow examination of the FROC data structure at a deeper level than is possible with formalism alone.

Since serious misunderstandings and confusion regarding the FROC paradigm exist, several key points are re-emphasized:

1. An important distinction is made between observable and unobservable events. Observable events, such as unmarked lesions, can safely be assigned the $-\infty$ rating. Negative infinity ratings cannot be assigned to unobservable events.
2. A location-level true negative is an unobservable event and usage of this term has no place in the FROC lexicon.
3. The FROC curve does not reach unit ordinate except in special cases when the lesions are easy to find.
4. The limiting end-point abscissa of the FROC, that is, what the observer would have reached had the observer marked *every* latent NL, is unconstrained to the range (0,1).
5. The inclusion of NLs on diseased cases introduces an undesirable dependence of the FROC curve on disease prevalence. A valid operating characteristic should be independent of disease prevalence.
6. The notion that maximum NLF is determined by the ratio of the image area to the lesion area is incorrect. This simplistic model is not supported by eye-movement data acquired on radiologists performing clinical tasks.
7. In contrast to the FROC, the limiting end-point of the AFROC is constrained, that is, both coordinates are in the range (0,1).
8. For the observer, who does not generate any marks, the operating point is (0,0) and the AFROC is the inaccessible line connecting (0,0) to (1,1), contributing empirical AUC = 0.5. This observer has unit specificity but zero sensitivity, which is better than chance level performance (AUC = 0). The corresponding ROC observer displays chance level performance and gets no credit for perfect performance on non-diseased cases.
9. The weighted AFROC curve is the preferred way to summarize performance in the FROC task. Usage of the FROC to derive measures of performance is strongly discouraged.
10. The highest NL rating carries more information about the other NLs on the case than the rating of a randomly selected NL. The implication is that the AFROC does not sacrifice much power relative to FROC curve based measures.
11. The highest rating method of inferring ROC data is justified; alternatives such as average and stochastically dominant rating do not appear to have substantive advantages.

It is unfortunate that the optimal way of summarizing FROC data, namely the AFROC, has been known for a long time, specifically 1977 in the Bunch et al. paper although they imply that it is not the preferred way. An 1989 paper by the author,[4] *stated unambiguously that the area under the AFROC is an appropriate figure of merit for the FROC paradigm.* Unfortunately, this recommendation has been largely ignored and CAD research, which would have benefited most from it, has proceeded, over more than two decades, entirely based on the FROC curve. Currently there is much controversy about CAD's effectiveness, especially for masses in breast cancer screening. The author believes that CAD's current poor performance is in part due to choosing the incorrect operating characteristic used to evaluate and optimize it.

If the author appears to have picked on other's mistakes, and on CAD, it is with the objective of learning from one's mistakes. The author has made his own share of mistakes,[15] which are unavoidable in science, and has contributed to some of the confusion, an example of which is the temporary recommendation of the AFROC1 noted above. Progress in science rarely proceeds in a straight line.

A legitimate concern at this point could be that most of the recommendations are based on the validity of the FROC data simulator. Details of the simulator and evidence for its validity are deferred to **Chapter 16** and **Chapter 17**.

Having defined various operation characteristics associated with FROC data, and how to compute the coordinates of operating points, it is time to turn to formula for figures of merit, e.g., AUCs, that can be derived from these plots without recourse to planimetry (i.e., without actually counting squares) and their physical meanings, the subject of the next chapter.

References

1. Duchowski AT. *Eye Tracking Methodology: Theory and Practice.* Clemson, SC: Clemson University; 2002.
2. Yoon HJ, Zheng B, Sahiner B, Chakraborty DP. Evaluating computer-aided detection algorithms. *Med Phys.* 2007;34(6):2024–2038.
3. Bunch PC, Hamilton JF, Sanderson GK, Simmons AH. A free-response approach to the measurement and characterization of radiographic-observer performance. *J Appl Photogr Eng.* 1978;4:166–171.
4. Chakraborty DP. Maximum Likelihood analysis of free-response receiver operating characteristic (FROC) data. *Med Phys.* 1989;16(4):561–568.
5. Song T, Bandos AI, Rockette HE, Gur D. On comparing methods for discriminating between actually negative and actually positive subjects with FROC type data. *Med Phys.* 2008;35(4):1547–1558.
6. Chakraborty DP, Zhai X. On the meaning of the weighted alternative free-response operating characteristic figure of merit. *Med Phys.* 2016;43(5):2548–2557.
7. Chakraborty DP, Berbaum KS. Observer studies involving detection and localization: Modeling, analysis and validation. *Med Phys.* 2004;31(8):2313–2330.
8. Chakraborty DP, Winter LHL. Free-response methodology: Alternate analysis and a new observer-performance experiment. *Radiology.* 1990;174:873–881.
9. Lockwood P, Piper K. AFROC analysis of reporting radiographer's performance in CT head interpretation. *Radiography.* 2015;21(3):e90–e95.
10. Popescu LM. Nonparametric signal detectability evaluation using an exponential transformation of the FROC curve. *Medi Phys.* 2011;38(10):5690–5702.
11. Chakraborty DP, Yoon HJ. JAFROC analysis revisited: Figure-of-merit considerations for human observer studies. *Paper presented at: Proceedings of the SPIE Medical Imaging: Image Perception, Observer Performance, and Technology Assessment;* 2009;7263:72630T. Lake Buena Vista (Orlando Area), FL.
12. DeSantis C, Siegel R, Bandi P, Jemal A. Breast cancer statistics, 2011. *CA: Cancer J Clin.* 2011;61(6):408–418.
13. Zhai X, Chakraborty DP. A bivariate contaminated binormal model for robust fitting of proper ROC curves to a pair of correlated, possibly degenerate, ROC datasets. *Med Phys.* 2017;44:2207–2222.
14. Chakraborty DP. Validation and statistical power comparison of methods for analyzing free-response observer performance studies. *Acad Radiol.* 2008;15(12):1554–1566.
15. Chakraborty DP. Problems with the differential receiver operating characteristic (DROC) method. *Paper presented at: Proceedings of the SPIE Medical Imaging 2004: Image Perception, Observer Performance, and Technology Assessment;* 2004;5372:138–143. San Diego, CA.
16. Chakraborty DP, Howard NS, Kundel HL. The Differential Receiver Operating Characteristic (DROC) Method: Rationale and Results of Recent Experiments. *Paper presented at: Proceedings of the SPIE Medical Imaging 1999: Image Perception and Performance;* 1999. San Diego, CA.
17. Chakraborty DP, Kundel HL, Nodine CF, Narayan TK, Devaraju V. The differential receiver operating characteristic (DROC) method. *Paper Presented at: Proceedings of the Medical Imaging 1998: Image Processing;* 1998, San Diego, CA. doi: 10.1117/12.310893
18. Chakraborty DP. A brief history of free-response receiver operating characteristic paradigm data analysis. *Acad Radiol.* 2013;20(7):915–919.
19. Gur D, Rockette HE. Performance assessment of diagnostic systems under the FROC paradigm: Experimental, analytical, and results interpretation issues. *Acad Radiol.* 2008; 15:1312–1315.

20. Chakraborty DP. Counterpoint to Performance assessment of diagnostic systems under the FROC paradigm by Gur and Rockette. *Acad Radiol.* 2009;16:507–510.
21. Gur D, Rockette HE. Performance assessments of diagnostic systems under the FROC paradigm: Experimental, analytical, and results interpretation issues. *Acad Radiol.* 2009; 16(6):770–771.
22. Hillis SL, Chakraborty DP, Orton CG. ROC or FROC? It depends on the research question. *Med Phys.* 2017;44:1603–1606.
23. Levine D, Bankier AA, Halpern EF. Submissions to radiology: Our top 10 list of statistical errors1. *Radiology.* 2009;253(2):288–290.
24. Vikgren J, Zachrisson S, Svalkvist A, et al. Comparison of chest tomosynthesis and chest radiography for detection of pulmonary nodules: Human observer study of clinical cases. *Radiology.* 2008;249(3):1034–1041.
25. Berbaum KS, Franken EA, Dorfman DD, et al. Satisfaction of search in diagnostic radiology. *Invest Radiol.* 1990;25(2):133–140.
26. Dobbins JT, McAdams HP, Sabol JM, et al. Multi-institutional evaluation of digital tomosynthesis, dual-energy radiography, and conventional chest radiography for the detection and management of pulmonary nodules. *Radiology.* 2016;282:236–250.
27. Zanca F, Hillis SL, Claus F, et al. Correlation of free-response and receiver-operating-characteristic area-under-the-curve estimates: Results from independently conducted FROC/ROC studies in mammography. *Med Phys.* 2012;39(10):5917–5929.
28. McEntee MF, Littlefair S, Pietrzyk MW. A comparison of ROC inferred from FROC and conventional ROC. *Paper presented at: Proceedings of the SPIE Medical Imaging*: Image Perception, Observer Performance, and Technology Assessment; 903718; 2014; San Diego, CA. doi: 10.1117/12.2044341; 2014.
29. Swensson RG. Unified measurement of observer performance in detecting and localizing target objects on images. *Med Phys.* 1996;23(10):1709–1725.

14

Computation and meanings of empirical FROC FOM-statistics and AUC measures

14.1 Introduction

The previous chapter focused on empirical *plots* possible with FROC data. Examples are the FROC, AFROC, wAFROC, inferred ROC, and so on. Expressions were given for computing operating points for each plot. Because of the same conundrum faced with quoting sensitivity and specificity pairs, section 3.8, operating points are rarely used as figures of merit. Rather, interest is in area measures derived from operating characteristics.

In this chapter, the empirical area under a plot is generically denoted A_{oc}, where the oc subscript denotes the applicable operating characteristic. For example, the area under the empirical weighted AFROC is denoted A_{wAFROC}. Calculating areas from operating points using planimetry or geometry is tedious. *Needed are formula for calculating them directly from ratings.* An example of this was encountered before in **Chapter 5**, where it was shown that the area under the empirical ROC A_{ROC} equaled the Wilcoxon statistic.

A distinction is being made between the empirical AUC under a plot, that is, an area, and a FOM-statistic, generically denoted θ, that is computed directly from the ratings. The FOM-statistic is a scalar function of the ratings that is intended to quantify performance. While any function of the ratings is a possible FOM-statistic, whether a proposed function is useful depends upon whether it can be related to the area under a meaningful operating characteristic. This chapter derives formula for FOM-statistics θ_{oc}, which yield the same values as the empirical areas A_{oc} under the corresponding operating characteristics. The meanings of these FOM-statistics are discussed.[1]

The following is the organization of this chapter. Expressions for the empirical AFROC FOM-statistic θ_{AFROC} and the empirical weighted AFROC FOM-statistic θ_{wAFROC} are presented and their limiting values for chance-level and perfect performances are explored. Two important theorems are stated, whose proofs are in Online Appendix 14.A. The first theorem proves the equality between the empirical wAFROC FOM-statistic θ_{wAFROC} and the area A_{wAFROC} under the empirical wAFROC plot. A similar equality applies to the empirical AFROC FOM-statistic θ_{AFROC} and the area A_{AFROC} under the empirical AFROC plot. The second theorem derives an expression for the area under the straight-line extension of the *wAFROC* (or AFROC) from the observed end-point to (1,1), and explains why it is essential to include this area. R code, detailed in Online Appendix B, explains, with a small simulated-dataset, how NL and LL ratings and lesion weights determine the AFROC and wAFROC empirical plots follows. The example shows explicitly that the wAFROC gives equal importance to all diseased cases, a desirable statistical characteristic, while the AFROC gives excessive importance to cases with more lesions. Online Appendix B creates a small simulated dataset to illustrate the non-weighted and weighted versions of the AFROC. The small number of cases makes the explanations easier to follow and permits hand-calculation of the operating points. This is followed by physical interpretation of the corresponding AUCs/FOM-statistics. It shows explicitly how the ratings comparisons implied in the FOM-statistic properly credit and penalize the observer for correct and incorrect decisions, respectively. This section gives probabilistic meanings to the AFROC and wAFROC AUCs.

Detailed derivations of FOM-statistics applicable to the areas under the empirical FROC plot, the AFROC1, and wAFROC1 plots are not given. Instead, the results for all plots are summarized in Online Appendix 14.C. Online Appendix 14.D shows that the definitions work, that is, the FOM-statistics yield the correct areas as determined by numerical integrations.

14.2 Empirical AFROC FOM-statistic

The empirical AFROC-AUC was defined in Section 13.4.2 as the area under the empirical AFROC. The corresponding FOM-statistic θ_{AFROC} is defined[2] in terms of a quasi-Wilcoxon statistic* by (the kernel function ψ was defined in Equation 5.10)

$$\theta_{AFROC} \triangleq \frac{1}{K_1 L_T} \sum_{k_1=1}^{K_1} \sum_{k_2=1}^{K_2} \sum_{l_2=1}^{L_{k_2}} \psi\left(\max_{l_1}\left(z_{k_1 1 l_1 1} \middle| z_{k_1 1 l_1 1} \geq \zeta_1\right), z_{k_2 2 l_2 2}\right) \tag{14.1}$$

FROC notation is summarized in Table 13.1. The conditioning on Z-samples greater than or equal to ζ_1 ensures that *only marked regions contribute to* θ_{AFROC}. In **Chapter 13**, the relevant formula ensured that only marked regions contributed to operating points. A similar requirement needs to be imposed in expressions for FOM-statistics.

θ_{AFROC} *achieves its highest value, unity, if and only if every lesion is rated higher than every mark on every non-diseased case,* for then the ψ function always yields unity, and the summations yield[†]

$$\theta_{AFROC} = \frac{1}{K_1 L_T} \sum_{k_1=1}^{K_1} \sum_{k_2=1}^{K_2} \sum_{l_2=1}^{L_{k_2}} 1 = \frac{1}{K_1 \sum_{k_2=1}^{K_2} L_{k_2}} \sum_{k_1=1}^{K_1} L_T = \frac{1}{K_1} \sum_{k_1=1}^{K_1} 1 = 1 \tag{14.2}$$

If, on the other hand, every lesion is rated lower than every mark on every non-diseased case, the ψ function yields zero, and the FOM-statistic is zero. Therefore,

$$0 \leq \theta_{AFROC} \leq 1 \tag{14.3}$$

The range of θ_{AFROC}, Equation 14.3, is twice that of A_{ROC}, that is, *0.5–1.* This has the important consequence that treatment or modality related differences between θ_{AFROC} (i.e., effect sizes) tend to be larger relative to the corresponding ROC effect sizes, explained further in **Section 19.3**. Equation 14.3 is another reason why the positive diagonal of the AFROC corresponding to AUC = 0.5, does not reflect chance-level performance. An area under the AFROC equal to 0.5 is actually a reasonable performance, being smack in the middle of the allowed range. An example of this was given in Section 13.4.2.2 for the case of an expert radiologist who does not mark any cases.

---

* Quasi: seemingly; apparently, but not really.
† Here is another example of how the maximum value carries more information than a randomly selected value. If the maximum rating over all NLs on a case is smaller than a specified value, then every NL on the case must be rated smaller than the specified value.

14.3 Empirical weighted AFROC FOM-statistic

The empirical weighted AFROC plot and lesion weights were defined in Section 13.7. The empirical weighted AFROC FOM-statistic[3] is defined in terms of a quasi-Wilcoxon statistic:

$$\theta_{wAFROC} \triangleq \frac{1}{K_1 K_2} \sum_{k_1=1}^{K_1} \sum_{k_2=1}^{K_2} \sum_{l_2=1}^{L_{k_2}} W_{k_2 l_2} \, \psi \left(\max_{l_1} \left(z_{k_1 1 l_1 1} \middle| z_{k_1 1 l_1 1} \geq \zeta_1 \right), z_{k_2 2 l_2 2} \right) \tag{14.4}$$

The weights obey the constraint:

$$\sum_{l_2=1}^{L_{k_2}} W_{k_2 l_2} = 1 \tag{14.5}$$

If not explicitly qualified as a weighted AFROC FOM-statistic, it is understood that one means the conventional (not-weighted) AFROC FOM-statistic defined by Equation 14.2. In the special case of one lesion per diseased case, the two definitions are identical. However, for multiple lesions per case, they are not, as one cannot set the weights equal to unity, for example, for *equally weighted lesions*.

$$W_{k_2 l_2} = \frac{1}{L_{k_2}} \tag{14.6}$$

For equally weighted lesions, for a case with three lesions, each weight equals one-third (1/3).

14.4 Two theorems

The area A_{wAFROC} under the wAFROC plot is obtained by summing the areas of individual trapezoids defined by drawing vertical lines from each pair of adjacent operating points to the x-axis. A sample plot is shown Figure 14.1. The points are labeled following the usual convention: the trivial point at the upper right corner is labeled 0, the next lower-left one is labeled 1, etc.. The labels increment by one as one moves down the curve. Shown shaded are two trapezoids, one defined by points 0 and 1 and the other defined by points i and $i + 1$.

The operating point labeled i has coordinates $(FPF_i, wLLF_i)$ given by Equations 13.11 and 13.17, respectively, reproduced here for convenience:

$$FPF_i \equiv FPF\left(\zeta_i\right) = \frac{1}{K_1} \sum_{k_1=1}^{K_1} I \left(\max_{l_1} \left(z_{k_1 1 l_1 1} \middle| z_{k_1 1 l_1 1} \geq \zeta_1 \right) \geq \zeta_i \right) \tag{14.7}$$

$$wLLF_i \equiv wLLF\left(\zeta_i\right) = \frac{1}{K_2} \sum_{k_2=1}^{K_2} \sum_{l_2=1}^{L_{k_2}} W_{k_2 l_2} I \left(z_{k_2 2 l_2 2} \geq \zeta_i \right) \tag{14.8}$$

As usual, the summation in Equation 14.7 is conditioned on marked NLs. The conditioning is unnecessary for LLs. Online Appendix 14.A proves the following theorems.

14.4.1 Theorem 1

The area A_{wAFROC} under the empirical *wAFROC* plot equals the weighted AFROC FOM-statistic θ_{wAFROC} defined by Equation 14.4.

$$\theta_{wAFROC} = A_{wAFROC} \tag{14.9}$$

This is the FROC counterpart of Bamber's Wilcoxon – empirical ROC-AUC equivalence theorem,[4] derived in **Chapter 5**.

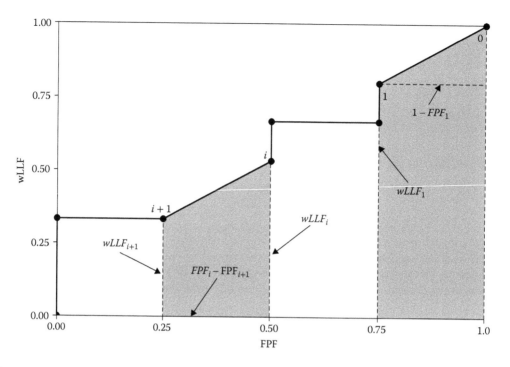

Figure 14.1 An example wAFROC plot; from left to right, the two shaded areas correspond to A_i and A_0, respectively, defined in Online Appendix 14.A. This plot was produced by **mainProofPlot.R**.

14.4.2 Theorem 2

The area A_0 under the straight-line extension of the wAFROC from the observed end-point $(FPF_1, wLLF_1)$ to $(1,1)$ is given by

$$A_0 = \frac{(1 - FPF_1)(1 + wLLF_1)}{2}$$ (14.10)

According to Equation 14.10, A_0 increases as FPF_1 decreases, that is, as more non-diseased cases are not marked and as $wLLF_1$ increases, that is, as more lesions, especially those with greater weights, are marked. Both observations are in keeping with the behavior of a valid FOM. Failing to include the area under the straight-line extension results in not counting the positive contribution to the FOM of unmarked cases, each of which represents a perfect decision. For a perfect observer whose operating characteristic is the vertical line from (0,0) to (0,1), the area under the straight-line extension comprises the entire AUC for the wAFROC. Excluding it would predict zero AUC for a perfect observer, which is illogical.

14.5 Understanding the AFROC and wAFROC empirical plots

The code for this example is in file **mainExamples.R** in Online Appendix 14.B. Parameters of the simulation are $\mu = 2$, $\lambda = 1$, $\nu = 1$, $\zeta_1 = -1$ and $L_{max} = 2$. It simulates a dataset consisting of four non-diseased cases and four diseased cases. The first two diseased cases have one lesion each, and the remaining two have two lesions each. Clicking **source** yields Section 14.5.1.

14.5.1 Code output

```
> source(...)
AFROC AUC =  0.7708333
wAFROC AUC =  0.7875
```

Use the commands in Section 14.5.2 to get the number of lesions per case and the ratings in the NL and LL arrays.

14.5.2 Code snippets

```
> Lk2
[1] 1 1 2 2
> frocData$NL[1,1,,]
               [,1]        [,2]
[1,]        -Inf        -Inf
[2,]    0.4874291       -Inf
[3,]    0.7383247  0.5757814
[4,]   -0.3053884       -Inf
[5,]    1.5117812       -Inf
[6,]        -Inf        -Inf
[7,]        -Inf        -Inf
[8,]        -Inf        -Inf
> frocData$LL[1,1,,]
               [,1]        [,2]
[1,]    0.8523430       -Inf
[2,]   -0.2146999       -Inf
[3,]    1.5884892       -Inf
[4,]    2.9438362  1.98381
```

The length of the third dimension of the NL array is eight (4 non-diseased + 4 diseased cases). The fifth sequential case corresponds to NLs on the first diseased case, and so on. The simulated Z-samples displayed in Section 14.5.2 are shown in Table 14.1 for non-diseased and in Table 14.2 for diseased cases. The columns labeled $k_t tl_s s$ list the case-location indexing subscripts, the columns labeled $z_{k_t tl_s s}$ list the corresponding Z-samples, when realized and otherwise NAs are listed. Column 5 in Table 14.1 illustrates the conversion of the NL Z-samples to FP Z-samples according to the highest-rating assumption (the first non-diseased case illustrates the rule that in the absence of any marks the FP rating is $-\infty$). The tables show that the simulator did not realize any Z-sample on the first non-diseased case (alternatively, if it did, the Z-sample(s) fell below ζ_1; one cannot tell the difference), and for the second lesion on the third diseased case. Because non-diseased cases have no lesions, all Z-samples listed in Table 14.1 are NLs. In contrast, in Table 14.2, each case can generate NLs and LLs. The second column of Table 14.2 lists the number of lesions per diseased case L_{k_2}. Columns 3 and 4 illustrate NL indexing and Z-samples and columns 5 and 6 illustrate LL indexing and Z-samples. Column 8 illustrates the conversion of the NL and LL Z-samples to TP Z-samples according to the highest-rating assumption.

Table 14.1 Simulated data used in example plots for non-diseased cases yielding the light FP circles in Figure 14.2 and 14.4

Non-diseased case	NLs		FPs		Operating points in figures 14.2 and 14.3
	$k_t tl_s s$	$z_{k_t tl_s s}$	$k_t t$	$FP_{k_t t}$	
1	1111	NA	11	$-\infty$	*
	1121	NA			
2	2111	0.487	21	0.487	F
	2121	NA			
3	3111	0.738	31	0.738	E
	3121	0.576			
4	4111	−0.305	41	−0.305	H
	4121	NA			

Note: * = unmarked; NA = not applicable.

Table 14.2 Simulated data used in example AFROC and wAFROC plots for diseased cases yielding the dark LL circles in Figures 14.2 and 14.4

Diseased case	L_{k_2}	NLs		LLs		TPs		Lesion weights $W_{k_2 l_2}$	Operating points in figures 14.2 and 14.3
		$k_t tl_s s$	$z_{k_t tl_s s}$	$k_t tl_s s$	$z_{k_t tl_s s}$	$k_t t$	$TP_{k_2 2}$		
1	1	1211	1.512	1212	0.852	12	1.512	1	D
		1221	NA	1222	NA			NA	*
2	1	2211	NA	2212	−0.215	22	−0.215	1	G
		2221	NA	2222	NA			NA	
3	2	3211	NA	3212	1.588	32	1.588	0.6	C
		3221	NA	3222	NA			0.4	
4	2	4211	NA	4212	2.944	42	2.944	0.4	A
		4221	NA	4222	1.984			0.6	B

Note: NA = not applicable; * = unmarked lesion; note that the Z-samples listed under NLs are ignored in AFROC and wAFROC analyses, but are used in (inferred) ROC analysis.

14.5.3 The AFROC plot

In Figure 14.2, FPs and LLs, represented by light and dark circles, respectively, are shown ordered from left to right, with higher Z-samples to the right (this is subsequently referred to as a *one-dimensional* depiction of the data). Each circle is labeled using the $k_t tl_s s$ notation. For example, the rightmost dark circle corresponds to the LL Z-sample originating from the first lesion in the fourth diseased case, that is, z_{4212}. Consistent with Table 14.1, there are three light circles (FPs) not counting FP_{11}, which occurs at $z = -\infty$ representing the first non-diseased case with no marks. Likewise, consistent with Table 14.2, there are five dark circles (LLs) not counting $z_{3222} = -\infty$ representing the unmarked second lesion on the third diseased case.

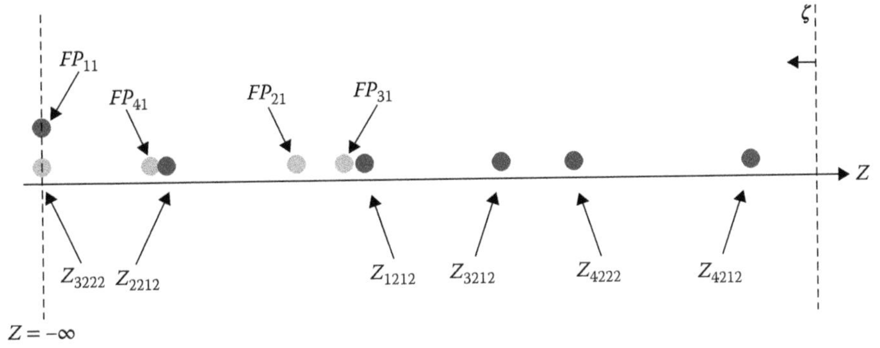

Figure 14.2 A one-dimensional depiction of the data in Tables 14.1 and 14.2, showing Z-samples used for plotting the AFROC shown in Fig. 14.3; the dark circles correspond to latent lesion localizations (LLs) and the light ones to latent false positives (FPs). As the sliding threshold ζ moves to the left, latent events "flip" to real events. This plot was generated by `main1DPlotAfroc.R`.

Starting from $+\infty$, moving a virtual threshold continuously to the left generates the AFROC plot (Figure 14.3). As each FP is crossed, the operating point moves to the right by

$$\frac{1}{K_1} = 0.25 \tag{14.11}$$

As each LL is crossed, the operating point moves up by

$$\frac{1}{\sum\limits_{k_2=1}^{K_2} L_{k_2}} = \frac{1}{6} \tag{14.12}$$

Since it has one lesion, crossing the Z-sample for the first case would result in an upward movement of 1/6, and likewise for the second case. Since the third case contains two lesions, crossing the corresponding Z-samples would result in a net upward movement by 1/3. *This behavior shows explicitly that the non-weighted method gives greater importance to diseased cases with more lesions*, that is, such cases make a greater contribution to AUC. The jumps from lesions in the same case need not be contiguous; they could be distributed with intervening jumps from lesions on other cases, but eventually the jumps will occur and contribute to the net upward movement. As an example, the jumps due to the two lesions on the fourth case are contiguous, see points A and B in Figure 14.3. However, the jumps due to the two lesions on the third case are not contiguous; the first lesion gives the point C, but the unmarked lesion on this case, indicated by the asterisk in Table 14.2, eventually contributes when the operating point moves diagonally from point H to (1,1). The last columns of Tables 14.1 and 14.2 label the correspondences of the z-samples to the operating points shown in Figures 14.3.

14.5.4 The weighted AFROC (wAFROC) plot

Figure 14.4, which is analogous to Figure 14.2, is a one-dimensional depiction of the data in Tables 14.1 and 14.2 but this time the lesion weights, shown in Table 14.2, are incorporated, as indicated by varying the size of each dark circle to indicate its weight (in Figure 14.4 all dark circles were of the same size). In addition, each lesion is labeled with its rating *and* weight. The listed weights in Table 14.2 were assigned arbitrarily, subject to the rule that the weights on a case must sum to unity.

Moving a virtual threshold continuously to the left generates the wAFROC plot, Figure 14.5. The movement of the operating point in response to crossing FPs is the same as before. However, as each LL is crossed the operating point moves up by an amount that *depends on the lesion*.

$$\frac{W_{k_2 l_2}}{K_2} = \frac{W_{k_2 l_2}}{4} \tag{14.13}$$

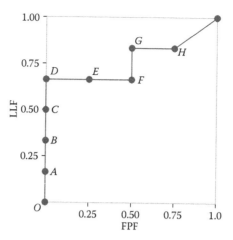

Figure 14.3 The empirical AFROC plot for the data shown in Tables 14.1 and 14.2. The labels correspond to the last columns of the tables. The corresponding one-dimensional depiction is Figure 14.2. This plot was generated by **mainPlotAfroc.R**.

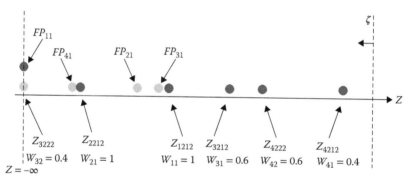

Figure 14.4 A one-dimensional depiction of the data in Tables 14.1 and 14.2, including the weights, showing z-samples used for plotting the weighted AFROC plot shown in Fig. 14.5; the sizes of the dark circles code the lesions weights (the lesion-specific weights are shown below each z-sample). This plot was generated by **main1DPlotwAfroc.R**.

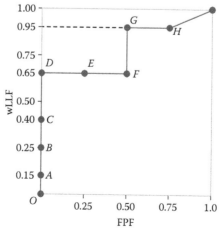

Figure 14.5 The empirical weighted AFROC (wAFROC) plot for the data shown in Tables 14.1 and 14.2. The operating point labels correspond to the last columns of the tables. The corresponding one dimensional plot is Figure 14.4. The area under the wAFROC is 0.7875. This plot was generated by **mainPlotwAfroc.R**.

Since the first two diseased cases have one lesion each (i.e., unit weights), crossing the corresponding Z-samples results in upward jumps of 0.25, Figure 14.5 (C to D and F to G). According to the weights in Figure 14.4, crossing the Z-sample of the first lesion in the third diseased case results in an upward jump of 0.6/4. That from the second lesion in the same case results in an upward jump of 0.4/4 for a net upward jump in the third case of 0.25, the same as for the first two diseased cases. Likewise, crossing the Z-samples of the two lesions in the fourth disease case results in an upward jump of 0.4/4 = 0.1 (O to A) for the first lesion and 0.6/4 = 0.15 (B to C) for the second lesion, for a net upward jump of 1/4, which is the same as for the first three diseased cases. *This shows explicitly that the weighting method gives each diseased case the same importance, regardless of the number of lesions in it; a property not shared by the area under the AFROC.*

14.6 Physical interpretation of AFROC-based FOMs

From the preceding sections, it is seen that the AFROC-based trapezoidal plots consist of upward and rightward jumps, starting from the origin (0,0) and ending at (1,1). This is true regardless of whether the Z-samples are binned or not, that is, at the microscopic level the jumps always exist. Each upward jump is associated with an *LL* rating exceeding a virtual threshold. Each rightward jump is associated with an FP rating exceeding the threshold. *Upward jumps tend to increase the area under the AFROC-based plots and rightward jumps tend to decrease it.* This makes physical sense in terms of correct decisions being rewarded and incorrect ones being penalized, and can be seen from two extreme-case examples. If there are only upward jumps, then the trapezoidal plot rises from the origin to (0,1), where all lesions are correctly localized without any generating FPs and performance is perfect—the straight-line extension to (1,1) ensures that the net area is unity. If there are only horizontal jumps, that takes the operating point from the origin to (1,0), where none of the lesions are localized and every non-diseased image has at least one NL mark, representing worst possible performance. Here, despite the straight-line extension to (1,1), the net area is zero.

14.6.1 Physical interpretation of area under AFROC

The area under the AFROC has the following physical interpretation: it is the fraction of LL versus FP Z-sample comparisons where the LL sample is equal (counting as half a comparison) or greater (counting as a full comparison) than the FP Z-sample. From Tables 14.1 and 14.2, there are four FPs and six LLs for 24 possible comparisons. Inspection of the tables reveals that there are $4 \times 4 = 16$ comparisons contributing 1s, two comparisons (from the second diseased case) contributing 1s, and one comparison (from the second lesion on the third diseased case) contributing 0.5, which sum to 18.5. Dividing by 24 yields 18.5/24 = 0.7708, the empirical AFROC-AUC, Section 14.5.1. In probabilistic terms:

> The area under the AFROC is the probability that a lesion is rated higher than any mark on a non-diseased case.

14.6.2 Physical interpretation of area under wAFROC

The area under the wAFROC has the following physical interpretation: *it is the lesion-weight adjusted fraction of diseased cases versus non-diseased case comparisons where LL Z-samples are equal (counting as half a comparison times the weight of the lesion in question) or greater (counting as a full comparison times the weight of the lesion) than FP Z-samples.* Note that there are still 24 LL versus FP comparisons but the counting proceeds differently. The fourth diseased case contributes

$0.4 \times 4 + 0.6 \times 4$, that is, 4 (compared to 8 in the preceding example). The third diseased case contributes $0.6 \times 4 + 0.4 \times 0.5$, that is, 2.6 (compared to 4.5 in the preceding example). The second diseased case contributes $1 \times 2 = 2$ (compared to 2 in the preceding example), and the first diseased case contributes $1 \times 4 = 4$ (compared to 4 in the preceding example). Summing these values and dividing by 16 (the total number of diseased cases versus non-diseased cases comparisons) one gets $12.6/16 = 0.7875$, which is the area under the wAFROC, Section 14.5.1. In probabilistic terms:

> The area under the weighted AFROC is the lesion-weight adjusted probability that a lesion is rated higher than any mark on a non-diseased case.

14.7 Discussion/Summary

The primary aim of this chapter was to develop expressions for FOMs (i.e., functions of ratings) and show their equivalences to the empirical AUCs under corresponding operating characteristics. Unlike the ROC, the AFROC and wAFROC figures of merit are represented by quasi-Wilcoxon like constructs, not the well-known Wilcoxon statistic.[5]

The author is aware from users of his software that their manuscript submissions have sometimes been held up with the critique that the meaning of the AFROC FOM-statistic is not intuitively clear.[6] This is one reason the author has tried to make the meaning clear, perhaps at the risk of making it *painfully* clear. Clinical interpretations do not always fit into convenient easy-to-analyze paradigms. Not understanding something is not a reason for preferring a simpler method. Use of the simpler ROC paradigm to analyze location specific tasks results in loss of statistical power and sacrifices better understanding of what is limiting performance. It is unethical to analyze a study with a method with lower statistical power when one with greater power is available.[7–9] The title of the paper by Halpern et al. is "The continuing unethical conduct of underpowered clinical trials." The AFROC FOM-statistic has been used, at the time of writing, in over 107 publications.*

The subject material of this chapter is not that difficult. However, it does require the researcher to be receptive an unbiased. Dirac addressed an analogous then-existing concern about quantum mechanics, namely it did not provide a satisfying picture of what is going on, as did classical mechanics.† To paraphrase him, the purpose of science (quantum physics in his case) is not to provide satisfying pictures but to explain data. FROC data is inherently more complex than the ROC paradigm and one should not expect a simple FOM-statistic. The detailed explanations given in this chapter should allow one to understand the wAFROC and AFROC FOMs.

A misconception regarding the wAFROC FOM-statistic is that the weighting may sacrifice statistical power and render the method equivalent to ROC analysis in terms of statistical power. Analysis of clinical datasets and simulation studies suggests that this is not the case; loss of power is minimal. As noted earlier, the highest rating carries more information than a randomly selected rating and likewise, the summed weighted lesions ratings carry more information than arbitrarily selected lesion ratings.

Bamber's equivalence theorem led to much progress in nonparametric analysis of ROC data. The proofs of the equivalences between the areas under the AFROC and wAFROC and the corresponding quasi-Wilcoxon statistics provide a starting point. To realize the full potential of these proofs, similar work, like that conducted by DeLong et al.,[10] is needed for the FROC paradigm.

---

* https://www.researchgate.net/publication/311102039_Publications_that_have_used_JAFROC_software_as_of_112916
† "In answer to the first criticism it may be remarked that the main object of physical science is not the provision of pictures, but is the formulation of laws governing phenomena and the application of these laws to the discovery of new phenomena. If a picture exists, so much the better; but whether a picture exists or not is a matter of only secondary importance." P.A.M. Dirac, Principles of Quantum Mechanics, 4th edition, Chapter 1. http://www.informationphilosopher. com/solutions/scientists/dirac/chapter_1.html

This work is not going to be easy, one reason being the relative dearth of researchers working in this area, but it is possible. Indeed work has been published by Popescu[11] on nonparametric analysis of the exponentially transformed FROC (EFROC) plot which, like the AFROC and wAFROC, is completely contained within the unit square. This work should be extended to the wAFROC. For reasons stated in **Chapter 13**, nonparametric analysis of FROC curves[12-14] is not expected to be useful—the only exception being if one is interested in comparing two observers with similar performances, whose FROC curves share a common range, and even here the neglect of unmarked cases is a serious concern.

Current terminology prefixes each of the AFROC-based FOMs with the letter J for jackknife. The author recommends dropping this prefix, which has to do with significance testing procedure rather than the actual definition of the FOM-statistic. For example, the correct way is to refer to the AFROC figure of merit, not the JAFROC figure of merit. For continuity, the software packages implementing the methods are still referred to as **JAFROC** (Windows) or **RJAfroc** (cross-platform, open-source).

To gain deeper insight into the FROC paradigm, it is necessary to look at methods used to measure visual search, the subject of the next chapter.

References

1. Chakraborty DP, Zhai X. On the meaning of the weighted alternative free-response operating characteristic figure of merit. *Med Phys*. 2016;43(5):2548–2557.
2. Chakraborty DP, Yoon HJ. JAFROC analysis revisited: Figure-of-merit considerations for human observer studies. Proceedings of the SPIE Medical Imaging: Image Perception, Observer Performance, and Technology Assessment; 2009;7263:72630T.
3. Chakraborty DP, Berbaum KS. Observer studies involving detection and localization: Modeling, analysis and validation. *Med Phys*. 2004;31(8):2313–2330.
4. Bamber D. The area above the ordinal dominance graph and the area below the receiver operating characteristic graph. *J Math Psychol*. 1975;12(4):387–415.
5. Wilcoxon F. Individual comparison by ranking methods. *Biometrics*. 1945;1:80–83.
6. Hillis SL, Chakraborty DP, Orton CG. ROC or FROC? It depends on the research question. *Med Phys*. 2017;44:1603–1606.
7. Lerman J. Study design in clinical research: Sample size estimation and power analysis. *Can J Anaesth*. 1996;43(2):184–191.
8. Halpern SD, Karlawish JH, Berlin JA. The continuing unethical conduct of underpowered clinical trials. *JAMA*. 2002;288(3):358–362.
9. Breau RH, Carnat TA, Gaboury I. Inadequate statistical power of negative clinical trials in urological literature. *J Urol*. 2006;176(1):263–266.
10. DeLong ER, DeLong DM, Clarke-Pearson DL. Comparing the areas under two or more correlated receiver operating characteristic curves: A nonparametric approach. *Biometrics*. 1988;44:837–845.
11. Popescu LM. Nonparametric signal detectability evaluation using an exponential transformation of the FROC curve. *Med Phys*. 2011;38(10):5690–5702.
12. Bandos AI, Rockette HE, Song T, Gur D. Area under the free-response ROC curve (FROC) and a related summary index. *Biometrics*. 2008;65(1):247–256.
13. Samuelson FW, Petrick N, Paquerault S. Advantages and examples of resampling for CAD evaluation. Paper presented at: 4th IEEE International Symposium on Biomedical Imaging 2007: Biomedical Imaging: From Nano to Macro, 2007. ISBI 2007; Arlington, VA: IEEE.
14. Samuelson FW, Petrick N. Comparing image detection algorithms using resampling. IEEE International Symposium on Biomedical Imaging: From Nano to Micro; 2006:1312–1315. Arlington, VA: IEEE.

15

Visual search paradigms

15.1 Introduction

To understand free-response data, specifically how radiologists interpret images, one must come to grips with visual search. Casual usage of everyday terms such as *search, recognition,* and *detection* in specific scientific contexts can lead to confusion, so in this chapter the author will attempt to carefully define them. Visual search, including the medical imaging search task, is defined in a broad sense as grouping and labeling parts of an image. Two experimental methods for studying search are described. The more common method, widely used in the non-medical imaging context, consists of showing observers known targets (i.e., known shapes and sizes but unknown locations) and known distractors (again, known shapes and sizes but unknown locations). One measures how rapidly the observer can perceive the presence of the target (reaction time) and their accuracy (fraction of correct decisions on target present versus target absent presentations). In the medical imaging paradigm, one does not know the sizes, shapes, and locations of the targets and distractors. Instead, one relies on eye-tracking measurements to determine where the observer is looking and for how long. A clustering algorithm is applied to determine regions that received prolonged examination (dwell time) representing sites where decisions were made. The focus in this chapter is on the second paradigm, which closely parallels the free-response ROC (FROC) paradigm. A schema of how radiologists find lesions, termed the *Kundel–Nodine search model*, is described. *The importance of this major conceptual model is not widely appreciated by researchers.* It is a two-stage model, where the first stage identifies suspicious regions. The second stage analyzes each suspicious region for level of suspicion, extracting a level of suspicion index, and if the level of suspicion is high enough, the region is marked. The Kundel–Nodine search model is the basis of the radiological search model (RSM) described in the next chapter. A section is devoted to a recently developed method for analyzing simultaneously acquired eye-tracking and FROC data.

The starting point is the definition of recognition/finding. The following sections draw heavily on work by Nodine and Kundel.[1-5] The author also wishes to acknowledge critical insights gained through conversations with Dr. Claudia Mello-Thoms.

15.2 Grouping and labeling ROIs in an image

Looking at and understanding an image involves *grouping and assigning labels* to different regions of interest (ROIs) in the image, where the labels correspond to entities that exist (or have existed in the examples to follow) in the real world. As an example, if one looks at Figure 15.1, one would label them (from left to right and top to bottom, in raster fashion): Franklin Roosevelt, Harry Truman, Lyndon Johnson, Richard Nixon, Jimmy Carter, Ronald Reagan, George H. W. Bush, and the presidential seal. The accuracy of the labeling depends on prior-knowledge, that is, expertise, of the observer. If one were ignorant about U.S. presidents, one would be unable to correctly label them.

Image interpretation in radiology is not fundamentally different. It involves assigning labels to an image by grouping and recognizing areas of the image that have correspondences to the radiologist's knowledge of the underlying anatomy, and, most importantly, deviations from the underlying anatomy. Most physicians, who need not be radiologists, can look at a chest x-ray and say *this is a clavicle, this is a rib, these are the lungs, this is the mediastinum, this is the diaphragm,* and so on (Figure 15.2). This is because they know (from their training) the underlying anatomy and have a basic understanding of the x-ray image formation physics that relates anatomy to the image.

Figure 15.1 This image consists of eight sub-images or ROIs. Understanding an image involves *grouping and assigning labels* to different ROIs, where the labels correspond to entities that exist in the real world. One familiar with U.S. history would label them, from left to right and top to bottom, in raster fashion: Franklin Roosevelt, Harry Truman, Lyndon Johnson, Richard Nixon, Jimmy Carter, Ronald Reagan, George H. W. Bush, and the presidential seal. Labeling accuracy depends on expertise of the observer. The row and column index of each ROI identifies its location.

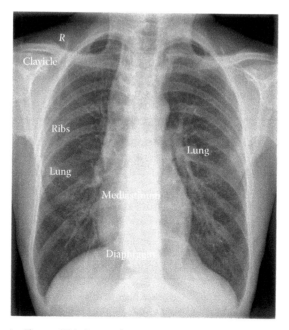

Figure 15.2 Analogous to Figure 15.1, image interpretation in radiology also involves assigning labels to an image by grouping and recognizing areas of the image that have correspondences to the radiologist's knowledge of the underlying anatomy. Most physicians, who do not have radiologists can look at a chest x-ray and say *this is the clavicle, this is a rib, these are the lungs, this is the mediastinum, this is the diaphragm,* and so on. This is because they know the underlying anatomy and have a basic understanding of x-ray image formation physics that relates anatomy to the image.

15.3 Recognition versus detection

The process of grouping and labeling parts of an image is termed *recognition*. This was illustrated with the pictures of the U.S. presidents in Figure 15.1. *Recognition* is distinct from *detection*, which is deciding about the presence of something that is unexpected or the absence of something that is expected, in other words, a *deviation*, in either direction, from what is expected. An example of detecting the *presence* of something that is *unexpected* would be a lung nodule and an example of detecting the *absence* of something that is *expected* would be an image of a patient with a missing rib (yes, it does occur, even excluding the biblical Adam).

The terms *expected* and *unexpected* are important. They imply expertise-dependent expectations regarding the true structure of the non-diseased image, which the author terms a *non-diseased template*, and therefore an ability to *recognize*** clinically relevant deviations or perturbations, in either direction, from this template, for example, a lung nodule that could be cancer. By *clinically relevant* the author means perturbations related to the patient's health outcome—recognizing scratches, dead pixels, artifacts of know origin, and lead patient ID markers do not count. There is a *location* associated with recognition[†], but not with detection. Detection is the presence or absence of something; the perturbation could be anywhere in the image. For example, in Figure 15.1, recognizing a face is equivalent to assigning a row and column index in the image. Specifically, recognizing George H.W. Bush implies pointing to row 2 and column 3. Detecting George H.W. Bush implies stating that George H.W. Bush is present in the image, but the location could be in any of the eight locations. Recognition is a FROC paradigm task, while detection is a receiver operating characteristic (ROC) paradigm task.

15.3.1 Search versus classification expertise

Since template perturbations can occur at different locations in the images, the ability to *selectively* recognize them is related to *search expertise*. The term *selectively* is important. A non-expert can trivially recognize all perturbations by claiming all regions in the image are perturbed. *Search expertise is the ability to find clinically relevant perturbations that are actually present while minimizing finding what appear to be clinically relevant perturbations that are actually not present.* In FROC terminology, search expertise is the ability to find latent LLs while minimizing the numbers of found latent NLs. Lesion-classification expertise is the ability to correctly classify a found suspicious region as malignant or benign.

The skills required to recognize a nodule in a chest x-ray are different from that required to recognize a low-contrast circular or Gaussian-shaped artificial nodule against a background of random noise. In the former instance, the skills of the radiologist are relevant, for example, the skilled radiologist knows not to confuse a blood vessel viewed end-on for a nodule, especially since the radiologist knows where to expect these vessels, for example, the aorta. In the latter instance, (i.e., viewing artificial nodules superposed on random noise) there are no expected anatomic structures, *so the skills possessed by the radiologist are rendered irrelevant.* This is the reason why having radiologists interpret random noise images and pretending that this makes it clinically relevant is a waste of reader resources and bad science. One might as well use undergraduates with good eyesight, motivation, and training. To quote Nodine and Kundel,[1] "Detecting an object that is hidden in a natural scene is not the same as detecting an object displayed against a background of random noise." This paragraph also argues against usage of phantoms as stand-ins for clinical images for clinical performance assessment. Phantoms are fine in the QC context, as in **Chapter 1**, but they do not allow radiologists the opportunity to exercise their professional skills.

---

* The author originally wrote, out of habit, *detect* instead of *recognize*, undermining his own message. Since commonly used terms are used in a scientific sense, in this field proper terminology is important.

† There could be additional descriptors associated with recognition, for example, *I see a lesion here and it is a microcalcification*, and/or *I see a lesion here and it is a fibroadenoma*.

15.4 Two visual search paradigms

There are two visual search paradigms: which the author terms the *conventional paradigm* and the *medical imaging paradigm*.

15.4.1 The conventional paradigm

In the conventional paradigm, one measures *reaction time* and *percent correct* in the following task. Images are shown briefly and followed, after an interstimulus interval, by a mask image (e.g., random noise, to wipe out memory of the preceding image). Each image may contain a *defined target* in a set of *defined distractors*. *Defined targets and defined distractors* mean that their presence and numbers are under the control of the experimenter and the observer, via training images, knows their characteristics (e.g., shapes, sizes, etc.). For example,[6] a target could be the letter T and a distractor could be the letter L. The observer's task is to discriminate between two conditions, (1) target and distractors present and (2) only distractors present, by pressing a *yes* (target is present) or *no* key. Also measured is the time it takes, from image onset, to make the target-present target-absent decision, termed *reaction time*. This is by far the most widely used paradigm[7,8] (see, for example, Ref. 7 and the literature cited therein; the paper, cited 2908 times as of 8/18/2016, is an excellent review of the conventional paradigm). Typically measured is the dependence of *percent correct* and *reaction time* on set size (defined as the number of distractors). The following example, adapted from Ref. 8, describes an actual study using this paradigm:

> *Stimuli (stimuli = items = distractors plus a possible target) were red and green Xs and Os on a black background. Individual items could be placed at any of 36 locations within a pre-defined square field. On each trial, items were presented at 8, 16, or 32 randomly chosen positions within the square field (thereby varying the set size, i.e., the number of distractors). The target was a red O (in target present images) and distractors were green Os and red Xs. On target present trials, one of the locations contained a target item. Targets were present on 50% of trials. Set size, positions of target and distractors, and presence or absence of a target was random across trials. Subjects responded by pressing one of two keys: a yes key if a target was detected and a no key if it was not. Reaction times were measured from stimulus onset. The stimuli remained visible until the subject responded and feedback was given on each trial.*

The results are used to test different models of visual search. In particular, there has been interest in determining if the items are processed in parallel or sequentially. There is, as stated above, a very large selection of literature on this paradigm, or variants of it, and this brief account is given simply to distinguish it from the medical imaging paradigm that follows.

15.4.2 The medical imaging visual search paradigm

The key difference is the use of eye-tracker technology.[1,9,10] This is not to imply that users of the conventional paradigm have not used eye-tracker technology. They have, but the medical imaging paradigm is crucially dependent on this technology whereas the conventional paradigm is not. Eye-tracker technology determines the location and duration of the axis of gaze (i.e., where and for how long the radiologist looks at different locations in the image). The difference between the two paradigms is necessitated by several factors.

Unlike the conventional paradigm, one does not know the numbers and precise shapes, sizes, contrasts, and so on, of the true lesions, the targets in the conventional paradigm. These are camouflaged in anatomic noise and are more complex than the Ts and Ls or Xs and Os of the conventional method.

One does not know the numbers and precise shapes, sizes, contrasts, and so on, of the distractors. In fact, the radiologist *perceives* these and what constitutes a latent NL to one radiologist may not be a latent NL to another. Two radiologists may not even agree on the number of latent NLs on a specific image. Unlike the conventional paradigm, the number of NLs in the medical imaging paradigm must be treated as a radiologist and case dependent random non-negative integer.

The medical image paradigm allows for zero latent marks (i.e., no distractors), which has no counterpart in the conventional method. These images are the unambiguous or easy non-diseased cases that do not generate any marks.

In medical imaging, one is more interested in objective performance measurements (does a radiologist find the lesion at high confidence?) than in reaction time.

In addition to eye-tracking data, one may acquire ratings data as in the ROC paradigm,[11] or more recently, mark-rating data, as in the FROC paradigm.[12,13] If using the ROC method the performance measure (e.g., AUC under ROC curve) is comparable to the percent correct obtained in the conventional paradigm, except that the ROC-ratings method is more efficient.[14] However, since location of the perceived target or lesion is ignored, the scoring is ambiguous in the sense originally noted by Bunch et al.,[15] that is, the observer may have missed the target and mistaken a distractor for the target on a target present image, and that event combination would be scored as a correct decision. The FROC paradigm requires correct localization, thereby ruling out this ambiguity.

Compared to the many papers using the conventional visual search paradigm, research in medical imaging visual search is relatively limited. Prof. Kundel, Prof. Nodine, Prof. Krupinski, and Dr. Claudia Mello-Thoms have made major contributions to this field. The following is an example of how data is collected in the medical imaging visual search paradigm.[16]

Eye-position data were recorded using a limbus reflection technique. (Limbus, Figure 15.3, is defined as the border between the cornea [the transparent layer making up the outermost front part of the eye, covering the iris and pupil] and sclera [opaque white of the eye].) Eye movements are measured by having the observer wear a specially designed spectacle frame containing infrared emitters and sensors that measure changes in light reflected from the border between the iris and sclera, Figure 15.3. The viewers were told they had 15 seconds to search the lung fields of each image for the presence of a nodule and additionally to remember the locations of regions suspected of containing a nodule but considered negative. Following the 15-second presentation, the viewers rated each image for confidence in the presence of disease.

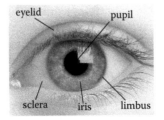

Figure 15.3 Meaning of terms relating to the anatomy of the eye.

Eye-position data is collected only during the initial 15 seconds while the radiologist searches the image. One issue with this way of collecting data is that during the reporting phase, a radiologist may discover something new and proceed to investigate this finding, but because eye-position recording has been terminated, that information is not captured. In the data collection methodology used in a recent study[13] searching and reporting occur simultaneously with eye-position collection.* The newer paradigm more closely resembles clinical practice, and potentially allows one to follow the perceptual and interpretative process entailed in case reading from beginning to end, without researcher-initiated interruptions.

15.5 Determining where the radiologist is looking

The eye-tracking (ET) device the author is familiar with[13] (ASL Model H6, Applied Sciences Laboratory, Bedford, MA) uses a magnetic head tracker to monitor head position, and this allows the radiologists to freely move their head. The ET system integrates eye position calculated from limbus-reflection, and head position, to calculate the intersection of the line of gaze and the display plane. The data stream (raw data) provided by the eye-tracker consists of several bytes of data at 60 Hz, containing the (x,y) coordinates of where the observer is looking plus various flag bits (e.g., indicating blinks). The eye moves in rapid jumps (*saccades*) with intervening longer pauses (*fixations*). The eye-movement induced reflectance changes are converted to display coordinates, which indicate the locations and durations of fixations. Fixations occurring in clusters indicate where attention is being directed and decisions are made. The raw data needs to be processed, Appendix 15.A, to determine regions where decisions were made; the processing, which is guided by models of human perception, depends on the researcher.

15.6 The Kundel–Nodine search model

The Kundel–Nodine model[1-5] is a schema of events that occur from the radiologist's first glance to the decision about the image. The model is similar to the guided search model[7,8,17] proposed by Prof. Jeremy Wolfe in the non-medical imaging context.

Assuming the task has been defined prior to viewing, based on eye-tracking recordings obtained on radiologists while they interpreted images, Kundel and Nodine proposed the following schema for the diagnostic interpretation process, consisting of two major† components: (1) *glancing or global impression* and (2) *scanning or feature analysis* (Figure 15.4).

15.6.1 Glancing/global impression

The colloquial term *glancing* is meant literally.‡ The glance is brief, typically lasting about 100–300 ms, too short for detailed foveal examination and cognitive interpretation. Instead, during this brief interval peripheral vision and reader expertise are the primary mechanisms responsible for the identification of latent marks. The glance results in a global impression, or *gestalt*, that identifies perturbations from the template defined earlier. Object recognition occurs at a holistic level, that is,

---

* The software manual for this study is available as additional online material in file `Data acquisition program. pdf`.
† The author's organization differs from the way Nodine and Kundel state it, but the essential concepts are the same.
‡ Glance: to take a brief or hurried look.

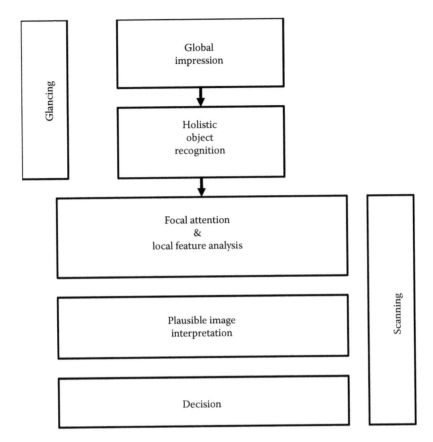

Figure 15.4 The Kundel–Nodine two-stage model of radiological search. The glancing/global stage identifies perturbations from the template of a generic non-diseased case. The scanning stage performs detailed analysis of the identified perturbations and calculates the probability that the perturbation is a true lesion. Only perturbations with sufficiently high probability are marked/reported.

in the context of the whole image, as there is insufficient time for detailed viewing and all of this is going on using peripheral vision. It is remarkable that radiologists can make reasonably accurate interpretations from information obtained in a brief glance (Figure 6 in Ref. 1). Suspicious regions, which are perturbations from the template, are flagged for subsequent detailed viewing, that is, *the initial glance tells the visual system where to look more closely*. See **Chapter 12**, the section on solar analogy, for further background on this important aspect of vision. Since eye-tracking technology does not measure peripheral vision, the locations of the perturbations need to be inferred from the scanning stage described next.

15.6.2 Scanning/local feature analysis

The global impression identifies suspicious regions that are tagged for detailed foveal viewing by the central vision.[18] During this process—termed *scanning or feature analysis*—the observer scrutinizes* and analyzes the suspicious regions for evidence of disease. In principle, for each suspicious region they calculate the probability of malignancy. For those readers more familiar

---

* Scrutinize: to examine or inspect closely and thoroughly.

with how CAD works, this corresponds to the feature analysis stage of CAD where regions found by the global search, termed *initial detections* in CAD, are analyzed for probability of malignancy. The scrutiny is conducted via clusters of closely spaced fixations. In the absence of closely spaced fixations, or *retinal jitter,* which can be induced in a laboratory condition known as *retinal stabilization,* perceptions tend to rapidly fade away.[19] Perception is sensitive to temporal changes. It is as if there is a high-pass temporal filter that suppresses stationary features, so that changes from it are quickly perceived, no doubt a result of biologic evolution. The probability of malignancy is used to decide whether to report the region. The corresponding locations are the big-clusters in Figure 15.A.1 (C) in the online appendix. After places identified during the global impression have been scrutinized, the viewer may follow the same scanning pattern aimed at discovering something that was missed, or, may simply scan at random while thinking about the image.

The fixations that cluster at perturbations are collecting data necessary to test for the presence of a lesion. If testing yields a sufficiently high confidence of a lesion, a decision is made to report the lesion. If testing is negative or inconclusive, the search continues. Thus, the report "normal chest" is an overall impression based on a series of local decisions that were needed because the relevant anatomic perturbations can only be resolved by foveal vision. The viewer is not aware of all of the decisions, positive and negative, made during scanning.* The eye-tracking record however, reveals where the eye lingered, providing indirect evidence about where conscious decisions were made. However, as noted earlier, the eye-tracking record does not include perturbations perceived by peripheral vision. It is believed[23–24] that prolonged or multiple fixations that cluster on image detail signal the testing and decision-making activity associated with the interpretation of anatomical perturbations that are potential lesions. This is the reason for the use, in Appendix 15.A, of a total dwell time of 800 ms to determine where decisions occurred. The value is somewhat arbitrary and investigator dependent.

The essential point that emerges is that *decisions are made at a finite, relatively small, number of regions.* Attention units are not uniformly distributed through the image, in raster-scan fashion, a common misconception; rather the global impression identifies a smaller set of regions that require detailed scanning.

Eye-tracker recordings for a two-view digital mammogram for two observers are shown in Figure 15.5, for an inexperienced observer (upper two panels) and an expert mammographer (lower two panels). The small circles indicate individual fixations (dwell time ~ 100 ms). The larger high-contrast circles indicate clustered fixations (cumulative dwell time ~ 1 s). The larger low-contrast circles indicate a mass visible on both views (this is clearer in the lower panels). The inexperienced observer finds many more suspicious regions than does the expert mammographer but misses the lesion in the MLO view (top-left image). In other words, the inexperienced observer generates many latent NLs but only one latent LL. The mammographer finds the lesion in the MLO view, which qualifies as a latent LL, without finding suspicious regions in the non-diseased parenchyma, that is, the expert generated zero latent NLs on this case and one latent LL. It is possible the observer was so confident in the malignancy found in the MLO view that there was no need to fixate the visible lesion in the other view-the decision had already been made to recall the patient for further imaging.

15.6.3 Resemblance of Kundel–Nodine model to CAD algorithms

It turns out that the designers of CAD algorithms independently arrived at a two-stage process remarkably similar to that described by Kundel–Nodine for radiologist observers. CAD algorithms are designed to emulate expert radiologists, and while this goal is not yet met, these algorithms are reasonable approximations to radiologists, and include the critical elements of search and lesion

---

* This is another reason why the term *location-level true negative* is so misleading, **Chapter 13**; it is an unmeasurable event. Trying to enumerate unmeasurable events is futile.

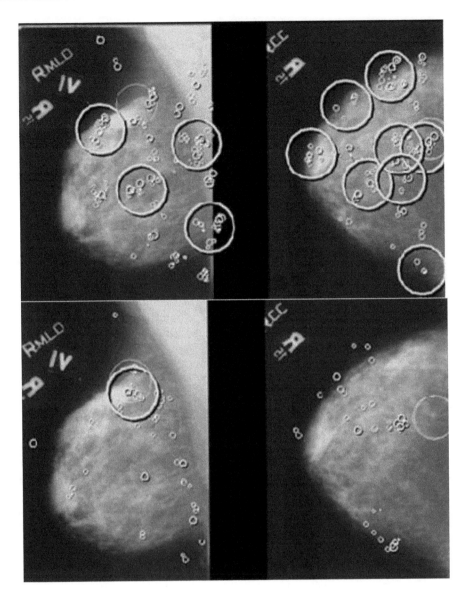

Figure 15.5 Eye-tracking recordings for a two-view digital mammogram display for two observers, an inexperienced observer (upper two panels) and an expert mammographer (lower two panels). The small circles indicate individual fixations (dwell time ~ 100 ms). The larger high-contrast circles indicate clustered fixations (cumulative dwell time ~ 1 sec). The latter correspond to the latent marks in the search-model. The larger low-contrast circles indicate a mass visible on both views. The inexperienced observer finds many more suspicious regions than does the expert mammographer but misses the lesion in the MLO view. In other words, the inexperienced observer generates many latent NLs but only one latent LL. The mammographer finds the lesion in the MLO view, which qualifies as a latent LL, without finding suspicious regions in the non-diseased parenchyma, that is, the expert generated zero latent NLs on this case and one latent LL. It is possible the observer was so confident in the malignancy found in the MLO view that there was no need to fixate the visible lesion in the other view-the decision had already been made to recall the patient for further imaging, which confirmed the finding.

classification that characterize clinical tasks. CAD algorithms involve two steps analogous to the holistic and cognitive stages of the Kundel–Nodine visual search model. In other words, CAD has a *perceptual* correspondence to human observers that, to the author's knowledge, is not shared by other methods of predicting how radiologists perform on clinical images.

In the first stage of CAD, termed *initial detections*,[20] the algorithm finds all reasonable regions that could possibly be a malignancy. The term *all reasonable* is used because an unreasonable CAD could trivially find every malignancy by marking all regions of the image. A reasonable CAD preferably marks lesions while minimizing marking other regions. Therefore, the idea of CAD's initial detection stage is to find many of the malignancies as possible while not finding too many non-diseased regions. This corresponds to the search stage of the Kundel–Nodine model. Unfortunately, CAD is rather poor at this task compared to expert radiologists. Progress in this area has been stymied by lack of understanding of search and how to measure performance in the FROC task. Indeed, a widely held misconception is that CAD is *perfect* (!) at search, because it looks at everything (Dr. Ron Summers, NIH, private communication, Dublin, ca. 2010). In giving equal attention units to all parts of the image, CAD will trivially find all cancers, but it will also find a large number of NLs. Expert radiologists are particularly good at giving more attention units to cancers than the surround, especially for the mass detection task (Figure 15.5) lower panels. Measuring search performance is addressed in **Chapter 17**.

CAD researchers are, in the author's opinion, at the forefront of those presuming to understand how radiologists interpret cases. They work with real images and real lesions and the CAD manufacturer's reputation is on the line, just like a radiologist's, and Medicare even reimburses CAD interpretations. While their current track record is not that good for breast masses compared to expert radiologists, with proper understanding of what is limiting CAD, namely the search process, there is no doubt that future generations CAD algorithms will approach and even surpass expert radiologists.

15.7 Analyzing simultaneously acquired eye-tracking and FROC data

Studies of medical image interpretation have focused on either assessing radiologists' performance using, for example, the receiver operating characteristic (ROC) paradigm, or assessing the interpretive process by analyzing eye-tracking (ET) data. Analysis of ET data has not benefited from threshold-bias independent figures of merit (FOMs) analogous to the area under the ROC curve. In essence, research in this field is restricted to sensitivity/specificity analysis, and ignoring the benefits of accounting for their anti-correlation (recall the study that showed large decrease in inter-reader variability when AUC was used as a figure of merit instead of sensitivity or specificity, Figure 3.6 and Table 3.3). A recent study[13] demonstrated the feasibility of such FOMs and measured agreement between figures of merit derived from free-response ROC (FROC) and ET data. A pre-publication copy, **Analysis of simultaneously acquired ET-FROC data.pdf**, is included in the online supplemental material. This section summarizes the salient points.

15.7.1 FROC and eye-tracking data collection

The data collection is shown schematically in Figure 15.6. A head-mounted eye-position tracking system was worn that used an infrared beam to calculate line of gaze by monitoring the pupil and the corneal reflection. A magnetic head tracker was used to monitor head position which allows the radiologists to freely move their head. The eye-tracker integrates eye position and head position to calculate the intersection of the line of gaze and the display plane.

The computer automatically captured the following information:

1. The (x,y) location of marks made by the radiologists. Each mark was compared to the locations of the actual lesion and classified as FROC lesion localization (FROC-LL) if it fell within $2.5°$ (the proximity criterion) of visual angle (roughly 200 pixels). Otherwise, it was classified as FROC non-lesion localization (FROC-NL).
2. The confidence level (rating) for each mark.

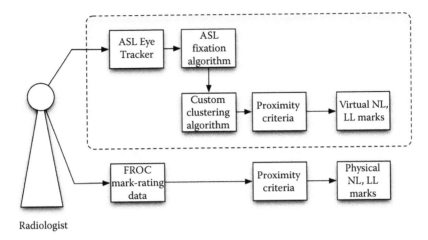

Figure 15.6 Schematic of the data collection and processing to obtain real and eye-tracking marks: the radiologists interpreted the images using a two-monitor workstation. Concurrently, and for the duration of the interpretation, an ASL eye-position tracking system determined the line of gaze. The ASL fixation and clustering algorithms are described in the text. The proximity criterion, defined as 2.5° of visual angle, is the maximum distance between a lesion center and a mark for the mark to be considered a LL (correct localization). Non-lesion localizations are all other marks. ASL = Applied Sciences Laboratory; NL = non-lesion localization; LL = lesion localization.

3. Time-stamped, raw eye-position data was collected during the entire time that the radiologist examined the case. This data, acquired at 60 frames per second, included flags to indicate when image manipulation activities (such as marking, rating or window/level adjustments) and blinks occurred. The flagged data frames were excluded from analysis.

Eight expert breast radiologists interpreted a case set of 120 two-view mammograms while eye-tracking (ET) data and FROC data were continuously collected during the interpretation interval. Regions that attract prolonged (>800 ms) visual attention, using the algorithm in Appendix 15.A, were considered to be *eye-tracking marks*. Based on the dwell and approach-rate (inverse of time-to-hit), *eye-tracking ratings* were assigned to each ET-mark. The ET-ratings were used to define threshold-bias independent FOMs in a manner analogous to the area under the trapezoidal alternative FROC (AFROC) curve (0 = worst, 1 = best). Agreement between ET FOM and FROC FOM was measured (0.5 = chance, 1 = perfect) using the jackknife, 95% confidence intervals (CI) for the FOMs, and agreement was estimated using the bootstrap.

15.7.2 Measures of visual attention

Online Appendix 15.A describes a fixation clustering algorithm that reduces the large number of fixations to big-clusters, where each big-cluster represents a region where a decision was made, i.e., a latent mark. At each big-cluster location, the following eye-position quantities were calculated:

1. *Dwell time (D):* this was defined as the cumulative gaze in seconds of all fixations that comprised the big-cluster with total dwell exceeding 800 ms.
2. *Approach-rate (A):* this was defined as the reciprocal (s^{-1}) of shortest time-to-hit a big-cluster with total dwell exceeding 800 ms, that is, approach the center of the big-cluster to within 2.5°. The reciprocal is taken to maintain a common directionality. In most cases, greater perceptual attention is expected to be accompanied by greater approach-rate and larger values of dwell (the exception to this occurs for very large lesions, which pop out from the surrounding background but do not need much cognitive processing to be resolved—in this case, dwell is not expected to be long).

Dwell time has been linked to the amount of cognitive processing at a given location, and a dwell threshold has been proposed to separate the different types of errors.[5] Approach-rate can be thought of as a perceptual measure of how much a perceived area pops out from the background, and it has been shown to be significantly related to the likelihood that a given breast cancer will be reported by radiologists,[21] with greater approach-rates being related to correct decisions.[22]

15.7.3 Generalized ratings, figures of merit, agreement, and confidence intervals

The eye-tracking paradigm is conceptually similar to the FROC paradigm in the sense that both yield decisions at locations found by the observer. In effect, the big-clusters can be regarded as eye-tracking marks. In the FROC paradigm the observer marks regions that are considered sufficiently suspicious for presence of a lesion, and the degree of suspicion is recorded as a conscious rating. Analogously, eye-tracking yields the locations of regions that attracted visual attention long enough to allow a decision to be made at the location (the big-clusters), and for each region, there is a dwell time and an approach-rate. Dwell and approach-rate can be regarded as generalized (unconscious) ratings. Just as a figure of merit can be defined from FROC mark-rating data, likewise, figures of merit can be defined from the eye-tracking marks and generalized ratings. Details are in Appendix 15.B, where three figures of merit are defined, $\theta_{R,j}$ $\theta_{D,j}$ and $\theta_{A,j}$, where R stands for ratings, D for dwell, and A for approach-rate, and j is the reader index. These are analogous to the AFROC-AUC (since the dataset contained only one lesion per diseased case, these are the same as the wAFROC-AUC). The range of each figure of merit is from zero to unity.

A jackknife-based method[25] for measuring individual case-level agreement between any pair of figures of merit is described in Appendix 15.C. Defined are Γ_{RD}, Γ_{DA}, and Γ_{RA}, which measure agreement between ratings and dwell, dwell and approach-rate, and ratings and approach-rate, respectively. Each agreement measure ranges from 0.5 (chance level agreement) to 1 (perfect agreement). A bootstrap-based method for obtaining confidence intervals for figures of merit and agreements is described in Appendix 15.D. The two-sided Wilcoxon signed rank test was used to measure the significance of differences between matched pairs of variables, one pair per reader, such as numbers of marks, ratings, figures of merit, and agreements.

The AFROC mark-ratings FOM was largest 0.734, CI = (0.65, 0.81) followed by the dwell 0.460 (0.34, 0.59) and then by the approach-rate FOM 0.336 (0.25, 0.46). The differences between the FROC mark-ratings FOM and the perceptual FOMs were significant ($p < 0.05$). All pairwise agreements were significantly better than chance: ratings versus dwell 0.707 (0.63, 0.88), dwell versus approach-rate 0.703 (0.60, 0.79), and rating versus approach-rate 0.606 (0.53, 0.68). The agreement between ratings versus approach-rate was significantly smaller than that between dwell versus. approach-rate (p = 0.008).

This brief description shows how methods developed for analyzing observer performance data can be leveraged to complement current ways of analyzing ET data and lead to new insights.

15.8 Discussion/Summary

This chapter has introduced the terminology associated with a search task: *recognition/finding, classification,* and *detection.* Search involves finding lesions and correctly classifying them, so two types of expertise are relevant. Search expertise is the ability to find (true) lesions without finding non-lesions, while classification accuracy is concerned with correct classification (benign versus malignant) of a suspicious region that has already been found. Quantification of these abilities is described in the next chapters. Two paradigms are used to measure search, one in the non-medical context and the other, the focus of this book, in the medical context. The second method is based on the eye-tracking measurements performed while radiologists perform quasi-clinical tasks (performing eye-tracking measurements in a true clinical setting is more difficult). A method for analyzing eye-tracking data using methods developed for FROC analysis has been described. It has the advantage of taking into account information present in eye-tracking data, such as dwell time and approach rate, in a quantitative manner, essentially by treating them as eye-tracking ratings to which modern FROC methods can be applied. The Kundel–Nodine model of visual search in diagnostic imaging was described. The next chapter describes a statistical parameterization of this model, termed the *radiological search model* (RSM).

References

1. Nodine CF, Kundel HL. Using eye movements to study visual search and to improve tumor detection. *RadioGraphics*. 1987;7(2):1241–1250.
2. Kundel HL, Nodine CF, Conant EF, Weinstein SP. Holistic component of image perception in mammogram interpretation: Gaze-tracking Study. *Radiology*. 2007;242(2):396–402.
3. Kundel HL, Nodine CF. Modeling visual search during mammogram viewing. *Proc SPIE*. 2004;5372:110–115.
4. Kundel HL, Nodine CF. A visual concept shapes image perception. *Radiology*. 1983;146:363–368.
5. Kundel HL, Nodine CF, Carmody D. Visual scanning, pattern recognition and decision-making in pulmonary nodule detection. *Invest Radiol*. 1978;13:175–181.
6. Horowitz TS, Wolfe JM. Visual search has no memory. *Nature*. 1998;394(6693):575–577.
7. Wolfe JM. Guided Search 2.0: A revised model of visual search. *Psychon Bull Rev*. 1994;1(2):202–238.
8. Wolfe JM, Cave KR, Franzel SL. Guided Search: An alternative to the feature integration model for visual search. *J Exp Psychol: Hum Percept Perform*. 1989;15(3):419.
9. Carmody DP, Kundel HL, Nodine CF. Performance of a computer system for recording eye fixations using limbus reflection. *Behav Res Meth Instrum*. 1980;12(1):63–66.
10. Duchowski AT. *Eye Tracking Methodology: Theory and Practice*. Clemson, SC: Clemson University; 2002.
11. Nodine C, Mello-Thoms C, Kundel H, Weinstein S. Time course of perception and decision making during mammographic interpretation. *AJR*. 2002;179:917–923.
12. Nodine CF, Kundel HL, Mello-Thoms C, et al. How experience and training influence mammography expertise. *Acad Radiol*. 1999;6(10):575–585.
13. Chakraborty DP, Yoon H-J, Mello-Thoms C. Application of threshold-bias independent analysis to eye-tracking and FROC data. *Acad Radiol*. 2012;19(12):1474–1483.
14. Burgess AE. Comparison of receiver operating characteristic and forced choice observer performance measurement methods. *Med Phys*. 1995;22(5):643–655.
15. Bunch PC, Hamilton JF, Sanderson GK, Simmons AH. A free-response approach to the measurement and characterization of radiographic-observer performance. *J Appl Photogr Eng*. 1978;4:166–171.
16. Kundel HL, Nodine CF, Krupinski EA. Searching for lung nodules: Visual dwell indicates locations of false-positive and false-negative decisions. *Invest Radiol*. 1989;24:472–478.
17. Wolfe JM. Visual search. In: Pashler H, ed. *Attention*. London: University College London Press; 1998.
18. Larson AM, Loschky LC. The contributions of central versus peripheral vision to scene gist recognition. *J Vis*. 2009;9(10):6–6.
19. Pritchard RM, Heron W, Hebb DO. Visual perception approached by the method of stabilized images. *Can J Psychol (Revue canadienne de psychologie)*. 1960;14(2):67.
20. Edwards DC, Kupinski MA, Metz CE, Nishikawa RM. Maximum likelihood fitting of FROC curves under an initial-detection-and-candidate-analysis model. *Med Phys*. 2002;29(12):2861–2870.

21. Kundel HL, Nodine CF, Krupinski EA, Mello-Thoms C. Using gaze-tracking data and mixture distribution analysis to support a holistic model for the detection of cancers on mammograms. *Acad Radiol.* 2008;15(7):881–886.
22. Mello-Thoms C, Hardesty LA, Sumkin JH, et al. Effects of lesion conspicuity on visual search in mammogram reading. *Acad Radiol.* 2005;12:830–840.
23. Nodine CF. Recording and analyzing eye-position data using a microcomputer workstation. *Behav Res Meth Instrum Comput.* 1992;24(3):475–485.
24. Hillstrom A. Repetition effects in visual search. *Percept Psychophys.* 2000;2:800–817.
25. Chakraborty DP. Measuring agreement between ratings interpretations and binary clinical interpretations of images: A simulation study of methods for quantifying the clinical relevance of an observer performance paradigm. *Phys Med Biol.* 2012;57:2873–2904.

16

The radiological search model (RSM)

16.1 Introduction

Brief summaries of the radiological search model (RSM) were presented earlier in connection with the simulator used to generate FROC data. This chapter describes the statistical model in more detail. The RSM embodies the essential ideas of the Kundel–Nodine model of visual search described in the previous chapter. *It turns out that all that is needed to model seemingly as complex a process as visual search, at least to first order, is one additional parameter.* All models of receiver operating characteristic (ROC) data involve two parameters (not counting thresholds). For example, the unequal variance binormal model, **Chapter 6,** requires the (a,b) parameters. Alternative ROC models described in **Chapter 20** also require two parameters. The model described below contains three parameters, μ, λ and ν. The μ parameter is the simplest to understand: it is the perceptual signal-to-noise ratio (*pSNR*) of latent LLs relative to latent NLs. The parameters λ and ν describe the search process, that is, the first stage of the Kundel–Nodine model (glancing or global impression). Essentially, they describe the ability of the observer to find latent LLs while not finding latent NLs. It turns out that it is easier to understand the search process via intermediate primed parameters, λ' and ν'. However, unlike λ, ν, the primed parameters depend on μ; that is, they are not intrinsic. So, the following introduces, in order, μ, λ', and ν', and explains their meaning via software examples. Physical meanings for these parameters are given as well as how one might measure them via eye-tracking measurements. Finally, a model reparameterization is proposed, which takes into account that λ' and ν' must depend on μ, and this is where the parameters λ and ν are introduced, which are expected to be intrinsic, that is, independent, of μ.

The online appendices explain Poisson and binomial sampling at a simple level. It is the author's experience that users of the author's can benefit from the simple examples.

16.2 The radiological search model (RSM)

The radiological search model (RSM) for the free-response paradigm is a statistical parameterization of the Kundel–Nodine model. It consists of two stages:

A *search stage,* corresponding to the initial glance in the Kundel–Nodine sense, in which suspicious regions, that is, the *latent marks,* are flagged for subsequent scanning. The total number of latent marks on a case is ≥ 0, so some cases may have zero latent marks, a fact that will turn out to have important consequences for the shapes of all RSM-predicted operating characteristics.

A *decision stage,* during which each latent mark is scanned, features are extracted and analyzed, and the observer realizes a decision variable (i.e., a Z-sample) at each latent mark. Typically radiologists spend >1 s per site and high-resolution foveal examination is necessary to extract relevant details of the region being examined and make a decision whether to mark it. The number of Z-samples equals the number of latent marks on the case.

A *marking stage* during which the degree of suspicion associated with each latent mark is compared to a minimum reporting threshold, and if exceeded, the corresponding location is marked and rated according to its degree of suspicion, with higher ratings assigned to suspicious regions with higher degrees of suspicion.

Latent marks can be either *latent NLs* (corresponding to non-diseased regions) or *latent LLs* (corresponding to diseased regions). The number of latent NLs on a case is denoted $l_1 \geq 0$. The number of latent LLs on a diseased case is denoted $l_2 \geq 0$. Latent NLs can occur on non-diseased and diseased cases, but latent LLs can only occur on diseased cases. Assume that every diseased case has L actual lesions. [Later this is extended to an arbitrary number of lesions per diseased case.] Since the number of latent LLs cannot exceed the number of lesions, $0 \leq l_2 \leq L$. (The symbol l_s is for *location* and the subscript is the site-level truth-state s)

Distributional assumptions are made for the numbers of latent NLs and latent LLs and the associated Z-samples. Since in this chapter one is dealing with a parametric model of search, one does not need to show explicitly case and location dependence as in **Chapter 13**. This allows for a simpler notation unencumbered by a plethora of subscripts.

16.2.1 RSM assumptions

The number of latent NLs, $l_1 \geq 0$, is an integer random variable sampled from the Poisson distribution with mean λ'.

$$l_1 \sim Poi(\lambda') \tag{16.1}$$

The reason for the prime will become clear in Section 16.4. The probability mass function (*pmf*) of the Poisson distribution is defined by

$$pmf_{Poi}(l_1;\lambda') = e^{-\lambda'}\frac{(\lambda')^{l_1}}{l_1!} \tag{16.2}$$

The number of latent LLs, l_2, $0 \leq l_2 \leq L$, is an integer random variable sampled from the binomial distribution with success probability ν' and trial size L.

$$l_2 \sim B(L,\nu') \tag{16.3}$$

The *pmf* of the binomial distribution is defined by

$$pmf_B(l_2;\nu',L) = \binom{L}{l_2}(\nu')^{l_2}(1-\nu')^{L-l_2} \tag{16.4}$$

Each latent mark is associated with a Z-sample. That for a latent NL is denoted z_{l_1} while that for a latent LL is denoted z_{l_2}.

For latent NLs, the Z-samples are obtained by sampling $N(0,1)$.

$$z_{l_1} \sim N(0,1) \tag{16.5}$$

For latent LLs, the Z-samples are obtained by sampling $N(\mu,1)$.

$$z_{l_2} \sim N(\mu,1) \tag{16.6}$$

In an FROC study with R-ratings, the observer adopts R ordered cutoffs ζ_r (r = 1, 2, ..., R). Defining $\zeta_0 = -\infty$ and $\zeta_{R+1} = \infty$, if $\zeta_r \leq z_{l_s} < \zeta_{r+1}; s = 1, 2$ the corresponding latent site is marked and rated in bin r, and if $z_{l_s} < \zeta_1$ the site is not marked.

The location of the mark is at the center of the latent site that exceeded the cutoff and an infinitely precise proximity criterion is adopted. Consequently, there is no possibility of confusing a mark made because of a latent LL Z-sample exceeding the cutoff with one made because of a latent NL Z-sample exceeding the cutoff, and vice-versa. Therefore, any mark made because of a latent NL Z-sample that satisfies $\zeta_r \leq z_{l_1} < \zeta_{r+1}$ will be scored as a non-lesion localization (NL) and rated r. Any mark made because of a latent LL Z-sample that satisfies $\zeta_r \leq z_{l_2} < \zeta_{r+1}$ will be scored as a lesion-localization (LL) and rated r. Unmarked LLs (latent or not) are assigned the negative infinity-rating. By *latent or not* the author means that even lesions that were not flagged by the search stage, and therefore do not qualify as latent LLs, are assigned the negative infinity-rating.

By choosing R large enough, the above discrete rating model becomes applicable to continuous Z-samples.

Summary of RSM

1st stage, initial glance, identifies latent NLs and latent LLs.
- # of NLs ~ Poisson with mean λ'
- # of LLs ~ binomial with success probability ν' and trial size L

2nd stage calculates Z-sample at each latent mark.
- Z-sample for latent NL $\sim N(0,1)$
- Z-sample for latent LL ~ $N(\mu,1)$

The latent mark is marked if $z \geq \zeta_1$.

The rating assigned to a mark is the index of the nearest threshold that is just equaled or exceeded by the Z-sample; unmarked latent NLs are unobservable events, but unmarked LLs, latent or not, are assigned the negative infinity-rating.

16.2.2 Summary of RSM defining equations

The RSM is summarized in Table 16.1.

16.3 Physical interpretation of RSM parameters

The parameters μ, λ', and ν' have the following meanings.

16.3.1 The μ parameter

The parameter $\mu \geq 0$ is the *perceptual signal-to-noise ratio, pSNR,* introduced in **Chapter 12**, between latent NLs and latent LLs. *It is not the perceptual signal-to-noise ratio of the latent LL relative to its immediate surround.* The immediate surround provides background context and is what makes the lesion conspicuous, that is, it is part of *pSNR*. The competition for latent marks is other regions that could be mistaken for lesions, that is, latent NLs. The immediate surround has no chance of being mistaken for a lesion. Calculating *pSNR* the conventional way using the immediate surround, as in computer analysis of mammography phantom images[1,2] (CAMPI), will yield an infinite value (because the contrast is measured relative to the surround, i.e., the latter has zero noise).

The μ parameter is identical to the separation parameter of two unit normal distributions required to achieve the observed probability of correct choice (PC) in a two-alternative forced

Table 16.1 Summary of RSM defining equations

Latent NL site sampling	$l_1 \sim Poi(\lambda')$ $pmf_{Poi}(l_1; \lambda') = e^{-\lambda'} \dfrac{(\lambda')^{l_1}}{l_1!}$	$\lambda' \geq 0;\ l_1 \geq 0$
Latent LL site sampling	$l_2 \sim B(L, \nu')$ $pmf_B(l_2; \nu', L) = \begin{pmatrix} L \\ l_2 \end{pmatrix}(\nu')^{l_2}(1-\nu')^{L-l_2}$	$0 \leq \nu' \leq 1;\ 0 \leq l_2 \leq L$
Z-sampling for latent NLs on case with truth state t	$z_{l_1} \sim N(0,1)$	Applies to NL Z-samples on non-diseased and diseased cases
Z-sampling for latent LLs	$z_{l_2} \sim N(\mu, 1)$	$\mu \geq 0$
Marking and binning rule for NLs	if $\zeta_r \leq z_{l_1} < \zeta_{r+1};\ r = 1, 2 ..., R \Rightarrow bin = r$	$\zeta_0 < \zeta_1 < ... < \zeta_{R+1}$
Marking and binning rule for LLs	if $\zeta_r \leq z_{l_2} < \zeta_{r+1};\ r = 0, 1 ..., R \Rightarrow bin = r$	$\zeta_0 = -\infty;\ \zeta_{R+1} = \infty$
Unmarked latent NL sites	if $z_{l_1} < \zeta_1 \Rightarrow bin = NA$	Unobservable events
Unmarked lesions, latent or not	if $z_{l_2} < \zeta_1 \Rightarrow bin = -\infty$	Inferred zero-rating for latent LLs

choice (2AFC) task between cued (i.e., pointed to by toggle-able arrows*) NLs and cued LLs. One measures the locations of the latent marks using eye-tracking apparatus[3] and clusters the data as described in **Chapter 15**, then runs a 2AFC study as follows. Pairs of images are shown, each with a cued location, one a latent NL (a big-cluster, as defined in **Chapter 15**) and the other a latent LL, where all locations were recorded in prior eye-tracking sessions for the specific radiologist. The radiologist's task is to pick the image with the latent LL. The probability correct (PC) in this task is related to the μ parameter by

$$\mu = \sqrt{2}\ \Phi^{-1}(PC) \tag{16.7}$$

It is essential that the radiologist on whom the eye-tracking measurements were performed and the one who performs the two-alternative forced choice tasks be the same. Two radiologists will not agree on latent marks. This will be more evident for the latent NL marks since greater disagreement is expected on what truly does not exist and was conjured in the radiologist's mind and this is expected to vary between radiologists. Disagreement is also expected, but to a lesser degree, on the latent LLs; one expects less disagreement on what truly exists (as the true lesions are independent of the radiologist). However, the set of latent LLs for the two observers are expected to be different. Specifically, a region perceived as a latent LL by one might be missed by the other. A complication in conducting such a study is that, because of memory effects, a lesion can only be shown once; this could result in a limited number of comparisons and a consequential imprecise estimate of μ (the author ran into this very problem in a preliminary attempt at this type of measurement).

16.3.2 The λ' parameter

The λ' parameter determines the tendency of the observer to generate latent NLs. The mean number of latent NLs per case estimates λ'; this is a property of the Poisson distribution. It can also be

---

* The presence of the arrow can distort perception of the lesion, so it needs to be toggleable.

measured via an eye-tracking apparatus. This time it is only necessary to cluster the marks and classify each mark as a latent NL or latent LL according to the adopted acceptance radius. An eye-tracking based estimate would be the total number of latent NLs (the big-clusters in the previous chapter) in the dataset divided by the total number of cases.

Consider two observers, one with $\lambda' = 1$ and the other with $\lambda' = 2$. While one cannot predict the number of latent NLs on any specific case, one can predict the *average* number of latent NLs on a given case set. Use **R** (copy and paste the commands into the **Console** window) to demonstrate this.

16.3.2.1 Code snippets

```
> seed <- 1;set.seed(seed)
> samples1 <- rpois(100,1)
> mean(samples1)
[1] 1.01
> seed <- 1;set.seed(seed)
> samples2 <- rpois(100,2)
> mean(samples2)
[1] 2.02
> samples1[1:10]
[1] 0 1 1 2 0 2 3 1 1 0
> samples2[1:10]
[1] 1 1 2 4 1 4 4 2 2 0
```

In this example, the number of samples has been set to 100 (the first argument to **rpois()**). For the first observer, $\lambda' = 1$, the first case generated zero latent NLs, the second and third generated one NL each, and so on. For the second observer, the first and second case generated one latent NL each, the third generated two, and so on. While one cannot predict what will happen on any specific case, one can predict that the average number of latent NL marks per case for the first observer will be close to one (the observed values is 1.01) and that for the second one will be close to two (the observed values is 2.02).

Estimates should be accompanied by confidence intervals. The code is in file **mainPoissonExample.R**, explained in Online Appendix 16.A.1, which illustrates Poisson sampling and estimation of an exact confidence interval for the mean parameter of the Poisson distribution. **Source** the code to get the following output.

16.3.2.2 Code output

```
> source(...)
K = 100,
lambdaP 1st reader = 1
lambdaP 2nd reader = 2
obs. mean, reader 1 = 1.01
obs. mean, reader 2 = 2.02
Rdr. 1: 95% CI = 0.8226616 1.227242
Rdr. 2: 95% CI = 1.751026 2.318599
```

This code uses 100 samples (**K**, line 5 in listing). For reader 1 the estimate of the Poisson parameter (the mean parameter of the Poisson distribution is frequently referred to as the Poisson parameter) is 1.01 with 95% confidence interval (0.823, 1.227); for reader 2 the corresponding estimate is 2.02 with 95% confidence interval (1.751, 2.319). *As the number of cases increases, the widths of the confidence intervals shrink; for* example, with 10,000 cases, that is, 100 times the value in the previous example:

16.3.2.3 Code output

```
> source(...)
K = 10000 ,
lambdaP 1st reader = 1
lambdaP 2nd reader = 2
obs. mean, reader 1 = 1.0055
obs. mean, reader 2 = 2.006
Rdr. 1: 95% CI =   0.9859414 1.025349
Rdr. 2: 95% CI =   1.978335 2.033955
```

This time for reader 1, the estimate of the Poisson parameter is 1.01 with 95% confidence interval (0.986, 1.025); for reader 2 the corresponding estimate is 2.01 with 95% confidence interval (1.978, 2.034). The width of the confidence interval is inversely proportional to the square root of the number of cases (the example below is for reader 1).

16.3.2.4 Code snippet

```
> -(0.8226616-1.2272416) # reader 1
[1] 0.40458
> -(0.9859414-1.0253490) # reader 1
[1] 0.0394076
```

Since the number of cases was increased by a factor of 100, the width decreased by a factor of 10.

16.3.3 The ν' parameter

The ν' parameter determines the ability of the observer to find lesions. Assuming the same number of lesions per diseased case, the mean fraction of latent LLs per diseased case is an estimate of ν'. It too can be measured via eye-tracking apparatus performed on a radiologist. An eye-tracking based estimate would be the total number of latent LLs (big-clusters, each of which localizes a lesion) in the dataset divided by the total number of lesions. Consider two observers, one with $\nu'=0.5$ and the other with $\nu'=0.9$. Again, while one cannot predict the precise number of latent LLs on any specific diseased case, or which specific lesions will be correctly localized, one can predict the average number of latent LLs. The code is in file **mainBinomialExample1.R** in Online Appendix 16.B.1. **Source** it to get the following output.

16.3.3.1 Code output

```
> source(...)
K2 =   100
nuP 1st reader = 0.5
nuP 2nd reader = 0.9
mean, reader 1 = 0.48
mean, reader 2 = 0.94
Rdr. 1: 95% CI = 0.3790055 0.5822102
Rdr. 2: 95% CI =   0.8739701 0.9776651
```

This code uses 100 samples (**K2**, line 5 in listing). The result shows that for reader 1 the estimate of the binomial success rate parameter is 0.48 with 95% confidence interval (0.38, 0.58); for reader 2 the corresponding estimate is 0.94 with 95% confidence interval (0.87, 0.98). As the number of diseased cases increases, the confidence interval shrinks in inverse proportion to the square root of cases. (Example, increasing the number of cases by a factor of 100, for reader 1 the CI width ratio is [0.4854532–0.5051496]/[0.3790055–0.5822102] = 0.097.)

As a more complicated but clinically realistic example, consider a dataset with 100 cases where 97 have one lesion per case, two have two lesions per case and one has three lesions per case (these are typical lesion distributions observed in screening mammography). The code is in file **mainBinomialExample2.R** in Online Appendix 16.C.1. **Source** it to get the following output.

16.3.3.2 Code output

```
> source(...)
K2[1] = 97
K2[2] = 2
K2[3] = 1
nuP1 = 0.5
nuP2 = 0.9
obsvd. mean, reader 1 = 0.4903846
obsvd. mean, reader 2 = 0.9326923
Rdr. 1: 95% CI = 0.3910217 0.5903092
Rdr. 2: 95% CI = 0.8662286 0.9725125
```

16.4 Model reparameterization

While the parameters μ, λ', and v' are physically meaningful, and can be estimated from eye-tracking measurements, a little thought reveals that they cannot be varied independently of each other. Rather, μ is the *intrinsic* parameter. Its value, together with two other *intrinsic* parameters, $v \geq 0$ and $\lambda \geq 0$, determine the physically more meaningful parameters λ' and v', respectively, according to the following reparameterization:

$$v' = 1-\exp(-v\mu) \tag{16.8}$$

$$\lambda' = \lambda/\mu \tag{16.9}$$

The parameterization is not unique, but the one adopted is relatively simple. The need for the first reparameterization (involving v) was foreseen (using different notation) in the original search model papers[4,5] but the need for the second reparameterization (involving λ) was discovered more recently. Since it determines v', the v parameter can be considered as the intrinsic (i.e., μ-independent) ability to find lesions; specifically, it is the rate of increase of v' with μ at small μ.

$$v = \frac{\partial v'}{\partial \mu}\bigg|_{\mu=0} \tag{16.10}$$

The dependence of v' on μ is consistent with the fact that higher contrast lesions will be easier to find by any observer. This is why v' is not an intrinsic property—any observer, even one without special expertise, can find a high contrast lesion. Conversely, lower contrast lesions will be more difficult to find even by expert observers. The colloquial term *find* is used as shorthand for *flagged for further inspection by the holistic first stage of the search mechanism, thus qualifying as a latent site*. In other words, finding a lesion means the lesion was perceived as a suspicious region, which makes it a latent site, independent of whether or not the region was actually marked. Finding refers

---

* Intrinsic: belonging naturally, essential.

to the search stage. Marking refers to the decision stage, where the region's Z-sample is determined and compared to a reporting threshold.

According to Equation 16.8, as $\mu \to \infty$, $\nu' \to 1$, and in the opposite limit as $\mu \to 0$, $\nu' \to 0$. Recall the analogy to finding the sun made in **Chapter 12**: objects with very high perceptual SNR are certain to be found and conversely, objects with zero perceptual SNR are found only by chance and are marked based on cost-benefit considerations.

According to Equation 16.9, the value of μ also determines λ as $\mu \to \infty$ $\lambda' \to 0$ and conversely, as $\mu \to 0$ then $\lambda' \to \infty$. This too is clear from the sun analogy of **Chapter 12**.

The reparameterization used here is not unique, but is simple and has the right limiting behaviors.

16.5 Discussion/Summary

This chapter has described a statistical parameterization of the Kundel–Nodine model. The 3-parameter model of search, in the medical imaging context, accommodates key aspects of the process. Search, the ability to find lesions while minimizing finding non-lesions, is described by two parameters, specifically, λ', ν'. The ability to correctly mark a found lesion (while not marking found non-lesions) is characterized by the third parameter of the model, μ. While the primed parameters have relatively simple physical meaning, they depend on μ. Consequently, it is necessary to define them in terms of intrinsic parameters. The modeled dependence on μ is not unique.

The next chapter explores the predictions of the radiological search model.

References

1. Chakraborty DP. Computer analysis of mammography phantom images (CAMPI): An application to the measurement of microcalcification image quality of directly acquired digital images. *Med Phys*. 1997;24(8):1269–1277.
2. Chakraborty DP, Eckert MP. Quantitative versus subjective evaluation of mammography accreditation phantom images. *Med Phys*. 1995;22(2):133–143.
3. Chakraborty DP, Yoon H-J, Mello-Thoms C. Application of threshold-bias independent analysis to eye-tracking and FROC data. *Acad Radiol*. 2012;19(12):1474–1483.
4. Chakraborty DP. ROC Curves predicted by a model of visual search. *Phys Med Biol*. 2006;51:3463–3482.
5. Chakraborty DP. A search model and figure of merit for observer data acquired according to the free-response paradigm. *Phys Med Biol*. 2006;51:3449–3462.

17

Predictions of the RSM

17.1 Introduction

The preceding chapter described the radiological search model (RSM) for free-response ROC (FROC) data. This chapter describes predictions of the RSM and how they square with evidence. The starting point is the inferred receiver operating characteristic (ROC) curve. While mathematically complex, the results are important because they are needed to derive the ROC-likelihood function, which is used to estimate RSM parameters from ROC data in **Chapter 19**. The preceding sentence should cause the inquisitive reader to ask: since the ROC paradigm ignores search, how is it possible to derive parameters of a model of search from the ROC curve? The answer is that the *shape* of the ROC curve contains information about the RSM parameters. The RSM-predicted shape is fundamentally different from predictions of all conventional ROC models (binormal,[1] contaminated binormal model,[2-4] bigamma,[5] and proper ROC[6,7]) as it has the constrained end-point property. All other models predict that the end-point, namely the uppermost non-trivial point on the ROC reached at an infinitely low reporting threshold, is (1,1), while the RSM predicts it does not reach (1,1). The nature of search is such that the limiting end-point is constrained to be below and to the left of (1,1). This key difference, which effectively compresses the observable range of the ROC curve in both horizontal and vertical directions, allows one to estimate search parameters from ROC data, **Chapter 19**. Next, the RSM is used to predict FROC and alternative FROC (AFROC) curves. Two following sections show how search performance and lesion-classification performance can be quantified from the location of the ROC end-point. Search performance is the ability to find lesions while *avoiding* finding non-lesions, and lesion-classification performance is the ability, having found a suspicious region, to correctly classify it. If classified as an NL, it would not be marked (in the mind of the observer every mark is a potential LL, albeit at different confidence levels). *Note that lesion-classification is different from classification between diseased and non-diseased cases, which is measured by the ROC-AUC.* Based on the ROC/FROC/AFROC curve predictions of the RSM, a comparison is presented between area measures that can be calculated from FROC data, and this leads to an important conclusion, namely, that the FROC curve is a poor descriptor of search performance and that the AFROC/wAFROC are preferred. This will come as a surprise (shock?) to most researchers somewhat familiar with this field, since the overwhelming majority of users of FROC methods, particularly in computer aided detection (CAD), have relied on the FROC curve. Finally, evidence for the validity of the RSM is presented.

Online Appendix 17.A describes implementation of the error function in R, as this enters the formula. The analytic expression for the RSM-predicted TPF is in Online Appendix 17.B. An expression for the *pdf* for diseased cases that demonstrates the proper property of the RSM predicted ROC is in Online Appendix 17.C. R code for generating RSM-predicted ROC curves and *pdf*'s is in Online Appendix 17.D. Detailed explanation of the code comparing FROC vs. AFROC vs. ROC is on

Online Appendix 17.E. Demonstration of the ability of the RSM, in certain circumstances, to mimic the widely successful binormal model is in Online Appendix 17.F. Finally, Online Appendix 17.G explains the code that shows that the RSM predicts ROC curves that are consistent with certain observations made by Swets et al in the late 1960s.

The starting point is inferring ROC data from FROC data.

17.2 Inferred integer ROC ratings

Consider a $R_{FROC} \geq 1$ rating FROC study with allowed ratings $r = 1, 2, ..., R_{FROC}$. In **Chapter 13**, the *inferred* ROC rating of a case was defined as the rating of the highest rated mark on a case, or $-\infty$, if the case has no marks. Since a $-\infty$ rating is an can be inconvenient when the other ratings are integers and the ratings are ordered integer labels, the corresponding integer rating is taken to be ROC:1. No ordering information is lost provided every other rating is also incremented by unity. Therefore, the integer inferred ROC scale extends from one to $R_{FROC} + 1$. Thus, an R_{FROC} rating FROC study formally corresponds to a $R_{FROC} + 1$ rating ROC study. Henceforth the word *inferred* will be implicit when referring to an RSM-predicted ROC curve.

In addition, instead of referring to false positive (FP) and true positive (TP) ratings, in this chapter it will be more convenient to use the symbol $h_{k_t t}$ to denote the rating of the highest rated Z-sample on case k_t with truth-state t. Thus $h_{k_1 1}$ refers to the highest rating on a non-diseased case $k_1 1$, and $h_{k_2 2}$ refers to the highest rating on diseased case $k_2 2$, etc. For non-diseased cases, the maximum is over all latent NLs on the case. For diseased cases, the maximum is over all latent NLs and latent LLs on the case. Reiterating, *the integer ROC rating is the one-incremented highest FROC rating of the case, or ROC:1 if the case has no marks.* As before, when there is a possibility of confusion, one precedes the rating with the acronym for the applicable paradigm. Consider a set of ordered thresholds $\zeta_r < \zeta_{r+1}$ and dummy thresholds defined by $\zeta_0 = -\infty$, $\zeta_{R_{FROC}+1} = \infty$, then, if $\zeta_r \leq h_{k_t t} < \zeta_{r+1}$ the case is rated $ROC : (r+1)$. As an example, if $h_{k_t t} < \zeta_1$ the case is rated ROC:1, corresponding to an unmarked case.

17.2.1 Comments

1. Since $r = 1, 2, ..., R_{FROC}$ the lowest allowed ROC rating on a case with at least one mark is ROC:2.
2. On a case with no latent marks *or* the highest rated latent site satisfies $h_{k_t t} < \zeta_1$ the observer gives the ROC:1 rating. From the analyst's point of view, one cannot distinguish between whether the ROC:1 rating was the result of the case not having any latent marks or the case had at least one latent mark, but none of the Z-samples exceeded ζ_1.

A consequence of the possibility that some cases have no marks is that all RSM-predicted operating characteristics share a *constrained end-point property*, which is the next topic.

17.3 Constrained end-point property of the RSM-predicted ROC curve

The full range of ROC space, that is, $0 \leq FPF(\zeta) \leq 1$ and $0 \leq TPF(\zeta) \leq 1$, is *not continuously accessible* to the observer. In fact, $0 \leq FPF(\zeta) \leq FPF_{max}$ and $0 \leq TPF(\zeta) \leq TPF_{max}$ where FPF_{max} and TPF_{max} are each generally less than (or, in special cases, equal to) unity. Therefore, (FPF_{max}, TPF_{max}) represents a constraint on the end-point; the abscissa of the end-point has to be smaller than or equal to FPF_{max} and the ordinate has to be smaller than or equal to TPF_{max}.

In the following explanation, for the sake of simplicity and without loss of generality, the effect of lowering only ζ_1 is considered, while the remaining thresholds are left unchanged. Lowering ζ_1

will flip ROC ratings from ROC:1 to ROC:2, but never to ROC:3. So all operating points will be unaffected with the exception of the highest non-trivial point, which will creep upwards and to the right. A detailed explanation follows.

As ζ_1 is lowered to $-\infty$, some of the previously ROC:1 rated cases that had at least one latent site will be marked and bumped-up to the ROC:2 bin, until eventually only cases with no latent marks remain in the ROC:1 bin. *These remaining cases will never be rated ROC:2.* A rational observer who finds no suspicious regions, literally nothing to report, will assign the lowest available bin to the case, which happens to be ROC:1. The *finite* number of cases in the ROC:1 bin at an infinitely low threshold has the consequence that the uppermost non-trivial continuously accessible operating point, obtained by cumulating ratings ROC:2 and above, is below and to the left of (1,1). The (1,1) point is reached trivially when the researcher cumulates the counts in all bins, i.e., ROC:1 and above. This behavior is distinct from traditional ROC models where the entire curve extending from (0, 0) to (1, 1) is continuously accessible to the observer. This is because in conventional models every case yields a finite decision variable, no matter how small. Lowering the lowest threshold to $-\infty$ eventually flips all cases in the previously ROC:1 bin to the ROC:2 bin, and one is eventually left with zero counts in the ROC:1 bin, and the operating point, obtained by cumulating bins ROC:2 and above, is (1,1).

Starting with a finite threshold, as the observer is encouraged to be more aggressive in reporting lesions, the ROC point moves continuously upwards and to the right from (0, 0) to (FPF_{max}, TPF_{max}) and no further. The ROC curve cannot just hang there since cumulating all cases yields the trivial (1,1) operating point. Therefore, the complete ROC curve is obtained by extending the end-point using a dashed line that connects it to (1,1). *The observer cannot operate along the dashed line. See further elaboration of this point, in particular, why guessing to operate along the dashed line, is not an option, in* Section 17.12.1.

How closely the observer approaches the limiting point (FPF_{max}, TPF_{max}) is unrelated to the number of bins, which determines the number of points on the continuous section of the ROC curve, not the position of the uppermost non-trivial point. Rather, it depends on the lowest threshold, i.e., ζ_1. As the latter is lowered, the observed end-point approaches (FPF_{max}, TPF_{max}) from below. How closely (FPF_{max}, TPF_{max}) approaches (1,1) depends on λ' and ν'. As λ' approaches infinity and as ν' approaches unity, the limiting point (FPF_{max}, TPF_{max}) approaches (1,1) from below, see Equation 17.1 and 17.2 to follow. These parameters determine the probability that a case has one or more marks, and depending on the truth-state of the case, non-diseased or diseased, these probabilities equal FPF_{max} or TPF_{max}, respectively, as discussed next.

17.3.1 The abscissa of the ROC end-point

One needs the probability that a non-diseased case has at least one latent NL. Such a case will generate a finite $h_{k_1 1}$ and with an appropriately low ζ_1 the case will be rated ROC:2 or higher. The probability of zero latent NLs (see Equation 16.2) is $pmf_{Poi}(0; \lambda') = e^{-\lambda'}$. Therefore, the probability that the case has at least one latent NL, which is the maximum continuously reachable abscissa of the ROC, is

$$FPF_{max} = 1 - e^{-\lambda'} = 1 - e^{-\lambda/\mu} \tag{17.1}$$

The second form on the right-hand side of Equation 17.1 expresses the result in terms of intrinsic RSM parameters (the μ independent λ, ν parameters, not the physical ones; see section on model re-parameterization in Section 16.4). As μ increases FPF_{max} moves to the left, reaching zero in the limit $\mu \to \infty$. Recall the by now familiar solar analogy in **Chapter 12**. Conversely, for fixed $\mu > 0$, increasing λ causes FPF_{max} to move to the right approaching one in the limit $\lambda \to \infty$, because in this limit every case has a latent NL.

17.3.2 The ordinate of the ROC end-point

A diseased case has no marks, even for very low ζ_1, if it has zero latent NLs, the probability of which is $e^{-\lambda'}$, *and* it has zero latent LLs, the probability of which is, Equation 16.4, $pmf_{Bin}(0; L, v') = (1 - v')^L$. Here L is the number of lesions in each diseased case, assumed constant.

Independence Assumption 1: occurrences of latent LLs are independent of each other, that is, the probability that a lesion is found is independent of whether other lesions were found on the same case.

Independence Assumption 2: occurrences of latent NLs are independent of each other; that is, the probability that a non-diseased region is found is independent of whether other non-diseased regions were found on the same case.

Independence Assumption 3: occurrence of a latent NL is independent of the occurrence of a latent LL on the same case.

Comment: Assumptions 1 and 3 are not consistent with satisfaction of search (SOS) effects.[8,9] The RSM is a first-order description of visual search; higher order effects need to be eventually included.

By the independence assumptions, the probability of zero latent NLs *and* zero latent LLs on a diseased case is the product of the two probabilities, namely $e^{-\lambda'}(1 - v')^L$. Therefore, the probability that there exists at least one latent site is

$$TPF_{max} = 1 - e^{-\lambda'}(1 - v')^L = 1 - e^{-\lambda/\mu}\exp(-v\mu L) \qquad (17.2)$$

The second expression on the right-hand side, in terms of intrinsic parameters, follows from Equations 16.8 and 16.9. As $\lambda \to \infty$, (FPF_{max}, TPF_{max}) approaches $(1,1)$ from below, because in this limit every case is assured to have a latent NL, therefore yields a finite Z-sample, and is marked at sufficiently low ζ_1. Conversely, as $0(TPF_{max})$ to $\mu \to \infty$, the product of the two exponents tends to zero and the ROC plot approaches perfect performance, i.e., a *vertical* line from the origin to $(0,1)$, which is the continuously accessible section, followed by the *inaccessible* horizontal line connecting $(0,1)$ to $(1,1)$.

17.3.3 Variable number of lesions per case

Define f_L *as the fraction of diseased cases with L lesions* and L_{max} as the maximum number of lesions per diseased case in the dataset, then

$$\sum_{L=1}^{L_{max}} f_L = 1 \qquad (17.3)$$

By restricting attention to the set of diseased cases with L lesions each, Equation 17.2 for $TPF_{max}(\mu,\lambda',\nu'\,|\,L)$ applies, and since TPF is a probability, and probabilities of independent processes add, it follows that

$$TPF_{max}\left(\mu,\lambda',\nu'\,|\,\overrightarrow{f_L}\right)=\sum_{L=1}^{L_{max}} f_L TPF_{max}\left(\mu,\lambda',\nu'\,|\,L\right) \tag{17.4}$$

In other words, the ordinate of the uppermost point is a weighted sum over the lesion distribution vector. It is seen that the ordinate is less than or equal to unity (as each term $TPF_{max}(\mu,\lambda',\nu'\,|\,L)$ in the weighted summation is less than or equal to unity). The expression for FPF_{max} is unaffected.

17.4 The RSM-predicted ROC curve

To predict the continuous ROC curve, one assumes the observer indicates the *actual value of the Z-sample* for each latent site. The ROC decision variable is the rating of the highest rated mark $h_{k_t t}$ for the case. Since one is on the continuous section of the curve, each case must have at least one site and one does not have to worry about cases with no marks and conditioning on the null set is not needed. Therefore, false positive fraction (FPF) is the probability that $h_{k_1 1}$ on a non-diseased case exceeds the virtual threshold ζ and true positive fraction (TPF) is the probability that $h_{k_2 2}$ on a diseased case exceeds ζ:

$$FPF(\zeta)=P\left(h_{k_1 1}\geq\zeta\right)$$
$$TPF(\zeta)=P\left(h_{k_2 2}\geq\zeta\right) \tag{17.5}$$

Varying the threshold parameter ζ from $+\infty$ to $-\infty$ sweeps out the continuous section of the theoretical* RSM-predicted ROC curve, extending from $(0,0)$ to (FPF_{max},TPF_{max}). Derivations of the RSM-predicted ROC coordinates are shown next.

17.4.1 Derivation of FPF

Independence Assumption 4: the Z-samples of NLs on the same case are independent of each other.

Consider the set of non-diseased cases with n latent NLs each, where $n>0$. According to the RSM, each latent NL yields a Z-sample from $N(0,1)$. The probability that a Z-sample from a latent NL is smaller than ζ is $\Phi(\zeta)$. By the independence assumption the probability that *all* n samples are smaller than ζ is $\left[\Phi(\zeta)\right]^n$. If all Z-samples are smaller than ζ, the highest Z-sample $h_{k_1 1}$ must be smaller than ζ. Therefore, the probability that $h_{k_1 1}$ exceed ζ is

$$FPF\left(\zeta\,|\,n\right)=P\left(h_{k_1 1}>\zeta\,|\,n\right)=1-\left[\Phi(\zeta)\right]^n \tag{17.6}$$

The conditioning notation in Equation 17.6 reflects the fact that this expression applies specifically to non-diseased cases with n latent NLs. To obtain $FPF(\zeta)$ one performs a Poisson pmf-weighted summation of $FPF(\zeta\,|\,n)$ over n from 1 to ∞ (the probabilities of independent processes can be added):

$$FPF(\zeta)=\sum_{n=0}^{\infty}\left[pmf_{Poi}\left(n;\lambda'\right)\left(1-\left(\Phi(\zeta)\right)^n\right)\right] \tag{17.7}$$

* *Theoretical* is used to distinguish it from the *empirical*, or, observed end-point. It is the end-point predicted by setting $\zeta=-\infty$.

The infinite summations (see below) are easier performed using symbolic algebra software such as Maple™ or Mathematica™. The author is more familiar with Maple™. Inclusion, in the summation, of the $n = 0$, which term evaluates to zero, is done to make it easier for Maple™ to evaluate the summation in closed form; otherwise one would need to simplify the Maple™-generated result. The result is shown below (Maple 17, Waterloo Maple Inc.), where **lambda** and **nu** refer to the primed quantities.

17.4.1.1 Maple code

```
restart;
phi := proc (t, mu) exp(-(1/2)*(t-mu)^2)/sqrt(2*Pi) end:
PHI := proc (c, mu) local t; int(phi(t, mu), t = -infinity .. c) end:
Poisson := proc (n, lambda) lambda^n*exp(-lambda)/factorial(n) end:
FPF := proc(zeta,lambda) sum(Poisson(n,lambda)*(1 - PHI(zeta,0)^n),
n=0..infinity);end:
FPF(zeta, lambda);
```

The Maple™ code yields the following result (the second line uses the physical [primed] to intrinsic transformation):

$$FPF\left(\zeta;\lambda'\right)=1-\exp\left(-\frac{\lambda'}{2}\left[1-erf\left(\frac{\zeta}{\sqrt{2}}\right)\right]\right)$$

$$FPF\left(\zeta;\lambda\right)=1-\exp\left(-\frac{\lambda}{2\mu}\left[1-erf\left(\frac{\zeta}{\sqrt{2}}\right)\right]\right)$$

(17.8)

The error function in Equation 17.8 is defined by[10]

$$erf\left(x\right)=\frac{2}{\sqrt{\pi}}\int_{0}^{x}e^{-t^2}\,dt$$

(17.9)

It is related, Online Appendix 17.A, to the normal CDF function Φ by

$$erf\left(x\right)=2\Phi\left(\sqrt{2}x\right)-1$$

(17.10)

The error function ranges from -1 to $+1$ as its argument ranges from $-\infty$ to $+\infty$. For $\zeta=-\infty$, Equation 17.8 yields the same expression for FPF_{max}, as does Equation 17.1:

$$1-\exp\left(-\frac{\lambda}{2\mu}\left[1-erf\left(\frac{\zeta\to-\infty}{\sqrt{2}}\right)\right]\right)=1-\exp\left(-\frac{\lambda}{2\mu}[1+1]\right)=1-\exp\left(-\frac{\lambda}{\mu}\right)$$

(17.11)

Therefore, *FPF* ranges from zero to FPF_{max} to zero as ζ ranges from $+\infty$ to $-\infty$, confirming the constrained property of the abscissa of the theoretical end-point.

17.4.2 Derivation of TPF

The derivation of the true positive fraction $TPF(\zeta)$ follows a similar line of reasoning except this time one needs to consider the highest of the latent NLs *and* latent LL Z-samples. Consider a diseased case with n latent NLs and l latent LLs. Each latent NL yields a decision variable sample from $N(0,1)$ and each latent LL yields a sample from $N(\mu,1)$. The probability that all n latent NLs have Z-samples less than ζ is $\left[\Phi(\zeta)\right]^{n}$. The probability that all l latent LLs have Z-samples less than ζ is $\left[\Phi(\zeta-\mu)\right]^{l}$. Using the independence assumptions, the probability that all latent marks have Z-samples less than ζ is the product of these two probabilities. The probability that $h_{k_2 2}$ (the highest Z-sample on diseased case $k_2 2$) is larger than ζ is the complement of the product probabilities, that is,

$$TPF_{n,l}(\zeta\,|\,\mu,n,l)=P\!\left(h_{k_2 2}>\zeta\,|\,\mu,n,l\right)=1-\left[\Phi(\zeta)\right]^{n}\left[\Phi(\zeta-\mu)\right]^{l} \tag{17.12}$$

One sums the Poisson and binomial pmf-weighted $TPF_{n,l}(\zeta\,|\,\mu n,l)$ over n and l to obtain the desired ROC-ordinate (as before, the probabilities of independent processes can be added):

$$TPF\!\left(\zeta\,|\,\mu,\lambda',\nu',L\right)=\sum_{n=0}^{\infty}f_{Poi}\left(n;\lambda'\right)\sum_{l=0}^{L}pmf_{B}\left(l;\nu',L\right)TPF_{n,l}\left(\zeta;\mu,n,l\right) \tag{17.13}$$

This can be evaluated using Maple™, Online Appendix 17.B, yielding:

$$TPF\!\left(\zeta;\mu,\lambda',\nu',L\right)=1-\left(1-\frac{\nu'}{2}+\frac{\nu'}{2}erf\!\left(\frac{\zeta-\mu}{\sqrt{2}}\right)\right)^{L}e^{\left(-\frac{\lambda'}{2}+\frac{\lambda'}{2}erf\left(\frac{\zeta}{\sqrt{2}}\right)\right)} \tag{17.14}$$

It can be confirmed that for $\zeta=-\infty$ Equation 17.14 yields the same expression for $TPF_{\max}$ as Equation 17.2:

$$TPF\!\left(-\infty;\mu,\lambda',\nu',L\right)=1-\left(1-\nu'\right)^{L}e^{-\lambda'}=1-e^{-\lambda/\mu}\exp(-\nu\mu L) \tag{17.15}$$

17.4.3 Extension to varying numbers of lesions

To extend the results to varying numbers of lesions per diseased case, one averages the right-hand side of Equation 17.14 over the *fraction* of diseased cases with L lesions, defined in Equation 17.3. The expression for TPF is

$$TPF\!\left(\zeta;\mu,\lambda',\nu',\overline{f_L}\right)=1-\sum_{L=1}^{L_{max}}f_{L}\left(1-\frac{\nu'}{2}+\frac{\nu'}{2}erf\!\left(\frac{\zeta-\mu}{\sqrt{2}}\right)\right)^{L}e^{\left(-\frac{\lambda'}{2}+\frac{\lambda'}{2}erf\left(\frac{\zeta}{\sqrt{2}}\right)\right)} \tag{17.16}$$

The right-hand side can be expressed in terms of intrinsic parameters, but the resulting expression is cumbersome and is not shown. The expression for *FPF* is, of course, unaffected.

17.4.4 Proper property of the RSM-predicted ROC curve

A proper ROC curve has the property that it never crosses the chance line and its slope decreases monotonically as the operating point moves up the ROC curve.[11] It is shown next that the continuously accessible portion of the ROC curve is proper. For convenience one abbreviates *FPF* and *TPF* to x and y, respectively, and suppresses the dependence on model parameters. From Equations 17.14 and 17.8 one can express the ROC coordinates as

$$x(\zeta) = 1 - G(\zeta)$$

$$y(\zeta) = 1 - F(\zeta)G(\zeta) \tag{17.17}$$

where

$$G(\zeta) = e^{\left(-\frac{\lambda'}{2} + \frac{\lambda'}{2}erf\left(\frac{\zeta}{\sqrt{2}}\right)\right)} \tag{17.18}$$

and

$$F(\zeta) = \left(1 - \frac{\nu'}{2} + \frac{\nu'}{2}erf\left(\frac{\zeta - \mu}{\sqrt{2}}\right)\right)^{L} \tag{17.19}$$

These equations have exactly the same structure as Swensson's[12] Equations 1 and 2 and the logic he used to demonstrate that ROC curves predicted by his model were proper also applies to the present situation. Specifically, since the error function ranges between -1 and 1 and $0 \le \nu' \le 1$, it follows that $F(\zeta) \le 1$. Therefore, $y(\zeta) \ge x(\zeta)$ and the ROC curve is constrained to the upper half of the ROC space, namely the portion above the chance diagonal. Additionally, the more general constraint shown by Swensson applies, namely the slope of the ROC curve at any operating point (x, y) cannot be less than the slope of the straight line connecting (x, y) and (x_{max}, y_{max}), the coordinates of the theoretical end-point. This implies that the slope decreases monotonically and rules out curves with hooks (the term "hook" is used to describe the inappropriate chance-line crossing and approach to $(1,1)$ with infinite slope that characterizes all binormal model fits to real datasets.

The results in this section are also valid for arbitrary numbers of lesions per case. Proving this is left as an exercise for the reader.

It is important to note that the proper ROC prediction applies to the continuous section of the ROC only. It is possible for the slope to change abruptly, even increasing, as one switches from the continuous section to the inaccessible part of the plot. Recall that the inaccessible part corresponds to cases that do not provide Z-samples, and are therefore not subject to the logic of this section, which applies to the continuous section of the curve.

17.4.5 The *pdfs* for the ROC decision variable

In **Chapter 6** the *pdf* functions for non-diseased and diseased cases for the unequal variance binormal ROC model[1] were derived. The procedure was to take the derivative of the appropriate cumulative distribution function (CDF) with respect to ζ. An identical procedure is used for the RSM. The *pdf* corresponding to non-diseased cases is given by (the *CDF* for non-diseased cases is the complement of *FPF* see Equation 17.8):

$$\mathrm{pdf}_N(\zeta) = \frac{\partial}{\partial \zeta}\exp\left(-\frac{\lambda'}{2}\left[1 - erf\left(\frac{\zeta}{\sqrt{2}}\right)\right]\right) \tag{17.20}$$

Using Maple™, this evaluates to

$$\mathrm{pdf}_N(\zeta) = \frac{\lambda' e^{-\frac{1}{2}\zeta^2} e^{-\frac{\lambda'}{2}\left[1 - erf\left(\frac{\zeta}{\sqrt{2}}\right)\right]}}{\sqrt{2\pi}} \tag{17.21}$$

Similarly, for the diseased cases,

$$\text{pdf}_D(\zeta) = \frac{\partial}{\partial \zeta}\left(1 - \frac{\nu'}{2} + \frac{\nu'}{2}erf\left(\frac{\zeta - \mu}{\sqrt{2}}\right)\right)^L e^{\left(-\frac{\lambda'}{2} + \frac{\lambda'}{2}erf\left(\frac{\zeta}{\sqrt{2}}\right)\right)} \tag{17.22}$$

Maple™ does evaluate the derivative, but it is cumbersome to display; it is shown in Online Appendix 17.C. The formula for the RSM-predicted ROC curves and the *pdfs* of non-diseased and diseased cases, are coded in **RJafroc** in function **PlotRsmOperatingCharacteristics()**. That the integrals of the *pdfs* are given by (non-diseased followed by diseased):

$$\int_{-\infty}^{\infty} \text{pdf}_N(\zeta)d\zeta = \exp\left(-\frac{\lambda'}{2}\left[1 - erf\left(\frac{\zeta}{\sqrt{2}}\right)\right]\right)\Bigg|_{-\infty}^{\infty} = 1 - \exp(-\lambda') = FPF_{\max} \tag{17.23}$$

$$\left.\begin{aligned} \int_{-\infty}^{\infty} \text{pdf}_D(\zeta)d\zeta &= \left(1 - \frac{\nu'}{2} + \frac{\nu'}{2}erf\left(\frac{\zeta - \mu}{\sqrt{2}}\right)\right)^L e^{\left(-\frac{\lambda'}{2} + \frac{\lambda'}{2}erf\left(\frac{\zeta}{\sqrt{2}}\right)\right)}\Bigg|_{-\infty}^{\infty} \\ &= 1 - \exp(-\lambda')(1 - \nu')^L \\ &= TPF_{\max} \end{aligned}\right\} \tag{17.24}$$

In other words, they evaluate to the coordinates of the theoretical end-point, each of which is less than or equal to one. The reason is that not every case is marked. The probability of no marks on non-diseased cases is $1 - FPF_{\max}$, and that on diseased cases is $1 - TPF_{\max}$. If one restricts to cases with at least one mark, then each of the *pdfs* would integrate to unity but this amounts, effectively, to not counting cases with no marks. One can avoid not counting cases with no marks *and* have *pdfs* that integrate to unity if one conditions on cases with at least one mark, and subsequently multiplies by the appropriate probability of the case having at least one mark. Alternatively, one can include the probability of having a mark in the definition of the *pdf*, as was done above, in which case the *pdfs* will not integrate to unity. The author is aware that pdfs that do not integrate to unity present problems to some researchers, and the original RSM publications[13,14] unnecessarily introduced Dirac delta functions to artificially force the integrals to be unity. Further comments on the constrained end-point property are to be found in the Discussion section, Section 17.12.1, in response to Dr. Wagner's review. The reason for the emphasis on the finite end-point property is that in the author's experience it presents a conceptual difficulty to researchers exposed to over six-decades of learning that ROC curves must extend to (1,1).

17.4.6 RSM-predicted ROC-AUC and AFROC-AUC

It is possible to perform the integration under the RSM-ROC curve to get RSM-ROC-AUC, denoted $AUC_{RSM}^{ROC}\left(\mu, \lambda, \nu, \overrightarrow{f_L}\right)$:

$$AUC_{RSM}^{ROC}\left(\mu, \lambda, \nu, \overrightarrow{f_L}\right) = \int_0^1 TPF\left(\zeta; \mu, \lambda, \nu, \overrightarrow{f_L}\right)d\left(FPF\left(\zeta; \lambda\right)\right) \tag{17.25}$$

Since the RSM predicts other operating characteristics, the superscript is needed to keep track of the specific operating characteristic. The right-hand side can be evaluated using numerical integration using the **integrate()** function of **R**. The code implementing it is in **UtilAucsRSM.R**. This function returns ROC-AUC and AFROC-AUC as a **list** object. The calculation of AFROC-AUC is described in Section 17.7. The procedure is illustrated in file **mainRsmAuc.R**, a listing of which follows.

17.4.6.1 Code listing

```
rm(list = ls()) # mainRsmAuc.R
library(RJafroc)

seed <- 1;set.seed(seed)
mu <- 3; lambda <- 1; nu <- 1
ret <- UtilConvertIntrinsic2PhysicalRSM(
    mu, lambda, nu)
lambdaP <- ret$lambdaP;nuP <- ret$nuP
Lmax <- 3;K2 <- 700
Lk2 <- floor(runif(K2, 1, Lmax + 1))
nLesPerCase <- unique(Lk2)
lesionDist <- array(dim = c(length(nLesPerCase), 2))
for (i in nLesPerCase)
    lesionDist[i, ] <- c(i, sum(Lk2 == i)/K2)
aucs <- UtilAucsRSM(
    mu, lambdaP = lambdaP,
    nuP = nuP, lesionDistribution = lesionDist)
cat("mu = ", mu,
    "\nlambda = ", lambda,
    "\nnu = ", nu,
    "\nLmax = ", Lmax,
    "\nRSM AUC/ROC = ", aucs$aucROC,
    "\nRSM AUC/AFROC = ", aucs$aucAFROC,
    "\n")
```

Line 5 initializes the intrinsic search model parameters, lines 6–7 converts them to physical parameters, and the number of diseased cases, line 9, is deliberately set to a large value so that small numbers do not limit the precision of the lesion distribution vector $\overline{f_L}$. This detail should not be fretted over, but to be clear, one is not simulating 700 diseased cases; after all, this chapter is a about theoretical predictions of the RSM, which are independent of sampling effects. **Source** the code. Lines 9–12 generate the histogram of the number of lesions per case, i.e., the lesion distribution vector defined in Section 17.3.3. For example, with **Lmax** = 3 (i.e., the maximum number of lesions per diseased case is 3) one gets: the following code.

```
> lesionDist
     [,1] [,2]
[1,]   1  0.3057143
[2,]   2  0.3685714
[3,]   3  0.3257143
```

This implies that fraction 0.3057 of diseased cases has one lesion, fraction 0.3686 of diseased cases has two lesions, and a fraction 0.3257 of diseased cases has three lesions. One can confirm that these fractions sum to unity. Lines 15–18 call the function **UtilAucsRSM()** which returns the two AUCs.

Source **mainRsmAuc.R** repeatedly with appropriately entered parameter values at line 5 to obtain the values listed in Table 17.1. The relevant AUCs are in the last two columns. Examination of Table 17.1 reveals the following points.

- Both AUCs are *decreasing* functions of λ, i.e., increasing λ results in more latent NLs and decreases performance. The main reason for this is that with increasing numbers of latent NLs, the mean of the non-diseased pdf moves to the right, thereby decreasing the effective difference between the two peaks. Since the pdf of the diseased distribution is also shifted to the right, this is a relatively weak effect (0.802–0.715, for ROC-AUC, as λ increases from 0.1 to 1).
- Both AUCs are *increasing* functions of ν, i.e., increasing ν results in more latent LLs, which increases performance (more lesions are *hit*, to use eye-tracking terminology, and therefore more likely to be marked, and the diseased distribution *pdf* moves to the right). This has a relatively strong effect on performance (0.575–0.794, for ROC-AUC, as ν increases from 0.25 to 2).
- Both AUCs are *increasing* functions of μ, i.e., increasing *perceptual signal-to-noise ratio* (*pSNR*) always leads to improved performance. For background on this fundamental dependence the reader is referred back to the solar analogy in **Chapter 12**. Increasing μ increases the separation between the two *pdfs*. Furthermore, the number of NLs decreases because λ' decreases: $\lambda' = \lambda/\mu$ and this increases performance. Finally, ν' increases and approaches unity: $\nu' = 1 - \exp(-\mu\nu)$. Because all three effects reinforce each other, a change in μ results in a large effect on performance (e.g., 0.715–0.970, for ROC-AUC, as μ increases from 1 to 3 for one lesion per case).
- ROC-AUC increases with L_{max}, since with more lesions per case there is increased probability that that at least one of them will be found, i.e., will be a latent LL and therefore more likely to be marked, and the diseased distribution *pdf* moves to the right (0.715 < 0.767 < 0.808, for ROC-AUC, for one to three lesions per case). However, AFROC-AUC is independent of L_{max}, because the y-axis is LLF, which is independent of the number of lesions in the dataset.
- ROC-AUC values are constrained to the range 0.5–1 while the AFROC-AUC values are constrained to the larger range 0–1 (try setting **mu** = 0.001 to be convinced of this; setting mu to zero will cause a divide by zero error). The reader should confirm that the difference in ROC-AUCs between any two rows in Table 17.1 is smaller than the corresponding difference in AFROC-AUCs. *The AFROC effect-size is always larger than the corresponding ROC effect-size.*

Table 17.1 Dependence of RSM predicted ROC and AFROC AUCs on model parameters.

μ	λ	ν	L_{max}	AUC_{RSM}^{ROC}	AUC_{RSM}^{AFROC}
1				0.71483	0.57789
2	1			0.90266	0.87402
3		1		0.97000	0.96282
	0.5		1	0.75511	0.67603
	0.1			0.80159	0.78375
		2		0.79386	0.72281
1		0.5		0.63372	0.42916
	1	0.25		0.57518	0.32180
		1	2	0.76710	0.57789
				0.80840	0.57789
			3		
3	1	1		0.98974	0.96282

Note: The last two columns list the RSM-predicted AUCs under the ROC and AFROC, respectively. The corresponding RSM parameters are listed in the first four columns.

For example, taking the differences between the first two rows, difference ROC-AUC = −0.188 while AFROC-AUC = −0.296. This is one reason why AFROC-AUC yields greater statistical power than ROC-AUC.

Exercise: The author leaves it as an exercise for the reader to *randomly* simulate pairs of values of sets of search model parameters (all three parameters have to be positive and L_{max} has to be an integer) and confirm that the AFROC effect-size is *always* larger than the ROC effect-size. Actually, this is not always true, but the exercise will help the reader better understand the RSM. Try counting the fraction of simulations where the AFROC effect-size is *smaller* than the ROC effect-size.

17.5 RSM-predicted ROC and *pdf* curves

So far the focus has been on the ROC and AFROC AUCs. The file **mainRsmPdfRoc.R**, a listing of which is in Online Appendix 17.D.2, displays *pdf* and ROC curves, in that order, for different values of μ. Line 5 defines **muArr** containing the six values of μ: 0.001, 1, 2, 3, 4, and 5. The remaining *intrinsic* RSM model parameters are defined as: $\lambda = \nu = L_{max} = 1$. The appropriate value of **mu** inside the for-loop is selected at line 7. The function **PlotRsmOperatingCharacteristics()** is used at lines 8 through 11. The option, line 10, **type = ALL** generates all plots (ROC, AFROC, wAFROC, FROC, and *pdfs*). In this section, only the *pdf* (lines 13–22) and ROC (lines 25–48) plots are used. These are continuous parametric plots, unlike the empirical plots generated from binned data in **Chapter 13**. The command **str(ret1)** reveals, near the top of the long output, that each plot contains 1301 closely spaced points. **Source** the file **mainRsmPdfRoc.R**. One sees the following output in the **Console** window, the ROC curves shown in Figure 17.1a through f and the *pdfs* shown in Figure 17.2a through f.

17.5.1 Code output

```
> source(...)
mu =   0.001
lambda =   1
nu =   1
AUC =   0.5000003
fpfMax =   1
tpfMax =   1
mu =   1
lambda =   1
nu =   1
AUC =   0.7148289
fpfMax =   0.6316236
tpfMax =   0.8644745
mu =   2
lambda =   1
nu =   1
AUC =   0.9026559
fpfMax =   0.3930598
tpfMax =   0.9178594
mu =   3
```

```
lambda =   1
nu =   1
AUC =   0.970004
fpfMax =   0.2831462
tpfMax =   0.96431
mu =   4
lambda =   1
nu =   1
AUC =   0.9902785
fpfMax =   0.2209363
tpfMax =   0.985731
mu =   5
lambda =   1
nu =   1
AUC =   0.9965944
fpfMax =   0.1810482
tpfMax =   0.9944819
```

In Section 17.5.1 **fpfMax** = FPF_{max} and **tpfMax** = TPF_{max} correspond to the theoretical end-point of the ROC. The graphs in the **Plots** window are ordered as (*pdf*, ROC) pairs, in order of increasing μ values (the left-most plot is the *pdf* corresponding to μ = 0.001, click right arrow to see the corresponding ROC, click right arrow to see the *pdf* for next higher value of μ, etc.).

In the *ROC* curves shown in Figure 17.1a through f the solid part is the continuous section and the dashed part, when shown, is the inaccessible part. The corresponding *pdf* curves are shown in Figure 17.2a through f for non-diseased (solid line) and diseased cases (dotted line).

Question for reader: for μ = 0.001, why is the ROC plot a straight line connecting the origin to (1,1); and why is there no dotted line on either the ROC or the *pdf*?

The first five members of **ret1**, line 8, contain the plots (**ROC, AFROC, wAFROC, FROC**, and **pdf**).* Each of these is a huge **ggplot** object, difficult to list, but each plot object is easily viewed by using the **print()** function. The remaining elements return AUCs under different operating characteristics. The following code snippet applies to mu = 0.001; insert a break point at line 13, source the code, stop debug mode and enter the shown commands in the Console window followed by Enter.

17.5.2 Code snippet

```
> ret1$aucROC
[1] 0.5000003
> ret1$aucAFROC
[1] 1.002317e-06
> ret1$aucwAFROC
[1] 1.002317e-06
> ret1$aucFROC
[1] 0.4986837
```

---

* Because each **ggplot** object is huge, this can be hard to see: try typing **ret1$** in the **Console** window—one should see the different plot options in a pop-up window provided by **RStudio**. The top five elements in the pop-up window are **ROCPlot, AFROCPlot, wAFROCPlot, FROCPlot**, and **PDFPlot**.

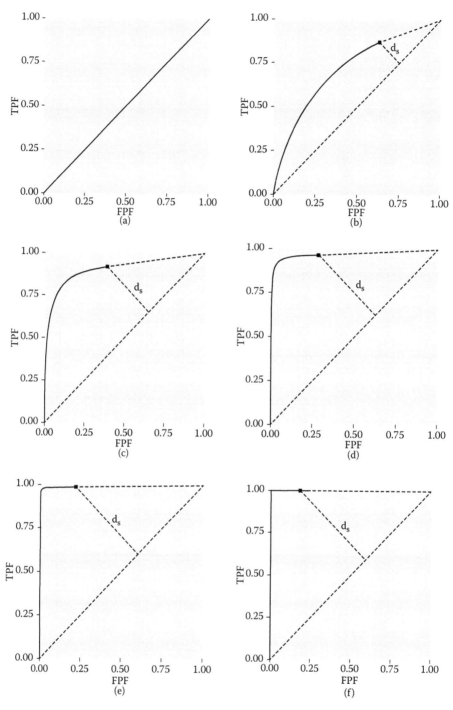

Figure 17.1 RSM-predicted ROC curves for following values of the μ parameter: (a) 0.001, (b) 1, (c) 2, (d) 3, (e) 4, (f) 5. Other parameters are $\lambda = \nu = L_{max} = 1$. The solid line is the continuously accessible portion of the ROC. Notice the transition, as μ increases, from chance level performance (a) to almost perfect performance (f) and the end-point moves from (1,1) towards (0,1). The area under the ROC curve includes that under the dashed line connecting the end-point to (1,1), which credits unmarked non-diseased cases. If this area is not included, a severe underestimate of performance can occur, especially for large μ. The length d_s of the perpendicular line from an end-point and ending on the chance diagonal is a measure of search performance, explained in Section 17.8. These plots were generated by `mainRsmPdfRoc.R`.

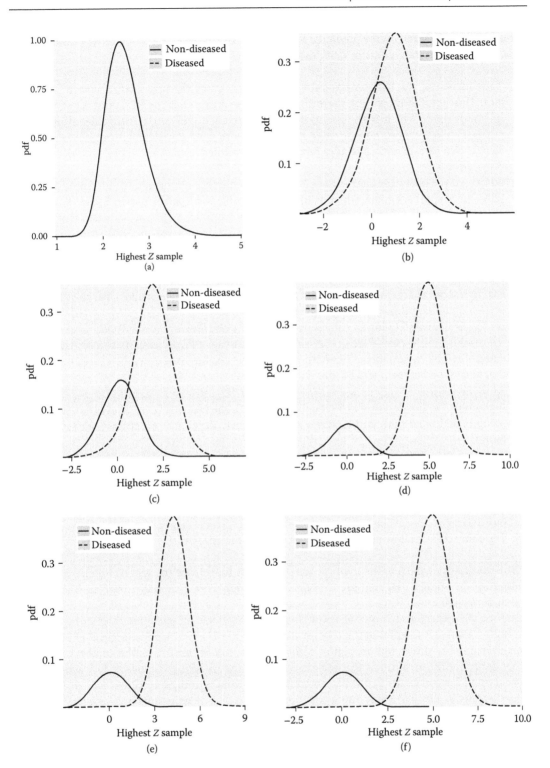

Figure 17.2 RSM-predicted ROC *pdf* curves for following values of the μ parameter: (a) 0.001, (b) 1, (c) 2, (d) 3, (e) 4, (f) 5. Other parameters: $\lambda = \nu = L_{max} = 1$. The solid line is the *pdf* of the highest rating on non-diseased cases; the dotted line is the *pdf* of the highest rating on diseased cases. Due to overlap, the dotted line in not visible in plot (a). Because of cases with no latent marks, the areas under each individual curve are less than unity; see text for explanation. These plots were generated by **mainRsmPdfRoc.R**.

For now, focus attention on the predicted ROC curves, in Figure 17.1a through f. Try different values for the parameters to confirm that the following statements are true.

1. As μ increases, the ROC curve more closely approaches the upper-left corner of the ROC plot. This signifies increasing performance and the area under the ROC and AFROC curves approach unity, which is the best possible performance. FPF_{max} decreases, denoting greater numbers of unmarked non-diseased cases, i.e., more good decisions on non-diseased cases. TPF_{max} increases, denoting smaller numbers of unmarked lesions, i.e., more good decisions on diseased cases.
2. For $\mu \rightarrow 0$ the operating characteristic approaches the chance diagonal and the area under the ROC curve approaches 0.5, which is the worst possible ROC performance, Figure 17.1a.
3. The area under the ROC increases monotonically from 0.5 to 1 as μ increases from zero to infinity.
4. For $\mu \rightarrow \infty$ the accessible portion of the operating characteristic approaches the vertical line connecting (0,0) to (0,1), the area under which is zero. The complete ROC curve is obtained by connecting this point to (1,1) by the dashed line and in this limit the area under the complete ROC curve approaches unity. *Omitting the area under the dashed portion of the curve will result in a severe underestimate of true performance.*
5. As L_{max} is increased (allowed values are 1, 2, 3, etc.), the area under the ROC curve increases, approaching unity and TPF_{max} approaches unity while FPF_{max} remains constant at a value determined by λ, Equation 17.1. With more lesions per diseased case, the chances are higher that at least one of them will be found and marked.
6. As λ decreases, FPF_{max} decreases to zero and TPF_{max} decreases approaching $1 - \exp(\mu \nu L)$, Equation 17.2. The decrease in TPF_{max} is in line with the fact that there is less chance of an NL being rated higher than an LL, and one is completely dependent on at least one lesion being found.
7. As ν increases FPF_{max} stays constant at the value determined by λ and μ, Equation 17.1, while TPF_{max} approaches unity. The corresponding physical parameter $\nu' = 1 - \exp(-\nu \mu)$ increases approaching unity, guaranteeing every lesion will be found.
8. The ROC curve is always proper (no matter what the parameter values, as long as each is positive).
9. The code will not allow negative values of the parameters μ, λ, ν.

The meanings of the lengths d_s of the perpendicular lines from the respective end-points and ending on the chance diagonal are explained in Section 17.8; essentially, they are measures of search performance.

Figure 17.2a through f shows *pdf* plots for the same values of parameters as in Figure 17.1a through f. Consider the plot of the *pdfs* for $\mu = 1$ in Figure 17.2b. Since the integral of a *pdf* function over a defined interval amounts to counting events occurring in the interval, it should be evident that the area under the non-diseased *pdf* equals FPF_{max}, Equation 17.1, which *is less than unity*. This can create problems understanding what is going on[16] (*how can a pdf not integrate to unity?*). The *pdf* would integrate to unity if each case generated a Z-sample. However, if the radiologist finds nothing suspicious in the case, what are they supposed to do—pick a location at random? They do not work that way and even designer level CAD algorithms[17] do not generate marks on every case.* Unit area under the *pdf* can happen only if every case generates a Z-sample, and for experts there is a substantial fraction of non-diseased cases not generating Z-samples.

For the chosen value $\lambda = 1$ one has $FPF_{max} = 1 - e^{-\lambda/\mu} = 1 - e^{-1} = 0.632$. The area under the *pdf* is less than unity because it is missing the contribution of non-diseased cases with zero marks,

---

* Here is the relevant quotation from the cited reference: "*In fact, about 12% of the images in our CAD data sets had no marks. This reflects the fact that these are expert algorithms, such as could be used in the clinic.*"

the probability of which is $e^{-1} = 0.368$. Likewise, the area under the diseased *pdf* equals TPF_{max}, Equation 17.2, which is also less than unity. For the chosen values of $\mu = \lambda = \nu = L_{max} = 1$, it equals $TPF_{max} = 1 - e^{-1}e^{-1} = 0.865$; this area is somewhat larger than that under the non-diseased *pdf*, as is evident from visual examination of Figure 17.2b. In other words, a greater fraction of diseased cases generates marks than do non-diseased cases, which is consistent with the presence of lesions in diseased cases. The complement of 0.865 is due to diseased cases with no marks, which account for a fraction 0.135 of diseased cases. To summarize, the *pdfs* do not integrate to unity for the reason that there are cases with zero marks that do not generate Z-samples.

The plot in Figure 17.2a may be surprising. Since it corresponds to a very small value of μ, one may expect both *pdfs* to overlap and to be centered at *zero*. Instead, while they do overlap, the shape is definitely non-Gaussian and it is centered at approximately 3.2. This is because the small value of μ results in a large value of the λ' parameter, since $\lambda' = \lambda / \mu$. The highest of a large number of samples from the unit normal distribution is not normal and is peaked at a value above zero. Fisher and Tippett[18,19] investigated this distribution in 1928, since generalized to extreme value distributions.[20]

17.6 The RSM-predicted FROC curve

The derivation of the FROC curve is much simpler. From the property of the Poisson distribution, namely, its mean is the λ' parameter of the distribution, it follows that the expected number of latent NLs per case is λ'. To obtain NLF, one multiplies this parameter by $P(Z > \zeta | Z \sim N(0,1))$, i.e., $1 - \Phi(\zeta)$. Therefore, the expected number of latent NLs per case that are actually marked is:

$$NLF(\lambda',\zeta) = \lambda'\left(1 - \Phi(\zeta)\right) = \frac{\lambda}{\mu}\left(1 - \Phi(\zeta)\right) \tag{17.26}$$

Calculation of LLF with variable numbers of lesions per case: diseased cases are separated into groups, each with a fixed number of lesions L per case, where L varies from one to L_{max}. For each group characterized by L, one seeks the fraction of the expected number of latent LLs per case divided by the total number of lesions in each case in the group (L). Since ν' is the probability that a lesion is found, it must equal the desired fraction. Next, one multiplies by $P(Z > \zeta | Z \sim N(\mu,1))$ i.e., $1 - \Phi(\zeta - \mu)$, to obtain the fraction that is actually marked. Finally, one performs a weighted summation over the different groups with f_L as the weighting fraction. Therefore,

$$LLF(\zeta;\mu,\nu',\overrightarrow{f_L}) = \sum_{L=1}^{L_{max}} f_L \nu'\left(1 - \Phi(\zeta - \mu)\right) = \nu'\left(1 - \Phi(\zeta - \mu)\right) = \left(1 - \exp(-\nu\mu)\right)\left(1 - \Phi(\zeta - \mu)\right) \tag{17.27}$$

Note that $LLF(\zeta;\mu,\nu',\overrightarrow{f_L})$ is independent of $\overrightarrow{f_L}$. Summarizing, the coordinates of the RSM-predicted point on the FROC curve are given by

$$\left.\begin{aligned} NLF(\zeta) &= \lambda'\left[1 - \Phi(\zeta)\right] = \lambda'\Phi(-\zeta) \\ LLF(\zeta) &= \nu'\left[1 - \Phi(\zeta - \mu)\right] = \nu'\Phi(\mu - \zeta) \end{aligned}\right\} \tag{17.28}$$

In terms of the intrinsic parameters, the FROC curve is given by

$$\left.\begin{aligned} NLF(\zeta) &= \frac{\lambda}{\mu}\left[1 - \Phi(\zeta)\right] \\ LLF(\zeta) &= \left(1 - \exp(-\nu\mu)\right)\left[1 - \Phi(\zeta - \mu)\right] \end{aligned}\right\} \tag{17.29}$$

The FROC curve starts at $(0,0)$ and ends at (λ', v'); the x-coordinate does not extend to arbitrarily large values and the y-coordinate does not approach unity (unless $v' = 1$, which implies easy to find lesions). The constrained end-point property, demonstrated before for the ROC curve, also applies to the FROC curve:

$$\left.\begin{aligned} NLF_{max} &= \lambda' = \lambda / \mu \\ LLF_{max} &= v' = 1 - \exp(-v\mu) \end{aligned}\right\} \tag{17.30}$$

Source the file **mainRsmPlotsFROC.R**. Note the change at line 7 with **type = FROC**, which generates *FROC* plots shown in Figure 17.3a through f for the following values of μ: 0.001, 1, 2, 3, 4 and 5.

17.6.1 Code listing

```
rm(list = ls()) # mainRsmPlotsFROC.R
library(RJafroc)
muArr <- c(0.001,1,2,3,4,5)
lambda <- 1;nu <- 1; Lmax <- 1
for (i in 1:length(muArr)) {
  mu <- muArr[i]
  ret <- PlotRsmOperatingCharacteristics(
    mu, lambda, nu,
    type = "FROC", lesionDistribution = Lmax,
    llfRange = c(0,1))
  frocPlot <- ret$FROCPlot +
    scale_color_manual(values = "black") +
    theme(axis.title.y = element_text(size = 25,face="bold"),
          axis.title.x = element_text(size = 30,face="bold"))
  frocPlot$layers[[1]]$aes_params$size <- 2 # line
  print(frocPlot)
  cat("mu = ", mu,
      "\nlambda = ", lambda,
      "\nnu = ", nu,
      "\nAUC = ", ret$aucFROC,
      "\nnlfMax = ", max(ret$FROCPlot$data$NLF),
      "\nllfMax = ", max(ret$FROCPlot$data$LLF),"\n")
}
```

There are other subtle changes in the code relative to the previous ROC/*pdf* code listing; at line 11 one uses **ret$FROCPlot** instead of **ret$ROCPlot**. The reasons for this, and the other changes can be appreciated by examining the structure of **ret**. This is left to the reader.

17.6.2 Code output

The listed AUC's are the total areas under the FROC curves. As μ increases, starting from just above zero, the total area under the FROC curve *decreases* from 0.5 to $\approx \lambda / \mu = 1/5 = 0.2$. The limiting values can be understood from the predicted end-point $(\lambda', v') = (\lambda / \mu, 1 - \exp(-v\mu))$. In the limit $\mu \to 0+$, the end-point is to the far-right bottom corner of the plot at $(\lambda / \mu, v\mu)$, that is, $(1000, 1/1000)$. The area of the triangle connecting the origin to this point is $0.5(\lambda / \mu)(v\mu) = 0.5\lambda v = 0.5$. The predicted value 0.5 is due to particular choices $\lambda = 1; v = 1$, that is, *there is nothing special about it*.

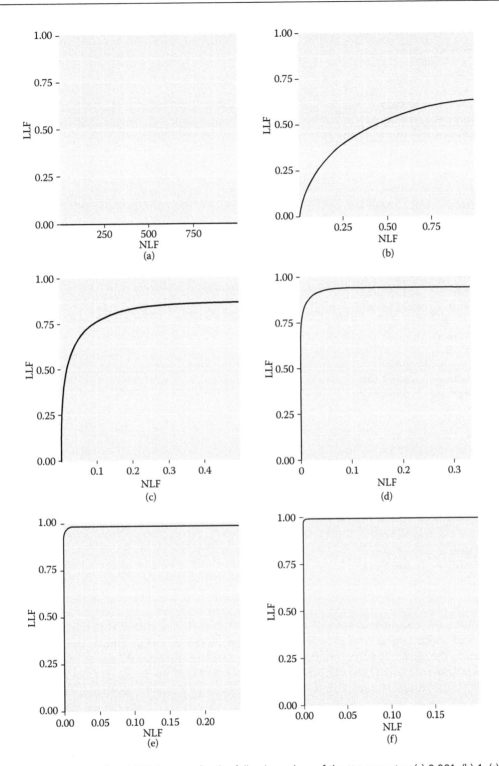

Figure 17.3 RSM-predicted FROC curves for the following values of the μ parameter: (a) 0.001, (b) 1, (c) 2, (d) 3, (e) 4, (f) 5. Other parameters: $\lambda = \nu = L = 1$. As μ increases the curve approaches the top left corner, in the limit it is the vertical line connecting the origin to (0,1). Notice the wide range of variation of the *x*-axis scaling; in (a) it ranges from 0 to 100 while in (f) it ranges from 0 to 0.2. The total area under the FROC curve actually decreases as μ increases. *Because it is not contained within the unit square, the FROC cannot be used as the basis of a meaningful figure of merit.* These plots were generated by **mainRsmPlotsFROC.R**.

```
> source(...)
mu =  0.001
lambda =  1
nu =  1
AUC =  0.4986837
nlfMax =  998.6501
llfMax =  0.0009981554
mu =  1
lambda =  1
nu =  1
AUC =  0.4797164
nlfMax =  0.9986501
llfMax =  0.6321005
mu =  2
lambda =  1
nu =  1
AUC =  0.397746
nlfMax =  0.4993251
llfMax =  0.8646645
mu =  3
lambda =  1
nu =  1
AUC =  0.3109422
nlfMax =  0.3328834
llfMax =  0.9502129
mu =  4
lambda =  1
nu =  1
AUC =  0.2445158
nlfMax =  0.2496625
llfMax =  0.9816844
mu =  5
lambda =  1
nu =  1
AUC =  0.1983476
nlfMax =  0.19973
llfMax =  0.9932621
```

For $\mu = 5$, the end-point is at $\left(\lambda/\mu, 1 - \exp(-\nu\mu)\right) \approx \left(\lambda/\mu, 1\right)$. The area of the rectangle with unit height and width λ/μ is $\lambda/\mu = 1/5 = 0.2$. Since the curve starts with infinite slope and is concave down, the rectangle will overestimate the true AUC. In the general case, as the μ parameter increases from zero to the final value μ, the area under the FROC curve ranges from $\lambda\nu/2$ to λ/μ. Since $\nu \geq 0$, the area increases with μ if $\lambda\nu/2 < \lambda/\mu$, that is, $\mu\nu < 2$, and otherwise, it will decrease. In the example just detailed, the product $\mu\nu$ happened to be five, which is why the area decreased. With a different choice of parameters, the area could have increased. *The parameter-dependent behavior of the area under the FROC curve, as μ increases, argues strongly that it is an inappropriate measure of performance.* Unlike FROC, as μ increases, the area under the predicted ROC increased from 0.5 to 1, independent of the other parameters in the model. A similar property is demonstrated below for the predicted area under the AFROC, except the range of variation is double that of the ROC.

One could argue that one should not be using the complete area under the predicted FROC. First, since observers do not reach the end-point, it cannot be estimated. Second, a partial area measure that only integrates up to a fixed value of NLF, which one denotes NLF_α, may be more appropriate.[21,22] One problem with this approach, proposed by FDA investigators, is that the choice of NLF_α is arbitrary, which opens the system to gaming. Moreover, the measure does not behave reasonably as μ increases. As μ increases, there will come a point at which one cannot interpolate the value of LLF @ NLF_α. This occurs for the theoretical curve when the following condition is satisfied (see Equation 17.30 for explanation):

$$\lambda / \mu = NLF_\alpha \tag{17.31}$$

As μ increases, the above equality will eventually be satisfied; for smaller values data exists to the right of the desired NLF value, making interpolation possible. For larger μ the FROC curve does not extend far enough to the right to allow the AUC calculation to be performed. This is actually a practical problem. Experts tend to not traverse as far to the right of the FROC as non-experts. Using a partial area measure, one would be unable to calculate AUC for experts (if NLF_α were chosen too large) or be unable to use all the data from non-experts (if NLF_α were chosen too small). However, perhaps the most important reason for not using the partial area under the FROC is the fact that this curve is defined by the marks. Therefore, unmarked non-diseased cases, which represent perfect decisions, are not properly accounted for.

Problems with the FROC are resolved if one uses the complete area under the AFROC curve, to which one now turns.

17.7 The RSM-predicted AFROC curve

The AFROC x-coordinate is the same as the ROC x-coordinate and Equation 17.8 applies.

$$FPF(\zeta) = 1 - \exp\left(-\frac{\lambda}{2\mu}\left[1 - erf\left(\frac{\zeta}{\sqrt{2}}\right)\right]\right) \tag{17.32}$$

The AFROC y-coordinate is identical to the FROC y-coordinate and the second Equation 17.29 applies.

$$LLF(\zeta) = v'[1 - \Phi(\zeta - \mu)] = [1 - \exp(-v\mu)][1 - \Phi(\zeta - \mu)] \tag{17.33}$$

The second expression on the right-hand side uses the intrinsic RSM parameters. Note that the expression is independent of the number of lesions in the dataset or their distribution (unlike the AFROC, the weighted AFROC does depend on the lesion distribution $\vec{f_L}$). Since $erf(-\infty) = -1$ the limiting coordinates of the AFROC are

$$\left.\begin{array}{l} FPF_{\max} = 1 - e^{-\lambda'} \\ \\ LLF_{\max} = v' \end{array}\right\} \tag{17.34}$$

In other words, unlike previous predictions,[12,23,24] the AFROC does not extend to (1,1). It too has the constrained end-point property. In terms of the intrinsic RSM parameters,

$$\left.\begin{array}{l} FPF_{\max} = 1 - e^{-\lambda/\mu} \\ \\ LLF_{\max} = 1 - \exp(-v\mu) \end{array}\right\} \tag{17.35}$$

As μ increases starting from 0+, $FPF_{\max}$ decreases starting from 1−, approaching 0+ as μ approaches +∞. In the same limit, $TPF_{\max}$ increases starting from 0, approaching 1−. (The notation 1− denotes a number just less than one.)

Source the file **mainRsmAFROC.R**. Note the change at line 9, **type = AFROC**, which generates *AFROC* plots shown in Figure 17.4a through f for the following values of μ: 0.001, 1, 2, 3, 4 and 5.

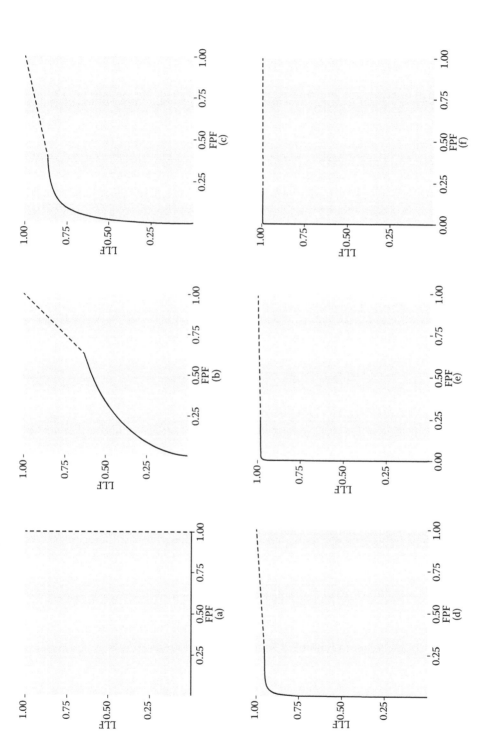

Figure 17.4 RSM-predicted AFROC curves for following values of the μ parameter: (a) 0.001, (b) 1, (c) 2, (d) 3, (e) 4, (f) 5. Other parameters: $\lambda = \nu = L_{max} = 1$. As μ increases, AFROC-AUC increases; the curve increasingly approaches the top left corner, followed by an inaccessible dashed linear extension to (1,1). Unlike FROC plots, these plots are completely contained within the unit square, which makes it easy to define figures of merit, for example, the area under the AFROC, including the area under the straight-line extension. These plots were generated by `mainRsmAFROC.R`.

17.7.1 Code listing

```
rm(list = ls()) # mainRsmPlotsAfroc.R
library(RJafroc)
muArr <- c(0.001,1,2,3,4,5)
lambda <- 1;nu <- 1; Lmax <- 10
for (i in 1:length(muArr)) {
  mu <- muArr[i]
  ret <- PlotRsmOperatingCharacteristics(
    mu, lambda, nu,
    type = "AFROC",
    lesionDistribution = Lmax)
  afrocPlot <- ret$AFROCPlot
  print(afrocPlot)
  cat("mu = ", mu,
      "\nlambda = ", lambda,
      "\nnu = ", nu,
      "\nAUC = ", ret$aucAFROC,
      "\nfpfMax = ", max(ret$AFROCPlot$data$FPF),
      "\nllfMax = ", max(ret$AFROCPlot$data$LLF),"\n")
}
```

The corresponding AFROC plots are shown in Figure 17.4a through f.

17.7.2 Code output

```
> source('~/onlinebookk21778/Ch17/software/mainRsmPlotsAFROC.R')
mu =  0.001
lambda =  1
nu =  1
AUC =  1.002317e-06
fpfMax =  1
llfMax =  0.0009981554
mu =  1
lambda =  1
nu =  1
AUC =  0.577889
fpfMax =  0.6316236
llfMax =  0.6321005
mu =  2
lambda =  1
nu =  1
AUC =  0.8740216
fpfMax =  0.3930598
llfMax =  0.8646645
mu =  3
lambda =  1
nu =  1
AUC =  0.9628192
fpfMax =  0.2831462
```

```
llfMax  =   0.9502129
mu  =   4
lambda  =   1
nu  =   1
AUC  =   0.9882477
fpfMax  =   0.2209363
llfMax  =   0.9816844
mu  =   5
lambda  =   1
nu  =   1
AUC  =   0.9959835
fpfMax  =   0.1810482
llfMax  =   0.9932621
```

As μ increases, the area under the AFROC increases monotonically from zero to one. The reader should check that this is true regardless of the choices of the other parameters in the model. This is expected of a well-behaved area measure that can be used as a figure of merit.

Experiment with different values for the parameters to confirm that the following statements are true:

1. The AFROC plot is independent of the number of lesions per case. Area under AFROC is independent of L_{max}. These statements are not true for the wAFROC. In contrast, the ROC ordinate increases with increasing numbers of lesions per case.
2. From Equations 17.2 and 17.30, it follows that $TPF_{max} \geq LLF_{max}$ with the equality holding in the limit $\mu \to \infty$, as shown in following Eqn. 17.36.

$$
\left.
\begin{aligned}
TPF_{max} &= 1 - e^{-\lambda/\mu}\exp(-\nu\mu L) \geq 1 - e^{-\lambda/\mu}\exp(-\nu\mu) \geq 1 - \exp(-\nu\mu) \\
LLF_{max} &= 1 - \exp(-\nu\mu) \\
\therefore\ TPF_{max} &\geq LLF_{max}
\end{aligned}
\right\}
\tag{17.36}
$$

Compare corresponding TPF_{max}, LLF_{max} values in Section 17.5.1 and Section 17.7.2: at $\mu = 0.001$, the corresponding values are 1 which is greater than 0.001 and at $\mu = 5$ they are 0.9945 which is greater than 0.9933.

The physical reason for $TPF_{max} \geq LLF_{max}$ is that the ROC gives credit for incorrect localizations on diseased cases, while the AFROC does not. This is the famous right for wrong reason argument[15] originally advanced by Bunch et al. For large μ, such events are less likely.

3. As μ increases the AFROC curve more closely approaches the upper left corner of the plot, denoting increasing performance and the area under the AFROC curve approaches one, which is the best possible performance: (a) FPF_{max} decreases and (b) LLF_{max} increases, denoting decreasing numbers of incorrect decisions on non-diseased and diseased cases, respectively.
4. For $\mu \to 0$ and non-zero λ the operating characteristic approaches the horizontal line extending from the origin to (1,0), which is the continuous section of the curve, followed by the vertical dashed line connecting (1,0) to (1,1) and AFROC-AUC approaches zero. In this limit,

none of the lesions are localized and every case has at least one NL mark, which implies worst possible performance.

5. For $\mu \to \infty$ the accessible portion of the operating characteristic approaches the vertical line connecting (0,0) to (0,1), the area under which is zero. The complete AFROC curve is obtained by connecting this point to (1,1) by the dashed line and in this limit the area under the complete ROC curve approaches one. *As with the ROC, omitting the area under the dashed portion of the curve will result in a severe underestimate of true performance.*

6. As λ decreases, FPF_{max} decreases to zero while LLF_{max} stays constant, Equation 17.35, as the latter is independent of λ.

7. As ν increases, FPF_{max} stays constant (it is independent of ν) while LLF_{max} approaches unity. As ν increases, the corresponding physical parameter $\nu' = 1 - \exp(-\nu\mu)$ increases, approaching unity, guaranteeing that every lesion is found. AFROC-AUC approaches one.

8. Over the range (0,0) to (FPF_{max}, LLF_{max}), the slope of the AFROC decreases monotonically. It is infinite at the origin and zero at the end-point.

17.7.3 Chance level performance on AFROC

There appears to be a misconception[25,26] that chance level performance on an AFROC corresponds to the positive diagonal of the plot, yielding AFROC-AUC of 0.5. Figure 17.4a shows that, in fact, chance level performance corresponds to the horizontal line connecting the origin to (1,0) and a vertical dashed line connecting (1,0) to (1,1) corresponding to AFROC-AUC of zero. If the lesion perceptual contrast is zero, then no lesions are found and all marks are NLs. The AFROC-AUC FOM, namely the probability that a lesion rating exceeds the ratings of NLs on non-diseased cases, (see Section 14.2), is zero.

17.7.4 The reader who does not yield any marks

Suppose the radiologist does not mark any case, as in Section 13.4.2.2, resulting in an empty data file. One possibility is that the radiologist did not interpret the cases and simply whizzed through them. In this situation, the radiologist is not performing the diagnostic task. The AFROC operating point is stuck at the origin and connecting the straight-line extension yields AFROC-AUC = 0.5, would be incorrect as it implies finite performance (any value greater than zero for AFROC-AUC implies some degree of expertise). All models of observer performance assume that the observer is behaving rationally,[10] so this possibility is not analyzable. On the other hand, there is the real possibility that the radiologist did not detect any lesions and did not mark any non-diseased cases. Assuming the radiologist is behaving rationally, one needs an explanation for AFROC-AUC = 0.5. It turns out that this observer is perfect at not generating NLs on non-diseased cases, so no patient is recalled incorrectly. The radiologist needs to get some credit for this ability, and this is the explanation of AFROC-AUC = 0.5 for a rational observer. Since this radiologist's LLF = 0 obviously the radiologist is far from perfect. The radiologist needs to be trained to find lesions. A suitable training set would consist of diseased cases only with the radiologist being tasked to locate the lesions. The FROC data from such a dataset could be analyzed using the AFROC1 figure of merit, which could be used to measure the improvement of the radiologist in finding lesions. There is no point wasting non-diseased cases on training this radiologist, as the radiologist has already proven perfect performance on them by not generating NL marks on any non-diseased case.

The following code, `mainNoMarks.R`, illustrates how this observer can be simulated.

17.7.4.1 Code listing

```
# mainNoMarks.R
rm(list = ls())
library(RJafroc)

Lmax <- 1;mu <- 0.001;lambda <- 0.000001;nu <- 1

retFroc <- PlotRsmOperatingCharacteristics(
  mu, lambda, nu,
  type = "FROC", lesionDistribution = Lmax,
  llfRange = c(0,1)
)
retRoc <- PlotRsmOperatingCharacteristics(
  mu, lambda, nu,
  type = "ROC", lesionDistribution = Lmax
)
retAfroc <- PlotRsmOperatingCharacteristics(
  mu, lambda, nu,
  type = "AFROC", lesionDistribution = Lmax
)

afrocPlot <- retAfroc$AFROCPlot
print(afrocPlot) # AFROC plot
cat("mu = ", mu,
    "\nlambda = ", lambda,
    "\nnu = ", nu,
    "\naucFroc = ", retFroc$aucFROC,
    "\naucRoc = ", retRoc$aucROC,
    "\naucAfroc = ", retAfroc$aucAFROC,"\n")
```

The corresponding AFROC plot, obtained by sourcing this code, is shown in Figure 17.5.

The explanation lies in the values of the chosen parameters. The μ parameter was set to 0.001 as setting it to zero would create a divide-by-zero error when $\lambda' = \lambda / \mu$ is calculated. Instead one sets λ to 0.000001, so $\lambda' = 0.001$. With such small λ' the probability is almost zero that any case will have an NL mark. According to the Poisson distribution, the probability of no mark is $\exp(-\lambda') \sim 1$. Of course, the fact that μ is close to zero means the $v' = 1 - \exp(-\mu v) \sim 0$, so no lesion is found.

The corresponding code output is shown below.

17.7.4.2 Code output

```
> source(...)
mu =  0.001
lambda =  1e-06
nu =  1
aucFroc =  4.986837e-07
aucRoc =  0.5004993
aucAfroc =  0.5
```

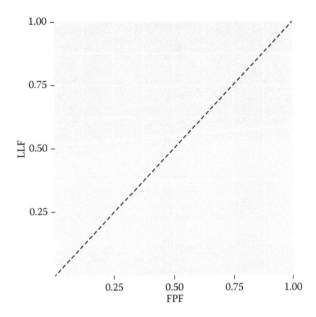

Figure 17.5 The case of the rational observer who does not mark any image, who operates at (0,0) on the AFROC and for whom the AFROC plot consists only of the straight-line extension connecting the origin to (1,1) and AFROC-AUC = 0.5. This observer has better performance (specifically unit case-level specificity) than the worst observer shown in Figure 17.4a who yielded AFROC-AUC = 0 (zero sensitivity and zero specificity). This figure was generated by sourcing file **mainNoMarks.R**.

This tells us that FROC-AUC = 0, ROC-AUC = 0.5, and AFROC-AUC = 0.5. FROC-AUC is meaningless as the operating point is stuck at the origin and one has no idea where it is supposed to end. ROC-AUC = 0.5 means that the observer is showing chance level performance at the task of separating non-diseased and diseased cases. *The ROC paradigm does not credit the observer for avoiding marking non-diseased cases.* AFROC-AUC = 0.5 credits the observer for not marking any non-diseased cases but there is no credit for marking lesions. The difference between the ROC and FROC paradigms, both predicting AUC = 0.5, but with different meanings, is because FROC is a location specific paradigm but ROC is not.

17.8 Quantifying search performance

In Figure 17.6, the line labeled (**a**) is a conventional model ROC curve ending at (1,1), the filled circle, while (**b**) shows a typical search model ROC curve ending at a point below and to the left of (1,1), the filled square. The location of the end-point of the RSM-predicted curve determines the *search performance* of the observer. The square root of two times the perpendicular distance d_S (the subscript s is for search) from the end-point to the chance diagonal in Figure 17.6, the line labeled (**c**), is defined as search performance, denoted S. For example, if $\lambda' = 0$ and $\nu' = 1$, then the end-point is (0,1) and $S = 1$. This observer has *perfect* search performance since no NLs were found *and* all lesions were found. The perpendicular distance from (0,1) to the chance diagonal is $1/\sqrt{2}$, which multiplied by $\sqrt{2}$ yields unity. Search performance ranges from 0 to 1. Using geometry, Equations 17.1 and 17.2, it follows that

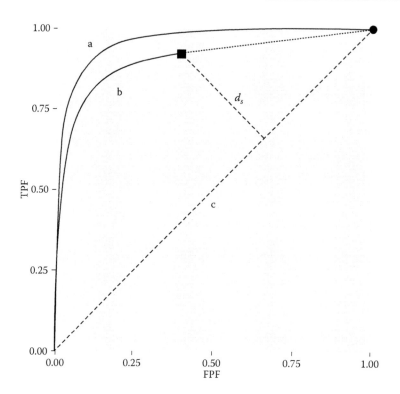

Figure 17.6 This figure shows a typical conventional model ROC curve labeled (**a**) predicted by ROC models that do not account for search performance. [Search performance is defined as the ability to find lesions while avoiding non-lesions]. The end-point of a typical conventional model curve is (1,1), shown by the filled circle. By adopting a sufficiently low reporting threshold the observer can continuously move the operating point from (0,0) to (1,1). The curve labeled (**b**) is a typical RSM-predicted ROC curve. The end-point is downward and leftward shifted relative to (1,1), as shown by the filled square. The observer cannot move the operating point continuously all the way from (0,0) to (1,1) because a finite fraction of images contain no marks. The fractions of unmarked non-diseased and diseased cases determine the location of the abscissa/ordinate of the end-point, respectively. The observer can move the operating point continuously from the origin to the end-point and no further. *The location of the end-point is a measure of search performance.* Higher search performance is characterized by the end-point moving upward and to the left, ideally to (0,1) which corresponds to perfect search performance. The perpendicular distance from the end-point to the chance diagonal (**c**) multiplied by $\sqrt{2}$, that is, $S = \sqrt{2}d_s$, is defined as a measure of search performance. Lesion-classification performance $A_C = \Phi\left(\dfrac{\mu}{\sqrt{2}}\right)$ is defined as the implied AUC of two unit variance normal distributions separated by the μ parameter of the search model. It measures the ability, having found a suspicious region, to correctly classify it as a true lesion. The code for this plot is in file **mainQuantifySearchPerformance.R**.

$$d_s = \frac{TPF_{\max} - FPF_{\max}}{\sqrt{2}} = \frac{1}{\sqrt{2}}\left(1 - e^{-\mu\nu}\right)\exp\left(-\frac{\lambda}{\mu}\right) \qquad (17.37)$$

Therefore, search performance is given by

$$\left.\begin{aligned} S = \sqrt{2}d_s = \left(1 - e^{-\mu\nu}\right)\exp\left(-\frac{\lambda}{\mu}\right) \\ S = \nu'\exp(-\lambda') \end{aligned}\right\} \qquad (17.38)$$

The second form in Equation 17.38 shows S in terms of the physical (i.e., primed) parameters; *it shows that search performance is the product of two terms: the probability v' of finding lesions times the probability $\exp(-\lambda')$ of avoiding finding any non-lesions.* This puts into a mathematical form the qualitative definition of search performance as the ability to find lesions while avoiding finding non-lesions. Since at least one parameter is needed to describe each of these probabilities, quantifying search requires at least two parameters.

Applying this definition to the case of the rational observer who does not generate any marks, one sees that the observer's search performance is zero, $S = v' \exp(-\lambda') \sim 0$. This emphasizes the point that not generating NLs is not enough; one must also be able to find lesions. It is also consistent with the fact that the origin lies on the positive diagonal of the ROC, implying zero perpendicular distance between them, that is, $d_s = 0$.

17.9 Quantifying lesion-classification performance

To avoid misunderstanding, *the author emphasizes that lesion-classification performance is being used in a different sense from that used in ROC methodology, where classification is between diseased and non-diseased cases, not between diseased and non-diseased regions,* i.e., latent NLs and latent LLs, as in the current context.

Having found a suspicious region, how good is the observer at correctly classifying true lesions and non-lesions? Lesion-classification performance C is determined by the μ parameter, and is defined by the implied AUC of unit variance normal distributions separated by μ.

$$A_C = \Phi\left(\frac{\mu}{\sqrt{2}}\right) \tag{17.39}$$

It ranges from 0.5 to 1. Only one parameter is needed for this, so one needs three parameters to quantify search and lesion-classification performance.

17.9.1 Lesion-classification performance and the 2AFC LKE task

Lesion-classification performance described above is similar but not identical to what is commonly measured in model-observer research using the *location-known-exactly* (LKE) paradigm. In this one uses 2AFC methods as in Figure 4.3. On diseased cases, the lesion is cued, but to control for false positives, one must also cue a control region on non-diseased cases, as in Figure 4.3. In that figure, the lesion, present in one of the two images, is always in the center of one of the two fields. The probability of a correct choice in the 2AFC task is AUC_{LKE}, that is, AUC conditioned on full knowledge of the (possible) position of the lesion being cued. One expects $A_C \leq AUC_{LKE}$. The reason for the inequality is that on a non-diseased case, the location being cued, in all likelihood, will not correspond to a latent NL, if any, found by the observer's search mechanism. Latent NLs are more suspicious for disease than other arbitrary locations in the case. A_C measures the separation parameter between latent NLs and LLs. The separation parameter between latent LLs and researcher chosen locations is expected to be larger. It is known that performance under the LKE condition exceeds that in a free-search 2AFC or ROC study, denoted AUC. Pointing to the possible location of the lesion takes out the need to search the rest of the image; searching introduces the possibility of not finding the lesion and/or finding non-lesions. One expects the following ordering: $AUC < A_C \leq AUC_{LKE}$. AUC is expected to be the least, as there is maximal uncertainty about possible lesion location. A_C is expected to be next in order, as now uncertainty has been reduced, and the observer's task is to pick between two cued locations, one a latent NL and the other a latent LL.

A_{LKE} is expected to be highest, as the observer's task is to pick between two cued locations, one a latent LL and the other a researcher chosen location, most likely not a latent NL. It should be clear that lowering uncertainty leads to an easier decision task and a correspondingly larger performance measure. Data supporting the expected inequality $AUC < A_C$ is presented in Section 19.5.4.6.

17.9.2 Significance of measuring search and lesion-classification performance from a single study

The ability to quantify search and lesion-classification performance from a ROC paradigm study is highly significant, going well-beyond modeling the ROC curve. ROC-AUC measures how well an observer is able to separate two groups of patients, a group of diseased patients from a group of non-diseased patients. While important, it does not inform us about how the observer goes about doing this and what is limiting the observer's performance (an exception is the CBM model which yields information about how good the observer is at finding lesions but does not account for the ability of the observer to avoid NLs on non-diseased cases). In contrast, the search and lesion-classification measures described above can be used as a diagnostic aid in determining what is limiting performance. If search performance is poor, it indicates that the observer needs to be trained on many non-diseased cases, learn the variants of non-diseased anatomy, and learn not to confuse them with lesions. On the other hand, if lesion-classification performance is poor, then one needs to train the observer using images where the location of a possible lesion is cued, and the observer's task is to determine if the cued location is a real lesion. The classic example here is breast CAD where the designer level ROC curve goes almost all the way to (1,1), implying poor search performance, while lesion-classification performance could actually be quite good because CAD has access to the pixel values and the ability to apply complex algorithms to properly classify lesions as benign or malignant.

Of course, before one can realize these benefits, one needs a way of estimating the end-point shown in Figure 17.6, plot (**b**). The observer will generally not oblige by reporting every suspicious region. RSM-based curve fitting is needed to estimate the end-point's location (**Chapter 19**).

17.10 The FROC curve is a poor descriptor of search performance

The basic reason is that the FROC curve is unconstrained in the x-direction.[27] Experts do not traverse as much along the positive x-direction as non-experts and partial area measures lose their meaning. Another reason is that the FROC depends only on the marks; unmarked non-diseased cases, representing perfect decisions, are not taken into account. The only meaningful comparison between two FROC curves occurs when they have a common NLF range, but this is rarely the case. As predicted by the RSM, a common range of NLF occurs when the two curves differ only in the ν parameter: if μ and λ are the same, then Equation 17.30 predicts the two curves will have identical $NLF_{max} = \lambda / \mu$. As shown below with numerical integration, this is the only situation where the area under the FROC tracks the area under the ROC, with the latter regarded as the gold standard.

The code in file **mainIsFrocGood.R**, explained in Online Appendix 17.E, calculates, by numerical integration, the areas under the *full* FROC, ROC, and AFROC curves. Each *full* curve consists of the continuously accessible part plus any straight-line extension to (1,1), if applicable. The code is divided into three parts:

- Part I, lines 16 through 83, calculates $AUC_{RSM}^{FROC}(\mu, \lambda, \nu)$, $AUC_{RSM}^{ROC}(\mu, \lambda, \nu, \overrightarrow{f_L})$, and $AUC_{RSM}^{AFROC}(\mu, \lambda, \nu)$ for varying μ, with $\lambda = \nu = 1$.
- Part II, lines 85 through 153, calculates the same AUCs for varying λ, with $\mu = 2$ and $\nu = 1$.
- Part III, lines 155 through 220, calculates the same AUCs for varying ν, with $\mu = 2$ and $\lambda = 1$.

The plots generated are shown in Figure 17.7a through f. The figure panels are arranged in two columns and three rows. The columns show variation of FROC-AUC and AFROC-AUC vs. ROC-AUC, respectively. In the first row, only μ is varied, in the second only λ is varied and in the third

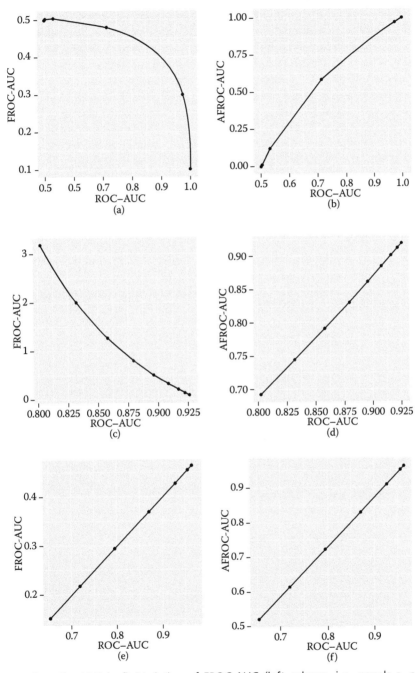

Figure 17.7 Plots Fig. 17.7 (a–f): Variation of FROC-AUC (left column, i.e., panels a, c and e) and AFROC-AUC (right column, i.e., panels b, d and f) vs. the gold standard, i.e., ROC-AUC. Each row of the plot corresponds to a different varied parameter. Plots (a) and (b) correspond to varying μ, plots (c) and (d) correspond to varying λ and plots (e) and (f) correspond to varying ν. If the quantity plotted along the y-axis is a valid FOM, it should be monotonic with the gold standard. This is always true for AFROC-AUC but not always true for FROC-AUC. See text for details. These plots were generated by `mainIsFrocGood.R`.

row only v is varied. $AUC_{RSM}^{ROC}\left(\mu,\lambda,v,\overrightarrow{f_L}\right)$ is plotted along the x-axis, $AUC_{RSM}^{FROC}\left(\mu,\lambda,v\right)$ is plotted along the y-axis in the left plot, and $AUC_{RSM}^{AFROC}\left(\mu,\lambda,v\right)$ is plotted along the y-axis in the right plot. $AUC_{RSM}^{ROC}\left(\mu,\lambda,v,\overrightarrow{f_L}\right)$ is the gold standard as it measures basic classification ability between diseased and non-diseased cases. So, for a valid figure of merit, the quantity plotted along the y-axis should monotonically increase with the gold standard, that is, the slope should be positive. This is always true for $AUC_{RSM}^{AFROC}\left(\mu,\lambda,v\right)$ but is not for $AUC_{RSM}^{FROC}\left(\mu,\lambda,v\right)$; it is only true when v is varied, which, as noted above, is the only situation when the range of integration along the NLF axis is constant.

The plots of the FROC-AUC in Figure 17.7a and c are non-linear and have negative slopes. In contrast, the AFROC-AUCs have a quasi-linear dependence on ROC-AUC. (The linear approximation slopes are printed by the code. For the plot in Figure 17.7b the slope is 2.00, for the plot in Figure 17.7d it is 1.84 and for the plot in Figure 17.7f is 1.42. The slopes indicate how much an ROC-AUC effect-size is amplified in the AFROC FOM. If only μ is different between two modalities, the amplification is almost exactly a factor of two. In the worst-case scenario, if only v is different, the amplification is a factor of 1.42. In general, all three quantities could be different; one expects an intermediate amplification of the effect-size, in the range 1.4–2.)

It is instructive to consider the extreme cases of a *perfect* observer and the *worst* observer to see how the two methods of plotting would deal with defining the *average* observer. To make the comparison easier, consider that the lesions are small compared to the image area, so that the chance of a random LL is very small.

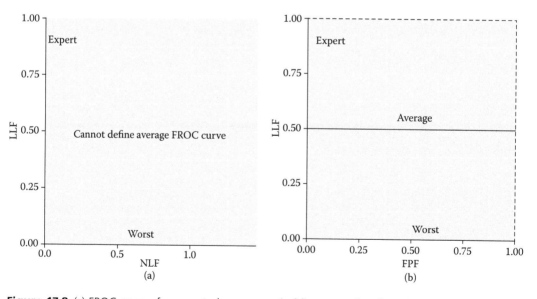

Figure 17.8 (a) FROC curves for expert observer: vertical line extending from (0,0) to (0,1) and worst observer: horizontal line over the indicated NLF range. It is not possible to define an average FROC curve as a common NLF range for the two observers does not exist. (b) Corresponding AFROC curves. AFROC-AUC for a perfect observer is unity (the area includes that under the dashed section extending from (0,1) to (1,1)). The corresponding area for the worst observer is zero, and the average AFROC curve is a straight line parallel to the x-axis at an ordinate of 0.5, so the area under the average AFROC-AUC is 0.5 (unlike the ROC-AUC, AFROC-AUC = 0.5 does not denote worst possible performance). This plot was generated by the code in **mainBestWorstObserver.R**.

The perfect observer (LLF = 1 @ NLF = 0) and the worst observer (LLF = 0 @ NLF < some constrained value) both yield identical areas (zero) under the FROC curves. It is not possible to define an average FROC curve, Figure 17.8a. Because the two plots do not share a common range of abscissa values one cannot define an average FROC curve. In contrast, the AFROC is contained to the unit square and the area under the AFROC curve, Figure 17.8b, is unity for the perfect observer. The corresponding area for the worst observer is zero, and the average AFROC curve is a straight line parallel to the x-axis at ordinate of 0.5, so the area under the average AFROC is 0.5 (as already noted, unlike the ROC area, AFROC area = 0.5 does not denote worst possible performance).

The FROC curve depends only on the marks. A valid FOM should reward correct decisions and penalize incorrect decisions on *all* cases (in the author's judgment, the use of partial area measures, widespread in the literature, needs to reconsidered). Unmarked non-diseased cases are perfect decisions, but these are not accounted for in the FROC curve (they indirectly affect the curve by the leftward movement of the uppermost point, all the way to NLF = 0 for a perfect observer, but these decisions are not accounted for in FROC curve based partial area measures). The area under the horizontal dashed curve in the AFROC curve shown in Figure 17.8b is due to unmarked images. See Section 14.4.2 for further discussion of the meaning of the area under the dashed portion of the AFROC plot.

Finally, FP marks on diseased cases don't have the same negative connotation as those on non-diseased cases since, following diagnostic workup, it is possible that the cancer will be found on the recalled cases but, unlike the AFROC, both contribute to the FROC x-axis.

The RSM is a first-order model and a lot of interesting science remains to be uncovered. It does not account for the satisfaction of search (SOS) effects[8,9] observed in medical imaging. It is as if the radiologist senses that an image is possibly diseased, without being able to pinpoint the specific reason, and therefore adopts a more cautious reporting style. They are more reluctant to mark NLs on diseased than on non-diseased cases. This means the probability of an LL rating exceeding the rating of an NL on *diseased* cases is not equal to the probability that an LL rating exceeding the rating of an NL on *non-diseased* cases:

$$P\left(Z_{k_2 2 l_2 2} > Z_{k_2 2 l_1 1}\right) \neq P\left(Z_{k_2 2 l_2 2} > Z_{k_1 1 l_1 1}\right) \tag{17.40}$$

Therefore, inclusion of inter-comparisons between LLs and NLs on diseased cases would make the figure of merit depend on disease prevalence, thereby violating a desirable property of a valid figure of merit. This is a reason for excluding such comparisons on diseased cases in the AFROC/wAFROC figures of merit.

17.10.1 What is the "clinically relevant" portion of an operating characteristic?

The reason for the quotes is that in the author's experience this term is used rather loosely in the literature. There is a serious misconception that the *clinically relevant* part of an operating characteristic is the steep portion emanating from the origin. The purpose of this section is to clarify. One needs to go back to the definition of the FROC, particularly the linear plot, Figure 14.2, showing how the raw plot is generated as a virtual threshold is moved from the far right to the far left. While this plot applies to the AFROC, the essential idea is the same. One orders the LL marks (dark dots) from left to right in increasing order according to their Z-samples. Likewise, the NL marks (light dots) are also ordered from left to right in increasing order according to their Z-samples. As the virtual threshold is moved to the left, starting from $\zeta = \infty$, mostly dark dots and occasional light dots are crossed. Each time a dark dot is crossed the operating point moves up by 1/(total number of lesions) and each time a light dot is crossed the operating point moves to the right by 1/(total number of cases). This causes the operating point to rise, starting from the origin and move upward and to the right. The steep portion of the plot corresponds to crossings by LL and NL marks with high Z-samples; it is the contribution of mostly easily visible lesions and the occasional NL. All observers are expected to localize the easy lesions, and there is nothing *clinically significant* about this. This argument applies to all operating characteristics. The clinical significance arises from the application. In a screening application, it is important to maintain high sensitivity at a reasonable specificity. In fact Jiang, Metz, and Nishikawa[28] had it right when they proposed the area *above* a preselected high sensitivity threshold TPF_0 should be divided by $(1 - TPF_0)$. Such a measure would emphasize the upper right corner of the ROC curve, not the steep portion near the origin. In the screening context, most of the Z-samples (99.5% to be more precise) are from non-diseased cases and only 0.5% are from diseased cases. This implies the *clinically relevant* part of the plot is near the upper right corner of the ROC plot. With the FROC, a normalized area above a preselected LLF_0 cannot be defined. On the other hand, the AFROC is amenable to such a partial area measure.

To do it right, one needs to include the costs and benefits of correct and incorrect decisions on diseased and non-diseased cases, the prevalence of disease and the actual population distribution of the Z-samples for non-diseased and diseased cases, and perform a weighted average over the entire ROC or AFROC curve. In the screening context, this would tend to weight the upper end of the curve. This is not an easy problem but it can be solved.

17.11 Evidence for the validity of the RSM

Listed below are some of the reasons arguing for the validity of the RSM.

1. Its correspondence to the Kundel—Nodine model of radiological search, which in turn is tied to actual eye-tracking measurement performed on radiologists. A model that is directly linked to physical measurements is compelling.
2. In special cases, the RSM is indistinguishable from the binormal model; even with its short-comings, the binormal model has been viewed, for over 6 decades, as highly successful at fitting ROC data.
3. It explains the following:
   a. The empirical observation[29] that most ROC datasets are characterized by b-parameter < 1.
   b. The empirical observation[30] that the b-parameter tends to decrease as contrast increases.
   c. The empirical observation,[30] that the difference in means of the two *pdf*s divided by the difference in standard deviations is roughly constant for a fixed set of experimental conditions.
4. It explains data degeneracy, that is, no interior operating points, which tend to be observed with expert observers.
5. It predicts FROC, AFROC, and LROC curves that fit real datasets.

As described in **Chapter 20**, the CBM model explains #3(a) and #4 while the bigamma model[5] explains #3(c).

17.11.1 Correspondence to the Kundel–Nodine model

Perhaps the strongest evidence for the validity of the RSM is its correspondence to the Kundel–Nodine model of radiological search,[31-35] which in turn is derived from eye-tracking measurements made on radiologists while they perform diagnostic tasks. These measurements show that radiologists identify suspicious regions in a short time and this ability corresponds to the λ, ν parameters of the RSM. Having found suspicious regions, the next activity uncovered by eye-tracking measurements is the detailed examination of each suspicious region in order to determine if it is a reportable finding. This is where the Z-sample is calculated, and this process is modeled by two unit normal distributions separated by the μ parameter of the RSM. Other ROC models do not share this close correspondence to actual physical measurements performed during diagnostic interpretations. The CBM model comes closest as it models the possibility that some lesions are not found, which is part of search performance, but the other part of search performance is the ability to find lesions while *avoiding* finding NLs. CBM does not model the possibility that latent non-lesions are also found in the process of finding lesions. Like other ROC models, it predicts that the point (1,1) is continuously accessible to the observer, which implies zero search performance (Figure 17.6).

17.11.2 The binormal model is a special case of the RSM

ROC models assume that every case provides a decision variable sample. This is what permits the observer to continuously move the operating point to (1,1). While in general the RSM allows cases with no decision samples, one way to ensure a decision sample on every case is to increase λ. In the limit of large λ every case has at least one latent NL. Under these conditions one expects the *pdfs* to integrate to unity and to be quasi-normal (they are not strictly normal, rather they are related to extreme-value distributions, but, as shown by Hanley,[36] the binormal model is quite robust to deviations from normality).

The point of the following exercise is to demonstrate that, in certain limits, RSM-generated ROC data are well fitted by the binormal model. Consider the code in **mainRsmVsEng.R**, Online Appendix 17.F.2, which simulates FROC data using the RSM, converts it to highest rating ROC data, bins the data into five bins, and prints out two sets of five integers. These are the bins count for non-diseased and diseased cases, respectively, analogous to the ROC counts data in Table 4.1 in **Chapter 4**. The counts data are analyzed by the Eng Java program,[37] which implements Metz's ROCFIT program, **Chapter 6**, yielding the binormal parameter values a, b, and the goodness of fit χ^2 statistic and a *p*-value.

Ensure that line 49 is not commented. Insert a break point at line 49 and click **Source** yielding the output shown in Section 17.11.2.1. The value of **Row**, on the first pass through the for-loop at line 10, is 1, corresponding to the same value of *Row* in Table 17.2, which lists RSM parameter values corresponding to this value of *Row.*, selected at line 12. Consult **R** help for the meaning of the **switch** statement.

17.11.2.1 Code output

```
> debugSource(...)
K1 =    500
K2 =    700
zeta1 =   -1
seed =   1
Lmax =   1
mu =    2
lambda =   10
nu =   1
RSM-ROC-AUC =   0.7874696
        [,1] [,2] [,3] [,4] [,5]
[1,]    196    88    72    72    72
[2,]     72    72    82    94   380
...  (debugger generated lines)
```

The last two lines contain the ROC counts table. Copy the numbers (not the square brackets) and paste them into the input window of the Eng program, and run the program with format option 3 selected. It yields the Row-1 values listed in the last six columns of Table 17.2. The other rows in

Table 17.2 Demonstration of RSM predicting binormal model equivalent data, showing results of simulating ROC ratings tables using RSM parameter values specified in columns two through six, the predicted RSM-AUC is in column 7, and fitting each ratings table using the binormal model, yielding the listed corresponding parameters, Az and goodness of fit statisics. In all cases the binormal fits to RSM generated data are excellent.

			R code						Eng Java program			
Row	seed	L_{max}	μ	λ	ν	AUC_{ROC}^{RSM}	a	b	A_z	χ^2	df	p-value
1	1	1	2.0	10	1	0.787	1.002	0.861	0.776	3.006	2	0.223
2	1	1	2.5	10	1	0.879	1.497	0.752	0.884	1.554	2	0.460
3	1	1	3.0	10	1	0.938	1.928	0.736	0.940	3.235	2	0.198
4	2	1	2.5	10	1	0.879	1.221	0.682	0.844	4.241	2	0.120
5	2	2	2.0	10	1	0.841	1.250	0.786	0.837	0.205	2	0.903
6	2	2	2.5	10	1	0.920	1.554	0.646	0.904	1.049	2	0.592
7	2	2	3.0	10	1	0.963	2.057	0.676	0.956	9.344	2	0.009
8	2	2	3.0	1	1	0.984	2.391	0.405	0.987	0.024	2	0.988
9	2	2	3.0	0.1	1	0.987	2.015	0.068	0.978		NA	

Note: The number of non-diseased cases was 500, the number of diseased cases was 700, and the reporting threshold was set to −1. The resulting ratings table, e.g., Section 17.11.2.1, was analyzed by Eng Java software to obtain the values listed in the remaining columns. Note the close correspondence between the RSM-AUC and the binormal fitted AUC, with the former being slightly larger. The proper RSM fits are expected to have a larger AUC. The last column lists the *p*-value for the chi-square goodness of fit statistic. Values greater than 0.001 are generally considered good fits. (NA: the chi-square goodness of fit statistics could not be calculated because some of the cell counts were less than five.)

Table 17.2 were generated by clicking **Next** and copying the output to the Eng Java program running the Eng program, which corresponds to repeating the above steps with other values (2, 3, ..., 9) of the **Row** variable.

The (*a*, *b*) binormal parameters were transferred to the appropriate location, between lines 57 through 65 in **mainRsmVsEng.R**. Once all values are populated, the plots shown in Figure 17.9 were obtained by stopping debug mode, clearing any breakpoints, commenting line 49 and sourcing the program. All this could have been automated using **RocfitR.R, Chapter 6**, but the author wanted a direct comparison of RSM prediction versus binormal fits using Metz's code. The plots in Figure 17.9 represent binormal model fitted ROC curves to ROC ratings data generated by the RSM.

These plots show that ROC data generated by the RSM are well fitted by the binormal model. The *p*-values of the goodness of fit statistic are all in the range of what is considered acceptable fits to a model.

Of interest in Table 17.2 is the observation that the qualities of the fits are quite good (p > 0.001 is generally considered acceptable, *Numerical Recipes*,[9] page 779). There is one exception: Row-9 shows what happens when $\mu = 3; \lambda = 0.1; \nu = 1$; this makes the data almost degenerate. ROCFIT-JAVA is still able to fit it, note the very small value (0.068) of the *b*-parameter, but it cannot calculate a goodness of fit statistic.

One expects AUC_{ROC}^{RSM} to exceed the binormal fitted value A_z. This has to do with the *proper* property of the RSM-ROC curve, which implies an ideal observer, while the binormal model predicts *improper* ROC curves; this is explained in **Chapter 20**. For Row = 2 and Row = 3, the expected orderings are reversed, although the discrepancies are small (about 1%). This is because RSM-*predicted* values are not subject to sampling variability as they are derived by numerical integration, Equation 17.25 and the numerical estimation error is very small

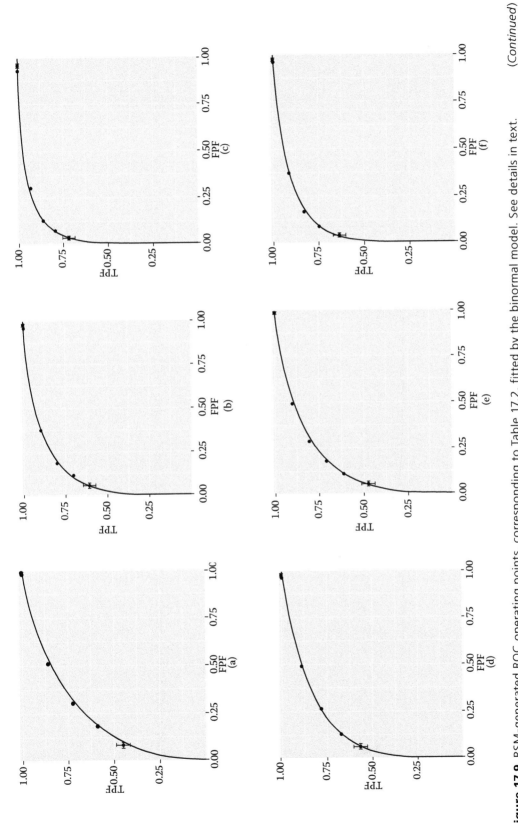

Figure 17.9 RSM-generated ROC operating points, corresponding to Table 17.2, fitted by the binormal model. See details in text.

(Continued)

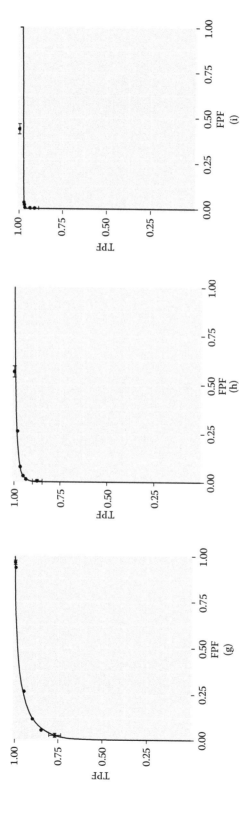

Figure 17.9 (Continued) RSM-generated ROC operating points, corresponding to Table 17.2, fitted by the binormal model. See details in text.

compared to the sampling error. In contrast, the estimates of A_z are subject to sampling variability, even though large numbers of cases were used (i.e., 500+700 = 1200, line 9). Row = 4 repeats Row = 2 with a different value of **seed:**, this time the expected ordering is observed (RSM-AUC > binormal-AUC).

Figure 17.9 shows *binormal model fitted curves for RSM-generated data*. The plots are labeled by the value of **Row** used to generate them. The hooks in the binormal ROC curves are not easily visible (the exception is 9) but they are nevertheless present as each of the b-parameters in Table 17.2 is less than unity. The error bars are exact 95% binomial confidence intervals on the operating points.

The plots in Figure 17.9 show RSM-generated ROC operating points using parameters and seeds specified in Table 17.2 and corresponding binormal model fitted curves. Plots a–i correspond to Row = 1–9 in Table 17.2. The binormal parameters were obtained by running lines 1 through 49 in **mainRsmVsEng.R**, transferring the ROC counts table to the Eng Java program, running the Eng program, and transferring the parameter values to the appropriate location, between lines 57 and 65. Once all values are populated, the plots were obtained by sourcing the file with different values for variable **Row**. Even with the large number of cases, sampling variability affects the binormal model fits, e.g., the binormal model curves in plots labeled b and d differ only in seed values. These plots show that over a wide range of parameters, the binormal model fits RSM-generated ROC data reasonably.

> Over a wide range of parameters, RSM-generated ROC data is fitted reasonably by the binormal model. As far as the binormal model-fitting software is concerned, the counts data arose from two normal distributions (apart from the arbitrary monotonic transformation uncertainty that applies to all such fitting). Since the binormal model has been used successfully for almost six decades, the ability of the RSM to mimic it is an important justification for validity of the RSM.

17.11.3 Explanations of empirical observations regarding binormal parameters

As summarized previously, there are three empirical observations: (1) $b < 1$, (2) b decreases as μ increases, and (3) $\Delta(mean)/\Delta(\sigma)$ is approximately constant for fixed experimental conditions. So as not to confuse with the RSM μ parameter, the difference in means is denoted $\Delta(mean)$, not $\Delta\mu$.

17.11.3.1 Explanation for empirical observation $b < 1$

The RSM-predicted ROC curves are consistent with empirical observations,[29] going back almost six decades, that observed ROC data, when fitted by the unequal variance binormal model, yield b-parameters < 1, implying the diseased case *pdf* is wider than the non-diseased case *pdf*. The RSM provides a natural explanation for this, rather than by imposing it empirically.[38] The reason the diseased distribution has a wider *pdf* is that diseased cases yield *two* types of Z-samples, variable numbers of NL Z-samples from a zero-centered unit variance normal distribution and variable numbers of LL Z-samples from a μ-centered unit variance normal distribution. The resulting *mixture* distribution is expected, when one attempts to fit it with a normal distribution, to yield standard deviation greater than 1, or, equivalently, b-parameter < 1.

Examples of RSM-generated datasets consistent with $b < 1$ binormal model fits were presented in Table 17.2 and Figure 17.9. The code in **mainRsmPlots.R**, Online Appendix 17.D.2, displays *pdf*s, shown in Figure 17.10a through d, for the following parameter values: (a) $\mu = 2; v = 0.15$; (b) $\mu = 2; v = 0.25$; (c) $\mu = 3; v = 0.15$; (d) $\mu = 3; v = 0.25$. In all cases $\lambda = 1$ and $L_{max} = 1$. In plots

(a) and (b) the dotted curves, corresponding to diseased cases, are wider than the solid ones, corresponding to non-diseased cases, and there is a hint of mixture behavior for the dotted curve in plot (d), and bimodality is prominent for the largest μ and smallest ν in plot (c). The mixture distribution is expected to lead to a larger estimate of standard deviation of the assumed normal distribution. The fit would not be ideal, but it is known that for relatively small numbers of cases, as is true with clinical data, it is difficult to detect deviations from binormality. The binormal model is

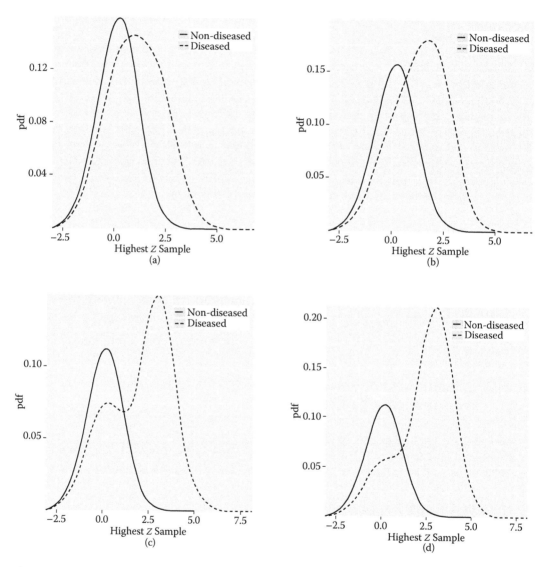

Figure 17.10 (a) $\mu = 2; \nu = 0.15$ (b) $\mu = 2; \nu = 0.25$ (c) $\mu = 3; \nu = 0.15$ (d) $\mu = 3; \nu = 0.25$. This figure explains the empirical observation that binormal model b-parameter <1. Shown are probability density functions (*pdf*) plots against the highest Z-sample for displayed values of RSM parameters μ and ν; the other parameters are $\lambda = 1$ and $L_{max} = 1$. The dotted curve corresponds to diseased cases and the solid curve to non-diseased cases. In (a) and (b) the dotted curve is noticeably broader. In (d) there is a hint of a peak at zero, which is quite prominent in (c). The broadening and/or mixture behavior is due to the two types of Z-samples occurring on diseased cases, those from NLs, which give the peak just above zero and those from LLs, which give the peak at μ. When fitted by a normal distribution, the dotted curve will appear to have larger standard deviation, relative to the non-diseased cases, or equivalently, binormal model b-parameter < 1. See Online Appendix 17.G for further discussion of this and another observation made by Swets et al.[30] These plots were generated by **mainRsmPlots.R**.

quite robust with respect to deviations from strict normality.[36] Several examples of this are evident in the goodness of fit p-values in Table 17.2, which show good binormal fits to RSM-generated data, even with 1200 cases.

17.11.3.2 Explanation of Swets et al. observations

More than 55 years ago, Swets et al.[30] noticed in non-medical imaging contexts:

1. The standard deviation of the non-diseased distribution divided by the standard deviation of the diseased distribution, tended to decrease as contrast increased.
2. The ratio $\Delta(mean) / \Delta(\sigma)$ is approximately constant for a fixed set of experimental conditions.

In the RSM μ is the perceptual signal-to-noise, $pSNR$, of found lesions. Swets' first empirical observation implies that the b-parameter should decrease with increasing μ. Testing this proposition over a wide range of μ using the preceding methods (i.e., based on binormal model maximum likelihood fitting, either using the Eng Java code or **RocfitR.R**) is direct but cumbersome and subject to failure. This is because it depends on convergence of the binormal model algorithm, which is problematic for larger values of μ which lead to degenerate datasets. Instead, the following method was used. The search model predicted $pdfs$ were normalized so that they individually integrated to unit areas. This was accomplished by dividing the non-diseased pdf by the x-coordinate of the end-point, and the diseased pdf by the y-coordinate of the end-point. The means and standard deviations of these distributions were calculated by numerical integration. If $f(x)$ is a unit-normalized pdf its mean $\langle x \rangle$ and variance σ_x^2 are defined by [DIS = diseased; NON = non-diseased]:

$$\left. \begin{aligned} \langle x \rangle &= \int_{-\infty}^{\infty} x f(x) \, dx \\[2mm] \sigma_x^2 &= \int_{-\infty}^{\infty} \left(x - \langle x \rangle \right)^2 f(x) \, dx \end{aligned} \right\} \tag{17.41}$$

$$\left. \begin{aligned} \Delta(mean) &= \langle x \rangle_{DIS} - \langle x \rangle_{NON} \\[2mm] \Delta(\sigma) &= \sigma_{DIS} - \sigma_{NON} \end{aligned} \right\} \tag{17.42}$$

The needed quantities are defined as

Varying experimental conditions, in the Swets et al sense[30], were simulated by individually varying two of the three parameters of the RSM under the constraint that the RSM-predicted AUC remained constant at a specified value. Without this constraint, variation of a single parameter, e.g., μ, would allow AUC to vary over the entire range 0.5 to 1, which is uncharacteristic of radiologists interpreting the same case set. The underlying assumption is that observers, characterized by different RSM parameters, nevertheless tend to have about the same RSM-AUCs. In other words, they tend to compensate for deficiencies in some area (e.g., finding too many NLs, large λ) with increased performance in other areas (e.g., finding more lesions, i.e., larger ν, and/or extracting greater $pSNR$ from found lesions, i.e., larger μ).

The code for this is in file **mainRsmSwetsObservations.R**, explained in Online Appendix 17.G, which was used to populate Table 17.3. The table is organized into parts A, B, and

Table 17.3 Demonstration of RSM's consistency with Swet's observations: (i) decrease in b as μ increases and (ii) near constancy of the mean to sigma ratio.

	A				B				C			
	$\lambda = 2$				$\nu = 1$				$\mu = 2$			
AUC_{RSM}^{ROC}	μ	ν	b	$\dfrac{\Delta(mean)}{\Delta(\sigma)}$	μ	λ	b	$\dfrac{\Delta(mean)}{\Delta(\sigma)}$	λ	ν	b	$\dfrac{\Delta(mean)}{\Delta(\sigma)}$
0.7	2.00	0.30	0.75	3.14	0.95	1.00	0.98	25.4	0.10	0.26	0.90	16.09
	2.21	0.26	0.71	2.81	0.99	1.20	0.98	25.6	0.15	0.26	0.86	11.27
	2.45	0.23	0.67	2.59	1.04	1.43	0.98	25.8	0.24	0.26	0.83	8.13
	2.71	0.20	0.62	2.43	1.09	1.71	0.98	26.0	0.37	0.26	0.79	6.08
	3.01	0.17	0.58	2.32	1.14	2.04	0.97	25.8	0.57	0.27	0.77	4.75
	3.33	0.16	0.54	2.24	1.19	2.45	0.97	25.0	0.88	0.28	0.75	3.90
	3.68	0.14	0.50	2.20	1.24	2.92	0.97	23.4	1.36	0.29	0.75	3.38
	4.08	0.13	0.46	2.17	1.29	3.50	0.97	20.9	2.10	0.31	0.76	3.12
	4.52	0.11	0.43	2.16	1.35	4.18	0.96	17.8	3.24	0.33	0.77	3.06
	5.00	0.10	0.40	2.17	1.40	5.00	0.95	14.5	5.00	0.37	0.79	3.16
0.8	2.00	0.58	0.81	5.95	1.35	1.00	0.94	18.0	0.10	0.46	0.95	34.85
	2.21	0.48	0.76	5.01	1.40	1.20	0.94	17.1	0.15	0.47	0.93	23.86
	2.45	0.41	0.72	4.44	1.45	1.43	0.94	16.4	0.24	0.47	0.90	16.74
	2.71	0.36	0.68	4.07	1.50	1.71	0.93	15.7	0.37	0.48	0.87	12.15
	3.01	0.32	0.64	3.84	1.56	2.04	0.93	15.2	0.57	0.49	0.85	9.21
	3.33	0.28	0.60	3.69	1.61	2.45	0.93	14.6	0.88	0.51	0.82	7.39
	3.68	0.25	0.56	3.60	1.67	2.92	0.92	13.9	1.36	0.54	0.81	6.34
	4.08	0.23	0.53	3.56	1.72	3.50	0.92	13.2	2.10	0.58	0.81	5.93
	4.52	0.20	0.50	3.54	1.78	4.18	0.91	12.3	3.24	0.65	0.82	6.11
	5.00	0.18	0.47	3.55	1.84	5.00	0.90	11.2	5.00	0.76	0.84	6.88

Note: The table is organized into parts A, B and C. The parameter of the RSM that is held constant, while the other two are varied to maintain the ROC-AUC listed in column 1, is shown in the second header. **Part A**: This show that b decreases as μ increases and the $\Delta(mean)$ to $\Delta(\sigma)$ ratio, i.e., non-diseased to diseased is approximately constant. A similar organization and conclusions apply to the other parts of the table. [All parameters in this table represent intrinsic values.]

C. Part A varies (μ, ν) for constant RSM-AUC for $\lambda = 2$. Part B varies (μ, λ) for constant RSM-AUC for $\nu = 1$. Part C varies (λ, ν) for constant RSM-AUC for $\mu = 2$. The first column lists AUC_{RSM}^{ROC}, set to either 0.7 or 0.8, the second and third columns list the parameters that were varied, subject to the AUC constraint, followed by the empirical b-parameter and $\Delta(mean)/\Delta(\sigma)$. A similar organization applies to part B, where only (μ, λ) are varied and to part C where only (λ, ν) are varied. The number of lesions per diseased case was set to one. The function **FindParamFixAuc()** finds the missing RSM parameter, indicated by initializing it with **NA**, prior to the function call, given the two other RSM parameters. The function **rsmPdfMeansAndStddevs()** calculates the means and standard deviations of the two distributions, after appropriately normalizing them to unit areas.

Examination of Table 17.3 reveals, for both values of AUC, an inverse relation between μ and b, parts A and B, *and* an inverse relation between v and b, part C. In part C, for fixed μ, the only way to improve performance is by increasing v, that is, by the observer getting better at finding lesions. Furthermore, Table 17.3 reveals an approximately constant value of $\Delta(mean)/\Delta(\sigma)$ especially in parts A and B.

The biggest deviation from approximate constancy is observed in part C. This could be due to unreasonably low values of λ, for example, 0.1, necessary in order to satisfy the AUC constraint. It should be noted that the Swets et al. observations are based on only two datasets and they further state that $\Delta(mean)/\Delta(\sigma) = 4$, their observed value, is probably not generally applicable.

The empirical observation that b decreases with increasing lesion detectability is likely more generally true and the results, Table 17.3, support it. It makes sense since with increasing μ one is averaging samples from two widely divergent distributions. This implies the existence of two types of Z-samples in diseased cases. Another explanation is the existence of heterogeneity in the distribution for diseased cases, for example, a mix of easy and hard lesions, which too would tend to broaden the diseased distribution and make $b < 1$ and b decreasing with increasing lesion detectability. Putting all this together, the author makes the following prediction: if the possible location of the lesion is prespecified *and* all lesions have the same detectability, then one expects $b = 1$, to within sampling variability, and no dependence of b on lesion detectability. This is a fairly simple experiment to conduct using simulated Gaussian noise backgrounds and superposed Gaussian disk lesions[39–41] with fixed contrast (using simulated clinical backgrounds is not recommended for this study as it could cause differences in lesion detectability depending on variations in the background in the immediate vicinity of the lesion).

17.11.4 Explanation of data degeneracy

An ROC dataset is said to be degenerate if the corresponding ROC plot does not have any interior data points. Data degeneracy is a significant problem faced by the binormal model,[1,6,42]. Degenerate datasets cannot be analyzed by the binormal model. The RSM provides a natural explanation for such datasets, and as shown in **Chapter 19**, such datasets are readily fitted by the RSM. The CBM[2–4] also provides an alternative explanation for the data degeneracy and a method for fitting such datasets.

The reason for degenerate datasets is the existence of cases not providing any decision variable samples. Such cases are always binned in the lowest ROC:1 bin. The possibility of data degeneracy can be appreciated by examining Figure 17.1 (in particular, plots in Figure 17.1d through f, i.e., the higher values of μ). As μ increases the accessible portion of the ROC curve shrinks and the curve increasingly approaches the top-left corner of the plot. The effect is particularly pronounced for observers characterized by large values of μ and small values of λ, that is, the experts. For them the operating points will be clustered near the initial near-vertical section of the ROC curve. It will be difficult to get such observers to generate appreciable numbers of false positives. Instructions such as *spread your ratings*[43] or the use *continuous* ratings[44] may not always work and furthermore, on general principles, interference with the radiologist's readings style to make the data easier to analyze is undesirable. To the experimenter it will appear that the observer is not cooperating, when in fact they are being rational. A similar issue affected Dr. Swensson's LROC analysis method[12] in which originally every case had to be assigned a *most-suspicious* region, even if the radiologist thought the case was perfectly normal; this met with resistance from radiologists. In later versions of his software, Dr. Swensson removed the forced localization requirement and instead did it in software by sampling a random number generator. Radiologists don't like to be told, "*Even if you believe the case is definitely normal, there must be some region that is <u>least</u> normal, or <u>most</u> suspicious*" All of these difficulties go away if one abandons the premise that every case generates a Z-sample.

17.11.5 Predictions of FROC/AFROC/LROC curves

Besides predicting ROC, FROC, and AFROC curves, as shown in this chapter, the RSM also predicts LROC curves.[16] Moreover, these are generally better fits to experimental data since they do not allow the AFROC and LROC curve to go continuously to FPF = 1, as do earlier models, two by the author[23,24] and one by the late Dr. Swensson.[12] As a historical note, the FROCFIT and AFROC software developed[23] by the author in 1989 was more successful at fitting microcalcification data than mass data (private communication, Prof. Heang-Ping Chan, ca. 1990). This is consistent with the premise that the microcalcification task is characterized by larger λ than the mass task. Radiologists literally use a magnifying glass (a physical one or a software implementation) to search each image for the much smaller specks, and this increases the potential for finding NLs, hence the larger λ. Mass detection is more a function of the global gestalt view described in the previous chapter. Larger λ yields a FROC curve that traverses more to the right than the corresponding mass curve. The FROCFIT program allows the FROC curve to go far to the right and reach unit ordinate, which is usually not observed with mass data, but could approximate microcalcification data.

17.12 Discussion/Summary

This chapter has described ROC, FROC, and AFROC curves predicted by the radiological search model (RSM). All RSM-predicted curves share the *constrained end-point property* that is qualitatively different from previous ROC models. In the author's experience, it is a property that most researchers in this field have difficulty accepting. There is too much history going back to the early 1940s, of the ROC curve extending from (0,0) to (1,1) that one has to let go of, and this can be difficult.

The author is not aware of any direct evidence that radiologists can move the operating point continuously in the range (0,0) to (1,1) in search tasks, so the existence of such a ROC is an assumption. Algorithmic observers that do not involve the element of search can extend continuously to (1,1). An example of an algorithmic observer not involving search is a diagnostic test that rates the results of a laboratory measurement, for example, the A1C measure of blood glucose* for presence of a disease. If A1C $\geq$ 6.5% the patient is diagnosed as diabetic. By moving the threshold from infinity to –infinity, and assuming a large population of patients, one can trace out the entire ROC curve from the origin to (1,1). This is because every patient yields an A1C value. Now imagine that some finite fraction of the test results are "lost in the mail;" then the ROC curve, calculated over all patients, would have the constrained end-point property, albeit due to an *unreasonable* cause.

The situation in medical imaging involving search tasks is qualitatively different. Not every case yields a decision variable. There is a *reasonable* cause for this: to render a decision variable sample the radiologist must find something suspicious to report, and if none is found, there is no decision variable to report. The ROC curve calculated over all patients would exhibit the constrained end-point property, even in the limit of an infinite number of patients. If calculated over only those patients that yielded at least one mark, the ROC curve would extend from (0,0) to (1,1) but then one would be ignoring the cases with no marks, which contain information. Unmarked non-diseased cases represent perfect decisions and unmarked diseased cases represent worst-case decisions.

RSM-predicted ROC, FROC, and AFROC curves were derived. These were used to demonstrate that the FROC is a poor descriptor of performance. Since almost all work to date, including some early work by the author,[45,46] has used FROC curves to measure performance, this is going to be difficult for some to accept. The examples in Figures 17.7 and 17.8 should convince the reader that the FROC curve is indeed a poor measure of performance. The only situation

---

* http://www.diabetes.org/diabetes-basics/diagnosis/

where one can safely use the FROC curve is if the two modalities produce curves extending over the same NLF range. This can happen with two variants of a CAD algorithm, but rarely with radiologist observers.

A unique feature is that the RSM provides measures of search and lesion-classification performance. It bears repeating that search performance is the ability to find lesions while avoiding finding non-lesions. Search performance can be determined from the position of the ROC end-point (this can be determined by RSM-based fitting of ROC data, **Chapter 19**). The perpendicular distance between the end-point and the chance diagonal is, apart from a factor of 1.414, a measure of search performance. All conventional ROC models that predict continuous curves extending to (1,1), imply zero search performance.

Lesion-classification performance is measured by the AUC value corresponding to the μ parameter. Lesion-classification performance is the ability to discriminate between LLs and NLs, not between diseased and non-diseased cases; the latter is measured by RSM-AUC. There is an analogy between the two ways of measuring classification performance (i.e., at lesion level or at case level) and usage of CAD to find lesions in screening mammography versus usage of CAD in the diagnostic context to determine if a lesion found at screening is actually malignant.

Search and lesion-classification performance can be used as diagnostic aids to optimize performance of a reader. For example, if search performance is low, then training using mainly non-diseased cases is called for, so the resident learns the different variants of non-diseased tissues that can appear to be true lesions. If lesion-classification performance is low then training with diseased cases only is called for, so the resident learns the distinguishing features characterizing true lesions from non-diseased tissues that appear to be true lesions.

Finally, evidence for the RSM is summarized. Its correspondence to the empirical Kundel–Nodine model of visual search that is grounded in eye-tracking measurements. It reduces in the limit of large λ to the binormal model. The predicted *pdfs* in this limit are not strictly normal, but deviations from normality would require very large sample size to demonstrate. Examples were given where even with 1200 cases the binormal model provides statistically good fits, as judged by the chi-square goodness of fit statistic, Table 17.2, to ROC data generated by the RSM. Since the binormal model has proven successful in describing a large body of data, it is satisfying that the RSM can mimic it. The RSM explains most empirical results regarding binormal model fits. The common finding that $b < 1$, that b decreases with increasing lesion *pSNR* (large μ and / or v), and the finding that the difference in means divided by the difference in standard deviations is fairly constant for a fixed experimental situation, Table 17.3. The RSM explains data degeneracy, especially for radiologists with high expertise.

The contaminated binormal model[2-4] (CBM), **Chapter 20**, which models the diseased distribution as having two peaks, one at zero and the other at a constrained value, also explains the empirical observation that b-parameter < 1 and data degeneracy. Because it allows the ROC curve to go continuously to (1,1), CBM does not completely account for search performance. It accounts for search when it comes to finding lesions, but not for avoiding finding non-lesions.

The author does not want to leave the impression that RSM is the ultimate model. The current model does not predict satisfaction of search (SOS) effects.[8-9] As stated earlier, the RSM is a first-order model and a lot of interesting science remains to be uncovered.

17.12.1 The Wagner review

The two RSM papers[13,14] were included in a list of 25 papers the "Highlights of 2006" in *Physics in Medicine and Biology*. As stated by the publisher: "I am delighted to present a special collection of articles that highlight the very best research published in Physics in Medicine and Biology in 2006. Articles were selected for their presentation of outstanding new research, receipt of the highest praise from our international referees, and the highest number of downloads from the journal website."

One of the reviewers was the late Dr. Robert ("Bob") Wagner. He had an open-minded approach to imaging science that is lacking these days and a unique writing style. The author reproduces one of his comments with minor edits, as it pertains to the most interesting and misunderstood prediction of the RSM, namely its constrained end-point property.

I'm thinking here about the straight-line piece of the ROC curve from the max to (1, 1).

> *This can be thought of as resulting from two overlapping uniform distributions (thus guessing) far to the left in decision space (rather than delta functions). Please think some more about this point--because it might make better contact with the classical literature.*
>
> *BTW -- it just occurs to me (based on the classical early ROC work of Swets & co.) -- that there is a test that can resolve the issue that I struggled with in my earlier remarks. The experimenter can try to force the reader to provide further data that will fill in the space above the max point. If the results are a straight line, then the reader would just be guessing -- as implied by the present model. If the results are concave downward, then further information has been extracted from the data. This could require a great amount of data to sort out--but it's an interesting point (at least to me).*

Dr. Wagner made two interesting points. With his passing, the author has been deprived of the penetrating and incisive evaluation of his ongoing work, which the author deeply misses. Here is the author's response (ca. 2006):

The need for delta functions at negative infinity can be seen from the following argument. Let us postulate two unit width *pdfs* with the same shapes but different areas, centered at a common value far to the left in decision space, but not at negative infinity. These *pdfs* would also yield a straight-line portion to the ROC curve. However, they would be inconsistent with the search model assumption that some images yield no decision variable samples and therefore cannot be rated in bin ROC:2 or higher. Therefore, if the distributions are as postulated above then choice of a cutoff in the neighborhood of the overlap would result in some of these images being rated 2 or higher, contradicting the RSM assumption. The delta function *pdfs* at negative infinity are seen to be a consequence of the search model.

One could argue that when the observer sees nothing to report then he starts guessing and indeed this would enable the observer to move along the dashed portion of the curve. This argument implies that the observer knows when the threshold is at negative infinity, at which point the observer turns on the guessing mechanism (the observer who always guesses would move along the chance diagonal). In the author's judgment, this is unreasonable. The existence of two thresholds, one for moving along the non-guessing portion and one for switching to the guessing mode would require abandoning the concept of a single decision rule. To preserve this concept, one needs the delta functions at negative infinity.

Regarding Dr. Wagner's second point, it would require a great amount of data to sort out whether forcing the observer to guess would fill in the dashed portion of the curve, but the author doubts it is worth the effort. Given the bad consequences of guessing (incorrect recalls) the author believes that in the clinical situation, *the radiologist will never guess*. If the radiologist sees nothing to report, nothing will be reported. In addition, the author believes that *forcing the observer into changing reporting style, to prove a research point, is not a good idea*. Finally, with the benefit of hindsight, there was no need for postulating delta functions at negative infinity. One simply accepts that some images do not generate Z-samples and therefore are not counted in the integrated pdfs. This explains why the pdfs do not integrate to unity.

References

1. Dorfman DD, Alf E. Maximum-Likelihood Estimation of Parameters of Signal-Detection Theory and Determination of Confidence Intervals—Rating-Method Data. *Journal of Mathematical Psychology*. 1969;6:487–496.
2. Dorfman DD, Berbaum KS, Brandser EA. A contaminated binormal model for ROC data: Part I. Some interesting examples of binormal degeneracy. *Acad Radiol*. 2000;7(6):420–426.
3. Dorfman DD, Berbaum KS. A contaminated binormal model for ROC data: Part II. A formal model. *Acad Radiol*. 2000;7(6):427–437.
4. Dorfman DD, Berbaum KS. A contaminated binormal model for ROC data: Part III. Initial evaluation with detection. *Acad Radiol*. 2000;7(6):438–447.
5. Dorfman DD, Berbaum KS, Metz CE, Lenth RV, Hanley JA, Abu Dagga H. Proper Receiving Operating Characteristic Analysis: The Bigamma model. *Acad Radiol*. 1997;4(2):138–149.
6. Metz CE, Pan X. Proper Binormal ROC Curves: Theory and Maximum-Likelihood Estimation. *J Math Psychol*. 1999;43(1):1–33.
7. Pan X, Metz CE. The "proper" binormal model: Parametric receiver operating characteristic curve estimation with degenerate data. *Academic Radiology*. 1997;4(5):380–389.
8. Berbaum KS, Franken EA, Dorfman DD, et al. Satisfaction of Search in Diagnostic Radiology. *Invest Radiol*. 1990;25(2):133–140.
9. Franken EA, Berbaum KS, Lu CH, et al. Satisfaction of search in the detection of plain-film abnormalities in abdominal contrast studies. *Invest Radiol*. 1994;29(4):403–409.
10. Press WH, Teukolsky SA, Vetterling WT, Flannery BP. Numerical Recipes: The Art of Scientific Computing. 3 ed. Cambridge: Cambridge University Press; 2007.
11. Macmillan NA, Creelman CD. Detection Theory: A User's Guide. New York: Cambridge University Press; 1991.
12. Swensson RG. Unified measurement of observer performance in detecting and localizing target objects on images. *Med Phys*. 1996;23(10):1709–1725.
13. Chakraborty DP. ROC Curves predicted by a model of visual search. *Phys Med Biol*. 2006;51:3463–3482.
14. Chakraborty DP. A search model and figure of merit for observer data acquired according to the free-response paradigm. *Phys Med Biol*. 2006;51:3449–3462.
15. Bunch PC, Hamilton JF, Sanderson GK, Simmons AH. A free-response approach to the measurement and characterization of radiographic-observer performance. *J Appl Photogr Eng*. 1978;4:166–171.
16. Chakraborty DP, Yoon HJ. Operating characteristics predicted by models for diagnostic tasks involving lesion localization. *Med Phys*. 2008;35(2):435–445.
17. Yoon HJ, Zheng B, Sahiner B, Chakraborty DP. Evaluating computer-aided detection algorithms. *Med Phys*. 2007;34(6):2024–2038.
18. Weisstein EW. Fisher-Tippett Distribution. MathWorld--A Wolfram Web Resource http://mathworld.wolfram.com/Fisher-TippettDistribution.html, 2005.
19. Fisher RA, Tippett LHC. Limiting forms of the frequency distribution of the largest and smallest member of a sample. *Proc Cambridge Phil Society*. 1928;24:180–190.
20. Kotz S, S. N. Extreme Value Distributions: Theory and Applications. Covent Garden, London: Imperial College Press; 2000.
21. Samuelson FW, Petrick N, Paquerault S. Advantages and examples of resampling for CAD evaluation. Paper presented at: Biomedical Imaging: From Nano to Macro, 2007. ISBI 2007. 4th IEEE International Symposium on Biomedical Imaging 2007; Arlington VA, USA.
22. Samuelson FW, Petrick N. Comparing image detection algorithms using resampling. *IEEE International Symposium on Biomedical Imaging: From Nano to Micro*; 2006:1312–1315.
23. Chakraborty DP. Maximum Likelihood analysis of free-response receiver operating characteristic (FROC) data. *Med Phys*. 1989;16(4):561–568.

24. Chakraborty DP, Winter LHL. Free-Response Methodology: Alternate Analysis and a New Observer-Performance Experiment. *Radiology*. 1990;174:873–881.

25. Lockwood P, Piper K. AFROC analysis of reporting radiographer's performance in CT head interpretation. *Radiography*. 2015;21(3):e90–e95.

26. Litchfield D, Donovan T. Worth a Quick Look? Initial Scene Previews Can Guide Eye Movements as a Function of Domain-Specific Expertise but Can Also Have Unforeseen Costs. 2016.

27. Popescu LM. Nonparametric signal detectability evaluation using an exponential transformation of the FROC curve. *Med Phys*. 2011;38(10):5690–5702.

28. Jiang Y, Metz CE, Nishikawa RM. A receiver operating characteristic partial area index for highly sensitive diagnostic tests. *Radiology*. 1996;201(3):745–750.

29. Green DM, Swets JA. *Signal Detection Theory and Psychophysics*. New York, NY: John Wiley & Sons; 1966.

30. Swets JA, Tanner Jr WP, Birdsall TG. Decision processes in perception. *Psycholog Rev*. 1961;68(5):301.

31. Kundel HL, Nodine CF, Conant EF, Weinstein SP. Holistic Component of Image Perception in Mammogram Interpretation: Gaze-tracking Study. *Radiology*. 2007;242(2):396–402.

32. Kundel HL, Nodine CF. Modeling visual search during mammogram viewing. *Proc SPIE*. 2004;5372:110–115.

33. Nodine CF, Kundel HL. Using eye movements to study visual search and to improve tumor detection. *RadioGraphics*. 1987;7(2):1241–1250.

34. Kundel HL, Nodine CF. A visual concept shapes image perception. *Radiology*. 1983;146:363–368.

35. Kundel HL, Nodine CF, Carmody D. Visual scanning, pattern recognition and decision-making in pulmonary nodule detection. *Invest Radiol*. 1978;13:175–181.

36. Hanley JA. The Robustness of the "Binormal" Assumptions Used in Fitting ROC Curves. *Med Decis Making*. 1988;8(3):197–203.

37. Eng J. ROC analysis: Web-based calculator for ROC curves, http://www.jrocfit.org. 2006.

38. Hillis SL. Simulation of Unequal-Variance Binormal Multireader ROC Decision Data: An Extension of the Roe and Metz Simulation Model. *Acad Radiol*. 2012;19(12):1518–1528.

39. Chakraborty DP. An alternate method for using a visual discrimination model (VDM) to optimize softcopy display image quality. *J Soc Inf Disp*. 2006;14(10):921–926.

40. Chakraborty DP. Predicting detection task performance using a visual discrimination model-II. *Paper presented at: Proceedings of SPIE Medical Imaging 2005: Image Perception, Observer Performance, and Technology Assessment*; 2005; San Diego, CA.

41. Chakraborty DP. Predicting detection task performance using a visual discrimination model. *Proceedings of SPIE Medical Imaging 2004: Image Perception, Observer Performance, and Technology Assessment*; 2004;5372:53–61.

42. Metz CE, Herman B, Shen J. Maximum-likelihood estimation of receiver operating characteristic (ROC) curves from continuously-distributed data. *Stat Med*. 1998;17:1033–1053.

43. Metz CE. Some practical issues of experimental design and data analysis in radiological ROC studies. *Invest Radiol*. 1989;24:234–245.

44. Metz CE, Herman BA, Shen J-H. Maximum likelihood estimation of receiver operating characteristic (ROC) curves from continuously-distributed data. *Stat Med*. 1998;17(9):1033–1053.

45. Chakraborty DP, Breatnach ES, Yester MV, Soto B, Barnes GT, Fraser RG. Digital and Conventional Chest Imaging: A Modified ROC Study of Observer Performance Using Simulated Nodules. *Radiology*. 1986;158:35–39.

46. Niklason LT, Hickey NM, Chakraborty DP, et al. Simulated Pulmonary Nodules: Detection with Dual-Energy Digital versus Conventional Radiography. *Radiology*. 1986;160:589–593.

Analyzing FROC data

18.1 Introduction

The focus of this chapter is on analyzing multiple-reader multiple-case (MRMC) free-response ROC (FROC) datasets. Sample size estimation for FROC studies is deferred to **Chapter 19**. Analyzing FROC data is, apart from a single crucial difference, similar to analyzing ROC data. The difference is the selection of the appropriate figure of merit. The reason is that the DBMH and ORH methods are applicable to any scalar figure of merit (this is not true of non-parametric methods that work only with the empirical ROC-AUC). Any appropriate FROC FOM reduces the mark-rating data for a single dataset (i.e., a specified treatment and a specified reader) to a single scalar figure of merit. No assumptions are made regarding independence of ratings on the same case—an often-misunderstood point. The author recommends usage of the weighted AFROC figure of merit, where the lesions should be equally weighted, the **RJafroc** default, unless there are strong clinical reasons for assigning unequal weights. Experience has shown that the main results, i.e., p-values, are, barring extreme examples, rather insensitive to the actual choice of weights.

18.2 Example analysis of a FROC dataset

The following is a listing of file **mainAnalyzewAFROC.R**. It performs both wAFROC and inferred ROC analyses of the same dataset and the results are saved to tables similar in structure to the Excel output tables shown for DBMH analysis of ROC data in Section 9.10.2.

18.2.1 Code listing

```
rm(list = ls()) #mainAnalyzewAFROC.R
library(RJafroc)

# included datasets
fileNames <-  c("TONY", "VD", "FR", "FED",
                "JT", "MAG", "OPT", "PEN",
                "NICO","RUS", "DOB1", "DOB2",
                "DOB3", "FZR")
f <- 4
fileName <- fileNames[f] # i.e., FED data
```

```
# the datasets already exist as R objects in RJafroc
theData <- get(sprintf("dataset%02d", f))
reportFileNamewAfroc <- paste0(fileName,"wAfroc.xlsx")
reportFileNamewHrRoc <- paste0(fileName,"HrAuc.xlsx")
if (!file.exists(reportFileNamewAfroc)){
  UtilOutputReport(
    dataset = theData,
    reportFormat = "xlsx",
    reportFile = reportFileNamewAfroc,
    renumber = "TRUE",
    method = "DBMH", fom = "wAFROC",
    showWarnings = "FALSE")
}
if (!file.exists(reportFileNamewHrRoc)){
  UtilOutputReport(
    dataset = theData,
    reportFormat = "xlsx",
    reportFile = reportFileNamewHrRoc,
    renumber = "TRUE",
    method = "DBMH", fom = "HrAuc",
    showWarnings = "FALSE")
}
```

The datasets are described in **Online Chapter 24,** and a preview was given in Section 9.1.5. Four of these are ROC datasets, one an LROC dataset and the rest (nine) are FROC datasets. For non-ROC datasets, the highest rating method was used to infer the corresponding ROC data. The datasets are identified in the above code by strings contained in the string-array variable **fileNames** (lines 5–8). Line 9 selects the dataset to be analyzed. In the example shown, the **FED** dataset has been selected. It is is a 5-treatment 4-radiologist FROC dataset[1] acquired by Dr. Federica Zanca. Line 12 accesses the dataset using the function **get()** which converts a string to an R object. Line 13 constructs the name of the wAFROC output file and line 14 does the same for the ROC output datafile. Lines 16–22 generates an output file by performing DBMH significance testing (**method = DBMH**) using **fom = wAFROC**, that is, the wAFROC figure of merit—*this is the critical change*. If one changes this to **fom = HrAuc**, line 30, then inferred ROC analysis occurs. In either case the default analysis, that is, **option = ALL** is used, that is, results of random-reader random-case (RRRC), fixed-reader random-case (FRRC), and random-reader fixed-case (RRFC) are reported. Results are shown below for random-reader random-case only.

The results of wAFROC analysis are saved to **FedwAfroc.xlsx** and that of inferred ROC analysis are saved to **FedHrAuc.xlsx**. The output file names need to be explicitly stated as otherwise they would overwrite each other (as a time-saver, checks are made at lines 15 and 24 to determine if the output files already exist; to run the analysis from scratch the reader should delete these files).

In the Excel data file, the readers are named 1, 3, 4, and 5—the software treats the reader names as labels. The author's guess is that for some reason complete data for reader 2 could not be obtained. The **renumber = TRUE** option has the effect of renumbering the readers 1 through 4. Without renumbering, the output would be aesthetically displeasing, but have no effect on the conclusions.

Figures of merit, empirical wAFROC-AUC, and empirical ROC-AUC, and the corresponding reader averages for both analyses are summarized in Table 18.1 The weighted AFROC results were obtained by copy and paste operations from worksheet **FOMs** in file **FedwAfroc.xlsx**. The

Table 18.1 Empirical wAFROC-AUC and ROC-AUC for all combinations of treatments and readers, and reader-averaged AUCs for each treatment

| Trt | wAFROC-AUC | | | | | ROC-AUC | | | | |
	Rdr-1	Rdr-2	Rdr-3	Rdr-4	Rdr. Avg.	Rdr-1	Rdr-2	Rdr-3	Rdr-4	Rdr. Avg.
1	0.77927	0.72489	0.70362	0.80509	0.75322	0.90425	0.79820	0.81175	0.86645	0.84516
2	0.78700	0.72690	0.72262	0.80378	0.76007	0.86425	0.84470	0.82050	0.87160	0.85026
3	0.72969	0.71576	0.67231	0.77266	0.72260	0.81295	0.81635	0.75275	0.85730	0.80984
4	0.81013	0.74312	0.69436	0.82941	0.76925	0.90235	0.83150	0.78865	0.87980	0.85058
5	0.74880	0.68227	0.65517	0.77125	0.71437	0.84140	0.77300	0.77115	0.84800	0.80839

Note: The weighted AFROC results were obtained from worksheet **FOMs** in file **FedwAfroc.xlsx**. The highest rating AUC results were obtained from worksheet **FOMs** in file **FedHrAuc.xlsx**. The wAFROC-AUCs are smaller than the corresponding ROC-AUCs. [Rdr. Avg. = reader averaged FOM]

Table 18.2 Results of random-reader random-case (RRRC) analysis with wAFROC-AUC FOM. The upper part of the table shows results for treatment FOM differences and the lower part applies to individual treatment FOMs. The overall F-test yielded $F(4,36.8) = 7.8$, p-value = 0.00012. For each paired difference standard error = 0.0122 and degrees of freedom of t-distribution = 36.8

95% CI's FOMs, treatment difference					
Difference	Estimate	t	Pr>t	Lower	Upper
1–2	−0.00686	−0.56265	0.57709	−0.03155	0.01784
1–3*	0.03061	2.51234	0.01651	0.00592	0.05531
1–4	−0.01604	−1.31592	0.19634	−0.04073	0.00866
1–5*	0.03884	3.18766	0.00292	0.01415	0.06354
2–3*	0.03747	3.07499	0.00396	0.01278	0.06217
2–4	−0.00918	−0.75328	0.45608	−0.03387	0.01552
2–5*	0.04570	3.75031	0.00061	0.02100	0.07040
3–4*	−0.04665	−3.82827	0.00048	−0.07135	−0.02195
3–5	0.00823	0.67532	0.50370	−0.01647	0.03292
4–5*	0.05488	4.50358	0.00007	0.03018	0.07957

95% CI's FOMs, each treatment					
Treatment	Estimate	StdErr	df	Lower	Upper
1	0.75322	0.02976	7.70845	0.68413	0.82231
2	0.76007	0.02843	10.69208	0.69727	0.82288
3	0.72260	0.02692	8.61918	0.66128	0.78392
4	0.76925	0.03572	5.24242	0.67870	0.85981
5	0.71437	0.03325	6.58542	0.63473	0.79402

Note: CI = confidence interval; * = Significantly different at alpha = 0.05; **Pr>t** is the p-value of the t- test; **StdErr** = standard error of FOM estimate; df = degrees of freedom of t-distribution.

highest rating AUC results were obtained by similar operations from worksheet **FOMs** in Excel file **FedHrAuc.xlsx**. As expected, each wAFROC-AUC is smaller than the corresponding ROC-AUC.

Table 18.2 shows results for RRRC analysis using the wAFROC-AUC FOM. The overall F-test of the null hypothesis that all treatments have the same reader-averaged FOM, rejected the NH: $F(4, 36.8) = 7.8, p = 0.00012$.* The numerator degree of freedom ndf is $I-1 = 4$. Since the null hypothesis is that *all* treatments have the same FOM, this implies that *at least* one pairing of treatments yielded a significant FOM difference. The control for multiple testing is in the formulation of the null hypothesis and no further Bonferroni-like[2] correction is needed. To determine which pairings are significantly different one examines the p-values (listed under `Pr>t`) in the **95% CI's FOMs, treatment difference** portion of the table. It shows that the following differences are significant at alpha = 0.05, namely 1–3, 1–5, 2–3, 2–5, 3–4, and 4–5; these are indicated by asterisks. The values listed under the **95% CI's FOMs, each treatment** portion of the table show that treatment 4 yielded the highest FOM (0.769) followed closely by treatments 2 and 1, while treatment 5 had the least FOM (0.714), slightly worse than treatment 3. This explains why the p-value for the difference 4-5 is the smallest (0.00007) of all the listed p-values in the **95% CI's FOMs, each treatment** portion of the table. Each instance where the p-value for the individual treatment comparisons yields a significant p-value is accompanied by a 95% confidence interval that does not include zero. The two statements of significance, one in terms of a p-value and one in terms of a CI, are equivalent. When it comes to presenting results for treatment FOM differences, the author prefers the 95% CI but some journals insist on a p-value, even when it is not significant. Note that *two sequential tests* are involved, an overall F-test of the NH that all treatments have the same performance and *only* if this yields significant results is one justified in looking at the p-values of individual treatment pairings.

Table 18.3 shows corresponding results for the inferred ROC-AUC FOM. Again, the null hypothesis was rejected: $F(4, 16.8) = 3.46, p = 0.032$. This means at least two treatments have significantly

Table 18.3 Results of random-reader random-case (RRRC) analysis with inferred ROC-AUC FOM. The upper part of the table shows results for treatment FOM differences and the lower part applies to individual treatment FOMs. The overall F-test yielded F(4,16.8) = 3.47, p-value = 0.03054. For each paired difference standard error = 0.01654 and degrees of freedom of t-distribution = 16.8

95% CI's FOMs, treatment difference					
Difference	Estimate	t	Pr>t	Lower	Upper
1–2	−0.00510	−0.30840	0.76157	−0.04002	0.02982
1–3*	0.03533	2.13611	0.04769	0.00040	0.07025
1–4	−0.00541	−0.32729	0.74749	−0.04033	0.02951
1–5*	0.03678	2.22379	0.04017	0.00185	0.07170
2–3*	0.04043	2.44450	0.02584	0.00550	0.07535
2–4	−0.00031	−0.01890	0.98515	−0.03523	0.03461
2–5*	0.04188	2.53218	0.02161	0.00695	0.07680
3–4*	−0.04074	−2.46340	0.02487	−0.07566	−0.00582
3–5	0.00145	0.08768	0.93117	−0.03347	0.03637
4–5*	0.04219	2.55108	0.02079	0.00727	0.07711
95% CI's FOMs, each treatment					
Treatment	Estimate	StdErr	df	Lower	Upper
1	0.84516	0.02858	5.45748	0.77351	0.91681
2	0.85026	0.01993	27.72258	0.80942	0.89110
3	0.80984	0.02665	7.03714	0.74689	0.87278
4	0.85058	0.02941	5.40326	0.77664	0.92451
5	0.80839	0.02576	6.77565	0.74707	0.86971

Note: CI = confidence interval; * = Significantly different at alpha = 0.05; **Pr>t** is the p-value of the *t*- test; **StdErr** = standard error of FOM estimate; df = degrees of freedom of t-distribution.

---

* This is the way results should be reported in publications.

different FOMs. Looking down the table, one sees that the same 6 pairs (as with wAFROC analysis) are significantly different, 1–3, 1–5, etc., as indicated by the asterisks. The last five rows of the table show that treatment 4 had the highest performance while treatment 5 had the lowest performance. At the 5% significance level, both methods yielded the same significant differences, but this is not always true. While it is incorrect to conclude from a single dataset that a smaller p-value is indicative of higher statistical power, simulation testing under controlled conditions has consistently shown higher statistical power for the wAFROC-AUC FOM[3,4] as compared to the inferred ROC-AUC FOM.

18.3 Plotting wAFROC and ROC curves

It is important to display empirical wAFROC/ROC curves, not just for publication purposes, but to get a better feel for the data. Since treatments 4 and 5 showed the largest difference, the corresponding wAFROC/ROC plots are displayed. The code is in file **mainwAfrocRocPlots.R**.

18.3.1 Code listing

```
rm(list = ls()) #mainwAfrocPlots.R
library(RJafroc);library(ggplot2)

# included datasets
fileNames <-  c("TONY", "VD", "FR", "FED",
                "JT", "MAG", "OPT", "PEN",
                "NICO","RUS", "DOB1", "DOB2",
                "DOB3", "FZR")

f <- 4
fileName <- fileNames[f] # i.e., FED data
# the datasets already exist as R objects in RJafroc
theData <- get(sprintf("dataset%02d", f))

# wAFROC plots
# plots for all treatment reader combinations
plotT <- c(4,5);plotR <- c(1,2,3,4)
p <- PlotEmpiricalOperatingCharacteristics(
  theData,trts = plotT, rdrs = plotR,
  opChType = "wAFROC")
print(p$Plot)

# create reader averaged curves in each modality
plotT <- list(4,5);plotR <- list(c(1:4),c(1:4))
p <- PlotEmpiricalOperatingCharacteristics(
  theData,trts = plotT, rdrs = plotR,
  opChType = "wAFROC")
print(p$Plot)

# ROC plots
# plots for all treatment reader combinations
plotT <- c(4,5);plotR <- c(1,2,3,4)
rocData <- DfFroc2Roc(theData)
p <- PlotEmpiricalOperatingCharacteristics(
  rocData,trts = plotT, rdrs = plotR,
```

```
    opChType = "ROC")
  print(p$Plot)

  plotT <- list(4,5);plotR <- list(c(1:4),c(1:4))
  p <- PlotEmpiricalOperatingCharacteristics(
    rocData,trts = plotT, rdrs = plotR,
    opChType = "ROC")
  print(p$Plot)
```

Sourcing this code yields Figure 18.1. Plot (a), originating from lines 17 through 21, shows individual reader wAFROC plots for treatment 4 (solid lines) and treatment 5 (dashed lines). Running the software on one's computer shows color-coded plots that are easier to discriminate. While difficult to see, close examination of this plot shows that all readers performed better in treatment 4 than in treatment 5 (i.e., for each reader the solid line is above the dashed line). Plot (b), originating from lines 24 through 28, shows reader-averaged wAFROC plots for treatments 4 (dark line, upper curve) and 5 (light line, lower curve). If one changes, for example, line 21 from **print (p$Plot)** to **print (p$Points)** the code will output the *coordinates* of the points describing the curve, which gives the user the option to copy and paste the operating points into alternative plotting software.

Lines 17 through 21 create plots for all specified treatment-reader combinations. The method to creating reader-averaged curves, such as in (b), is defining two **list** variables, **plotT** and **plotR**, at line 24, the first containing the treatments to be plotted, **list(4,5)**, and the second, a **list** of equal length, containing the arrays of readers to be averaged over, **list(c(1:4), c(1:4))**. More examples can be found in the help page for **PlotEmpiricalOperatingCharacteristics()**.

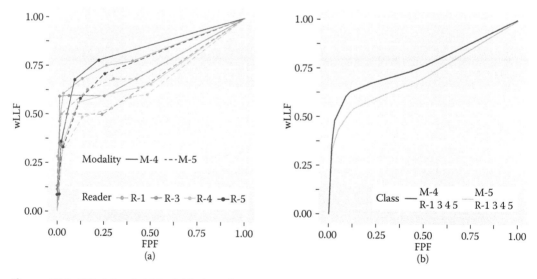

Figure 18.1 FED dataset. (a) Individual reader wAFROC plots for treatments 4 and 5. All readers performed better in treatment 4 as indicated by each solid line being above the corresponding dashed lines. (b) Reader-averaged wAFROC plots for treatments 4 and 5. The performance superiority of treatment 4 is obvious in this curve. The difference is significant, $p = 0.00012$.

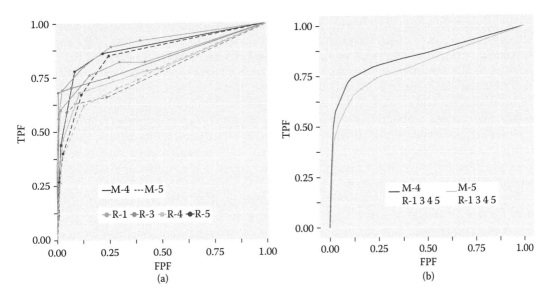

Figure 18.2 FED dataset. (a) Individual reader ROC plots for treatments 4 and 5. While difficult to see, all readers performed better in treatment 4. (b) Reader-averaged ROC plots for treatments 4 and 5. The performance superiority of treatment 4 is fairly obvious in this curve. The difference is significant, $p = 0.03054$.

> Meaningful operating points on the reader-average curves cannot be defined. This is because ratings are treatment and reader specific *labels*, so one cannot, for example, sum the bin counts over all readers to construct a table like ROC Table 4.1 or its AFROC counterpart, Table 13.3.

Instead, the following procedure is used internal to `PlotEmpiricalOperating Characteristics()`. The reader-averaged plot for a specified treatment is obtained by dividing the *FPF* range from 0 to 1 into finely spaced steps. For each *FPF* value the *wLLF* values for that treatment are averaged over all readers, yielding the reader-averaged ordinate. Calculating confidence intervals on the reader-averaged curve is possible but cumbersome and unnecessary in the author's opinion. The relevant information, namely the 95% confidence interval on the difference in reader-averaged AUCs, is already contained in the program output, see Table 18.2, row labeled **4–5***. The difference is 0.05488 with a 95% confidence interval (0.03018, 0.07957) which does not include zero.

Inferred ROC plots corresponding to Figure 18.2 are generated by lines 32 through 43, that is, by changing `opChType = wAFROC` to `opChType = ROC` resulting in Figure 18.3. From Table 18.3 it is seen that the difference in reader-averaged AUCs is 0.04219 with a 95% confidence interval (0.00727, 0.07711). The observed wAFROC effect size, 0.05488, is larger than the corresponding inferred ROC effect size, 0.04219. This is a common observation, but sampling variability compounded with small differences, could give different results in some situations; see suggested simulation study following Table 17.1.

18.3.2 Reporting a study

The methods section of a report should make it clear exactly how the study was conducted. The information should be enough to allow someone else to replicate the study. How many readers, how many cases, how many treatments were used? How was ground truth determined and if the FROC paradigm was used, how were true lesion locations determined? The instructions to the readers should be clearly stated and put in writing. Precautions to minimize reading order effects should be stated—usually this is accomplished by interleaving cases from different treatments so that the chances that cases from a particular treatment is always seen first by every reader are minimized. Additionally, images from the same case, but in different treatments, should not be viewed in the same reading session. Reading sessions are usually an hour, and the different sessions should ideally be separated by at least one day. Users generally pay minimal attention to training sessions. It is recommended that at least 25% of the total number of interpretations be training cases and cases used for training should not be used in the main study. Feedback should be provided during the training session to allow the reader to become familiar with the range of difficulty levels regarding diseased and non-diseased cases in the dataset. Deception, for example, stating a higher prevalence than is actually used, is usually not a good idea. The user-interface should be explained carefully. The best user interface is intuitive, minimizes keystrokes, and requiring the least explanation.

In publications, the paradigm used to collect the data (ROC, FROC, etc.) and the figure of merit used for analysis should be stated. If FROC, the proximity criterion should be stated. The analysis should state the NH and the alpha of the test, and the desired generalization. The software used and appropriate references should be cited. The results of the overall F-test, the p-value, the observed F-statistic and its degrees of freedom should be stated. If the NH is not rejected, one should cite the observed inter-treatment FOM differences, confidence intervals, and p-values and, ideally, provide preliminary sample size estimates. This information could be useful to other researchers attempting to conduct a larger study. If the NH is rejected, a table of inter-treatment FOM differences such as Table 18.3 should be summarized. Reader-averaged plots of the relevant operating characteristics for each treatment should be provided. In FROC studies it is recommended to vary the proximity criterion, perhaps increasing it by a factor of 2, to test if the final conclusions (is NH rejected and if so which treatment is highest) are unaffected.

Assuming the study has been done properly and with sufficiently large number of cases, the results should be published in some form, even if the NH is not rejected. The dearth of datasets to allow reasonable sample size calculations is a real problem in this field. The dataset should be made available, perhaps on **ResearchGate**, or if communicated to the author, they will be included in future updates to **RJafroc**. Datasets acquired via NIH or other government funding must be made available upon request, in an easily decipherable format. Subsequent users of these datasets must cite the original source of the data. Given the high cost of publishing excess pages in some journals, an expanded version, if appropriate for clarity, should be made available using online posting avenues such as **ResearchGate**.

18.4 Single fixed-factor analysis

So far focus has been in comparing two or more treatments, i.e., whether the reader-averaged inter-treatment FOM differences are significantly different from zero. What does one do if there is only data for a single treatment? Now interest could be in whether the readers, regarded as a fixed effect, have significantly different performances in this treatment. Alternatively, one could have a single reader interpreting a common set of cases in multiple treatments, in which case, interest is in whether the reader's performances on the treatments are significantly different. Both of these problems fall under the category of single fixed-factor analysis. In the first example the reader is treated as a single fixed-factor and in the second example treatment is regarded as a single fixed-factor. (If an effect is regarded as random, which is possible with readers, but not with treatments, then it would *not* make sense to ask whether the random reader samples are significantly different from each other. For a random factor, all conclusions apply to its population parameters, such as mean and variance, not to individual levels of the factor.)

Assume J (> 1) readers interpret the same cases in a single treatmentand and interest is in determining if the performances of the readers are significantly different. The null hypothesis is that the reader performances are identical, and if it can be rejected then at least one pair of readers have significantly different performances. This is illustrated for the same dataset used earlier, except this time one needs to create a single treatment dataset from a multiple treatment dataset. The function **DfExtractDataset()** serves that purpose. The code is in file **mainAnalyzeSingleModality.R**, which implements the single factor analysis method described in Ref. 5.

18.4.1 Code listing

```
rm(list = ls()) # mainAnalyzeSingleModality.R
library(RJafroc)

# included datasets
fileNames <-  c("TONY", "VD", "FR", "FED",
                "JT", "MAG", "OPT", "PEN",
                "NICO","RUS", "DOB1", "DOB2",
                "DOB3", "FZR")

f <- 4
fileName <- fileNames[f] # i.e., FED data
# the datasets already exist as R objects in RJafroc
theData <- get(sprintf("dataset%02d", f))

singleModData <- DfExtractDataset(
  theData, trts = 1)
singleModDataRes <-
  StSignificanceTestingSingleFixedFactor(
    singleModData, fom = "wAFROC")
print(singleModDataRes)
```

At line 15–16 the data for treatment 1 is extracted to the dataset object **singleModData**. Line 17–19 applies single fixed-factor analysis, **StSignificanceTestingSingleFixedFactor()**, and the next line prints the results. The output of sourcing the above code is shown in Section 18.4.2.

18.4.2 Code output

```
source(...)
$f
[1] 4.061743

$ddf
[1] 597

$pValue
[1] 0.007138494

$fomStats
Reader      Area     stdErr    DF   CILower   CIUpper
1          0.7793   0.02908   199   0.7219    0.8366
3          0.7249   0.02893   199   0.6678    0.7819
4          0.7036   0.03284   199   0.6389    0.7684
5          0.8051   0.02737   199   0.7511    0.8591
```

```
$diffFomStats
          DifferenceFom      t      DF    Pr > t    CILower CIUpper
1 - 3            0.054     1.90    199    0.058    -0.0020    0.11
1 - 4            0.076     2.09    199    0.038     0.0042    0.15
1 - 5           -0.026    -0.790   199    0.431    -0.0903    0.04
3 - 4            0.021     0.589   199    0.557    -0.0500    0.09
3 - 5           -0.080    -2.731   199    0.007    -0.138    -0.02
4 - 5           -0.101    -2.97    199    0.003    -0.169    -0.03
```

The F-statistic is 4.06 with 597 denominator degrees of freedom. The numerator degree of freedom, not shown, is the number of readers minus 1, that is, 3.* The corresponding p-value is 0.0071 = 1 - $\mathbf{pf}$(4.06, 3 ,597), which is significant at alpha = 0.05. This means that one can reject the null hypothesis that all readers have identical performance in treatment 1. The **$diffFomStats** values show that the differences between readers 1 and 4, 3 and 5, and 4 and 5 are significant. From the **$fomStats values** one sees that reader FOMs are ordered, in increasing order of performance, as follows: 4 < 3 < 1 < 5 (using original reader numbers; there is no reader 2 in this dataset).

18.5 Crossed-treatment analysis

This analysis was developed for a particular application[6] in which nodule detection in an anthropomorphic chest phantom in computed tomography (CT) was evaluated as a function of tube charge (mAs) and reconstruction method. The phantom was scanned at 4 values of mAs and images were reconstructed with adaptive iterative dose reduction 3D (AIDR[3D]) and filtered back projection (FBP). Thus, there are *two treatment* factors and the factors are *crossed* since for each value of the mAs factor there were two values of the reconstruction algorithm factor. Interest was in determining if whether performance depends on the mAs and/or reconstruction method.

In a typical analysis of MRMC ROC or FROC study, treatment is considered as a single factor with I levels, where I is usually small. The figure of merit for treatment i ($i = 1,2,...,I$) and reader j ($j = 1,2,...,J$) is denoted θ_{ij}. MRMC analysis compares the observed magnitude of the difference in reader-averaged figures of merit between treatments i and i', $i \neq i'$, to the estimated standard deviation of the difference. For example, the reader-averaged difference in figures of merit is $\theta_{i\bullet} - \theta_{i'\bullet}$, where the dot symbol represents the average over the corresponding (reader) index. The standard deviation of the difference is estimated using the DBMH or the ORH method. With I levels, the number of distinct i versus i' comparisons is $I(I-1)/2$. If the current study were analyzed in this manner, where $I = 8$ (4 levels of mAs and two image reconstruction methods), then this would imply 28 comparisons. The large number of comparisons does not inform the researcher regarding the main questions of interest, that is, whether performance depends on the mAs and/or the reconstruction method. For example, in standard analysis the two reconstruction algorithms might be compared at different mAs levels, and one is in the dark as to which factor (algorithm or mAs) caused an observed significant difference.

Unlike conventional ROC type studies, the images in this study are defined by two factors. The first factor, *tube charge*, had four levels: 20, 40, 60, and 80 mAs. The second factor, *reconstruction method*, had two levels: FBP and AIDR[3D]. The figure of merit is represented by $\theta_{i_1 i_2 j}$, where i_1 ($i_1 = 1,2,...,I_1$) represents the levels of the first factor (mAs), $I_1 = 4$ and i_2 ($i_2 = 1,2,...,I_2$) represents the levels of the second factor (reconstruction method), $I_2 = 2$. Two sequential analyses were performed: (1) *mAs analysis*, where the figure of merit was averaged over i_2 (the reconstruction index), and (2) *reconstruction analysis*, where the figure of merit was averaged over i_1 (the mAs index). For example, the *mAs analysis* figure of merit is $\theta_{i_1 \bullet j}$, where the dot represents the average over the

---

* A future update of the software will include this value.

reconstruction index, and the corresponding reconstruction analysis figure of merit is $\theta_{\bullet i_2 j}$, where the dot represents the average over the mAs index. Thus, in either analysis, *the figure of merit is dependent on a single treatment factor,* and therefore standard DBMH or ORH methods apply.

The *mAs analysis* determines whether tube charge is a significant factor and in this analysis the number of possible comparisons is only six. The *reconstruction analysis* determines whether AIDR[3D] offers any advantage over FBP and in this analysis the number of possible comparisons is only one. Multiple testing on the same dataset increases the probability of Type-I error, therefore a Bonferroni correction is applied by setting the threshold for declaring significance at 0.025; this is expected to conservatively maintain the overall probability of a Type-I error at $\alpha = 0.05$. *Crossed-treatment* analysis is used to describe this type of analysis of ROC/FROC data, which yields clearer answers on which of the two factors effects performance. The averaging over the other treatment has the effect of increasing the power of the study in detecting differences in each of the two factors.

Since the phantom is unique, and conclusions are only possible that are specific to this one phantom, the case (or image) factor was regarded as fixed. For this reason, only results of random-reader fixed-case analyses are reported.

18.5.1 Example of crossed-treatment analysis

The dataset is contained in file **CrossedModalitiesDataFile.xlsx**.[*3] Examination of this file, Figure 18.3, reveals that it has two treatment ID fields, **Modality _ ID1**, the reconstruction algorithm with two levels, *I* and *F*, corresponding to AIDR[3D] and FBP, respectively, and **Modality _ ID2**, the mAs factor with four levels, 20, 40, 60, and 80.

The code for this analysis is in file **MainCrossedModalities.R**, a listing of which follows.

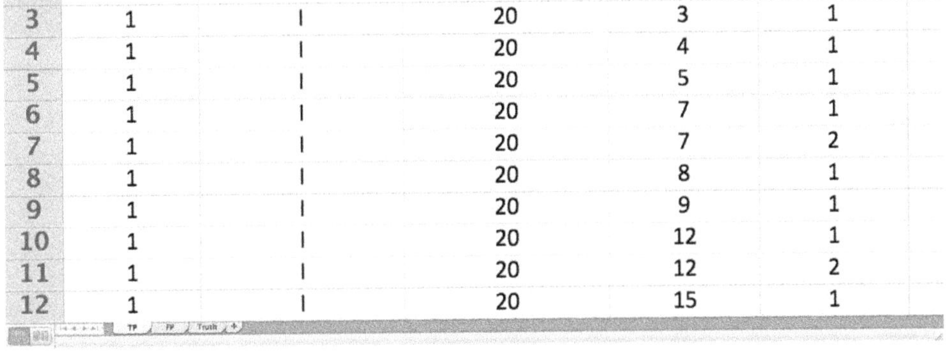

	A	B	C	D	E	T
1	Reader_ID	Modality_ID1	Modality_ID2	Case_ID	Lesion_ID	
2	1	I	20	1	1	
3	1	I	20	3	1	
4	1	I	20	4	1	
5	1	I	20	5	1	
6	1	I	20	7	1	
7	1	I	20	7	2	
8	1	I	20	8	1	
9	1	I	20	9	1	
10	1	I	20	12	1	
11	1	I	20	12	2	
12	1	I	20	15	1	

Figure 18.3 Structure of the **TP** worksheet for a crossed modality dataset. The I in column B denotes iterative reconstruction; the other possibility if F for filtered back-projection. In column C are listed the mAs values. A similar structure applies to the **FP** worksheet. The only change from a standard JAFROC format data file is the presence of the additional modality column.

---

* The author thanks Dr. John Thompson and Prof. Peter Hogg, University of Salford, for the dataset.

18.5.2 Code listing

```
rm(list=ls(all=TRUE)) # MainCrossedModalities.R
library(RJafroc)

InputFilenameBase <- "CrossedModalitiesDataFile"
InputFilename <- paste0(InputFilenameBase,".xlsx")
fom <- "wAFROC"

cat("***Data file is ", InputFilenameBase, '***\n')
cat("fom = ", fom, '\n')

dataset <- DfReadCrossedModalities(InputFilename)
res1 <- StSignificanceTestingCrossedModalities(
  dataset,avgIndx = 1, fom = fom, option = "RRFC")
res2 <- StSignificanceTestingCrossedModalities(
  dataset,avgIndx = 2, fom = fom, option = "RRFC")
cat("**** Averaging over reconstruction index  *****\n")
print(res1)
cat("**** Averaging over mAs index  *****\n")
print(res2)
```

Line 6 defines **fom**, that is, the figure of merit, as **wAFROC**. Line 11 uses **DfReadCrossedModalities()** to read the crossed modality dataset. Lines 12–13 perform the mAs analysis using function **StSignificanceTestingCrossedModalities**. The first argument to this function is the **dataset**. The argument **avgIndx = 1** tells the function to average the FOM over the first modality index, that is, over the reconstruction index. The third argument passes the FOM argument and the **option** argument requests RRFC analysis, as this is a phantom study. The results are in Section 18.5.3, where due to page-width constraints, some of the output is suppressed; the reader should run the code to view the full output.

18.5.3 Code output, partial

```
> source(...)
***Data file is  CrossedModalitiesDataFile ***
fom =  wAFROC
**** Averaging over reconstruction index  *****
$fomArray
... (suppressed)

$msT
[1] 0.006286043

$msTR
[1] 0.0003938919

$varComp
             varCov
Var(R)       2.419778e-03
Var(T*R)     6.212086e-05
COV1         5.838726e-04
```

```
COV2        4.686406e-04
COV3        3.880314e-04
Var(Error)  9.962529e-04


$fRRFC
[1]  15.9588


$ddfRRFC
[1]  30


$pRRFC
[1]  2.185176e-06
```

This shows that for random-reader fixed-case analysis (RRFC), the F-test yields F(3,30) = 15.96; p = 2.18e-06 (since two sequential tests are involved, one uses the Bonferroni corrected alpha = 0.025 for the overall test to be valid at the 5% level). All mAs differences are significant except 40–60 and 60–80. The reader-averaged FOMs show that FOM, averaged over both reconstruction methods, increases with mAs.

Lines 14–15 perform the reconstruction algorithm analysis. The argument **avgIndx = 2** tells the function to average the FOM over the second modality index, that is, over the mAs index. The abbreviated results are in Section 18.5.4.

18.5.4 Code output, partial

```
**** Averaging over mAs index  *****
$fomArray
...  (suppressed)


$msT
[1]  1.268641e-05


$msTR
[1]  0.0001676095


$varComp
                 varCov
Var(R)       2.426193e-03
Var(T*R)     1.823035e-05
COV1         6.091068e-04
COV2         4.113551e-04
COV3         4.050123e-04
Var(Error)   7.648287e-04


$fRRFC
[1]  0.07569028


$ddfRRFC
[1]  10


$pRRFC
[1]  0.7888258
```

Unlike mAs, the reconstruction method does not have a significant effect. The FOM difference, -0.00152, is quite small, $F(1,10) = 0.076$, and $p = 0.79$. The reader is referred to the referenced publication for full details of the study. Finally, the reader should compare the above results to those obtained using the ROC figure of merit, which uses the trapezoidal ROC-AUC, and confirm that they are consistent with the wAFROC FOM (the p-values for the mAs analysis is slightly larger).* Because the mAs effect was so large, both wAFROC and ROC analyses declared the same significant differences for levels of mAs while not finding reconstruction algorithm to have a significant effect.

18.6 Discussion/Summary

An IDL (Interactive Data Language, currently marketed by Exelis Visual Information Solutions, www.exelisvis.com) version of JAFROC was first posted to a now obsolete website on 4/16/2004. This software required a license for IDL, which most users did not have. Subsequently (9/27/2005) a version was posted which allowed analysis using the freely downloadable IDL Virtual Machine software (a method for freely distributing compiled IDL code). On 1/11/2011 the standalone Windows-compatible version was posted (4.0) and the current version is 4.2. JAFROC is windows compatible (XP, Vista and Windows 7, 8, and 10).

To our knowledge JAFROC is the only easily accessible software currently available that can analyze FROC data. Workstation software for acquiring ROC and FROC data is available from several sources.[7-9] The Windows version is no longer actively supported (bugs, if pointed out, will be corrected). Current effort to conduct research and distribute software uses the **R** platform.[10] There are several advantages to this. **R** is an open-source platform—the author has already benefited from a bug pointed out by a user.† **R** runs on practically any platform (Windows, OSX, Linux, etc.). Also, developing an **R** package benefits from other contributed R-packages, which allow easy computation of probability integrals, random number generation, and parallel computing to speed up computations, to name just a few. The drawback with **R**, and this has to do with its open source philosophy, is that one cannot readily integrate existing ROC code, developed on other platforms and other programming languages (specifically, **DLL**s are not allowed in **R**). So useful programs like CORROC2 and CBM were coded in C++; **R** allows C++ programs to be compiled and included in a package provided the source code is included.

Due to the random number of marks per image, data entry in the FROC paradigm is inherently more complicated and error-prone than in ROC analysis, and consequently, and in response to feedback from users, much effort has gone into error-checking. The users have especially liked the feature where the program indicates the Excel sheet name and line-number where an error is detected. User-feedback has also been very important in detecting program bugs and inconsistencies in the documentation and developing additional features (e.g., crossed-modality analysis).

Interest in the FROC paradigm is evidenced by the fact that Ref. 3 describing the JAFROC method has been cited over 289 times. Over 25,000 unique visitors have viewed the author's website, at least 73 have downloaded the software, and over 107 publications using JAFROC have appeared. The list is available on the author's website. JAFROC has been applied to magnetic resonance imaging, virtual computerized tomography colonoscopy, digital tomosynthesis (chest and breast), mammography dose and image processing optimization, computer aided detection (CAD), computerized tomography, and other applications.

Since confusion still appears to exist, especially among statisticians, regarding perceived neglect of intra-image correlations of ratings and how true negatives are handled in FROC analysis,[11] we

---

* One should not blindly expect the p-value for ROC analysis to be always larger than the corresponding wAFROC analysis p-value. After all, the observed p-value is a realization of a random variable. See Online Appendix 18.A for a recent instructive correspondence with a new user of the software.

† The author thanks Dr. Lucy D'Agostino McGowan for pointing out an error and the fix in an early version of the software posted to CRAN.

close with a quote from respected sources:[12] "(Chakraborty and Berbaum) have presented a solution to the FROC problem using a jackknife resampling approach that respects the correlation structure in the images ... their paradigm successfully passes a rigorous statistical validation test." Since 2005 the National Institutes for Health (NIH) has been generous with supporting the research and users of JAFROC have been equally generous with providing their datasets, which have resulted in several collaborations.

It is time to address a crucial issue—how does one estimate RSM parameters from FROC or ROC data and how do RSM predicted ROC fits compare to other methods of fitting ROC data? Besides allowing estimation of factors limiting performance, it turns out the ROC curve fitting ability is needed for sample size estimation for FROC studies. These are the topics of the next chapter.

References

1. Zanca F, Jacobs J, Van Ongeval C, et al. Evaluation of clinical image processing algorithms used in digital mammography. *Med Phys.* 2009;36(3):765–775.
2. Bland JM, Altman DG. Multiple significance tests: The Bonferroni method. *BMJ.* 1995;310(6973):170.
3. Chakraborty DP, Berbaum KS. Observer studies involving detection and localization: Modeling, analysis and validation. *Med Phys.* 2004;31(8):2313–2330.
4. Chakraborty DP. Validation and statistical power comparison of methods for analyzing free-response observer performance studies. *Acad Radiol.* 2008;15(12):1554–1566.
5. Hillis SL, Obuchowski NA, Schartz KM, Berbaum KS. A comparison of the Dorfman-Berbaum-Metz and Obuchowski-Rockette methods for receiver operating characteristic (ROC) data. *Stat Med.* 2005;24(10):1579–1607.
6. Thompson JD, Chakraborty DP, Szczepura K, et al. Effect of reconstruction methods and x-ray tube current-time product on nodule detection in an anthropomorphic thorax phantom: A crossed-modality JAFROC observer study. *Med Phys.* 2016;43(3):1265–1274.
7. Thompson J, Hogg P, Thompson S, Manning D, Szczepura K. ROCView: Prototype software for data collection in jackknife alternative free-response receiver operating characteristic analysis. *Br J Radiol.* 2012;85(1017):1320–1326.
8. Håkansson M, Svensson S, Zachrisson S, Svalkvist A, Båth M, Månsson LG. ViewDEX: An efficient and easy-to-use software for observer performance studies. *Radiat Prot Dosimetry.* 2010;139(1–3):42–51.
9. Jacobs J, Zanca F, Bosmans H. A novel platform to simplify human observer performance experiments in clinical reading environments. *Proc SPIE.* 2011;7966:79660B.
10. *R: A language and environment for statistical computing. R Development Core Team,* http://www.R-project.org/; 2011.
11. Levine D, Bankier AA, Halpern EF. Submissions to radiology: Our top 10 list of statistical errors1. *Radiology.* 2009;253(2):288–290.
12. Wagner RF, Metz CE, Campbell G. Assessment of medical imaging systems and computer aids: A tutorial review. *Acad Radiol.* 2007;14(6):723–748.

19

Fitting RSM to FROC/ROC data and key findings

19.1 Introduction

The radiological search model (RSM) is based on what is known, via eye-tracking measurements, about how radiologists interpret medical images.[1] The ability of this model to predict search and lesion-classification expertise was described in **Chapter 17**. If one could estimate the three parameters of the RSM from clinical datasets, then one would know which component of expertise is limiting performance. This would provide insight into the decision-making efficiency of observers that goes beyond ideal observer models limited to simplistic backgrounds and simulated lesions at known locations.[2,3], the parameters of the RSM would provide insight into the decision making efficiency of observers as measured on actual clinical images. For this potential to be realized, one has to be able to reliably estimate parameters of the RSM from data, and this turned out to be a difficult problem.

To put progress in this area in context a brief historical background is needed. The author has worked on and off on the FROC estimation problem since 2002, and two persons (Dr. Hong-Jun Yoon and Xuetong Zhai) can attest to the effort, the missteps, and so on, that are part of research. Initial attempts focused on fitting the free-response ROC (FROC) curve, in the (subsequently shown to be mistaken) belief that this was using all the data, so it must be the optimal approach. In fact, unmarked non-diseased cases are not taken into account in the FROC plot. In addition, there are degeneracy issues (see Section 19.2.3 below), which make parameter estimation difficult except in special uninteresting situations. Accordingly, early work involved maximization of the FROC log-likelihood function (see Equation 19.6 below). The FROC curve based method was applied to seven designer-level computer aided detection (CAD) datasets. With CAD data, one has a large number of marks and unmarked cases are relatively rare (they did make up about 12.5% of the cases).[4] Only the CAD designer knows of their existence since in the clinic only a small fraction of the marks, that is, those whose Z-samples exceed a manufacturer-selected threshold, are actually shown to the radiologist. In other words, the full raw FROC curve, extending to the end-point, is available to the CAD algorithm designer, which makes estimation of the end-point defining parameters (λ', ν') almost trivial, see Equation 19.8. Estimating the remaining parameter of the RSM is then also relatively easy.

It was gradually recognized that the FROC curve-based method worked only for designer-level CAD data, and not for human observer data. Consequently, subsequent efforts focused on ROC curve-based fitting, and this proved successful at fitting radiologist datasets where a detailed definition of the ROC curve is not available, i.e., the number of marks is much smaller than with designer-level CAD, and furthermore, human observers may not reach the ROC end-point, that is, they cannot be assumed to report *every* suspicious region. A preliminary account of this work can be found in a conference proceeding.[5]

> *The reader may be surprised to read that the research eventually turned to ROC curve-based fitting, which implies that one does not even need FROC data in order to estimate RSM parameters. The author has previously stated that the ROC paradigm ignores search, so how can one estimate search-model parameters from ROC data? The reason is that the shape of the ROC curve, in particular, the position of the uppermost observed operating point, depends on the RSM parameters, and this information can be used for a successful fitting method that is not as susceptible to degeneracy (it turns out degeneracy is not entirely eliminated). In addition, the algorithm to be described below can fit degenerate datasets, defined as those that do not provide any interior data points, that is, all operating points lie on the edges of the ROC square.*

The chapter starts with fitting FROC curves. This is partly for historical reasons and to make contact with a method used by CAD designers. Then focus shifts to fitting ROC curves and comparing the RSM-based method to existing methods, namely the proper ROC[6,7] (PROPROC) and the contaminated binormal model[8-10] (CBM) methods, both of which are proper ROC fitting models. These are described in more detail in **Chapter 20**. The comparison is based on a large number of interpretations, namely, 14 MRMC datasets comprising 43 modalities, 80 readers, and 2012 cases, most of which are from the author's international collaborations. The datasets are described in **Online Chapter 24**. Besides providing further evidence for the validity of the RSM, the estimates of search and lesion-classification performance derived from the fitted parameters demonstrate that there is information in ROC data that is currently ignored by analyses that do not account for search performance. *Specifically, it shows that search performance is the bottleneck that is currently limiting radiologist performance. This is the scientific significance of the ability to fit the RSM to ROC data.*

The ability to fit RSM to clinical datasets is critical to sample size estimation—*this was the practical reason why the RSM fitting problem had to be solved.* Sample size estimation requires relating the wAFROC-AUC FOM to the corresponding ROC-AUC FOM in order to obtain a physically meaningful effect size. Lacking a mathematical relationship between them, comparing the effect sizes in the two units would be like comparing apples and oranges. A mathematical relation is only possible if one has a parametric model that predicts both ROC and wAFROC curves, as does the RSM, **Chapter 17**. Therefore, this chapter concludes with sample size estimation for FROC studies using the wAFROC FOM. However, as long as one can predict the appropriate operating characteristic using RSM parameters, the method is readily extended to other paradigms,[11] for example, the location ROC (LROC) and the region of interest (ROI) paradigms.

A word of advice to the reader: this chapter may not be the easiest to assimilate. The author has spent considerable effort in the organization and writing of this chapter to make it as accessible as possible, but it will need some effort from the reader to understand the material. Since this is a key chapter, the author believes the effort will be worth it.

19.2 FROC likelihood function

Recall that the likelihood function is the probability of observing the data as a function of the parameter values. FROC notation was summarized in Table 13.1. Thresholds $\vec{\zeta} \equiv \zeta_r$; $r = 0, 1, 2, ..., (R_{FROC} + 1)$ were defined, where R_{FROC} is the number of FROC bins, with $\zeta_0 = -\infty$ and $\zeta R_{FROC+1} = \infty$. Since each Z-sample is obtained by sampling an appropriately centered unit variance normal distribution, the probability p_r that a latent NL is marked and rated in FROC bin r, and the corresponding probability q_r that a latent LL is marked and rated in FROC bin r, are given by (see row #7 in Table 13.1):

$$\left. \begin{array}{l} p_r = \Phi(\zeta_{r+1}) - \Phi(\zeta_r) \\[2mm] q_r = \Phi(\zeta_{r+1} - \mu) - \Phi(\zeta_r - \mu) \end{array} \right\} \tag{19.1}$$

Understanding these equations is easy. The CDF function evaluated at a threshold is the probability that a Z-sample is less than the threshold, that is, $\Phi(\zeta_r) = P(z_1 \le \zeta_r | z_1 \sim N(0,1))$; the 1-subscript corresponds to the non-diseased location-level truth-state ($s = 1$). The first equation is the difference between the CDF functions of a unit normal distribution evaluated at the two thresholds. This is the probability that the NL Z-sample falls in bin FROC: r. Likewise, since $\Phi(\zeta_r - \mu) = P(z_2 \le \zeta_r | z_2 \sim N(\mu,1))$, the second equation gives the probability that the LL Z-sample falls in bin FROC: r; the 2-subscript corresponds to the diseased location-level truth-state ($s = 2$). The probabilities p_r, q_r each sum to unity when all bins are included.

Since NL and LL events are assumed independent, the contributions to the log-likelihood function can be separated, and one need not enumerate counts at the *individual* case-level. Instead, in the description that follows, one enumerates NL and LL counts in the various bins over the *whole* dataset.

19.2.1 Contribution of NLs

Define n (*an unknown random non-negative integer*) as the total number of latent NLs in the dataset; the observed NL counts vector is $\vec{n} \equiv \{n_r\}; r = 0,1,2,..., R_{FROC}$. Here n_r is the total number of NL counts in FROC ratings bin r, $n_0 = n - \sum_{r=1}^{R_{FROC}} n_r = n - N$, is the (*unknown*) number of unmarked latent NLs and N is the total number of *observed* NLs in the dataset. The probability $P(\vec{n} | n, \vec{\zeta})$ of observing the *NL* counts vector $\vec{n}$ is (the factorials come from the multinomial distribution):

$$P(\vec{n} | n, \vec{\zeta}) = n! \prod_{r=0}^{R_{FROC}} \frac{[p_r]^{n_r}}{n_r!} \tag{19.2}$$

Since n is a random integer, the above probability needs to be averaged over the Poisson distribution of n, yielding:

$$P(\vec{n}; \lambda', \vec{\zeta}) = \sum_{n=N}^{\infty} pmf_{Poi}(n; K\lambda') P(\vec{n} | n, \vec{\zeta}) \tag{19.3}$$

In this expression $K = K_1 + K_2$ is the total number of cases. The pmf_{Poi} of the Poisson distribution yields the probability of n counts from a Poisson distribution with mean $K\lambda'$; the multiplication by the total number of cases is required because one is counting the total number of latent NLs over the entire dataset, *not per case*. The lower limit on n is needed because n cannot be smaller than N, the total number of observed NL counts. To summarize, Equation 19.3 is the probability of observing the NL counts vector $\vec{n}$ as a function of RSM parameters. Not surprisingly, since NLs are sampled from a zero-mean normal distribution, the μ parameter does not enter the above expression.

19.2.2 Contribution of LLs

Likewise, define l (a non-negative random integer) the total number of latent LLs in the dataset and the LL counts vector is $\vec{l} \equiv \{l_r\}$. Here l_r is the number of LL counts in FROC ratings bin r, $l_0 = l - \sum_{r=1}^{R} l_r = l - L$ is the (known) number of unmarked latent LLs, and L is the total number of observed LLs in the dataset. The probability $P\left(\vec{l} \mid l, \mu, \vec{\zeta}\right)$ of observing the LL counts vector $\vec{l}$ is

$$P\left(\vec{l} \mid l, \mu, \vec{\zeta}\right) = l! \prod_{r=0}^{R_{FROC}} \frac{[q_r]^{l_r}}{l_r!} \tag{19.4}$$

The above probability needs to be averaged over the binomial distribution of l.

$$P\left(\vec{l}; \mu, v', \vec{\zeta}\right) = \sum_{l=L}^{L_T} pmf_B\left(l; L_T, v'\right) P\left(\vec{l} \mid l, \mu, \vec{\zeta}\right) \tag{19.5}$$

In this expression L_T is the total number of lesions in the dataset and the lower limit on l is needed because l cannot be smaller than L, the total number of observed LLs. Performing the two summations using Maple™, multiplying the two probabilities, Equations 19.3 and 19.5, and taking the logarithm yields the final expression[4] for the log-likelihood function, where terms not involving parameters have been dropped:

$$
\begin{aligned}
LL_{FROC} \equiv LL_{FROC}\left(\vec{n}, \vec{l} \mid \mu, \lambda', v'\right) = \sum_{r=1}^{R_{FROC}} \left\{ n_r \log(p_r) + l_r \log(q_r) \right\} \\
+ N \log(\lambda') + L \log(v') - K\lambda'(1 - p_0) + (L_T - L)\log(1 - v' + v'q_0)
\end{aligned}
\tag{19.6}
$$

19.2.3 Degeneracy problems

The product $\lambda'(1 - p_0) = \lambda'\Phi(-\zeta_1)$ reveals degeneracy*; the effect of increasing λ' can be counterbalanced by increasing ζ_1; increasing λ' yields more latent NLs but increasing ζ_1 results in fewer of them being marked. The two possibilities cannot be distinguished. A similar degeneracy occurs in the term involving the product $-v' + v'q_0 = -v'(1 - q_0) = -v'\Phi(\mu - \zeta_1)$, where increasing v' can be counterbalanced by decreasing $(\mu - \zeta_1)$, that is, by increasing ζ_1. Again, the effect of increasing v' is to produce more latent LLs, but increasing ζ_1 results in fewer of them being marked. This is the fundamental problem with fitting RSM FROC curves to radiologist FROC data.

19.3 IDCA likelihood function

In the limit $\zeta_1 \to -\infty$, $p_0 \to 0$, $q_0 \to 0$, and Equation 19.6 reduces to (notice that in this limit the degeneracies described above vanish):

$$
\begin{aligned}
LL_{FROC}^{IDCA} = \sum_{r=1}^{R_{FROC}} \left\{ n_r \log(p_r) + l_r \log(q_r) \right\} + N \log(\lambda') + L \log(v') \\
- K\lambda' + (L_T - L)\log(1 - v')
\end{aligned}
\tag{19.7}
$$

---

* Degeneracy is used in the sense that two quantities appear in combination, that is, the product, so that they cannot be individually separated.

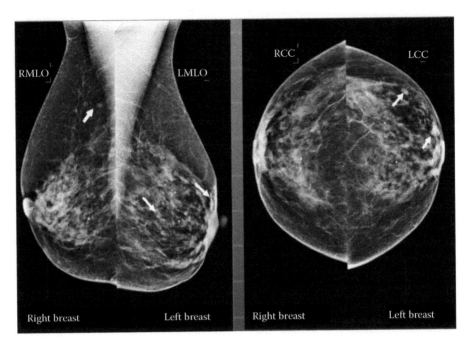

Figure 19.1 A typical 4-view display of a patient mammogram with the CAD cues (the arrows) turned on. In practice, the CAD cues are initially turned off, the radiologist interprets the case without CAD, and subsequently the cues are shown and the radiologist is free to use or ignore the CAD marks in revising their interpretation. For this case the algorithm actually found many more candidate lesions but only the ones with probability of malignancy exceeding the CAD manufacturer's threshold were shown.

The superscript IDCA[12] comes from *initial detection and candidate analysis*. All CAD algorithms consist of an *initial detection* stage, which identifies possible lesion candidates. In the second stage, the algorithm analyzes each candidate lesion and calculates a probability of malignancy. If the probability of malignancy exceeds a threshold value selected by the CAD manufacturer, and this is selected based on a compromise between sensitivity and specificity, the location of the candidate lesion is *cued* (i.e., shown) to the radiologist, Figure 19.1. The reader may have noticed a resemblance between CAD and the Kundel–Nodine model; the latter is parameterized in the RSM—all use a two-stage sampling model. Good ideas generally tend to get adopted in different applications, often without full appreciation of the theoretical underpinnings; if the underpinnings were fully appreciated by CAD developers, they would not be claiming that CAD is perfect at search.

According to Equations 17.28–17.30, in the limit $\zeta_1 = -\infty$ the observed end-point x, y coordinates of the designer-level FROC curve represent estimates of (λ', ν'), respectively:

$$\left.\begin{array}{l} \lambda' = NLF_{\max} \\ \nu' = LLF_{\max} \end{array}\right\} \tag{19.8}$$

In other words, if $\zeta_1 \to -\infty$ then two of the parameters of the RSM are trivially determined from the location of the observed end-point. Suppressing the two already determined known parameters the log-likelihood function, Equation 19.6, effectively reduces to

$$LL_{FROC}^{IDCA} = \sum_{r=1}^{R_{FROC}} \left\{ n_r \log(p_r) + l_r \log(q_r) \right\} + \dots \tag{19.9}$$

The above equation contains only *one* parameter, namely μ, which is implicit in the definition of q_r, Equation 19.1.

Equation 19.9 resembles the log-likelihood function for the binormal model, since, according to Equation 6.37, the LL function for the binormal model with R_{FROC} bins* is

$$LL = \sum_{r=1}^{R_{FROC}} K_{1r} \log\Big(\big(\Phi(\zeta_{r+1}) - \Phi(\zeta_r)\big)\Big) + \sum_{r=1}^{R_{FROC}} K_{2r} \log\Big(\big(\Phi(b\zeta_{r+1} - a) - \Phi(b\zeta_r - a)\big)\Big) \qquad (19.10)$$

In this equation K_{tr} is the number of counts in bin r of case-level truth state t of a ROC study consisting of R_{FROC} bins. Define the unequal variance binormal model versions of Equation 19.1 as follows:

$$\left.\begin{aligned} p_r' &= \Phi(\zeta_{r+1}) - \Phi(\zeta_r) \\ q_r' &= \Phi(b\zeta_{r+1} - a) - \Phi(b\zeta_r - a) \end{aligned}\right\} \qquad (19.11)$$

Here (a,b) are the parameters the unequal variance binormal model. Equation 19.10 becomes

$$LL = \sum_{r=1}^{R_{FROC}} K_{1r} \log(p_r') + \sum_{r=1}^{R_{FROC}} K_{2r} \log(q_r') \qquad (19.12)$$

- With the identifications $K_{1r} \rightarrow n_r$ and $K_{2r} \rightarrow l_r$, Equation 19.12 looks exactly like Equation 19.9. This implies that the binormal ROC fitting method can be used to determine a and b. Notice that instead of fitting an *equal* variance binormal model to determine the remaining *single* remaining μ parameter of the RSM, one is using an *unequal* variance binormal model with *two* parameters, a and b. It turns out that the extra parameter helps. It gives some flexibility to the fitting curve to match the data.
- Here is the idea: regard the NL marks as *non-diseased cases* ($K_{1r} \rightarrow n_r$) and the LL marks as *diseased cases* ($K_{2r} \rightarrow l_r$). Construct a pseudo-ROC counts table, analogous to Table 4.1, where n_r is defined as the pseudo-FP counts in ratings bin r, and likewise, l_r is defined as the pseudo-TP counts in ratings bin r. The pseudo-ROC counts table has the same structure as the ROC counts table, Table 4.1, and can be fitted by the binormal model or other alternatives.
- The pseudo-FP and pseudo-TP counts can be used to define pseudo-FPF and pseudo-TPF in the usual manner; the respective denominators are the total number of NL and LL counts, respectively. These probabilities define the pseudo-ROC plot.
- The prefix *pseudo* is needed because one is regarding localized regions in a case as *independent* cases. Since the fitting algorithm assumes each rating is from an independent case, one is violating a basic assumption, but with CAD data it appears one can get away with it, because the method yields good fits, especially with the extra parameter!
- *The fitted FROC curve is obtained by scaling (i.e., multiplying) the ROC curve along the y-axis by* LLF_{max} *and along the x-axis by* NLF_{max}. The method is illustrated in Figure 19.2.
- This method of fitting FROC data was well known to CAD researchers but was first formalized in the referenced paper.[12]

Assuming binormal fitting is employed, yielding parameters a and b, the equations defining the IDCA fitted FROC curve are (see Equations 6.19 and 6.20)

$$\left.\begin{aligned} NLF(\zeta) &= \lambda' \Phi(-\zeta) \\ LLF(\zeta) &= \nu' \Phi(a - b\zeta) \end{aligned}\right\} \qquad (19.13)$$

---

* For example, ROC bin 1 is bounded by negative infinity and ζ_1.

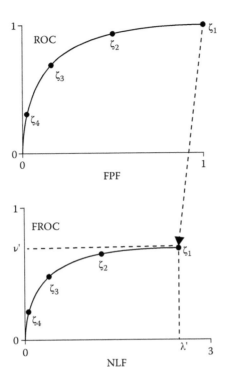

Figure 19.2 This figure illustrates the IDCA method of fitting designer-level CAD FROC data. In the upper half of the figure, the y-axis of the pseudo-ROC is pseudo-TPF and the x-axis is pseudo-FPF. The method is illustrated for a dataset with four FROC bins $R_{FROC} = 4$, but in practice CAD yields many more bins. Regarding the NLs and LLs as non-diseased and diseased cases, respectively, one constructs a table similar to Table 4.1, but this time with only four ROC bins (i.e., three non-trivial operating points). This defines the four operating points, the filled circles, including the trivial one at the upper right corner, shown in the upper half of the plot, which is the pseudo-ROC. One fits the ratings counts data using, for example, the binormal model, yielding the continuous line (based on experience the unequal variance binormal model is needed; the equal variance model does not fit as well). In practice, the operating points will not fall exactly on the fitted line. Finally, one scales (or stretches, or multiplies) the y-axis by v'. Likewise, the x-axis is scaled by λ'. This yields the continuous line shown in the lower half of the figure. Upon adding the FROC operating points one finds that they are magically fitted by the line, which is a scaled replica of the ROC fit in the upper curve.

The RSM-predicted FROC curve is repeated as follows for convenience:

$$\left. \begin{array}{l} NLF(\zeta) = \lambda' \Phi(-\zeta) \\ LLF(\zeta) = v' \Phi(\mu - \zeta) \end{array} \right\} \tag{19.14}$$

IDCA uses the *unequal variance binormal model* to fit the pseudo-ROC, which of course opens up the possibility of an inappropriate chance-line crossing and a predicted FROC curve that is non-monotonically increasing with NLF (this is always present with IDCA fits, but one would need to examine the curve near the end-point very closely to see it). In practice, based on the author's experience with several datasets, the unequal variance model gives visually good fits.

In fact, IDCA yields excellent fits to some designer-level FROC datasets. However, the issue is not with the quality of the fits, rather the appropriateness of the FROC curve as a measure of performance, especially for human observers. For CAD, the method works, so if one wished one could use IDCA to fit designer-level CAD FROC data. However, with closely spaced operating points, the empirical FROC would also work and it does not involve any fitting assumptions. The issue is not fitting designer-level CAD data but comparing standalone performance of designer-level CAD to radiologists and this is not solved by IDCA, which works for designer-level CAD, but not for human observers. The latter do not report all suspicious regions so the IDCA assumption $\zeta_1 = -\infty$ is invalid. The problem of analyzing standalone performance of CAD against a group of radiologists interpreting the same cases is addressed in **Chapter 22**.

19.4 ROC likelihood function

The second attempt at determining RSM parameters used the ROC likelihood function. In **Chapter 17** expressions were derived for the coordinates (x, y) of the ROC curve predicted by the RSM, see Equations 17.8 and 17.16.

$$x(\zeta;\lambda') = 1 - \exp\left(-\frac{\lambda'}{2} + \frac{\lambda'}{2}erf\left(\frac{\zeta}{\sqrt{2}}\right)\right) \tag{19.15}$$

$$y(\zeta\,|\,\mu,\lambda',\nu',f_L) = 1 - \sum_{L=1}^{L_{max}} f_L\left(1 - \frac{\nu'}{2} + \frac{\nu'}{2}erf\left(\frac{\zeta-\mu}{\sqrt{2}}\right)\right)^L e^{\left(-\frac{\lambda'}{2}+\frac{\lambda'}{2}erf\left(\frac{\zeta}{\sqrt{2}}\right)\right)} \tag{19.16}$$

Let (F_r, T_r) denote the number of false positives and true positives, respectively, in ROC rating bin r defined by thresholds $[\zeta_r, \zeta_{r+1}); r = 0, 1, 2, ..., R_{FROC}.$[*] The range of r shows explicitly that R_{FROC} FROC ratings correspond to $R_{FROC} + 1$ ROC bins, as expected, the extra bin is due to unmarked cases. Specifically, (F_0, T_0) represent the *known* numbers of non-diseased and diseased cases, respectively, with no marks, (F_1, T_1) represent the numbers of non-diseased and diseased cases, respectively, with highest rating equal to one, etc. The probability P_{1r} of a count in non-diseased ROC bin r is

$$P_{1r} = x(\zeta_r) - x(\zeta_{r+1}) \tag{19.17}$$

(One needs to subtract the CDF evaluated at $r+1$ from that evaluated at r; the CDF is the complement of x, which results in the reversal. It should also make sense because the higher indexed FPF is to the left of the lower indexed one. Recall that the operating points are numbered starting from the top-right and working down.) Likewise, the probability P_{2r} of a count in diseased ROC bin r is

$$P_{2r} = y(\zeta_r) - y(\zeta_{r+1}) \tag{19.18}$$

The likelihood function is ignoring combinatorial factors that do not depend on parameters.

$$\prod_{r=0}^{R_{FROC}} (P_{1r})^{F_r}(P_{2r})^{T_r}$$

The log-likelihood function is

$$LL_{ROC}\left(\mu,\lambda',\nu',\overrightarrow{f_L}\right) = \sum_{r=0}^{R_{FROC}} \left[F_r \log(P_{1r}) + T_r \log(P_{2r})\right] \tag{19.19}$$

The area AUC_{RSM} under the parametric RSM-ROC curve was obtained by numerical integration (x and y were defined above):

$$AUC_{ROC}^{RSM}\left(\mu,\lambda',\nu',\overrightarrow{f_L}\right) = \int_{x=0}^{1} y\left(\mu,\lambda',\nu',\overrightarrow{f_L}\right)dx \tag{19.20}$$

---

[*] The rating bookkeeping can be confusing. Basically, $r = 0$ corresponds to unmarked cases, $r = 1$ corresponds to cases where the highest rated FROC mark was rated 1, and so on, and $r = R_{FROC}$ corresponds to cases where the highest rated FROC mark was rated R_{FROC}.

The total number of parameters to be estimated, including the R_{FROC} thresholds, is $3 + R_{FROC}$. Maximizing the likelihood function defined by Equation 19.19 yields parameter estimates. The Broyden–Fletcher–Goldfarb–Shanno (BFGS)[13–18] minimization algorithm, as implemented as function **mle2()** in the R-package *bbmle*,[19] was used to minimize the negative of the likelihood function. Since the BFGS algorithm varies each parameter in an unrestricted range $(-\infty, +\infty)$, which would cause problems (e.g., the parameters cannot be negative and thresholds need to be properly ordered), appropriate variable transformations (both *forward* and *inverse*) were used so that parameters supplied to the log-likelihood function were always in the valid range, irrespective of values chosen by the BFGS algorithm. The goodness of fit statistic is calculated using the method described in Section 6.4.2. Because of the additional parameter, the degrees of freedom (*df*) of the chi-square goodness of fit statistic is $(R_{FROC} - 3)$. One can appreciate that calculating goodness of fit for the RSM can fail in situations, where the corresponding statistic can be calculated for the binormal model, for example, three (nontrivial) ROC operating points, corresponding to $df = 1$.* With FROC data one needs at least four (nontrivial) ROC operating points, each defined by bins with at least five counts in both non-diseased and diseased categories.

19.5 RSM versus PROPROC and CBM, and a serendipitous finding

The RSM fitting algorithm was applied, along with PROPROC and CBM, both of which are conventional (i.e., non-search) proper ROC fitting methods described in **Chapter 20**, to a number of datasets. Comparing RSM against the binormal model would be inappropriate, as the latter does not predict proper ROCs.

> *The RSM and CBM, implemented in* **RJafroc** *and PROPROC, implemented in software available from the University of Iowa ROC website, were applied to fourteen (14) MRMC datasets described, comprising 43 modalities, 80 readers and 2012 cases. The total number of individual modality-reader combinations is 236, in other words, there are 236 datasets to which each of the three algorithms was applied. A description of the datasets is generated by sourcing file* **mainDataSummary.R**.

The RSM has three parameters μ, λ', and ν' (excluding thresholds). CBM has two parameters α and μ_{CBM}, detailed in **Chapter 20**. Both RSM and CBM fitting methods are implemented in **RJafroc**. PROPROC has two-parameters c, d_a, detailed in **Chapter 20**. The PROPROC parameters were determined by running Windows software[20] **OR-DBM-MRMC 2.5** (Sept. 04, 2014, Build 4) with PROPROC selected as the curve fitting method. The relevant results are saved in files that end with **proproc norm area pooled.csv** contained in **~/RJafroc/inst/MRMCRuns**. This directory is part of the **RJafroc** installation. It, and its contents, becomes visible when the package is installed. The files can be opened by **RStudio** and the values c and d_a are in the last two columns, see screen-shot Figure 19.3. For example, for the **"VD"** dataset, for modality 2 and reader 1, $c = -0.321$ and $d_a = 2.35$.

The file **mainRSM.R**, listed in Section 19.5.1, explains the fitting procedure. The capabilities of this program are implemented in **RJafroc**. For example, one can view the results by simply running **ret <- ExampleCompare3ProperRocFits()** at the **Console** prompt >. However, for now one is interested in how the program works. Briefly, it extracts the dataset from the **RJafroc** installation, converts it to a highest rating inferred ROC dataset, bins the data to 5 yield 5 ROC points, if possible, applies RSM and CBM fitting, reads the appropriate PROPROC parameters from **OR-DBM-MRMC 2.5** generated files, e.g., Fig. 19.3, and saves all results, and plots, to an object

---

* With three operating points, each defined by bins with at least five counts in both non-diseased and diseased categories, the number of usable ROC bins is four. Subtracting three one gets $df = 1$, and the ROC goodness of fit statistic can be calculated. However, because of the extra RSM parameter, the corresponding $df = 0$.

```
 VD_MRMC proproc area pooled.csv
 1  T,  R,  area      , numCAT, adjPMean ,    c      ,   d_a
 2  1,  1,  0.934040362,  5,  0.934040362 ,-0.298007234, 2.125541232
 3  1,  2,  0.891071412,  4,  0.891071412 ,-0.280900426, 1.731472469
 4  1,  3,  0.907832135,  5,  0.907832135 ,-0.745509798, 0.000118696
 5  1,  4,  0.977459481,  4,  0.977459481 ,-0.931580746, 0.000523679
 6  1,  5,  0.84055976,   5,  0.84055976  ,-0.507426932, 0.895863826
 7  2,  1,  0.951935931,  5,  0.951935931 ,-0.321235433, 2.348149498
 8  2,  2,  0.925992584,  4,  0.925992584 ,-0.790867615, 4.09947E-05
 9  2,  3,  0.930431758,  5,  0.930431758 ,-0.329982208, 2.07854334
10  2,  4,  1,           3,  1           , 1          , 0
11  2,  5,  0.942687414,  4,  0.942687414 ,-0.55308191 , 2.019659866
```

Figure 19.3 Screen-shot showing the contents of the **"VD"** dataset file with name ending with **proproc area pooled.csv**. The last two columns contain the c, d_a PROPROC parameters for this dataset. For example, for modality 2 and reader 1, $c = -0.321$ and $d_a = 2.35$. Since there are five readers and two modalities, the file has 10 data rows. Row 10, corresponding to modality 2 and reader 5, yielded $c = 1$ and $d_a = 0$, i.e., perfect performance, AUC = 1. [T = treatment; R = reader; area = PROPROC AUC; **numCAT** = # ROC bins; **adjPMean** = undocumented value; c = c-parameter; **d_a** = d_a-parameter; the PROPROC parameters are defined in Ref. 6.]

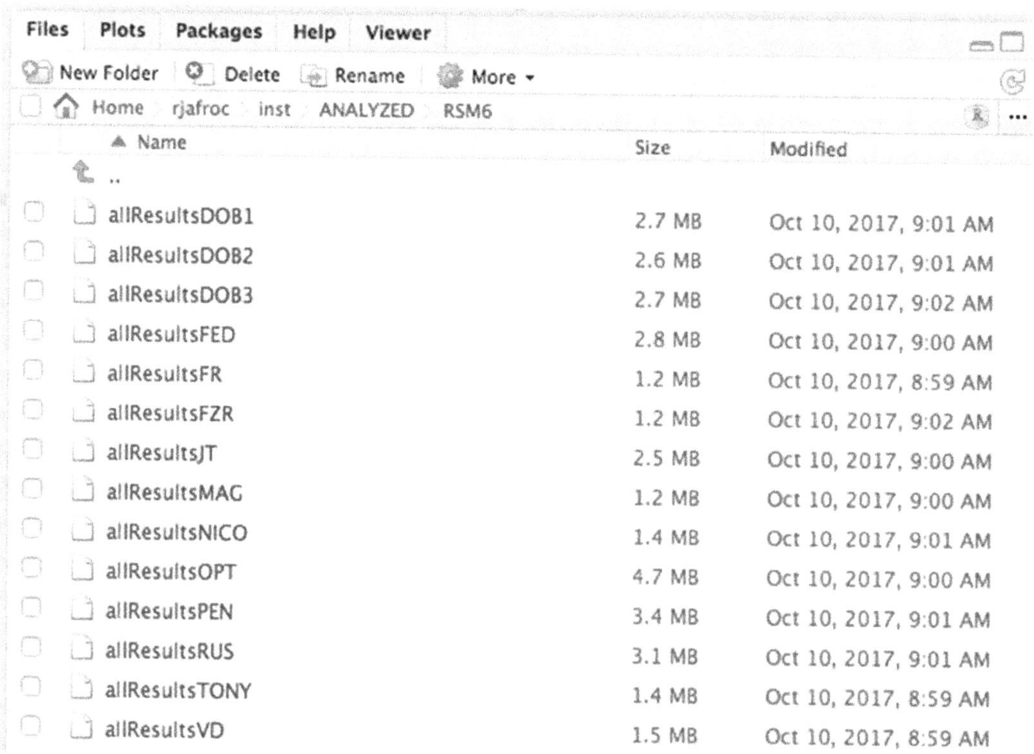

Files Plots Packages Help Viewer		
New Folder Delete Rename More ▾		
Home rjafroc inst ANALYZED RSM6		
▲ Name	Size	Modified
..		
allResultsDOB1	2.7 MB	Oct 10, 2017, 9:01 AM
allResultsDOB2	2.6 MB	Oct 10, 2017, 9:01 AM
allResultsDOB3	2.7 MB	Oct 10, 2017, 9:02 AM
allResultsFED	2.8 MB	Oct 10, 2017, 9:00 AM
allResultsFR	1.2 MB	Oct 10, 2017, 8:59 AM
allResultsFZR	1.2 MB	Oct 10, 2017, 9:02 AM
allResultsJT	2.5 MB	Oct 10, 2017, 9:00 AM
allResultsMAG	1.2 MB	Oct 10, 2017, 9:00 AM
allResultsNICO	1.4 MB	Oct 10, 2017, 9:01 AM
allResultsOPT	4.7 MB	Oct 10, 2017, 9:00 AM
allResultsPEN	3.4 MB	Oct 10, 2017, 9:01 AM
allResultsRUS	3.1 MB	Oct 10, 2017, 9:01 AM
allResultsTONY	1.4 MB	Oct 10, 2017, 8:59 AM
allResultsVD	1.5 MB	Oct 10, 2017, 8:59 AM

Figure 19.4 Contents of **~/RJafroc/inst/ANALYZED/RSM6** containing pre-analyzed datasets. Each file contains the results of RSM, PROPROC and CBM fits to the dataset named following the string **allResults**. Plots for RSM and CBM ROC operating characteristics are also saved in each file, which explains their size.

allResults in an appropriately named disk file to prevent overwriting, Fig. 19.4. For example, **allResultsDOB1** contains the saved **allResults** object corresponding to dataset **"DOB1"**. The saved files are shown in the screen-shot Fig. 19.4. The structure of **allResults** is explained in Online Appendix 19.A. The listing of **mainRSM.R** follows.

19.5.1 Code listing

```
# mainRSM.R
# RSM fits vs. PROPROC and CBM fits;
# Windows proproc results must be saved to MRMCRuns
# directory prior to running this
rm(list = ls())
library(RJafroc)
library(ggplot2)
library(bbmle)
library(stats)
library(binom)

options(digits = 3)
reAnalyze <- FALSE;showPlot <- TRUE;saveProprocLrcFile <- FALSE

# included datasets
fileNames <-   c("TONY", "VD", "FR",
                 "FED", "JT", "MAG",
                 "OPT", "PEN", "NICO",
                 "RUS", "DOB1", "DOB2",
                 "DOB3", "FZR")

f <- 1 # selected dataset
fileName <- fileNames[f]
# the datasets already exist as R objects
theData <- get(sprintf("dataset%02d", f))
# RSM ROC fitting needs to know lesionDistribution
lesionDistribution <- lesionDistribution(theData)

rocData <- DfFroc2Roc(theData)

if (saveProprocLrcFile) {
  DfSaveDataFile(rocData,
                 fileName =
                 paste0(fileName,".lrc"),
                 format = "MRMC")
}
I <- length(rocData$modalityID);J <- length(rocData$readerID)
K <- dim(rocData$NL)[3];K2 <- dim(rocData$LL)[3];K1 <- K - K2

## retrieve PROPROC parameters
csvFileName <- paste0(fileName, " proproc area pooled.csv")
sysCsvFileName <- system.file(
  paste0(
    "MRMCRuns/",fileName), csvFileName, package = "RJafroc")
if (!file.exists(sysCsvFileName))
  stop("Run Windows PROPROC for this dataset using VMwareFusion")
proprocRet <- read.csv(sysCsvFileName)
c1 <- matrix(data =
               proprocRet$c,
             nrow = length(unique(proprocRet$T)),
             ncol = length(unique(proprocRet$R)), byrow = TRUE)
da <- matrix(data =
               proprocRet$d_a,
             nrow = length(unique(proprocRet$T)),
             ncol = length(unique(proprocRet$R)), byrow = TRUE)
```

```
retFileName <- paste0("allResults", fileName)
sysAnalFileName <- system.file(
  "ANALYZED/RSM6", retFileName, package = "RJafroc")

if (fileName %in% c("JT", "NICO", "DOB1", "DOB3")){
  binnedRocData <- DfBinDataset(rocData, desiredNumBins = 5)
}else{
  binnedRocData <- rocData
}

if (reAnalyze || !file.exists(sysAnalFileName)){
  allResults <- list()
  AllResIndx <- 0
  for (i in 1:I){
    for (j in 1:J){
      AllResIndx <- AllResIndx + 1
      cat("f, i, j:", f, i, j, "\n")
      # fit to CBM
      retCbm <- FitCbmRoc(
        binnedRocData, trt = i, rdr = j)
      # fit to RSM, need lesionDistribution matrix
      retRsm <- FitRsmRoc(
        binnedRocData, trt = i, rdr = j, lesionDistribution)
      if (showPlot) {
        x <- allResults[[AllResIndx]]
        lesionDistribution <- x$lesionDistribution
        empOp <- UtilBinCountsOpPts(binnedRocData, trt = i, rdr = j)
        fpf <- empOp$fpf; tpf <- empOp$tpf
        compPlot <- gpfPlotRsmPropCbm(
          which(fileNames == fileName),
          x$retRsm$mu, x$retRsm$lambdaP, x$retRsm$nuP,
          lesionDistribution, c1[i, j], da[i, j],
          x$retCbm$mu, x$retCbm$alpha,
          fpf, tpf, i, j, K1, K2, c(1, length(fpf)))
        print(compPlot)
      }
    }
  }
  # safety comments
  # sysSavFileName <-
  # paste0("/Users/Dev/rjafroc/inst/ANALYZED/RSM6/", retFileName)
  # save(allResults, file = sysSavFileName)
}else{
  load(sysAnalFileName)
  AllResIndx <- 0
  # cat(fileName,    "i, j, mu, lambdaP, nuP, c,      da,    alpha,")
  # cat("muCbm,      AUC-RSM, AUC-PROPROC, AUC-CBM, chisq, p-value,
  df\n")
  for (i in 1:I){
    for (j in 1:J){
      AllResIndx <- AllResIndx + 1
      x <- allResults[[AllResIndx]]
      if (showPlot) {
        empOp <- UtilBinCountsOpPts(binnedRocData, trt = i, rdr = j)
        fpf <- empOp$fpf; tpf <- empOp$tpf
        compPlot <- gpfPlotRsmPropCbm(
```

```
            which(fileNames == fileName),
            x$retRsm$mu, x$retRsm$lambdaP, x$retRsm$nuP,
            lesionDistribution = lesionDistribution, cl[i, j], da[i, j],
            x$retCbm$mu, x$retCbm$alpha,
            fpf, tpf, i, j, K1, K2, c(1, length(fpf)))
        compPlot <- compPlot +
          theme(
            axis.text=element_text(size=10),
            axis.title=element_text(size=28,face="bold"))
        print(compPlot)
      }
      # follows same format as RSM6 Vs. Others.xlsx
      cat(fileName, i, j, x$retRsm$mu, x$retRsm$lambdaP,
          x$retRsm$nuP,
          cl[i,j], da[i,j],
          x$retCbm$alpha, x$retCbm$mu,
          x$retRsm$AUC, x$aucProp, x$retCbm$AUC,
          x$retRsm$ChisqrFitStats[[1]], x$retRsm$ChisqrFitStats[[2]],
          x$retRsm$ChisqrFitStats[[3]],"\n")
    }
  }
}
```

Strings contained in the array variable **fileNames** (lines 16–20) identify the 14 datasets. They are index by **f** (for filename) currently assigned 1, line 22, corresponding to the **"TONY"** dataset[21]. Line 23–25 extracts this pre-installed dataset, which is assigned the name **theData**. Line 27 calculates the **lesionDistribution** of the dataset, i.e., how many cases contain one lesion, how many contain 2 lesions, etc. RSM fitting needs this information; see lines 78–79. Line 29 converts **theData** to a highest-rating inferred ROC dataset using function **DfFroc2Roc**; **"TONY"** happens to be an FROC dataset with two modalities and five readers. The PROPROC parameters are extracted from saved installation files, as in Fig. 19.3, at line 40–55. Since **c()** is reserved for the concatenation operator, line 48 uses **cl** to denote the PROPROC c-parameter. Line 57–59 constructs the name of the system file, **sysAnalFileName**, containing the results of the analysis. Data binning is performed, if necessary, at lines 61–65, resulting in the binned dataset object **binnedRocData**. If the **reAnalyze** flag, line 13, is **TRUE** or the **sysAnalFileName** file does not exist, then the **for-loop** at line 67 is entered. Lines 75–76 perform CBM fitting for modality **i** and reader **j**. Lines 78–79 perform RSM fitting. If **showPlot**, line 13, is **TRUE** then lines 81–91 displays the 3 fits using **gpfPlotRsmPropCbm()**, which is a general purpose function to plot RSM, PROPROC and CBM curves, with superposed operating points and error bars. The black line fit is the RSM fit, the dark-gray line is the PROPROC fit and the light-gray line is the CBM fit. Examples of these plots are shown in Fig. 19.5. The dotted-line extension, when shown, applies to the RSM fit. [On a computer monitor the RSM fits are in red and the CBM fits are in blue.]

Since all 14 datasets have already been analyzed and the results saved, the preceding analysis is bypassed (see test at line 67 for existence of a pre-analyzed results file **sysAnalFileName**) and program execution starts effectively at line 100, which loads the saved results using the **load()** function, which is the converse of the **save()** function used at line 98 for saving a new results file (as a precaution against unintentional overwriting, lines 96–98 are commented). Line 100 creates the **allResults** object described previously and in Online Appendix 19.A. Line 107 extracts the appropriate results for each dataset-modality-reader combination and saves it to variable **x**. RSM and CBM parameter extraction from **x** is illustrated in lines 124–129. For example, **x$retRsm$mu** is the RSM mu-parameter for the selected dataset, modality and reader. **Source** the code to get the code output listed in 19.5.1.1 and 10 plots in the **Plots** window, one of which is shown in Fig. 19.5a.

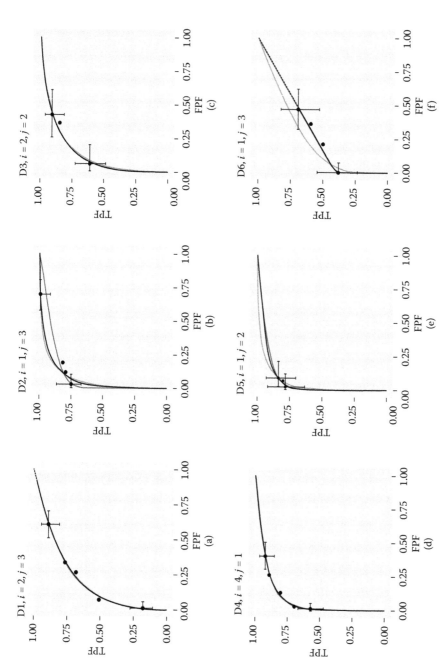

Figure 19.5 This figure shows a sampling of 12 plots, roughly one from each of the datasets. The labels at the top identify the dataset f, the modality i and the reader j. Each panel shows RSM (black), CBM (light-gray) and PROPROC (dark-gray) curves fitted to the same ROC dataset. Because of overlap, all curves may not be visible on some of the plots. Operating points are shown as filled circles. The plots show that the three fitting methods give comparable fits to the data, given that the data points are subject to sampling errors, as indicated by the confidence intervals. In some cases, e.g., plots a, c, d, g and i, the fits are very similar.

(Continued)

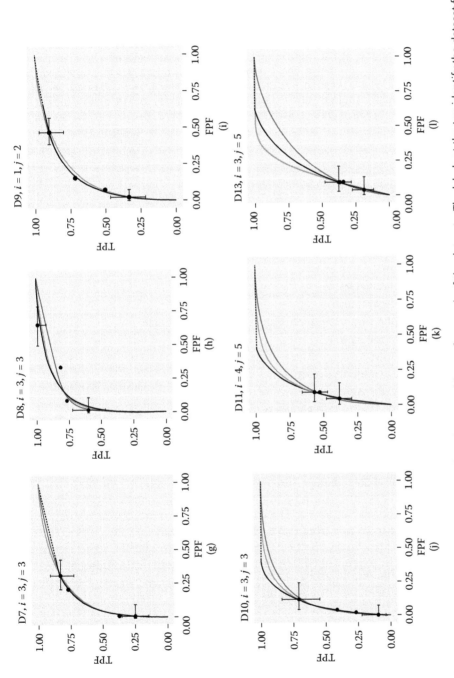

Figure 19.5 (Continued) This figure shows a sampling of 12 plots, roughly one from each of the datasets. The labels at the top identify the dataset f, the modality i and the reader j. Each panel shows RSM (black), CBM (light-gray) and PROPROC (dark-gray) curves fitted to the same ROC dataset. Because of overlap, all curves may not be visible on some of the plots. Operating points are shown as filled circles. The plots show that the three fitting methods give comparable fits to the data, given that the data points are subject to sampling errors, as indicated by the confidence intervals. In some cases, e.g., plots a, c, d, g and i, the fits are very similar.

19.5.1.1 Code output

```
> source(...)
TONY 1 1 1.78 1.04 0.729 -0.132 1.2 0.775 1.83 0.813 0.801 0.812 1.96 0.161 1
TONY 1 2 1.97 0.841 0.867 -0.087 1.77 0.924 2.04 0.891 0.895 0.893 NA NA NA
TONY 1 3 2.85 19.7 0.939 -0.144 1.48 0.855 1.87 0.853 0.853 0.849 0.305 0.581 1
TONY 1 4 1.22 1.06 0.999 0.0805 1.51 0.999 1.49 0.865 0.858 0.853 6.23 0.0126 1
TONY 1 5 1 0.575 0.999 0.223 1.74 0.999 1.65 0.896 0.891 0.878 4.6 0.032 1
TONY 2 1 1.21 1.06 0.478 -0.0817 0.628 0.598 1.17 0.678 0.672 0.676 2.1 0.147 1
TONY 2 2 1.05 2.35 0.998 0.0498 0.974 0.998 0.951 0.758 0.754 0.749 NA NA NA
TONY 2 3 2.03 2.42 0.729 -0.133 1.16 0.754 1.72 0.793 0.793 0.792 1.36 0.507 2
TONY 2 4 1.03 0.653 0.999 0.118 1.62 0.999 1.51 0.888 0.874 0.857 NA NA NA
TONY 2 5 0.777 1.89 0.998 0.0781 0.893 0.998 0.845 0.741 0.736 0.725 0.912 0.34 1
```

The first column is the dataset name, the second column is the modality index and the third column is the reader index. These are followed by the following parameters:

- **mu** = μ = RSM perceptual SNR parameter
- **lambdaP** = λ' = RSM Poisson parameter
- **nuP** = ν' = RSM binomial parameter
- **c** = c = PROPROC c-parameter
- **da** = d_a = PROPROC d-parameter
- **alpha** = α = CBM alpha parameter, probability that disease is visible
- **muCbm** = μ_{CBM} = CBM mu parameter
- **aucRsm** = AUC_{RSM} = RSM-AUC
- **aucProproc** = AUC_{PRO} = PROPROC-AUC
- **aucCbm** = AUC_{CBM} = CBM-AUC
- **chisqr statistic** = χ^2 = RSM goodness of fit statistic
- **chisqr(p-value)** = p = RSM goodness of fit statistic p-value
- **df** = df = RSM goodness of fit degrees of freedom

As an example, for the **TONY** dataset, for modality 2 and reader 3, one has: $\mu = 2.03$, $\lambda' = 2.42$, $\nu' = 0.729$, $c = -0.133$, $d_a = 1.16$, $\alpha = 0.754$, $\mu_{CBM} = 1.72$, $AUC_{RSM} = 793$, $AUC_{PRO} = 0.793$, $AUC_{CBM} = 0.792$, $\chi^2 = 1.36$, $p = 0.507$, $df = 2$. The corresponding plot (use the arrow buttons in the **Plots** window to get to it; it is labeled **D1, i=2, j=3**) is shown in Fig. 19.5 a. For this example the three AUCs are almost identical (and close to 0.793) and the plots are practically indistinguishable. The p-value (0.507) informs us that the statistical validity of the RSM fit cannot be rejected at the 5% significance level. The chisquare goodness of fit statistics are listed if they can be calculated; this explains some of the NA entries in Section 19.5.1.1. [The plot in question has 5 operating points, which corresponds to 6 ROC bins. The degrees of freedom for a conventional ROC model would be 3, but because of the extra parameter, RSM df = 2.]

Source the code repeatedly, with the appropriate dataset selected, yielding the plots in Fig. 19.5 (a–l). The figure shows a sampling of 12 plots from the 236 plots yielded by the 14 datasets. A dataset number, a modality index and a reader index label each plot. For example, plot Fig. 19.5 k, labeled **D11, i=4, j=5** corresponds to the 11th dataset identified in lines 16–20, modality index 4 and reader index 5. The plots show that the three fitting methods give comparable fits to the data. In some cases, e.g., plots a, c, d, g and i, the fits are very close. In fact 67 of the 236 fits were visually very similar, which should impress the reader, considering the differences between the fitting models.

Detailed results for three datasets (**"TONY"**, **"VD"** and **"FR"**) are reported here. Results for all 14 datasets can be viewed by sourcing the code with appropriately selected datasets at line 22.

19.5.1.2 Application of RSM/PROPROC/CBM to three datasets

Sourcing **mainRSM.R** two more times, each time with the appropriate dataset specified, yields the results summarized in Table 19.1 for dataset **"TONY"**, in Table 19.2 for (Van Dyke) dataset[22] **"VD"**, and in Table 19.3 for (Franken) dataset[23] **"FR"**. For the RSM, PROPROC and CBM fitted AUCs, the last row, labeled AVG, lists the averages of the corresponding AUCs. Note that the three averages are very close to each other. Results of bootstrap analysis, described in Section 19.5.2, averaged over all datasets, are summarized in Table 18.4. Since the RSM parameter values have clear physical meanings, their values should be meaningful. Examination of the values reveals that for the most part they behave reasonably but there are exceptions. For example, for some datasets the λ' parameter reached unreasonably high values; an example is in Table 19.1 for modality 1 and reader 3, where $\lambda' = 19.7$, which effectively "clipped" the upper limit 20 of the allowed range. Likewise, for other datasets the v' parameter effectively reached its upper limit of unity; when it did, the CBM α parameter also reached the identical upper limit. Clearly the degeneracy problem is not entirely resolved and work needs to be done to improve the RSM fitting algorithm. In the interim, the agreement of the AUCs with PROPROC and CBM estimates, gives some confidence that while the parameters may not be quite right yet, the AUC predictions are probably accurate.

The serendipitous finding, alluded to in the title to this section, is that all three proper ROC fitting methods yield almost identical AUCs. Results of bootstrap analysis, averaged over all datasets, are summarized in Table 19.4. In this table columns 5–8 show that, when averaged over all datasets, PROPROC-AUC was 1% larger than RSM-AUC, while CBM-AUC was 0.6% lower. The near equality of proper ROC AUCs, as estimated by different methods, has not been noted in the literature except in a proceedings paper by the author and a collaborator.[5] The explanation is deferred to Section 19.6.

Table 19.1 *The serendipitous finding:* for the **"TONY"** dataset *the three methods yield near identical estimates of average AUC*

		RSM			PROPROC		CBM		AUCs		
i	j	μ	λ'	v'	c	d_a	α	$μ_{BCM}$	RSM	PROPROC	CBM
1	1	1.7812	1.0396	0.7291	−0.1323	1.1972	0.7749	1.8306	0.8132	0.8014	0.8117
	2	1.9674	0.8415	0.8670	−0.0870	1.7712	0.9239	2.0446	0.8909	0.8948	0.8935
	3	2.8476	19.7416	0.9391	−0.1444	1.4819	0.8554	1.8738	0.8525	0.8527	0.8485
	4	1.2187	1.0565	0.9990	0.0805	1.5138	0.9990	1.4863	0.8647	0.8578	0.8530
	5	1.0023	0.5750	0.9990	0.2226	1.7402	0.9990	1.6535	0.8957	0.8909	0.8785
2	1	1.2061	1.0606	0.4780	−0.0817	0.6281	0.5978	1.1656	0.6785	0.6717	0.6764
	2	1.0475	2.3523	0.9984	0.0498	0.9739	0.9984	0.9508	0.7579	0.7545	0.7489
	3	2.0346	2.4186	0.7289	−0.1326	1.1559	0.7535	1.7173	0.7926	0.7932	0.7921
	4	1.0330	0.6533	0.9989	0.1182	1.6202	0.9989	1.5127	0.8878	0.8740	0.8572
	5	0.7774	1.8885	0.9984	0.0781	0.8929	0.9984	0.8453	0.7412	0.7361	0.7246
				AVG					0.8175	0.8127	0.8084

Note: Listed are RSM parameters (3 columns) followed by PROPROC parameters (2 columns) and CBM parameters (2 columns). The last three columns list the respective AUCs, and the last row, labeled AVG, lists the average AUCs. [i = modality index; j = reader index; for this dataset there were two modalities and 5 readers.]

Table 19.2 This is similar to Table 19.1, but for the **"VD"** dataset. This was a two-modality five-reader ROC dataset. Again, the average AUCs are almost identical.

		RSM			PROPROC		CBM		AUCs		
i	*j*	μ	λ'	ν'	*c*	d_a	α	μ_{BCM}	RSM	PROPROC	CBM
1	1	2.6893	0.3836	0.8718	−0.2980	2.1255	0.8831	3.1128	0.9270	0.9340	0.9293
	2	2.1731	0.2435	0.7514	−0.2809	1.7315	0.7931	2.7458	0.8650	0.8911	0.8759
	3	3.3863	19.9224	0.9989	−0.7455	0.0001	0.9988	2.1424	0.9328	0.9078	0.9346
	4	3.9998	0.1197	0.9301	−0.9316	0.0005	0.9487	3.9997	0.9648	0.9775	0.9721
	5	3.7839	14.1633	0.7017	−0.5074	0.8959	0.6676	3.0509	0.8305	0.8406	0.8235
2	1	3.8772	19.7009	0.9678	−0.3212	2.3481	0.9608	2.7603	0.9544	0.9519	0.9559
	2	3.9997	2.5508	0.8155	−0.7909	0.0000	0.8070	3.9997	0.9036	0.9260	0.9016
	3	2.8323	1.1514	0.9031	−0.3300	2.0785	0.9068	2.7989	0.9319	0.9304	0.9317
	4	4.0000	0.0100	1.0000	1.0000	0.0000	1.0000	4.0000	1.0000	1.0000	1.0000
	5	3.9901	1.4266	0.8761	−0.5531	2.0197	0.8749	3.9848	0.9353	0.9427	0.9353
					AVG				0.9245	0.9302	0.9260

Table 19.3 This is similar to Table 19.1 but for the **"FR"** dataset. This was a two-modality four-reader ROC dataset. Again, the average AUCs are almost identical.

		RSM			PROPROC		CBM		AUCs		
i	*j*	μ	λ'	ν'	*c*	d_a	α	μ_{BCM}	RSM	PROPROC	CBM
1	1	2.5395	1.5164	0.8149	−0.2899	1.5702	0.8179	2.4125	0.8732	0.8699	0.8730
	2	2.3647	1.2497	0.8705	−0.2546	1.7670	0.8797	2.3241	0.8953	0.8949	0.8957
	3	2.9812	19.5488	0.9751	−0.1354	1.6497	0.9229	1.8742	0.8780	0.8783	0.8760
	4	2.5928	9.2706	0.8974	−0.1350	1.3940	0.8286	1.8297	0.8380	0.8379	0.8332
2	1	2.3743	1.7007	0.8160	−0.2161	1.5346	0.8266	2.1653	0.8612	0.8615	0.8613
	2	2.8729	19.6322	0.9979	−0.1200	1.5972	0.9162	1.8124	0.8716	0.8706	0.8665
	3	2.6458	2.1552	0.7847	−0.3552	1.4018	0.7860	2.3407	0.8552	0.8541	0.8545
	4	3.4828	18.9016	0.7310	−0.4420	0.9837	0.6893	2.5337	0.8244	0.8253	0.8194
					AVG				0.8621	0.8616	0.8600

19.5.1.3 Validating the RSM fits

This topic was addressed in Section 19.4. The goodness of fit *p*-value is calculated in the **RJafroc** implementations of RSM and CBM fitting. A valid chi-square goodness-of-fit statistic, satisfying the requirement of at least 5 counts in each bin in each truth state, could be calculated in 42 out of the 236 RSM fits. In each instance where a valid statistic could be calculated, the *p*-value exceeded 0.05. A case for validity of the RSM can be made from the observed consistency of its ROC-curve based AUC predictions when compared to PROPROC and CBM over a large number of fits. Other arguments for validity were presented in Section 17.11.

19.5.2 Inter-correlations between different methods of estimating AUCs and confidence intervals

Correlations are expected between parameters of different ROC models estimated from the same dataset. The reason for calling these *inter-correlations* is because they measure correlations between *different* methods of estimating AUCs. If one accepts the proposition that the AUCs of the three methods are identical, then the following relations should hold with each m, the slope, close to unity.

$$\left.\begin{array}{c} AUC_{PRO} = m_{PR} AUC_{RSM} \\ AUC_{CBM} = m_{CR} AUC_{RSM} \end{array}\right\} \tag{19.21}$$

For example, a plot of PROPROC versus RSM AUCs should be linear with *zero* intercept and slope m_{PR} close to unity (PR = PROPROC versus RSM; CR = CBM versus RSM). The reason for the zero intercept is if one of the AUCs indicates zero performance (actually perfect performance but the decision variable is reversed), the other AUC must also be zero. Likewise, chance-level performance (AUC = 0.5) must be common to all methods of estimating AUC. Finally, perfect performance must be common to all methods. All of these conditions dictate a zero-intercept fitting model. An analysis was conducted to determine the average slopes (i.e., averaged over all datasets) in Equation 19.21 and a bootstrap analysis was conducted to determine the corresponding confidence intervals.

Plots of PROPROC-AUC versus RSM-AUC and CBM-AUC versus RSM-AUC, where each plot has the constrained linear fit superposed on the data points (each data point corresponds to a distinct modality-reader combination in the dataset) are generated by `mainInterCorrelations.R`, listed in Online Appendix 19.B.1. The code also prints the average slopes and R^2 values (R^2 is the fraction of variance explained by the straight line fit, see Section 19.5.2.1). Sourcing the code yielded the values used to populate columns 5–8 (excepting for the last 3 rows) in Table 19.4. The code also produces 28 plots, two per dataset, two of which, corresponding to dataset #7, are shown in Figure 19.6 (a–b). This dataset had 5 modalities and 7 readers (as indicated in the labels at the top), so there are 35 points on each plot. In plot (a) the constrained straight-line fit testing the relation $AUC_{PRO} = m_{PR} AUC_{RSM}$ yielded slope $m_{PR} = 1.0158$ and $R^2 = 0.9996$, see Table 19.4. Plot (b) is the corresponding CBM-AUC versus RSM-AUC plot. The constrained straight-line fit yielded slope $m_{CR} = 1.0044$ and $R^2 = 0.9998$. Figure 19.6 (c–d) shows the histograms of the corresponding bootstrap values averaged over all datasets, as explained in Section 19.5.2. The row in Table 19.4 labeled AVG lists the grand averages of the corresponding columns and the next two rows list the lower and upper bounds of the 95% bootstrap confidence intervals. The grand average PROPROC-AUC versus RSM-AUC slope is 1.0092 (1.0066, 1.0124). The average CBM-AUC versus RSM-AUC slope is 0.9938 (0.9918, 0.9963). The three fitted AUCs, averaged over 14 datasets are in close agreement, with PROPROC AUC being 1% larger than RSM-AUC while CBM-AUC is 0.6% smaller. These observations confirm the shaded emphasized text at the end of Section 19.5.1.2. The explanation of the last two columns of Table 19.4 is deferred to Section 19.5.3. [It is left as an exercise to the reader to calculate the corresponding slopes of empirical AUC versus RSM, PROPROC and CBM.]

19.5.2.1 A digression on regression through the origin

For regression through the origin of the traditional formula, e.g., the Excel implementation, does not apply and **R** has the correct implementation,[24] defined as follows:

$$R^2 = 1 - \frac{\displaystyle\sum_{i=1}^{N} r_i^2}{\displaystyle\sum_{i=1}^{N} (y_i - y^*)^2} \tag{19.22}$$

Table 19.4 Results of bootstrap analysis, described in Section 19.5.2, of PROPROC, CBM and RSM fitting of 236 common ROC datasets. On the average the PROPROC AUC was 1% higher than RSM while CBM was 0.6% lower (see values and confidence intervals in columns 5 and 7).

Dataset #	AVG. AUCs			$AUC_{PRO} = m_{PR} AUC_{RSM}$		$AUC_{CBM} = m_{CR} AUC_{RSM}$		$Cor(\mu, \mu_{CBM})$	$Cor(\nu', \alpha)$
	PROPROC	CBM	RSM	m_{PR}	R^2	m_{CR}	R^2		
1	0.8127	0.8084	0.8175	0.9942	1.0000	0.9887	0.9999	0.7304	0.9707
2	0.9302	0.9260	0.9245	1.0058	0.9998	1.0016	1.0000	0.5919	0.9824
3	0.8616	0.8600	0.8621	0.9993	1.0000	0.9975	1.0000	0.1070	0.9181
4	0.8587	0.8467	0.8484	1.0117	0.9998	0.9978	0.9999	0.8391	0.9750
5	0.8911	0.8798	0.8813	1.0106	0.9996	0.9984	0.9999	0.8385	0.7762
6	0.7252	0.6968	0.6992	1.0342	0.9985	0.9962	0.9999	0.9818	0.9805
7	0.8561	0.8463	0.8422	1.0158	0.9996	1.0044	0.9999	0.6815	0.8303
8	0.8357	0.8270	0.8292	1.0072	0.9996	0.9973	0.9998	0.8608	0.9192
9	0.8566	0.8547	0.8551	1.0014	0.9999	0.9994	1.0000	0.7454	0.9483
10	0.7877	0.7832	0.7885	0.9986	0.9999	0.9930	0.9999	0.7989	0.9635
11	0.7299	0.7028	0.7179	1.0153	0.9990	0.9774	0.9991	0.9212	0.9484
12	0.6972	0.6738	0.6913	1.0073	0.9985	0.9732	0.9975	0.9431	0.9731
13	0.6354	0.6152	0.6206	1.0240	0.9992	0.9892	0.9987	0.9683	0.8198
14	0.8837	0.8810	0.8813	1.0027	0.9999	0.9996	1.0000	0.9930	0.9914
Grand average over datasets									
GR. AVG	0.8116	0.8001	0.8042	1.0092	0.9995	0.9938	0.9996	0.7858	0.9284
BS CI, Lower	0.803	0.7918	0.7955	1.0066		0.9918		0.6752	0.8714
BS CI, Upper	0.820	0.8088	0.8128	1.0124		0.9963		0.8466	0.9499

Note: AUCs, averaged over all datasets, for the three fitting methods (columns 2–4); columns 5–8 show slopes and R^2 of assumed zero-intercept linear relationships between AUCs estimaed by different methods. The last two columns show correlation between similar RSM and CBM parameters, Section 19.5.3. [*GR. AVG* = grand average over all datasets; PRO = PROPROC; BS = bootstrap; CI = 95% confidence interval; AVG = average] This table was generated by sourcing `mainInterCorrelationsCI.R`.

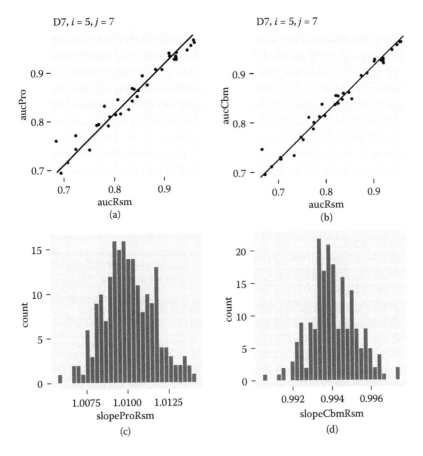

Figure 19.6 Examples of the slope (ratio of conventional model AUC to RSM-AUC) and bootstrap analysis to determine confidence intervals of the slopes between RSM and conventional model parameters. Panel (a) shows PROPROC-AUC versus RSM-AUC for dataset #7 (Online Table 23.1), which has five modalities and seven readers (as indicated in the label at the top), for a total of 35 points. The constrained straight-line fit yielded $m_{PR} = 1.0158$ and $R^2 = 0.9996$, Table 19.4. Panel (b) is the corresponding CBM-AUC versus RSM-AUC plot; the constrained straight-line fit yielded $m_{CR} = 1.0044$ and $R^2 = 0.9998$. Panel (c) shows the histogram of the 200 bootstrap PROPROC-AUC versus RSM-AUC slopes, where each slope represents an average over all 14 datasets. The mean slope was 1.0092 and the 95% confidence interval was (1.0066, 1.0124). Panel (d) shows the corresponding histogram for the CBM-AUC versus RSM-AUC slopes, with mean slope 0.9938 and 95% confidence interval (0.9918, 0.9963). These plots were generated by **mainInterCorrelations.R**. (aucPro = AUC_{PRO}; aucCbm = AUC_{CBM}; aucRsm = AUC_{RSM}; avgSlopeProRSM = m_{PR}; avgSlopeCbmRSM = m_{CR}; see Equation 19.21.)

Here, y_i is the residual (observed value minus fitted value) and y *is the mean of* y_i *if there is an intercept and zero otherwise.*

19.5.2.2 The bootstrap algorithm

Returning to the problem of how to determine slopes and bootstrap confidence intervals for the slopes, the code in **mainInterCorrelations.R** treats readers as random, Online Appendix 19.B.1. One could also have treated cases as random, but that would greatly increase computation time as all (!) fitting (RSM, PROPROC and CBM) would have to be repeated for each bootstrapped case sample (this would be especially difficult, as PROPROC cannot be run in batch mode). Given the large number of cases (i.e., over two thousand) it is likely that addition of case-level bootstrapping would not greatly alter the overall conclusions.

A detailed explanation of the code is given in Online Appendix 19.B.2. Briefly, for each dataset one extracts, from previously saved `allResults`, and for each modality-reader combination, three AUC values (RSM, PROPROC and CBM). One performs a constrained linear fit of the $I \times J$ PROPROC versus RSM AUC values, yielding the slope m_{PR} and an R^2 value. A similar constrained linear fit of CBM versus RSM AUC values was performed yielding the slope m_{CR} and an R^2 value. All relevant values are collected into a `list`, which is used by subsequent bootstrap code. The values populate each row labeled Dataset # 1–14 in Table 19.4. The last two columns in Table 19.4 are explained in Section 19.5.3.

Within the bootstrap loop, for each dataset one bootstraps readers and calculates average AUCs (over $I \times J$ values per dataset) for the three methods, RSM, PROPROC and CBM, using the bootstrapped (not the original) readers. The slopes m_{PR} and m_{CR} are calculated. The values are averaged over all datasets. This is repeated 200 times. The cluster-code returns 200 values for each slope m_{PR} and m_{CR} and each AUC: RSM, PROPROC and CBM. For each variable, empirical 95% confidence intervals are calculated from the bootstrap samples. The output of **sourcing** the code is shown in Online Appendix 19.B.3.

Fig. 19.6 panel (c) shows the histogram of the 200 m_{PR} slopes. The 95% confidence interval was calculated from the lower and upper 2.5% quantile of this distribution. Panel (d) shows the corresponding histogram for the m_{CR} slopes. The final bootstrap results are listed in the last 3 rows of Table 19.4 (within each column, excepting those labeled R^2, the last three rows are, in order, the average over all datasets—with no bootstrapping—the lower and the upper 95% confidence levels). The grand average $m_{PR} = 1.0092$ with an empirical bootstrap 95% confidence interval of (1.0066, 1.0124). The corresponding values for m_{PR} were = 0.9938 and confidence interval (0.9918, 0.9963). In other words, grand average PROPROC AUC was 1% larger than grand average RSM AUC, while grand average CBM AUC was 0.6% smaller.

In Tables 19.1 through 19.3, the straight-line fits in Figure 19.6, and the near unity slopes and near unit R^2 values in Table 19.4, the results collectively show that the three methods of estimating AUC, RSM, PROPROC, and CBM give almost identical results despite the fact that the shapes of the predicted curves are qualitatively different, as can be appreciated from Figure 19.5. There is strong evidence that the AUCs of the three proper ROC fitting methods are almost identical. Relative to RSM, PROPROC-AUCs were about 1% larger, while CBM-AUCs were about 0.6% smaller.

So far correlations between AUCs predicted by the three fitting methods were studied. As noted earlier, one also expects correlations between (μ, μ_{CBM}) and between (ν', α), as demonstrated in the last two columns of Table 19.4.

19.5.3 Inter-correlations between RSM and CBM parameters

The title of this section refers to *inter-correlations* as it investigates the correlations between parameters obtained by different fitting methods for the same dataset. The slopes m_{PR} and m_{CR} listed in Table 19.4 represent examples of inter-correlations between AUCs estimated by different methods. Here we examine examples of correlations between parameters. The PROPROC parameters do not have intuitive meanings but the RSM and CBM parameters do, and when their meanings are similar one expects correlations between them. The meanings of the RSM parameters were described in Section 16.3. The CBM α parameter, **Chapter 20**, is the fraction of diseased cases on which the disease is visible, so one expects it to correlate with the RSM ν' parameter. Also, the CBM μ_{CBM} parameter is the separation of two unit variance normal distributions, so it should correlate with the RSM μ parameter. The correspondences can perhaps be discerned in the tables, but a careful analysis requires bootstrapping to determine the correlations averaged over all datasets and corresponding 95% confidence intervals. (There is no counterpart to the RSM λ' parameter in PROPROC or CBM.)

The code for this is also in file **mainInterCorrelations.R**, ensure that line 29 is commented. Lines 44 through 50 extract the three pre-saved AUC values, followed by the two relevant RSM parameters and the two CBM parameters. Lines 114 and 115 calculate the correlations $Cor(\mu,\mu_{CBM})$ and $Cor(v',\alpha)$, listed in the last two columns of Table 19.4. The notation in the code file should be fairly transparent. For example, **nupRsm** is v'. The grand averages and bootstrap confidence intervals are determined as in the preceding section. Example plots for dataset #7 are shown in Figure 19.7 (a–b). Panel (a) shows a representative scatter plot

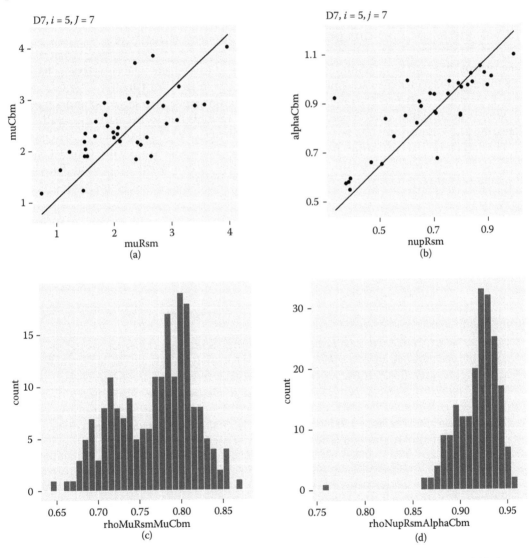

Figure 19.7 Examples of the expected correlations between CBM and RSM parameters. Panel (a) shows, for dataset #7, a scatter plot of μ_{CBM} versus the RSM μ parameter; the slope is 0.6815. Panel (b) shows, for the same dataset, a scatter plot of the CBM α parameter versus the RSM v' parameter; the slope is 0.8303. Panel (c) is the histogram of 200 correlations between μ_{CBM} and RSM μ, where each correlation is an average over all datasets. Panel (d) is the corresponding histogram for α versus v'. These plots were generated by **mainInterCorrelations.R**. (**muCbm** = μ_{CBM}; **muRsm** = μ; **alphaCbm** = α; **nupRsm** = v'; **rhoMuRsmMuCbm** = $Cor(\mu,\mu_{CBM})$; **rhoNupRsmAlphaCbm** = $Cor(v',\alpha)$; see Section 19.5.3.)

of `muCbm` versus `muRsm` while panel (b) shows a representative scatter plot of `alphaCbm` versus `nupRsm`. Panel (c) is the bootstrap histogram of the average correlation, over all datasets, corresponding to panels like (a), while panel (d) is the bootstrap histogram corresponding to panels like (b).

The details should not obscure the fact that the observed correlations are strong. The final values, including 95% confidence intervals are $Cor(\mu, \mu_{CBM}) = 0.7858$ (0.675, 0.847) and $Cor(v', \alpha) = 0.9284$ (0.871, 0.950).

19.5.4 Intra-correlations between RSM derived quantities

The title of this section refers to *intra-correlations* as it investigates the correlations between the three performance measures predicted by RSM parameters. In **Chapter 17** it was shown that RSM parameters predict a measure of search performance (Section 17.8) and a measure of lesion-classification performance (Section 17.9) in addition to predicting overall AUC, Equation 17.25. Equation 17.38 relates search performance to RSM parameters λ', v'. Equation 17.39 relates lesion-classification performance to RSM parameter μ. The three model parameters collectively predict RSM-AUC, which is calculated by numerical integration. For clarity, the equations are reproduced here.

$$S = v' \exp(-\lambda') \tag{19.23}$$

$$A_C = \Phi(\mu / \sqrt{2}) \tag{19.24}$$

*Consider search and lesion-classification performances. Since high values for each of them is a sign of expertise, one expects either of them to be positively correlated with AUC. However, assuming that overall ROC-AUC performance is somehow intrinsically limited (see below), then it is reasonable to assume that each radiologist tends to compensate for deficiencies in one area (e.g., search performance) by perfecting one's strengths (e.g., lesion classification performance), the desired end-point being to maximize performance. **In this situation one expects an inverse correlation between search and lesion classification performance.***

Signal detection theory, which considers the statistical distribution of signal and noise, places a limitation on the performance of the observer.[3] The limiting observer is termed an **ideal observer**. If both search and lesion-classification performance can be optimized *independently*, then ideal observer performance is in principle achievable. Human observers usually do not achieve ideal observer performance—this is the *intrinsic limit* mentioned in the previous paragraph. Papers by Pelli, Burgess, and others, have studied human observer efficiency in location known exactly ROC tasks.[2,25–30] Ideal observer performance in the FROC paradigm is currently an unsolved problem.

Preliminary data for inverse correlation between search and lesion-classification was presented at a conference.[31] *The purpose of this section is to show that this expectation holds for every one of the 14 datasets that were since analyzed.*

The analysis is similar to that presented for inter-correlations. The essential difference is that this time all quantities are RSM-related (i.e., PROPROC and CBM do not enter). For each dataset, modality and reader, one extracts RSM parameters from the pre-saved object `allResults`. The parameters are used to calculate search performance S, lesion classification performance C and ROC area A. For each dataset one calculates the three pairs of correlations between the performance measures S, C and A. The three performance measures and the three correlations are averaged over all datasets. This yields the base-line values. To estimate confidence intervals around the baseline values, for each dataset one bootstraps readers and recomputes the 3 performance measures and the 3 correlations and averages over all datasets. This is repeated 200 times. The empirical distributions of the quantities are used to estimate 95% confidence intervals. The code is in file `mainIntraCorrelations.R` which is listed in in Online Appendix 19.C.1 and explained in Online Appendix 19.C.2. The output of sourcing this code is in Online Appendix 19.C.3 and summarized in Table 19.5. In addition, the code produces scatter plots of C versus S, A versus S and A versus C for each dataset. Results for dataset #11 are shown in Figure 19.8 (a–c). Finally, the code produces histograms of the bootstrap distributions of S, C and A, shown in Figure 19.9 (a–c) and histograms of the bootstrap distributions of the correlations between S and C, between A and C and between S and A, Figure 19.9 (d–f).

We start with a discussion of the values of lesion classification performance C and non-diseased versus diseased classification performance A in Table 19.5.

Table 19.5 The results for individual datasets (i.e., original data, no bootstrapping) and bootstrap confidence intervals.

Dataset # *f*	S	C	A	$\rho(S,C)$	$\rho(S,A)$	$\rho(CA)$
1	0.254	0.833	0.817	−0.394	0.636	0.246
2	0.356	0.986	0.925	−0.246	0.479	0.400
3	0.084	0.970	0.862	−0.815	0.601	−0.503
4	0.266	0.931	0.848	−0.681	0.112	−0.417
5	0.253	0.919	0.881	−0.547	0.209	0.313
6	0.091	0.928	0.699	−0.384	0.336	−0.490
7	0.312	0.917	0.842	−0.333	0.391	0.296
8	0.160	0.968	0.829	−0.311	0.074	0.203
9	0.185	0.933	0.855	−0.718	−0.303	0.226
10	0.150	0.858	0.789	−0.616	0.215	0.243
11	0.130	0.749	0.718	−0.637	0.865	0.516
12	0.215	0.764	0.691	−0.481	0.581	0.078
13	0.039	0.759	0.621	−0.383	0.675	−0.253
14	0.150	0.969	0.881	−0.438	−0.445	0.580
Grand average over datasets and BS CIs						
GR. AVG	0.189	0.892	0.804	−0.499	0.316	0.018
BS CI, Lower	0.160	0.877	0.795	−0.608	0.125	0.114
BS CI, Upper	0.218	0.907	0.813	−0.350	0.437	0.266

Note: All search performance versus lesion classification performance correlations are negative. On the average, there is a negative correlation between search performance S and lesion classification performance C −0.499 (−0.61, −0.35). On the average, there is a positive correlation 0.316 (0.13, 0.44) between search and AUC but not between lesion classification and AUC 0.02 (−0.11, 0.27). Lesion-classification performance C is always larger than A and the difference is larger when search performance is smaller; see, for example datasets #6 and #13 [S = search performance; C = lesion classification performance; A = RSM-AUC. AVG = average; BS = bootstrap; CI = 95% confidence interval].

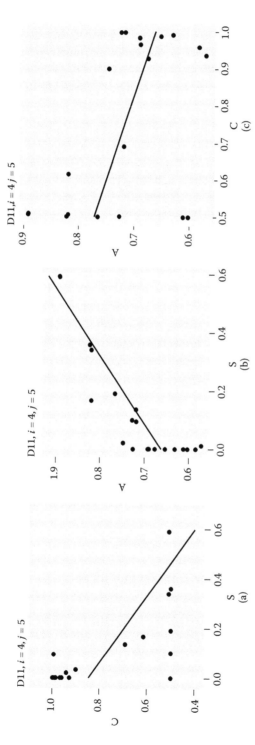

Figure 19.8 Scatter plots of pairings of search (S), lesion classification (C) and AUC (A) performance for dataset #11. Panel (a) is a scatter plot of lesion-classification performance C versus search-performance S, demonstrating the inverse correlation between them; the observed correlation is −0.637, which is typical. Panel (b) is a scatter plot between AUC and S; the observed correlation is 0.868, which is atypically high. Panel (c) is a scatter plot between AUC and C; the observed correlation is −0.516, which is atypically low. Modality-reader averaged search, lesion classification and AUC, for this dataset, are 0.130, 0.748 and 0.718, respectively. Note the low value of search relative to lesion classification and AUC performances. These plots were generated by `mainIntraCorrelations.R` with analysis restricted to dataset #11. [C = lesion-classification performance; S = search performance; A = RSM-AUC; rho = correlation.]

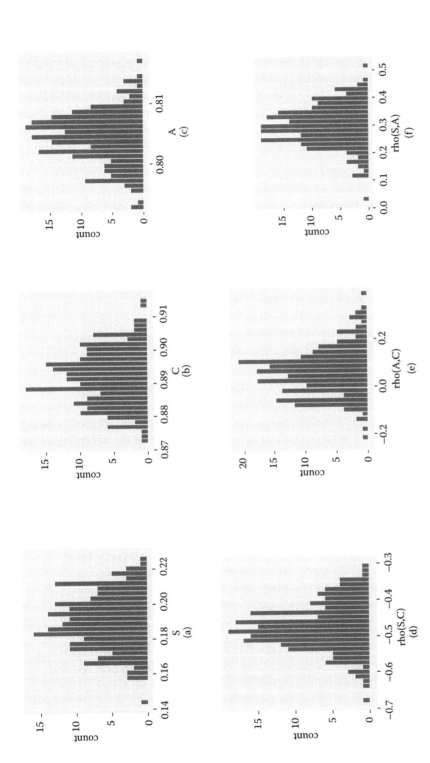

Figure 19.9 Bootstrap distributions of search performance S, lesion classification performance C and AUC performance A, and correlations between them. Panels (a–c) show bootstrap histograms, averaged over all datasets, to determine confidence intervals for search, lesion-classification and ROC-AUC performances. Average S is 0.189 (0.160, 0.218), average C is 0.892 (0.877, 0.907), and average AUC is 0.804 (0.795, 0.813). These values demonstrate that search expertise is the bottleneck limiting overall performance. Panels (d–f) show bootstrap distributions of correlations between the indicated pairs of measures. Panel (d) shows a histogram of 200 bootstrapped Pearson correlations between S and C, (grand) averaged over all datasets. Panel (e) is the corresponding histogram of correlations between A and C. Panel (f) is between S and A. (d) Average correlation between S and C is −0.499, (0.61, 0.35)—the evidence of inverse correlation between search and lesion-classification performance is strong. (e) Average correlation between S and A is 0.316, (0.13, 0.44). (f) Average correlation between C and A is 0.018, (−0.11, 0.27); one cannot rule out zero for this correlation. These plots were generated by `mainIntraCorrelations.R`. [C = lesion-classification performance; S = search performance; A = AUC; rho = correlation.]

19.5.4.1 Lesion classification performance versus AUC

It was shown in Section 17.9.1 that one expects the ordering $AUC < A_C \leq AUC_{LKE}$. Focusing on the first part of the inequalities, and using abbreviated notation: $A < C$. Here A is ROC-AUC measured under "normal" reading conditions, when there is uncertainty about lesion location, and the task is to separate non-diseased from diseased cases. C is the performance measured in a special 2AFC study in which on one image a latent LL is cued and on another a latent NL is cued and the task is to choose the latent LL. The LL and NL cued locations would need to be determined by a prior eye-tracking study, as described in Section 16.3.1. Table 19.5 confirms the expectation $A < C$ and furthermore shows that the difference tends to increase as search performance decreases. For example, see results for datasets #6 and #13 in the second and third columns of Table 19.5. It should be obvious why lesion classification performance under the cued condition is higher than when possible lesion locations are not cued (the latter is sometimes termed "free-search"[32,33] which is the "normal" condition under which images are interpreted in the clinic). If one knows where to look, the possibility of mistaking non-diseased regions as lesions is eliminated thereby enhancing performance (there is no problem locating the sun on a cloudy day if one has the correct locat time and GPS coordinates). One needs to explain why with poorer search performance, the expected difference is greater. If search performance is perfect (i.e., unity), the observer sees the lesion even if it is not cued, so in this situation $C = A$, in other words cuing has no effect, and the difference is zero. As search performance degrades, the probability increases that a non-diseased region will be higher rated than a lesion, which causes A to decrease, and the difference $C - A$ to increase. Formally, since search performance is determined by ν' and λ' parameters, one can hold the RSM μ parameter constant, which fixes C, but A decreases as search performance is degraded by decreasing ν' and / or increasing λ'. Since C is fixed but A decreases, the difference increases.

19.5.4.2 Summary of search and lesion classification performances versus AUC and intra-correlations of RSM parameters and a key finding

The plots in Figure 19.8 are noisy, with much evidence of parameter "clipping" referred to earlier and firm conclusions regarding the slopes are difficult. It turns out that plots (b) and (c) are atypical. It is necessary to average over all datasets to arrive at firmer conclusions and perform bootstrap analysis to determine confidence intervals.

Base line values for all quantities are listed in the row labeled GR. AVG. (for grand average over all original datasets). The following two rows list bootstrap 95% confidence limits, lower and upper, respectively.

A key result is that average search performance, for the datasets studied, was only 0.189 (0.16, 0.22), while average lesion-classification performance was 0.892 (0.877, 0.907) and average AUC was 0.804 (0.795, 0.813). One expects lesion-classification performance to be larger than AUC. *The surprising finding that search performance is only about 19% while lesion-classification is about 89% means that the <u>bottleneck in reader performance is suboptimal search</u>.*

Another, finding, consistent with the arguments presented in Section 19.5.4, is the relatively strong inverse correlation between S and C: −0.499 (−0.61, −0.35). Moreover, there is a positive correlation between S and A: 0.316 (0.13, 0.44) and weak correlation between C and A, which does not exclude zero: 0.018 (−0.11, 0.27).

19.6 Reason for serendipitous finding

Recall the serendipitious finding that all three proper ROC methods yield almost the same AUC. The reason is that the proper ROC is a consequence of using a decision variable that is equivalent (in the arbitrary monotonic increasing transformations sense, see below) to a *likelihood ratio*.

Proper ROC fitting is discussed in **Chapter 20**. For now, it is sufficient to note that the likelihood ratio $l(z)$ is defined as the ratio of the diseased *pdf* to the non-diseased *pdf*.

$$l(z) = \frac{pdf_D(z)}{pdf_N(z)} \tag{19.25}$$

There is a theorem,[3] that an observer who uses the likelihood ratio $l(z)$, or any monotone increasing transformation of it, as the decision variable, has optimal performance, that is, maximum AUC. To use the terminology of Barrett and Myers,[3] any observer using the likelihood ratio as the decision variable is an *ideal observer* (Section 13.2.6 *ibid.*). The *pdfs* depend on the assumed model for the Z-samples, which could be very different. *Different ideal observers interpreting the same cases must yield the same AUC as otherwise, one observer would be more ideal than another.* Different models of fitting proper ROCs represent different approaches to modeling the decision variable and the likelihood ratio, but while the curves can have different shapes (a second theorem is that the slope of the ROC curve at a given point equals the likelihood ratio calculated at that point) their AUCs must agree. This would explain the empirical observation that RSM, PROPROC, and CBM all yield the same AUCs, as summarized in Table 19.4.

19.7 Sample size estimation: wAFROC FOM

The author will end this chapter by turning to sample size estimation for FROC studies using the wAFROC AUC as the FOM. FROC sample size estimation is not fundamentally different from the procedure outlined in **Chapter 11** for the ROC paradigm. To recapitulate, based on analysis of a pilot dataset and using a specified FOM, for example, the ROC-AUC, and either the DBMH or the ORH method for significance testing, one estimates the intrinsic variability of the data expressed in terms of variance components. For DBMH analysis, these are the pseudovalue variance components, while for ORH analysis these are the FOM treatment-reader variance component and the FOM covariances. The second step is to *postulate a clinically realistic effect size*, for example, the AUC difference between the two modalities. Given these values, the sample size functions implemented in **RJafroc** allow one to estimate the number of readers and cases necessary to detect (i.e., reject the null hypothesis) at specified Type II error rate, typically chosen to be 20% (corresponding to 80% statistical power) and specified Type I error rate, typically chosen to be 5%.

In FROC analysis the only difference, indeed the critical difference, is the choice of FOM; for example, the wAFROC-AUC instead of the inferred ROC-AUC. The FROC dataset is analyzed using either the DBMH or the ORH method. This yields the necessary variance components or the covariance matrix corresponding to the wAFROC-AUC. The next step is to specify an effect size in wAFROC-AUC units, and therein is the rub. What value does one use? The ROC-AUC has a historically well-known interpretation: the classification ability at separating diseased patients from non-diseased patients. *Needed is a way of relating the effect size in ROC-AUC units to one in wAFROC-AUC units.*

1. One chooses a ROC-AUC effect size that is realistic, one that clinicians understand and can therefore participate in, in the effect size postulation process.
2. One converts the ROC effect size to a wAFROC-AUC effect size. The method for this is described in the next section.
3. One uses the sample size tools in in **RJafroc**, that is, functions with names beginning with **Ss**, to determine the necessary sample size.

It is important to recognize is that all quantities have to be in the same "units." The quotes are used since the units are not like units in physics, e.g., the MKS versus CGS systems, or Rad versus Gray. When doing ROC analysis, everything (variance components and effect size) has to be in units of the selected FOM, for example, ROC-AUC. When doing wAFROC analysis, everything has to be in units of the wAFROC-AUC. The variance components and effect size in wAFROC-AUC units will be different from their corresponding ROC counterparts. In particular, as shown next, a ROC-AUC effect size of 0.05 generally correspond to a larger effect size in wAFROC-AUC units. The reason for this is that the range over which wAFROC-AUC can vary, namely 0–1, is twice the corresponding ROC-AUC range.

The next section explains the steps used to implement #2 above.

19.7.1 Relating an ROC effect size to a wAFROC effect size

For each modality-reader (*ij*) dataset, the inferred ROC data is fitted by the RSM using procedure described above, yielding estimates of the parameters $\mu_{ij}, \lambda_{ij}, \nu_{ij}$ (notice the usage of intrinsic RSM parameters, not the primed values; the latter are easily converted to intrinsic values). The pilot study represents an *almost* null hypothesis dataset; if a significant difference was observed one would not be going through the exercise of sample size estimation. In any case, the author recommends taking the *median* of the three sets of parameters, over *all indices*, as representing the *average* NH dataset. The reason for this is that the median is less sensitive to outliers than the average.

$$\left.\begin{array}{l} \mu_{NH} = median(\mu_{ij}) \\ \lambda_{NH} = median(\lambda_{ij}) \\ \nu_{NH} = median(\nu_{ij}) \end{array}\right\} \tag{19.26}$$

Using these values ROC-AUC and wAFROC-AUC, for the NH condition, denoted x_{NH} and y_{NH} respectively, are calculated by numerical integration of the RSM-predicted ROC and wAFROC curves (**Chapter 17**). As before, f_L is the lesion distribution vector.

$$x_{NH} = ROC_{AUC}(\mu_{NH}, \lambda_{NH}, \nu_{NH}, \vec{f}_L)$$

$$y_{NH} = wAFROC_{AUC}(\mu_{NH}, \lambda_{NH}, \nu_{NH}, \vec{f}_L) \tag{19.27}$$

To induce the alternative hypothesis condition one increments μ_{NH} by $\Delta\mu$. The resulting ROC-AUC and wAFROC-AUC are calculated, again by numerical integration of the RSM-predicted ROC and wAFROC curves, leading to the corresponding effect sizes (note that in each equation below one takes the difference between the AH value minus the NH value).

$$ES_{ROC}(\Delta\mu) = ROC_{AUC}(\mu_{NH}+\Delta\mu, \lambda_{NH}, \nu_{NH}, \vec{f}_L) - ROC_{AUC}(\mu_{NH}, \lambda_{NH}, \nu_{NH}, \vec{f}_L)$$

$$ES_{wAFROC}(\Delta\mu) = wAFROC_{AUC}(\mu_{NH}+\Delta\mu, \lambda_{NH}, \nu_{NH}, \vec{f}_L) - wAFROC_{AUC}(\mu_{NH}, \lambda_{NH}, \nu_{NH}, \vec{f}_L) \tag{19.28}$$

Equation 19.28, evaluated for different values of $\Delta\mu$, provides a calibration curve between the effect sizes expressed in the two units in Figure 19.10a. This allows one to interpolate the appropriate wAFROC effect size corresponding to any postulated ROC effect size.

19.7.2 Example using DBMH analysis

The following example uses the first two modalities of the **FED** dataset, which is a five-modality four-radiologist FROC dataset acquired by Dr. Federica Zanca.[34] The detailed explanation is in Online Appendix 19.C and the following is a summary. The datafile is read and a dataset object corresponding to modalities 1 and 2, only, is extracted. These were *not* found to be significantly different,[35] so it makes sense to regard them as *almost* NH modalities. As before, the **allResults** object contains preanalyzed data—it is a large list with 20 elements. If the data is not preanalyzed, one should be prepared for a short wait while the RSM algorithm calculates the needed values.

In this example, the DBMH method is used to get the needed pseudovalue variance components. The code is in **mainwAFROCPowerDBMH.R**, detailed in Online Appendix 19.C.1. Ensure that is set to 4 at line 13. **Source** it to get the abbreviated output (details are in online file **PowerComparison.xlsx**) summarized in Table 19.6 and Figure 19.10. The critcal part is how the sample size estimation function is called, as it is rather unforgiving at the time of writing. An example follows in the next section.

19.7.2.1 Code listing (partial)

```
varCompROC <- StSignificanceTesting(
  frocData,
  fom = "HrAuc",
  method = "DBMH",
  option = "RRRC")$varComp
varCompwAFROC <- StSignificanceTesting(
  frocData,
  fom = "wAFROC",
  method = "DBMH",
  option = "RRRC")$varComp

...

for (i in 1:length(effectSizeROC)) {
  varYTR <- varCompROC$varComp[3]
  varYTC <- varCompROC$varComp[4]
  varYEps <- varCompROC$varComp[6]
  powerROC[i] <- SsPowerGivenJK(
    JTest,
    KTest,
    alpha = 0.05,
    effectSize = effectSizeROC[i],
    option = "RRRC",
    method = "DBMH",
    varYTR = varYTR,
    varYTC = varYTC,
    varYEps = varYEps)$powerRRRC

  varYTR <- varCompwAFROC$varComp[3]
  varYTC <- varCompwAFROC$varComp[4]
  varYEps <- varCompwAFROC$varComp[6]
  powerwAFROC[i] <- SsPowerGivenJK(
    JTest, KTest,
    alpha = 0.05,
    effectSize = effectSizewAFROC[i],
    option = "RRRC",
    method = "DBMH",
    varYTR = varYTR,
```

```
    varYTC = varYTC,
    varYEps = varYEps)$powerRRRC

  cat("ROC effect-size = ,", effectSizeROC[i],
     "wAFROC effect-size = ,", effectSizewAFROC[i],
     "Power ROC, wAFROC:", powerROC[i], ",", powerwAFROC[i], "\n")
  }
```

Table 19.6 Sample size estimation for FROC studies

ES calibration data		Power table for 5 readers and 100 cases			
$ES_{ROC}(\Delta\mu)$	$ES_{wAFROC}(\Delta\mu)$	ES-ROC	ES-wAFROC	Power ROC	Power wAFROC
0.00076	0.00152	0.01	0.02024	0.0644	0.1165
0.00151	0.00303	0.02	0.04047	0.1088	0.3216
0.00226	0.00452	0.03	0.06071	0.1847	0.6092
0.00300	0.00601	0.04	0.08095	0.2908	0.8448
...	...	0.05	0.10118	0.4195	0.9597
0.01139	0.02306	0.06	0.12142	0.5574	0.9934
0.01205	0.02442	0.07	0.14166	0.6882	0.9993
0.01270	0.02577	0.08	0.16189	0.7984	1.0000
0.01335	0.02711	0.09	0.18213	0.8810	1.0000
0.01399	0.02844	0.10	0.20237	0.9361	1.0000

Note: Shown in the first two columns is a partial listing of the effect size calibration data. The remaining columns list different ROC effect sizes, the corresponding AFROC effect sizes, and statistical power for ROC and corresponding power for wAFROC FOM. Note the substantial statistical power penalty when using the ROC FOM.

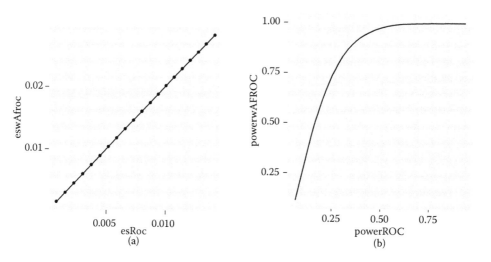

Figure 19.10 (a) Effect size calibration plot generated by computing the ROC and wAFROC effect sizes for different values of $\Delta\mu$, Equation 19.28. The points are well fitted by a straight line, which makes it easy to interpolate the wAFROC effect size for any specified ROC effect size. In this example, the slope of the fitting line is two, which translates to a doubling of the effect size in wAFROC units relative to ROC units. (b) Comparison of statistical power between ROC and wAFROC FOMs, for 100 cases and five readers. The increased power of the wAFROC is due in main part to the larger effect size. (eswAFROC = effect size using wAFROC FOM; esROC = effect size using ROC FOM; powerwAFROC = statistical power using wAFROC FOM; powerROC = statistical power using ROC FOM.)

In the first call to **SsPowerGivenJK()**, line 133, **method** is specified as **"DBMH"** and it is followed by three arguments. The Y symbol designates pseudovalue-based variance components and the order and names of the arguments are important; they stand for $\sigma^2_{Y;\tau R}$, $\sigma^2_{Y;\tau C}$ and $\sigma^2_{Y;\tau RC+\epsilon}$ as calculated by a previous call to **StSignificanceTesting()** with **fom** set to **"HrAuc"** and **effectSize** specified as **effectSizeROC[i]**. The second call, line 147, looks very similar but note the usage of variance components calculated by a previous call to **SignificanceTesting()** with **fom** set to **"wAFROC"** and **effectSize** specified as **effectSizewAFROC[i]**.

Shown in the first two columns of Table 19.6 is a partial listing of the effect size calibration data. In other words, the values of the two effect sizes listed in the left-hand side of Equation 19.28 for 20 values of $\Delta\mu$ ranging from 0.01 to 0.2 in steps of 0.01. These are fitted to a straight-line constrained to go through the origin, Figure 19.10a. The slope of the fit (2.02, i.e., the wAFROC effect size is about twice the ROC effect size) was used to convert the ROC effect sizes listed in the third column of Table 19.6 (ranging from 0.01 to 0.10) to the corresponding wAFROC effect sizes listed in the fourth column. DBMH analysis of the inferred ROC data yields pseudovalue variance components, which, when used with the ROC effect size, yields the predicted ROC power for five readers and 50/50 cases listed under *Power ROC*. Likewise, DBMH analysis using the wAFROC FOM yields pseudovalue variance components, which, when used with the *wAFROC effect size*, yields the predicted wAFROC power for five readers and 50/50 cases listed under *Power wAFROC*.

Because the effect size enters the sample size formulae as the square, the effect is dramatic. For an ES-ROC = 0.03, the power of the ROC method is about 18.5% while that of the wAFROC method is about 60.9%.

The observed wAFROC effect size was −0.00686 (this is a very small effect size and the corresponding ROC effect size is about −0.00343; the sign does not effect the calculations), see red font near bottom of Online Appendix 19.C.8, which was too small to reach 80% power, even with the number of cases equal to 200 and the number of readers equal to four, the actual conditions of the study,[35] see results of sourcing code with these values in Online Appendix 19.C.7. It is not surprising that the study did not find a significant difference between these two modalities.

While demonstrated for the wAFROC FOM, sample size estimation can be extended to other figures of merit. The only change is to use the same FOM for both variance components/covariances and effect size specification. Specifying the new FOM and running DBMH or ORH analysis determines the variance components. Next, one specifies the effect size in ROC-AUC units and converts it to the effect size in the new units—this is where parametric curve fitting is needed, currently unimplemented for LROC and ROI paradigms.

19.7.3 Example using ORH analysis

An example using the ORH method to get the needed FOM variance components is in code **mainwAFROCPowerORH.R**. **Source** it to get essentially the same output as in Table 19.6 and Figure 19.10.

19.7.3.1 Code snippet

```
varCompROC <- StSignificanceTesting(
  frocData,
  fom = "HrAuc",
  method = "ORH",
  option = "RRRC")$varComp
varCompwAFROC <- StSignificanceTesting(
  frocData,
  fom = "wAFROC",
  method = "ORH",
```

```
    option = "RRRC")$varComp
...
 for (i in 1:length(effectSizeROC)) {
   cov1 <- varCompROC$varCov[3]
   cov2 <- varCompROC$varCov[4]
   cov3 <- varCompROC$varCov[5]
   varTR <- varCompROC$varCov[2]
   varEps <- varCompROC$varCov[6]
   powerROC[i] <- SsPowerGivenJK(
     JTest,
     KTest,
     alpha = 0.05,
     effectSize = effectSizeROC[i],
     option = "RRRC",
     method = "ORH",
     cov1 = cov1,
     cov2 = cov2,
     cov3 = cov3,
     varTR = varTR,
     varEps = varEps, KStar = KStar)$powerRRRC

   cov1 <- varCompwAFROC$varCov[3]
   cov2 <- varCompwAFROC$varCov[4]
   cov3 <- varCompwAFROC$varCov[5]
   varTR <- varCompwAFROC$varCov[2]
   varEps <- varCompwAFROC$varCov[6]
   powerwAFROC[i] <- SsPowerGivenJK(
     JTest,
     KTest,
     alpha = 0.05,
     effectSize = effectSizewAFROC[i],
     option = "RRRC",
     method = "ORH",
     cov1 = cov1,
     cov2 = cov2,
     cov3 = cov3,
     varTR = varTR,
     varEps = varEps, KStar = KStar)$powerRRRC

   cat("ROC effect-size = ,", effectSizeROC[i],
     "wAFROC effect-size = ,", effectSizewAFROC[i],
     "Power ROC, wAFROC:", powerROC[i], ",", powerwAFROC[i], "\n")
 }
```

In the first call, **method** is specified as **"ORH"** and it is followed by five arguments and K^*, the number of cases in the pilot study. The passed variance components are, in order, Cov_1, Cov_2, Cov_3, $\sigma^2_{\tau R}$, and *Var*. They were calculated by a previous call to **StSignificanceTesting()** with **method** set to **"ORH"**, **fom** set to **"HrAuc"** and **effectSize** is specified as **effectSizeROC[i]**. The second call looks very similar but note the usage of variance components calculated by a previous call to **StSignificanceTesting()** **method** set to **"ORH"**, with **fom** set to **"wAFROC"** and **effectSize** is specified as **effectSizewAFROC[i]**.

For an ES-ROC of 0.03, the power of the ROC method is about 18.5% while that of the wAFROC method is about 60.9%, the same as obtained by *DBMH* analysis. The reader should experiment with other datasets to become more familiar with the method.

19.8 Discussion/Summary

Over the years, there have been several attempts at fitting FROC data. Prior to the RSM-based ROC curve approach described in this chapter, all methods were aimed at fitting FROC curves, in the mistaken belief that this approach was using all the data. The earliest was the author's FROCFIT software.[35] This was followed by Swensson's approach,[36] subsequently shown to be equivalent to the author's earlier work, as far as predicting the FROC curve was concerned.[11] In the meantime, CAD developers, who relied heavily on the FROC curve to evaluate their algorithms, developed an empirical approach that was subsequently put on a formal basis in the IDCA method.[12]

This chapter describes an approach to fitting ROC curves, instead of FROC curves, using the RSM. Fits were described for 14 datasets, comprising 236 distinct modality-reader combinations. All fits and parameter values are viewable by sourcing the cited code file **mainRSM.R**. Validity of fit was assessed by the chi-sare goodness of fit p-value; in 42 out of 236 fits, where a valid chi-square statistic could be calculated, the p-value was greater than 0.05. This provides strong evidence for validity of the RSM and the highest-rating method for inferring ROC data from FROC data, which is the basis of the RSM-predicted ROC curves, Section 19.4.

Two conventional, i.e., non-search, proper ROC methods, namely PROPROC and CBM, were fitted to the same datasets, yielding further insights. One of the insights was the finding that the AUCS were almost identical, with PROPROC yielding, on the average, a slightly higher value, by approximately 1% relative to the RSM, and CBM yielding a slightly lower value, by about 0.6%. The PROPROC-AUC and CBM-AUC versus RSM-AUC straight-line fits, constrained to go through the origin, had near unit slopes and R^2 values greater than 0.999, indicative of excellent fits. The equality can be understood based on the uniqueness of the performance of the ideal observer implied by a proper ROC curve. While the curves are model dependent and not unique, their AUCs are unique.

The reader may wonder why the author chose not to fit the wAFROC. After all, it is the recommended figure of merit for FROC studies. While the methods described in this chapter are readily adapted to the wAFROC, they are more susceptible to degeneracy issues. The reason is that the y-axis is defined by LL-events, in other words, by the μ, ν' parameters, while the x-axis is defined by the highest rated NL on non-diseased cases, in other words, by the λ' parameter. The consequent decoupling of parameters leads to degeneracy of the type described in Section 19.2.3. This is somewhat alleviated, but not eliminated, in ROC fitting because the y-axis is defined by LLs and NLs, in other words all parameters of the RSM are involved. The situation with the wAFROC is not quite as severe as with fitting FROC curves but it is expected to have a problem with degeneracy. There are some ideas but for now this remains an open research subject. Empirical wAFROC, which is the current method implemented in **RJafroc**, is expected to have the same issues with variability of thresholds between modalities as the empirical ROC-AUC, as discussed in Section 5.9. So, the fitting problem needs to be solved.

The current ROC-based effort led to some interesting findings. The near equality of the AUCs predicted by the three proper ROC fitting methods, summarized in Table 19.4, has been noted, which is explained by the fact that proper ROC fitting methods represent different approaches to realizing an ideal observer, and the ideal observer must be unique, Section 19.6.

This chapter explores what is termed *inter-correlations*, between RSM and CBM parameters. Since they have similar physical meanings, the RSM and CBM separation parameters were found to be significantly correlated as were the RSM and CBM parameters corresponding to the fraction of lesions that was actually visible. This type of correspondence between two different models can be interpreted as evidence of mutually reinforcing validity of each of the models.

An important finding is the inverse correlation between search performance and lesion-classification performance, which suggest there could be tradeoffs in attempts to optimize them. As a simplistic illustration, a low-resolution gestalt-view of the image,[1] such as seen by the peripheral viewing mechanism, is expected to make it easier to rapidly spot deviations from the expected normal template described in **Chapter 15**. However, the observer may not be able to switch effectively between this and the high-resolution viewing mode necessary to correctly classify found suspicious region.

The main scientific conclusion of this chapter is that search-performance is the bottleneck in limiting observer performance (S = 19%). It is unfortunate that search is ignored in current analytical approaches using the ROC paradigm, and influential researchers still continue to advocate ignoring localizaton information. Search is also ignored in approaches using the empirical AUC and in current model observer methods. Actually, the search information is present in the ROC curve—one just needs a model, such as the RSM, that includes search and predicts ROC curves. Finally, researchers need to reconsider the focus of their investigations, most of which is currently directed at improving lesion classification performance, which is not to the bottleneck (C = 89%). A general rule in systems optimization, namely focusing on the weakest link, is currently being largely ignored. Finally, work remains to be done in improving the RSM fitting algorithm and possibly extending it to include higher-order effects such as satisfaction of search.

This concludes **Part C** of the book. It is time to move on to a few advanced topics in **Part D**.

References

1. Kundel HL, Nodine CF, Conant EF, Weinstein SP. Holistic component of image perception in mammogram interpretation: Gaze-tracking study. *Radiology*. 2007;242(2):396–402.
2. Burgess AE, Wagner RF, Jennings RJ, Barlow HB. Efficiency of human visual signal discrimination. *Science*. 1981;214(2):93–94.
3. Barrett HH, Myers K. *Foundations of Image Science*. Hoboken, NJ: John Wiley and Sons; 2003.
4. Yoon HJ, Zheng B, Sahiner B, Chakraborty DP. Evaluating computer-aided detection algorithms. *Med Phys*. 2007;34(6):2024–2038.
5. Chakraborty DP, Svahn T. Estimating the parameters of a model of visual search from ROC data: An alternate method for fitting proper ROC curves. *Proc SPIE 7966*. 2011;7966. Lake Buena Vista (Orlando), FL.
6. Metz CE, Pan X. Proper binormal ROC curves: Theory and maximum-likelihood estimation. *J Math Psychol*. 1999;43(1):1–33.
7. Pan X, Metz CE. The "proper" binormal model: Parametric receiver operating characteristic curve estimation with degenerate data. *Acad Radiol*. 1997;4(5):380–389.
8. Dorfman DD, Berbaum KS, Brandser EA. A contaminated binormal model for ROC data: Part I. Some interesting examples of binormal degeneracy. *Acad Radiol*. 2000;7(6):420–426.
9. Dorfman DD, Berbaum KS. A contaminated binormal model for ROC data: Part III. Initial evaluation with detection. *Acad Radiol*. 2000;7(6):438–447.
10. Dorfman DD, Berbaum KS. A contaminated binormal model for ROC data: Part II. A formal model. *Acad Radiol*. 2000;7(6):427–437.
11. Chakraborty DP, Yoon HJ. Operating characteristics predicted by models for diagnostic tasks involving lesion localization. *Med Phys*. 2008;35(2):435–445.
12. Edwards DC, Kupinski MA, Metz CE, Nishikawa RM. Maximum likelihood fitting of FROC curves under an initial-detection-and-candidate-analysis model. *Med Phys*. 2002;29(12):2861–2870.
13. Shanno DF, Kettler PC. Optimal conditioning of quasi-Newton methods. *Math Comput*. 1970;24(111):657–664.
14. Shanno DF. Conditioning of quasi-Newton methods for function minimization. *Math Comput*. 1970;24(111):647–656.
15. Goldfarb D. A family of variable-metric methods derived by variational means. *Math Comput*. 1970;24(109):23–26.
16. Fletcher R. A new approach to variable metric algorithms. *Comput J*. 1970;13(3):317–322.

17. Broyden CG. The convergence of a class of double-rank minimization algorithms 1. general considerations. *IMA J Appl Math*. 1970;6(1):76–90.

18. Fletcher R. *Practical Methods of Optimization*. John Wiley & Sons; 1987. 2nd Edition; Chichester, UK.

19. Ben B and R Development Core Team bbmle: Tools for General Maximum Likelihood Estimation. R package version. 2017. https://CRAN.R-project.org/package=bbmle.

20. Berbaum KS, Metz CE, Pesce LL, Schartz KM. DBM MRMC User's Guide, DBM-MRMC 2.1 Beta Version 2, http://www-radiology.uchicago.edu/cgi-bin/roc_software.cgi and http://perception.radiology.uiowa.edu, Accessed 28 December 2009.

21. Svahn T, Andersson I, Chakraborty D, et al. The diagnostic accuracy of dual-view digital mammography, single-view breast tomosynthesis and a dual-view combination of breast tomosynthesis and digital mammography in a free-response observer performance study. *Radiat Prot Dosimetry*. 2010;139:113–117.

22. Van Dyke CW, White RD, Obuchowski NA, Geisinger MA, Lorig RJ, Meziane MA. Cine MRI in the diagnosis of thoracic aortic dissection. *79th RSNA Meetings*. 1993. Chicago, IL.

23. Franken EA, Jr., Berbaum KS, Marley SM, et al. Evaluation of a digital workstation for interpreting neonatal examinations: A receiver operating characteristic study. *Invest Radiol*. 1992;27(9):732–737.

24. Eisenhauer JG. Regression through the origin. *Teach Stat*. 2003;25(3):76–80.

25. Burgess AE. Visual signal detection. I. Ability to use phase information. *J Opt Soc Am A*. 1984;1(8):900–905.

26. Burgess AE, Ghandeharian H. Visual signal detection. II. Signal-location identification. *J Opt Soc Am*. 1984;1(8):906–910.

27. Burgess AE. Visual signal detection. III. On Bayesian use of prior knowledge and cross correlation. *J Opt Soc Am*. 1985;2(9):1498–1507.

28. Burgess AE, Colborne B. Visual signal detection. IV. Observer inconsistency. *J Opt Soc Am*. 1988;5(4):617–627.

29. Pelli D, Blakemore C. *The Quantum Efficiency of Vision*. Chapter 1 in Vision: coding and efficiency, ed. by C. Blakemore. Cambridge: Cambridge University Press 1990.

30. Burgess AE. Visual perception studies and observer models in medical imaging. *Semin Nucl Med*. 2011;41(6):419–436.

31. Chakraborty DP, Yoon H-J, Mello-Thoms C. Inverse dependence of search and classification performances in lesion localization tasks. *Proc SPIE*. 2012;8318:83180H.

32. Burgess AE, Li X, Abbey CK. Visual signal detectability with two noise components: Anomalous masking effects. *J Opt Soc Am A*. 1997;14(9):2420–2442.

33. Burgess AE, Humphrey K. Density dependence of signal detection in radiographs. *Med Phys*. 1981;8(5):646–651.

34. Zanca F, Jacobs J, Van Ongeval C, et al. Evaluation of clinical image processing algorithms used in digital mammography. *Med Phys*. 2009;36(3):765–775.

35. Chakraborty DP. Maximum likelihood analysis of free-response receiver operating characteristic (FROC) data. *Med Phys*. 1989;16(4):561–568.

36. Swensson RG. Unified measurement of observer performance in detecting and localizing target objects on images. *Med Phys*. 1996;23(10):1709–1725.

PART D

Selected advanced topics

20

Proper ROC models

20.1 Introduction

So far, two methods have been described for fitting curves to receiver operating characteristic (ROC) operating points, namely the (unequal variance) binormal model, **Chapter 6** and the radiological search model (RSM), **Chapter 19**. The binormal model has been widely used to fit ROC operating points in many studies, dating to the early 1960s, and in a wide range of applications that are not just limited to medical imaging. It is a two-parameter model, that is, (a, b), excluding threshold parameters. However, binormal model fits almost invariably lead to ROC curves that inappropriately cross the chance diagonal, leading to a prediction of a region of the ROC curve where performance is worse than chance, even for expert observers. By convention, such curves are termed *improper*. The chance line crossing near the upper right corner, which occurs when $b < 1$, and the fact that the ROC curve must eventually reach $(1, 1)$ implies the curve must turn upward as one approaches $(1, 1)$, thereby displaying a *hook*. (The hook occurs near the origin if $b > 1$.) Since b is treated as a continuous-variable parameter, there is *zero probability that an estimate of b will be exactly equal to one*, the only condition under which there is no hook. *Therefore, every fitted binormal model ROC curve is improper.*

The improper behavior is often not readily visible. One may need to zoom in on the upper right corner to see it. This has been used as an argument for minimizing its importance/relevance.[1] In an influential publication[1] Metz has stated that the hook may be an *extrapolation artifact*. The late Professor Charles ("Charlie") E. Metz has been a seminal force in this field, but science does not care how right one has been in the past; every scientist, including Einstein, makes mistakes.* If Charlie really believed in the artifact excuse, he would not have put a major effort into developing the proper ROC model and implementing it in PROPROC software.[2,3] Some have termed the hook a quirk (defined in the dictionary as a strange chance occurrence) of the binormal model when in fact it is present in every binormal fit to observer data. Arguments, bordering on semantics, have been made[4], one example of which is that proper ROC behavior is only expected of *ideal observers* and *real observers are not ideal*. The difference between real and ideal observers is taken into account in model observer theory[5] by allowing an observer's efficiency to be less than unity. Efficiency is the ability of the observer to extract relevant information from the image, relative to a theoretical maximum, and there is variability between observers' efficiencies, i.e., the familiar between-reader variability. For a given observer, efficiency is defined as the square of the ratios, that of the observer to that of an

---

* Einstein famously fought against quantum mechanics, coming up with clever thought experiments that in turn kept other quantum mechanics researchers busy refuting each argument. He famously inserted a fudge parameter into his general relativity equations—now called the cosmological constant—to stop the universe from expanding; "the worst mistake of my life" as subsequently acknowledged by him.

ideal observer optimally extracting all information. This is usually measured by 2AFC experiments, and is expected to be less than unity. Since 2AFC derived AUCs are model independent, efficiency has nothing to do with the shape of the ROC curve, in particular whether it is proper or not. A real observer, who cannot extract all available image information can nevertheless yield proper curves if the observer's decision variable is equivalent to a likelihood ratio (this point is explained shortly). So, while there is confusion about levels of ideality, *in the author's judgment, any fit that predicts worse than chance level performance anywhere on the ROC plot, visible or not, is scientifically indefensible.*

Much effort has gone into developing models that always predict proper ROC curves, that is, those that do not drop below the chance diagonal. These are called proper ROC curves (some of the literature use quotes and some do not). There are at least four methods for fitting proper ROC curves, and these are listed in reverse chronological order. The most recent (2016) one[6,7] is based on the radiological search model (RSM), which was described in **Chapter 19**. Next, (2000) is a method[8-10] based on the contaminated binormal model (CBM) Next, (1999) is the binormal model-based proper ROC fitting algorithm[2,3] developed by Metz and Pan, implemented in PROPROC software. Next, (1997) is the bigamma model[11] fitting algorithm, for which no software currently exists, as far as the author is aware. Finally, (1996) is LROCFIT, based on the location ROC (LROC) paradigm,[12] which also predicts proper ROC curves, but the data has to be acquired according to the LROC paradigm, while all of the other methods described in this chapter, and in **Chapter 19**, work with ROC data.

A related issue is data degeneracy that occurs when a reader fails to provide an interior ROC point, i.e., a point that lies *inside* the ROC unit-square, as distinct from points *on* the axes. The binormal model is unable to yield reasonable fits to such data. The CBM and RSM methods are able to fit degenerate datasets. It has been claimed incorrectly, as shown below, that PROPROC can fit degenerate datasets—it cannot.

There is a theorem relating the slope of a general ROC curve, independent of model assumptions, to a quantity called the *likelihood ratio*. This will lead to the distinction between *proper* and *improper* ROC curves. Needed first is an understanding of the *likelihood ratio* and demonstration (not a proof) of an important theorem, namely the slope of the ROC curve always equals the likelihood ratio, independent of parametric assumptions.

20.2 Theorem: Slope of ROC equals likelihood ratio

Using the (a, b) notation, the equation for the operating point on a binormal model based ROC curve is (Equations 6.19 and 6.20)

$$\left. \begin{array}{l} FPF(z) = \Phi(-z) \\[2mm] TPF(z) = \Phi(a - bz) \end{array} \right\} \tag{20.1}$$

The slope m of the ROC curve is

$$m(z|a,b) = \frac{\partial(TPF)}{\partial(FPF)} = \frac{\partial(TPF)/\partial z}{\partial(FPF)/\partial z} = \frac{b\phi(a-bz)}{\phi(-z)} = b\exp\left(-\frac{1}{2}\left[(bz-a)^2 - z^2\right]\right) \tag{20.2}$$

The probability density functions (*pdfs*) for the binormal model were defined in Equations 6.23 and 6.25 and for convenience are reproduced here (N = non-diseased, D = diseased):

$$pdf_N(z) = \frac{1}{\sqrt{2\pi}} \exp\left(-\frac{z^2}{2}\right)$$

(20.3)

$$pdf_D(z) = \frac{b}{\sqrt{2\pi}} \exp\left(-\frac{(bz-a)^2}{2}\right)$$

(20.4)

The *likelihood ratio* $l_{BIN}(z|a,b)$ is defined by (the subscript BIN is for binormal model)

$$l_{BIN}(z|a,b) = \frac{pdf_D(z)}{pdf_N(z)} = b\exp\left(-\frac{1}{2}\left[(bz-a)^2 - z^2\right]\right)$$

(20.5)

It is seen from Equations 20.2 and 20.5 that the likelihood ratio equals the slope of the ROC curve.

$$l_{BIN}(z|a,b) = m(z|a,b)$$

(20.6)

While illustrated for the binormal model, this theorem is actually independent of any distributional assumptions.[13] The reason for the name *likelihood ratio* is that early researchers in the field, mostly psychophysicists, used the term *likelihood function* for what statisticians term the *probability density function*. Hence, the ratio of two likelihood functions was termed a *likelihood ratio*, when in fact it is a ratio of two *pdfs*. The *likelihood ratio* should never be confused with the *likelihood function* defined in **Chapter 6**, which is the probability of an observed dataset, given a set of parameter values, and which is the starting point of maximum likelihood estimation methods.

20.3 Theorem: Likelihood ratio observer maximizes AUC

An observer who maximizes *TPF* for a given *FPF*, for all values of FPF in the range (0,1), is called a *Neyman-Pearson observer*.[14] It can be shown that the Neyman-Pearson observer uses the likelihood ratio as the decision variable.[15] The proof is a little involved and will not be repeated here. A mathematical description is found in the book[16] by Barrett and Myers, Section 13.2.6. A practical consequence of the theorem is that AUC for a likelihood ratio observer represents an upper limit on performance, and such an observer is called an *ideal observer*. A final theorem is that an observer using the likelihood ratio as the decision variable always yields proper ROC curves.

Since this topic confused the author in his learning curve, a more detailed explanation follows. So far, the decision rule has been (**Chapter 3**) that if the Z-sample for a case is equal to or greater than threshold ζ the case is diagnosed as diseased, and otherwise it is diagnosed as non-diseased. The likelihood ratio involves a deeper concept. It is assumed that *the observer is aware of the distribution of the Z-samples for diseased and non-diseased cases*, that is, the observer knows the *pdfs* as functions of z. Then, for an observed value of z for a particular case, the observer computes the *pdf* for the diseased distribution at that value of z, namely $pdf_D(z)$ and does the same for the non-diseased distribution, $pdf_N(z)$. Assuming equal numbers of diseased and non-diseased cases, if $pdf_D(z) \geq pdf_N(z) \Rightarrow l(z) \geq 1$ it is equally or more probable that the case is diseased, and the reverse is true if $pdf_D(z) < pdf_N(z) \Rightarrow l(z) < 1$. In other words, the likelihood ratio observer does not use the Z-sample to make decisions. Instead, the observer uses the observed Z-sample and knowledge of the value of the two *pdfs* at the observed value of z to compute the likelihood ratio $l(z) = pdf_D(z) / pdf_N(z)$ and only if the ratio is equal or greater than some preset value λ is the case diagnosed as diseased (otherwise the case is diagnosed as non-diseased).

A word on notation: the author is attempting to be consistent. When the decision variable was z, he used the corresponding Greek character to denote a threshold on that axis. Since likelihood ratio is denoted l, the corresponding Greek character is λ. This is not to be confused with RSM parameter with the same symbol.

So, in effect, the observer has transformed the decision variable z to a new variable l defined by $l = pdf_D(z) / pdf_N(z)$, and the new variable is used for making decisions. Varying λ allows the observer to move along the resulting likelihood ratio ROC curve. The value of λ used by the observer is determined by the disease prevalence and the costs and benefits of incorrect (FN and FP) and correct decisions (TP and TN). Assuming equal costs, benefits, and disease prevalence of 0.5, the observer will select $\lambda = 1$ threshold, on the likelihood scale, for declaring a case as diseased. If on the other hand disease-prevalence is 0.1, the observer will adopt a higher value of λ, corresponding to moving down the ROC curve, to decrease FPF.

If decision variable z results in improper ROC curves, as does the Z-sample in the binormal model, the transformation will not be monotonic.* If the transformation were monotonic, that would imply that z was already equivalent to the likelihood ratio, where by *equivalent* is meant that a monotonic transformation exists that relates the two. From earlier chapters, the ROC curve is invariant to monotonic transformations of the decision variable. If the Z-sample from a different model, for example, the highest rating z_h sample of the RSM, always yields proper ROC curves, **Chapter 17**, then it must already be equivalent to the likelihood ratio.

To summarize these concepts in equations, the *likelihood ratio observer* transforms z to l:

$$l(z) \equiv pdf_D(z) / pdf_N(z) \tag{20.7}$$

The decision variable $M(l(z))$ will yield a proper ROC curve. Here, $M(x)$ is an arbitrary monotone increasing function of its argument, that is, if $x_2 > x_1$:

$$M(x_2) \geq M(x_1) \tag{20.8}$$

Strictly speaking, *increasing* should be replaced by *non-decreasing*, that is, the transformation allows flat portions; examples of monotonic transformations were shown in Fig. 6.2.

20.4 Proper versus improper ROC curves

If $b = 1$,

$$m(z | b = 1) = \exp\left(-\frac{1}{2}(a^2 - 2az)\right) \tag{20.9}$$

At $z = -\infty$ the slope is zero and at $z = +\infty$ the slope is infinity; all this assumes the a-parameter is greater than zero; if the a-parameter is negative, performance is worse than chance meaning the

---

* For a positive-directed decision variable, as in this book, the term *monotonic* is always meant to be *monotonic increasing*. It is possible for a transformation to be monotonic decreasing, as in $l = 1/z$.

observer has the decision variable reversed. This is reasonable behavior; the ROC curve starts out at the origin with infinite slope and gradually bends down and eventually approaches (1, 1) with zero slope. This is an example of a proper ROC. *However, if $0 < b < 1$, which is true for most data-sets (radiological or not[16]), the slope is infinite at the origin, which is fine, and at (1, 1), which spells trouble.* For the ROC curve to have started out at the origin with infinite slope and then approach (1, 1), again with infinite slope, means that at some intermediate point, it must have crossed the chance diagonal. This behavior is termed the *hook* and it generally occurs just below the upper right corner (Figure 20.1a). Such curves are termed *improper*. If $b \neq 1$ the binormal model predicts a portion of the curve in which performance is worse than chance-level. If $b > 1$, which is relatively less common, then the binormal model predicts that the slope is *zero* near the origin, which means the ROC curve starts out below the chance diagonal. In other words, the hook is still there, the only

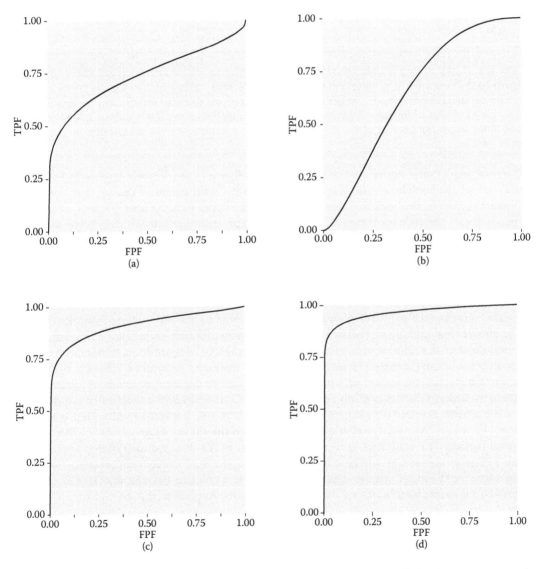

Figure 20.1 (a) through (d): Improper binormal model ROC plots for the following parameter values, shown as (a,b) pairs: a: (0.7,0.5); b: (0.7,1.5); c: (1.5,0.5) and d: (2,0.5). If $b \neq 1$, the hook is always present, near the upper right corner if $0 < b < 1$, or near the origin if $b > 1$. The code for this figure is in **mainImproperRocs.R**. An improper ROC curve implies a region where predicted performance is worse than chance-level.

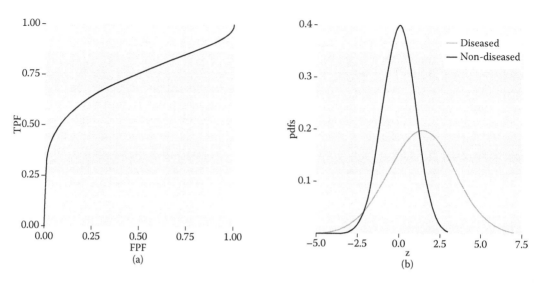

Figure 20.2 Plots of the ROC curve and *pdfs* for $a = 0.7, b = 0.5$. (a) Plot of ROC showing hook at upper right corner and (b) Plots of *pdfs*. Notice the wide spread of the diseased *pdf* causing the diseased *pdf* value to exceed that of the non-diseased *pdf* at both very high and very low values of z, leading to the hook at the upper right corner of the ROC plot. The code for this figure is in **MainPdfs.R**.

difference being that it occurs just above the origin (Figure 20.1b). Usually (Figure 20.1c and d), with $0 < b < 1$ the hook is not visible unless one zooms in on the part near the top right corner. To summarize, if $b \neq 1$, the hook is always present, near the upper right corner if $0 < b < 1$, or near the origin if $b > 1$. (The code for Figure 20.1 is in **mainImproperRocs.R**, which is listed in Online Appendix 20.A.1.)

Restricting to the more common $0 < b < 1$ situation, the reason for the hook can be appreciated from Figure 20.2b showing the *pdf* functions for non-diseased and diseased cases as functions of z. The plot corresponds to the same parameter values that generated Figure 20.1a. Since $0 < b < 1$, the diseased *pdf* is *broader* and has a lower peak (since the integrals under each distribution are unity) than the non-diseased *pdf*. Sliding an imaginary threshold to the left, starting from the extreme right, one sees that initially, just below $z = 7$, the diseased distribution starts being "picked up" while the non-diseased distribution is not picked up, causing the ROC, Figure 20.2a, to start with infinite slope near the origin (because TPF increases while FPF does not). Around $z = 2.5$ the non-diseased distribution starts being picked up, causing the ROC slope to decrease. Around $z = -3$, almost all of the non-diseased distribution has been picked up which means FPF is near unity, but since not all of the broader diseased distribution has been picked up, TPF is less than unity. Here is a region where TPF < FPF, meaning the operating point is below the chance diagonal. As the threshold is lowered further, TPF continues to increase, as the rest of the diseased distribution is picked up while FPF stays almost constant at unity. In this region, the ROC curve is approaching the upper right corner with almost infinite slope (because TPF is increasing but FPF is not). The code for Figure 20.2 is in **mainPdfs.R**, and the listing is in Online Appendix 20.A.2.

20.5 Degenerate datasets

Metz[2] defined binormal *degenerate* data sets as those that result in exact-fit binormal ROC curves of inappropriate shape consisting of a series of horizontal and/or vertical line segments in which the ROC curve crosses the chance line. The crossing of the chance line occurs because the degenerate datasets can be fitted exactly by infinite or zero values for the model slope parameter b, and infinite values for the decision thresholds, or both. To understand this, consider that the non-diseased distribution is a Dirac delta function centered at zero (by definition such a function integrates to unity), and the unit variance diseased distribution is centered at 0.6744898. In other words, this binormal model is characterized by $a = 0.6744898$ and $b = 0$ and the two distributions consist of a delta function at $z = 0$ and a unit normal centered at $z = a$. What is the expected ROC curve? As the threshold ζ is moved from the far right, gradually to the left, TPF will increase but FPF will be stuck at zero until the threshold reaches zero. Just before reaching this point, the coordinates of the ROC operating point are (0, 0.75). The 0.75 is due to the fact that $z = 0$ is -0.6744898 units relative to the center of the diseased distribution, so the area under the diseased distribution below $z = 0$ is $\Phi(-0.6744898) = 0.25$.* Since **pnorm** is giving the probability *below* the threshold, TPF must be its

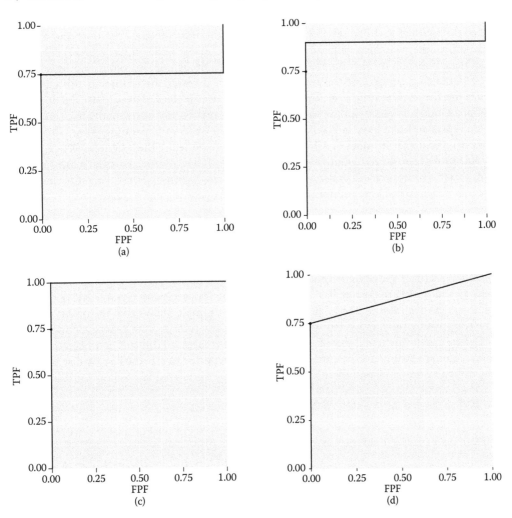

Figure 20.3 Examples of exact-fits to a degenerate data set yielding a single non-interior operating point (0, 0.75). Plots (a) through (c) represent exact binormal model fits, with the following parameter values shown as (a,b) pairs: (a) corresponds to (0.674, 0), (b) corresponds to (1.28, 0), and (c) corresponds to (infinity, 0). The respective AUCs are 0.75, 0.9, and 1. Fits (a) and (b) cross the chance diagonal and fit (c) obviously overestimates performance. The fits are unreasonable and not even unique. Fit (d) contaminated binormal model (CBM) fit, for the following values of CBM parameters. infinite mu parameter and alpha parameter equal to 0.75 and AUC = 0.875. The RSM yields an identical fit as CBM, as does the empirical AUC. These figures were generated by sourcing **mainDegenerate.R**.

---

* **pnorm**$(-0.6744898) = 0.25$.

complement, namely 0.75. This explains the operating point (0,0.75), which lies on the y-axis. As the threshold crosses the zero-width delta function, FPF shoots up from 0 to 1, but TPF stays constant. Therefore, the operating point has jumped from (0, 0.75) to (1, 0.75). When the threshold is reduced further, the operating point moves up vertically, along the right side of the ROC plot, until the threshold is so small that virtually all of the diseased distribution exceeds it and the operating point reaches (1, 1). The ROC curve is illustrated in Figure 20.3a. This is an extreme example of a ROC curve with a hook. If the data is such that the only operating point provided by the observer is (0,0.75) then this curve will be an exact fit to the operating point. An example of such a dataset is shown in Table 20.1.

Actually, given one operating point (0, 0.75), the preceding fit is not even unique. If the diseased distribution is shifted appropriately to the right of its previous position, and one can determine* the necessary value of a, then the ROC curve will shoot upward through the operating point (0, 0.75) to (0, 0.9), as in Figure 20.3b, before proceeding horizontally to (1, 0.9) and then completing the curve to (1, 1). If the diseased distribution is shifted well to the right; that is, a is very large, then the ROC curve will shoot upward past the operating point, as in Figure 20.3c, all the way to (0,1) before proceeding horizontally to (1, 1). All of these represent exact maximum likelihood fits to the observed operating point, albeit with b = zero, and different values of a. Not one of them is reasonable. These figures were generated by **mainDegenerate.R**, Online Appendix 20.A.3. [If b is infinite, one can rescale the decision variable so that the pdf of the diseased distribution corresponds to a delta function while the non-diseased pdf is a unit normal and the single operating point is now on the x-axis.]

20.5.1 Comments on degeneracy

Degeneracy occurs if the observer does not provide any interior operating points. So why should one worry about it? Perhaps one has a non-cooperating observer who is not heeding the instructions[2] to *spread the ratings, use all the bins*. A simple example shows that the observer could in fact be cooperating fully and is still unable to provide any interior data points. Consider that the 100 diseased cases shown in Table 20.1 consist of 75 easy cases, 25 difficult cases and the 100 non-diseased cases are assumed to be easy. The observer is expected to rate the 75 easy diseased cases as 5s, the difficult ones as 1s, and the 100 non-diseased cases are rated 1s, as shown in Table 20.1. No

Table 20.1 A degenerate data set: The bins counts are listed separately for actually non-diseased and actually diseased cases

	Counts in ratings bins				
	Rating = 1	**Rating = 2**	**Rating = 3**	**Rating = 4**	**Rating = 5**
$K_1 = 100$	100	0	0	0	0
$K_2 = 100$	25	0	0	0	75
	Operating points				
	≥ 5	≥ 4	≥ 3	≥ 2	≥ 1
FPF	0	0	0	0	0
TPF	0.75	0.75	0.75	0.75	1

Note: There are 100 non-diseased cases and 100 diseased cases in this dataset. The lower half of the table lists the corresponding FPF and TPF values, that is, the abscissa and ordinate, respectively, of the operating points.

---

* **qnorm(0.1)** $= -1.281552$; therefore $a = 1.28$.

amount of coaxing—*please, please spread your ratings*—is going to convince this observer to rate with 2s, 3s, and 4s any of the 75 easy diseased cases. If the cases are obviously diseased, and that is what the author means by easy cases, they are supposed to be rated 5s: *definitely diseased*. Forcing them to rate some of them as *probably diseased* or *possibly diseased* would be irrational and guilty of bending the reading paradigm to fit the convenience of the researcher (early in his research career, the author used to believe in the existence of non-cooperating observers, so Metz's advice to "spread the ratings" did not seem irrational at that time).

Faced with a degenerate dataset the binormal model is unable to yield a reasonable and unique ROC curve (Figure 20.3a through c). Essentially, the analyst has to discard the reader's dataset (as the author was forced to do, albeit in an FROC study, with Dr. Fraser's data, in 1986)—always a bad idea. Discarding a dataset just because the model is unable to fit it is a failing of the model, and an expensive failing at that, because radiologist-generated operating points do not come cheap. Furthermore, degenerate datasets are more likely with *expert* observers. What is an obvious diseased case to an expert may not be obvious to a less experienced observer. The latter observer is more likely to assign intermediate ratings to some of the cases, i.e., to this observer the cases are not easy or difficult; rather, there is a non-binary gradation to them. Therefore, the less experienced observer is more likely to provide a number of interior data points, and the binormal model would fit their data without encountering degeneracy problems. Therefor, the effect of the binormal model is to screen-out expert radiologists in favor of those with less expertise.

One could argue that the choice of dataset was incorrect; a better dataset would have consisted of a mixture of easy and difficult cases. A robust model should be able to deal with the data as it comes, and not be restricted to datasets that are convenient to the model. *As a general scientific principle, the model should accommodate the data, not vice-versa.*

If the dataset yields a single operating point (0, 0.75), what is a reasonable ROC plot? There is a theorem that given an observed operating point, the line connecting that point to (1, 1) represents a *lower bound* on achievable performance by the observer. The lower bound is achieved by the observer using a *guessing* mechanism to classify the remaining cases. Here is an explanation of this theorem. Having rated the 75 easy diseased cases as fives, the observer is left with 25 diseased cases and 100 non-diseased cases, all of which appear definitely non-diseased to the observer. Suppose the observer randomly rates 20% of the remaining cases as 4s. This would pick up five of the actually diseased cases and 20 non-diseased ones. Therefore, the total number of diseased cases rated 4 or higher is 80, and the corresponding number of non-diseased cases is 20. The new operating point of the observer is (0.20, 0.80). Now, one has two operating points, the original one on the *y*-axis at (0, 0.75) and an interior point (0.20, 0.80). Next, instead of randomly rating 20% of the remaining cases as 4s, the observer rates 40% of them as 4s, then the interior point would have been (0.40, 0.85). The reader can appreciate that simply by increasing the fraction of remaining cases that are randomly rated 4s, the observer can move the operating point along the straight line connecting (0, 0.75) and (1, 1), as in Figure 20.3d. Since a guessing mechanism is being used, this must represent a lower bound on performance. The resulting ROC curve is proper and the net AUC = 0.875. This is in fact the fit yielded by the CBM and RSM models for this dataset. The empirical AUC also yields this result.

Why should one select the lowest possible performance consistent with the data? Because it yields a unique value for performance; any higher performance is not unique, Figure 20.3a through d.

20.6 The likelihood ratio observer

An observer using the likelihood ratio, or any monotonic transformation of it, as the decision variable, is termed a likelihood ratio observer. This is worth some illustrations (Figure 20.4a through f) all of which apply to a = 0.7 and b = 0.5. The y-axis is labeled slope, but one knows by now that that is the same as likelihood ratio. The code for Figure 20.4 is in **mainSlopes.R**, *listed in Online Appendix 20.A.4.*

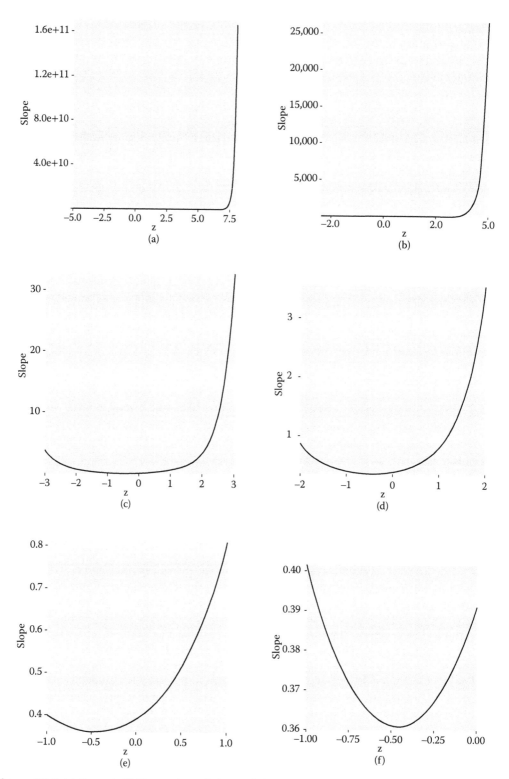

Figure 20.4 (a) through (f) These plots of slope of the ROC curve, equivalent to the likelihood ratio, all apply to $a = 0.7$, $b = 0.5$. The plots correspond to different ranges along the x-axis, intended to show a birds-eye view (a), and gradually focusing in on the region near the minimum (f). The code for this Figure is in **mainSlopes.R**.

For now, focus on Figure 20.4a in which the plot is over the range z: $(-5, 8)$. Imagine a sliding threshold λ, *moving down along the ordinate*, starting from very high values, $\sim 1.5 \times 10.^{11}$ The decision rule is to diagnose the case as diseased if $l \geq \lambda$, and otherwise the case is diagnosed as non-diseased. In Figure 20.4a, the likelihood ratio function seems, for all practical purposes, to be monotonically related to z, so on the surface it makes no difference which decision rule is used, one based on $z \geq \zeta$ or the one based on $l \geq \lambda$. The same conclusion is reached if one views Figure 20.4b, which shows the range z: $(-3, 5)$. Figure 20.4c reveals the first hint of trouble. It shows the range z: $(-3, 3)$; now, the ordinate l is not monotonically related to z, rather the function $l(z|a,b)$ has a *minimum*, so for values of λ below approximately 1 (the value depends on the choice of the a and b parameters) *there are two values of ζ corresponding to a single value of λ. This behavior is amplified in the remaining plots, as one homes in on the minimum. Taking the clearest example, Figure 20.4f, for $\lambda = 0.37$ the two values are $(-0.7277494065, -0.2055839269)$. (The author achieved this extraordinary precision by solving the quadratic equation in **Maple**™.) The two values are denoted $\zeta_L = -0.728$ and $\zeta_U = -0.206$. The likelihood ratio observer, using threshold $\lambda = 0.37$, declares all cases with $l \geq 0.37$ as diseased, and non-diseased otherwise. Therefore, the corresponding Z-sample binormal model observer must declare cases with $z \geq \zeta_U$ *and* cases with $z \leq \zeta_L$ as diseased, and non-diseased otherwise. The Z-sample based decision rule is non-intuitive: *why should cases with very low Z-samples be declared diseased?* The reason is that the model assumed by the observer (as noted earlier, the analysis assumes that the observer knows or learns the two distributions and makes rational decisions based on this knowledge) is itself unphysical. According to the that is, binormal model, both very high and very low values of z are consistent with the case being diseased. If the observer is indeed rational, then he will abandon making decisions based on z and instead use the transformation $l \equiv l_{BIN}(z|a,b)$ to make decisions based on l. The l-based decision rule is much simpler: *if $l \geq \lambda$ diagnose the case as diseased, otherwise diagnose it as non-diseased.*

Since likelihood ratio (= slope) decreases monotonically as λ is reduced (i.e., one is moving *down* the y-axis in Figure 20.4, starting from a very high value), the slope of the resulting ROC curve *decreases* monotonically.* In other words, *the likelihood ratio observer generates a proper ROC curve*. Another theorem,[13,15,17] is that the likelihood ratio observer achieves highest AUC as compared to an observer using any other decision rule, e.g., one based on the binormal model Z-sample exceeding threshold ζ. Since likelihood ratio is the ratio of two probabilities, it is always positive. More accurately, it is non-negative. The slope of the proper ROC at the upper right corner is non-negative. Unlike traditional ROC curves, the slope of proper ROC fits do not generally decrease monotonically to zero as one moves up the curve. Rather, it approaches a positive constant. In special cases, the limiting slope can be zero, as in the $b = 1$ binormal model example shown earlier, see the following Equation 20.9, so it is more accurate to say that the limiting slope is non-negative.

For the binormal model based likelihood ratio observer, the slope will not decrease to zero, rather it will decrease to the limiting value, 0.3606620722, suggested by Figure 20.4f. For values of λ below the minimum, there are no solutions for z.

The observer can use a decision rule that is based on *any* monotonic increasing transformation $M(l_{BIN}(z|a,b))$ of the likelihood ratio and the resulting ROC curve will be identical to that based on the likelihood ratio $l_{BIN}(z|a,b)$; the slope of the resulting curve always equals the likelihood ratio (before the monotonic transformation). For example, the monotonic transformation could yield a variable ranging from $-\infty$ to $+\infty$, but the slope of the resulting ROC curve will still be that for the likelihood ratio observer prior to the transformation. The limiting value of the slope is model dependent; specifically, if the binormal model is used, the limiting slope depends on the values of (a,b).

A final theorem not explicitly stated in the literature states that since AUC of the likelihood ratio observer is optimal with respect to other observers using different decision rules, or different underlying

---

* One needs to think carefully about this sentence to eventually get it.

models of the decision variable, the AUC achieved by the likelihood ratio observer represents a unique value, regardless of the model used to fit the proper ROC curve. *In other words, the shapes of the predicted proper ROC curves are model-dependent and could differ from each other, but the AUCs will be the same.* This was demonstrated in **Chapter 19**, in connection with RSM, PROPROC, and CBM fitted AUCs.

20.7 PROPROC formalism

An algorithm[2,3,18] based on the binormal model based likelihood ratio observer has been implemented by Metz, Pan, and Pesce. The software is called **PROPROC**, for proper ROC. (The 1999 publication[2] is a difficult read but worth it.) The method uses two parameters (c, d_a) defined as follows:

$$c = \frac{b-1}{b+1}; d_a = \frac{\sqrt{2}a}{\sqrt{1+b^2}} \tag{20.10}$$

Allowed values of the parameters are as follows:

$$\left.\begin{array}{l} -1 < c < 1 \\ 0 \le d_a \le \infty \end{array}\right\} \tag{20.11}$$

Equation 20.10 can be solved for the (a, b) parameters as functions of the (c, d_a) parameters.

$$a = \frac{d_a}{\sqrt{2}} \sqrt{1 + \left(\frac{c+1}{c-1}\right)^2} \tag{20.12}$$

$$b = -\frac{c+1}{c-1} \tag{20.13}$$

Since $b < 1$ with most clinical datasets, one expects to find $c < 0$,* Equation 20.10. The proper ROC curve is defined by[3]

$$FPF(v) = \Phi\left(-[1-c]v - \frac{d_a}{2}\sqrt{1+c^2}\right) + \left\{\Phi\left(-[1-c]v + \frac{d_a}{2c}\sqrt{1+c^2}\right) - H(c)\right\} \tag{20.14}$$

$$TPF(v) = \Phi\left(-[1+c]v + \frac{d_a}{2}\sqrt{1+c^2}\right) + \left\{\Phi\left(-[1+c]v + \frac{d_a}{2c}\sqrt{1+c^2}\right) - H(c)\right\} \tag{20.15}$$

The (Heaviside) step function $H(x)$ is defined by

$$\left.\begin{array}{l} H(x < 0) = 0 \\ H(x > 0) = 1 \end{array}\right\} \tag{20.16}$$

The function is discontinuous, but its value at $x =$ zero is irrelevant because $c =$ zero implies $b =$ one, in which case the equal variance binormal model applies, which predicts proper ROCs. Depending on the value of c, the threshold variable v in Equations 20.14 and 20.15 has different ranges.

---

* Of the 236 datasets described in **Chapter 19**, 180 (76%) yielded negative values of c; see file **RSM Vs. Others.xlsx** in Online Supplementary material to that chapter.

$$\frac{d_a}{4c}\sqrt{1+c^2} \le v \le \infty \text{ if} \qquad c<0$$

$$-\infty \le v \le \infty \qquad \text{if} \qquad c=0 \qquad (20.17)$$

$$-\infty \le v \le \frac{d_a}{4c}\sqrt{1+c^2} \qquad \text{if} \qquad c>0$$

PROPROC software implements a maximum likelihood method to estimate the (c,d_a) parameters from ratings data. The 1999 publication[2] states, without proof, that the area under the proper ROC is given by

$$A_{prop} = \Phi\left(\frac{d_a}{\sqrt{2}}\right) + 2F\left\{-\frac{d_a}{\sqrt{2}}, 0; -\frac{1-c^2}{1+c^2}\right\} \qquad (20.18)$$

Here $F(X,Y;\rho)$ is the bivariate standard-normal (i.e., zero means and unit variances) cumulative distribution function (CDF) with correlation coefficient ρ. In the notation of **Chapter 21**:

$$F(X,Y;\rho) = \int_{x=-\infty}^{X}\int_{y=-\infty}^{Y} dx\,dy\, f\left(\begin{pmatrix} x \\ y \end{pmatrix} \middle| \begin{pmatrix} 0 \\ 0 \end{pmatrix}, \begin{pmatrix} 1 & \rho \\ \rho & 1 \end{pmatrix}\right) \qquad (20.19)$$

Here f is the *pdf* of a standard-normal bivariate distribution with correlation ρ, that is, the mean is the zero-column vector of length 2 and the 2×2 covariance matrix has 1s along the diagonal and ρ along the off-diagonal.

The first term in Equation 20.18 is equal to the area under the binormal model ROC curve, **Chapter 6** Equation 6.48:

$$A_z = \Phi\left(\frac{a}{\sqrt{1+b^2}}\right) \qquad (20.20)$$

Therefore,

$$A_{prop} = A_z + 2F\left\{-\frac{d_a}{\sqrt{2}}, 0; -\frac{1-c^2}{1+c^2}\right\} \qquad (20.21)$$

Since F (a CDF, which is a probability) is non-negative,

$$A_{prop} \ge A_z \qquad (20.22)$$

This reinforces the earlier-stated, more general result (more general, as it is not restricted to any distributional assumptions) that performance of a likelihood ratio, or ideal, observer is greater than or equal to that of any other observer using a different decision rule.

20.7.1 Example: Application to two readers

The analytic expressions stated above are coded in **R**, listed in Online Appendix 20.B.1, in file **mainProprocRocs.R**. This yields the proper ROC predicted by the specified values of the PROPROC model parameters (c,d_a) and plots of the slopes as a function of the binormal model decision variable z. **Source** the file, yielding Figure 20.5a through d. Plot (a) is the ROC

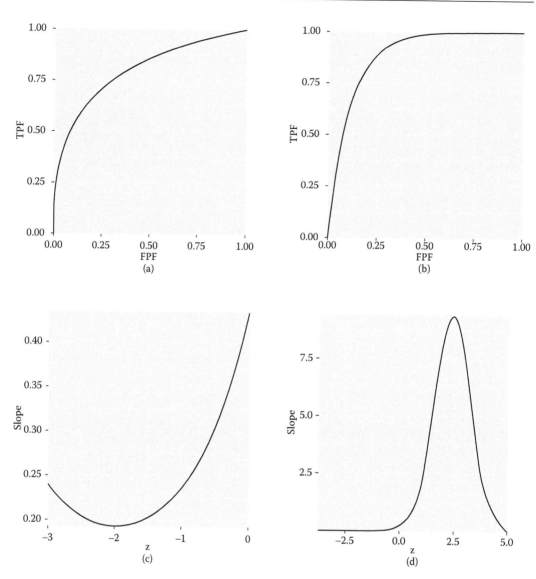

Figure 20.5 Plot (a) is for $c = -0.1322804$, $\mathbf{da} = 1.197239$ while plot (b) is for $c = 0.2225588$, $\mathbf{da} = 1.740157$. Plots (c) and (d) are the corresponding slope plots as functions of the binormal model Z-sample. In plot (a), the slope is infinite near the origin and the curve approaches the upper right corner with finite slope. The situation is reversed in plot (b) where the slope is finite near the origin and the curve approaches the upper right corner with zero slope. Plot (c) shows the slope as functions of binormal model z for the reader in (a), and plot (d) does the same for the reader in (b). The code for these figures is in file **mainProprocRocs.R**.

for **c1** $= -0.132$, $\mathbf{da} = 1.197$ while plot (b) is the ROC for **c1** $= 0.223$, $\mathbf{da} = 1.74$ (because **c** is used as the concatenation operator in **R**, it is not a good idea to use it as a variable name). Plots (c) and (d) are the corresponding slope plots.

These two readers are from a clinical dataset,[6,19] specifically, breast tomosynthesis modality, readers 1 and 5, see Table 19.1. Highest rating inferred ROC data from original FROC data were analyzed by PROPROC and the resulting parameter values were coded in **mainProprocRocs.R**. They were chosen as they demonstrate key differences in the shapes of proper ROC plots. Figure 20.5a corresponds to a negative value of c, which implies $b < 1$, which is the usual observation with clinical cases. In this situation, the slope of the proper ROC is *infinite* near the origin (similar to the very large values of slope in Figure 20.4a for large z) and approaches a *positive constant* near the upper right corner of the ROC (corresponding to the positive minimum slope in Figure 20.4f). Figure 20.5b is for a positive value of c, that is, for $b > 1$, which is an unusual situation. Now the slope of the proper ROC is *finite* near the origin and approaches *zero* near the upper right corner (i.e., the minimum in Figure 20.4f is at zero; the minimum cannot be negative, as it is a likelihood ratio).

Figure 20.5c and d shows the slopes as functions of binormal model z for the readers shown in Figure 20.5a and b, respectively. Figure 20.5c is similar in shape to Figure 20.4f and explains the finite slope near the upper right corner of the corresponding ROC. Considering Figure 20.5d, as one cuts the slope axis horizontally with a sliding threshold λ, starting with very high values and moving downwards, the slope of the ROC curve starts at the origin with a large but finite value. This corresponds to the peak in Figure 20.5d. Above the peak, there are no solutions for z. The slope decreases monotonically to zero, corresponding to the flattening out of the slope at zero for $z \sim -2$.

The two values of z corresponding to each cut implies, of course, that the binormal model based proper algorithm has to do a lot of bookkeeping, since each horizontal cut splits the decision axis into three regions. One can think of shrinking each of plots Figure 20.5c and d to zero width, and all that remains is the slope axis with a thick vertical line superimposed on it, corresponding to the horizontally collapsed curves. In Figure 20.5c, this vertical line extends from positive infinity down to about 0.1, and represents the range of decision variable samples encountered by the observer on the likelihood ratio scale. In Figure 20.5d this vertical line extends from a finite value (~9.4) to zero. Values outside of these ranges are not possible for the stated binormal model parameters.

20.7.2 Example: Check of Equation 36 in the Metz–Pan paper

The full dataset[6,19], embedded in **RJafroc** as **dataset01**, contains FROC data for two modalities and five readers. Modality 1 is breast tomosynthesis and modality 2 is conventional digital mammograms. The highest rating inferred ROC data was analyzed by PROPROC to get the (c, d_a) parameters, Online Appendix 20.B.2 and 20.B.3. Online Appendix 20.B.4 describes code that uses these values to generate proper ROC curves, and performs a numerical check of Equation 20.21. The code is in file **mainMetzPanEqn36Check.R**. Line 22, that plots the ROC curve, is commented just so one does not get overwhelmed with 10 ROC plots (if one uncomments this line, the plots are produced in the following order, modality 1, reader 1, modality 1, reader 2, ..., modality 2, reader 1, modality 2, reader 2, etc.). The code also yields a numerical estimate of AUC and it evaluates Equation 20.21, which is the same as Equation 36 in the Metz-Pan paper.[3] Sourcing this file yields the results shown in tabular form in Table 20.2. The last column lists the normalized difference, i.e., *the analytical value minus the numerical estimate divided by the analytical value*. Note the close correspondence between the Metz formula, Equation 20.21, Equation 36 in the Metz-Pan paper, and the numerical estimate.

As a historical note, Equations 31 and 36 (they differ only in parameterizations) in the referenced publication[3,4] are provided without proof—it was probably obvious to Prof. Metz or he wanted to leave it to us mere mortals to figure it out as a final parting gesture of his legacy. The author once put a significant effort into proving it and even had a bright graduate student from the biostatistics department work on it to no avail. The author has observed that these equations always yield very close to the numerical estimates, to within numerical precisions, so the theorem is correct *empirically*, but he has been unable to *prove* it analytically. It is left as an exercise for a gifted reader to prove/disprove these equations.

20.7.3 Issue with PROPROC

The author recalls a conversation ca. 2010 when Dr. Robert (Bob) Wagner made the comment that "PROPROC is bulletproof." At the time, the author agreed with that statement. However, recent studies has identified an issue with how PROPROC handles degenerate datasets. Comparisons were

Table 20.2 Comparison of results of numerical integration of ROC curve with Metz's Equation 36

i	j	c	da	Num. AUC	Equation 36	Norm. diff
1	1	−0.1322804	1.197239	0.8014164	0.8014164	0.000E+00
1	2	−0.08696513	1.771176	0.8947898	0.8947898	0.000E+00
1	3	−0.1444419	1.481935	0.8526604	0.8526605	1.173E-07
1	4	0.08046016	1.513757	0.8577776	0.8577776	0.000E+00
1	5	0.2225588	1.740157	0.8909392	0.8909392	0.000E+00
2	1	−0.08174248	0.6281251	0.6716573	0.6716574	1.489E-07
2	2	0.04976448	0.9738786	0.7544738	0.7544739	1.325E-07
2	3	−0.1326126	1.155871	0.7931786	0.7931787	1.261E-07
2	4	0.1182226	1.620176	0.8740273	0.8740274	1.144E-07
2	5	0.0781033	0.8928816	0.7360989	0.7360989	0.000E+00

Note: The last column lists the normalized difference, that is, the analytical value minus the numerical estimate divided by the analytical value. Note the close agreement between the numerical estimates and the analytical formula. [i is modality, j is reader, **c** and **da** are the PROPROC parameters, Num. AUC is the numerical estimate, and Equation 36 is identical to Equation 20.21 in this book.]

presented in **Chapter 19**, showing that PROPROC yields AUC estimates that are within 1% of those yielded by RSM and CBM, when averaged over all datasets (236 comparisons).

However, there is an issue with how PROPROC handles degenerate datasets. A publication[3] claiming that PROPROC can fit degenerate datasets actually uses a different definition of degeneracy, one which allows an interior operating point, Figure 2 *ibid*. Since it is based on the binormal model, it seemed unreasonable to the author that the method could yield valid fits to truly degenerate datasets, that is, ones with no interior data points. To test this hypothesis, the author created degenerate datasets with 100 non-diseased and 100 diseased cases yielding a single operating point at (0, TPF), where 0 < TPF < 1. In all cases PROPROC yielded AUC = 1, regardless of how *small* the value of TPF was, for example, TPF = 0.01. Likewise, given a dataset yielding a single operating point at (FPF, 0), where 0 < FPF < 1, PROPROC still yielded AUC = 1, regardless of how *large* the value of FPF was (e.g., FPF = 0.99). These statements are demonstrated upon sourcing the file `mainProprocDegeneracyIssue.R`.

> Stated simply, consider 100 non-diseased and 100 diseased patients. An operating point at (0.99, 0) implies 199 incorrect decisions on 200 patients (100 on diseased cases and 99 on non-diseased cases). PROPROC predicts perfect performance for this dataset. Likewise, an operating point at (0, 0.01) implies 99 incorrect decisions on 100 diseased patients. PROPROC still predicts perfect performance for this dataset.

The problem is basic: the only way to deal with a degenerate dataset within the binormal model is to set $b = 0$, which results in PROPROC parameter $c = -1$, Equation 20.10, which is outside the allowed range, Equation 20.11. Imposing the proper property on the binormal model for $b = 0$ results in an unresolvable mismatch.* The problem is easily fixed: one performs a preliminary check for degeneracy and if the data is degenerate, PROPROC should default to the CBM algorithm. Unfortunately, as stated on the University of Chicago ROC website, the software is no longer supported. If the software were open-source, Metz's seminal effort at putting together this complex code would have lived on, and perhaps even been improved upon (the current software does not report the covariance matrix or a goodness of fit statistic).

---

* It is analogous to trying to put a square peg into a round hole.

20.7.4 The role of simulation testing in validating curve fitting software

PROPROC was tested with millions of simulations[18]. It is not the *quantity* but the *quality* of simulations, which ultimately determines whether a method has been properly vetted. The adopted approach was apparently random sampling of model parameters, simulating datasets assuming equally spaced and fixed thresholds, applying the algorithm to estimate the parameters, and comparing estimates to population values. Given sufficiently large numbers of cases any MLE method is guaranteed to converge to the population values. The authors concluded that, "the new algorithm never failed to converge and produced good fits for all of the several million datasets on which it was tested." They arrived at this conclusion without a single reported measure of goodness of fit; PROPROC and CBM do not report goodness of fit statistics.

The author is emphasizing this example because, unfortunately, this is how validation work is *currently* done—by simulations. The lessons of history and basic statistics have apparently been forgotten. Notable *past* researchers in this field, and the author includes in this select group Dorfman,[20] Chakraborty,[21] Swensson[12] and the authors of CBM,[9] did not include simulations in the original publication reporting a new fitting method (in order, RSCORE, FROCFIT, LROCFIT, and CBM). Rather, in each instance, the researcher(s) reported goodness of fit statistics. For CBM, the likelihood-ratio G^2 statistic was reported.[9]

In the author's experience, simulations are useful to detect programming errors. Given a large sample size, if the method does not yield the correct parameters, there is something wrong at the programming level. An example of this occurred in a recent publication[22] where the authors were strongly encouraged by reviewers to include simulation testing of the newly proposed curve fitting method. *Generating data after from an assumed sampling model and fitting the data to an algorithm that assumes the same sampling model is circular logic.* Figure 20.6a through d shows this for new software, recently published by Zhai and the author.[22] The software is called CORCBM, for correlated CBM. It extends the current method of fitting a pair of correlated ROC datasets, namely the bivariate binormal model, **Chapter 21**, as implemented in CORROC2 (for correlated ROC). Correlated ROC data occurs when a single reader interprets the same cases in two modalities. The extension consists of replacing the bivariate binormal model with the bivariate contaminated binormal model, which, unlike the binormal model, is able to fit degenerate datasets. All this will make more sense in **Chapter 21**. Each plot below yields two fits to the paired operating points, one per modality, as indicated by the circles and triangles in Figure 20.6. As the numbers of cases is increased, the observed operating points line up on the theoretical curves, as expected, barring programming errors.

The real test of a fitting method is showing that it works with real datasets, as revealed by goodness of fit statistics. Complications such as degeneracy, which are easily overlooked in simulations, then become too obvious to be ignored.

20.7.5 Validity versus robustness

Validity of a fitting model means that the probability model assumed by the fitting model is consistent with the observed cell counts. In addition, one needs to be cognizant of model *robustness*, that is, stability of the model to deviations from the assumed distributions. This is what Hanley showed[23] when he examined the robustness of the binormal model to deviations from binormal distributed ratings. His main conclusion was that it would take large datasets to demonstrate

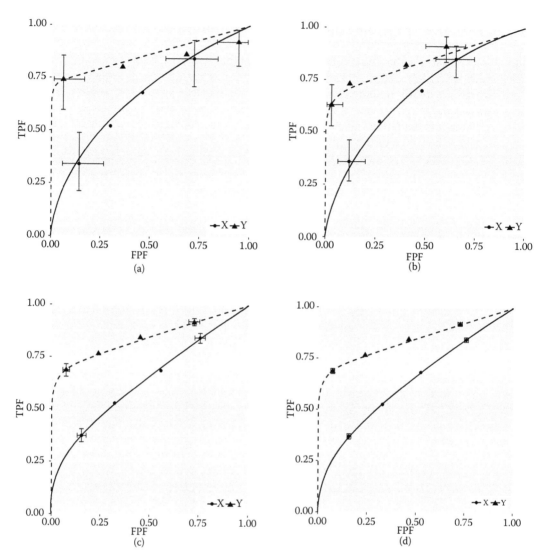

Figure 20.6 Results of simulation study: shown are simulated operating points, filled circles and triangles, and exact 95% confidence intervals for the two conditions, and the corresponding CORCBM fits. As the number of cases increases, the fits increasingly match the simulated operating points. The simulation parameters were set at the values specified in the first row of Table 5 in (From Zhai X, Chakraborty DP, *Med Phys.*, 44, 2017. With permission) Plot (a) corresponds to 50 non-diseased and 50 diseased cases: the uppermost operating point lies below the chance diagonal and cannot be fitted by CORCBM which yields fits restricted to lie above the chance diagonal. Plot (b) corresponds to 100 non-diseased and 100 diseased cases and now the fits are noticeably better. Plot (c) corresponds to 1000 non-diseased and 1000 diseased cases. Plot (d) corresponds to 5000 non-diseased and 5000 diseased cases; here the operating points are almost exactly fitted by CORCBM and the confidence intervals are comparable in size to the filled circles and triangles.

deviations from the binormal model. We arrived at a similar conclusion when the latency property of the rating was examined in Section 6.2.2. The binormal is quite robust, provided the data is not degenerate. It just cannot handle that type of data, which unfortunately, tends to occur with expert observers. As one example, Vannier et al.[24] have reported a study where a binormal model based software (CORROC2, to be discussed in **Chapter 20**) was unable to fit 14 out of 16 datasets.

The author has a historical anecdote. Around 1984 he got involved in performing a FROC study comparing Picker International's conventional chest machine versus a prototype digital chest unit made by the same manufacturer. One of the reasons that the University of Alabama at Birmingham was chosen to evaluate Picker's new machine was the reputation of the late Dr. Robert F. Fraser, the author of a famous book* on chest radiography,[25] and Dr. Gary Barnes, an eminent medical physicist. The FROC curves generated by Dr. Fraser in the digital modality did not contain any interior points. The plots were degenerate, and Dr. Fraser's data could not be used by the author. This is just one instance of expert data that could not be analyzed because of lack of interior points.

20.8 The contaminated binormal model (CBM)

The contaminated binormal model,[9–11] or CBM, is an alternate model for decision variable sampling in an ROC study. Like the binormal model, it too is a two-parameter model, excluding cutoffs. The first parameter is μ_{CBM}, which is the separation of two *unit-variance* normal distributions, so the model does not allow unequal variances, which, as we have seen, is the basic cause for improper ROC curves (see Figure 20.2b). CBM assumes that sampling for non-diseased cases is from the $N(0,1)$ distribution, while sampling for diseased cases is from the $N(\mu_{CBM},1)$ distribution, *provided the disease is visible*, otherwise it is from the $N(0,1)$ distribution. In other words, for diseased cases the sampling is from a *mixture distribution* of two unit variance normal distributions, one μ-centered and the other zero-centered, with mixing fraction α; α, the second parameter, is the fraction of diseased cases where the abnormality is actually visible. The binning is accomplished, as usual, by the cutoff vector $\vec{\zeta} = (\zeta_1, ..., \zeta_{R_{roc}-1})$, where R_{ROC} is the number of ROC bins. Defining dummy cutoffs $\zeta_0 = -\infty$ and $\zeta_{R_{ROC}} = \infty$, the binning rule is as before (Equation 4.12).

$$if\ \zeta_{r-1} \le z < \zeta_r \Rightarrow rating = r \tag{20.23}$$

Therefore, CBM is characterized by the parameters (μ, α, and ζ_r, $r = 1, 2, ..., R_{ROC} - 1$). The parameters μ_{CBM} and α can be used to predict the ROC curve and the area under the curve (AUC). (The CBM μ_{CBM}-parameter should not be confused with the binormal model parameter $\mu \equiv a/b$ or the RSM μ-parameter.)

20.8.1 CBM formalism

For non-diseased cases, the *pdf* is the same as for the binormal model, Equation 6.23.

$$pdf_N(z) = \phi(z) \tag{20.24}$$

For diseased cases, the *pdf* is a mixture of two unit variance distributions, one centered at zero and the other at μ, with mixing fractions $(1-\alpha)$ and α, respectively.

$$pdf_D(z) = (1-\alpha)\ \phi(z) + \alpha\ \phi(z-\mu) \tag{20.25}$$

The likelihood ratio for the CBM model is given by the ratio of the two *pdfs*.

$$l_{CBM}(z|a,b) = \frac{(1-\alpha)\ \phi(z) + \alpha\ \phi(z-\mu)}{\phi(z)} = (1-\alpha) + \alpha \exp\left(-\frac{\mu^2}{2} + z\mu\right) \tag{20.26}$$

---

* Over $300 on Amazon, for the 4-volume set!

The likelihood function increases monotonically as z increases; this proves that the predicted ROC curve is proper, that is, its slope decreases monotonically as the operating point moves up the curve (causing both z and the likelihood ratio to decrease). The slope at the origin (infinite z) is infinite and the slope at the upper right corner is $(1-\alpha)$. The predicted ROC coordinates are

$$FPF(\zeta) = 1 - \Phi(\zeta) = \Phi(-\zeta) \tag{20.27}$$

$$TPF(\zeta) = (1-\alpha)(1-\Phi(\zeta)) + \alpha(1-\Phi(\zeta-\mu)) = (1-\alpha)\Phi(-\zeta) + \alpha\,\Phi(\mu-\zeta) \tag{20.28}$$

Since on non-diseased cases the sampling behaviors are identical, the expression for $FPF(\zeta)$ is identical to that for the binormal model, Equation 6.12. Equation 20.28 can be understood as follows. $TPF(\zeta)$ is the probability that a Z-sample exceeds ζ. There are two possibilities: the Z-sample arose from the 0-centered distribution, which occurs with probability $(1-\alpha)$, or it arose from the μ-centered distribution, which occurs with probability α. In the former case, the probability that the Z-sample exceeds ζ is $\Phi(-\zeta)$. In the latter case, the probability that the Z-sample exceeds ζ is $\Phi(\mu-\zeta)$. To obtain the net probability one sums these component probabilities in proportion to the probabilities $(1-\alpha)$ and α that the samples arose from the zero-centered or μ-centered distributions, respectively. A similar logic can be used to derive the AUC under the CBM fitted ROC curve.

$$AUC_{CBM} = 0.5\,(1-\alpha) + \alpha\;\Phi\left(\frac{\mu}{\sqrt{2}}\right) \tag{20.29}$$

In the limit $\mu \to \infty$, $AUC_{CBM} \to 0.5\,(1+\alpha)$ and in the limit $\mu \to 0$, $AUC_{CBM} \to 0.5$ and the bounds $0.5 \le AUC_{CBM} \le 1$ is always true.

The following figures should further clarify these equations. Figure 20.7a through d shows ROC curves predicted by the CBM model; the corresponding values of the (α,μ) parameters are indicated in the legend. For small μ and/or α the curve approaches the chance diagonal, consistent with the notion that if the lesion is not visible, performance can be no better than chance-level.

As one might expect, as μ and/or α increases, performance gets better and the curve approaches the top left corner. Try as one might, by varying the parameters, one can never get the predicted curve to cross the chance diagonal, confirming viscerally that one is dealing with a proper ROC model. A pedagogic approach to proving this property is Equation 20.26 and the paragraph following it.

Figure 20.8a through d are the corresponding slope (or likelihood ratio) plots for the same parameter values as in Figure 20.7a through d. The important point to note is that all plots are monotonic with z: as z increases, the slope increases. This makes the reader using z, under the CBM model, a likelihood ratio observer. In addition, according to Equation 20.26, the slope approaches $1-\alpha$ as z approaches –infinity; that is, once again we encounter a proper ROC curve, which approaches the upper right corner with *non-zero* slope (except for the special case $\alpha = 1$).

The *pdf* functions, which resemble Figure 20.4a, are not shown. They can be viewed by running the code. Close examination of the region near the flat part shows it does not plateau at zero; rather, the minimum is at $1-\alpha$, explaining the non-zero limiting slope of the predicted curve near (1, 1).

20.9 The bigamma model

The gamma function is defined by[26] (as of writing, this book has been cited over 85,005 times! Now, *that* is an impact statement!)

$$\Gamma(x) = \int_0^\infty t^{x-1} e^{-t}\, dt \tag{20.30}$$

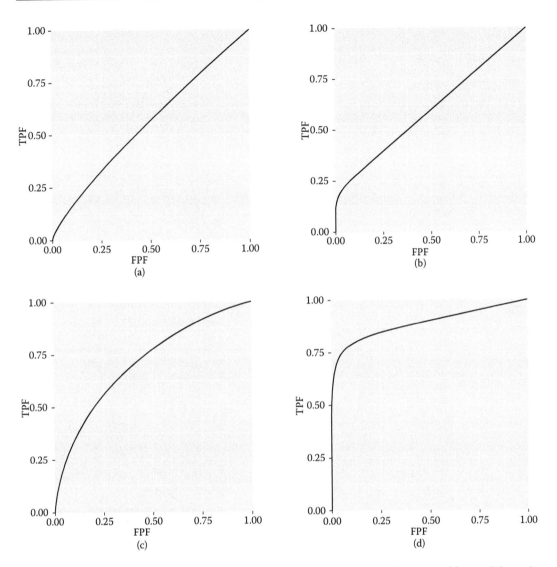

Figure 20.7 CBM predicted ROC curves. (a) $\alpha = 0.2; \mu = 1$; (b) $\alpha = 0.2; \mu = 3$; (c) $\alpha = 0.8; \mu = 1$; (d) $\alpha = 0.8; \mu = 3$. The figure was generated by code **mainCbmPlots.R** described in Online Appendix 20.C.1.

A random variable X is said to have a gamma distribution if its *pdf* is[26,27]

$$f\left(x\middle|r,\lambda\right) = \frac{\lambda^r}{\Gamma(r)} x^{r-1} e^{-\lambda x} \qquad\qquad x \geq 0, r > 0, \lambda > 0 \qquad (20.31)$$

It has the following interpretation: for integer r, assuming events are Poisson distributed with rate parameter λ, the *pdf* of the rth event follows Equation 20.31. However, in this equation, r is not restricted to integers. Alternatively, $1/\lambda$ is called the *scale* parameter and r is called the *shape* parameter of the gamma distribution.

The mean and variance of this distribution are given by

$$\left.\begin{aligned} E(X) &= \frac{r}{\lambda} \\[2mm] Var(X) &= \frac{r}{\lambda^2} \end{aligned}\right\} \qquad (20.32)$$

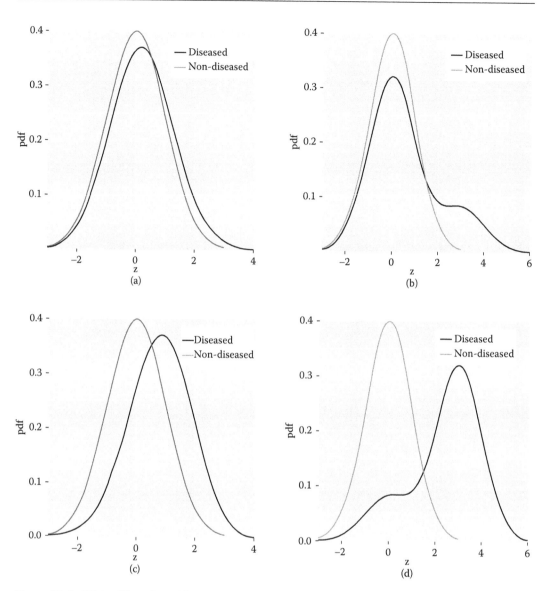

Figure 20.8 CBM *pdf* functions. (a) $\alpha = 0.2; \mu = 1$; (b) $\alpha = 0.2; \mu = 3$; (c) $\alpha = 0.8; \mu = 1$; (d) $\alpha = 0.8; \mu = 3$. The μ labels in these plots correspond to the CBM μ parameter. The figure was generated by code **mainCbmPlots.R** described in Online Appendix 20.C.1.

The constant-shape bigamma model is defined[11] by the following *pdfs* (note $\lambda = 1$ for the non-diseased distribution):

$$\text{pdf}_N(x) = f(x|r,1) = \frac{1}{\Gamma(r)} x^{r-1} e^{-x} \qquad x \geq 0 \tag{20.33}$$

$$\text{pdf}_D(x) = f(x|r,\lambda) = \frac{\lambda^r}{\Gamma(r)} x^{r-1} e^{-\lambda x}, \qquad x \geq 0 \tag{20.34}$$

The reason for the name *bigamma* is that it involves two gamma distributions (just as binormal refers to two normal distributions).

If the variance of the diseased distribution is to be larger than that of the non-diseased one, then, from Equations 20.32 through 20.34, the following constraint follows that

$$0 < \lambda \leq 1 \qquad (20.35)$$

The limitation $x \geq 0$ suggests that x is equivalent to a likelihood ratio (the ratio of two *pdfs* must be non-negative), which will be shown shortly to be true. The model assumes the same shape parameter r for non-diseased and diseased cases, which is the basic reason why it predicts proper ROC curves. The likelihood ratio function $l_{BG}(x; r, \lambda)$ is given by

$$l_{BG}(x|r, \lambda) \equiv \frac{\text{pdf}_D(x)}{\text{pdf}_N(x)} = \lambda^r e^{-(\lambda-1)x}, \qquad x \geq 0 \qquad (20.36)$$

The subscript BG is for bigamma model. Since the exponential is a monotonic function of its argument, the likelihood ratio is a monotonic function of x, which implies that an observer[11] using knowledge of this distribution and basing decisions on observed values of x exceeding a threshold, is a likelihood ratio observer. Notice the usage of x rather than z; the latter is used to denote samples from normal distributions, and there is even a term z-deviate, or z-score, to denote samples from $N(0,1)$. *The bigamma model observer bases decisions on the x-sample, not the Z-sample.* In addition, for notational consistency one uses the symbol ξ to denote a threshold on the x-axis. If $x \geq \xi$ the case is diagnosed as diseased and otherwise, it is non-diseased.

Since $0 < \lambda < 1$, the slope at the upper right corner (corresponding to $x = 0$) is λ^r, which is less than one, but not zero, and the slope is infinite at the origin (corresponding to $x =$ infinity). As with PROPROC and CBM, once again one encounters a proper ROC curve approaching the upper right corner with *finite* slope. (If $\lambda > 1$ then the slope is zero at the origin—which would cause the ROC curve to start out below the chance diagonal. Therefore, constraint Equation 20.35 is needed not just to yield a wider distribution for diseased cases but to also assure the ROC curve is proper. If $\lambda = 1$ then regardless of the value of r, the ROC curve is the chance diagonal; the reason is that the two *pdfs* Equations 20.33 and 20.34 then become identical.)

The cumulative distribution function (*CDF*) of the gamma distribution is given by

$$F(x; r, \lambda) = \int_0^x f(u; r, \lambda) du = \frac{\gamma(r, \lambda x)}{\Gamma(r)} \qquad (20.37)$$

Here the lower incomplete gamma function[26,27] $\gamma(s, x)$ is defined by

$$\gamma(s, x) = \int_0^x t^{s-1} e^{-t} dt \qquad (20.38)$$

(Since **R** implements the *CDF* of the gamma distribution directly, as **pgamma()**, see explanation in Online Appendix 20.D, one actually has no use for Equation 20.38; it is included for pedagogic completeness.) The coordinates of the predicted ROC at threshold ξ are (the complementary probability is needed as FPF and TPF are the probabilities of exceeding a threshold, i.e., $P(x \geq \xi)$ for non-diseased and diseased cases, respectively):

$$FPF(\xi|r) = 1 - \frac{\gamma(r, \xi)}{\Gamma(r)} = \int_\xi^\infty f(u; r, 1) du \qquad (20.39)$$

$$TPF(\xi|r, \lambda) = 1 - \frac{\gamma(r, \lambda \xi)}{\Gamma(r)} = \int_\xi^\infty f(u; r, \lambda) du \qquad (20.40)$$

Intimidating as all these equations appear, life gets simpler when one implements them in code. The relevant code, **mainBigammaPlots.R**, listed in Online Appendix 20.D. It generates ROC plots, Figure 20.9a through d, and plots of the *pdfs* for non-diseased and diseased cases, Figure 20.10a through d, for specified values of the two bigamma model parameters. The parameter values are shown as pairs of (r, λ) values: a: $(1,1)$; b: $(4.39, 0.439)$; c: $(5, 0.3)$; d: $(10, 0.1)$. Plot (a) is for $\lambda = 1$ when the ROC becomes the chance diagonal regardless of the value of r. The values for plot (b) are identical to those used for simulations in the Dorfman et al. publication[12]. Plots (c) and (d) illustrate that increasing the ratio r / λ, that is, the mean of the diseased distribution (the mean of the non-diseased distribution is r), increases performance. The plots in Figure 20.9 are consistent with the *pdfs* shown in Figure 20.10. For example, for plot (d), the mean of the diseased distribution is at 100, relative to 10 for the non-diseased distribution and performance is close to perfect (AUC = 1).

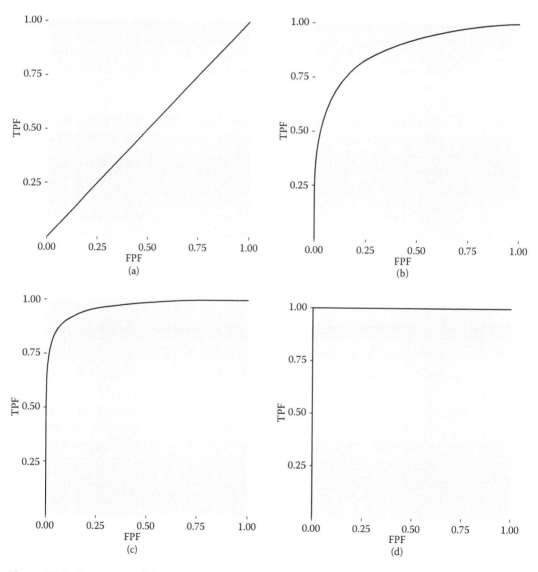

Figure 20.9 Bigamma model ROC curves. The parameter values shown as (r, λ) values; (a) (1, 1); (b) (4.391, 0.439); (c) (5,0.03); (d) (10,0.1). The corresponding AUCs are (a) 0.5; (b) 0.8790038; (c) 0.9645528; (d) 0.9999984. These plots were generated by **mainBigammaPlots.R**.

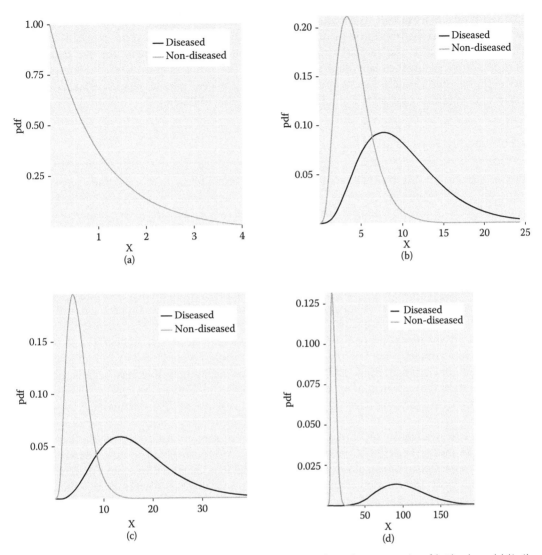

Figure 20.10 Bigamma model slope curves. The parameter values shown as pairs of (r, λ) values; (a) (1, 1); (b) (4.391, 0.439); (c) (5,0.03); (d) (10,0.1). The pair of values in (b) is from the cited publication. These plots were generated by **mainBigammaPlots.R**.

The author is unaware of an analytic expression for AUC of the bigamma model. However, as illustrated in Online Appendix 20.D, the ROC curve can be numerically integrated almost trivially. Using the code, one can confirm the following values of AUC: (a) AUC = 0.5, (b) AUC = 0.8790038, (c) AUC = 0.9645528, (d) AUC = 0.9999984. The reader is referred to Online Appendix 20.D for details on how the code works.

One should confirm that values of $\lambda > 1$ yield ROC curves below the chance diagonal and AUC less than 0.5. The ROC curves are actually reflected versions of proper ROC curves; for each such curve, the slope starts out at zero near the origin and increases to infinity near the top right corner.

There used to be software called BIGAMMA (formerly MAXGAM[12]) implementing the bigamma model, but the University of Iowa ROC website now states "BIGAMMA is not available for download at this time. Users who desire a proper ROC model should use PROPROC or CBM" (http://perception.radiology.uiowa.edu, accessed 4/29/ 2017). However, constructing one is relatively easy. The author leaves it as an exercise to the reader to write **R** code implementing the method. The integrals of the *pdfs* are already implemented in **R**, namely **pgamma()**, so this should be a relatively easy and productive exercise. The author can understand the de-emphasis of this method, as the proposed bigamma distribution is inconsistent with expectations based on the central limit theorem, namely, when several random processes contribute to a measurement, the net distribution tends to be normal, independent of the distributions of the individual processes. The latency property of the decision variable does not help here—the existence of a monotonic transformation that transforms a bigamma model to a binormal model is highly unlikely.

20.10 Discussion/Summary

Most researchers in this field have avoided the subject of modeling ROC data. The exceptions are Dorfman, Metz, Swensson, Berbaum, and the author. There are a few reasons for this. First, the empirical AUC is easy to calculate. There is no degeneracy issue such as encountered with the binormal model. One can always calculate the empirical AUC, even when one should not, that is, when a close look at the plot suggests there might be a problem with the dataset. Second, some prefer nonparametric analysis, in the mistaken belief that it is always better to make the analysis free of assumptions, especially normality assumptions. Good science is driven by testable assumptions. When the assumptions are valid, they add value to the analysis. As an example, assuming the validity of the RSM, one can tease out measures of search and lesion-classification performance that say much more about what is limiting performance than a trivial calculation of empirical AUC. Due to the central limit theorem, the normal distribution plays a key role in data modeling in all branches of science. To abandon it and use the empirical AUC could be construed as lack of intellectual curiosity and / or rigidity of thinking. Third, nonparametric methods that have been proposed are applicable only to the empirical AUC. Usage of these methods is limited to empirical ROC analysis. One cannot even use them with fitted ROC curves, nor with other FOMs that measure localization performance.

With the easy availability of CBM and RSM, there is now no excuse for not using them. These programs are indeed bulletproof. Based on their very design, they can fit any dataset that one cares to throw at them. For reasons explained in the previous chapter, the author's preference is to use RSM as it yields information about search and lesion-localization performance, which should impact how radiologists and computer aided detection algorithms are trained and evaluated. The Windows version of CBM needs to be extended to calculate a goodness of fit statistic (this is done in the **RJafroc** implementation). Finally, the time has come to let go of the binormal model. It has outlived its 60 years of usage.

References

1. Metz CE. Some practical issues of experimental design and data analysis in radiological ROC studies. *Invest Radiol.* 1989;24:234–245.
2. Metz CE, Pan X. Proper binormal ROC curves: Theory and maximum-likelihood estimation. *J Math Psychol.* 1999;43(1):1–33.
3. Pan X, Metz CE. The "proper" binormal model: Parametric receiver operating characteristic curve estimation with degenerate data. *Acad Radiol.* 1997;4(5):380–389.
4. Pesce LL, Metz CE, Berbaum KS. On the convexity of ROC curves estimated from radiological test results. *Acad Radiol.* 2010;17(8):960–968. e964.
5. Burgess A, Wagner R, Jennings R, Barlow H. Efficiency of human visual signal discrimination. *Science.* 1981;214(4516):93–94.
6. Chakraborty DP, Svahn T. Estimating the parameters of a model of visual search from ROC data: An alternate method for fitting proper ROC curves. *Proc SPIE 7966.* 2011;7966. Lake Buena Vista (Orlando), Florida. doi: 10.1117/12.878231; http://dx.doi.org/10.1117/12.878231

7. Chakraborty DP, Yoon H-J, Mello-Thoms C. Inverse dependence of search and classification performances in lesion localization tasks. *Proc SPIE*. 2012;8318:83180H.

8. Dorfman DD, Berbaum KS, Brandser EA. A contaminated binormal model for ROC data: Part I. Some interesting examples of binormal degeneracy. *Acad Radiol*. 2000;7(6):420–426.

9. Dorfman DD, Berbaum KS. A contaminated binormal model for ROC data: Part III. Initial evaluation with detection. *Acad Radiol*. 2000;7(6):438–447.

10. Dorfman DD, Berbaum KS. A contaminated binormal model for ROC data: Part II. A formal model. *Acad Radiol*. 2000;7(6):427–437.

11. Dorfman DD, Berbaum KS, Metz CE, Lenth RV, Hanley JA, Abu Dagga H. Proper receiving operating characteristic analysis: The bigamma model. *Acad Radiol*. 1997;4(2):138–149.

12. Swensson RG. Unified measurement of observer performance in detecting and localizing target objects on images. *Med Phys*. 1996;23(10):1709–1725.

13. Egan JP. *Signal Detection Theory and ROC Analysis*. 1st ed. New York, NY: Academic Press; 1975.

14. Neyman J, Pearson ES. On the problem of the most efficient tests of statistical hypotheses. In Kotz S, Johnson NL, eds. *Breakthroughs in Statistics Foundations and Basic Theory*, pp. 73–108. New York, NY: Springer-Verlag; 1992.

15. Green DM, Swets JA. *Signal Detection Theory and Psychophysics*. New York, NY: John Wiley & Sons; 1966.

16. Barrett HH, Myers K. *Foundations of Image Science*. Hoboken, NJ: John Wiley and Sons; 2003.

17. Macmillan NA, Creelman CD. *Detection Theory: A User's Guide*. New York, NY: Cambridge University Press; 1991.

18. Pesce LL, Metz CE. Reliable and computationally efficient maximum-likelihood estimation of proper binormal ROC curves. *Acad Radiol*. 2007;14(7):814–829.

19. Andersson I, Ikeda DM, Zackrisson S, et al. Breast tomosynthesis and digital mammography: A comparison of breast cancer visibility and BIRADS classification in a population of cancers with subtle mammographic findings *Eur Radiol*. 2008;18(12):2817–2825.

20. Dorfman DD, Alf E. Maximum-likelihood estimation of parameters of signal-detection theory and determination of confidence intervals—rating-method data. *J Math Psychol*. 1969;6:487–496.

21. Chakraborty DP. Maximum likelihood analysis of free-response receiver operating characteristic (FROC) data. *Med Phys*. 1989;16(4):561–568.

22. Zhai X, Chakraborty DP. A bivariate contaminated binormal model for robust fitting of proper ROC curves to a pair of correlated, possibly degenerate, ROC datasets. *Med Phys*. 2017;44(3):2207–2222.

23. Hanley JA. The robustness of the "binormal" assumptions used in fitting ROC curves. *Med Decis Making*. 1988;8(3):197–203.

24. Vannier MW, Pilgram TK, Marsh JL, et al. Craniosynostosis: Diagnostic imaging with three-dimensional CT presentation. *Am J Neuroradiol*. 1994;15(10):1861–1869.

25. Fraser RS, Müller N, Colman N, Pare P. *Fraser and Paré's Diagnosis of Diseases of the Chest*. Vol. 1–4. Philadephia, PA: WB Saunders; 1999.

26. Abramowitz M, Stegun IA. *Handbook of Mathematical Functions: With Formulas, Graphs, and Mathematical Tables*. Vol. 55. US Dept. of Commerce, National Bureau of Standards. Applied Mathematics Series V55, US Govt. Printing Office, Washington, DC; 1964.

27. Larsen RJ, Marx ML. *An Introduction to Mathematical Statistics and Its Applications*. 3rd ed. Upper Saddle River, NJ: Prentice-Hall; 2001.

21

The bivariate binormal model

21.1 Introduction

Until now, the focus for the receiver operating characteristic (ROC) paradigm has been on the single rating per case scenario. This applies when a single reader interprets a set of cases in a single modality using the ROC paradigm. Of greater clinical interest is a study where a group of readers interprets a common set of cases in multiple modalities, termed the *MRMC scenario*. This is because such a study, properly analyzed, yields results that are expected to generalize to the populations of readers and cases from which the study readers and cases were sampled. However, before one can understand *several* readers interpreting a common case-set in multiple modalities, a good place to start is *two* readers interpreting a common set of cases. Historically, the field progressed from this starting point (the DBMH and ORH analyses methods came after the method to be described here was developed, ca. 1980s). One might wonder why even bother describing the older approach, since methods for analyzing MRMC datasets have already been described. The answer is that before one can truly understand multiple-treatment multiple-reader studies, which permit a large number of pairings, without at least learning the history of analysis of data consisting of a single pairing.

From the measurement point of view the analyst is dealing with *two* ratings per case, one per reader, a key difference from the ROC paradigm we have been considering so far, which yielded *one* rating per case. An analogous situation applies when one has a single reader interpreting cases in two modalities. From the measurement point of view the analyst is again faced with *two* ratings per case, one per modality.

In **Chapter 6** the *univariate* binormal model was described, which is appropriate for modeling the single rating per case situation (*binormal* comes from the two *truth* states, not two *modalities*). From the previous discussion, it is evident that this needs to be extended to a *bivariate* model, which would describe studies where each case yields two ratings. Because one has dual interpretations of the same case, the two ratings are *correlated*. On especially easy non-diseased cases, the two ratings will both tend to be low (e.g., 1's) and on especially easy diseased cases, the two ratings will both tend to be high (e.g., 5's) and on difficult cases, both ratings are likely to be in the ambiguous range (e.g., 3's). This type of tandem behavior is unlikely to be observed if the two interpretations did not come from the same case.

Here is the plan for this chapter. First, the bivariate extension of the univariate binormal model is described. Since most of the readers of this book are not expected to be experts in statistics, bivariate sampling will be described at a relatively simple level. Visualization of the resulting probability density functions (*pdfs*) is demonstrated, which allows the reader to *interactively view them*, which should allow a better understanding of this relatively more complex distribution. Estimation of parameters of the bivariate model from ratings data is addressed. The usage of free software implementing this procedure, that is, CORROC2, is then described. The data format used by CORROC2 is described in some detail as well as how to interpret the output of the program. Since this software runs on **Windows**, material is provided to allow the user to run it on a MAC (most of this necessary digression, since the author is using OS X, is in an Online Appendix). This digression is not needed in one is using a **Windows** platform. Finally, the application of the software to clinical datasets is described. A recent advance that solves degeneracy problems associated with CORROC2 is described, namely CORCBM, for correlated CBM. We conclude with a discussion of the applications of CORROC2.

21.2 The Bivariate binormal model

In a single-modality single-reader ROC study the modality and reader indices are superfluous. Each z-sample can be thought of as arising from a binormal model with mean parameter μ_t, where t the truth index ($t = 1, 2$). The Z-samples of diseased cases are distributed $N(\mu, \sigma^2)$. Non-diseased case Z-samples are distributed $N(0, 1)$. In addition, for a ratings study, one needs thresholds ζ_r. Recall, from **Chapter 04**, that for an R rating task with allowed ratings $r = 1, 2, ..., R$, one needs $R - 1$ thresholds $\zeta_1, \zeta_2, ..., \zeta_{R-1}$. Defining ζ_0 and ζ_R as negative and positive infinity, respectively, the decision rule is to label a case with rating r if the realized z-sample satisfies $\zeta_{r-1} \leq z < \zeta_r$.

In the two-modality scenario, each case yields two Z-samples, z_i, where $i = 1, 2$ represent the two modalities, or generically, two "conditions." Additionally, the threshold vectors have condition dependencies,: i.e., there exist two sets of thresholds $\zeta_{11}, \zeta_{12}, ..., \zeta_{1(R-1)}$ and $\zeta_{21}, \zeta_{22}, ..., \zeta_{2(S-1)}$, assuming R ratings in one condition and S ratings in the other. The pairs (ζ_{10}, ζ_{20}) and (ζ_{1R}, ζ_{2S}) are defined as negative and positive infinities, respectively. The rating depends on the condition. The decision rule is to label a case in condition 1 with rating r if $\zeta_{1(r-1)} \leq z_1 < \zeta_{1r}$, and the same case in condition 2 is labeled with rating s if $\zeta_{2(s-1)} \leq z_2 < \zeta_{2s}$.

In the bivariate binormal model the decision variable is $Z_{ik_t t}$, where the i-subscript corresponds to the two modalities and the t subscript corresponds to the two truth states. The correlated ratings pairs $(Z_{1k_1 1}, Z_{2k_1 1})$ and $(Z_{1k_2 2}, Z_{2k_2 2})$, corresponding to Z-samples from non-diseased ($t = 1$) and diseased cases ($t = 2$), respectively, are abbreviated, using vector notation, to $\overrightarrow{Z_{k_t t}}$:

$$\overrightarrow{Z_{k_t t}} \equiv \begin{pmatrix} Z_{1k_t t} \\ Z_{2k_t t} \end{pmatrix} \tag{21.1}$$

According to the bivariate binormal model, $\overrightarrow{Z_{k_t t}}$ is sampled from $N_2(\overrightarrow{\mu_t}, \Sigma_t)$, the bivariate normal distribution with mean $\overrightarrow{\mu_t}$ and covariance matrix Σ_t.

$$\overrightarrow{Z_{k_t t}} \sim N_2(\overrightarrow{\mu_t}, \Sigma_t) \quad t = 1, 2 \tag{21.2}$$

In Equation 21.2, the symbol $\sim N_2(\overrightarrow{\mu_t}, \Sigma_t)$ realises a (column) vector $\overrightarrow{Z_{k_t t}}$ of length 2 from the bivariate normal distribution with mean $\overrightarrow{\mu_t}$ and covariance matrix Σ_t. Each case yields two samples, that is, the vector has two rows, corresponding to the two modalities. The subscript 2 on N distinguishes the *bivariate* distribution from the corresponding univariate distribution N (which has an implicit 1 subscript). The vectors $\overrightarrow{\mu_1}$ and $\overrightarrow{\mu_2}$ in Equation 21.2 are defined by

$$\vec{\mu_1} = \begin{pmatrix} \mu_{11} \\ \mu_{21} \end{pmatrix} \equiv \begin{pmatrix} 0 \\ 0 \end{pmatrix}$$

$$\vec{\mu_2} = \begin{pmatrix} \mu_{12} \\ \mu_{22} \end{pmatrix} \equiv \begin{pmatrix} \mu_1 \\ \mu_2 \end{pmatrix}$$

(21.3)

The indices are *it*, that is, modality followed by the truth index. The equations state that the two non-diseased distributions, one per modality, are both centered on zero, while the diseased distribution for modality 1 is centered at μ_1, shorthand for μ_{12}, while that for modality 2 is centered at μ_2, which is shorthand for μ_{22}.

The covariance matrices Σ_1 and Σ_2 in Equation 21.2 are defined by (also called variance-covariance matrices, but the author will stick with the shorter term)

$$\Sigma_1 = \begin{pmatrix} 1 & \rho_1 \\ \rho_1 & 1 \end{pmatrix}$$

$$\Sigma_2 = \begin{pmatrix} \sigma_1^2 & \rho_2\sigma_1\sigma_2 \\ \rho_2\sigma_1\sigma_2 & \sigma_2^2 \end{pmatrix}$$

(21.4)

The quantities in Equation 21.4 are defined as follows. For non-diseased cases, the Pearson correlation coefficient of the pairs of Z-samples is ρ_1 and the corresponding value for diseased cases is ρ_2. The ones along the diagonal of Σ_1 confirm that the variances of the two samples for non-diseased cases are individually equal to unity. The corresponding variances for diseased cases are σ_1^2 and σ_2^2.

The sampling model depicted in Equation 21.2 is a natural extension to two dimensions of the univariate binormal model described in **Chapter 6**, namely,

$$Z_{k_t t} \sim N\left(\mu_t, \sigma_t^2\right) \quad t = 1, 2$$

(21.5)

where,

$$\mu_1 = 0, \mu_2 = \mu$$

$$\sigma_1^2 = 1, \sigma_2^2 = \sigma^2$$

(21.6)

The univariate model yields *one* Z-sample per case: samples for non-diseased cases have mean zero and unit variance; samples for diseased cases have mean μ and variance σ^2. In contrast, the bivariate model yields *two* Z-samples per case where, for either modality, the samples for non-diseased cases have zero mean and unit variance, and the correlation coefficient of the two samples per non-diseased case is ρ_1. For diseased cases, the samples for the first modality have mean μ_1 and variance σ_1^2 while, for the second modality, the samples have mean μ_2 and variance σ_2^2 and the correlation coefficient of the two samples per diseased case is ρ_2. To summarize, the parameters describing the bivariate model, not including the modality dependent thresholds needed for binning the data, are $\vec{\mu_t}, \Sigma_t (t = 1, 2)$. The total number of parameters is six: two means for the diseased cases, two variances for the diseased cases, and two correlation coefficients.

In Online Appendix 21.A, file **mainBivariateSampling.R**, illustrates the bivariate sampling model in **R**. Specific values, selected arbitrarily, are assigned to the parameters as per Equation 21.7. The function **mvrnorm()** (to be understood as *multivariate version of* $rnorm()$) is used to generate the bivariate samples. The code generates **N** = 10,000 samples the bivariate distribution and the corresponding means and standard deviations were calculated.

$$\mu_1 = 1.5; \mu_2 = 2.0; \sigma_1 = 1.0; \sigma_2 = 1.5; \rho_1 = 0.3; \rho_2 = 0.6 \tag{21.7}$$

Sourcing the file yields output 21.2.1 and Figure 21.1a.

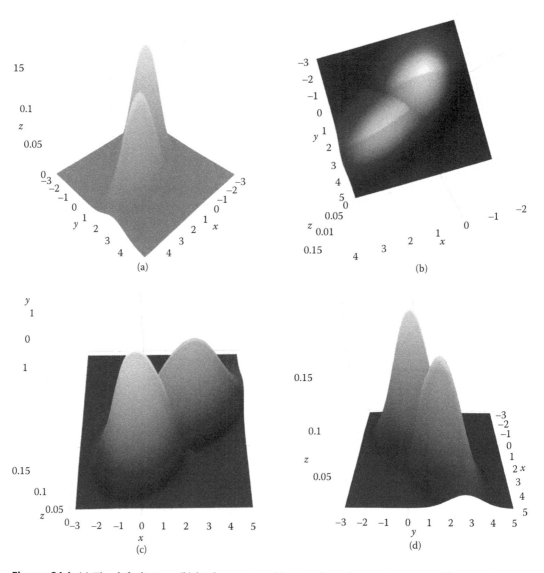

Figure 21.1 (a) The default view; (b) birds-eye view; (c) view along the y-axis at $x = 0$; (d) view along the x-axis at $y = 0$. These plots were generated by sourcing **mainBivariateSampling.R**.

21.2.1 Code output

```
> source (…)
expected means of non-diseased cases =  0 0
expected correlation of non-diseased cases =  0.3
expected covariance of non-diseased cases =
        [,1] [,2]
[1,]   1.0  0.3
[2,]   0.3  1.0


observed means of non-diseased cases =  -0.002791453 -0.007749206
observed correlation of non-diseased cases =  0.3194812
observed covariance of non-diseased cases =
         [,1]      [,2]
[1,] 1.0051009 0.3225885
[2,] 0.3225885 1.0143726


expected means of non-diseased cases =  1.5 2
expected correlation of non-diseased cases =  0.6
expected covariance of non-diseased cases =
        [,1] [,2]
[1,]   1.0 0.90
[2,]   0.9 2.25


observed means of diseased cases =  1.509351 2.009172
observed correlation of diseased cases =  0.6100278
observed covariance of diseased cases =
          [,1]      [,2]
[1,] 1.0247098 0.9295477
[2,] 0.9295477 2.2659128
```

For ease of comparison, the output lists the *expected* values (i.e., based on the known parameter values) for non-diseased cases, followed by the *observed* values, and similarly for the diseased cases. Agreement between expected and observed values is good—this is due to the very large number of samples (10,000) over which the estimates are calculated. This exercise should give one some understanding of the bivariate sampling, the parameters needed to describe the samples and estimation of the parameters of the bivariate model from the observed samples. Section 21.4 below, on visualizing the bivariate *pdf*, will reinforce the reader's understanding.

21.3 The multivariate density function

The multivariate (specifically *p*-variate) probability density function (*pdf*) in *p*-dimensions is defined by

$$f(\vec{x}|\vec{\mu},\Sigma) = \frac{1}{(2\pi)^{\frac{p}{2}}|\Sigma|^{\frac{1}{2}}} \exp\left(-\frac{(\vec{x}-\vec{\mu})^T \Sigma^{-1}(\vec{x}-\vec{\mu})}{2}\right) \tag{21.8}$$

Here $\vec{x}$ is a length p column vector at which one wishes to evaluate the *pdf*; its mean is $\vec{\mu}$ (also a length p column vector), the $p \times p$ covariance matrix is Σ, $|\Sigma|$ is the determinant of the covariance matrix, and T is the transpose operator. The reader should be convinced that the numerator of the exponent is a scalar. Equation 21.8 breaks down if the determinant of the covariance matrix (appearing in the denominator of the factor outside the exponent) is zero. This is consistent with the fact that the inverse matrix, appearing in the numerator of the exponent, only exists if the determinant is non-zero.

For the special case of a bivariate normal distribution this is equivalent to[1]

$$f\left(x_1, x_2 \,|\, \sigma_1, \sigma_2, \rho\right) = \frac{1}{2\pi\sigma_1\sigma_2\sqrt{1-\rho^2}} \exp\left(-\frac{t}{2\left(1-\rho^2\right)}\right) \tag{21.9}$$

where,

$$t = \frac{\left(x_1 - \mu_1\right)^2}{\sigma_1^2} - \frac{2\rho\left(x_1 - \mu_1\right)\left(x_2 - \mu_2\right)}{\sigma_1\sigma_2} + \frac{\left(x_2 - \mu_2\right)^2}{\sigma_2^2} \tag{21.10}$$

In the univariate case (i.e., $p = 1$), Σ in Equation 21.8 is replaced by σ^2, and the above expression reduces, as expected, to the well-known *pdf* of the normal distribution $N\left(\mu, \sigma^2\right)$.

$$f\left(x \,|\, \mu, \sigma^2\right) = \frac{1}{\sqrt{2\pi}\sigma} \exp\left(-\frac{\left(x - \mu\right)^2}{2\sigma^2}\right) \tag{21.11}$$

With some difficulty, if one is new to **R**, one could code Equation 21.8 (along the way, inverting the matrix will be but one minor challenge), but as one might expect, someone has already done it. The function is called **dmvnorm()**, to be understood as *density of multivariate normal distribution*, and it is available via a package **mvtnorm()**. Online Appendix 21.B.2 illustrates calculation of the *pdf* of the multivariate normal distribution at a specified point. The reader should view and source file **MainBivariateDensity.R**.

21.3.1 Code output

```
> source (…)
rho = 0.5 , var1 = 1.5 , var2 = 1.5
density at 0.1, 0.2 =  0.1208948
```

21.4 Visualizing the bivariate density function

In understanding the bivariate distribution, it is helpful to visualize the *pdfs*. Since one is dealing with a vector $\vec{x}$ of length 2, one needs two axes to select $\vec{x}$ and a third axis to depict the *pdf* at the selected $\vec{x}$, that is, one needs a 3D plot on 2D paper. Various visualization techniques have been developed and there is a sophisticated one available for **R**. The package is called **plotly**. The last two lines of the code in **mainBivariateSampling.R** display an interactive 3D plot in the **Viewer** panel, which uses color to code the third dimension. It is *interactive* in the sense that by dragging the cursor over the plot one can visualize it from different viewpoints. Since the book will not appear in color, one needs to run the code and get familiar with it. Feel free to change the parameters (within reasonable limits). The examples shown below are for the following choices:

$$\mu_1 = 1.5; \mu_2 = 2.0; \sigma_1 = 1.0; \sigma_2 = 1.5; \rho_1 = 0.3; \rho_2 = 0.6 \qquad (21.12)$$

Figure 21.1a is the default view. Figure 21.1b is the birds-eye view, with the *pdf*-axis literally coming at the reader at right angles to the paper or screen. Figure 21.1c is the view looking along the *y*-axis at $x = 0$. Figure 21.1d is the view looking along the *x*-axis at $y = 0$. If the cursor is placed inside the plot area, `plotly` displays the three coordinates (*x*, *y*, and the *pdf*) and slices along the two primary axes, as well as contours connecting all points with the same *pdf* value. *These capabilities cannot be appreciated from the printed Figure 21.1 and the above description—one has to experience them.*

21.5 Estimating bivariate binormal model parameters

In **Chapter 6**, a method for estimating the parameters of the univariate binormal model for a single-reader single-modality dataset was described. The method involved maximizing the likelihood function, that is, the probability of the observed data as a function of the model parameters. Specifically, for a 5-rating study, there were four thresholds. The likelihood was a function of six variables: two basic parameters, namely the separation of the two normal distributions μ and the standard deviation σ of the diseased distribution relative to the non-diseased distribution. In addition, there are the four thresholds. The threshold parameters are sometimes referred to, in the literature, as *nuisance* parameters but, nuisance or not, they have to be estimated in order to get to the basic parameters. To estimate the parameters the likelihood function was maximized with respect to these parameters. The values of the parameters yielding the maximum are the maximum-likelihood estimates (*MLEs*).

With a bivariate model, one is dealing with six basic parameters $\vec{\mu_t}, \Sigma_t (t = 1,2)$ *and threshold parameters corresponding to each modality.* The estimation procedure is not fundamentally different, but the presence of the second dimension does complicate the notation. Again, the starting point is the likelihood function, that is, the probability of the observed data as a function of the parameter values. Since there is a possibility of confusion between the truth index t and the modality index i, each of which ranges from 1 to 2, the non-diseased counts in bin r of the first modality and bin s of the second modality is denoted by K_{rs1} (i.e., the two modality indices occur *before* the truth index). The corresponding diseased counts are denoted K_{rs2}. One needs two indices rs because each case yields two integer ratings, r and s. Assume R ratings in the first modality and S ratings in the second modality. To construct the matrix K_{rs1} one starts with a zero-initialized $R \times S$ matrix, increment the cell at row r and column s of this matrix by unity for each non-diseased case that received an r rating in the first modality and an s rating in the second modality. When all non-diseased cases are exhausted, the count in row r and column s of this matrix is K_{rs1}; the procedure is repeated for the diseased cases yielding K_{rs2}.

For non-diseased cases, the probability of a Z-sample in bin r of the first modality and bin s of the second modality is determined by the volume under the bivariate distribution $N_2(\vec{\mu_1}, \Sigma_1)$ between modality 1 thresholds $\zeta_{1(r-1)}$ and ζ_{1r}, *and* between modality 2 thresholds $\zeta_{2(s-1)}$ and ζ_{2s}. Complicated as this may seem, and prior to Alan Genz's work[2] showing how to evaluate such multidimensional integrals, this was near impossible, but now it is quite easy, especially in **R**. The required probability is given by the **pmvnorm()** function—this is the generalization to the multivariate case of the **pnorm()** function, which gives the area under the normal distribution from negative infinity to a specified value. Specifically, **pmvnorm()** gives the probability under the multivariate distribution between specified thresholds, as illustrated next.

21.5.1 Examples of bivariate normal probability integrals

Needed is the bivariate extension of the univariate cumulative distribution function Φ defined in Equation 3.7. Specifically, one needs a way of calculating the integral.

$$p_{rst} = \int\limits_{x_1 = \zeta_{1(r-1)}}^{\zeta_{1r}} \int\limits_{x_2 = \zeta_{2(s-1)}}^{\zeta_{2s}} dx_1\, dx_2\, f\left(\overrightarrow{X}\middle|\overrightarrow{\mu}_t, \Sigma_t\right) \tag{21.13}$$

Online Appendix 21.C.1 shows how to do this in **R**. The code file **mainBivariateIntegrals.R** implements the code equation:

$$p_{rst} = pmvnorm\left(c\left(\zeta_{1(r-1)}, \zeta_{1(s-1)}\right), c\left(\zeta_{2r}, \zeta_{2s}\right), mean = \overrightarrow{\mu}_t, sigma = \Sigma_t\right) \tag{21.14}$$

Here **pmvnorm()** is the integral of the bivariate normal distribution with specified **mean** and covariance matrix **sigma** between limits $\zeta_{1(r-1)} \le x_1 < \zeta_{1r}$ and $\zeta_{2(s-1)} \le x_2 < \zeta_{2s}$. The limits are specified by two length-2 arrays that are supplied as the first two arguments to this function. *The first argument to the function represents the concatenated **lower** thresholds in the two modalities. The second argument represents the concatenated **upper** thresholds in the two modalities.* These are followed by the separation parameter vector and the covariance matrix. The function is not limited to two-dimensional covariance matrices. With a $p \times p$ covariance matrix, the lower and upper limits would each be length p vectors. In lines 27–28, Online Appendix 21.C.1, $c\left(\zeta_{1(r-1)}, \zeta_{1(s-1)}\right) = c(0.3, 0.4)$ and $c\left(\zeta_{2r}, \zeta_{2s}\right) = c(0.4, 0.5)$. Sourcing the code yields the following.

21.5.1.1 Examples of bivariate normal probability integrals

```
> source(...)
Integrals of the bivariate normal
Over the entire space =  1
Over the full space in one dimension
and the -ve half space in the other dimension =  0.5
Between specified ctff. values =  0.001136472
```

The facts that the integral over the entire space yields unity, i.e., $c\left(\zeta_{1(r-1)}, \zeta_{1(s-1)}\right) = c(-\infty, -\infty)$ and $c\left(\zeta_{2r}, \zeta_{2s}\right) = c(\infty, \infty)$, and that over each half space yields 0.5, i.e., $c\left(\zeta_{1(r-1)}, \zeta_{1(s-1)}\right) = c(-\infty, -\infty)$ and $c\left(\zeta_{2r}, \zeta_{2s}\right) = c(\infty, 0)$, should make intuitive sense.

21.5.2 Likelihood function

The probability of observing K_{rs1} and K_{rs2} non-diseased and diseased counts, respectively, in bin r in the first modality and bin s in the second modality is

$$\prod_{t=1}^{2} \left(p_{rst}\right)^{K_{rst}} \tag{21.15}$$

Including all bins and taking the logarithm, the logarithm of the likelihood function is given by (neglecting a combinatorial factor that does not depend on the parameters)

$$LL\left(\overrightarrow{\mu_2}, \Sigma_1, \Sigma_2, \overrightarrow{\zeta_1}, \overrightarrow{\zeta_2}\right) \equiv \sum_{t=1}^{2}\sum_{s=1}^{S}\sum_{r=1}^{R} K_{rst}\log\left(p_{rst}\right) \tag{21.16}$$

The maximum likelihood estimates of the parameters are obtained by maximizing the LL function. This was solved[3,4] by Metz and colleagues more than three decades ago and implemented in software CORROC2. The Metz software is particularly useful as it measures correlations at the underlying Z-sample level, which can properly be used in designing a data simulator that is matched to a clinical dataset, the subject of the next chapter. Much as the author has emphasized that ratings are not numerical quantities, and that the observed ratings are not hard numbers, the measured correlations according to this algorithm are valid, because CORROC2 models the ratings as continuous variables and estimates the correlation based on the model.

21.6 CORROC2 software

Around 1980 Prof. Charles E. Metz and Helen B. Kronman developed software implementing maximum likelihood estimation of the parameters of the bivariate normal model outlined above from data corresponding to two correlated ROC ratings per case. The software is called **CORROC2**, for *correlated ROC*. Subsequent revisions to the program were made by Pu-Lan Wang and Jong-Her Shen and more recently by Ben Herman.[3–6] *CORROC2 is used to analyze paired ratings that result when a common set of cases is interpreted under two **conditions**.* The pairing can be simple, as when a single reader interprets a common set of cases in two modalities, or more general, as when two readers interpret a common set of cases in one modality, or when two different readers interpret a common set of cases in two different modalities. The generality makes the method quite useful in a way that, in the judgment of the author, has not been fully exploited to date.*

CORROC2 software allows selection of three methods of testing for differences between two conditions.

1. Testing for simultaneous differences between the (*a, b*) parameters corresponding to two ROC curves.
2. Testing the difference between the binormal-fitted AUCs, under two correlated ROCs.
3. Testing the difference between TPFs on two ROC curves at a specified FPF.

To understand the difference between methods 1 and 2, consider that is that it is possible to have two ROC curves with different shapes but the same fitted AUCs. If the difference in shapes is large enough, method 1 would declare the two conditions different, but method 2 would not. In the examples below, we use method 2.

This section is one of the few in this book that uses **Windows** code. If using **OS X**, one needs to download virtual machine software as described in Online Appendix 21.D. *If using a* **Windows** *machine, one does not need to do anything extra. Therefore, in following this chapter it is important that one has either a Windows computer or a Mac computer with a virtual machine installed.*

---

* The author recalls a conversation with the late Prof. Richard (Dick) Swensson at an SPIE, where Dick enthusiastically explained a project he was working on that utilized CORROC2 repeatedly.

21.6.1 Practical details of CORROC2 software

The Windows executable code is named **CORROC2.exe**. The author downloaded the software, including the **Fortran** source code, many years ago, ca. 1997, from the University of Chicago website, but it is currently unavailable and unsupported on that site. The author is fortunate that he held onto the downloaded code. The reason is that the **FORTRAN** source code, with appropriate modifications, can be interfaced to the **R** code. The currently distributed equivalent code named **ROC-kit** cannot be run in command-line mode (i.e., without having to use the mouse to select the input file, specify the output file, etc.). The author has confirmed with Mr. Ben Herman, who is familiar with the code, that by design it cannot be used in command-line mode. The ability to run in command-line mode is needed if one is to call a program from **R**.

For simplicity, it is assumed that the number of allowed ROC ratings bins, *R, is the same in each modality* (the software can handle different numbers of ratings in the two modalities). For two modalities, the total number of parameters to be estimated is $6 + 2(R - 1)$. The six comes from the two means and two variances of the diseased distributions, plus the two correlations (one per truth-state) and $2(R - 1)$ from the two sets of thresholds, each with $R - 1$ values. In addition to providing *estimates of all parameters* of the model, the algorithm provides estimates of the *covariance matrix of the parameters*. This, not to be confused with Σ_t, is needed for the same reason it was needed in the univariate case, see **Chapter 6**, that is, to estimate variances of the six estimates and variances of functions of estimates, for example, AUC or differences in AUCs.

21.6.2 CORROC2 data input

Here is a partial listing of how the paired ratings data is input to the program (the data file is **DataFileInp.txt** in the software distribution under **software/corrocii/Debug**).

```
TYPICAL EXAMPLE OF 5-CATEGORY DATA --X
5
5
TYPICAL EXAMPLE OF 5-CATEGORY DATA --Y
5
5
4  4
3  5
5  3
1  1
1  1
1  3
...
```

The first line, whose content is immaterial to the analysis, is a description of the data in the first modality. The second line (5), is the total number of ratings bins in the first modality. The third line (5) is the rating with the highest evidence of disease in the first modality (5 in the current example). The fourth line is a description of data in the second modality, again its content is immaterial. The next two 5s are the number of bins in the second modality, and the bin number denoting highest evidence of disease in the second modality.

The actual ratings data starts on line 7, which states that the first non-diseased case was rated ROC: 4 in the first modality and ROC: 4 in the second modality. *The two ratings for each case, separated by spaces, are input on the same line.* Line eight states that the second non-diseased case was rated ROC: 3 in the first modality and ROC: 5 in the second modality, and so on. An asterisk after the last non-diseased case signals the end of the non-diseased cases. The next line has the two ratings corresponding to the first diseased case. Here is a partial example:

```
4  4
*
4  5
5  5
```

This says that the first diseased case was rated ROC: 4 in the first modality and ROC: 5 in the second modality, the second diseased case was rated ROC: 5 in both modalities, and so on. The last diseased case is followed by another asterisk, and the last line is the word **area**.

```
5  5
2  5
*
area
```

Other choices are available for the last line, but the author will stick to **area**, which means one wishes to use the area under the binormal-fitted ROC curve A_z as the figure of merit.

21.6.3 CORROC2 output

The **CORROC2** output associated with this software for the above dataset, 347 lines in all, is in the file **DataFileOut.txt**, Figure 21.2 (use **RStudio** to go down the directory structure and click on the appropriate file; it is displayed, see Figure 21.2 below, with line numbers).

Condition X and **Condition Y** denote modalities 1 and 2, respectively.* Line 10 of the output[†] summarizes the statistical test to be employed **AREA (A SUB Z) TEST**. This is because we ended the reader data file **DataFileInp.txt** with **area**. Lines 13 through 22 summarize the dataset, how many ratings, which rating represents greatest evidence of disease, and so on. Lines 26 through 37 summarize the observed ratings matrix for non-diseased cases (labeled **ACTUALLY NEGATIVE**), that is, the matrix K_{rs1}. For example, there were three cases that were rated $s = 3$ in **CONDITION X** and $r = 5$ in **CONDITION Y**. The matrix is labeled by its row index, followed by its column index and last with the truth index. The output lists the 5-rating bin as the first row, the 4-rating bin as the second row, etc. Likewise, lines 40 through 51 summarize the ratings matrix K_{rs2} for diseased cases. Lines 53 through 59 list the observed operating points in the two modalities, followed by the initial and final estimates of the parameters. The program lists the parameters using the (a,b) notation, not the (μ, σ) notation, but one knows how to transform between them, **Chapter 6**, Equation 6.16, and see exercise on conversion below. Here is the output showing the final parameter values (14 values in all starting at line 74):

```
            FINAL ESTIMATES OF THE PARAMETERS:
   AX= 1.3006      BX= 0.4919      AY= 1.6049      BY= 0.8704
   R(NEGATIVE CASES) = 0.2599      R(POSITIVE CASES) = 0.4544
   T(I)-0.281   0.221   0.687   1.941
   U(J)-0.791  -0.417   0.175   0.990
```

---

* *Condition* is a more general term than modalities; as already noted the pairing could be reader 1 versus reader 2 in modality 1, or even reader 1 in modality 1 versus reader 2 in modality 2; *all that is required for a valid pairing is interpretations of a common set of cases.*

† All this will not make sense unless one views the output; obviously the full output cannot be shown in the book, but one can view it using the downloaded software.

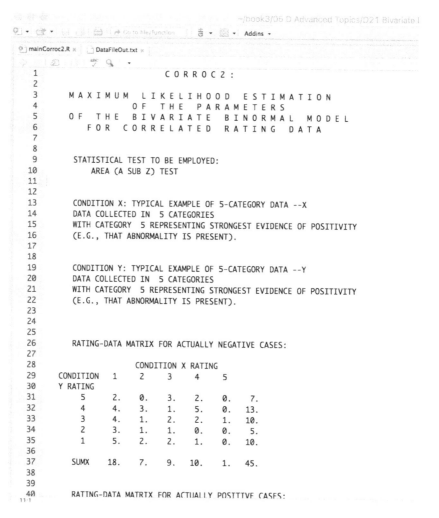

Figure 21.2 Lines of `CORROC2` output in `DataFileOut.txt` that are associated with the dataset in `DataFileInp.txt`.

Here **X** refers to condition X and **Y** to condition Y. **AX** is the a-parameter in condition **X** (what the author calls a_1). **BX** is the b-parameter in condition **X** (our b_1). **AY** is the a-parameter in condition **Y** (i.e., a_2). **BY** is the b-parameter in condition **Y** (i.e., b_2). **R(NEGATIVE CASES)** is the correlation coefficient for the non-diseased cases (i.e., ρ_1). **R(POSITIVE CASES)** is the correlation coefficient for the diseased cases (i.e., ρ_2). Finally, **T(I)** and **U(J)** are the thresholds in the two conditions, corresponding to ζ_{1r} and ζ_{2s} in our notation. Since there are five ratings bins, there are four thresholds in each modality.

21.6.3.1 Exercise: Convert the above parameter values to (μ, σ) notation

Using the transformations, Equation 6.16, $\mu_1 = 1.3006 / 0.4919 = 2.644$, $\mu_2 = 1.844$, $\sigma_1 = 1 / .4919 = 2.033$, $\sigma_2 = 1.149$, $\rho_1 = 0.2599$ and $\rho_2 = 0.4544$ (the correlations are unaffected by linear transformations). Regarding the thresholds, for modality 1, $\zeta_{11} = -0.281 / .4919 = -0.571$, $\zeta_{12} = 0.221 / .4919 = 0.449$, and so on. Similarly, the thresholds for modality 2 are $\zeta_{21} = -0.791 / .8704 = -0.909$, $\zeta_{22} = -0.417 / .8704 = -0.479$, etc. Question for the reader: why were the thresholds divided by the corresponding b-values?

21.6.4 Covariance matrix

For convenience of the yet-to-be-described **R**-program that does all this automatically, the FORTRAN source code was modified to print the elements of the covariance matrix *sequentially*, that is, one per line. The output corresponding to lines 83 through 280 in file **DataFileOut.txt** is (shown is a partial listing)

```
        VARIANCE-COVARIANCE MATRIX
              14
       0.621003E-01
       0.205564E-01
       0.171881E-01
       0.294564E-02
       0.104652E-03
      -0.684784E-02
       0.144067E-01
       0.135140E-01
       0.127711E-01
       0.393710E-02
       0.285687E-02
       0.285592E-02
       0.285248E-02
       0.284967E-02
       0.205564E-01
```

The number 14 is the dimension of each edge of the 2D covariance matrix, that is, there are 14^2 (196) elements in all. The 14 comes from the six basic parameters plus the four thresholds in the two conditions. The elements are listed in *row-major* order, that is, all the elements of the first row, followed by all elements of the second row, and so on. One can anticipate some difficulty displaying it, so only the upper left 6 × 6 corner of the full matrix is shown below, Table 21.1; that is, *any covariance involving a threshold is not shown*. For convenience, the numbers have been rounded to fit and since the matrix is symmetric, there is no need to show numbers below the diagonal. The diagonal elements, which represent variances, are all positive.

21.6.5 Rest of CORROC2 output

Lines 286 through 320 allow one to plot the ROC curves for the two modalities, but one does not need them, as one knows by now how to generate binomial model predicted curves. Lines

Table 21.1 Covariance matrix of the parameters for the dataset in **datafileinp.txt**

	a_1	b_1	a_2	b_2	ρ_1	ρ_2
a_1	.062	.021	.017	.003	.000	−.007
b_1		.022	.002	.002	.001	−.004
a_2			.089	.045	.000	−.010
b_2				.078	.003	−.008
ρ_1					.026	.000
ρ_2						.035

Note: Since the matrix is symmetric, values below the diagonal are not shown. Also, not shown are covariance-elements associated with threshold parameters.

325 through 327 lists the areas under the respective fitted ROC curves, and their correlation (the is the AUC-level correlation, which could be estimated by the bootstrap of jackknife methods explained in **Chapter 10**):

```
      AREA(X) =   0.8784       AREA(Y) =   0.8870
      STD DEV OF AREA(X) =   0.0392     STD DEV OF AREA(Y) =
0.0376
      CORRELATION OF AREA(X) AND AREA(Y)  = 0.2569
```

The remaining lines, 333 through 341, list the p-value.

```
            STATISTICAL SIGNIFICANCE OF THE DIFFERENCE
        BETWEEN THE TWO CORRELATED ROC CURVES ACCORDING TO THE
   SELECTED TEST:
      ****************************************************************
   **********
        THE COMPUTED CORRELATED 'AREA TEST' STATISTIC VALUE IS   -0.1829
        WITH A CORRESPONDING TWO-TAILED P-LEVEL OF 0.8549
        AND ONE-TAILED P-LEVEL OF 0.4274.
```

The famed p-value is 0.8549 (it is customary to quote the two-tailed value, see **Chapter 8** for details on this). This means that the difference in areas (0.8784− 0.8870 = −0.0086) is not statistically significant at Type-I error probability $\alpha = 0.05$. The calculation of the test statistic is described in code **mainDiffAzTestStatistic.R**. It uses methods described in **Chapter 6** to calculate the standard deviation of the difference of two AUCs.

$$\sigma^2_{X-Y} = \sigma^2_X + \sigma^2_Y - 2\rho_{XY}\sigma_X\sigma_Y \tag{21.17}$$

The test–statistic for determining the significance of the AUC difference is

$$(X-Y)/\sigma_{X-Y} \sim N(0,1) \tag{21.18}$$

21.7 Application to a real dataset

Assuming one has set up the connection between the Mac side and the Windows side properly, as described in Online Appendix 21.D, one can apply **CORROC2** to real ROC datasets. Open the **R** file **mainCorroc2.R** in the **software** folder corresponding to this chapter, described in detail in Online Appendix 21.E. **Source** the file, yielding the following code output.

21.7.1 Code output

```
> source(...)
...

The 6 parameters are  1.72 0.55 1.87 0.527 0.332 0.798
The 2 sided pValue is  0.482
The covariance matrix is:
```

	[,1]	[,2]	[,3]	[,4]	[,5]	[,6]
[1,]	0.120075	0.05096	0.050939	0.01054	6.58e-04	-1.05e-02
[2,]	0.050956	0.03667	0.013392	0.00550	2.01e-03	-4.87e-03
[3,]	0.050939	0.01339	0.111921	0.04078	2.28e-04	-1.16e-02
[4,]	0.010545	0.00550	0.040777	0.02908	1.51e-03	-4.56e-03
[5,]	0.000658	0.00201	0.000228	0.00151	2.09e-02	3.11e-06
[6,]	-0.010453	-0.00487	-0.011566	-0.00456	3.11e-06	8.67e-03

What the **R** code does is detailed in Online Appendix 21.D. It reads the file containing the data in the above example, extracts the ratings for reader 1 in both modalities, writes the data in the appropriate format for **CORROC2**, runs the latter program in a Windows environment, reads the resulting output, and prints out model parameters and the covariance matrix.

Like most ROC software, the output uses the (a,b) convention, **Chapter 6**. Recall that in this convention the diseased distribution has unit standard deviation, the non-diseased distribution has standard deviation b, and the separation of the two means is a. Since there are two modalities, there are two a-parameters and two b-parameters. In the CORROC2 output, the two modalities are referred to as X (1 in our notation) and Y (2 in our notation), respectively, and the ordering of the six parameters is $a_1, b_1, a_2, b_2, \rho_1, \rho_2$. The ordering also determines the ordering of the covariance matrix. We start with the meanings of the header row and first column of the matrix shown in Section 21.7.1. The notation [,1] means the first column of the matrix, likewise [,2] is the second column, and so on. Likewise, [1,] means the first row of the matrix, [2,] means the second row of the matrix, and so on. For example, the above table tells us that the second row and third column of the covariance matrix is 0.013392. This is the covariance between b_1 and a_2, i.e., $cov(b_1, a_2)$. The diagonal elements of the matrix are the variances. For example, the variance of ρ_1 is 2.09e–02, whose square root is the standard deviation, equal to 0.145.

In the **Files** panel find the folder named **corrocii** and open it (the reader needs to click on the name, not the picture of the folder; watch how the folder name is underlined when the cursor is above it, which is **RStudio**'s way of telling you *yes, it is OK to click here, something will happen*). You should see a number of files with the **.FOR** extension (these are FORTRAN source code files) and a folder named **Debug**; click on **Debug**. Figure 21.3 is a screenshot showing the contents of the **Debug** window. Table 21.2 summarizes the meanings of the files listed in this figure.

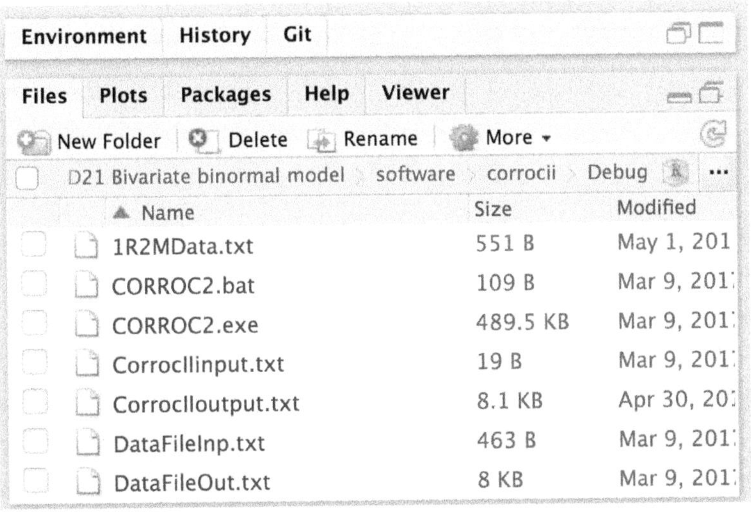

Figure 21.3 Contents of the **Debug** window, as viewed from the Mac side.

Table 21.2 This table summarizes the meanings of the files listed in Figure 21.3

File name	Meaning	Comment
DataFileInp.txt	Input Ratings data in CORROC2 format	Read-only file used to illustrate data input format
DataFileOut.txt	Output file corresponding to DataFileInp.txt	Read-only file used to illustrate the output
1R2MData.txt	Ratings data in CORROC2 format	Created by **RunCorrocOnPairedData.R**, line 25
CORROC2.bat	Batch file that runs CORROC2.exe	
CORROC2.exe	The actual CORROC2 program	See below
CorrocIIinput.txt	Input file to the batch file	
CorrocIIoutput.txt	Output file created by batch file	

The Windows software is located in **software/CORROC2/Debug**, Figure 21.3. **CORROC2.bat** is a batch file, whose listing follows.

21.7.2 Contents of CORROC2.bat

```
cd CORROCII
cd Debug
del CorrocIIoutput.txt
CORROC2.exe < CorrocIIinput.txt > CorrocIIoutput.txt
cd ..
cd ..
```

The batch file runs on the Windows side. The first two lines change the current directory to **corrocii/Debug**. The next line deletes the output file, **CorrocIIoutput.txt**, if it exists. The next line pipes the contents of **CorrocIIinput.txt** to the executable file **CORROC2.exe** and pipes the output to **CorrocIIoutput.txt**. The last two lines restore the starting directory. The contents of **CorrocIIinput.txt** are

```
Y
1R2MData.txt
N
Q
```

21.7.3 Contents of CorrocIIinput.txt

These answer the four questions asked by **CORROC2**. Try running **CORROC2** from the Windows side, Figure 21.4.

The first line of this file is the answer (yes) to the first question asked by **CORROC2.exe** (you can check these out by clicking on **CORROC2.exe** in your physical Windows machine or virtual Windows machine): *Do you want to use data from a previously created input file for the next run?*

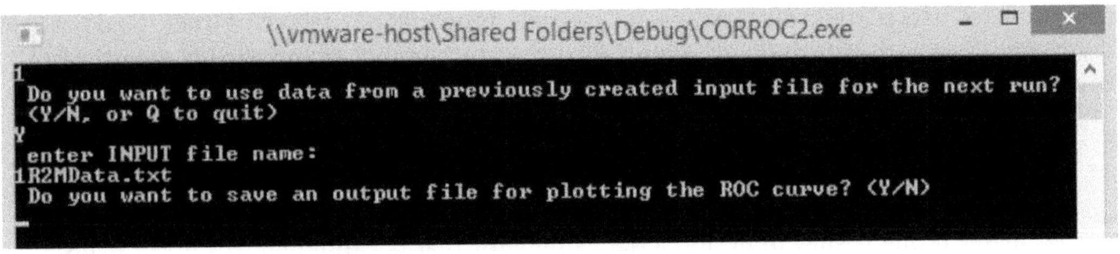

Figure 21.4 Result of running CORROC2 at the Windows 8 command prompt.

The second line is the name of the input file. The third line is the answer (no) to the next question: *Do you want to save an output file for plotting the ROC curve?* The reason for (no) is that we are more interested in the screen output, which contains all the estimates, including the covariance matrix, rather than plotting ROC curves. The last line is the answer (Q) to the next question asked by **CORROC2.exe**: *Do you want to use data from a previously created input file for the next run? (Y/N, or Q for quit).* Now, *in a Windows environment*, click on the batch file **CORROC2.bat**. The latest version of the file **CorrocIIoutput.txt** should appear in your Explorer window, with the correct date and time stamps. Open this file. The data file is named **1R2MData.txt** and if you open it, you will recognize that it is similar to the file used in the section illustrating the data input format for **CORROC2**. It was created by the **R**-code.

21.8 Discussion/Summary

CORROC2 is one of the relatively under-utilized tools developed by Prof. Charles E. Metz. There are only four publications describing it[3-5,7] and two of them are difficult to find,[4,7] nor has the software been maintained on a level comparable to ROCFIT. For example, a method for assessing the goodness of the fit is currently not implemented. With R × S cells, it is almost impossible to maintain greater than five counts per cell, so an alternate method of combining bins needs to be implemented.

One reason for the relative underutilization of CORROC2 could be that it is not designed to analyze multiple readers interpreting a common set of cases in two or more modalities, that is, MRMC datasets. For example, in a study in which CORROC2 was used,[8] four readers interpreted cases in four modalities. With four modalities, there are six possible combinations (4 × 3/2), so a conservative Bonferroni type correction for the *p*-value would be to divide 0.05 by 6. For each reader CORROC2 yields a *p*-value for the difference between the chosen pairs of modalities. If all readers agree, then there is an unambiguous answer, but the method does not allow for correlations between different readers, which undoubtedly leads to loss of power. More fundamentally, it does not address the question of interest: *when averaged over all four readers, are the figures of merit in the four modalities different?*

It is possible to analyze multiple-reader multiple-modality MRMC-ROC datasets without actually using CORROC2, as was done in **Chapters 09 and 10**. The correlations in the resulting AUC values, that is, *FOM correlations* as distinct from *ratings correlations*, are explicitly accounted for in the DBMH/ORH analyses, i.e., the described methods are not concerned with ratings correlations. One possible use of CORROC2 would be if one had to compare two CAD algorithms applied to a common dataset, but even here, it is of limited utility since, as a ROC study, location is ignored.

CORROC2 was developed ca. 1980 by Metz, Wang, and Kronman.[4,10] Subsequent revisions to the program were made by Jong-Her Shen and more recently by Benjamin Herman.[5,6] There are in excess of 116 citations to this software. One reason is that, at that time, it was the only software allowing analysis of paired datasets. However, no advances have been made in the intervening three decades, which would allow fitting, for example, proper ROC curves to paired—and possibly degenerate—datasets. Part of the reason for this neglect is the shift in emphasis to empirical AUC based analysis, which does not require parametric modeling or curve fitting.

The real use of CORROC2 is if one is interested in *ratings correlations*. In fact, until recently, it was the only way to estimate the ratings correlations. Interest in the correlations of ratings arises if one wishes to design a data simulator that is calibrated to a given dataset. Roe and Metz performed some work along this line[9] when a set of ratings simulators were suggested as representative of real ROC datasets. These simulators have been widely used in the methodology literature to validate analysis methods. While simulators are of primary use to methodology developers, their impact is much greater than their seemingly limited applicability. Part of the unfinished business of this book is to develop a calibrated simulator. Initial progress in this direction was stymied by the fact that binormal model based CORROC2 cannot handle degenerate datasets. Progress in using a correlated version of CBM is currently underway. It should be robust to degenerate datasets and yield proper ROC fits in each modality.

At the time of writing, 4/30/17, an extension of CORROC2 has been published.[11] It is called CORCBM, for correlated CBM. Details are in a document CORCBM.pdf in the online supplementary material directory corresponding to this chapter. The title of the publication is: "A bivariate contaminated binormal model for robust fitting of proper ROC curves to a pair of correlated, possibly degenerate, ROC datasets." It replaces the bivariate binormal model with the bivariate contaminated binormal model; hence its name CORCBM, for correlated CBM. Since CBM was designed to fit practically any *single* dataset, including degenerate ones, CORCBM is likewise able to fit practically any *paired* dataset. An application of CORCBM to calibrating a simulator to a data set containing single-modality multiple-reader ratings is described in **Chapter 23**.

References

1. Tong YL. *The Multivariate Normal Distribution*. 1st ed. New York, NY: Springer-Verlag; 1990.
2. Genz A. Numerical computation of multivariate normal probabilities. *J Comput Graph Stat*. 1992;1(2):141–149.
3. Metz CE, Kronman H. Statistical significance tests for binormal ROC curves. *J Math Psychol*. 1980;22(3):218–242.
4. Metz CE, Wang P-L, Kronman HB. A new approach for testing the significance of differences between ROC curves measured from correlated data. In: Deconinck F, ed. *Information Processing in Medical Imaging*. The Hague: Nijhoff; 1984.
5. Metz C, Kronman H, Shen J, Wang P. *CORROC2: A Program for ROC Analysis of Correlated, Inherently Categorical, Rating-Scale Data*. Chicago, IL: Department of Radiology and the Franklin McLean Research Institution, University of Chicago; 1989.
6. Metz CE, Herman BA, Roe CA. Statistical comparison of two ROC-curve estimates obtained from partially-paired datasets. *Med Decis Making*. 1998;18(1):110–121.
7. Metz CE, Kronman H. A Test for the statistical significance of differences between ROC curves. *INSERM*. 1979;88:647–660.

8. Farquhar TH, Llacer J, Hoh CK, et al. ROC and localization ROC analyses of lesion detection in whole-body FDG PET: Effects of acquisition mode, attenuation correction and reconstruction algorithm. *J Nucl Med.* 1999;40:2043–2052.

9. Roe CA, Metz CE. Dorfman-Berbaum-Metz method for statistical analysis of multireader, multimodality receiver operating characteristic data: Validation with computer simulation. *Acad Radiol.* 1997;4:298–303.

10. Metz CE. *ROCKIT Beta version,* http://www-radiology.uchicago.edu/krl/University of Chicago; [updated 27 February 2007; cited October 2008]. 1998.

11. Zhai X, Chakraborty DP. A bivariate contaminated binormal model for robust fitting of proper ROC curves to a pair of correlated, possibly degenerate, ROC datasets. *Med Phys.* 2017;44(3):2207–2222.

22

Evaluating standalone CAD versus radiologists

22.1 Introduction

This chapter is focused on computer aided detection (CAD) in screening mammography. On a mammogram, there are four signs associated with breast cancer.[1] The two most common are masses (56%) and microcalcifications (29%). Less common are asymmetries (12%) and architectural distortions (4%). The majority (95%) of cases presenting as masses are invasive cancers, i.e., have spread from the original location of the cancer and are associated with greater mortality, and the majority (68%), presenting as microcalcifications, are associated with ductal carcinoma in-situ (i.e., have not spread from the milk ducts). Mammography screening is the a proven technique to reduce mortality from breast cancer,[2] annually about[3] 40,000 patients in the United States, but mammographic screening, particularly detecting masses, is one of the most difficult tasks in radiology. There are a few reasons for this: the (fortunate) low incidence of breast cancer, about five per 1000, the lack of anatomic landmarks in the breast, the difficulty of imaging near the chest wall, and the camouflaging effect of dense (glandular) breast tissues. Typical sensitivity is about 80% with wide variability,[4,5] as much as 40%. A similar variability is observed for specificity, with experts achieving about 90% specificity. *As a result, a woman's chance of a correct diagnosis depends on the radiologist who interprets her images.*

Novel imaging treatments like breast computerized tomography[6] and breast tomosynthesis[7] have become realities. A major effort has been underway for almost three decades to develop CAD algorithms to analyze images for signs of cancer.[8–10] CAD complements technological innovations because it can be applied to images from the newer modalities where performance, while better than conventional 2D images, is still far from perfect.[11] Almost all major imaging device manufacturers provide CAD as part of their imaging workstations. In the United States, the majority of digital mammograms are now analyzed by CAD.[12] In the United Kingdom CAD is not widely used; instead the radiologists depend on double reading.

A typical CAD algorithm consists of two stages: (1) a *search stage*, where image processing techniques are used to identify suspicious regions and (2) a *lesion-classification stage* where features (quantitative descriptors of the suspicious regions, such as size, contrast, shape, texture, etc.), are analyzed by a binary classifier (i.e., benign versus malignant) to yield a *malignancy index* that an identified suspicious region is an actual cancer. If the index exceeds a preselected threshold, set by the manufacturer based on a compromise between sensitivity and specificity, the suspicious region in question is marked (or annotated or cued). After an initial unaided interpretation of the mammogram, the radiologist reviews the annotations and reinterprets the mammogram; this is the current FDA-approved *second-reader* mode for usage of CAD.

While there is consensus that CAD works for microcalcifications (regions containing small high contrast speck-like lesions), where it relieves the radiologist from the tedium of having to exhaustively search images with a magnifying glass, for masses (larger and relatively lower contrast lesions) its performance is controversial as evidenced by dueling editorials on the merits of CAD.[13,14] Two very large clinical studies[15,16] (222,135 and 684,956 women, respectively) showed that CAD can actually have a detrimental effect on patient outcome. These publications led to a flurry[17-24] of rebuttals from the CAD community, mainly questioning the methodology used in these studies, one of which, relating to the use of clinical ratings for ROC studies was discussed in Section 4.10. A more recent study has confirmed the negative view of the efficacy of CAD[25] and there has even been a call for ending Medicare reimbursement for CAD interpretations.[26]

Why the poor performance of CAD, even after 25 years of research? One reason has already been identified: excessive reliance on a poor FOM, namely FROC curve based measures. Another reason is lack of focus on the most obvious measure of the relative performance of CAD, namely a direct comparison of CAD to that of a group of experts. Clinical CAD systems sometimes simply report the locations of suspicious regions, that is, it may not provide ratings. However, a (continuous variable) *malignancy index* for every CAD-found suspicious region is available to the CAD algorithm designer.[27] *Standalone performance, that is, performance of CAD by itself, regarded as an algorithmic reader, versus a group of expert radiologists, is rarely measured.* The author knows of only one study[28] in breast CAD where standalone CAD performance was measured; other examples are to be found in CT colonography CAD and other modalities.[29-33]

The argument made for not measuring standalone performance is that since CAD will be used *only* as a second reader, one need only measure performance of a radiologist with and without CAD. Indeed, a CAD expert has stated:[34] "High stand-alone performance is neither a necessary nor a sufficient condition for CAD to be truly useful clinically." Unfortunately, measuring performance with and without CAD has set a low bar for CAD to be considered useful. For example, CAD is not penalized for missing lesions as long as the radiologist finds them and CAD is not penalized for excessive FPs as long as the radiologist ignores them. There are also serious methodological problems, with excessive reliance on the low statistical power ROC paradigm and wide usage of the FROC curve as a measure of performance.

The purpose of this chapter is to describe a method to compare standalone performance of CAD to a group of radiologists interpreting the cases. The paradigm used to collect the data (e.g., ROC, FROC, LROC, or ROI) is irrelevant—all that is needed is a scalar performance measure, or figure of merit (FOM), for the actual paradigm used. Before proceeding, in this chapter all references to CAD are to *standalone* CAD with designer-level data available. The author is grateful to Prof. Nico Karssemeijer for providing a dataset on standalone performance of CAD, without which this chapter would not be possible.

22.2 The Hupse–Karssemeijer et al. study

The study[28] (henceforth referred to as the *Hupse–Karssemeijer study*) compared standalone performance of a CAD device to that of nine radiologists interpreting the same cases (120 non-diseased and 80 with a single malignant mass per diseased case) using the location receiver operating characteristic (LROC) paradigm,[35-38] see **Online Chapter 24** for summaries of all included datasets. In LROC, for each case the radiologist gives an overall rating for presence of disease, that is, a ROC rating, and indicates the location of the most suspicious region. On non-diseased cases, the rating is classified as a false positive (i.e., the location is ignored), but on a diseased case the rating is classified as a *correct localization* if the mark is sufficiently close to the lesion and otherwise it is classified as an *incorrect localization*.

For a given threshold, the total number of correct localizations divided by the total number of diseased cases is an estimate of the probability of correct localization on a diseased case, denoted *PCL*. On non-diseased cases, the total number of false positives divided by the number of non-diseased cases is an estimate of the probability of a false positive, that is, FPF. In the following equation, *CL* denotes correct localization, *t* denotes the truth status, ζ denotes the decision threshold, and $Z_{k_t t}$ denotes the (random) decision variable for case $k_t t$.

$$PCL(\zeta) = P\left(Z_{k_2 2} > \zeta \mid CL\right)$$

$$FPF(\zeta) = P\left(Z_{k_1 1} > \zeta\right)$$

(22.1)

The plot of *PCL* (ordinate) versus *FPF* defines the LROC plot. In the Hupse–Karssemeijer study "quasi-continuous" 0–100 integer ratings were used and empirical LROCs were used to summarize performance.

Hupse–Karssemeijer used as FOMs the interpolated PCL at two values of FPF, namely 0.05 and 0.2. The two FOMs are denoted by $PCL_{FPF=0.05}$ and $PCL_{FPF=0.2}$, respectively. Note that these figures of merit are point measures, as they depend on a single point on the LROC curve. In contrast, full-area measures, such as AUC, depend on the entire curve (partial-area measures are also possible, for example, the area under the curve to the left of some prespecified value of FPF). The t-test between the observed radiologist *PCL* values and that of CAD was used to compute the (two-sided) *p*-value for rejecting the NH of equal performance. Hupse–Karssemeijer reported *p*-value = 0.17 for $PCL_{FPF=0.05}$ and *p*-value < 0.001, with CAD being inferior, for $PCL_{FPF=0.2}$. Table 22.1 summarizes relevant results from the Hupse–Karssemeijer publication (Table 1 *ibid.*). In the published table, the ordinate of the LROC is labeled TPF, whereas in LROC notation[36] it is customary to label it PCL.

The next section presents **R** code that duplicates the Hupse–Karssemeijer analysis.

Table 22.1 Hupse–Karssemeijer analysis with cases treated as a fixed factor

FOM	⟨RAD⟩	CAD	⟨RAD⟩ − CAD	CI⟨RAD⟩	*p*-value
$PCL_{FPF=0.05}$	0.518	0.487	0.031	(0.45,0.59)	0.17
$PCL_{FPF=0.2}$	0.736	0.620	0.116	(0.69,0.78)	<0.001

Note: Different conclusions, depending on the choice of FPF value at which to evaluate the partial curve measure; <> denotes an average over readers.

The paired t-test applied by Hupse–Karssemeijer treats the case sample as fixed. In other words, the analysis is not accounting for case-sampling variability but it is accounting for reader variability. As a historical/personal note, the author recalls a discussion (ca. 1986) with the late Prof. Charles E. Metz where Prof. Metz patiently explained to the author and Prof. Gary Barnes, his mentor at University of Alabama at Birmingham, *that the (paired) t-test we were using in our Radiology publication[39] was accounting for reader variability but not case variability.* To paraphrase his comment: *I am comfortable with the claim that the observed significant difference generalizes to readers but it is specific to the one case sample that you studied.* Since the 1986 study was a phantom study, the case sample happened to be the phantom with superposed simulated lesions. So, the finding was actually specific to the one phantom that was used. Since a population of phantom is not meaningful, therefore, even in retrospect, the paired t-test analysis used in the 1986 study, which disregarded case-variability, was appropriate. However, in the context of the Hupse–Karssemeijer study, it is necessary to account for case-sampling variability, that is, patient-to-patient variability (to make the point explicit, patients are not phantoms; there is variability between them and a population of patients *does* exist).

22.2.1 Random-reader fixed-case analysis

The code and explanation to implement the Hupse–Karssemeijer analysis is in file **mainAnalysisFixed.R**, described in Online Appendix 22.A. The word *fixed* in the file name emphasizes that case is regarded as a fixed factor in the analysis. At line 11, the variable named **FOM** is set to **"PCL"**, which means the figure of merit is interpolated *"PCL"* at the value of *FPF* specified at line 21, that is, **FPF** = 0.05 on the first iteration of the for-loop starting at line 20. In other words, the figure of merit for the first iteration is $PCL_{FPF=0.05}$. Insert a break point at line 53 and click on **Source** to get Code Output 22.2.1.1 and Figure 22.1.

22.2.1.1 Code output

```
> debugSource(...)
Hupse-Karssemeijer analysis
: random readers fixed cases
FOM = PCL
FPF = 0.05
FomCad = 0.4625
AVG radiologist performance = 0.5063657
95%CI = 0.4368152 0.5759163
AVG diff. performance = 0.04386574
95%CI =   -0.02568478 0.1134163
t-statistic = 1.454404
df = 8
p-value = 0.1839158
```

Code Output 22.2.1.1 shows that FOM CAD, namely $PCL_{FPF=0.05}$ is smaller than the average of the radiologists (0.463 < 0.506), but the *p*-value is 0.18, so the difference is not significant at $\alpha = 0.05$. This result is mirrored by the fact that the 95% CI for the radiologists (0.44, 0.58) includes the figure of merit of CAD (0.463). The output lists the value of the t-statistic (1.45) and the degrees of freedom (8), which is one less than the number of radiologists.

Use the first command in the code snippet below to examine **thetajc**, the figure of merit array (*j* is the reader index and *c* is the case sample index; see Equation 22.2 below). The first reader is standalone CAD and the subsequent nine readers correspond to the radiologists. The second command shows the difference between CAD and the average of the radiologists (the negative index removes CAD), with CAD being worse (–0.0439).

22.2.1.2 Code snippet

```
Browse[2]> thetajc
 [1] 0.4625 0.4250 0.5156 0.6750 0.5292 0.4479 0.5979 0.5375 0.4250 0.4042
Browse[2]> mean(thetajc[-1]-thetajc[1])
 [1] 0.04387
```

The radiologists are slightly (0.044) better as a group than CAD. The question one wishes to answer is: is the difference statistically significant? Related questions are: what is the p-value and what is the confidence interval for the FOM difference?

Figure 22.1 shows empirical LROC plots for the 9 radiologists and standalone CAD (heavy black line). Unlike ROC curves, the LROC curves flatten out at a value of PCL < 1. This behavior should be fairly obvious because even as the threshold is lowered, lesions that are not found by the search mechanism will not be correctly localized, especially since the reader is only allowed one mark.

Exit debug mode, remove any breakpoints and **source** the code. This time all iterations in the for-loop are executed, where the iterations correspond to different FPF values at which to evaluate PCL. The code output was used to populate the rows labeled "*fixed*" in Table 22.2. These correspond to the values of FPF defined in the array at line 12, namely 0.05, 0.2, 0.5, 1. The table is split into four rows corresponding to different choices of FPF and each row is further split into two rows, corresponding to treating cases as fixed and random, respectively. All rows labeled "fixed" were obtained by sourcing **mainAnalysisFixed.R**, while rows labeled "random" were obtained by sourcing

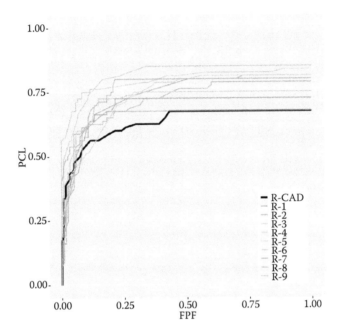

Figure 22.1 Empirical LROC plots; the thick black plot applies to standalone CAD while the rest apply to the nine radiologists. This figure, which shows the entire operating characteristic, leads to the inescapable conclusion that the radiologists as a group outperformed CAD. Whether the difference is significant is determined by the analysis.

Table 22.2 Comparison of fixed-case and random-case analyses

FPF	Cases	CAD	$\langle RAD \rangle$	$CI_{\langle RAD \rangle}$	$\langle RAD \rangle - CAD$	$CI_{\langle RAD \rangle - CAD}$	p-value
0.05	fixed	0.463*	0.506*	(0.44,0.58)*	0.044	(−0.026,0.11)	0.18*
	random			(0.36,0.65)		(−0.100 0.188)	0.55
0.2	fixed	0.592*	0.711*	(0.67,0.75)*	0.119``	(0.079, 0.16)	0.0001*
	random			(0.60,0.83)		(0.004, 0.235)	0.042
0.5	fixed	0.675	0.775	(0.73,0.82)	0.100	(0.059, 0.141)	0.0005
	random			(0.67,0.88)		(−0.004, 0.204)	0.059
1	fixed	0.675	0.783	(0.74,0.83)	0.108	(0.065, 0.152)	0.0004
	random			(0.68,0.89)		(0.005, 0.212)	0.041

Note: All rows labeled *fixed* were obtained by sourcing **mainAnalysisFixed.R**, while rows labeled *random* were obtained by sourcing **mainAnalysisRandom.R**. The figure of merit (FOM) = PCL @ specified FPF. An asterisk indicates instances where direct comparison with Hupse–Karssemeijer analyses are possible. Because it is accounting for an additional source of variability, each of the lines labeled *random* yields a larger p-value and a wider confidence interval than the corresponding line labeled *fixed*.

`mainAnalysisRandom.R`, to be described below. Instances where direct comparisons with the Hupse–Karssemeijer analysis are possible are indicated with asterisks.

At the two instances where direct comparison with the Hupse–Karssemeijer analyses is possible, i.e., fixed case analysis with PCL @ FPF = 0.05 and PCL @ 0.2, respectively, the agreement between 95% CIs and *p*-values is quite good. For example, @ FPF = 0.05, the CIs are (0.45, 0.59) in Table 22.1 versus (0.44, 0.58) in Table 22.2. The minor differences in FOMs (Table 22.2 yields consistently smaller values than those in the Hupse–Karssemeijer publication) could be due to differences between the dataset actually used for the reported study and that received by us and / or methods for interpolating the data.

It should be fairly obvious from these plots that the radiologists, as a group, outperformed CAD. The difference is especially striking for the third radiologist (the color coded figure makes this clearer). However, this straightforward conclusion can be obfuscated by focusing on the very low end of the plots, for example, FPF = 0.05, where the plots are very steep, resulting in a not-significant *p*-value (i.e., 0.18). Some CAD developers use a point measure at the low end of FPF in the mistaken belief that this is the "clinically relevant" portion of the curve; this issue was discussed at length in Section 17.10.1. Appropriate analysis, using whole area measures of performance, is described in Section 22.5.

The output of the analysis for the **FPF** = 0.2 loop is shown below.

22.2.1.3 Code output

```
FOM  =   PCL
FPF  =   0.2
FomCad  =   0.5917
AVG radiologist performance  =   0.7111
95%CI =   0.6702 0.7519
AVG diff. performance  =   0.1194
95%CI =   0.07858 0.1603
t-statistic  =   6.742
df  =   8
p-value  =   0.0001463
```

As expected from Figure 22.1, at this FPF the difference (between the average of the radiologists and CAD) is significant, *p*-value = 0.00015. The average difference in FOMs is much larger, 0.119 versus 0.044 in the previous example, and the 95% confidence interval for the radiologists (0.67, 0.75) now does not include the FOM for CAD (0.59). However, as emphasized previously, claiming that CAD is inferior based on a point FOM would be incorrect as the choice of FPF is arbitrary.

The Hupse–Karssemeijer analysis treats the case sample as fixed. The following section extends the analysis to random cases.

22.3 Extending the analysis to random cases

Standalone CAD is an algorithmic reader, not a different treatment. Therefore, a method for analyzing readers in a single treatment,[40,41] as described in Section 10.7, is appropriate, where the figure of merit for the radiologists was modeled as shown in Equation 22.2. (*j* is the reader index, excluding CAD).

$$\theta_{j\{c\}} = \mu + R_j + \varepsilon_{j\{c\}} \tag{22.2}$$

$\theta_{j\{c\}}$ is the figure of merit for radiologist j ($j = 1, 2, ..., J$) interpreting case sample $\{c\}$; R_j is the random effect of radiologist j and $\varepsilon_{j\{c\}}$ is the error term described in **Chapter 10** and below. (Equation 22.6).

The formula presented in Section 10.7, allow one to test the NH: $\mu = \mu_0$ where μ_0 is a prespecified *constant*. One could set μ_0 equal to the standalone performance of CAD, but that would ignore the fact that the performance of CAD is itself a random variable, not a constant, whose case-sampling variability needs to be accounted for.

Instead, the following model is used for the figure of merit fluctuations of the radiologists and CAD:

$$\theta_{j\{c\}} = \theta_{0\{c\}} + \Delta\theta + R_j + \varepsilon_{j\{c\}} \qquad (j=1,2,...,J) \tag{22.3}$$

$\theta_{0\{c\}}$ is the CAD figure of merit estimate for case sample $\{c\}$, that is, $j = 0$ is used to denote the CAD reader; $\Delta\theta$ is the average figure of merit increment of the radiologists over CAD. Subtract the CAD figure of merit from the radiologist figure of merit for the same case sample, and define this as the *difference (radiologist minus CAD) figure of merit* $\psi_{j\{c\}}$, that is,

$$\psi_{j\{c\}} \equiv \theta_{j\{c\}} - \theta_{0\{c\}} \tag{22.4}$$

Then Equation 22.3 can be written as

$$\psi_{j\{c\}} = \Delta\theta + R_j + \varepsilon_{j\{c\}} \tag{22.5}$$

Equation 22.5 is homologous (i.e., identical in form) to Equation 22.2 with the subtle difference that the figure of merit on the left-hand side of Equation 22.5 is a *difference FOM*, between the radiologist's FOM and that of CAD. Equation 22.5 describes a model for J *difference radiologists* interpreting a common case set $\{c\}$, each of whose performances is a difference from that of CAD (the difference is positive if the radiologist is better). Under the NH the expected difference is zero: $NH : \Delta\theta = 0$. The method described in Section 10.7 for single-treatment multiple-reader analysis is now directly applicable to the model described by Equation 22.5. One assumes that the error term is distributed as

$$\varepsilon_{j\{c\}} \sim N_J\left(\vec{0}, \Sigma\right) \tag{22.6}$$

The $J \times J$ covariance matrix Σ is defined by two parameters, *Var* and *Cov₂*, as follows:

$$\Sigma_{jj'} = Cov\left(\varepsilon_{j\{c\}}, \varepsilon_{j'\{c\}}\right) = \begin{cases} Var & j = j' \\ Cov_2 & j \neq j' \end{cases} \tag{22.7}$$

The terms *Var* and *Cov₂* can be estimated using resampling methods, for example, the jackknife, see **Chapter 07** and **Chapter 10**. Denote the difference FOM with case k removed by $\psi_{j(k)}$ (the index in parenthesis (k) denotes deleted case k; i.e., it is the FOM with case k removed from case set $\{1\}$. For notational conciseness, and since one is restricted to $c = 1$, the case set index is henceforth suppressed. The covariance matrix is estimated using (the variance inflation factor is shown explicitly):

$$\left.\Sigma_{jj'}\right|_{jack} = \frac{(K-1)^2}{K}\left[\frac{1}{K-1}\sum_{k=1}^{K}\left(\psi_{j(k)} - \psi_{j(\bullet)}\right)\left(\psi_{j'(k)} - \psi_{j'(\bullet)}\right)\right] \tag{22.8}$$

The final estimates of *Var* and Cov_2 are averaged (indicated by the angular brackets below) over all pairings of radiologists satisfying the relevant equalities/inequalities shown just below the closing angular bracket:

$$Var = \left\langle \Sigma_{jj'}\big|_{jack} \right\rangle_{j=j'}$$

$$Cov_2 = \left\langle \Sigma_{jj'}\big|_{jack} \right\rangle_{j \neq j'} \tag{22.9}$$

Under the null hypothesis $\Delta\theta = 0$ the observed value of the statistic $t_{I=1}$ defined below follows; (compare to Equation 10.70) the *t*-distribution with $df_H^{I=1}$ degrees of freedom, where $df_H^{I=1}$ is defined in Equation 22.12, (compare to Equation 10.72)

$$t_{I=1} \equiv \psi_\bullet \sqrt{\frac{J}{MS_R + \max(JCov_2, 0)}} \sim t_{df_H^{I=1}} \tag{22.10}$$

Here $\psi_\bullet$ is the *observed effect size*, that is, the average, over radiologists, of ψ_j, Equation 22.4, and *MSR* is defined by (compare to Equation 10.68)

$$MSR = \frac{1}{J-1} \sum_{j=1}^{J} (\psi_j - \psi_\bullet)^2 \tag{22.11}$$

The degrees of freedom of the *t*-statistic is defined by (compare to Equation 10.72)

$$df_H^{I=1} = \left[\frac{MSR + \max(JCov_2, 0)}{MSR} \right]^2 (J-1) \tag{22.12}$$

The 2-sided *p*-value for rejecting the NH: $\Delta\theta = 0$ is

$$p = P\left(|t| > |t_{I=1}| \big| t \sim t_{df_H^{I=1}} \right) \tag{22.13}$$

A $(1-\alpha) \times 100$ percent confidence interval for the observed effect size $\psi_\bullet$ is given by[40] (compare to Equation 10.74)

$$\psi_\bullet \pm t_{\alpha/2; df_H^{I=1}} \sqrt{\frac{MSR + \max(JCov_2, 0)}{J}} \tag{22.14}$$

Here $t_{\alpha/2; df_H^{I=1}}$ is the upper $\alpha/2$ quantile of $t_{df_H^{I=1}}$, that is, the value such that $\alpha/2$ of the specified distribution lies above it.

Open the file **mainAnalysisRandomBrief.R**, a listing of which follows (*Brief* in the file name signifies that this file is a brief version, a more general version handles other FOMs). *Random* in the file name implies that cases are treated as random, as in the analysis presented in Equations 22.3 through 22.14. In the brief version, the **FOM** is set to **AUC**, the empirical area under the ROC curve; **FPF** is then irrelevant, but it is needed to use other FOMs. In case one wonders how to get ROC data from LROC, one uses the maximum rating, which could be a correct or an incorrect localization event (see the **pmax()** function in line 17 of the following code listing; this stands for *parallel maximum* over the correct and incorrect localization arrays, each of length K_2).

22.3.1 Code listing

```
# MainAnalysisRandomBrief.R
rm(list = ls())
library(RJafroc)
library(ggplot2)

alpha <- 0.05
FOM <- "AUC"
FPF <- 0.0
cat("FOM = ", FOM, "\n")
if (FOM == "PCL") cat("FPF = ", FPF, "\n")
cat("Random-reader random-case analysis")
cat("\nof Hupse Karssemeijer radiologist data:\n")
retNico <- DfReadLrocDataFile()
zjk1 <- retNico$NL[1,,,1]
zjk2Cl <- retNico$LLCl[1,,,1]
zjk2Il <- retNico$LLIl[1,,,1]
zjk2 <- pmax(zjk2Cl,zjk2Il)
K1 <- length(zjk1[1,])
K2 <- length(zjk2Cl[1,]);K <- c(K1,K2)
ret_nh <- DiffFomAnal2007Hillis53 (
  alpha,
  zjk1,
  zjk2,
  FOM,
  FPF)
cat(str(ret_nh))
```

RJafroc has an embedded function **DiffFomAnal2007Hillis53()**, which implements the analysis described in Equations 22.3 through 22.14 and gives due credit to Section. 5.3 in a Hillis publicatioctin.[41] Depending on the choice of FOM, this funon is called at line 27, 34 or 41 and the returned **list** is printed. **Source** the file to get the following output.

22.3.2 Code output

```
> source(...)
FOM = AUC
Random-reader random-case analysis
of Hupse Karssemeijer radiologist data:
List of 9
 $ Var    : num 0.0014
 $ Cov2   : num 0.000924
 $ PsiMean: num 0.0317
 $ CI     : num [1:2(1d)] -0.031 0.0945
 $ reject : num 0
 $ ddfH   : num 878
 $ Tstat  : num 0.993
 $ p_val  : num 0.321
 $ thetajc: num [1:10(1d)] 0.817 0.842 0.841 0.9 0.838 ...
```

The lines starting with **$Var** and **$Cov2** show the two elements of the covariance matrix: $Var = 0.0014$ and $Cov_2 = 0.000924$. Covariance is smaller than variance, which should make sense

514 Evaluating standalone CAD versus radiologists

since covariance equals the variance times a correlation, and the latter is less than unity. The next line lists the mean value of the difference figure of merit, **PsiMean** which stands for ψ• = 0.0317 (i.e., the difference in empirical AUCs of the radiologists minus that of CAD, averaged over radiologists). The next line lists the 95% confidence interval for ψ•, which is [−0.031, 0.0945]. Since this includes zero, the NH, that average differential performance is zero, is not rejected, emphasized by the next line **reject** = 0. The next line lists the degrees of freedom of the t-test, = 878, the $t_{I=1}$ statistic is 0.993, the p-value is 0.321, and the last line prints **thetatjc** ($\theta_{j\{c\}}$): the first value corresponds to CAD; the second value corresponds to the first radiologist, and so on.

22.3.3 Extension to more FOMs

The file **mainAnalysisRandom.R**, listed in Online Appendix 22.B.1, extends **mainAnalysisRandomBrief.R** to other FOMs. Ensure that FOM is set to **"PCL"** at line 10. **Source** the code to get the random-case results summarized in Table 22.2. Because it is accounting for an additional source of variability, each of the lines labeled *random* yields a larger p-value and a wider confidence interval than the corresponding line labeled *fixed*.

22.4 Ambiguity in interpreting a point-based FOM

Both analyses, i.e., fixed-case or random-case, in Table 22.2 declare *not significant difference at FPF = 0.05 and significant difference at FPF = 0.2*. The difference in conclusions between FPF = 0.05 and FPF = 0.2 illustrates the ambiguity when using point-based measures, which do not use all the data. For example, at FPF = 0.05, most of the data contributing to the plots to the right of this operating point, Figure 22.1, are not used in the analysis. An even more important reason is the steepness of the curves at this low FPF, which introduces greater variability in the interpolated estimates.

The variable results of point measures, such as those used in Tables 22.1 and 22.2, invite gaming the system; that is, the researcher is under temptation to report performance at a FPF value where the algorithm "looks good." Using FPF = 0.05 the designer could claim (incorrectly as it turns out, because inability to reject the NH does not mean the NH is true) that their CAD algorithm is no worse than experts. Another investigator using FPF = 0.2 could claim that the algorithm is worse than radiologists (a correct conclusion as it turns out, but incorrect method, because the p-value depends on the choice of FPF, which is arbitrary).

22.5 Results using full-area measures

Table 22.3 presents results for choices of FOM, *each of which is a full-area measure*. A full-area measure removes arbitrariness inherent in any partial-area or point-based measure. Shown are results for the empirical area under the LROC plot (**FOM** = **"ALroc"** at line 10 in **mainAnalysisRandom.R**) and the empirical area under the ROC plot (**FOM** = **"AUC"**).

Table 22.3 Result of using area empirical FOM measures. Two FOMs are used, as indicated in the first column

FOM	Cases	CAD	$\langle RAD \rangle$	$CI_{\langle RAD \rangle}$	$\langle RAD \rangle - CAD$	$CI_{\langle RAD \rangle - CAD}$	p-value
A_{LROC}	fixed	0.628	0.734	(0.69,0.77)	0.105	(0.065, 0.146)	0.0003**
	random			(0.64,0.83)		(0.008, 0.203)	0.0349
A_{ROC}	fixed	0.817	0.849	(0.83,0.87)	0.0317	(0.009, 0.055)	0.0124**
	random			(0.79,0.91)		(−0.031, 0.094)	0.321

Legend: **Significant at alpha corrected for multiple testing, that is, test alpha = 0.025.

Before getting into a discussion of these results, it is appropriate to note that a correction for multiple testing is needed when applying multiple analyses to the same dataset. One way of accomplishing this is by setting the alpha for multiple testing at the nominal alpha divided by the number of tests. In other words, to maintain the overall probability of a Type-I error at 0.05, one needs to adopt the multiple testing alpha as 0.05/2, that is, 0.025. This is often referred to as the Bonferroni correction.[42,43] The Bonferroni correction is conservative. It is accurate when the data are independent but in the presence of correlations it would lead to inability to reject the NH more often than is appropriate (i.e., loss of statistical power). It has the advantage of being the simplest method for accounting for multiple comparisons.

Using the total area under the empirical LROC (first results row in Table 22.3) yields a not-significant difference for random cases ($p = 0.0349 > 0.025$), but a significant difference for fixed cases ($p = 0.0003 < 0.025$; the use of 0.025 for alpha is explained below), with CAD being inferior, consistent with the visual impression of the plots in Figure 22.1. Similarly, the corresponding analysis using the empirical area under the ROC yields a significant difference only if cases are regarded as fixed ($p = 0.0124 < 0.025$). Both full-area measures, LROC and ROC, are in agreement, yielding a significant difference only if the case is regarded as a fixed factor. However, the p-value for LROC is much smaller ($0.0003 \ll 0.0124$). This is consistent with greater statistical power using LROC, which takes location into account, as compared to ROC. The increased statistical power has been noted by Swensson.[38,44,45]

If prior to conducting the analysis one had committed to either FOM, not both, then one would be justified in not applying the Bonferroni correction. In the current situation, based on the expectation of higher power, the author would have chosen the full area under the LROC, in which case the difference is significant even when case is regarded as a random factor ($p = 0.0349 < 0.05$). Of course, having committed, scientific integrity requires that one does not change course after the study is completed just because the results look better using an alternate FOM. If one does change course, then the Bonferroni correction is needed. It is there to keep one honest.

22.6 Discussion/Summary

This chapter has described an extension of the Hupse–Karssemeijer analysis that accounts for case variability. The method extends the single treatment analysis of **Chapter 10** to a situation where one of the readers is a special reader, and the desire is to compare performance of this reader against the average of the rest of the readers. The method was used to analyze, using different figures of merit, the LROC data set kindly provided by Prof. Karssemeijer. It is shown that point-based measures of performance, such as PCL @ specified FPF, lead to different results, depending on the choice of FPF. Similar conclusions apply to partial-area based FOMs. The only way of avoiding this is to use full-area based measures, such as the area under the ROC or the LROC. An added advantage of doing so is the increased statistical power due to the fact that one is using all the data; point-based measures or partial-area measures do not use all the data. The currently encouraged focus on the low FPF end of the ROC curve is misleading, as one is trying to measure performance on the initial steep area of the curve, where there is little data, and most of the data that exists comes from highly visible lesions, which both CAD and the radiologists are expected to find easily, leading to further loss of statistical power. Using a full-area measure takes all lesions in the dataset into account, and the findings extrapolate to the population of cases, not just to the subset of cases with highly visible lesions.

While demonstrated using ROC and LROC data, the method is applicable to any scalar figure of merit. It is recommended that future CAD studies use FROC methodology in conjunction with the empirical wAFROC-AUC figure of merit. The advantage of using the FROC paradigm lies specifically in higher statistical power and the diagnostics yielded by RSM-based ROC curve fitting; specifically, the RSM parameters yield information on what is limiting CAD performance, i.e., search or classification performance. Based on the author's experience current CAD algorithms are much worse than experts in search performance (e.g., CAD generates 30 times the number of NLs per 4-view case than experts), but CAD's classification performance might actually be better (explained by the fact that it has access to the digital pixel values and the ability to implement sophisticated feature-detection and lesion-classification methods). The author has been told by Dr. Ronald Summers that the CAD community believes in precisely the opposite—namely, that CAD is perfect at searching because it "looks at everything."

References

1. Shetty MK, Editor. *Breast Cancer Screening and Diagnosis: A Synopsis*. New York, NY: Springer Science+Business Media; 2015.
2. Kerlikowske K, Grady D, Rubin SM, Sandrock C, Ernster VL. Efficacy of screening mammography: A meta-analysis. *JAMA*. 1995;273(2):149–154.
3. ACS. *American Cancer Society: Cancer Facts & Figures 2015 Atlanta*. Reston, VA: American Cancer Society; 2015.
4. Beam CA, Layde PM, Sullivan DC. Variability in the interpretation of screening mammograms by US radiologists. Findings from a national sample. *Arch Intern Med*. 1996;156(2):209–213.
5. Elmore JG, Jackson SL, Abraham L, et al. Variability in interpretive performance at screening mammography and radiologists' characteristics associated with accuracy. *Radiology*. 2009;253(3):641–651.
6. Boone JM, Kwan ALC, Yang K, Burkett GW, Lindfors KK, Nelson TR. Computed tomography for imaging the breast. *J Mammary Gland Biol Neoplasia*. 2006;11:103–111.
7. Niklason LT, Christian BT, Niklason LE, et al. Digital tomosynthesis in breast imaging. *Radiology*. 1997;205(2):399–406.
8. Giger ML, Chan H-P, Boone J. Anniversary paper: History and status of CAD and quantitative image analysis: The role of medical physics and AAPM. *Med Phys*. 2008;35(12):5799–5820.
9. Nishikawa RM. Current status and future directions of computer-aided diagnosis in mammography. *Comput Med Imaging Graph*. 2007;31(4–5):224–235.
10. Li Q, Nishikawa RM, eds. *Computer-Aided Detection and Diagnosis in Medical Imaging, Imaging in Medical Diagnosis and Therapy*. Boca Raton, FL: CRC Press, Taylor & Francis; 2012.
11. Chan H-P, Wei J, Sahiner B, et al. Computer-aided detection system for breast masses on digital tomosynthesis mammograms: Preliminary experience. *Radiology*. 2005;237(3): 1075–1080.
12. Rao VM, Levin DC, Parker L, Cavanaugh B, Frangos AJ, Sunshine JH. How widely is computer-aided detection used in screening and diagnostic mammography? *J Am Coll Radiol*. 2010;7:802–805.
13. Philpotts LE. Can computer-aided detection be detrimental to mammographic interpretation? *Radiology*. 2009;253(1):17–22.
14. Birdwell RL. The preponderance of evidence supports computer-aided detection for screening mammography1. *Radiology*. 2009;253(1):9–16.
15. Fenton JJ, Taplin SH, Carney PA, et al. Influence of computer-aided detection on performance of screening mammography. *N Engl J Med*. 2007;356(14):1399–1409.
16. Fenton JJ, Abraham L, Taplin SH, et al. Effectiveness of computer-aided detection in community mammography practice. *J Natl Cancer Inst*. 2011;103(15):1152–1161.

17. Ciatto S, Houssami N. Computer-aided screening mammography. *N Engl J Med*. 2007;357(1):83–85.
18. Feig SA, Birdwell RL, Linver MN. Computer-aided screening mammography. *N Engl J Med*. 2007;357(1):83–85.
19. Fenton JJ, Barlow WE, Elmore JG. Computer-aided screening mammography. *N Engl J Med*. 2007;357(1):83–85.
20. Gur D. Computer-aided screening mammography. *N Engl J Med*. 2007;357(1):83–85.
21. Nishikawa RM, Schmidt RA, Metz CE. Computer-aided screening mammography. *N Engl J Med*. 2007;357(1):83–85.
22. Ruiz JF. Computer-aided screening mammography. *N Engl J Med*. 2007;357(1):83–85.
23. Berry DA. Computer-assisted detection and screening mammography: Where's the beef? *J Natl Cancer Inst*. 2011;103(15):1139–1141.
24. Nishikawa RM, Giger ML, Jiang Y, Metz CE. Re: Effectiveness of computer-aided detection in community mammography practice. *J Natl Cancer Inst*. 2012;104(1):77.
25. Lehman CD, Wellman RD, Buist DS, Kerlikowske K, Tosteson AN, Miglioretti DL. Diagnostic accuracy of digital screening mammography with and without computer-aided detection. *JAMA Intern Med*. 2015;175(11):1828–1837. doi:10.1001/jamainternmed.2015.5231.
26. Fenton JJ. Is it time to stop paying for computer-aided mammography? *JAMA Intern Med*. 2015;175(11):1837–1838. doi:10.1001/jamainternmed.2015.5319.
27. Edwards DC, Kupinski MA, Metz CE, Nishikawa RM. Maximum likelihood fitting of FROC curves under an initial-detection-and-candidate-analysis model. *Med Phys*. 2002;29(12):2861–2870.
28. Hupse R, Samulski M, Lobbes M, et al. Standalone computer-aided detection compared to radiologists' performance for the detection of mammographic masses. *Eur Radiol*. 2013;23(1):93–100.
29. Hein PA, Krug LD, Romano VC, Kandel S, Hamm B, Rogalla P. Computer-aided detection in computed tomography colonography with full fecal tagging: Comparison of standalone performance of 3 automated polyp detection systems. *Can Assoc Radiol J*. 2010;61(2):102–108.
30. Summers RM, Handwerker LR, Pickhardt PJ, et al. Performance of a previously validated CT colonography computer-aided detection system in a new patient population. *Am J Roentgenol*. 2008;191(1):168–174.
31. Taylor SA, Halligan S, Burling D, et al. Computer-assisted reader software versus expert reviewers for polyp detection on CT colonography. *Am J Roentgenol*. 2006;186(3):696–702.
32. De Boo DW, Uffmann M, Weber M, et al. Computer-aided detection of small pulmonary nodules in chest radiographs: An observer study. *Acad Radiol*. 2011;18(12):1507–1514.
33. Tan T, Platel B, Huisman H, Sánchez C, Mus R, Karssemeijer N. Computer-aided lesion diagnosis in automated 3-D breast ultrasound using coronal spiculation. *IEEE Trans Med Imaging*. 2012;31(5):1034–1042.
34. Nishikawa RM, Pesce LL. Fundamental limitations in developing computer-aided detection for mammography. *Nucl Instrum Methods Phys Res A*. 21 August 2011;648(Supp 1),S251–S254.
35. Metz CE, Starr SJ, Lusted LB. Observer performance in detecting multiple radiographic signals. Prediction and analysis using a generalized ROC approach. *Radiology*. 1976;121(2):337–347.
36. Starr SJ, Metz CE, Lusted LB, Goodenough DJ. Visual detection and localization of radiographic images. *Radiology*. 1975;116:533–538.
37. Swensson RG, Judy PF. Detection of noisy visual targets: Models for the effects of spatial uncertainty and signal-to-noise ratio. *Percept Psychophys*. 1981;29(6):521–534.
38. Swensson RG. Unified measurement of observer performance in detecting and localizing target objects on images. *Med Phys*. 1996;23(10):1709–1725.
39. Chakraborty DP, Breatnach ES, Yester MV, Soto B, Barnes GT, Fraser RG. Digital and conventional chest imaging: A modified ROC study of observer performance using simulated nodules. *Radiology*. 1986;158:35–39.

40. Hillis SL. A marginal-mean ANOVA approach for analyzing multireader multicase radiological imaging data. *Stat Med.* 2014;33(2):330–360.
41. Hillis SL. A comparison of denominator degrees of freedom methods for multiple observer ROC studies. *Stat Med.* 2007;26:596–619.
42. Perneger TV. What's wrong with Bonferroni adjustments. *BMJ.* 1998;316(7139):1236.
43. Bland JM, Altman DG. Multiple significance tests: The Bonferroni method. *BMJ.* 1995;310 (6973):170.
44. Swensson RG, King JL, Good WF, Gur D. Observer variation and the performance accuracy gained by averaging ratings of abnormality. *Med Phys.* 2000;27(8):1920–1933.
45. Swensson RG. Using localization data from image interpretations to improve estimates of performance accuracy. *Med Decis Making.* 2000;20(2):170–184.
46. Nishikawa R.M. (2010) Computer-aided Detection and Diagnosis. In: Bick U., Diekmann F. (eds) Digital Mammography. *Medical Radiology*. Springer, Berlin, Heidelberg.

23

Validating CAD analysis

23.1 Introduction

In **Chapter 22** a method was described for analyzing standalone performance of computer aided detection (CAD) versus radiologists interpreting the same cases. The analysis accounted for radiologist and case variability. Any proposed analysis method should be validated to ensure that the relevant variability estimates, used internally by the method, are accurate. A statistically valid method will reject the NH about 5% of the time (assuming nominal α is 0.05). If it rejects more often than 5% then variability (of the FOM) is being underestimated,* and conversely if it rejects less often, then variability is being overestimated. Validation involves determining the empirical (i.e., observed) NH rejection rate when the NH is true. This cannot be done with patient data as one does not know the truth and moreover one needs many (thousands) of independent datasets. Consequently, a data simulator is needed to generate artificial datasets under known conditions of truth. It can be shown that about 2000 simulations are needed to constrain the 95% confidence for the nominal NH rejection rate to the interval (0.04, 0.06), **Chapter 8**.

In what follows a decision variable simulator is described that generates CAD and radiologist receiver operating characteristic (ROC) ratings data. This is followed by the method used to calibrate the simulator to a real dataset. Finally, the simulator is used to validate the analysis.

A brief overview of the simulator calibration and validation follows. It uses a recent advance,[1] namely the bivariate contaminated binormal model (BCBM) fitting method, implemented in **RJafroc** software as function **FitCorCbm()**, to fit pairs of ratings data for readers interpreting the same cases. CORCBM replaces the bivariate binormal model[2-4] used in CORROC2 with the contaminated binormal model (CBM).[5-7] Unlike CORROC2, which was described in **Chapter 21**, CORCBM is robust. It inherits the robustness property from CBM. It can fit practically any dataset, including degenerate ones. Each fit yields two normal distribution separation parameters (μ_X, μ_Y), two disease visibility parameters (α_X, α_Y), two correlations (ρ_1, ρ_2), and a 6×6 covariance matrix of the parameters $\Sigma_{\mu_X \alpha_X \mu_Y \alpha_Y \rho_1 \rho_2}$. Here X and Y refer to the two arms of the pairing. Performing the fitting over all pairs of readers in the dataset, and using appropriate averaging for model parsimony, yields the final set of *calibrated parameters and the average covariance matrices* $\Sigma_{\mu_X \alpha_X \mu_Y \alpha_Y \rho_1 \rho_2}$ for CAD versus radiologist pairing, and radiologist versus radiologist pairings. One imposes the null hypothesis condition by equating certain parameters. The calibrated values are used to sample new parameters sets, that is, 6-parameters $(\mu_X, \alpha_X, \mu_Y, \alpha_Y, \rho_1, \rho_2)$ per simulation. Each distinct set of 6-parameters is used to generate ratings for each non-diseased and each diseased case in the case

---

* This can be appreciated from the proverbial extreme case—if the variability estimate is zero, the F-statistic is infinite and the NH is always rejected.

sample. The CAD versus radiologists significance testing described in **Chapter 22** is applied, using empirical AUC as the figure of merit. This yields Cov_2 and Var, and the p-value for rejecting the NH that CAD performance equals the average performance of the radiologists. If the p-value is smaller than 0.05, an instance of NH rejection is recorded. For each simulation one saves Cov_2, Var and an integer 1 if the NH is rejected and zero otherwise. When the 2000 simulations are over, one calculates averages for Cov_2, Var and the rejection rate.

The simulator model is considered valid if average Cov_2 and Var are contained within the corresponding bootstrap-determined 95% confidence intervals derived from the original dataset. If the rejection rate is contained within the interval (0.04, 0.06) one concludes that the analysis method described in **Chapter 22** is valid.

We start with an abbreviated description of the bivariate contaminated binormal model (BCBM) model; for details the reader should consult the cited publication.[1] The fitting method, which uses standard maximum likelihood estimation, is not described here. This is followed by a visual demonstration of what the bivariate CBM *pdfs* look like. Described next is the single-modality multiple-reader decision variable simulator, which is based on the BCBM, and the calibration process, that is, choosing the parameters of the model to optimally match the clinical dataset. This is followed by the method used to simulate ratings. To facilitate understanding of the method, an example using CAD paired with three radiologists is given. This is followed by **R** software implementing the example. Finally, the method is applied to the clinical dataset consisting of CAD and nine radiologists interpreting the same dataset.

The starting point is gaining an understanding of the BCBM model.

23.2 Bivariate contaminated binormal model (BCBM)

The aim is to analyze data sets consisting of paired interpretations of cases. One possible pairing is CAD versus radiologists interpreting the same cases. Another possible pairing is two radiologists interpreting the same cases. A generic pairing is defined by two *arms* of the pairing, X and Y. For notational compactness, X and Y will be used to also denote random decision variables in the corresponding arms (the meaning will be clear from the context).

BCBM is defined by two bivariate CBM distributions, one per truth state $t; t = 1, 2$. Each bivariate distribution yields two simultaneous samples (X and Y) for each case.

23.2.1 Non-diseased cases

Let X_1 and Y_1 denote the random variables for a non-diseased case, $t = 1$, in the two arms. According to the BCBM these are sampled as follows:

$$\begin{pmatrix} X_1 \\ Y_1 \end{pmatrix} \sim N_2(\mu_1, \Sigma_1) \tag{23.1}$$

In Equation 23.1 the parameters of the bivariate distribution are defined by

$$\mu_1 = \begin{pmatrix} 0 \\ 0 \end{pmatrix}$$

$$\Sigma_1 = \begin{pmatrix} 1 & \rho_1 \\ \rho_1 & 1 \end{pmatrix} \tag{23.2}$$

Here ρ_1 is the decision-variable correlation between the two arms of the pairing for non-diseased cases.

23.2.2 Diseased cases

For diseased cases, the bivariate extension of CBM needs to account for the possibility that the mixing fractions* can be different in the two arms. These are modeled by α_X and α_Y, and likewise the Z-sample means for cases where the disease is visible can be different in the two arms, these are modeled by μ_X and μ_Y. All distributions are unit variance but can have different correlations as described next. Four types of mixings, and corresponding bivariate distributions, need to be accounted for:

1. The disease is invisible in both arms. Therefore, the appropriate distribution, denoted $N_{2;00}$, an abbreviation for a bivariate N_2 mixture distribution centered at $(0,0)$, has correlation $\rho_{2;00}$ and mixing fraction $(1-\alpha_X)(1-\alpha_Y)$. The subscript 2 on the correlation denotes the truth state, $t = 2$. Since it is a mixture distribution, the net probability equals the mixing fraction.
2. The disease is visible in arm X but invisible in arm Y. Therefore, the appropriate distribution, denoted $N_{2;X0}$, an abbreviation for a bivariate N_2 mixture distribution centered at $(\mu_X,0)$, has correlation $\rho_{2;X0}$ and mixing fraction $\alpha_X(1-\alpha_Y)$.
3. The disease is invisible in arm X but is visible in arm Y. Therefore, the appropriate distribution, denoted $N_{2;0Y}$, an abbreviation for a bivariate N_2 mixture distribution centered at $(0,\mu_Y)$, has correlation $\rho_{2;0Y}$ and mixing fraction $(1-\alpha_X)\alpha_Y$.
4. The disease is visible in both arms X and Y. Therefore, the appropriate distribution, denoted $N_{2;XY}$, an abbreviation for a bivariate N_2 mixture distribution centered at (μ_X,μ_Y), has correlation $\rho_{2;XY}$ and mixing fraction $\alpha_X\alpha_Y$.

The net probability under the four mixture distributions equals unity. Let X_2 and Y_2 denote the random variables for a diseased case, $t = 2$, in arms X and Y, respectively. The means vectors and covariance matrices for above-mentioned four types of mixings are defined by

$$
\begin{aligned}
&\mu_{2;00} = \begin{pmatrix} 0 \\ 0 \end{pmatrix}; \quad \Sigma_{2;00} = \begin{pmatrix} 1 & \rho_{2;00} \\ \rho_{2;00} & 1 \end{pmatrix} \\[2mm]
&\mu_{2;X0} = \begin{pmatrix} \mu_X \\ 0 \end{pmatrix}; \quad \Sigma_{2;X0} = \begin{pmatrix} 1 & \rho_{2;X0} \\ \rho_{2;X0} & 1 \end{pmatrix} \\[2mm]
&\mu_{2;0Y} = \begin{pmatrix} 0 \\ \mu_Y \end{pmatrix}; \quad \Sigma_{2;0Y} = \begin{pmatrix} 1 & \rho_{2;0Y} \\ \rho_{2;0Y} & 1 \end{pmatrix} \\[2mm]
&\mu_{2;XY} = \begin{pmatrix} \mu_X \\ \mu_Y \end{pmatrix}; \quad \Sigma_{2;XY} = \begin{pmatrix} 1 & \rho_{2;XY} \\ \rho_{2;XY} & 1 \end{pmatrix}
\end{aligned}
\tag{23.3}
$$

23.2.3 Assumptions regarding correlations

The above model has four correlations. For parsimony, the following assumptions are made:

■ For diseased cases where the disease is *invisible* in both arms, the correlation is the same as that for non-diseased cases, ρ_1.
■ For diseased cases where the disease is visible in only one arm, the correlation is the mean of (a) the non-diseased cases correlation and (b) the correlation where disease is visible in both arms.

---

* The meanings of the parameters for the univariate CBM were explained in **Chapter 20**. The bivariate CBM essentially doubles the basic parameters and adds two correlation terms.

By these assumptions:

$$\rho_{2;00} = \rho_1$$

$$\rho_{2;X0} = \rho_{2;0Y} = \frac{1}{2}\left(\rho_1 + \rho_{2;XY}\right)$$ \qquad (23.4)

23.2.4 Visual demonstration of BCBM *pdfs*

Visualizing the probability density functions (*pdfs*) is helpful toward understanding the BCBM.

23.2.4.1 Non-diseased cases

For non-diseased cases, the *pdf* is identical to that for the non-diseased arm of the bivariate binormal model; see for example the *pdfs* centered in (0,0) in Figure 21.1 (a,b). A 3D-visualization (not shown) would be a single peak centered at (0,0) with isopleth eccentricity determined by ρ_1; if $\rho_1 = 0$ the isopleths are circles and if $\rho_1 = 1$ the isopleths are straight lines at 45°.

23.2.4.2 Diseased cases

The diseased case pdf is a mixture distribution of four bivariate distributions, each contributing a single peak. The net volume under the four peaks will equal unity. Figure 23.1a and b shows 3D-visualizations, from different viewing perspectives, of diseased-case *pdfs* of the bivariate CBM. It was generated by sourcing file **main3DPlot.R**, with the following parameter values: $\mu_X = 3$, $\mu_Y = 3$, $\alpha_X = 0.5$, $\alpha_Y = 0.6$, $\rho_1 = 0.3$, and $\rho_{2;XY} = 0.8$, lines 8–13. [For comparison, for diseased cases, the *pdf* of the bivariate binormal model (not shown) would be a single peak at (3,3); see for example diseased *pdf* in Figure 23.1]

In Figure 23.1a, a generic view, there are four peaks, corresponding to the four possibilities outlined in Section 2.2.2. The (0, 0) centered distribution is for cases where the disease is *invisible* in both arms. The (3, 3) centered distribution is for cases where the disease is *visible* in both arms. The (3, 0) and (0, 3) centered distributions correspond to cases where the disease is visible *only* in arm

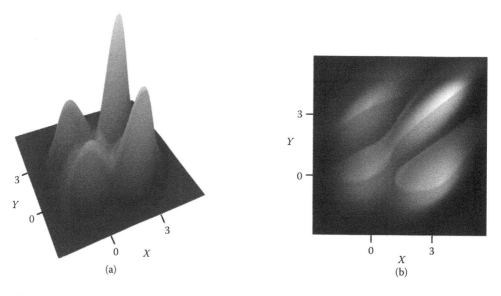

(a)   (b)

Figure 23.1 3D-visualizations of the *pdf* of the mixture distribution for diseased cases. (a) A generic view showing the peaks of the four bivariate normal distributions implicit in Section 23.2.2. (b) The bird's-eye view: the effect of stronger correlations when the disease is visible in both modalities (upper-right quadrant) can be best appreciated in this view. These figures are best visualized by running the code file and interacting with the 3D plots. These figures were created by sourcing **Main3DPlot.R**.

X or Y, respectively. Figure 23.1b shows a bird's eye view of the *pdfs*, which clearly shows the effect of the differing correlations. The isopleths approach narrow ellipses inclined at 45 degrees, whose minor-axis widths decrease as the correlation increases, approaching a straight line in the limit of unit correlation (unlike the bivariate binormal model, the variances are equal to unity in both directions—a great simplification). Perhaps evident in this view is that the (3, 3) centered distribution has the largest correlation, 0.8.

23.3 Single-modality multiple-reader decision variable simulator

CAD and a group of radiologists interpreting the same cases is a *single-modality multiple-reader paradigm*. It is necessary to *calibrate* the simulator to an actual clinical dataset, that is, assign realistic values to the parameters of the simulation model. A calibrated simulator will yield samples not *identical*, but *statistically* identical to the original dataset. The criteria used to assess statistical identity, is described later. The traditional Roe-Metz simulator, involving *univariate* sampling, that has been used in all ROC methodology validation work to date, is described in Online Appendix 23.A. This chapter describes a new method, involving *multivariate* sampling.

23.3.1 Calibrating the simulator to a specific dataset

The multivariate simulator was calibrated to the dataset used to illustrate the analysis described in Chapter 22 of the Hupse–Karssemeijer dataset[8]. The study used the location receiver operating characteristic (LROC) paradigm[9]. For the purpose of the current demonstration, the indicated location was ignored, i.e., only the ROC rating is used.

23.3.1.1 Calibrating parameters to individual radiologists

For each radiologist j, $j = 1, 2, ..., J$, one fits the clinical data using (univariate) CBM*, yielding μ_j, α_j, and a 2×2 covariance matrix $\Sigma_{\mu_j \alpha_j}$. In the study in question $J = 9$. Threshold related covariance terms are excluded from the following description. Since each radiologist is an equally valid sample from the population, for parsimony, the values were averaged over j, resulting in $\mu_\bullet, \alpha_\bullet, \Sigma_{\mu_\bullet \alpha_\bullet}$, where, as usual, the dot represents an average over the replaced index. For compactness, these are henceforth denoted $\mu_R, \alpha_R, \Sigma_{\mu\alpha R}$, where the "$R$" subscript denotes single radiologist-based estimates, i.e.,

$$\mu_\bullet, \alpha_\bullet, \Sigma_{\mu_\bullet \alpha_\bullet} \Rightarrow \mu_R, \alpha_R, \Sigma_{\mu\alpha R} \tag{23.5}$$

In Eqn. (23.5), μ_R is the CBM separation parameter when readers are analyzed individually using the CBM method and the parameters are averaged, and similarly for the parameter α_R. *These are the calibrated CBM parameters for the individual readers.*

23.3.1.2 Calibrating parameters to paired radiologists

Index 0 is reserved for CAD. For each pairing of different radiologists, $j \neq j'$; $jj' > 0$, one fits the data using CORCBM[1] yielding $\mu_{jj'}, \alpha_{jj'}, \mu_{jj}, \alpha_{jj}, \rho_{1;jj'}, \rho_{2;jj'}$, and the 6×6 covariance matrix $\Sigma_{\mu_{jj'} \alpha_{jj'} \mu_{jj} \alpha_{jj} \rho_{1;jj'} \rho_{2;jj'}}$.

> The first index on each ρ is case-truth. The parameter-indexing notation is explained for the separation parameter μ; similar notation is used for the visibility parameter α. $\mu_{jj'}$ represents the CBM separation parameter for the reader j arm of the pairing, when this reader is paired

---

* See description in **Chapter 20**. The CBM is implemented in `RJafroc`.

with reader j'. Its primary dependence is the intrinsic performance of reader j, which is the reason j is shown as a subscript; the secondary dependence, indicated as a *subscript on a subscript*, arises based on which reader constitutes the other arm of the pairing (this is a subtle point and the reader would be justified in mulling it over; basically, because of the paired analysis the primary dependence is affected by the other arm of the pairing). Likewise, μ_{jj} represents the separation parameter for the reader j' arm of the pairing, when this reader is paired with reader j. It is primarily dependent on the intrinsic performance of reader j', which could be quite different from that of reader j. Since for each pairing jj' there exists a pairing $j'j$, upon averaging over all pairings the result must be symmetric with respect to j and j', yielding $\mu_{\bullet\bullet}, \alpha_{\bullet\bullet}, \mu_{\bullet\bullet}, \alpha_{\bullet\bullet}, \rho_{1;\bullet\bullet}, \rho_{2;\bullet\bullet}$ and the average covariance matrix $\Sigma_{\mu_{\bullet\bullet}\alpha_{\bullet\bullet}\mu_{\bullet\bullet}\alpha_{\bullet\bullet}\rho_{1;\bullet\bullet}\rho_{2;\bullet\bullet}}$. For covariances there is no distinction between primary and secondary dependencies.

23.3.1.3 Reconciling single radiologist and paired radiologist parameters

To simultaneously simulate all readers according to a multivariate distribution, see below, one needs a *unique* value for the radiologist separation parameter. This is accomplished by averaging the two estimates, single-reader and paired-reader, as in the first equation below (a similar equation for the visibility parameter is not shown):

$$
\left.
\begin{aligned}
\mu_R &= \left(\mu_{\bullet} + \mu_{\bullet\bullet} \right)/2 \\
\Sigma_{RR} &\equiv \Sigma_{\mu_{\bullet\bullet}\alpha_{\bullet\bullet}\mu_{\bullet\bullet}\alpha_{\bullet\bullet}\rho_{1;\bullet\bullet}\rho_{2;\bullet\bullet}}
\end{aligned}
\right\}
\tag{23.6}
$$

Since different algorithms are used, one does not, for example, expect estimate $\mu_{\bullet}$ to equal $\mu_{\bullet\bullet}$.* The second equation introduces shorthand for the covariance matrix, where RR denotes radiologist-radiologist pairings.

23.3.1.4 Calibrating parameters to CAD

Like individual radiologists, the individual performance of CAD (regarded as reader "0") can be estimated using the CBM model, yielding:

$$
\mu_0, \alpha_0, \Sigma_{\mu_0,\alpha_0}
\tag{23.7}
$$

23.3.1.5 CAD-Radiologist pairings

Upon pairing CAD with radiologist $j; j > 0$, CORCBM yields $\mu_{0j}, \alpha_{0j}, \mu_{j_0}, \alpha_{j_0}, \rho_{1;0j}, \rho_{2;0j}$ and $\Sigma_{\mu_{0j}\alpha_{0j}\mu_{j_0}\alpha_{j_0}\rho_{1;0j}\rho_{2;0j}}$. Averaging over the radiologist index yields $\mu_{0\bullet}, \alpha_{0\bullet}, \mu_{\bullet 0}, \alpha_{\bullet 0}, \rho_{1;0\bullet}, \rho_{2;0\bullet}$ and $\Sigma_{\mu_{0\bullet}\alpha_{0\bullet}\mu_{\bullet 0}\alpha_{\bullet 0}\rho_{1;0\bullet}\rho_{2;0\bullet}}$. The CAD-only based estimates of the parameters are averaged with the average, over radiologists, of the CAD arm of CAD-radiologist pairings, yielding the first equation below (a similar equation for the visibility parameter is not shown):

$$
\left.
\begin{aligned}
\mu_C &= \left(\mu_0 + \mu_{0\bullet} \right)/2 \\
\Sigma_{CR} &\equiv \Sigma_{\mu_{0\bullet}\alpha_{0\bullet}\mu_{\bullet 0}\alpha_{\bullet 0}\rho_{1;0\bullet}\rho_{2;0\bullet}}
\end{aligned}
\right\}
\tag{23.8}
$$

The second equation introduces shorthand for the relevant averaged covariance.

---

* For the dataset in question the difference is about 10%

Under the null hypothesis the CAD and the radiologist parameters must be identical:

$$NH: \quad \begin{Bmatrix} \mu_C = (\mu_R + \mu_{\bullet_0})/2 \\ \alpha_C = (\alpha_R + \alpha_{\bullet_0})/2 \end{Bmatrix}$$
(23.9)

On the right hand side of the above equations, for each parameter one has the average of (a) the solo-radiologist based estimate (Eqn. 23.5) with (b) the average of the radiologist as the first arm, paired with CAD.

23.3.2 Simulating data using the calibrated simulator

Let J, K_1, K_2 denote the number of radiologists, non-diseased and diseased cases, respectively, to be simulated, and CAD is reader "0." For each pair of different readers jj', $j \neq j'$, where the range of either j or j' is 0 to J, each sample from the multivariate normal distribution yields a row-vector of 6—parameters:

$$[\mu_X, \alpha_X, \mu_Y, \alpha_Y, \rho_1, \rho_2]_{jj'} \sim \begin{cases} \left[N_6\left((\mu_C, \alpha_C, \mu_{\bullet_0}, \alpha_{\bullet_0}, \rho_{1;0\bullet}, \rho_{2;0\bullet}), \Sigma_{CR}\right) \Big| jj' = 0 \right] \\ \left[N_6\left((\mu_R, \alpha_R, \mu_R, \alpha_R, \rho_{1;\bullet\bullet}, \rho_{2;\bullet\bullet}), \Sigma_{RR}\right) \Big| jj' > 0 \right] \end{cases}$$
(23.10)

The two parameters, characterizing the first (i.e., "X") arm of the pairing, are μ_X, α_X and the corresponding ones characterizing the second (i.e., "Y") arm are μ_Y, α_Y. The last two parameters ρ_1, ρ_2 are the decision variable correlations for non-diseased and diseased cases, respectively, assuming, for diseased cases, that the disease is visible in both X- and Y-arms to the radiologist (intermediate visibility conditions are accommodated below).

- In Equation (23.10), the upper line on the right hand side describes the sampling when one of the readers is CAD, i.e., $j = 0$ or $j' = 0$. In this situation the other reader is the radiologist and one samples $N_6\left((\mu_C, \alpha_C, \mu_{\bullet_0}, \alpha_{\bullet_0}, \rho_{1;0\bullet}, \rho_{2;0\bullet}), \Sigma_{CR}\right)$. If $j = 0$ the first two samples, μ_X, α_X, correspond to CAD and the next two correspond to the radiologist paired with CAD; if $j' = 0$ the orderings are reversed.
- In Equation (23.10), the lower line on the right hand side describes the sampling when none of the readers is CAD, $j > 0$ and $j' > 0$. In this situation both readers are radiologist and one samples $N_6\left((\mu_R, \alpha_R, \mu_R, \alpha_R, \rho_{1;\bullet\bullet}, \rho_{2;\bullet\bullet}), \Sigma_{RR}\right)$. The first two samples, μ_X, α_X, correspond to one radiologist and the next two correspond to the other radiologist.
- Because of the nature of random sampling, the sampled value for pairing jj' was not necessarily identical to that for $j'j$; the different values are replaced by their average.

An average (and therefore more stable) value μ_j for reader j, $j = 0, 1, ..., J$, is obtained by averaging $[\mu_X]_{j_1 j_2}$ over j_2, when $j = j_1$ or averaging $[\mu_Y]_{j_1 j_2}$ over j_1, when $j = j_2$. This can be expressed by

$$\mu_j = (\mu_{X;j\bullet} + \mu_{Y;\bullet j})/2$$
(23.11)

A similar expression applies to average value α_j:

$$\alpha_j = \left(\alpha_{X;j\bullet} + \alpha_{Y;\bullet j}\right)/2 \tag{23.12}$$

One may wonder why the correlations do not have to be likewise averaged. This is because the correlation is between the two arms of the pairing, while the other parameters are specific to one or the other arm of the pairing.

23.3.2.1 Example with J = 3

Following is an example of generating Z-samples for CAD and three radiologists, $J = 3$ (four readers total).

23.3.2.1.1 Non-diseased cases

For non-diseased cases, one need only construct the covariance matrix $\Sigma_{1;J+1}$. The subscript 1 indicates its truth-state, and $J + 1$ indicates its dimension along one edge of the matrix. *All diagonal elements are set to unity.*[*] The off-diagonal elements are filled as follows:

$$\Sigma_{1;J+1}\left(j, j'\right) = \rho_{1;jj'}; j \neq j'; j, j' = 0,1,...,J \tag{23.13}$$

The covariance matrix is shown below. The correlations were sampled in Equation 23.9. One is using the non-diseased case correlation ρ_1 as the intent is to sample new non-diseased case Z-samples.

$$\Sigma_{1;4} = \begin{bmatrix} 1 & \rho_{1;01} & \rho_{1;02} & \rho_{1;03} \\ & 1 & \rho_{1;12} & \rho_{1;13} \\ & & 1 & \rho_{1;23} \\ & & & 1 \end{bmatrix} \tag{23.14}$$

23.3.2.1.2 Diseased cases

For diseased cases, define the *visibility-condition* $\overrightarrow{V_c}$ (row) vector indexed by $c; c = 1,2,...,2^{(J+1)}$. Each value of c contains $(J+1)$ elements, each of which is either zero or one, indicating whether disease was visible to the reader specified by the column number in the row-vector. If the disease was visible, the corresponding element is 1 and otherwise it is zero. For example, if the $(j+1)th$ element of $\overrightarrow{V_c}$ is one, then the disease was visible to reader j, $j = 0,1,...,J$.

For $J = 3$, there are 2^4 visibility conditions, for which the visibility-condition vectors are shown in Table 23.1 under $\overrightarrow{V_c}$. For example, $\overrightarrow{V_1} = (1,1,1,1)$, means that in visibility-condition $c = 1$, disease was *visible* to CAD (the first 1 in the array) and to all radiologists. Likewise, $\overrightarrow{V_2} = (0,1,1,1)$ means disease was *invisible* to CAD but was visible to all radiologists, $\overrightarrow{V_3} = (0,0,1,1)$ means disease was invisible to CAD and the first radiologist, but was visible to the other radiologists, and finally, $\overrightarrow{V_{16}} = (0,0,0,0)$ means the disease was invisible to CAD or any of the radiologists.

The corresponding $\overrightarrow{\mu_c}$ vectors follow from the second column in Table 23.1. For example, since $\overrightarrow{V_1} = (1,1,1,1)$, it follows that $\overrightarrow{\mu_1} = (\mu_0, \mu_1, \mu_2, \mu_3)$. Since the disease was visible, the first value is the separation parameter for CAD, that is, μ_0, the second value is the separation parameter for the first radiologist, that is, μ_1, etc. The μ_j were calculated via Equation 23.10. As another example, from $\overrightarrow{V_3} = (0,0,1,1)$, it follows that $\overrightarrow{\mu_3} = (0,0,\mu_2,\mu_3)$. Since the disease was invisible to CAD and the first radiologist, the first two values are zeroes and the third value is the separation parameter for the second radiologist, that is, μ_2, and the last value is the separation parameter for the third radiologist, that is, μ_3.

---

[*] This is the great simplification afforded by CBM; all distributions are unit variance.

Table 23.1 Shown in this table, for $J = 3$, are 16 visibility conditions

c	$\vec{V_c}$	$\vec{\mu_c}$	$\Sigma_{2;4;c}$	$\vec{P}$
1	$\vec{V_1} = (1,1,1,1)$	$\vec{\mu_1} = (\mu_0, \mu_1, \mu_2, \mu_3)$	$\Sigma_{2;4;1} = \begin{bmatrix} 1 & \rho_{2;01} & \rho_{2;02} & \rho_{2;03} \\ & 1 & \rho_{2;12} & \rho_{2;13} \\ & & 1 & \rho_{2;23} \\ & & & 1 \end{bmatrix}$	$P_1 = \alpha_0\alpha_1\alpha_2\alpha_3$
2	$\vec{V_2} = (0,1,1,1)$	$\vec{\mu_2} = (0, \mu_1, \mu_2, \mu_3)$	$\Sigma_{2;4;2} = \begin{bmatrix} 1 & \rho_{\bullet;01} & \rho_{\bullet;02} & \rho_{\bullet;03} \\ & 1 & \rho_{2;12} & \rho_{2;13} \\ & & 1 & \rho_{2;23} \\ & & & 1 \end{bmatrix}$	$P_2 = (1-\alpha_0)\alpha_1\alpha_2\alpha_3$
3	$\vec{V_3} = (0,0,1,1)$	$\vec{\mu_3} = (0,0,\mu_2,\mu_3)$	$\Sigma_{2;4;3} = \begin{bmatrix} 1 & \rho_{1;01} & \rho_{\bullet;02} & \rho_{\bullet;03} \\ & 1 & \rho_{\bullet;12} & \rho_{\bullet;13} \\ & & 1 & \rho_{2;23} \\ & & & 1 \end{bmatrix}$	$P_3 = (1-\alpha_0)(1-\alpha_1)\alpha_2\alpha_3$
...	...	...	...	
16	$\vec{V_{16}} = (0,0,0,0)$	$\vec{\mu_{16}} = (0,0,0,0)$	$\Sigma_{2;4;16} = \begin{bmatrix} 1 & \rho_{1;01} & \rho_{1;02} & \rho_{1;03} \\ & 1 & \rho_{1;12} & \rho_{1;13} \\ & & 1 & \rho_{1;23} \\ & & & 1 \end{bmatrix}$	$P_{16} = (1-\alpha_0)(1-\alpha_1)(1-\alpha_2)(1-\alpha_3)$

Note: The visibility-condition row-vectors are shown in column 2. the corresponding μ vector is in column = 2. The covariance matrices are listed in column 3 and the last column lists the probability of occurrence of each condition. All matrices are symmetric.

In Table 23.1, the covariance matrix $\Sigma_{2;4;1}$ has 1s along the diagonal (this is true of all covariance matrices in Table 23.1), the value at row 1 (CAD, $j = 0$) and column 2 (radiologist 1, $j' = 1$) is $\rho_{2;01}$. One is using ρ_2 because the disease was visible to *both* readers implied by row 1 and column 2, and the 01 indexing follows from the readers specified by the row and column indices. As another example, the value at row 2 (radiologist 1, $j = 1$) and column 3 (radiologist 2, $j' = 2$) is $\rho_{2;12}$. One is using ρ_2 because the disease was visible to both readers implied by the specified row and column indices. The 12 indexing follows from the readers implied by the specified row and column indices.

As a more complicated example, consider $\Sigma_{2;4;3}$. Since $\vec{V_3} = (0,0,1,1)$, the disease was invisible to CAD or the first radiologist. This implies that the correlation at row 1 ($j = 0$) and column 2 ($j' = 1$) must be ρ_1, subscripted with the appropriate reader indices, that is, 01. The value at row 1 and column 3 is the average of ρ_1 and ρ_2, that is, $\rho_\bullet$, subscripted with the appropriate reader indices, $\rho_{\bullet;02}$. As a final example, the value at row 3 and column 4 is $\rho_{2;23}$, because disease was visible to the second and third radiologists ($j = 2$, $j' = 3$), and these are the appropriate indices.

The last column of Table 23.1 lists the probabilities associate with each visibility-condition. For example, for visibility-condition $c = 1$, $P_1 = \alpha_0\alpha_1\alpha_2\alpha_3$ because disease was visible to CAD, the probability of which is α_0, it was visible to radiologist 1, the probability of which is α_1, it was visible to radiologist 2, the probability of which is α_2 and it was visible to radiologist 3, the probability of which is α_3. Because the radiologists are independent, the probability of the visibility-condition $c = 1$ is the product of the component probabilities, which is $P_1 = \alpha_0\alpha_1\alpha_2\alpha_3$. Similarly, the probability of observing visibility-condition $c = 2$ is $P_2 = (1 - \alpha_0)\alpha_1\alpha_2\alpha_3$, because disease was invisible to CAD, the probability of which is $(1 - \alpha_0)$, and so on. One can confirm that the probabilities listed in the last column of Table 23.1 sum to unity.

To determine the number of diseased cases in each visibility-condition one samples the multinomial distribution with trial size K_2 and cell probabilities specified by $\vec{P}$.

23.3.2.1.3 Illustration using R code

The software implementing the method described above is implemented in a file, in the **software** directory, named **mainCadVsRadCalibValidate.R**.

1. Highlight lines 1–29 and click **Run**. Echoed lines are not shown below and warnings are ignorable. Enter **q <- 1** in **Console** window and hit **Enter**. Highlight lines 31–32 and click **Run**. Highlight lines 37–93 and click **Run**. Highlight **simuMu**, line 87, and click **Run**. Highlight **simuAlpha**, line 88, and click **Run**. Highlight **rhoNorMatr**, line 90, and click **Run**. Highlight **rhoAbn2Matr**, line 91, and click **Run**. This should yield the values shown below, Section 23.3.2.1.2.1 under Part A. The rest of the example is under part B.
2. Highlight lines 95–110 and click **Run**. Enter **l <- 16** in **Console** window and hit **Enter**. Highlight lines 112–130 and click **Run**. [Highlight **conditionArray**, line 99, and click **Run**. Highlight **probVector**, line 100, and click **Run**.] Highlight **muTemp**, line 113, and click **Run**. Highlight **sigmaTemp**, line 115, and click **Run**. This should yield all values shown below under Part B, corresponding to $c = 1$ in Table 23.1. None of the elements of **muTemp** are zero, consistent with the fact that in this condition disease was visible to CAD and all radiologists.
3. Enter **l <- 14** and repeat #2 except for the square bracketed part. This corresponds to $c = 2$ in Table 23.1. The first element of **muTemp** is zero, consistent with the fact that in this condition disease was invisible to CAD.
4. Enter **l <- 13** and repeat #2 except for the square bracketed part. This corresponds to $c = 3$ in Table 23.1. The first two elements of **muTemp** are zero, consistent with the fact that in this condition disease was invisible to CAD and the first radiologist.
5. Enter **l <- 1** and repeat #2 except for the square bracketed part. This corresponds to $c = 16$ in Table 23.1. Note that all elements of **muTemp** are zero, consistent with the fact that in this condition disease was invisible to CAD and all radiologists.

23.3.2.1.3.1 Code output (partial)

```
#Part A
seed = 1 , binning= TRUE , J= 4 , K1= 120 , K2= 80
> simuMu
[1] 2.01 2.06 2.04 2.09
> simuAlpha
[1] 0.888 0.873 0.882 0.853
```

```
> rhoNorMatr
       [,1]   [,2]   [,3]   [,4]
[1,]  1.000  0.243  0.109  0.180
[2,]  0.243  1.000  0.515  0.568
[3,]  0.109  0.515  1.000  0.472
[4,]  0.180  0.568  0.472  1.000
> rhoAbn2Matr
       [,1]   [,2]   [,3]   [,4]
[1,]  1.000  0.754  0.451  0.745
[2,]  0.754  1.000  0.730  0.765
[3,]  0.451  0.730  1.000  0.786
[4,]  0.745  0.765  0.786  1.000

#Part B
> conditionArray
       [,1]  [,2]  [,3]  [,4]
 [1,]    0     0     0     0
 [2,]    0     1     0     0
 [3,]    1     0     0     0
 [4,]    1     1     0     0
 [5,]    0     0     1     0
 [6,]    0     1     1     0
 [7,]    1     0     1     0
 [8,]    1     1     1     0
 [9,]    0     0     0     1
[10,]    0     1     0     1
[11,]    1     0     0     1
[12,]    1     1     0     1
[13,]    0     0     1     1
[14,]    0     1     1     1
[15,]    1     0     1     1
[16,]    1     1     1     1

> probVector
 [1] 0.000247 0.001693 0.001955 0.013421 0.001844 0.012663 0.014622
0.100387 0.001433 0.009837 0.011359 0.077984 0.010717 0.073578
0.084962
[16] 0.583296

# l <- 16
> muTemp
[1] 2.01 2.06 2.04 2.09
> sigmaTemp
       [,1]   [,2]   [,3]   [,4]
[1,]  1.000  0.754  0.451  0.745
[2,]  0.754  1.000  0.730  0.765
[3,]  0.451  0.730  1.000  0.786
[4,]  0.745  0.765  0.786  1.000

# l <- 14
> muTemp
[1] 0.00 2.06 2.04 2.09
```

```
> sigmaTemp
       [,1]    [,2]    [,3]    [,4]
[1,]  1.000  0.499  0.280  0.463
[2,]  0.499  1.000  0.730  0.765
[3,]  0.280  0.730  1.000  0.786
[4,]  0.463  0.765  0.786  1.000

# 1 <- 13
> muTemp
[1]  0.000000  0.000000  2.035396  2.092945
> sigmaTemp
          [,1]        [,2]        [,3]        [,4]
[1,]  1.0000000  0.2430901  0.2802590  0.4625262
[2,]  0.2430901  1.0000000  0.6220517  0.6666261
[3,]  0.2802590  0.6220517  1.0000000  0.7859280
[4,]  0.4625262  0.6666261  0.7859280  1.0000000

# 1 <- 1
> muTemp
[1]  0 0 0 0
> sigmaTemp
       [,1]    [,2]    [,3]    [,4]
[1,]  1.000  0.243  0.109  0.180
[2,]  0.243  1.000  0.515  0.568
[3,]  0.109  0.515  1.000  0.472
[4,]  0.180  0.568  0.472  1.000
```

23.3.2.2 General case

Let $\vec{V_c}(j)$ denote column j of the row-vector (consisting of zeroes and ones) specified by $\vec{V_c}$. The rule for calculating the elements of the mean vector for visibility-condition c is

$$\overrightarrow{\mu_c}(j) = \mu_j \overrightarrow{V_c}(j) \tag{23.15}$$

The rule for calculating the covariance matrix for visibility-condition c is

$$\left. \begin{array}{l} \Sigma_{2;J+1;c}(j, j') = \rho_{1;jj'}; \text{if } \overrightarrow{V_c}(j) + \overrightarrow{V_c}(j') = 0 \\[2mm] \Sigma_{2;J+1;c}(j, j') = \rho_{\bullet;jj'}; \text{if } \overrightarrow{V_c}(j) + \overrightarrow{V_c}(j') = 1 \\[2mm] \Sigma_{2;J+1;c}(j, j') = \rho_{2;jj'}; \text{if } \overrightarrow{V_c}(j) + \overrightarrow{V_c}(j') = 2 \end{array} \right\} \tag{23.16}$$

The rule for calculating the probability vector $\vec{P} = (P_1, P_2, ..., P_{2^{J+1}})$ of the different visibility-conditions is

$$P_c = \prod_0^J p_j; p_j = \begin{cases} \alpha_j; \text{if } \overrightarrow{V_c}(j) = 1 \\[2mm] (1 - \alpha_j); \text{if } \overrightarrow{V_c}(j) = 0 \end{cases} \tag{23.17}$$

23.3.2.3 Using the simulator
The K_1 non-diseased ratings are generated as follows:

$$Z_1 \sim N_{J+1}\left(\vec{0}, \Sigma_{1;J+1}\right) \tag{23.18}$$

For diseased cases, one samples the multinomial distribution K_2 times with cell probabilities as specified in $\bar{P}$. This yields the number of diseased cases in each visibility-condition c. Let $K_{2;c}$ denote the number of diseased cases in visibility-condition c. One generates $K_{2;c}$ samples from the multivariate normal distribution:

$$Z_{2;c} \sim N_{J+1}\left(\overrightarrow{\mu_c}, \Sigma_{2;J+1;c}\right) \tag{23.19}$$

This is repeated for all values of c. This completes the simulation of continuous ratings for a single-modality and $(J + 1)$ readers ROC dataset, of which the first reader is CAD. Since the original dataset was binned into six bins, the simulated datasets were likewise binned into six bins.

23.4 Calibration, validation of simulator, and testing its NH behavior

Tables 23.2 through 23.4, summarize the results of the calibration process.

23.4.1 Calibration of the simulator

Table 23.2 This table shows, for the clinical dataset, the calibrated values for the parameter vectors needed in Equation 23.10

	μ_X	α_X	μ_Y	α_Y	ρ_1	ρ_2
$jj' = 0$	2.2000	0.7239	1.9654	0.8804	0.1847	0.6906
$jj' > 0$	1.9751	0.8747	1.9751	0.8747	0.4790	0.7751

Table 23.3 This table shows, for the clinical dataset, the calibrated values for Σ_{CR} needed in Equation 23.10

	μ_X	α_X	μ_Y	α_Y	ρ_1	ρ_2
μ_X	0.1030	−0.0120	0.0128	−0.0008	−0.0029	0.0062
α_X	−0.0120	0.0069	0.0013	0.0004	0.0002	−0.0032
μ_Y	0.0128	0.0013	0.0835	−0.0163	−0.0015	−0.0003
α_Y	−0.0008	0.0004	−0.0163	0.0099	−0.0002	−0.0027
ρ_1	−0.0029	0.0002	−0.0015	−0.0002	0.0126	0.0001
ρ_2	0.0062	−0.0032	−0.0003	−0.0027	0.0001	0.0237

Table 23.4 This table shows, for the clinical dataset, the calibrated values for Σ_{RR} needed in Equation 23.10

	μ_X	α_X	μ_Y	α_Y	ρ_1	ρ_2
μ_X	0.0716	−0.0095	0.0242	0.0008	−0.0034	0.0005
α_X	−0.0095	0.0061	0.0008	0.0007	0.0002	−0.0024
μ_Y	0.0242	0.0008	0.0716	−0.0095	−0.0034	0.0005
α_Y	0.0008	0.0007	−0.0095	0.0061	0.0002	−0.0024
ρ_1	−0.0034	0.0002	−0.0034	0.0002	0.0105	0.0002
ρ_2	0.0005	−0.0024	0.0005	−0.0024	0.0002	0.0097

Table 23.5 This table summarizes results of validation of the simulation method and results of NH testing

Row #	BIN	$J / K_1 / K_2$	$Cov_2^{\bullet}$ (95% CI)	$Var^{\bullet}$ (95% CI)	Reject rate
1	TRUE	5/25/25	0.00023	0.00085	0.04
2	TRUE	10/25/25	0.00025	0.00086	0.041
3	TRUE	5/50/50	0.00021*	0.00077	0.046
4	TRUE	10/50/50	0.00023	0.00078	0.047
5	TRUE	5/120/80	0.00023	0.00082	0.0455
6	TRUE	10/120/80	0.00025	0.00082	0.0575
7	TRUE	10/100/100	0.00022	0.00074	0.05
8	FALSE	10/100/100	0.00022	0.00068	0.058

Note: For the original dataset $Cov_2^{org} = 0.00033(0.000216, 0.000573)$ an $Var^{org} = 0.00087(0.000636, 0.00109)$. In the table $Cov_2^{\bullet}$ = average of 2000 values of Cov_2^{s}, $Var^{\bullet}$ = average of 2000 values of Var^{s}, where s is the simulation index $s = 1, 2, ..., 2000$. Instances where the 95% confidence interval for the original dataset did not include the corresponding simulation averaged estimate are indicated with an asterisk. The NH rejection rates were within that expected for 2000 simulations, i.e., they are all in the range (0.04, 0.06).

23.4.2 Validation of simulator and testing its NH behavior

The code was run with different values of J, K_1, K_2, as indicated in Table 23.5. The seed variable was set to **NULL**, line 28, which generates random seeds[*]. Data binning to six bins was used, line 26 and lines 138 through140. When the total number of cases is different from that in the clinical dataset, the values of Cov_2^{s} and Var^{s} corresponding to simulated dataset $s; s = 1, 2, ..., 2000$, need to be multiplied by (K_{NEW} / K_{ORG}), line 143. Here *NEW* refers to the simulated datasets and *ORG* refers to the original dataset. Confidence intervals for Var^{org} and Cov_2^{org} were obtained by bootstrapping readers and cases 2000 times.

Results of the evaluation summarized in Table 23.5 show that, with one exception, as indicated by the asterisk, the estimates of $Cov_2^{\bullet}$ and $Var^{\bullet}$ are contained within the 95% confidence interval of the corresponding values for the original data. The estimates of $Var^{\bullet}$ (average of rows 1 through 7 yields 0.00081) are close to that for the original dataset: $Var^{org} = 0.00087(0.000636, 0.00109)$. The

---

[*] For NH testing and validation, seed should not be set to a numeric value; the latter is only done for demonstration and debugging purposes.

estimates of Cov_2^* (corresponding average = 0.00023) are smaller by about 30% than that for the original dataset: $Cov_2^{org} = 0.00033(0.000216, 0.000573)$. Row 8 in Table 23.5 is identical to row 7, except that the data was not binned. The variance is smaller, suggesting that binning introduces additional noise, which seems intuitively reasonable.

23.5 Discussion/Summary

This chapter describes a method for designing a ratings simulator that is statistically matched (calibrated) to the single-modality multiple-reader ROC Hupse–Karssemeijer dataset. Showing that it yields, upon analysis with the ORH method, a figure of merit variance structure that is consistent with that of the original dataset validates the method. Furthermore, when the NH condition was imposed, the analysis method described in **Chapter 22** rejects the NH consistent with the nominal α, thereby validating the analysis method.

The Roe and Metz (RM) simulator[10] is outdated and moreover, there does not exist a systematic way of estimating its parameters. The online appendix to this chapter details the calibration of the Roe and Metz model to the Hupse–Karssemeijer dataset (it is specific to CAD versus radiologists). When calibrated and the null hypothesis imposed, simulations yield the correct rejection rate, consistent with 5%. The method yielded $Cov_2 = 0.00048$, consistent with the original data, but $Var = 0.00133$, which is outside the 95% CI of the original data estimate.

The method described in this chapter is currently being extended to multiple modalities. It will then be used to test the NH behavior of DBMH and ORH analyses for simulators calibrated to different clinical datasets.

References

1. Zhai X, Chakraborty DP. A bivariate contaminated binormal model for robust fitting of proper ROC curves to a pair of correlated, possibly degenerate, ROC datasets. *Med Phys.* 2017;44(3):2207–2222.
2. Metz CE, Kronman H. A test for the statistical significance of differences between ROC curves. *INSERM.* 1979;88:647–660.
3. Metz CE, Kronman H. Statistical significance tests for binormal ROC curves. *J Math Psychol.* 1980;22(3):218–242.
4. Metz CE, Wang P-L, Kronman HB. A new approach for testing the significance of differences between ROC curves measured from correlated data. In: Deconinck F, ed. *Information Processing in Medical Imaging.* The Hague: Nijhoff; 1984.
5. Dorfman DD, Berbaum KS. A contaminated binormal model for ROC data: Part II. A formal model. *Acad Radiol.* 2000;7(6):427–437.
6. Dorfman DD, Berbaum KS. A contaminated binormal model for ROC data: Part III. Initial evaluation with detection. *Acad Radiol.* 2000;7(6):438–447.
7. Dorfman DD, Berbaum KS, Brandser EA. A contaminated binormal model for ROC data: Part I. Some interesting examples of binormal degeneracy. *Acad Radiol.* 2000;7(6):420–426.
8. Hupse R, Samulski M, Lobbes M, et al. Standalone computer-aided detection compared to radiologists' performance for the detection of mammographic masses. *Eur Radiol.* 2013;23(1):93–100.
9. Swensson RG. Unified measurement of observer performance in detecting and localizing target objects on images. *Med Phys.* 1996;23(10):1709–1725.
10. Roe CA, Metz CE. Dorfman-Berbaum-Metz Method for Statistical Analysis of Multireader, Multimodality Receiver Operating Characteristic Data: Validation with Computer Simulation. *Acad Radiol.* 1997;4:298–303.

Index